Lecture Notes in Bioii

Subseries of Lecture Notes in Computer Science

Alberto Apostolico   Concettina Guerra
Sorin Istrail   Pavel Pevzner
Michael Waterman (Eds.)

Research in Computational Molecular Biology

10th Annual International Conference, RECOMB 2006
Venice, Italy, April 2-5, 2006
Proceedings

 Springer

Series Editors

Sorin Istrail, Brown University, Providence, RI, USA
Pavel Pevzner, University of California, San Diego, CA, USA
Michael Waterman, University of Southern California, Los Angeles, CA, USA

Volume Editors

Alberto Apostolico
Concettina Guerra
University of Padova, Department of Information Engineering
Via Gradenigo 6/a, 35131 Padova, Italy
E-mail: {axa, guerra}@dei.unipd.it

Sorin Istrail
Brown University, Center for Molecular Biology and Computer Science Department
115 Waterman St., Providence, RI 02912, USA
E-mail: sorin@cs.brown.edu

Pavel Pevzner
University of California at San Diego
Department of Computer Science and Engineering
La Jolla, CA 92093-0114, USA
E-mail: ppevzner@cs.ucsd.edu

Michael Waterman
University of Southern California
Department of Molecular and Computational Biology
1050 Childs Way, Los Angeles, CA 90089-2910, USA
E-mail: msw@usc.edu

Library of Congress Control Number: 2006922626

CR Subject Classification (1998): F.2.2, F.2, E.1, G.2, H.2.8, G.3, I.2, J.3

LNCS Sublibrary: SL 8 – Bioinformatics

ISSN   0302-9743
ISBN-10   3-540-33295-2 Springer Berlin Heidelberg New York
ISBN-13   978-3-540-33295-4 Springer Berlin Heidelberg New York

Springer is a part of Springer Science+Business Media

springer.com

© Springer-Verlag Berlin Heidelberg 2006
Printed in Germany

Typesetting: Camera-ready by author, data conversion by Scientific Publishing Services, Chennai, India
Printed on acid-free paper   SPIN: 11732990   06/3142   5 4 3 2 1 0

Preface

This volume contains the papers presented at the 10th Annual International Conference on Research in Computational Molecular Biology (RECOMB 2006), which was held in Venice, Italy, on April 2–5, 2006. The RECOMB conference series was started in 1997 by Sorin Istrail, Pavel Pevzner and Michael Waterman. The table on p. VIII summarizes the history of the meetings. RECOMB 2006 was hosted by the University of Padova at the Cinema Palace of the Venice Convention Center, Venice Lido, Italy. It was organized by a committee chaired by Concettina Guerra. A special 10th Anniversary Program Committee was formed, by including the members of the Steering Committee and inviting all Chairs of past editions. The Program Committee consisted of the 38 members whose names are listed on a separate page.

From 212 submissions of high quality, 40 papers were selected for presentation at the meeting, and they appear in these proceedings. The selection was based on reviews and evaluations produced by the Program Committee members as well as by external reviewers, and on a subsequent Web-based PC open forum. Following the decision made in 2005 by the Steering Committee, RECOMB Proceedings are published as a volume of *Lecture Notes in Bioinformatics* (LNBI), which is co-edited by the founders of RECOMB. Traditionally, the *Journal of Computational Biology* devotes a special issue to the publication of archival versions of selected conference papers.

RECOMB 2006 featured seven keynote addresses by as many invited speakers: Anne-Claude Gavin (EMBL, Heidelberg, Germany), David Haussler (University of California, Santa Cruz, USA), Ajay K. Royyuru (IBM T.J. Watson Research Center, USA), David Sankoff (University of Ottawa, Canada), Michael S. Waterman (University of Southern California, USA), Carl Zimmer (Science Writer, USA), Roman A. Zubarev (Uppsala University, Sweden). The Stanislaw Ulam Memorial Computational Biology Lecture was given by Michael S. Waterman. A special feature presentation was devoted to the 10th anniversary and is included in this volume.

Like in the past, an important ingredient for the success of the meeting was represented by a lively poster session.

RECOMB06 was made possible by the hard work and dedication of many, from the Steering to the Program and Organizing Committees, from the external reviewers, to Venice Convention, Venezia Congressi and the institutions and corporations who provided administrative, logistic and financial support for the conference. The latter include the Department of Information Engineering of the University of Padova, the Broad Institute of MIT and Harvard (USA), the College of Computing of Georgia Tech. (USA), the US Department of Energy, IBM Corporation (USA), the International Society for Computational Biology

(ISCB), the Italian Association for Informatics and Automatic Computation (AICA), the US National Science Foundation, and the University of Padova.

Special thanks are due to all those who submitted their papers and posters and who attended RECOMB 2006 with enthusiasm.

April 2006   Alberto Apostolico
                                               RECOMB 2006 Program Chair

Organization

Program Committee

Tatsuya Akutsu	(Kyoto University, Japan)
Alberto Apostolico	Chair (Accademia Nazionale Dei Lincei, Italy, and Georgia Tech., USA)
Gary Benson	(Boston University, USA)
Mathieu Blanchette	(McGill, Canada)
Philip E. Bourne	(University of California San Diego, USA)
Steve Bryant	(NCBI, USA)
Andrea Califano	(Columbia University, USA)
Andy Clark	(Cornell University, USA)
Gordon M. Crippen	(University of Michigan, USA)
Raffaele Giancarlo	(University of Palermo, Italy)
Concettina Guerra	(University of Padova, Italy, and Georgia Tech., USA)
Dan Gusfield	(University of California, Davis, USA)
Sridhar Hannenhalli	(University of Pennsylvania, USA)
Sorin Istrail	(Brown University, USA)
Inge Jonassen	(University of Bergen, Norway)
Richard M. Karp	(University of California, Berkeley, USA)
Simon Kasif	(Boston University, USA)
Manolis Kellis	(MIT, USA)
Giuseppe Lancia	(University of Udine, Italy)
Thomas Lengauer	(GMD Sant Augustin, Germany)
Michael Levitt	(Stanford, USA)
Michal Linial	(The Hebrew University in Jerusalem, Israel)
Jill Mesirov	(Broad Institute of MIT and Harvard, USA)
Satoru Miyano	(University of Tokyo, Japan)
Gene Myers	(HHMI, USA)
Laxmi Parida	(IBM T.J. Watson Research Center, USA)
Pavel A. Pevzner	(University of California San Diego, USA)
Marie-France Sagot	(INRIA Rhone-Alpes, France)
David Sankoff	(University of Ottawa, Canada)
Ron Shamir	(Tel Aviv University, Israel)
Roded Sharan	(Tel Aviv University, Israel)
Steve Skiena	(State University of New York at Stony Brook, USA)
Terry Speed	(University of California, Berkeley, USA)
Jens Stoye	(University of Bielefeld, Germany)
Esko Ukkonen	(University of Helsinki, Finland)
Martin Vingron	(Max Planck Institute for Molecular Genetics, Germany)
Michael Waterman	(University of Southern California, USA)
Haim J. Wolfson	(Tel Aviv University, Israel)

Steering Committee

Sorin Istrail	RECOMB General Vice-chair (Brown, USA)
Thomas Lengauer	(GMD Sant Augustin, Germany)
Michal Linial	(The Hebrew University of Jerusalem, Israel)
Pavel A. Pevzner	RECOMB General Chair (University of California, San Diego, USA)
Ron Shamir	(Tel Aviv University, Israel)
Terence P. Speed	(University of California, Berkeley, USA)
Michael Waterman	RECOMB General Chair (University of Southern California, USA)

Organizing Committee

Alberto Apostolico	(Accademia Nazionale dei Lincei, Italy, and Georgia Tech., USA)
Concettina Guerra	Conference Chair (University of Padova, Italy, and Georgia Tech., USA)
Eleazar Eskin	Chair, 10th Anniversary Committee (University of California, San Diego)
Matteo Comin	(University of Padova, Italy)
Raffaele Giancarlo	(University of Palermo, Italy)
Giuseppe Lancia	(University of Udine, Italy)
Cinzia Pizzi	(University of Padova, Italy, and Univ. of Helsinki, Finland)
Angela Visco	(University of Padova, Italy)
Nicola Vitacolonna	(University of Udine, Italy)

Previous RECOMB Meetings

Date/Location	Hosting Institution	Program Chair	Conference Chair
January 20-23, 1997 Santa Fe, NM, USA	Sandia National Lab	Michael Waterman	Sorin Istrail
March 22-25, 1998 New York, NY, USA	Mt. Sinai School of Medicine	Pavel Pevzner	Gary Benson
April 22-25, 1999 Lyon, France	INRIA	Sorin Istrail	Mireille Regnier
April 8-11, 2000 Tokyo, Japan	University of Tokyo	Ron Shamir	Satoru Miyano
April 22-25, 2001 Montréal, Canada	Université de Montréal	Thomas Lengauer	David Sankoff
April 18-21, 2002 Washington, DC, USA	Celera	Gene Myers	Sridhar Hannenhalli
April 10-13, 2003 Berlin, Germany	German Federal Ministry for Education & Research	Webb Miller	Martin Vingron
March 27-31, 2004 San Diego, USA	UC San Diego	Dan Gusfield	Philip E. Bourne
May 14-18, 2005 Boston, MA, USA	Broad Institute of MIT and Harvard	Satoru Miyano	Jill P. Mesirov and S. Kasif

The RECOMB06 Program Committee gratefully acknowledges the valuable input received from the following external reviewers:

Josep F. Abril	Oranit Dror	Juha Kärkkäinen
Mario Albrecht	Bjarte Dysvik	Hans-Michael
Gabriela Alexe	Nadia El-Mabrouk	Kaltenbach
Julien Allali	Rani Elkon	Simon Kasif
Miguel Andrade	Sean Escola	Klara Kedem
Brian Anton	Eleazar Eskin	Alon Keinan
Sebastian Böcker	Jean-Eudes Duchesne	Wayne Kendal
Vineet Bafna	Jay Faith	Ilona Kifer
Melanie Bahlo	David Fernandez-Baca	Gad Kimmel
Nilanjana Banerjee	Vladimir Filkov	Jyrki Kivinen
Ali Bashir	Sarel Fleishman	Mikko Koivisto
Amir Ben-Dor	Kristian Flikka	Rachel Kolodny
Asa Ben-Hur	Menachem Former	Vered Kunik
Yoav Benjamini	Iddo Friedberg	Vincent Lacroix
Chris Benner	Menachem Fruman	Quan Le
Gyan Bhanot	Irit Gat-Viks	Soo Lee
Trond Hellem Bø	Gad Getz	Celine Lefebvre
Elhanan Borenstein	Apostol Gramada	Hadas Leonov
Guillaume Bourque	Alex Gray	Jie Liang
Frederic Boyer	Steffen Grossmann	Chaim Linhart
Dan Brown	Jenny Gu	Zsuzsanna Lipták
Trevor Bruen	Roderic Guigo	Manway Liu
Renato Bruni	Matthew Hahn	Aniv Loewenstein
David Bryant	Yonit Halperin	Claudio Lottaz
Jeremy Buhler	Tzvika Hartman	Claus Lundegaard
Nello Cristianini	Christoph Hartmann	Hannes Luz
Jo-Lan Chung	Nurit Haspel	Aaron Mackey
Barry Cohen	Greg Hather	Ketil Malde
Inbar Cohen-Gihon	Morihiro Hayashida	Kartik Mani
Matteo Comin	Trond Hellem Bø	Thomas Manke
Ana Teresa Freitas	D. Hermelin	Yishay Mansour
Miklos Csuros	Katsuhisa Horimoto	Adam Margolin
Andre Dabrowski	Moseley Hunter	Florian Markowetz
Alessandro Dal Palú	Seiya Imoto	Setsuro Matsuda
Sanjoy DasGupta	Yuval Inbar	Alice McHardy
Gianluca Della Vedova	Nathan Intrator	Kevin Miranda
Greg Dewey	David Jaffe	Leonid Mirny
Zhihong Ding	Martin Jambon	Stefano Monti
Atsushi Doi	Shane Jensen	Sayan Mukherjee
Dikla Dotan	Euna Jeong	Iftach Nachman
Agostino Dovier	Tao Jiang	Masao Nagasaki

Rei-ichiro Nakamichi
Tim Nattkemper
Ilya Nemenman
Sebastian Oehm
Arlindo Oliveira
Michal Ozery-Flato
Kimmo Palin
Kim Palmo
Paul Pavlidis
Anton Pervukhin
Pierre Peterlongo
Kjell Petersen
Nadia Pisanti
Gianluca Pollastri
Julia Ponomarenko
Elon Portugaly
John Rachlin
Sven Rahmann
Jörg Rahnenführer
Daniela Raijman
Ari Rantanen
Ramamoorthi Ravi
Marc Rehmsmeier
Vicente Reyes
Romeo Rizzi
Estela Maris Rodrigues
Ken Ross
Juho Rousu
Eytan Ruppin
Walter L. Ruzzo
Michael Sammeth
Oliver Sander
Zack Saul

Simone Scalabrin
Klaus-Bernd Schürmann
Michael Schaffer
Stefanie Scheid
Alexander Schliep
Dina Schneidman
Russell Schwartz
Paolo Serafini
Maxim Shatsky
Feng Shengzhong
Tetsuo Shibuya
Ilya Shindyalov
Tomer Shlomi
A. Shulman-Peleg
Abdur Sikder
Gordon Smyth
Yun Song
Rainer Spang
Mike Steel
Israel Steinfeld
Christine Steinhoff
Kristian Stevens
Aravind Subramanian
Fengzhu Sun
Christina Sunita Leslie
Edward Susko
Yoshinori Tamada
Amos Tanay
Haixu Tang
Eric Tannier
Elisabeth Tillier
Wiebke Timm
Aristotelis Tsirigos

Nobuhisa Ueda
Igor Ulitsky
Sandor Vajda
Roy Varshavsky
Balaji Venkatachalam
Stella Veretnik
Dennis Vitkup
Yoshiko Wakabayashi
Jianyong Wang
Junwen Wang
Kai Wang
Li-San Wang
Lusheng Wang
Tandy Warnow
Arieh Warshel
David Wild
Virgil Woods
Terrence Wu
Yufeng Wu
Lei Xie
Chen Xin
Eric Xing
Zohar Yakhini
Nir Yosef
Ryo Yoshida
John Zhang
Louxin Zhang
Degui Zhi
Xianghong J. Zhou
Joseph Ziv Bar
Michal Ziv-Ukelson

RECOMB Tenth Anniversary Venue: il Palazzo Del Cinema del Lido di Venezia

Sponsors

Table of Contents

Integrated Protein Interaction Networks
for 11 Microbes

Balaji S. Srinivasan[1,2], Antal F. Novak[3], Jason A. Flannick[3],
Serafim Batzoglou[3], and Harley H. McAdams[2]

[1] Department of Electrical Engineering
[2] Department of Developmental Biology
[3] Department of Computer Science, Stanford University,
Stanford, CA 94305, USA

Abstract. We have combined four different types of functional genomic data to create high coverage protein interaction networks for 11 microbes. Our integration algorithm naturally handles statistically dependent predictors and automatically corrects for differing noise levels and data corruption in different evidence sources. We find that many of the predictions in each integrated network hinge on moderate but consistent evidence from multiple sources rather than strong evidence from a single source, yielding novel biology which would be missed if a single data source such as coexpression or coinheritance was used in isolation. In addition to statistical analysis, we demonstrate via case study that these subtle interactions can discover new aspects of even well studied functional modules. Our work represents the largest collection of probabilistic protein interaction networks compiled to date, and our methods can be applied to any sequenced organism and any kind of experimental or computational technique which produces pairwise measures of protein interaction.

1 Introduction

Interaction networks are the canonical data sets of the post-genomic era, and more than a dozen methods to detect protein-DNA and protein-protein interactions on a genomic scale have been recently described [1, 2, 3, 4, 5, 6, 7, 8, 9]. As many of these methods require no further experimental data beyond a genome sequence, we now have a situation in which a number of different interaction networks are available for each sequenced organism. However, though many of these interaction predictors have been individually shown to predict experiment[6], the networks generated by each method are often contradictory and not superposable in any obvious way [10, 11]. This seeming paradox has stimulated a burst of recent work on the problem of network integration, work which has primarily focused on *Saccharomyces cerevisiae*[12, 13, 14, 15, 16, 17]. While the profusion of experimental network data [18] in yeast makes this focus understandable, the objective of network integration remains general: namely, a summary network

A. Apostolico et al. (Eds.): RECOMB 2006, LNBI 3909, pp. 1–14, 2006.

for each species which uses all the evidence at hand to predict which proteins are functionally linked.

In the ideal case, an algorithm to generate such a network should be able to:

1. Integrate evidence sets of various types (real valued, ordinal scale, categorical, and so on) and from diverse sources (expression, phylogenetic profiles, chromosomal location, two hybrid, etc.).
2. Incorporate known prior information (such as individually confirmed functional linkages), again of various types.
3. Cope with statistical dependencies in the evidence set (such as multiple repetitions of the same expression time course) and noisy or corrupted evidence.
4. Provide a decomposition which indicates the evidence variables which were most informative in determining a given linkage prediction.
5. Produce a unified probabilistic assessment of linkage confidence given all the observed evidence.

In this paper we present an algorithm for network integration that satisfies all five of these requirements. We have applied this algorithm to integrate four different kinds of evidence (coexpression[3], coinheritance[5], colocation[1], and coevolution[9]) to build probabilistic interaction networks for 11 sequenced microbes. The resulting networks are undirected graphs in which nodes correspond to proteins and edge weights represent interaction probabilities between protein pairs. Protein pairs with high interaction probabilities are not necessarily in direct contact, but are likely to participate in the same functional module [19], such as a metabolic pathway, a signaling network, or a multiprotein complex. We demonstrate the utility of network integration for the working biologist by analyzing representative functional modules from two microbes: the eukaryote-like glycosylation system of *Campylobacter jejuni* NCTC 11168 and the cell division machinery of *Caulobacter crescentus*. For each module, we show that a subset of the interactions predicted by our network recapitulate those described in the literature. Importantly, we find that many of the novel interactions in these modules originate in moderate evidence from multiple sources rather than strong evidence from a single source, representing hidden biology which would be missed if a single data type was used in isolation.

2   Methods

2.1   Algorithm Overview

The purpose of network integration is to systematically combine different types of data to arrive at a statistical summary of which proteins work together within a single organism.

For each of the 11 organisms listed in the Appendix[1] we begin by assembling a training set of known functional modules (Figure 1a) and a battery of different predictors (Figure 1b) of functional association. To gain intuition for what our

---

[1] Viewable at http://jinome.stanford.edu/pdfs/recomb06182_appendix.pdf

algorithm does, consider a single predictor E defined on a pair of proteins, such as the familiar Pearson correlation between expression vectors. Also consider a variable L, likewise defined on pairs of proteins, which takes on three possible values: '1' when two proteins are in the same functional category, '0' when they are known to be in different categories, and '?' when one or both of the proteins is of unknown function.

We note first that two proteins known to be in the same functional module are more likely to exhibit high levels of coexpression than two proteins known to be in different modules, indicated graphically by a right-shift in the distribution of $P(E|L = 1)$ relative to $P(E|L = 0)$ (Figure 1b). We can invert this observation via Bayes' rule to obtain the probability that two proteins are in the same functional module as a function of the coexpression, $P(L = 1|E)$. This posterior probability increases with the level of coexpression, as highly coexpressed pairs are more likely to participate in the same functional module.

If we apply this approach to each candidate predictor in turn, we can obtain valuable information about the extent to which each evidence type recapitulates known functional linkages – or, more precisely, the efficiency with which each predictor *classifies* pairs of proteins into the "linked" or "unlinked" categories. Importantly, benchmarking each predictor in terms of its performance as a binary classifier provides a way to compare previously incomparable data sets, such as matrices[6] of BLAST[20] bit scores and arrays of Cy5/Cy3 ratios[3]. Even more importantly, it suggests that the problem of network integration can be viewed as a high dimensional binary classifier problem. By generalizing the approach outlined above to the case where E is a vector rather than a scalar, we can calculate the summary probability that two proteins are functionally linked given all the evidence at hand.

2.2   Training Set and Evidence Calculation

It is difficult to say *a priori* which predictors of functional association will be the best for a given organism. For example, microarray quality is known to vary widely, so coexpression correlations in different organisms are not directly comparable. Thus, to calibrate our interaction prediction algorithm, we require a training set of known interactions.

To generate this training set, we used one of three different genome scale annotations: the COG functional categories assigned by NCBI[21], the GO[22] annotations assigned by EBI's GOA project[23], and the KEGG[24] metabolic annotations assigned to microbial genomes. In general, as we move from COG to GO to KEGG, the fraction of annotated proteins in a given organism decreases, but the annotation quality increases. In this work we used the KEGG annotation for all organisms other than *Bacillus subtilis*, for which we used GO as KEGG data was unavailable.

As shown in Figure 1a, for each pair we recorded ($L = 1$) if the proteins had overlapping annotations, ($L = 0$) if both were in entirely nonoverlapping categories, and ($L = ?$) if either protein lacked an annotation code or was marked as unknown. (For the GO training set, "overlapping" was defined as overlap

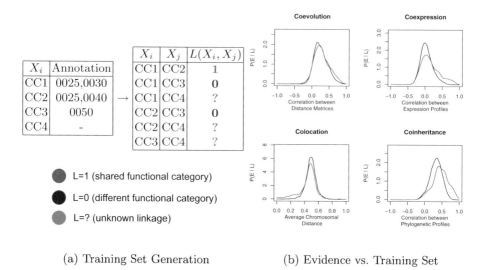

	X_i		X_i	X_j	$L(X_i, X_j)$
X_i	Annotation		CC1	CC2	1
CC1	0025,0030		CC1	CC3	0
CC2	0025,0040	→	CC1	CC4	?
CC3	0050		CC2	CC3	0
CC4	-		CC2	CC4	?
			CC3	CC4	?

● L=1 (shared functional category)

● L=0 (different functional category)

● L=? (unknown linkage)

(a) Training Set Generation   (b) Evidence vs. Training Set

Fig. 1. Training Sets and Evidence. (a) Genome-scale systematic annotations such as COG, GO or KEGG give functions for proteins X_i. As described in the text and shown on example data, we use this annotation to build an initial classification of protein pairs (X_i, X_j) with three categories: a relatively small set of likely linked (red) pairs and unlinked (blue) pairs, and a much larger set of uncertain (gray) pairs. (b) We observe that proteins which share an annotation category generally have more significant levels of evidence, as seen in the shifted distribution of linked (red) vs. unlinked (blue) pairs. Even subtle distributional differences contribute statistical resolution to our algorithm.

of specific GO categories beyond the 8th level of the hierarchy.) This "matrix" approach (consider all proteins within an annotation category as linked) is in contrast to the "hub-spoke" approach (consider only proteins known to be directly in contact as linked) [25]. The former representation produces a nontrivial number of false positives, while the latter incurs a surfeit of false negatives. We chose the "matrix" based training set because our algorithm is robust to noise in the training set so long as enough data is present.

Note that we have used an annotation on individual proteins to produce a training set on *pairs* of proteins. In Figure 1b, we compare this training set to four functional genomic predictors: coexpression, coinheritance, coevolution, and colocation. We include details of the calculations of each evidence type in the Appendix. Interestingly, despite the fact that these methods were obtained from raw measurements as distinct as genomic spacing, BLAST bit scores, phylogenetic trees, and microarray traces, Figure 1b shows that each method is capable of distinguishing functionally linked pairs ($L = 1$) from unlinked pairs ($L = 0$).

2.3 Network Integration

For clarity, we first illustrate network integration with two evidence types (corresponding to two Euclidean dimensions) in *C. crescentus*, and then move to the N-dimensional case.

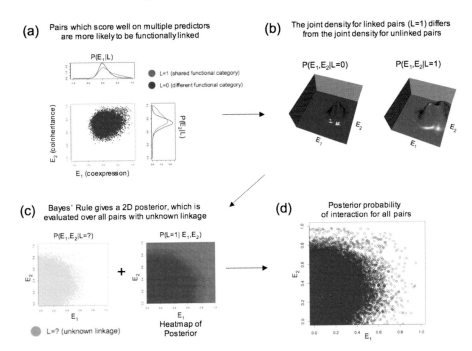

Fig. 2. 2D Network Integration in *C. crescentus*. (a) A scatterplot reveals that functionally linked pairs (red, $L = 1$) tend to have higher coexpression and coinheritance than pairs known to participate in separate pathways (blue, $L = 0$). (b) We build the conditional densities $P(E_1, E_2|L = 0)$ and $P(E_1, E_2|L = 1)$ through kernel density estimation. Note that the distribution for linked pairs is shifted to the upper right corner relative to the unlinked pair distribution. (c) We can visualize the classification process by concentrating on the decision boundary, corresponding to the upper right quadrant of the original plot. In the left panel, the scatterplot of pairs with unknown linkage status (gray) are the inputs for which we wish to calculate interaction probabilities. In the right panel, a heatmap for the posterior probability $P(L = 1|E_1, E_2)$ is depicted. This function yields the probability of linkage given an input evidence vector, and increases as we move to higher levels of coexpression and coinheritance in the upper right corner. (d) By conceptually superimposing each gray point upon the posterior, we can calculate the posterior probability that two proteins are functionally linked.

2D Network Integration. Consider the set of approximately 310000 protein pairs in *C. crescentus* which have a KEGG-defined linkage of $(L = 0)$ or $(L = 1)$. Setting aside the 6.6 million pairs with $(L = ?)$ for now, we find that $P(L = 1) = .046$ and $P(L = 0) = .954$ are the relative proportions of known linked and unlinked pairs in our training set.

Each of these pairs has an associated coexpression and coinheritance correlation, possibly with missing values, which we bundle into a two dimensional vector $E = (E_1, E_2)$. Figure 2a shows a scatterplot of E_1 vs. E_2, where pairs with $(L = 1)$ have been marked red and pairs with $(L = 0)$ have been marked blue.

We see immediately that functionally linked pairs aggregate in the upper right corner of the plot, in the region of high coexpression and coinheritance.

Crucially, the linked pairs (red) are more easily distinguished from the unlinked pairs (blue) in the 2-dimensional scatter plot than they are in the accompanying 1-dimensional marginals. To quantify the extent to which this is true, we begin by computing $P(E_1, E_2|L = 0)$ and $P(E_1, E_2|L = 1)$ via kernel density estimation[26, 27], as shown in Figure 2b. As we already know $P(L)$, we can obtain the posterior by Bayes' rule:

$$P(L = 1|E_1, E_2) = \frac{P(E_1, E_2|L = 1)P(L = 1)}{P(E_1, E_2|L = 1)P(L = 1) + P(E_1, E_2|L = 0)P(L = 0)}$$

In practice, this expression is quite sensitive to fluctuations in the denominator. To deal with this, we use M-fold bootstrap aggregation[28] to smooth the posterior. We find that $M = 20$ repetitions with resampling of 1000 elements from the $(L = 0)$ and $(L = 1)$ training sets is the empirical point of diminishing returns in terms of area under the receiver-operator characteric (ROC), as detailed in Figure 4.

$$P(L = 1|E_1, E_2) = \frac{1}{M} \sum_{i=1}^{M} \frac{P_i(E_1, E_2|L = 1)P(L = 1)}{P_i(E_1, E_2|L = 1)P(L = 1) + P_i(E_1, E_2|L = 0)P(L = 0)}$$

Given this posterior, we can now make use of the roughly 6.6 million pairs with $(L = ?)$ which we put aside at the outset, as pictured in Figure 2c. Even though these pairs have unknown linkage, for most pairs the coexpression (E_1) and coinheritance (E_2) are known. For those pairs which have partially missing data (e.g. from corrupted spots on a microarray), we can simply evaluate over the non-missing elements of the E vector by using the appropriate marginal posterior $P(L = 1|E_1)$ or $P(L = 1|E_2)$. We can thus calculate $P(L = 1|E_1, E_2)$ for every pair of proteins in the proteome, as shown in Figure 2d. Each of the formerly gray pairs with $(L = ?)$ is assigned a probability of interaction by this function; those with bright red values in Figure 2d are highly likely to be functionally linked.

In general, we also calculate $P(L = 1|E_1, E_2)$ on the training data, as we know that the "matrix" approach to training set generation produces copious but noisy data. The result of this evaluation is the probability of interaction for every protein pair.

N-dimensional Network Integration. The 2 dimensional example in *C. crescentus* immediately generalizes to N-dimensional network integration in an arbitrary species, though the results cannot be easily visualized beyond 3 dimensions. Figure 3 shows the results of calculating a 3D posterior in *C. crescentus* from coexpression, coinheritance, and colocation data, where we have once again applied M-fold bootstrap aggregation.

We see that different evidence types interact in nonobvious ways. For example, we note that high levels of colocation (E_2) can compensate for low levels of coexpression (E_1), as indicated by the "bump" in the posterior of Figure 3c. Biologically speaking, this means that a nontrivial number of *C. crescentus* proteins with shared function are frequently colocated yet not strongly coexpressed. This is exactly the sort of subtle statistical dependence between predictors that is crucial for proper classification. In fact, a theoretically attractive property of

$P(E_1,E_2,E_3|L=0)$ $P(E_1,E_2,E_3|L=1)$ $P(L=1|E_1,E_2,E_3)$

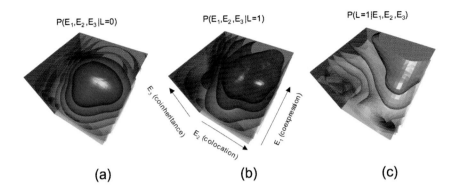

(a)   (b)   (c)

Fig. 3. 3D Network Integration in *C. crescentus*. (a)-(b) We show level sets of each density spaced at even volumetric increments, so that the inner most shell encloses 20% of the volume, the second shell encloses 40%, and so forth. As in the 2D case, the 3D density $P(E|L = 1)$ is shifted to the upper right corner. (c) For the posterior, we show level sets spaced at probability deciles, such that a pair which makes it past the upper right shell has $P(L = 1|E) \in [.9, 1]$, a pair which lands in between the upper two shells satisfies $P(L = 1|E) \in [.8, .9]$, and so on.

our approach is that the use of the conditional joint posterior produces the minimum possible classification error (specifically, the Bayes error rate [29]), while bootstrap aggregation protects us against overfitting[30].

Until recently, though, technical obstacles made it challenging to efficiently compute joint densities beyond dimension 3. Recent developments[26] in efficient kernel density estimation have obviated this difficulty and have made it possible to evaluate high dimensional densities over millions of points in a reasonable amount of time within user-specifiable tolerance levels. As an example of the calculation necessary for network integration, consider a 4 dimensional kernel density estimate built from 1000 sample points. Ihler's implementation[27] of the Gray-Moore dual-tree algorithm[26] allowed the evaluation of this density at the $\binom{3737}{2} \approx 7,000,000$ pairs in the *C. crescentus* proteome in only 21 minutes on a 3GHz Xeon with 2GB RAM. Even after accounting for the $2M$ multiple of this running time caused by evaluating a quotient of two densities and using M-fold bootstrap aggregation, the resulting joint conditional posterior can be built and evaluated rapidly enough to render approximation unnecessary.

Binary Classifier Perspective. By formulating the network integration problem as a binary classifier (Figure 4), we can quantify the extent to which the integration of multiple evidence sources improves prediction accuracy over a single source. As our training data is necessarily a rough approximation of the true interaction network, these measures are likely to be conservative estimates of classifier performance.

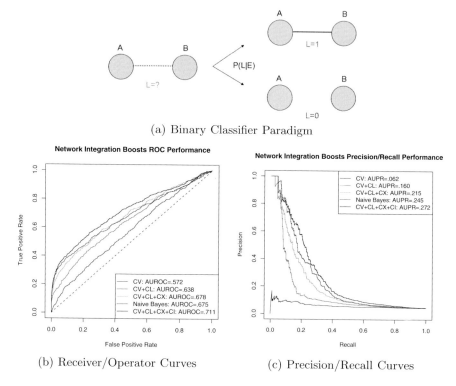

(a) Binary Classifier Paradigm

(b) Receiver/Operator Curves   (c) Precision/Recall Curves

Fig. 4. Network Integration as Binary Classifier. (a) We regard the network integration problem as a binary classifier in a high dimensional feature space. The input features are a set of evidences associated with a protein pair (A, B), and the output is the probability that a pair is assigned to the $(L = 1)$ category. (b) The area under the receiver operator characteristic (AUROC) is a standard measure[29] of binary classifier performance, shown here for several different ways of doing *C. crescentus* network integration. Here we have labeled data types as CV (coevolution), CL (colocation), CX (coexpression), and CI (coinheritance) and shown a successive series of curves for the integration of 1,2,3, and finally 4 evidence types. Classifier performance increases monotonically as more data sets are combined. Importantly, the true four dimensional joint posterior $P(L = 1|CV, CL, CX, CI)$ outperforms the Naive Bayes approximation of the posterior, where the conditional density $P(CV, CL, CX, CI|L = 1)$ is approximated by $P(CV|L = 1)P(CL|L = 1)P(CX|L = 1)P(CI|L = 1)$, and similarly for $L = 0$. For clarity we have omitted the individual curves for the CL (AUROC=.612), CX (AUROC=.619), and CV (AUROC=.653) metrics. Again, it is clear that the integrated posterior outperforms each of these univariate predictors. (c) Precision/recall curves are an alternate way of visualizing classifier performance, and are useful when the number of true positives is scarce relative to the number of false negatives. Again the integrated posterior outperforms the Naive Bayes approximation as a classifier. Note that since the "negative" pairs from the KEGG training set are based on the supposition that two proteins which have no annotational overlap genuinely do *not* share a pathway, they are a more noisy indicator than the "positive" pairs. That is, with respect to functional interaction, absence of evidence is not always evidence of absence. Hence the computed values for precision are likely to be conservative underestimates of the true values.

3   Results

3.1   Global Network Architecture

Applying the posterior $P(L = 1|E)$ to every pair of proteins in a genome gives the probability that each pair is functionally linked. If we simply threshold this result at $P(L = 1|E) > .5$, we will retain only those linkages which are more probable than not. This decision rule attains the Bayes error rate[29] and minimizes the misclassification probability. We applied our algorithm with this threshold to build 4D integrated networks for the 11 microbes and four evidence types listed in the Appendix. Figure 5 shows the global protein interaction networks produced for three of these microbes, where we have retained only those edges with $P(L = 1|E) > .5$.

To facilitate use of these protein interaction networks, we built an interactive netbrowser, viewable at http://jinome.stanford.edu/netbrowser. As a threshold of $P(L = 1|E) > .5$ tends to be somewhat stringent in practice, we allow dynamic, user-specified thresholds to produce module-specific tradeoffs between specificity and sensitivity in addition to a host of other customization options.

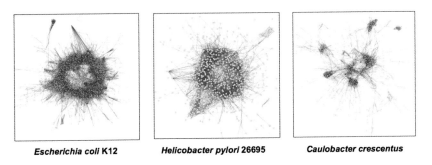

Escherichia coli K12   **Helicobacter pylori 26695**   **Caulobacter crescentus**

Fig. 5. Global visualization of integrated networks for *Escherichia coli* K12, *Helicobacter pylori* 26695, and *Caulobacter crescentus*. Only linkages with $P(L = 1|E_1, E_2, E_3, E_4) > .5$ are displayed.

3.2   *Campylobacter jejuni*: N-Linked Protein Glycosylation

N-linked protein glycosylation is one of the most frequent post-translational modifications applied to eukaryotic secretory proteins. Until recently[31] this process was thought to be absent from most microbes, but recent work[32] has shown that an operational N-linked glycosylation system does exist in *C. jejuni*. As the entire glycosylation apparatus can be successfully transplanted to *E. coli* K12, this system is of much biotechnological interest[33].

Figure 6a shows the results of examining the integrated network for *C. jejuni* around the vicinity of Cj1124c, one of the proteins in the glycosylation system. In addition to the reassuring recapitulation of several transferases and epimerases experimentally linked to this process[33], we note four proteins which are to our knowledge not known to be implicated in N-linked glycosylation (Cj1518,

Cj0881c, Cj0156c, Cj0128c). Importantly, all of these heretofore uncharacterized linkages would have been missed if only univariate posteriors had been examined, as they would be significantly below our cutoff of $P(L = 1|E) > .5$. As this system is still poorly understood – yet of substantial biotechnological and pathogenic[34] relevance – investigation of these new proteins may be of interest.

3.3   *Caulobacter crescentus*: Bacterial Actin and the Sec Apparatus

Van den Ent's[36] discovery that the ubiquitous microbial protein MreB was a structural homolog to actin spurred a burst of interest[37, 38, 39] in the biology of the bacterial cytoskeleton. Perhaps the most visually arresting of these recent findings is the revelation that MreB supports the cell by forming a tight spiral[37]. Yet many outstanding questions in this field remain, and prime among them is the issue of which proteins communicate with the bacterial cytoskeletal apparatus[40].

Figure 6b shows the proteins from the *C. crescentus* integrated network which have a 50% chance or greater of interacting with MreB, also known as CC1543. As a baseline measure of validity, we once again observe that known interaction partners such as RodA (CC1547) and MreC (CC1544) are recovered by network integration. More interesting, however, is the subtle interaction between MreB and the preprotein translocase CC3206, an interaction that would be missed if data sources were used separately. This protein is a subunit of the Sec machinery,

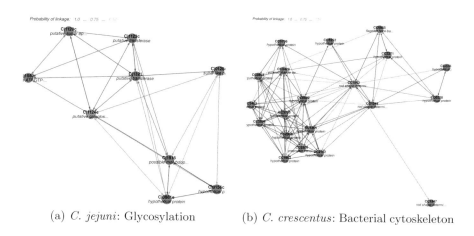

(a) *C. jejuni*: Glycosylation   (b) *C. crescentus*: Bacterial cytoskeleton

Fig. 6. Case Studies. (a) Network integration detects new proteins linked to glycosylation in *Campylobacter jejuni* NCTC 11168. High probability linkages are labeled in red and generally recapitulate known interactions, while moderately likely linkages are colored gray. Moderate linkages are generally not found by any univariate method in isolation, and represent the new biological insight produced by data integration. (b) In *Caulobacter crescentus*, data integration reveals that the Sec apparatus is linked to MreB, a prediction recently confirmed by experiment[35]. Again, moderate linkages revealed by data integration lead us to a conclusion that would be missed if univariate data was used.

and like MreB is an ancient component of the bacterial cell[41]. Its link to MreB is of particular note because recent findings[35] have shown that the Sec apparatus – like MreB – has a spiral localization pattern. While seemingly counterintuitive, it seems likely from both this finding and other work[42] that the export of cytoskeleton-related proteins beyond the cellular membrane is important in the process of cell division. We believe that investigation of the hypothetical proteins linked to both MreB and Sec by our algorithm may shed light on this question.

4   Discussion

4.1   Merits of Our Approach

While a number of recent papers on network integration in *S. cerevisiae* have appeared, we believe that our method is an improvement over existing algorithms.

First, by directly calculating the joint conditional posterior we require no simplifying assumptions about statistical dependence and need no complex parametric inference. In particular, removing the Naive Bayes approximation results in a better classifier, as quantified in Figure 4. Second, our use of the Gray-Moore dual tree algorithm means that our method is arbitrarily scalable in terms of both the number of evidence types and the number of protein pairs. Third, our method allows immediate visual identification of dependent or corrupted functional genomic data in terms of red/blue separation scatterplots – an important consideration given the noise of some data types [43]. Finally, because the output of our algorithm is a rigorously derived set of interaction probabilities, it represents a solid foundation for future work.

4.2   Conclusion and Future Directions

Our general framework presents much room for future development. It is straightforward to generalize our algorithm to apply to discrete, ordinal, or categorical data sets as long as appropriate similarity measures are defined. As our method readily scales beyond a few thousand proteins, even the largest eukaryotic genomes are potential application domains. It may also be possible to improve our inference algorithm through the use of statistical techniques designed to deal with missing data[44].

Moving beyond a binary classifier would allow us to predict different kinds of functional linkage, as two proteins in the same multiprotein complex have a different kind of linkage than two proteins which are members of the same regulon. This would be significant in that it addresses one of the most widely voiced criticisms of functional genomics, which is that linkage predictions are "one-size-fits-all". It may also be useful to move beyond symmetric pairwise measures of association to use metrics defined on protein triplets[8] or asymmetric metrics such that $E(P_i, P_j) \neq E(P_j, P_i)$.

While these details of the network construction process are doubtless subjects for future research, perhaps the most interesting prospect raised by the availability of a large number of robust, integrated interaction networks is the possibility of comparative modular biology. Specifically, we would like to *align* subgraphs of interaction networks on the basis of conserved interaction as well as conserved sequence, just as we align DNA and protein sequences. A need now exists for a network alignment algorithm capable of scaling to large datasets and comparing many species simultaneously.

Acknowledgments

We thank Lucy Shapiro, Roy Welch, and Arend Sidow for helpful discussions. BSS was supported in part by a DoD/NDSEG graduate fellowship, and HHM and BSS were supported by NIH grant 1 R24 GM073011-01 and DOE Office of Science grant DE-FG02-01ER63219. JAF was supported in part by a Stanford Graduate Fellowship, and SB, AFN, and JAF were funded by NSF grant EF-0312459, NIH grant UO1-HG003162, the NSF CAREER Award, and the Alfred P. Sloan Fellowship.

Authors' Contributions

BSS developed the network integration algorithm and wrote the paper. AFN designed the web interface with JAF under the direction of SB and provided useful feedback on network quality. HHM and SB provided helpful comments and a nurturing environment.

References

1. Overbeek, R., Fonstein, M., D'Souza, M., Pusch, G.D., Maltsev, N.: The use of gene clusters to infer functional coupling. Proc Natl Acad Sci U S A **96** (1999) 2896–2901
2. McAdams, H.H., Srinivasan, B., Arkin, A.P.: The evolution of genetic regulatory systems in bacteria. Nat Rev Genet **5** (2004) 169–178
3. Schena, M., Shalon, D., Davis, R.W., Brown, P.O.: Quantitative monitoring of gene expression patterns with a complementary DNA microarray. Science **270** (1995) 467–470
4. Enright, A.J., Iliopoulos, I., Kyrpides, N.C., Ouzounis, C.A.: Protein interaction maps for complete genomes based on gene fusion events. Nature **402** (1999) 86–90
5. Pellegrini, M., Marcotte, E.M., Thompson, M.J., Eisenberg, D., Yeates, T.O.: Assigning protein functions by comparative genome analysis: protein phylogenetic profiles. Proc Natl Acad Sci U S A **96** (1999) 4285–4288
6. Srinivasan, B.S., Caberoy, N.B., Suen, G., Taylor, R.G., Shah, R., Tengra, F., Goldman, B.S., Garza, A.G., Welch, R.D.: Functional genome annotation through phylogenomic mapping. Nat Biotechnol **23** (2005) 691–698
7. Yu, H., Luscombe, N.M., Lu, H.X., Zhu, X., Xia, Y., Han, J.D.J., Bertin, N., Chung, S., Vidal, M., Gerstein, M.: Annotation transfer between genomes: protein-protein interologs and protein-DNA regulogs. Genome Res **14** (2004) 1107–1118
8. Bowers, P.M., Cokus, S.J., Eisenberg, D., Yeates, T.O.: Use of logic relationships to decipher protein network organization. Science **306** (2004) 2246–2249

9. Pazos, F., Valencia, A.: Similarity of phylogenetic trees as indicator of protein-protein interaction. Protein Eng **14** (2001) 609–614 Evaluation Studies.

10. Gerstein, M., Lan, N., Jansen, R.: Proteomics. Integrating interactomes. Science **295** (2002) 284–287 Comment.

11. Hoffmann, R., Valencia, A.: Protein interaction: same network, different hubs. Trends Genet **19** (2003) 681–683

12. Jansen, R., Yu, H., Greenbaum, D., Kluger, Y., Krogan, N.J., Chung, S., Emili, A., Snyder, M., Greenblatt, J.F., Gerstein, M.: A Bayesian networks approach for predicting protein-protein interactions from genomic data. Science **302** (2003) 449–453 Evaluation Studies.

13. Troyanskaya, O.G., Dolinski, K., Owen, A.B., Altman, R.B., Botstein, D.: A Bayesian framework for combining heterogeneous data sources for gene function prediction (in Saccharomyces cerevisiae). Proc Natl Acad Sci U S A **100** (2003) 8348–8353

14. Lee, I., Date, S.V., Adai, A.T., Marcotte, E.M.: A probabilistic functional network of yeast genes. Science **306** (2004) 1555–1558

15. Tanay, A., Sharan, R., Kupiec, M., Shamir, R.: Revealing modularity and organization in the yeast molecular network by integrated analysis of highly heterogeneous genomewide data. Proc Natl Acad Sci U S A **101** (2004) 2981–2986

16. Wong, S.L., Zhang, L.V., Tong, A.H.Y., Li, Z., Goldberg, D.S., King, O.D., Lesage, G., Vidal, M., Andrews, B., Bussey, H., Boone, C., Roth, F.P.: Combining biological networks to predict genetic interactions. Proc Natl Acad Sci U S A **101** (2004) 15682–15687

17. Lu, L.J., Xia, Y., Paccanaro, A., Yu, H., Gerstein, M.: Assessing the limits of genomic data integration for predicting protein networks. Genome Res **15** (2005) 945–953

18. Friedman, A., Perrimon, N.: Genome-wide high-throughput screens in functional genomics. Curr Opin Genet Dev **14** (2004) 470–476

19. Hartwell, L.H., Hopfield, J.J., Leibler, S., Murray, A.W.: From molecular to modular cell biology. Nature **402** (1999) 47–52

20. Schaffer, A.A., Aravind, L., Madden, T.L., Shavirin, S., Spouge, J.L., Wolf, Y.I., Koonin, E.V., Altschul, S.F.: Improving the accuracy of PSI-BLAST protein database searches with composition-based statistics and other refinements. Nucleic Acids Res **29** (2001) 2994–3005

21. Tatusov, R.L., Fedorova, N.D., Jackson, J.D., Jacobs, A.R., Kiryutin, B., Koonin, E.V., Krylov, D.M., Mazumder, R., Mekhedov, S.L., Nikolskaya, A.N., Rao, B.S., Smirnov, S., Sverdlov, A.V., Vasudevan, S., Wolf, Y.I., Yin, J.J., Natale, D.A.: The COG database: an updated version includes eukaryotes. BMC Bioinformatics **4** (2003) 41

22. Ashburner, M., Ball, C.A., Blake, J.A., Botstein, D., Butler, H., Cherry, J.M., Davis, A.P., Dolinski, K., Dwight, S.S., Eppig, J.T., Harris, M.A., Hill, D.P., Issel-Tarver, L., Kasarskis, A., Lewis, S., Matese, J.C., Richardson, J.E., Ringwald, M., Rubin, G.M., Sherlock, G.: Gene ontology: tool for the unification of biology. The Gene Ontology Consortium. Nat Genet **25** (2000) 25–29

23. Camon, E., Magrane, M., Barrell, D., Lee, V., Dimmer, E., Maslen, J., Binns, D., Harte, N., Lopez, R., Apweiler, R.: The Gene Ontology Annotation (GOA) Database: sharing knowledge in Uniprot with Gene Ontology. Nucleic Acids Res **32** (2004) 262–266

24. Kanehisa, M., Goto, S., Kawashima, S., Okuno, Y., Hattori, M.: The KEGG resource for deciphering the genome. Nucleic Acids Res **32** (2004) 277–280

25. Bader, G.D., Hogue, C.W.V.: Analyzing yeast protein-protein interaction data obtained from different sources. Nat Biotechnol **20** (2002) 991–997
26. Gray, A.G., Moore, A.W.: 'n-body' problems in statistical learning. In: NIPS. (2000) 521–527
27. Ihler, A., Sudderth, E., Freeman, W., Willsky, A.: Efficient multiscale sampling from products of gaussian mixtures. In: NIPS. (2003)
28. Breiman, L.: Bagging predictors. Machine Learning **24** (1996) 123–140
29. Duda, R., Hart, P., Stork, D.: Pattern Classification. Wiley-Interscience Publication (2000)
30. Bauer, E., Kohavi, R.: An empirical comparison of voting classification algorithms: Bagging, boosting, and variants. Machine Learning **36** (1999) 105–139
31. Szymanski, C.M., Logan, S.M., Linton, D., Wren, B.W.: Campylobacter–a tale of two protein glycosylation systems. Trends Microbiol **11** (2003) 233–238
32. Wacker, M., Linton, D., Hitchen, P.G., Nita-Lazar, M., Haslam, S.M., North, S.J., Panico, M., Morris, H.R., Dell, A., Wren, B.W., Aebi, M.: N-linked glycosylation in Campylobacter jejuni and its functional transfer into E. coli. Science **298** (2002) 1790–1793
33. Linton, D., Dorrell, N., Hitchen, P.G., Amber, S., Karlyshev, A.V., Morris, H.R., Dell, A., Valvano, M.A., Aebi, M., Wren, B.W.: Functional analysis of the Campylobacter jejuni N-linked protein glycosylation pathway. Mol Microbiol **55** (2005) 1695–1703
34. Karlyshev, A.V., Everest, P., Linton, D., Cawthraw, S., Newell, D.G., Wren, B.W.: The Campylobacter jejuni general glycosylation system is important for attachment to human epithelial cells and in the colonization of chicks. Microbiology **150** (2004) 1957–1964
35. Campo, N., Tjalsma, H., Buist, G., Stepniak, D., Meijer, M., Veenhuis, M., Westermann, M., Muller, J.P., Bron, S., Kok, J., Kuipers, O.P., Jongbloed, J.D.H.: Subcellular sites for bacterial protein export. Mol Microbiol **53** (2004) 1583–1599
36. van den Ent, F., Amos, L.A., Lowe, J.: Prokaryotic origin of the actin cytoskeleton. Nature **413** (2001) 39–44
37. Gitai, Z., Dye, N., Shapiro, L.: An actin-like gene can determine cell polarity in bacteria. Proc Natl Acad Sci U S A **101** (2004) 8643–8648
38. Kurner, J., Frangakis, A.S., Baumeister, W.: Cryo-electron tomography reveals the cytoskeletal structure of Spiroplasma melliferum. Science **307** (2005) 436–438
39. Gerdes, K., Moller-Jensen, J., Ebersbach, G., Kruse, T., Nordstrom, K.: Bacterial mitotic machineries. Cell **116** (2004) 359–366
40. Cabeen, M.T., Jacobs-Wagner, C.: Bacterial cell shape. Nat Rev Microbiol **3** (2005) 601–610
41. Vrontou, E., Economou, A.: Structure and function of SecA, the preprotein translocase nanomotor. Biochim Biophys Acta **1694** (2004) 67–80
42. Kruse, T., Bork-Jensen, J., Gerdes, K.: The morphogenetic MreBCD proteins of Escherichia coli form an essential membrane-bound complex. Mol Microbiol **55** (2005) 78–89
43. Vidalain, P.O., Boxem, M., Ge, H., Li, S., Vidal, M.: Increasing specificity in high-throughput yeast two-hybrid experiments. Methods **32** (2004) 363–370
44. McLachlan, G., Krishnan, T.: The EM Algorithm and Extensions. John Wiley and Sons (1996)

Hypergraph Model of Multi-residue Interactions in Proteins: Sequentially–Constrained Partitioning Algorithms for Optimization of Site-Directed Protein Recombination

Xiaoduan Ye[1], Alan M. Friedman[2], and Chris Bailey-Kellogg[1]

[1] Department of Computer Science, Dartmouth College,
6211 Sudikoff Laboratory, Hanover NH 03755, USA
{ye, cbk}@cs.dartmouth.edu
[2] Department of Biological Sciences and Purdue Cancer Center,
Purdue University, West Lafayette, IN 47907, USA
afried@purdue.edu

Abstract. Relationships among amino acids determine stability and function and are also constrained by evolutionary history. We develop a probabilistic hypergraph model of residue relationships that generalizes traditional pairwise contact potentials to account for the statistics of multi-residue interactions. Using this model, we detected non-random associations in protein families and in the protein database. We also use this model in optimizing site-directed recombination experiments to preserve significant interactions and thereby increase the frequency of generating useful recombinants. We formulate the optimization as a sequentially-constrained hypergraph partitioning problem; the quality of recombinant libraries wrt a set of breakpoints is characterized by the total perturbation to edge weights. We prove this problem to be NP-hard in general, but develop exact and heuristic polynomial-time algorithms for a number of important cases. Application to the beta-lactamase family demonstrates the utility of our algorithms in planning site-directed recombination.

1 Introduction

The non-random association of amino acids, as expressed in pairwise potentials, has been usefully applied in a number of situations. Such pairwise contact potentials [1, 2] play a large role in evaluating quality of models in protein structure prediction [3, 4, 5, 6]. It has been suggested, however, that "it is unlikely that purely pairwise potentials are sufficient for structure prediction" [7, 8].

To better model evolutionary relationships that determine protein stability and functionality, it may be necessary to capture the higher-order interactions that are ignored in simple pairwise models (Fig. 1(a)). Researchers have begun to demonstrate the importance of accounting for higher-order terms. A statistical pseudo-potential based on four-body nearest neighbor interactions (as

A. Apostolico et al. (Eds.): RECOMB 2006, LNBI 3909, pp. 15–29, 2006.

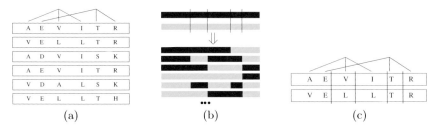

A	E	V	I	T	R
V	E	L	L	T	R
A	D	V	I	S	K
A	E	V	I	T	R
V	D	A	L	S	K
V	E	L	L	T	H

(a)   (b)   (c)

Fig. 1. Hypergraph model of evolutionary interactions, and effects of site-directed protein recombination. (a) Higher-order evolutionary interactions (here, order-3) determining protein stability and function are observed in the statistics of "hyperconservation" of mutually interacting positions. The left edge is dominated by Ala,Val,Ile and Val,Leu,Leu interactions, while the right is dominated by Glu,Thr,Arg and Asp,Ser,Lys ones. The interactions are modeled as edges in a hypergraph with weights evaluating the degree of hyperconservation of an interaction, both generally in the protein database and specific to a particular family. (b) Site-directed recombination mixes and matches sequential fragments of homologous parents to construct a library of hybrids with the same basic structure but somewhat different sequences and thus different functions. (c) Site-directed recombination perturbs edges that cross one or more breakpoints. The difference in edge weights derived for the parents and those derived for the hybrids indicates the effect of the perturbation on maintenance of evolutionarily favorable interactions.

determined by Delaunay tessellations) has successfully predicted changes in free energy caused by hydrophobic core mutations [8]. Similar formulations have been used to discriminate native from non-native protein conformations [9]. Geometrically less restricted higher-order interactions have also been utilized for recognition of native-like protein structures [10]. Recent work on correlated mutation analysis has moved from identifying pairwise correlations [11] to determining clusters or cliques of mutually-dependent residues that identify subclasses within a protein family and provide mechanistic insights into function [12, 13].

This paper develops a rigorous basis for representing multi-order interactions within a protein family. We generalize the traditional representations of sequence information in terms of single-position conservation and structural interactions in terms of pairwise contacts. Instead, we define a hypergraph model in which edges represent pairwise and higher-order residue interactions, while edge weights represent the degree of "hyperconservation" of the interacting residues (Sec. 2). Hyperconservation can reveal significant residue interactions both within members of the family (arising from structural and functional constraints) and generally common to all proteins (arising from general properties of the amino acids). We then combine family-specific and database-wide statistics with suitable weighting (Sec. 2.1), ensure non-redundancy of the information in super- and sub-edges with a multi-order potential score (Sec. 2.2), and derive edge weights by mean potential scores (Sec. 2.3). Application of our approach to beta-lactamases (Sec. 4) shows that the effect of non-redundant higher-order terms is significant and can be effectively handled by our model.

Protein recombination *in vitro* (Fig. 1(b)) enables the design of protein variants with favorable properties and novel enzymatic activities, as well as the exploration of protein sequence-structure-function relationships (see *e.g.* [14, 15, 16, 17, 18, 19, 20, 21, 22]). In this approach, libraries of hybrid proteins are generated either by stochastic enzymatic reactions or intentional selection of breakpoints. Hybrids with unusual properties can either be identified by large-scale genetic screening and selection, or many hybrids can be evaluated individually to determine detailed sequence-function relationships for understanding and/or rational engineering. We focus here on site-directed recombination, in which parent genes are recombined at specified breakpoint locations, yielding hybrids in which different sequence fragments (between the breakpoints) can come from different parents. Both screening/selection and investigational experiments benefit from recombination that preserves the most essential structural and functional features while still allowing variation. In order to enhance the success of this approach, it is necessary to choose breakpoint locations that optimize preservation of these features.

The labs of Mayo and Arnold [18, 23] have established criteria for non-disruption of contacting residue pairs and demonstrated the relationship between non-disruption and functional hybrids [18]. There is an on-going search for algorithms to select breakpoints for recombination based on non-disruption [23, 24], although none has yet been experimentally validated. Optimizing multi-order interactions after recombination (Fig. 1(c)) should help identify the best recombinants and thus the best locations for breakpoints. In support of this optimization, we develop criteria to evaluate the quality of hybrid libraries by considering the effects of recombination on edge weights (Sec. 2.4). We then formulate the optimal selection of breakpoint locations as a sequentially-constrained hypergraph partitioning problem (Sec. 3), prove it to be NP-hard in general (Sec. 3.1), develop exact and heuristic algorithms for a number of important cases (Secs. 3.2–3.5), and demonstrate their practical effectiveness in design of recombination experiments for members of the beta-lactamase family (Sec. 4).

2   A Hypergraph Model of Evolutionary Interactions

In order to more completely model statistical interactions in a protein, it is necessary to move beyond single-position sequence conservation and pairwise structural contact. We model a protein and its reference structure with a weighted hypergraph $G = (V, E, w)$, where vertices $V = \{v_1, v_2, \cdots, v_{|V|}\}$ represent residue positions in sequential order on the backbone, edges $E \subseteq 2^V$ represent mutually interacting sets of vertices, and weight function $w : E \rightarrow \mathbb{R}$ represents the relative significance of edges. We construct an order-c edge $e = \langle v_1, v_2, \cdots, v_c \rangle$ for each set of residues (listed in sequential order for convenience) that are in mutual contact; this construction can readily be extended to capture other forms of interaction, *e.g.* long-range interaction of non-contacting residues due to electrostatics. Note that subsets of vertices associated with a higher-order edge form lower-order edges. When we need to specify the exact order c of edges in a hypergraph, we use notation $G_c = (V, E_c, w)$. Since lower-order edges can be regarded as a special kind of higher-order ones, G_c includes "virtual" lower-order edges.

The definition of the edge weight is key to effective use of the hypergraph model. In the case where the protein is a member of a family with presumed similar structures, edge weights can be evaluated both from the general database and specific to the family. There are many observed residue values (across the family or database) for the vertices of any given edge. We thus build up to an edge weight by first estimating the probability of the residue values, then decomposing the probability to ensure non-redundant information among multi-order edges for the same positions. Finally we determine the effect on the pattern of these values due to recombination according to a set of chosen breakpoint locations.

2.1   Distribution of Hyperresidues in Database and Family

Let $R = \langle r_1, r_2, \cdots, r_c \rangle$ be a "hyperresidue," a c-tuple of amino acid types (e.g. $\langle \mathrm{Ala}, \mathrm{Val}, \mathrm{Ile} \rangle$). Intuitively speaking, the more frequently a particular hyperresidue occurs in functional proteins, the more important it is expected to be for their folding and function. We can estimate the overall probability p of hyperresidues from their frequencies in the database $\mathcal{D}$ of protein sequences and corresponding structures:

$$p(R) = (\#R \text{ in } \mathcal{D}) / |\mathcal{D}| , \tag{1}$$

where $|\mathcal{D}|$ represents the number of tuple instances in the database. When considering a specific protein family $\mathcal{F}$ with a multiple sequence alignment and shared structure, we can estimate position-specific (i.e., for edge e) probability of a hyperresidue:

$$p_e(R) = (\#R \text{ at } e \text{ in } \mathcal{F}) / |\mathcal{F}| , \tag{2}$$

where $|\mathcal{F}|$ is the number of tuple instances at specific positions in the family MSA, i.e. the number of sequences in the family MSA.

Estimation of probabilities from frequencies is valid only if the frequencies are large. Thus the general probability estimated from the whole database (Eq. 1) is more robust than the position-specific from a single family (Eq. 2). However, family-specific information is more valuable as it captures the evolutionarily-preserved interactions in that family. To combine these two aspects, we adopt the treatment of sparse data sets proposed by Sippl [25]:

$$q_e(R) = \omega_1 \cdot p(R) + \omega_2 \cdot p_e(R) , \tag{3}$$

but employing weights suitable for our problem:

$$\omega_1 = 1/(1 + |\mathcal{F}|\rho) \quad \text{and} \quad \omega_2 = 1 - \omega_1 , \tag{4}$$

where ρ is a user-specified parameter that determines the relative contributions of database and family. Note that when $\rho = 0$, $q_e(R) = p(R)$ and the family-specific information is ignored; whereas when $\rho = \infty$, $q_e(R) = p_e(R)$ and the database information is ignored. Using a suitable value of ρ, we will obtain a probability distribution that is close to the overall database distribution for a small family but approximates the family distribution for a large one.

2.2   Multi-order Potential Score for Hyperresidues

Since we have multi-order edges, with lower-order subsets included alongside their higher-order supersets, we must ensure that these edges are not redundant. In other words, a higher-order edge should only include information not captured by its lower-order constituents. The inclusion-exclusion principle ensures non-redundancy in a probability expansion, as Simons *et al.* [10] demonstrated in the case of protein structure prediction. We define an analogous multi-order potential score for hyperresidues at edges of orders 1, 2, and 3, respectively, as follows:

$$\phi_{v_i}(r_\alpha) = \log q_{v_i}(r_\alpha) , \tag{5}$$

$$\phi_{v_i v_j}(r_\alpha r_\beta) = \log \frac{q_{v_i v_j}(r_\alpha r_\beta)}{q_{v_i}(r_\alpha) \cdot q_{v_j}(r_\beta)} , \tag{6}$$

$$\phi_{v_i v_j v_k}(r_\alpha r_\beta r_\gamma) = \log \frac{q_{v_i v_j v_k}(r_\alpha r_\beta r_\gamma) \cdot q_{v_i}(r_\alpha) \cdot q_{v_j}(r_\beta) \cdot q_{v_k}(r_\gamma)}{q_{v_i v_j}(r_\alpha r_\beta) \cdot q_{v_i v_k}(r_\alpha r_\gamma) \cdot q_{v_j v_k}(r_\beta r_\gamma)} . \tag{7}$$

Here, $\phi_{v_i}(r_\alpha)$ captures residue conservation at v_i; $\phi_{v_i v_j}(r_\alpha r_\beta)$ captures pairwise hyperconservation and is zero if v_i and v_j are not in contact or their residue types are completely independent; $\phi_{v_i v_j v_k}(r_\alpha r_\beta r_\gamma)$ captures 3-way hyperconservation and is zero if v_i, v_j, and v_k are not in contact or their residue types are completely independent. The potential score of higher-order hyperresidues can be defined similarly. The potential score of a higher-order hyperresidue contains no information redundant with that of its lower-order constituents.

2.3   Edge Weights

In the hypergraph model, edge weights measure evolutionary optimization of higher-order interactions. For a protein or a set of proteins $\mathcal{S} \subseteq \mathcal{F}$, we can evaluate the significance of an edge as the average potential score of the hyperresidues appearing at the positions forming the edge:

$$w(e) = \sum_R \frac{\#R \text{ at } e \text{ in } \mathcal{S}}{|\mathcal{S}|} \cdot \phi_e(R) . \tag{8}$$

2.4   Edge Weights for Recombination

A particular form of edge weights serves as a guide for breakpoint selection in site-directed recombination. Suppose a set $\mathcal{S} \subseteq \mathcal{F}$ of parents is to be recombined at a set $X = \{x_1, x_2, \cdots, x_n\}$ of breakpoints, where $x_t = v_i$ indicates that breakpoint x_t is between residues v_i and v_{i+1}. We can view recombination as a two-step process: *decomposing* followed by *recombining*. In the decomposing step, each protein sequence is partitioned into $n + 1$ intervals according to the breakpoints, and the hypergraph is partitioned into $n + 1$ disjoint subgraphs by removing all edges spanning a breakpoint. The impact of this decomposition can be individually assessed for each edge, using Eq. 8 for the parents $\mathcal{S}$.

In the recombining step, edges removed in the decomposing step are reconstructed with new sets of hyperresidues according to all combinations of parent fragments. The impact of this reconstruction can also be individually assessed for each edge, yielding a breakpoint-specific weight:

$$w(e, X) = \sum_{R} \frac{\#R \text{ at } e \text{ in } \mathcal{L}}{|\mathcal{L}|} \cdot \phi_e(R) \ . \tag{9}$$

In this case, the potential score of hyperresidue R is weighted by the amount of its representation in the library $\mathcal{L}$. Note that we need not actually enumerate the set of hybrids (which can be combinatorially large) in order to determine the weight, as the frequencies of the residues at the positions are sufficient to compute the frequencies of the hyperresidues.

The combined effect of the two-step recombination process on an individual edge, the *edge perturbation*, is then defined as the change in edge weight:

$$\Delta w(e, X) = w(e) - w(e, X) \ . \tag{10}$$

If all vertices of e are in one fragment, we have $w(e) = w(e, X)$ and $\Delta w(e, X) = 0$. The edge perturbation thus integrates essential information from the database, family, parent sequences, and breakpoint locations.

3   Optimization of Breakpoint Locations

Given parent sequences, a set of breakpoints determines a hybrid library. The quality of this hybrid library can be measured by the total perturbation to all edges due to the breakpoints. The hypothesis is that the lower the perturbation, the higher the representation of folded and functional hybrids in the library. We formulate the breakpoint selection problem as follows.

Problem 1. c-RECOMB. Given $G_c = (V, E_c, w)$ and a positive integer n, choose a set of breakpoints $X = \{x_1, x_2, \cdots, x_n\}$ minimizing $\sum_{e \in E_c} \Delta w(e, X)$.

Recall from Sec. 2 that G_c represents a hypergraph with edge order uniformly c (where edges with order less than c are also represented as order-c edges).

This hypergraph partitioning problem is significantly more specific than general hypergraph partitioning, so it is interesting to consider its algorithmic difficulty. As as we will see in Sec. 3.1, *c-RECOMB* is NP-hard for $c = 4$ (and thus also for $c > 4$), although we provide polynomial-time solutions for $c = 2$ in Sec. 3.2 and $c = 3$ in Sec. 3.4.

A special case of *c-RECOMB* provides an efficient heuristic approach to minimize the overall perturbation. By minimizing the total weight of all edges E_X removed in the decomposing step, fewer interactions need to be recovered in the recombining step.

Problem 2. c-DECOMP. Given $G_c = (V, E_c, w)$ and a positive integer n, choose a set of breakpoints $X = \{x_1, x_2, \cdots, x_n\}$ minimizing $\sum_{e \in E_X} w(e)$.

c-DECOMP could also be useful in identifying modular units in protein structures, in which case there is no recombining step.

3.1   NP-Hardness of *4-RECOMB*

4-RECOMB is combinatorial in the set X of breakpoints and the possible configurations they can take relative to each edge. The number of possible libraries could be huge even with a small number of breakpoints (*e.g.* choosing 7 breakpoints from 262 positions for beta-lactamase results in on the order of 10^{13} possible configurations). The choices made for breakpoints are reflected in whether or not there is a breakpoint between each pair of sequentially-ordered vertices of an edge, and thus in the perturbation to the edge. We first give a decision version of *4-RECOMB* as follows and then prove that it is NP-hard. Thus the related optimization problem is also NP-hard. Our reduction employs general hypergraphs; analysis in the geometrically-restricted case remains interesting future work.

Problem 3. 4-RECOMB-DEC. Given $G_4 = (V, E_4, w)$, a positive integer n, and an integer W, does there exist a set of breakpoints $X = \{x_1, x_2, \cdots, x_n\}$ such that $\sum_{e \in E_4} \Delta w(e, X) \leq W$.

Theorem 3.1. 4-RECOMB-DEC is NP-hard.

Proof. We reduce from *3SAT*. Let $\phi = C_1 \wedge C_2 \wedge \cdots \wedge C_k$ be a boolean formula in 3-CNF with k clauses. We shall construct a hypergraph $G_4 = (V, E_4, w)$ such that ϕ is satisfiable iff there is a *4-RECOMB-DEC* solution for G_4 with $n = 3k$ breakpoints and $W = -|E_4|$. (See Fig. 2.). For clause $C_i = (l_{i,1} \vee l_{i,2} \vee l_{i,3})$ in ϕ, add to V four vertices in sequential order $v_{i,1}$, $v_{i,2}$, $v_{i,3}$, and $v_{i,4}$. Elongate V with $3k$ trivial vertices (v'_j in Fig. 2), where we can put trivial breakpoints that cause no perturbation. Let us define predicate $b(i, s, X) = v_{i,s} \in X$ for $s \in \{1, 2, 3\}$, indicating whether or not there is a breakpoint between $v_{i,s}$ and $v_{i,s+1}$. We also use indicator function I to convert a boolean value to 0 or 1. We construct E_4 with three kinds of edges: (1) For the 4-tuple of vertices for clause C_i, add an edge $e = \langle v_{i,1}, v_{i,2}, v_{i,3}, v_{i,4} \rangle$ with $\Delta w(e, X) = -I\{b(i, 1, X) \vee b(i, 2, X) \vee b(i, 3, X)\}$. (2) If two literals $l_{i,s}$ and $l_{j,t}$ are identical, add an edge $e = \langle v_{i,s}, v_{i,s+1}, v_{j,t}, v_{j,t+1} \rangle$ with $\Delta w(e, X) = -I\{b(i, s, X) = b(j, t, X)\}$. (3) If two literals $l_{i,s}$ and $l_{j,t}$ are complementary, add an edge $e = \langle v_{i,s}, v_{i,s+1}, v_{j,t}, v_{j,t+1} \rangle$ with $\Delta w(e, X) = -I\{b(i, s, X) \neq b(j, t, X)\}$.

There are $7k$ vertices and at most $k + 3\binom{k}{2} = O(k^2)$ edges, so the construction takes polynomial time. It is also a reduction. First, if ϕ has a satisfying assignment, choose breakpoints $X = \{v_{i,s} | l_{i,s}$ is TRUE$\}$ plus additional breakpoints between the trivial vertices to reach $3k$ total. Since each clause is satisfied, one of its literals is true, so there is a breakpoint in the corresponding edge e and its perturbation is -1. Since literals must be used consistently, type 2 and 3 edges also have -1 perturbation. Thus *4-RECOMB-DEC* is satisfied with $n = 3k$ and $W = -|E_4|$. Conversely, if there is a *4-RECOMB-DEC* solution with breakpoints X, then assign truth values to variables such that $l_{i,s} = b(i, s, X)$ for $s \in \{1, 2, 3\}$ and $i \in \{1, 2, \cdots, k\}$. Since perturbation to type 1 edges is -1, there must be at least one breakpoint in each clause vertex tuple, and thus a true literal in the clause. Since perturbation to type 2 and 3 edges is -1, literals are used consistently.

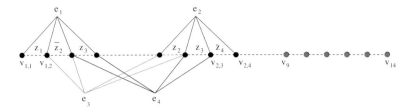

Fig. 2. Construction of hypergraph $G_4 = (V, E_4, w)$ from an instance of *3SAT* $\phi = (z_1 \vee \bar{z}_2 \vee z_3) \wedge (z_2 \vee z_3 \vee \bar{z}_4)$. Type 1 edges e_1 and e_2 ensure the satisfaction of clauses (-1 perturbation iff there is a breakpoint iff the literal is true and the clause is satisfied), while type 3 edge e_3 and type 2 edge e_4 ensure the consistent use of literals (-1 perturbation iff the breakpoints are identical or complementary iff the variable has a single value).

We note that *4-RECOMB-DEC* is in NP, since given a set of breakpoints X for parents S we can compute $\Delta w(e, X)$ for all edges in polynomial time ($O(S^4 E)$), and then must simply sum and compare to a provided threshold.

3.2 Dynamic Programming Framework

Despite the NP-hardness of the general sequentially-constrained hypergraph partitioning problem *c-RECOMB*, the structure of the problem (*i.e.* the sequential constraint) leads to efficient solutions for some important cases. Suppose we are adding breakpoints one by one from left to right (N- to C-terminal) in the sequence. Then the additional perturbation to an edge e caused by adding breakpoint x_t given previous breakpoints $X_{t-1} = \{x_1, x_2, \cdots, x_{t-1}\}$ can be written:

$$\Delta\Delta w(e, X_{t-1}, x_t) = \Delta w(e, X_t) - \Delta w(e, X_{t-1}) , \tag{11}$$

where $X_0 = \emptyset$ and the additional perturbation caused by the first breakpoint is $\Delta\Delta w(e, X_0, x_1) = \Delta w(e, X_1)$. Reusing notation, we indicate the total additional perturbation to all edges as $\Delta\Delta w(E, X_{t-1}, x_t)$. Now, if the value of $\Delta\Delta w(E, X_{t-1}, x_t)$ can be determined by the positions of x_{t-1} and x_t, independent of previous breakpoints, then we can adopt the dynamic programming approach shown below. When the additional perturbation depends only on x_{t-1} and x_t, we write it as $\Delta\Delta w(E, x_{t-1}, x_t)$ to indicate the restricted dependence.

Let $d[t, \tau]$ be the minimum perturbation caused by t breakpoints with the rightmost at position τ. If, for simplicity, we regard the right end of the sequence as a trivial breakpoint that causes no perturbation, then $d[n + 1, |V|]$ is the minimum perturbation caused by n breakpoints plus this trivial one, *i.e.* the objective function for Problem 1. We can compute d recursively:

$$d[t, \tau] = \begin{cases} \Delta w(E, \{\tau\}), & \text{if } t = 1 ; \\ \min\limits_{\lambda \leq \tau - \delta} \{d[t - 1, \lambda] + \Delta\Delta w(E, \lambda, \tau)\}, & \text{if } t \geq 2 . \end{cases} \tag{12}$$

where δ is a user-specified minimum sequential distance between breakpoints. The recurrence can be efficiently computed bottom-up in a dynamic programming style, due to its optimal substructure. In the following, we instantiate this

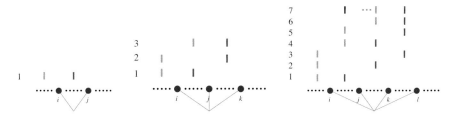

Fig. 3. All breakpoint configurations that cause additional perturbation to an edge as breakpoints in a c-$RECOMB$ problem are added one by one from left to right in the sequence. The dynamic programming formulation requires that we be able to distinguish the configurations from each other and from configurations with no additional perturbation. For an order-2 edge $\langle v_i, v_j \rangle$, there is additional perturbation if and only if the current breakpoint (red) is added between v_i and v_j and the previous breakpoint (green) is to the left of v_i. Similarly, the configurations on an order-3 edge $\langle v_i, v_j, v_k \rangle$ can be distinguished by the positions of the current breakpoint (red) and the preceding one (green) with respect to the intervals $[v_i, v_j]$ and $[v_j, v_k]$. However, for an order-4 edge, configurations 6 and 7 are ambiguous with respect to the intervals of $\langle v_i, v_j, v_k, v_l \rangle$. We cannot be certain about the existence of a potential breakpoint between v_i and v_j (blue) without potentially looking back at all previous breakpoints (green ellipses).

dynamic programming formulation with different forms of $\Delta\Delta w$ for different cases of c-$RECOMB$ and c-$DECOMP$. Due to space limitations, time complexity analyses are omitted.

The special case of 2-$DECOMP$ (disruption of pairwise interactions) has been previously solved as a shortest path problem [24]. A complexity analysis accounting for both the edge weight calculation and dynamic programming shows that the total time is $O(S^2 E + VE + nV^2)$.

The instantiation for 2-$RECOMB$ is as follows. Each order-2 edge $\langle v_i, v_j \rangle$ has two states: either there is breakpoint between v_i and v_j or not (Fig. 3). The state of e is changed by adding breakpoint x_t iff $x_{t-1} < v_i < x_t < v_j$. Thus the additional perturbation caused by adding x_t can be determined by the positions of x_{t-1} and x_t, and is independent of previous breakpoints. Our dynamic programming framework Eq. 12 is therefore applicable to 2-$RECOMB$; the time complexity is $O(S^2 E + VE + nV^2)$.

3.3   Reduction from c-$DECOMP$ to 2-$DECOMP$

A significant property of our multi-order potential score (Sec. 2.2) is that the score of a higher-order edge captures only higher-order hyperconservation and contains no information about its lower-order constituents. Thus in the decomposition phase, a higher-order edge is broken if there is a breakpoint *anywhere* in the set of residue positions it spans. The lack of breakpoints between any adjacent pair of its vertices will be captured by the weight of the appropriate lower-order constituent edge. By this reasoning, we can reduce the c-$DECOMP$ problem to the 2-$DECOMP$ problem: given hypergraph $G_c = (V_c, E_c, w_c)$, construct graph $G_2 = (V_2, E_2, w_2)$ such that $V_2 = V_c$ and each edge $e_c = \langle v_1, v_2, \cdots, v_c \rangle \in E_c$ is mapped to an edge $e_2 = \langle v_1, v_c \rangle \in E_2$ connecting the first and last vertex of e_c, putting weight $w_c(e_c)$ on $w_2(e_2)$. There is a breakpoint decomposing e_c in G_c iff there is one decomposing e_2 in G_2. G_2 can be constructed in $O(V + E)$ time, and optimal solutions for c-$DECOMP$ on G_c correspond to optimal solutions for 2-$DECOMP$ on G_2. Under this reduction (which adds only $O(E)$ computation), the total time complexity for c-$DECOMP$ is $O(\mathcal{S}E + VE + nV^2)$. Thus protein modules can be computed under c-$DECOMP$ in polynomial time for any order of edge.

3.4   Dynamic Programming for 3-$RECOMB$

We have seen that the c-$RECOMB$ problem is NP-hard when $c \geq 4$ (Sec. 3.1) and solvable in polynomial time when $c = 2$ (Sec. 3.2). In this section, we instantiate our dynamic programming framework to give a polynomial-time solution when $c = 3$.

An order-3 edge has four possible states, according to whether or not there is at least one breakpoint between each pair of its vertices listed in sequential order. As Fig. 3 illustrates, given only x_{t-1} and x_t, all breakpoint configurations that cause additional perturbation can be uniquely determined, and the additional perturbation can be computed as in Eq. 11. This edge perturbation calculation meets the restriction required for our dynamic programming framework, and Eq. 12 and be used to optimize 3-$RECOMB$ in $O(\mathcal{S}^3 E + VE + nV^2)$ time.

3.5   Stochastic Dynamic Programming for 4-$RECOMB$

Tetrahedra are natural building blocks of 3D structures, and Delaunay tetrahedra in the protein core have been shown to capture interactions important for protein folding [8]. Our results below show significant information in general order-4 hyperconservation. In order to solve 4-$RECOMB$ problems, we develop here a heuristic approach based on stochastic dynamic programming. Unlike 2-$RECOMB$ and 3-$RECOMB$, the additional perturbation of a breakpoint cannot always be determined by reference just to the current and previous breakpoint locations. As Fig. 3 shows, given x_{t-1} and x_t, there is ambiguity only between configurations 6 and 7.

We can still employ the dynamic programming framework if we move from a deterministic version, in which both the additional perturbation and next state are known, to a stochastic version, in which they are predicted as expected values. In the ambiguous case of configurations 6 and 7 with $t \geq 2$, let us assume that breakpoints before x_{t-1} are uniformly distributed in the sequence. Then the probability of finding no breakpoint between v_i and v_j, i.e. being in configuration 6 rather than 7, is

$$p = (1 - \frac{v_j - v_i}{x_{t-1}})^{t-2} , \tag{13}$$

since $\frac{v_j - v_i}{x_{t-1}}$ is the probability of a breakpoint being located between v_i and v_j and $t - 2$ is the number of breakpoints before position x_{t-1}. Thus for the ambiguous cases, the expected additional perturbation to e caused by adding x_t is

$$\Delta \Delta w(e, x_{t-1}, x_t, t) = p \cdot \Delta \Delta w_6(e, x_{t-1}, x_t) + (1 - p) \cdot \Delta \Delta w_7(e, x_{t-1}, x_t) , \tag{14}$$

where the subscript indicates the configuration. Note that, unlike our previous formulations, the additional perturbation depends on the number of previous breakpoints. Thus the time complexity of this stochastic dynamic programming is increased to $O(S^4 E + nVE + nV^2)$. This stochastic dynamic programming technique can also be applied to $c > 4$ c-RECOMB problems, but the effectiveness of the approximation is expected to decrease with an increasing number of ambiguous states.

4   Results and Discussion

We demonstrate our hypergraph model and recombination planning algorithms in analysis of the beta-lactamase protein family, since previous site-directed recombination experiments have employed beta-lactamase parents TEM-1 and PSE-4 [23]. We identified 123 beta-lactamases for $\mathcal{F}$, including TEM-1 and PSE-4, with no more than 80% sequence identity, and constructed a multiple sequence alignment with at most 20% gaps in any sequence. PDB file 1BTL was used as the representative family structure. Vertices were considered as located at the average position of non-hydrogen side-chain atoms, and edges formed for sets of vertices whose positions were within 8 Å of each other.

For the database $\mathcal{D}$, we started with a subset of sequences culled from the protein data bank according to structure quality (R-factor less than 0.25) and mutual sequence identity (at most 60%) by PISCES [26]. To minimize the effect of structural errors on statistical results, chains with nonconsecutive residue numbers, gaps (C^α-C^αdistance greater than 4.2 Å between consecutive residues), or incorrect atom composition of residues were excluded [9]. This left 687 chains. Contact maps were constructed as with the family.

We first considered the information content in higher-order interactions. Fig. 4 shows the distributions of hyperresidue potential scores in both the database and family, for increasing hyperresidue order. By the non-redundant decomposition, a higher-order potential score would be 0 if the lower-order terms were independent. Non-zero $\phi(R)$ scores represent positive and negative correlation. The figure shows that there is clearly information in the sets of higher-order terms. Note that the family distributions are biased (μ not at zero), presumably because many sets of amino acid types are not observed in the MSA. Family distributions are also more informative than database ones (larger σ for all orders). Dicysteine pairs are expected to be particularly informative (i.e. cysteines in disulfides are not independent), as reflected in the clear outliers marked in the $c = 2$ database histogram; there are no disulfides in the beta-lactamase family.

A limited amount of data is currently available for evaluating the experimental effectiveness of a recombination plan. Here, we use the beta-lactamase hybrid library of [23]. For each hybrid in the library, we computed both the total potential score and the mutation level. The total potential score is the sum, over all edges up to order-4, of the edge potential (Eq. 5– 7) for the residues in the hybrid sequence. The mutation level is the number of residues in the hybrid that differ from the closest parent. While hybrids with small mutation levels are expected

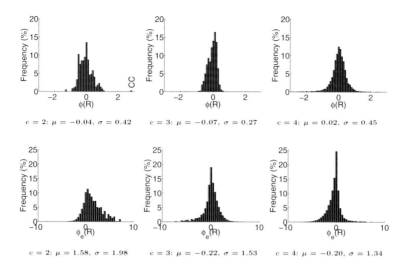

Fig. 4. Multi-order potential scores, derived from the database (top) and the beta-lactamase family (bottom). For each order c of hyperresidues, the distribution of potential scores is shown (pooled over all edges for the family version).

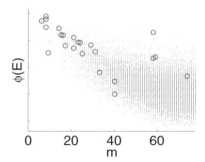

Fig. 5. Potential score $\phi(E)$ (sum over all interactions up to order-4) vs. mutation level m (# residues different from the closest parent) for all hybrids in a beta-lactamase library. Blue dots indicate hybrids, and red circles those determined to be functional [23].

to be functional, our potential yields high scores for the functional hybrids at high mutation levels (Fig. 5).

Next we applied our dynamic programming algorithms to optimize 7-break point sets for different beta-lactamase parents (Fig. 6), using minimum effective fragment length $\delta = 10$, database/family weight $\rho = 0.01$, and maximum order of edges $c = 3$. We found the results to be insensitive to ρ, beyond very small values placing all the emphasis on the database (data not shown). In the 1-parent case, the plan amounts to decomposing the protein (PDB file 1BTL as representative family structure) into modules preserving multi-order interactions. The 2-parent and 12-parent cases illustrated here would be useful in site-directed recombination experiments. We note that some locations can "float" due to parent

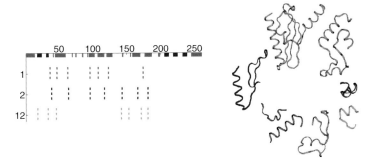

Fig. 6. Beta-lactamase breakpoint optimization. (Left) Optimized breakpoint locations when planning with 1, 2, or 12 parents. The sequence is labeled with residue index, with helices in red and β-sheets in blue. (Right) 3D structure fragments (PDB id: 1BTL) according to optimized breakpoint locations for the 1-parent case.

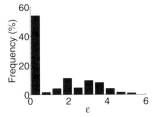

Fig. 7. Distribution of differences in edge perturbations between the two ambiguous configurations (6 and 7) in *4-RECOMB* for the beta-lactamases TEM-1 and PSE-4. Differences (ε) are expressed in standard deviations units.

sequence identity (*e.g.* in positions 17–20 with 2 parents). These all represent viable experiment plans, optimizing multi-order interactions according to sequence characteristics of different parents.

Finally, we considered the error that could be caused by the stochastic approximation in solving *4-RECOMB*. Fig. 7 shows the distribution, over all order-4 edges, of differences in perturbations between the ambiguous states. The differences are expressed in terms of perturbation standard deviations $\varepsilon = \frac{|\Delta\Delta w_6 - \Delta\Delta w_7|}{(\text{std}(\Delta\Delta w_6) + \text{std}(\Delta\Delta w_7))/2}$. Edges with identical residues at v_i or v_j are excluded, since the perturbation is necessarily the same. Even so, in a majority of cases the heuristic would lead to no or very small error. Thus the stochastic dynamic programming will provide a near optimal solution, which makes it reasonable to include 4-way interactions in practice.

5   Conclusion

We have developed a general hypergraph model of multi-order residue interactions in proteins, along with algorithms that optimize site-directed recombination experiments under the model. The model has a number of other potential applications for which multi-order interactions are significant, after suitable parameterization, including prediction of $\Delta G°$ of unfolding, $\Delta\Delta G°$ of mutagenesis, and modularity of protein structures. The algorithms likewise can be employed using potentials that incorporate additional information (*e.g.* weighted for active sites). Interesting future work includes selection of parent sequences, separation of stability-critical and functionality-critical multi-residue interactions, interpretation of experimental data, and feedback of experimental results to subsequent rounds of planning.

Acknowledgments

We thank Dr. Bruce Craig (Statistics, Purdue), Shobha Potluri and John Thomas (CS, Dartmouth), and Michal Gajda (IIMCB, Poland) for stimulating discussion. This work was supported in part by an NSF CAREER award to CBK (IIS-0444544) and an NSF SEIII grant to CBK, AMF, and Bruce Craig (IIS-0502801).

References

1. Tanaka, S., Scheraga, H.: Medium and long range interaction parameters between amino acids for predicting three dimensional strutures of proteins. Macromolecules **9** (1976) 945–950
2. Miyazawa, S., Jernigan, R.: Estimation of effective interresidue contact energies from protein crystal structures: Quasi-chemical approximation. Macromolecules **18** (1985) 531–552
3. Maiorov, V., Crippen, G.: Contact potential that recognizes the correct folding of globular proteins. J. Mol. Biol. **227** (1992) 876–88
4. Simons, K.T., Kooperberg, C., Huang, E., Baker, D.: Assembly of protein tertiary structures from fragments with similar local sequences using simulated annealing and Bayesian scoring functions. J. Mol. Biol. **268** (1997) 209–225
5. Kihara, D., Lu, H., Kolinski, A., Skolnick, J.: TOUCHSTONE: an *ab initio* protein structure prediction method that uses threading-based tertiary restraints. PNAS **98** (2001) 10125–10130
6. Godzik, A.: Fold recognition methods. Methods Biochem. Anal. **44** (2003) 525–546
7. Betancourt, M., Thirumalai, D.: Pair potentials for protein folding: Choice of reference states and sensitivity of predictive native states to variations in the interaction schemes. Protein Sci. **8** (1999) 361–369
8. Carter Jr., C., LeFebvre, B., Cammer, S., Tropsha, A., Edgell, M.: Four-body potentials reveal protein-specific correlations to stability changes caused by hydrophobic core mutations. J. Mol. Biol. **311** (2001) 621–638
9. Krishnamoorthy, B., Tropsha, A.: Development of a four-body statistical pseudo-potential to discriminate native from non-native protein conformations. Bioinformatics **19** (2003) 1540–1548
10. Simons, K., Ruczinski, I., Kooperberg, C., Fox, B., Bystroff, C., Baker, D.: Improved recognition of native-like protein structures using a combination of sequence-dependent and sequence-independent features of proteins. PROTEINS: Structure, Function, and Genetics **34** (1999) 82–95

11. Gobel, U., Sander, C., Schneider, R., Valencia, A.: Correlated mutations and residue contacts in proteins. PROTEINS: Structure, Function, and Genetics **18** (1994) 309–317
12. Lockless, S., Ranganathan, R.: Evolutionarily conserved pathways of energetic connectivity in protein families. Science **286** (1999) 295–299
13. Thomas, J., Ramakrishnan, N., Bailey-Kellogg, C.: Graphical models of residue coupling in protein families. In: 5th ACM SIGKDD Workshop on Data Mining in Bioinformatics (BIOKDD). (2005)
14. Stemmer, W.: Rapid evolution of a protein *in vitro* by DNA shuffling. Nature **370** (1994) 389–391
15. Ostermeier, M., Shim, J., Benkovic, S.: A combinatorial approach to hybrid enzymes independent of DNA homology. Nat. Biotechnol. **17** (1999) 1205–9
16. Lutz, S., Ostermeier, M., Moore, G., Maranas, C., Benkovic, S.: Creating multiple-crossover DNA libraries independent of sequence identity. PNAS **98** (2001) 11248–53
17. Sieber, V., Martinez, C., Arnold, F.: Libraries of hybrid proteins from distantly related sequences. Nat. Biotechnol. **19** (2001) 456–60
18. Voigt, C., Martinez, C., Wang, Z., Mayo, S., Arnold, F.: Protein building blocks preserved by recombination. Nat. Struct. Biol. **9** (2002) 553–558
19. O'Maille, P., Bakhtina, M., Tsai, M.: Structure-based combinatorial protein engineering (SCOPE). J. Mol. Biol. **321** (2002) 677–691
20. Aguinaldo, A., Arnold, F.: Staggered extension process (StEP) in vitro recombination. Methods Mol. Biol. **231** (2003) 105–110
21. Coco, W.: RACHITT: Gene family shuffling by random chimeragenesis on transient templates. Methods Mol. Biol. **231** (2003) 111–127
22. Castle, L., Siehl, D., Gorton, R., Patten, P., Chen, Y., Bertain, S., Cho, H.J., Duck, N., Wong, J., Liu, D., Lassner, M.: Discovery and directed evolution of a glyphosate tolerance gene. Science **304** (2004) 1151–4
23. Meyer, M., Silberg, J., Voigt, C., Endelman, J., Mayo, S., Wang, Z., Arnold, F.: Library analysis of SCHEMA-guided protein recombination. Protein Sci. **12** (2003) 1686–93
24. Endelman, J., Silberg, J., Wang, Z.G., Arnold, F.: Site-directed protein recombination as a shortest-path problem. Protein Eng., Design and Sel. **17** (2004) 589–594
25. Sippl, M.: Calculation of conformational ensembles from potentials of mean force. an approach to the knowledge-based prediction of local structures in globular proteins. J. Mol. Biol. **213** (1990) 859–883
26. Wang, G., R. L. Dunbrack, J.: Pisces: a protein sequence culling server. Bioinformatics **19** (2003) 1589–1591

Biological Networks: Comparison, Conservation, and Evolutionary Trees

(Extended Abstract)

Benny Chor and Tamir Tuller*

School of Computer Science, Tel Aviv University
{bchor, tamirtul}@post.tau.ac.il

Abstract. We describe a new approach for comparing cellular-biological networks, and finding conserved regions in two or more such networks. We use the length of describing one network, given the description of the other one, as a distance measure. We employ these distances as inputs for generating phylogenetic trees. Our algorithms are fast enough for generating phylogenetic tree of more than two hundreds metabolic networks that appear in KEGG. Using KEGG's metabolic networks as our starting point, we got trees that are not perfect, but are surprisingly good. We also found conserved regions among more than a dozen metabolic networks, and among two protein interaction networks. These conserved regions seem biologically relevant, proving the viability of our approach.

Keywords: Biological networks, tree reconstruction, relative description length, compression, metabolic networks, Conserved regions, networks' comparison, network evolution.

1 Introduction

With the advent of bio technologies, huge amounts of genomic data have accumulated. This is true not only for biological sequences, but also with respect to biological *networks*. Prominent examples are metabolic networks, protein-protein interaction networks, and regulatory networks. Such networks are typically fairly large, and are known for a number of species. On the negative side, they are error prone, and are often partial. For example, in the KEGG database [17] there are over 250 metabolic networks of different species, at very different levels of details. Furthermore, some networks are directly based on experiments, while others are mostly "synthesized" manually.

The goal in this study is to devise a quantitative and efficient method for local and global comparisons of such networks, and to examine their evolutionary signals. Our method of comparing two networks is based on the notion of *relative description length*. Given two labeled network A and B, we argue that the more similar they are, the fewer bits are required to describe A given B (and vice

---

* Corresponding author.

A. Apostolico et al. (Eds.): RECOMB 2006, LNBI 3909, pp. 30–44, 2006.

versa). Mathematically, this can give rise to Kolmogorov complexity-like measures, which are incomputable and inapproximable. Other approaches, based on labeled graph alignment, subgraph isomorphism, and subgraph homeomorphism, are computationally intractable [12].

By way of contrast, our algorithm is efficient: Comparing the man-mouse metabolic networks takes 10 seconds on a 3 years old PC (996 MHZ, 128 MB RAM, Pentium 3), and all $(240 \times 239)/2$ pairwise comparisons of the KEGG database took less than three days on the same machine. We extend the relative description length approach to *local* comparison of two or multiple networks. For every label of the nodes (describing a metabolic substrate), we identify if that label exists in the various networks, and build local neighborhoods of equal radius around these labels. Neighborhoods with high similarity, according to our criteria, are likely to be conserved. We seek a method that is efficient not only for one pair of networks, but for all $\binom{n}{2}$ pairs. Our global comparison produces a matrix for expressing the pairwise distances between networks. To test its quality we have built an evolutionary tree, based on the distance matrix constructed from KEGG's metabolic networks. To the best of our knowledge, this is the first time evolutionary trees are constructed based on biological networks. The results are surprisingly good. For example, the tree for 20 taxa with large networks (more than 3000) in the KEGG database perfectly clusters the taxa to Eukaryotes, Prokaryots and Archea, and clusters almost perfectly sub-partitions within each type. Neither the 20 taxa tree nor another KEGG based tree for 194 taxa are perfect, but this is hardly surprising given the huge disparity in detail between KEGG's metabolic networks, where some have more than 3000 nodes (metabolites) while as many as 10% of species have metabolic networks with fewer than 10 nodes. Bio networks are still at a state where the available data is much more fragmented and less accessible than biological sequences data. But network information certainly goes beyond sequence information, and our work makes some preliminary steps at the fascinating questions of network comparison and evolution.

Relative description length proved to be a useful parameter for measuring the disparity between biological sequences, such as genomes. In [21], Li *et al.* describe a distance based on compression [4] that was used for generating phylogenetic trees. In [3] Burstain *et. al* present a simple method based on string algorithms (average common substring) for generating phylogenetic trees. The main innovation in the present work is the use of the paradigm of relative description length in the domain of biological networks, which is very different than the one dimensional domain of biological sequences. The different domain necessitates a different approach. Our is based on the reasonable assumption that homologous nodes in close taxa will share more similar neighborhood, as compared to remote taxa.

To the best of our knowledge, this is the first time relative description length is used for comparing networks and constructing evolutional signals (trees). Ogata *et al.* [25] developed a heuristic for finding similar regions in two metabolic pathways. Their method is based on comparing the distances between pairs of

nodes in two metabolic pathways. Schreiber [28] developed a tool for visualization of similar subgraphs in two metabolic pathways. Tohsato *et al.* [31] deals
with alignment of metabolic pathways, where the topology of the pathways is
restricted to chains. Kelley *et al.* [18] data-mine chains in protein-protein networks by searching paths with high likelihood in a global alignments graph,
where each node represents a pair of proteins (one from each network). This
last work was generalized to identify conserved paths and clusters of protein-
protein interaction networks inmultiple organisms, by Sharan *at al.* [29]. They
build a graph with a node for each set of homologue proteins (one protein for
each organism). Two nodes are connected by an edge if all the pairs of proteins interact in each organism respectively. The second step is searching paths
and clusters in this graph. Koyuturk *et al.* [19] used a bottom up algorithm for
finding frequent subgraphs in biological networks. Pinter *et. al* [26] suggested
an $O(n^3/\log(n))$ algorithm for the alignment of two trees. While related, this
does not solve our problem as it is restricted to trees, and is not efficient enough
for multiple species. Another problem with the alignment approach is to define
the costs of deletion, mismatches. This problem is true for both sequences and
graphs' alignment. Chung, and Matula [5,23] suggest algorithms for a similar
problem of subgraph isomorphism on trees.

The rest of the paper is organized as follows: In section 2 we discuss the
general problem of comparing directed labelled graphs. Then we describe our
approach, the relative description length (RDL) method. In section 3 we describe
the properties of our measure. In section 4 we describe a method based on
the relative description measure for finding conserved regions in network. In
section 5 we demonstrate the method, where the inputs are metabolic networks
from KEGG. Section 6 contain concluding remarks and suggestions for further
research.

2   Distances and Phylogeny from Biological Networks

In this section we discuss the problem of comparing labeled, directed graphs. We
then describe our RDL method for computing distances between networks. The
"design criteria" is to find measures that accurately reflects biological disparity,
while concurrently be efficiently computable. The networks in this paper are
directed graphs with uniquely labeled nodes. Specifically we used the format
of Jeong *et al.* [16] for representing a metabolic networks, only the nodes have
labels, the edges have no labels. But our algorithms apply, *mutatis mutandis*, to
other types of networks with such representation. All metabolic substrates are
represented by graph nodes, and the reaction links in the pathway, associated
with enzymes, are represented by directed graph edges.

The basic measure we are interested in is the amount of bits needed to describe a network G_2, given the network G_1. The natural measure to consider
here is *Kolmogorove complexity* defined as follows $k(x) = k_U(x)$ is the length
of a shortest string, z that when given as an input to U, an *Universal Tur-
ing Machine* (TM) [30], U emits x and halts, namely $U(z) = x$ [22]. One may

consider *relative* Kolmogorov complexity. Given two strings x and y, $k(x|y)$ is defined as the length of the shortest string z one need to add to the the the string y as an input to a universal TM, U, such that $U(z, y) = x$. A variant of this measure is known to be a metric [22], *i. e.* it is symmetric and it satisfies the triangle inequality. Unfortunately, it is well known that Kolmogorov complexity, in its unconditional and conditional forms, is incomputable. Furthermore, there is a non constant function $f(x)$, a function that increases with x, such that even an $f(x)$ approximation of $k(x)$, and thus of $k(x|y)$ is incomputable. We now turn to the definition of the relative description length measure. Let pa_i denote the set of nodes that are parents of i in the network. A directed graph or network, G, with n labelled nodes can be encoded by using $\log(n)$ bits to denote the number of parents of each node, and $\log \binom{n}{|pa_i|}$ bits to name x_i's parents (for *sparse* networks, this is more succinct than the n bits per node of the naive description). Let $DL(G)$ denote the description length of G. Then for an n node network $DL(G) = \sum_{i=1}^{n} \left(\log(n) + \log \binom{n}{|pa_i|} \right)$. Suppose now we have a collection $\{G_i\}$ of labelled directed graphs, and let n_i denote the number of nodes in G_i. Let $n_{i,j}$ denote the number of labelled nodes that appear *both* in G_i and G_j. Let $pa_v(G)$ denote the number of parents of node v in the graph G. For encoding a subset T of a known set S, one needs $\log(|T|) + \log \binom{|S|}{|T|}$ bits. The first expression describes the size of the group T, and the second is for describing the subset out of $\binom{|S|}{|T|}$ possible subsets. We denote the number of bits encoding sub-set T of a known set S by $Enc(T|S)$. Two assumptions underly our procedure for describing one graph given the other:

1. The distance among corresponding pairs of nodes in networks of closely related species are similar.

2. It is possible that two nodes, corresponding to different species, have the same role even if their labeling is not identical.

The procedure for describing the graph G_2, given the graph G_1 was defined as follows:

DL$(G_2|G_1)$

1. There are $n_1 - n_{1,2}$ nodes that appear in G_1 and do not appear in G_2. Given G_1, they can be encoded using $Enc(n_1 - n_{1,2}|n_1)$ bits.

2. For each node v common to G_1 and G_2:

   (a) The node v has $|pa_v(G_1) \cap pa_v(G_2)|$ parents, which appear both in G_1 and G_2. We encode these nodes by $Enc(pa_v(G_1) \cap pa_v(G_2)|n_1)$ bits.

   (b) The node v has $|pa_v(2) \setminus (pa_v(G_1) \cap pa_v(G_2))|$ parents which appear in G_2 but not in G_1. We encode these nodes by $Enc(|pa_v(G_2) \setminus (pa_v(G_1) \cap pa_v(G_2))| |n_2 - n_{1,2})$ bits.

(c) The rest of the parents of the node v in G_2 appear in both G_1 and G_2, but are not parents of v in G_1. Denote the size of this set by n_v. Let d denote the minimal bidirectional radius of a ball around the node v in G_1 that contains all these parents. Let $n^{v,d}$ denote the number of nodes in this ball. We encode these parents using $\log(d) + \log(n_v) + \log\binom{n^{v,d}}{n_v}$ bits.

3. For each node v that appears in G_2 and not in G_1: Let c_v denote the number of bits need to describe the parents of node v by other node that appear both G_1 and G_2 using steps 1, 2. We encode the parents of the node by $1 + min(\log(n_1) + c_v, \log(n_2) + \log\binom{n_2}{|pa_v|})$ bits.

Definition 1. *Given two labelled, directed networks G_i and G_j, we define their relative description length "distance", $RDL(G_i, G_j)$, as follows: $RDL(G_i, G_j) = DL(G_i|G_j)/DL(G_i) + DL(G_j|G_i)/DL(G_j)$.*

The first term in this expression is the ratio of the number of bits needed to describe G_i when G_j is given and the number of bits needed to describe G_i without additional information. The second term is the dual. In general, $D(G_1, G_2)$ is larger when the two networks are more dissimilar, and $0 \le D(G_1, G_2) \le 2$. The extreme cases are $G_1 = G_2$, where $D(G_1, G_2)$ is $O(1/|V_1| + 1/|V_2|)$, and when G_1, G_2 have no nodes in common, where $D(G_1, G_2) = 2$.

In the preprocessing stage we first calculate the distances between all pairs of nodes in G_1 and G_2 by Dijkstra algorithm [6] or Johnson algorithm [6], we ignore directionality. The running time of these algorithms is $O(|E| \cdot |V| + |V|^2 \log(|V|))$. In all metabolic networks, the input degree of each node is bounded (in all the network in KEGG no one have more than 40 parents, usually it was much less, between 1 and 3 parents), thus $E = \Theta(V)$, and the time complexity is $O(|V|^2 \log(|V|))$ for all pairs. Note that there are algorithms of time complexity $O(|V|^{2.575})$ for finding distances between all pairs of nodes without any assumptions on the graphs structure [32]. We now sort the distance vector of each node in $O(|V| \log(|V|))$ time, so the total time is $O(|V|^2 \log(|V|))$. In the next stage, we sort the node names in each net in lexicographic order in $O(|V| \log(|V|))$ time. Then we sort each parent list in lexicographic order, this is done in $O(|V| \log(|V|))$ time.

Stage 1. in the procedure $DL(G_2|G_1)$ is done in linear time given a lexicographic ordering of the nodes in the two networks. The total of stages 2.(a) for all the nodes is done in linear time given a lexicographically ordered list of all the parent list. The total of stages 2.(b) for all the nodes is done in $O(|V| \log(|V|))$ time given a lexicographic sort of the nodes in G_1. The total of stages 2.(c) for all the nodes is done in $O(|V| \log(|V|))$ time given the sorted distances matrix of the network. Stage 3 done in total time of $O(|V|^2 \log(|V|))$ for all the nodes.

Thus the total time complexity of the pairwise network comparison algorithm is $O(|V|^2 \log(|V|))$.

We used neighbor joining algorithm [27] for generating a tree from the distance matrix. Recent variants of NJ run in in $O(N^2)$, where N is the number of taxa [9]. Thus the total time complexity of our method for generating a phylogenetic tree for N networks of up to $|V|$ nodes each, is $O(N^2 \cdot |V|^2 \log |V|)$. We discovered empirically that by skipping stage 3., the precision decreases by a few percentage points, while the time complexity becomes close to linear. Such shortcut may be suitable for larger inputs.

3   Properties of the RDL Networks Comparison Measure

It is easy to see that the measure $D(G_i, G_j)$ is symmetric. While $D(G, G) > 0$, it is small for large graphs. In general, our measure does not satisfy the triangle inequality. For example the distance of the following three networks in KEGG do not satisfy the triangle inequality. The networks are the bacteria Aquifex aeolicus (*aae*), the archea Archaeoglobus fulgidus (*afu*), and the bacteria Bacteroides fragilis YCH46 (*bfr*). The distance between *aae* and *bfr* is 4.7, while the distance between *aae* and *afu* is 0.7 and the distance between *afu* and *bfr* is 3.92. However, by empirically checking all the triplets in a distance matrix generated for all the 240 networks in KEGG we found that only a very small fraction of all triplets do not satisfy the triangle inequality - 363 triplets out of $2,257,280$ possible triplets. Usually these triplets involve very partial nets. For example the *bfr* network mentioned above includes only four nodes. After removing all the networks with less than 100 nodes, we got 194 networks left. For this set of species, all the triplets satisfy the triangle inequality.

We performed preliminary empirical studies, showing that our measure increases linearly as a function of the "evolutionary time". We used the following simple minded model: At each time period there is a probability p_1 of adding a new node to a net, probability p_2 of removing a node from a net (all nodes have the same probability to be removed), probability p_3 of adding a directed edge between any two vertices, probability p_4 of removing a directed existing edge between any two vertices (all edges have the same probability to be removed). We chose $p_1 = p_2$ in order to maintain the expected number of nodes in the graph, and choose $p_3 = p_4$ in order to maintain the average number of edges in the graph.

In the resultant graphs the growth was close to linear, suggesting that for networks with similar sizes, our method for generating phylogenetic trees using distances based methods, such as neighbor jointing, is justified. Furthermore, our method can also be used to estimate branch lengths of phylogenetic trees. These consequences do not necessarily apply to networks of different sizes. Of course, the preliminary simulation used a very simplistic model. More sophisticated ones, including unequal grows and elimination rates, may give a better indication for more realistic instances.

4   Finding Conserved Regions in Networks

In this section we describe our method for finding conserved regions in two or more networks, and the rationale behind it. The method is based on the RDL measure described in section 2. Consider a ball of bidirectional distance at most d from node v in the directed graph G. The d conservations score of the node v in two is ∞ if it is not appear in the two networks, if it appear in the two networks it defined as follows:

Definition 2. *A (d, c) conserved node:*
Let v be a shared node among G_1 and G_2. Let B_1 and B_2 be the balls of bidirectional radius d around v in G_1 and G_2, respectively. We say that v is (d, c) conserved in G_1, G_2 if $D(B_1, B_2) \leq c$.

The (d, c)–conservated region of the two network $G_1 = (V_1, E_1)$ and $G_2 = (V_2, E_2)$ is defined as the intersection of the two subgraphs of G_1, G_2 induced by the (d, c) conserved nodes with respect to G_1, G_2. Algorithmically, we get it as follows

---

Find the (d, c) conserved region of G_2, G_1:

1. For each node common to G_1 and G_2, compute its d-conservations score.
2. Generate a graph $G'_1 = (V'_1, E'_1)$ where V'_1 includes the nodes in G_1 that are (d, c) conserved with respect to G_1, G_2. The edge e is an directed edge in G'_1 if its two endpoints are in V'_1, and it is a directed edge in E.
3. The graph $G'_2 = (V'_2, E'_2)$ is defined analogously.

The parameters d (radius) and c (RDL score), determine the two conserved regions G'_1, G'_2. It is easy to see that decreasing c decreases the sizes of G'_1, G'_2. Increasing d may either increase or decrease the sizes of the conserved graphs.

In a similar way we now define a conservation score for a node with respect to more than two network.

Definition 3. *(d, c, k) conservation node:*
Let k satisfy $1 \leq k \leq \binom{N}{2}$. A node v is (d, c, k) conserved with respect to the N networks, $G_1, G_2,.., G_N$, if v is (d, c) conserved in at least k out of the $\binom{N}{2}$ networks pairs.

We adjusted the parameters d, c, k to our input graphs, by choosing parameters such that a random node is picked as conserved with probability smaller than p, where p is a pre-defined threshold (usually $p = 0.05$). The rational behind our approach is that the probability of mutations in "more important" parts of the network is smaller (just like for sequences). We filter noise by finding subgraphs that are conserved for sufficiently many pairs (k) of networks. Since every node in the network is a part of a process (*e.g.* a metabolic pathway, or a protein signaling pathway in a protein interaction network), we expect an "important" node to share "important" pathways and thus have a conserved neighborhood, which our definition is supposed to capture.

5 Experimental Results

In this section we describe the results of running our algorithms on the metabolic networks in the KEGG database. First, we describe the phylogenetic trees our method generated (for two different subsets of species), and discuss the similarity of these trees to the common taxonomy [8]. Then, we describe the results of applying our method for finding conserved regions and discuss the biological relevance of the results.

5.1 Phylogenetic Trees

We started with a relatively small subset, containing 19 taxa: 9 eukaryotes, 5 prokaryotes, and 5 archea. We chose species whose networks in KEGG have more than 900 nodes. We generated a distance matrix based on RDL, and finally constructed a tree, using the Phylip [11] implementation of NJ algorithm [27]. The tree with the true edges' length is depicted in figure 1. The resulting tree is reasonably close to the common accepted taxonomy of these species [8]. The five archea, the five prokaryotes, and the nine eukaryotes form a clade each. Within the eukaryotes, the three mammals (rat, mouse, and human) are clustered together. The fruit fly and the worm C. elegance, both from the *Bilateria* super family, are clustered together. The three yeasts (S. Scerevisiae, A. Gossyppi, and S. Pombe) are clustered together. One example of inaccuracy in our tree is the split inside the mammals, putting the human and mouse together and the rat as an outgroup. One possible explanation is that mouse is a much more popular model animal than rat (it indeed have about 30% more nodes in KEGG), consequently its investigated pathways are closer to human and this is reflected in KEGG. The length of the branches are reasonable, compared to analog methods for phylogeny that are based on sequences' compression [3, 21].

In the next step we generated a tree for all the 194 networks having more than 100 nodes in KEGG (KEGG has additional 56 species with smaller metabolic networks). The resulting tree is depicted in figure 2. Of the 194 taxa in the tree 13 are eukaryotes, 17 archea, and 164 are prokaryotes. This subset includes about 50 species with networks of a few hundreds nodes, and about 80 species with thousands nodes, the largest network (for example human or the bacteria Bardyrhizobium Japonicum - a gram negative bacteria that develops a symbiosis with the soybean plant) has more than 3000 nodes. The names of the taxa are their code name in KEGG. We colored eukaryotes blue, archea grin, and prokaryotes red.

All the archea formed a clade and so did the prokaryotes. All the eukaryotes but one, *plasmodium falciparum (pfa)*. Plasmodium is placed among the bacteria. One possible explanation is the loss of genes and mtabolic pathways that plasmodium, the malaria parasite, went through [13, 20]. The dataset we used has two super-families of archea. The first is *Euryarchaeota*, which contains

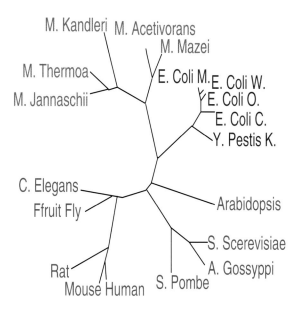

Fig. 1. A small phylogenetic tree, built upon distances of metabolic networks as computed using our method (tree topology and edges' length from NJ algorithm)

the species *pab, pho, hal, mja, afu, hma, pto, mth, tac, tyo,* and *mma.* The other is *Crenachaeota,* containing the species *pai, sto, sso, ape.* The only archea that "jumps family" from the second super family to the first is *Pyrobaculum aerophilum (pai),* which an extremely thermoacidophilic anaerobic taxa [1]. The partitioning within the eukaryotes kingdom is similar to its partition in the tree for the small dataset (figure 1). Most of the prokaryotes families are clustered together: For example the gamma proteobacteria *vvu, vvy, vpa, ppr, vch, son* form a clade. Most of the alpha bacteria are clustered together: *Mlo, Sme, Atu, Atc, Bme, Bms, Bja, Rpa,* and *Sil.* With the exception of *Ehrlichnia ruminantium Welgevonden (Eru)* that joined to the malaria parasite *pfa,* and of Caulobacter Crescentus *(ccr)* that is close (few splits away) but not in the same main cluster alpha bacteria. The two *Bartonella Bhe* and *Bqu* are clustered together, *Zmo* and *gox* are clustered close together but not in the main cluster of alpha bacteria. Considering the large variability in the sizes of the networks and the noisy inputs, we view the results as very good.

5.2 Conserved Regions in Metabolic Networks

In this section we describe the results of our algorithm for finding conserved regions on few dataset. The first contains two species: A bacteria and human, the second contains nine eukaryotes, and the last dataset has ten species, including four eukaryotes, three prokaryotes, and three archea. We also discuss another dataset of three species (Human, E. Coli and yeast) whose their pathways in KEGG are known to be constructed independently. For a lack of space we

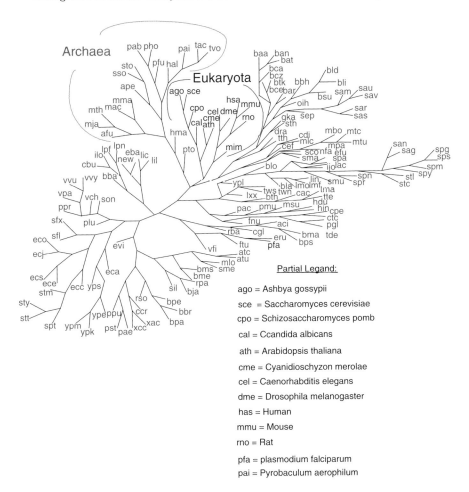

Fig. 2. Phylogenetic tree for 194 based on metabolic networks, all with more than 100 nodes in KEGG

describe here only the stoichiometric formulas of the compounds and very small fraction of the graphs we got, full details of the compounds can be found in KEGG. We describe here few of the subgraphs we found in the results conserved subgraphs. Note that even the relatively short subgraphs described here have since by our definition they are surrounded by a relatively conserved neighborhood.

Our first set contains two very far species: Human and the Gamma Enterobacteria - Yersinia Pestis KIM. Since these two species were split billions of years ago, we expect that the conserved regions found are common to many other taxa. The thresholds to our algorithm was diameter d of 20 nodes, and relative description score 0.9. From our experience, a threshold of 0.9 or lower is fairly strict.

KEGG's metabolic network of human includes more than 3000 nodes, while the metabolic network of the bacteria includes more than 2000 nodes. The resulting conserved networks includes 160 nodes, that are common to the two species. We describe here few of many other results we found: One of long simple paths in the conserved graph represents the metabolic pathway $C_{10}H_{14}N_2O_5$ ($C00214$) $\rightarrow$ $C_{10}H_{15}N_2O_8P$ ($C00364$) $\leftrightarrow$ $C_{10}H_{16}N_2O_{11}P_2$ ($C00363$) $\leftrightarrow$ $C_{10}H_{17}N_2O_{14}P_3$ ($C00459$), which is a part of the pyrimidine metabolism [14]. It includes the last four nodes at the end of the pathway Pyrimidine synthesis. Pyrimidine are the nucleotides T and C, which are building blocks of DNA.

Another simple path of length four represent the sub metabolic pathway $C_5H_{11}O_8P$ ($C00620$) $\leftrightarrow$ $C_5H_{11}O_8P$ ($C00117$) $\rightarrow$ $C_5H_{13}O_{14}P_3$ ($C00119$) $\leftarrow$ $C_{27}H_{46}O_3$ ($C01151$). This is a part of the the pentose phosphate pathway [24]. One of the functions of this anabolic pathway is to utilizes the 6 carbons of glucose to generate 5 carbon sugars, necessary for the synthesis of nucleotides and nucleic acids. This pathway is also part of purine synthesis metabolism, again - one of the building blocks of DNA.

In the next stage we checked for conserved regions in nine Eukaryotes. We chose Eukaryotes with networks larger than 2000 nodes in KEGG. We generated the $(20, 0.7, 6)$ conserved graph for this set of species.

The resulting nine conserved metabolic networks includes between 84 to 106 nodes, while each of the input networks has more than 2000 nodes. We describe here few of the results we found, some ultra conserved regions: The first subgraph $C_6H_9NO_2S_2R_2$ ($C00342$) $\leftrightarrow$ $C_6H_7NO_2S_2R_2$ ($C00343$) is shared by all nine subnetworks. It is part of the pyrimidine synthesis metabolism.

The second pathway is part of the Riboflavin (the left node in the pathway) synthesis metabolism: $C_{27}H_{33}N_9O_{15}P_2$ ($C00016$) $\leftrightarrow$ $C_{17}H_{21}N_4O_9P$ ($C00061$) $\leftrightarrow$ $C_{17}H_{20}N_4O_6$ ($C00255$) Riboflavin is a vitamin that supports energy meta bolism and biosynthesis of a number of essential compounds in eukaryotes, such as human, mouse, fruit fly, rat, S. Cerevisiae, and more [17]. The following ultra conserved subgraph is part of the Cysteine synthesis metabolism:

$C_6H_{12}N_2O_4S_2$ ($C00491$) $\leftrightarrow$ $C_3H_7NO_2S_2$ ($C01962$). Cysteine (the right node in the pathway above) is an amino acid with many important physiological functions in eukaryotes. It is part of Glutathione and is a precursor in its synthesis, which is found in almost all the eukaryotes tissues and has many functions such as activating certain enzymes, and degrading toxic compounds and chemical that contain oxygen.

The last dataset we includs four eukaryotes, three archea, and three bacteria. From each class, we chose species with a large number of nodes in KEGG, the input networks include between 1500 and 3000 nodes. We generated the $(20, 0.7, 6)$ conserved graph for this set of species. The resulting ten conserved metabolic networks include between 58 to 93 nodes. We describe here few of the interesting results. We found a ultra conserved sub-networks, related to nucleotides metabolism, this is the same part of the pyrimidine synthesis metabolism described above. Another path is part of the Bile acid biosynthesis metabolism: $C_{27}H_{48}N_2O_3$ ($C05444$) $\leftrightarrow$ $C_{27}H_{46}O_3$ ($C05445$). Bile acid is essential for fat digestion, and for eliminating wastes from the body. It is also generated by bacteria in the intestine [15].

An unexpected ultra conserved path, the subnetwork $C_2Cl(4)$ $(C06789)$ $\rightarrow$ C_2HCl_3 $(C06790)$ $\rightarrow (C_2H_2Cl_2$ $(C06791), C_2H_2Cl_2$ $(C06792)) \rightarrow C_2H_3Cl$ $(C06793)$ is the first part of the Tetrachloroethene degradation pathway. Tetrachloroethene is a toxin (also known as PCE). Different organisms have developed different processes for degrading PCE [2, 10, 7]. However, the part of this pathway we find here is shared by to many species (and in nine out of ten species in our dataset).

There are few species whose pathway in KEGG were reconstructed independently. Three such species are Human, E. Coli, S. cerevisiae (yeast). We implemented our method for finding conserved regions on these three species which have between 2000 to 3000 nodes in KEGG. We generated the $(20, 0.9, 3)$ conserved graph for this set of species. The conserved graphs of the Human, E. Coli, S. cerevisiae respectively included 79, 79, and 101 nodes respectively. Major fraction of the pathways found for other sets of species are also found here. One such example is the sub-graph of Pyrimidine synthesis.

In all the above results we noticed that conserved node, $i.$ $e.$ nodes that are part of the plotted resulting graphs, tend to be with a relative high in- and out-degrees, $i.$ $e.$ at least four, in the original networks. Note that in our graph representation of metabolic networks the edges (enzymes names) were unlabelled. However, in the case of the conserved sub-graphs described here the edges were also conserved.

5.3   Conserved Regions in Protein Interaction Networks

In addition to the metabolic networks, we have preliminary results on finding conserved regions in two protein interaction networks. In this subsection we report an initial study of finding conserved regions in the protein interaction networks of yeast and drosophila (7164 and 4737 nodes, respectively). We emphasis that these are preliminary results, which mainly establish the application of our approach to networks whose characteristics differ from metabolic networks. In contrast to the metabolic networks, protein interaction networks do not have labels that are shared across species. To identify corresponding nodes, we used Blast results. Two protein were declared identical if the drosophila's protein have the best blast score for the yeast protein, and the score were $< e^{-10}$. We now ran our algorithm. The two nodes with the highest conservation score the first node is the protein $YML064C$ in yeast and his homolog in drosophila (the protein $CG2108$). This protein catalyzed the basic reaction $GTP + H_2O \rightarrow GDP + phosphate$, and as such it is expected a-priori to be conserve. The second protein is $YLR447C$ in yeast (the protein $CG2934$ in drosophila) also involve in "basic" activities such as hydrogen-exporting ATPeas activity, catalyzing the reaction: $ATP + H_2O + H^+(in) \rightarrow ADP + phosphate + H^+(out)$.

6   Concluding Remarks and Further Research

We presented a novel method for comparing cellular-biological networks and finding conserved regions in two or more such networks. We implemented our method, and produced a number of preliminary biological results. It is clear that networks contains information, which is different than sequence information, and also differ from information in gene content. This work opens up a number of algorithmic and biological questions. The various networks in KEGG were not built independently. This biases the results, especially those of conserved regions. Interestingly, despite this fact, the our results seem surprisingly good.

The experimental work here concentrated mainly on metabolic networks taken from the KEGG database. Of course, there is no reason to consider only KEGG, and only metabolic networks. More importantly, we plan to examine our methods on more protein interaction networks, regulatory networks, and possibly a mixture thereof.

Our representation of the networks followed that of Jeong *et al.* [16] and ignored the edge labels (enzyme names). As shown in the conserved regions, identical node labels (substrates) seem to determine the enzymes involved. Yet, it is desireable to include edge labels explicitly. Indeed, the RDL approach allows such modification at relative ease. A more meaningful extension is to consider labels not just as equal or unequal. A continuous scale of similarity, as implied for example from the chemical description of substrates, can be used. Different representations of the directed graph (*e.g.* children instead of parents) are also possible. Other algorithms, based on variants of labeled subgraph isomorphism, can be considered as well. However, their efficiency should be carefully analyzed.

When dealing with biological networks, we should always keep in mind that they are still in their infancy. They are noisy due to experimental conditions, and they are partial, due to budgetary limitations and biases of the researchers. Thus the precision of the results is likely to evolve and improve, as more reliable data are gathered.

Finally, it will be of interest to combine different sources of data, for example sequence data (proteins and genes) and network data, to construct trees and find conserved regions. Of special interest are regions where the signals from the various sources are either coherent or incoherent. Of course, this work is only a first step, and calls for possible improvements.

Acknowledgements

We would like to thank Nadir Ashkenazi, Nathan Nelson, Eytan Ruppin, Roded Sharan, and Tomer Shlomi for helpful discussions.

References

1. S. Afshar, E. Johnson, S. Viries, and I. Schroder. Properties of a thermostable nitrate reductase from the hyperthermophilic archaeon pyrobaculum aerophilum. *Journal of Bacteriology*, pages 5491–5495, 2001.

2. D. M. Bagly and J. M. Gossett. Tetrachloroethene transformation to trichloroethene and cis-1,2-dichloroethene by sulfate-reducing enrichment cultures. *Appl. Environ. Microbio*, 56(8), 1990.

3. D. Burstein, I. Ulitsky, T. Tuller, and B. Chor. Information theoretic approaches to whole genome phylogenies. *RECOMB05*, pages 296–310, 2005.

4. X. Chen, S. Kwong, and M. Li. A compression algorithm for dna sequences and its applications in genome comparison. *RECOMB*, pages 107–117, 2000.

5. M. J. Chung. $o(n^{2.5})$ time algorithms for subgraph homeomorphisem problem on trees. *J. Algorithms*, 8:106–112, 1987.

6. T. H. Cormen, C. E. Leiserson, and R. L. Rivest. *Introduction to Algorithms*. The MIT Press, Cambridge, 1990.

7. J. Damborsky. Ttetrachloroethene–dehalogenating bacteria. *Folia. Microbiol*, 44(3), 1999.

8. NCBI Taxonomy Database. http://www.ncbi.nlm.nih.gov/entrez/linkout/tutorial/taxtour.html.

9. I. Elias and J. Lagergren. Fast neighbor joining. In *Proc. of the 32nd International Colloquium on Automata, Languages and Programming (ICALP'05)*, volume 3580 of *Lecture Notes in Computer Science*, pages 1263–1274. Springer-Verlag, July 2005.

10. B. Z. Fathepure and S. A. Boyd. Dependence of tetrachloroethylene dechlorination on methanogenic substrate consumption by methanosarcina sp. strain dcm. *Appl. Environ. Microbio*, 54(12), 1988.

11. J. Felsenstein. Phylip (phylogeny inference package) version 3.5c. *Distributed by the author. Department of Genetics, University of Washington, Seattle*, 1993.

12. M. R. Garey and D. S. Johnson. *Computers and Intractability*. Bell Telephone Laboratories, 1979.

13. R. Hernandez-Rivas, D. Mattei, Y. Sterkers, D. S. Peterson, T. E. Wellems, and A. Acherf. Expressed var genes are found in plasmodium falciparum subtelomeric regions. *Mol. Cell. Biol*, pages 604–611, 1997.

14. P. A. Hoffee and M. E. Jones. *Purin and pyrimidine nucleotide metabolism*. Academic Press, 1978.

15. A. F. Hofmann. Bile acids: The good, the bad, and the ugly. *News Physiol. Sci.*, 14, 1999.

16. H. Jeong, B. Tombor, R. Albert, Z. N. Oltvai, and A. L. Barabasi. The large-scale organization of metabolic networks. *Nature*, 407(6804):651–4, 2000.

17. M. Kanehisa and S. Goto. Kegg: Kyoto encyclopedia of genes and genomes. *Nucleic Acids Res*, 28:27–30, 2000.

18. B. P. Kelley, R. Sharan, R. M. Karp, T. Sittler, D. E. Root, B. R. Stock-well, and T. Ideker. Conserved pathways within bacteria and yeast as revealed by global protein network alignment. *PNAS*, 100:11394–11399, 2003.

19. M. Koyuturk, A. Grama, and W. Szpankowski. An efficient algorithm for detecting frequent subgraphs in biological networks. *Bioinformatics*, 20:i200–i207, 2004.

20. D. M. Krylov, Y. I. Wolf, I. B. Rogozin, and E. V. Koonim. Gene loss, protein sequence divergence, gene dispensability, expression level, and interactivity are correlated in eukaryotic evolution. *Genome Research*, 13:2229–2235, 2003.

21. M. Li, J. Badger, X. Chen, S. Kwong, P. Kearney, and H. Zhang. An information-based sequence distance and its application to whole mitochondrial genome phylogeny. *Bioinformatics*, 17(2):149–154, 2001.

22. M. Li and P. Vitanyi. *An Introduction to Kolmogorov Complexity and Its Applications*. Springer Verlag, 2005.

23. D. W. Matula. Subtree isomorphisem in $o(n^{5/2})$ time. *Ann. Discrete Math*, 2:91–106, 1978.

24. G. Michal. *Biochemical pathways*. John Wiley and Sons. Inc, 1999.

25. H. Ogata, W. Fujibuchi, S. Goto, and M. Kanehisa. A heuristic graph comparison algorithm and its application to detect functionally related enzyme clusters. *Nucleic Acids Research*, 28:4021–4028, 2000.
26. R. Y. Pinter, O. Rokhlenko, E. Yeger-Lotem, and M. Ziv-Ukelson. Alignment of metabolic pathways. *Bioinformatics*, 2005.
27. N. Saitou and M. Nei. The neighbor-joining method: a new method for reconstructing phylogenetic trees. *Mol. Biol. Evol.*, 4:406–425, 1987.
28. F. Schreiber. Comparison of metabolic pathways using constraint graph drawing. *Proceedings of the Asia-Pacific Bioinformatics Conference (APBC'03), Conferences in Research and Practice in Information Technology*, 19:105–110, 2003.
29. R. Sharan, S. Suthram, R. M. Kelley, T. Kuhn, S. McCuine, P. Uetz, T. Sittler, R. M. Karp, and T. Ideker. Conserved patterns of protein interaction in multiple species. *Proc. Natl. Acad. Sci. USA 102*, pages 1974 – 1979, 2005.
30. M. Sipser. *Introduction to the Theory of Computation*. PSW Publishing Company, 1997.
31. Y. Tohsato, H. Matsuda, and A. Hashimoto. A multiple alignment algorithm for metabolic pathway analysis using enzyme hierarchy. *Proc. 8th International Conference on Inelligent Systems for Molecular Biology (ISMB2000)*, pages 376–383, 2000.
32. U. Zwick. All pairs shortest paths using bridging sets and rectangular matrix multiplication, 2000.

Assessing Significance of Connectivity and Conservation in Protein Interaction Networks

Mehmet Koyutürk, Ananth Grama, and Wojciech Szpankowski

Dept. of Computer Sciences, Purdue University, West Lafayette, IN 47907
{koyuturk, ayg, spa}@cs.purdue.edu

Abstract. Computational and comparative analysis of protein-protein interaction (PPI) networks enable understanding of the modular organization of the cell through identification of functional modules and protein complexes. These analysis techniques generally rely on topological features such as connectedness, based on the premise that functionally related proteins are likely to interact densely and that these interactions follow similar evolutionary trajectories. Significant recent work in our lab, and in other labs has focused on efficient algorithms for identification of modules and their conservation. Application of these methods to a variety of networks has yielded novel biological insights. In spite of algorithmic advances, development of a comprehensive infrastructure for interaction databases is in relative infancy compared to corresponding sequence analysis tools such as BLAST and CLUSTAL. One critical component of this infrastructure is a measure of the statistical significance of a match or a dense subcomponent. Corresponding sequence-based measures such as E-values are key components of sequence matching tools. In the absence of an analytical measure, conventional methods rely on computer simulations based on ad-hoc models for quantifying significance. This paper presents the first such effort, to the best of our knowledge, aimed at analytically quantifying statistical significance of dense components and matches in reference model graphs. We consider two reference graph models – a $G(n, p)$ model in which each pair of nodes has an identical likelihood, p, of sharing an edge, and a two-level $G(n, p)$ model, which accounts for high-degree hub nodes generally occurring in PPI networks. We argue that by choosing conservatively the value of p, the $G(n, p)$ model will dominate that of the power-law graph that is often used to model PPI networks. We also propose a method for evaluating statistical significance based on the results derived from this analysis, and demonstrate the use of these measures for assessing significant structures in PPI networks. Experiments performed on a rich collection of PPI networks show that the proposed model provides a reliable means of evaluating statistical significance of dense patterns in these networks.

1 Introduction

Availability of high-throughput methods for identifying protein-protein interactions has resulted in a new generation of extremely valuable data [2, 38]. Effective analysis of the interactome holds the key to functional characterization, phenotypic mapping, and identification of pharmacological targets, among other important tasks. Computational infrastructure for supporting analysis of the interactome is in relative infancy, compared

A. Apostolico et al. (Eds.): RECOMB 2006, LNBI 3909, pp. 45–59, 2006.
© Springer-Verlag Berlin Heidelberg 2006

to its sequence counterparts [36]. A large body of work on computational analysis of these graphs has focused on identification of dense components (proteins that densely interact with each other) [3, 6, 19, 20, 23, 28]. These methods are based on the premise that functionally related proteins generally manifest themselves as dense components in the network [33]. The hypothesis that proteins performing together a particular cellular function are expected to be conserved across several species along with their interactions is also used to guide the process of identifying conserved networks across species. Based on this observation, PPI network alignment methods superpose PPI networks that belong to different species and search for connected, dense, or heavy subgraphs on these superposed graphs [11, 15, 16, 17, 26, 27].

There are two critical aspects of identifying meaningful structures in data – the algorithm for the identification and a method for scoring an identified pattern. In this context, the score of a pattern corresponds to its significance. A score is generally computed with respect to a reference model – i.e., given a pattern and a reference model, how likely it is to observe the pattern in the reference model that often is a probabilistic measure for scoring patterns. The less likely such an occurrence is in the reference model, the more interesting it is, since it represents a significant deviation from the reference (expected) behavior. One such score, in the context of sequences is the E-value returned by BLAST matches [37]. This score broadly corresponds to the likelihood that a match between two sequences is generated by a random process. The lower this value, the more meaningful the match. It is very common in a variety of applications to use a threshold on E-values to identify homologies across sequences. It is reasonable to credit E-value as one of the key ingredients of the success of sequence matching algorithms and software.

While significant progress has been made towards developing algorithms on graphs for identifying patterns (motifs, dense components), conservation, alignment, and related problems, analytical methods for quantifying the significance of such patterns are limited. In a related effort, Itzkovitz et al. [12] analyze the expected number of occurrences of certain *topological motifs* in a variety of random networks. On the other hand, existing algorithms for detecting generalized patterns generally adopt simple ad-hoc measures (such as frequency or relative density) [3, 15], compute z-scores for *the* observed pattern based on simplifying assumptions [16, 26, 27], or rely on Monte-Carlo simulations [26] to assess the significance of identified patterns. This paper represents the first such effort at analytically quantifying the statistical significance of the *existence* of a pattern of observed property, with respect to a reference model. Specifically, it presents a framework for analyzing the occurrence of dense patterns in randomly generated graph-structured data (based on the underlying model) with a view to assessing the significance of a pattern based on the statistical relationship between subgraph density and size.

The selection of an appropriate reference model for data and the method of scoring a pattern or match, are important aspects of quantifying statistical significance. Using a reference model that fits the data very closely makes it more likely that an experimentally observed biologically significant pattern is generated by a random process drawing data from this model. Conversely, a reference model that is sufficiently distinct from observed data is likely to tag most patterns as being significant. Clearly, neither extreme is desirable for good coverage and accuracy. In this paper, we consider two reference models (i) a $G(n, p)$ model of a graph with n nodes, where each pair of nodes has an identical probability, p, of sharing an edge, and (ii) a two level $G(n, p)$ model in which the graph is modeled as two separate $G(n, p)$ graphs with intervening edges. The latter model captures the heavy nodes corresponding to hub proteins. For these models, we analytically quantify the behavior of the largest dense subgraph and use this to derive a measure of significance. We show that a simple $G(n, p)$ model can be used to assess the significance of dense patterns in graphs with arbitrary degree distribution, with a conservative adjustment of parameters so that the model stochastically dominates a graph generated according to a given distribution. In particular, by choosing p to be maximal we assure that the largest dense subgraph in our $G(n, p)$ model stochastically dominates that of a power-law graph. Our two-level $G(n, p)$ model is devised to mirror key properties of the underlying topology of PPI graphs, and consequently yields a more conservative estimate of significance. Finally, we show how existing graph clustering algorithms [10] can be modified to incorporate statistical significance in identification of dense patterns. We also generalize these results and methods to the comparative analysis of PPI networks and show how the significance of a match between two networks can be quantified in terms of the significance of the corresponding dense component in a suitable specified product graph.

Our analytical results are supported by extensive experimental results on a large collection of PPI networks derived from BIND [2] and DIP [38]. These results demonstrate that the proposed model and subsequent analysis provide reliable means for evaluating the statistical significance of highly connected and conserved patterns in PPI networks. The framework proposed here can also be extended to include more general networks that capture the degree distribution of PPI networks more accurately, namely power-law [35, 39], geometric [21], or exponential [8] degree distributions.

2   Probabilistic Analysis of Dense Subgraphs

Since proteins that are part of a functional module are likely to densely interact with each other while being somewhat isolated from the rest of the network [33], many commonly used methods focus on discovering dense regions of the network for identification of functional modules or protein complexes [3, 6, 19, 23, 28]. Subgraph density is also central for many algorithms that target identification of conserved modules and complexes [11, 16, 26]. In order to assess the statistical significance of such dense patterns, we analyze the distribution of the largest "dense" subgraph generated by an underlying reference model. Using this distribution, we estimate the probability that an experimentally observed pattern will occur in the network by chance. The reference model must mirror the basic characteristics of experimentally observed networks in order to capture the underlying biological process correctly, while being simple enough to facilitate feasible theoretical and computational analysis.

2.1   Modeling PPI Networks

With the increasing availability of high-throughput interaction data, there has been significant effort on modeling PPI networks. The key observation on these networks is that a few central proteins interact with many proteins, while most proteins in the network have few interacting partners [13, 22]. A commonly accepted model that confirms this observation is based on power-law degree distribution [4, 34, 35, 39]. In this model, the number of nodes in the network that have d neighbors is proportional to $d^{-\gamma}$, where γ is a network-specific parameter. It has also been shown that there exist networks that do not possess a power-law degree distribution [9, 32]. In this respect, alternative models that are based on geometric [21] or exponential [8] degree distribution have been also proposed.

While assessing the statistical significance of identified patterns, existing methods that target identification of highly connected or conserved patterns in PPI networks generally rely on the assumption that the interactions in the network are independent of each other [14, 16, 26]. Since degree distribution is critical for generation of interesting patterns, these methods estimate the probability of each interaction based on the degree distribution of the underlying network. These probabilities can be estimated computationally by generating many random graphs with the same degree distribution via repeated edge swaps and counting the occurrence of each edge in this large collection of random graphs [26]. Alternately, they can be estimated analytically, by relying on a simple random graph model that is based on a given degree distribution [7]. In this model, each node $u \in V(G)$ of graph $G = (V, E)$ is associated with expected degree d_u and the probability of existence of an edge between u and v is defined as $P(uv \in E(G)) = d_u d_v / \sum_{u \in V(G)} d(u)$. In order for this function to be a well-defined probability measure for simple graphs, we must have $d_{\max}^2 \leq \sum_{u \in V(G)} d(u)$, where $d_{\max} = \max_{u \in V(G)} d_u$. However, available protein interaction data generally does not confirm this assumption. For example, based on the PPI networks we derive from BIND [2] and DIP [38] databases, yeast *Jsn1* protein has 298 interacting partners, while the total number of interactions in the *S. cerevisiae* PPI network is 18193. Such problems complicate the analysis of the significance of certain structures for models that are based on arbitrary degree distribution.

While models that assume power-law [35, 39], geometric [21], or exponential [8] degree distributions may capture the topological characteristics of PPI networks accurately, they require more involved analysis and may also require extensive computation for assessment of significance. To the best of our knowledge, the distribution of dense subgraphs, even maximum clique, which forms a special case of this problem, has not been studied for power-law graphs. In this paper, we first build a framework on the simple and well-studied $G(n, p)$ model and attempt to generalize our results to more complicated models that assume heterogeneous degree distribution.

2.2   Largest Dense Subgraph

Given graph G, let $F(U) \subseteq E(G)$ be the set of edges in the subgraph induced by node subset $U \subseteq V(G)$. The density of this subgraph is defined as $\delta(U) = |F(U)|/|U|^2$. Note here that we assume directed edges and allow self-loops for simplicity. PPI networks are undirected graphs and they contain self-loops in general, but any undirected network can be easily modeled by a directed graph and this does not affect the asymptotic correctness of the results. We define a ρ-dense subgraph to be one with density *larger* than pre-defined threshold ρ, i.e., U induces a ρ-dense subgraph if $F(U) \geq$

$\rho|U|^2$, where $\rho > p$. For any ρ, we are interested in the number of nodes in the largest ρ-dense subgraph. This is because any ρ-dense subgraph in the observed PPI network with size larger than this value will be "unusual", *i.e.*, statistically significant. Note that maximum clique is a special case of this problem with $\rho = 1$.

We first analyze the behavior of the largest dense subgraph for the $G(n, p)$ model of random graphs. We subsequently generalize these results to the piecewise degree distribution model in which there are two different probabilities of generating edges. In the $G(n, p)$ model, a graph G contains n nodes and each edge occurs independently with probability p.

Let random variable R_ρ be the size of the maximum subset of vertices that induce a ρ-dense subgraph, *i.e.*,

$$R_\rho = \max_{U \subseteq V(G):\delta(U)\geq\rho} |U|. \tag{1}$$

The behavior of R_1, which corresponds to maximum clique, is well studied on $G(n, p)$ model and its typical value is shown to be $O(\log_{1/p} n)$ [5]. In the following theorem, we present a general result for the typical value of R_ρ for any ρ.

Theorem 1. *If G is a random graph with n vertices, where every edge exists with probability p, then*

$$\lim_{n\to\infty} \frac{R_\rho}{\log n} = \frac{1}{\kappa(p, \rho)} \qquad (pr.), \tag{2}$$

where

$$\kappa(p, \rho) = -H_p(\rho) = \rho \log \frac{\rho}{p} + (1 - \rho) \log \frac{1 - \rho}{1 - p}. \tag{3}$$

Here, $H_p(\rho)$ denotes weighted entropy. More precisely,

$$P(R_\rho \geq r_0) \leq O\left(\frac{\log n}{n^{1/\kappa(p,\rho)}}\right), \tag{4}$$

where

$$r_0 = \frac{\log n - \log\log n + \log \kappa(p, \rho) - \log e + 1}{\kappa(p, \rho)} \tag{5}$$

for large n.

The proof of this theorem is presented in Section 4. Observe that, if n is large enough, the probability that a dense subgraph of size r_0 exists in the subgraph is very small. Consequently, r_0 may provide a threshold for deciding whether an observed dense pattern is statistically significant or not.

For a graph of arbitrary distribution, let d_{max} denote the maximum expected degree as defined in Section 2.1. Let $p_{max} = d_{max}/n$. It can be easily shown that the largest dense subgraph in the $G(n, p)$ graph with $p = p_{max}$ stochastically dominates that in the random graph generated according to the given degree distribution (e.g., power-law graphs). Hence, by estimating the edge probability conservatively, we can use the above result to determine whether a dense subgraph identified in a PPI network of arbitrary degree distribution is statistically significant. Moreover, the above result also provides a means for quantifying the significance of an observed dense subgraph. For a subgraph with size $\hat{r} > r_0$ and density $\hat{\rho}$, let $\epsilon = \frac{\hat{r} - \log n / \kappa(\hat{\rho}, p)}{\log n / \kappa(\hat{\rho}, p)}$. Then, as we show (cf. (12)) in the proof of Theorem 1 in Section 4, the probability of observing this subgraph in a graph generated according to the reference model is bounded by

$$P(R_{\hat{\rho}} \geq (1 + \epsilon) \log n / \kappa(\hat{\rho}, p)) \leq \frac{\sqrt{1 - \rho}}{2\pi\sqrt{\rho}} \frac{(1 + \epsilon) \log n}{n^{\epsilon(1+\epsilon) \log n / \kappa(\hat{\rho}, p)}}. \tag{6}$$

While these results on $G(n, p)$ model provide a simple yet effective way of assessing statistical significance of dense subgraphs, we extend our analysis to a more complicated model, which takes into account the degree distribution to capture the topology of the PPI networks more accurately.

2.3   Piecewise Degree Distribution Model

In the piecewise degree distribution model, nodes of the graph are divided into two classes, namely high-degree and low-degree nodes. More precisely, we define random graph G with node set $V(G)$ that is composed of two disjoint subsets $V_h \subset V(G)$ and $V_l = V(G) \setminus V_h$, where $n_h = |V_h| \ll |V_l| = n_l$ and $n_h + n_l = n = |V(G)|$. In the reference graph, the probability of an edge is defined based on the classes of its incident nodes as:

$$P(uv \in E(G)) = \begin{cases} p_h & \text{if } u, v \in V_h \\ p_l & \text{if } u, v \in V_l \\ p_b & \text{if } u \in V_h, v \in V_l \text{ or } u \in V_l, v \in V_h \end{cases} \tag{7}$$

Here, $p_l < p_b < p_h$. This model captures the key lethality and centrality properties of PPI networks in the sense that a few nodes are highly connected while most nodes in the network have low degree [13, 22]. Observe that, under this model, G can be viewed as a superposition of three random graphs G_l, G_h, and G_b. Here, G_h and G_l are $G(n, p)$ graphs with parameters (n_h, p_h) and (n_l, p_l), respectively. G_b, on the other hand, is a random bipartite graph with node sets V_l, V_h, where each edge occurs with probability p_b. Hence, we have $E(G) = E(G_l) \cup E(G_h) \cup E(G_b)$. This facilitates direct employment of the results in the previous section for analyzing graphs with piecewise degree distribution.

Note that the random graph model described above can be generalized to an arbitrary number of node classes to capture the underlying degree distribution more accurately. Indeed, with appropriate adjustment of some parameters, this model will approximate power-law or exponential degree distribution at the limit with increasing number of node classes. In order to get a better fit, we need to introduce three or four classes in our piecewise model.

We now show that the high-degree nodes in the piecewise degree distribution model contribute a constant factor to the typical size of the largest dense subgraph as long as n_h is bounded by a constant.

Theorem 2. *Let G be a random graph with piecewise degree distribution, as defined by (7). If $n_h = O(1)$, then*

$$P(R_\rho \geq r_1) \leq O\left(\frac{\log n}{n^{1/\kappa(p_l, \rho)}}\right), \tag{8}$$

where

$$r_1 = \frac{\log n - \log\log n + 2n_h \log B + \log \kappa(p_l, \rho) - \log e + 1}{\kappa(p_l, \rho)} \quad (9)$$

and $B = \frac{p_b q_l}{p_l} + q_b$, *where* $q_b = 1 - p_b$ *and* $q_l = 1 - p_l$.

Note that the above result is based on asymptotic behavior of r_1, hence the $\log n$ term dominates as $n \to \infty$. However, if n is not large enough, the $2n_h \log B$ term may cause over-estimation of the critical value of the largest dense subgraph. Therefore, the application of this theorem is limited for smaller n and the choice of n_h is critical.

A heuristic approach for estimating n_h is as follows. Assume that the underlying graph is generated by a power-law degree distribution, where the number of nodes with degree d is given by $nd^{-\gamma}/\zeta(\gamma)$ [1]. Here, $\zeta(.)$ denotes the Riemann zeta-function. If we divide the nodes of this graph into two classes where high-degree nodes are those with degree $d \geq (n/\zeta(\gamma))^{1/\gamma}$ so that the expected number of nodes with degree d is at most one, then $n_h = \sum_{d=(n/\zeta(\gamma))^{1/\gamma}}^{\infty} nd^{-\gamma}/\zeta(\gamma)$ is bounded, provided the above series converges.

2.4 Identifying Significant Dense Subgraphs

We use the above results to modify an existing state-of-the-art graph clustering algorithm, HCS [10], in order to incorporate statistical significance in identification of interesting dense subgraphs. HCS is a recursive algorithm that is based on decomposing the graph into dense subgraphs by repeated application of min-cut partitioning. The density of any subgraph found in this recursive decomposition is compared with a pre-defined density threshold. If the subgraph is dense enough, it is reported as a highly-connected cluster of nodes, else it is partitioned again. While this algorithm provides a strong heuristic that is well suited to the identification of densely interacting proteins in PPI networks [20], the selection of density threshold poses an important problem. In other words, it is hard to provide a biologically justifiable answer to the question "How dense must a subnetwork of a PPI network be to be considered biologically interesting?". Our framework provides an answer to this question from a statistical point of view by establishing the relationship between subgraph size and density as a stopping criterion for the algorithm.

For any subgraph encountered during the course of the algorithm, we estimate the critical size of the subgraph to be considered interesting by plugging in its density in (5) or (9). If the size of the subgraph is larger than this probabilistic upper-bound, then we report the subgraph as being statistically significant. Otherwise, we continue partitioning the graph. Note that this algorithm only identifies disjoint subgraphs, but can be easily extended to obtain overlapping dense subgraphs by greedily growing the resulting graphs until significance is lost. The Cytoscape [25] plug-in implementing the modified HCS algorithm is provided as open source at http://www.cs.purdue.edu/homes/koyuturk/sds/.

2.5 Conservation of Dense Subgraphs

Comparative methods that target identification of conserved subnets in PPI networks induce a cross-product or superposition of several networks in which each node

or preprocess the PPI networks to standardize the topological representation of protein complexes in the network model.

The behavior of largest dense subgraph size with respect to density threshold is shown in Figure 2 for *S. Cerevisiae* and *H. Sapiens* PPI networks and their intersection. It is evident from the figure that the observed size of the largest dense subgraph follows a similar trajectory with the theoretical values estimated by both models. Moreover, in both networks, the largest dense subgraph turns out to be significant for a wide range of density thresholds. For lower values of ρ, the observed subgraphs are either not significant or they are marginally significant. This is a desirable characteristic of significance-based analysis since identification of very large sparse subgraphs should be avoided while searching for dense patterns in PPI networks. Observing that the $G(n, p)$ model becomes more conservative than the piecewise degree distribution model for lower values of ρ, we conclude that this model may facilitate fine-grain analysis of modularity in PPI networks.

We implement the modified HCS heuristic described in Section 2.4 using a simple min-cut algorithm [29]. A selection of most significant dense subgraphs discovered on *S. cerevisiae* PPI network are shown in Table 1. In the table, as well as the size, density and significance of identified subgraphs, we list the GO annotations that are significantly shared by most of the proteins in the dense subgraph. The GO annotations may refer to function [F], process [P], or component [C]. The p-value for the annotations is estimated as the probability of observing at least the same number of proteins with the corresponding annotation if the proteins were selected uniformly at random.

Table 1. Seven most significant dense subgraphs identified in *S. cerevisiae* PPI network by the modified HCS algorithm and the corresponding functions, processes, and compartments with significant enrichment according to the GO annotation of the proteins in the subnet

# Prot	# Int	$p <$	GO Annotation
24	165	10^{-175}	[C] nucleolus (54%, $p < 10^{-7}$)
20	138	10^{-187}	[P] ubiquitin-dependent protein catabolism (80%, $p < 10^{-21}$)
			[F] endopeptidase activity (50%, $p < 10^{-11}$)
			[C] proteasome regulatory particle, lid subcomplex (40%, $p < 10^{-12}$)
16	104	10^{-174}	[P] histone acetylation (62%, $p < 10^{-15}$)
			[C] SAGA complex (56%, $p < 10^{-15}$)
			[P] chromatin modification (56%, $p < 10^{-14}$)
15	90	10^{-145}	[F] RNA binding (80%, $p < 10^{-12}$)
			[C] mRNA cleavage & polyadenylation spec fac comp (80%, $p < 10^{-24}$)
			[P] mRNA polyadenylylation (80%, $p < 10^{-21}$)
14	79	10^{-128}	[P] mRNA catabolism (71%, $p < 10^{-16}$)
			[F] RNA binding (64%, $p < 10^{-6}$)
			[P] nuclear mRNA splicing, via spliceosome (57%, $p < 10^{-7}$)
10	45	10^{-200}	[P] ER to Golgi transport (90%, $p < 10^{-14}$)
			[C] TRAPP complex (90%, $p < 10^{-23}$)
7	20	10^{-30}	[C] mitochondrial outer memb transloc comp (100%, $p < 10^{-20}$)
			[F] protein transporter activity (100%, $p < 10^{-14}$)
			[P] mitochondrial matrix protein import (100%, $p < 10^{-16}$)

This probability is upper-bounded using Chernoff's bound for the binomial tail; namely $P(S_{r,\hat{p}} \geq k) \leq \exp\{rH_{\hat{p}}(k/r)\}$, where r denotes the number of proteins in the subgraph, k denotes the number of proteins among these with the particular annotation, and $\hat{p}$ is the estimated probability that a random protein will carry this annotation [5].

For most of the significant dense subgraphs, most of the proteins that induce the subgraph are involved in the same cellular process. As an extreme case, the algorithm also identifies proteins that share a common function or that are part of a particular complex. For example, the dense subgraph of 7 proteins in the last row corresponds to the mitochondrial outer membrane translocase (TOM) complex, which mediates recognition, unfolding, and translocation of preproteins [18]. On the other hand, some dense subgraphs correspond to proteins that are involved in a range of processes but localize in the same cellular component, such as the largest dense subgraph identified by modified HCS, which contains 24 proteins.

The significant dense subgraphs that are conserved in *S. cerevisiae* and *H. sapiens* PPI networks are shown in Table 2. Most of these dense components are involved in fundamental processes and the proteins that are part of these components share a particular function. Among these, the 7-protein conserved subnet that consists of 6 Exosomal 3'-5' exoribonuclease complex subunits and Succinate dehydrogenase is interesting. As in the case of dense subgraphs in a single network, the conserved dense subgraphs provide an insight on the crosstalk between proteins that perform different functions. For example, the largest conserved subnet of 11 proteins contains Mismatch repair proteins, Replication factor C subunits, and RNA polymerase II transcription initiation/nucleotide excision repair factor TFIIH subunits, which are all involved in DNA

Table 2. Seven most significant conserved dense subgraphs identified in *S. cerevisiae* and *H. sapiens* PPI networks by the modified HCS algorithm and their functional enrichment according to COG functional annotations

# Prot	# Cons Int	$p <$	COG Annotation
10	17	10^{-68}	RNA polymerase (100%)
11	11	10^{-26}	Mismatch repair (33%)
			RNA polymerase II TI/nucleotide excision repair factor TFIIH (33%)
			Replication factor C (22%),
7	7	10^{-25}	Exosomal 3'-5' exoribonuclease complex (86%)
4	4	10^{-24}	Single-stranded DNA-binding replication protein A (50%)
			DNA repair protein (50%)
5	4	10^{-12}	Small nuclear ribonucleoprotein(80%)
			snRNP component (20%)
5	4	10^{-12}	Histone (40%)
			Histone transcription regulator (20%)
			Histone chaperone (20%)
3	3	10^{-9}	Vacuolar sorting protein (33%)
			RNA polymerase II transcription factor complex subunit (33%)
			Uncharacterized conserved protein (33%)

repair. The conserved subnets identified by the modified HCS algorithm are small and appear to be partial, since we employ a strict understanding of conserved interaction here. In particular, limiting the ortholog assignments to proteins that have a COG assignment and considering only matching direct interactions as conserved interactions limits the ability of the algorithm to identify a comprehensive set of conserved dense graphs. Algorithms that rely on sequence alignment scores and consider indirect or probable interactions [17, 26, 27] coupled with adaptation of the statistical framework in this paper have the potential of increasing the coverage of identified patterns, while correctly evaluating the interestingness of observed patterns.

4   Proof of Theorems

In this section we prove Theorems 1 and 2.

Proof 1. We first prove the upper-bound. Let $X_{r,\rho}$ denote the number of subgraphs of size r with density at least ρ, i.e., $X_{r,\rho} = |\{U \subseteq V(G) : |U| = r \wedge |F(U)| \geq \rho r^2\}|$. From first moment method, we obtain $P(R_\rho \geq r) \leq P(X_{r,\rho} \geq 1) \leq \mathbf{E}[X_{r,\rho}]$.

Let Y_r denote the number of edges induced by r vertices. Then, $\mathbf{E}[X_r, \rho] = \binom{n}{r} P(Y_r \geq \rho r^2)$. Moreover, since Y_r is a Binomial r.v. $B(r^2, p)$ and $\rho > p$, we have

$$P(Y_r \geq \rho r^2) \leq (r^2 - \rho r^2) P(Y_r = \rho r^2) \leq \binom{r^2}{\rho r^2}(r^2 - \rho r^2) p^{\rho r^2}(1-p)^{r^2 - \rho r^2}. \tag{10}$$

Hence, we get $P(R_\rho \geq r) \leq \binom{n}{r}\binom{r^2}{\rho r^2}(r^2 - \rho r^2) p^{\rho r^2}(1-p)^{r^2 - \rho r^2}$.

Using Stirling's formula, we find the following asymptotics for $\binom{n}{r}$:

$$\binom{n}{r} \sim \begin{cases} \frac{1}{\sqrt{2\pi r}} \frac{n^r}{r^r} e^{-r} & \text{if } r = o(\sqrt{n}) \\ \frac{1}{\sqrt{2\pi\alpha(1-\alpha)n}} 2^{nH(\alpha)} & \text{if } r = \alpha n \end{cases} \tag{11}$$

where $H(\alpha) = -\alpha \log \alpha - (1-\alpha)\log(1-\alpha)$ denotes the binary entropy.

Let $Q = 1/p^\rho(1-p)^{1-\rho}$. Plugging the above asymptotics into (4), we obtain

$$P(R_\rho \geq r) \leq \frac{r\sqrt{1-\rho}}{2\pi\sqrt{\rho}} \exp_2(-r^2 \log Q + r \log n - r \log r + r^2 H(\rho) - r \log e\}) \tag{12}$$

Defining $\kappa(p, \rho) = \log Q - H(\rho)$, we find $P(R_\rho \geq r_0) \leq \frac{r_0\sqrt{1-\rho}}{2\pi\sqrt{\rho}} \exp_2(f(r_0))$, where $f(r_0) = -r_0(r_0\kappa(p, \rho) - \log n + \log r + \log e)$. Plugging in (5) and working out the algebra, we obtain $f(r_0) = -r_0\left(1 - O\left(\frac{\log\log n}{\log n}\right)\right)$. Hence, $P(R_\rho \geq r_0) \leq O\left(2^{-r_0}\right) = O\left(\frac{\log n}{n^{1/\kappa(p,\rho)}}\right)$. This completes the proof for the upper-bound.

The lower-bound is not of a particular interest in terms of statistical significance, but we provide a sketch of the proof for completeness. By the second moment method [30], we have

$$P(R_\rho < r) \leq P(X_{r,\rho} = 0) \leq \frac{\mathbf{Var}[X_{r,\rho}]}{\mathbf{E}[X_{r,\rho}]^2} = \frac{1}{\mathbf{E}[X_{r,\rho}]} + \frac{\sum_{U_r \neq V_r} \mathbf{Cov}[X_\rho^{U_r}, X_\rho^{V_r}]}{\mathbf{E}[X_{r,\rho}]^2},$$

where $X_\rho^{U_r}$ is the indicator r.v. for the subgraph induced by the vertex set U_r being ρ-dense. Letting $r = (1 - \epsilon) \log n / \kappa(\rho)$, we observe that $\frac{1}{\mathbf{E}[X_{r,\rho}]} \to 0$ as $n \to \infty$. We split the sum $\sum_{U_r, V_r} \mathbf{Cov}[X_\rho^{U_r}, X_\rho^{V_r}] = g(r) + h(r)$, where $g(r)$ spans the set of node subsets U_r, V_r with intersection of cardinality at most $O(\rho r^2)$. Observe that when U_r overlaps with V_r on l vertices, then for $m = \rho r^2$

$$\mathbf{Cov}[X_\rho^{U_r}, X_\rho^{V_r}] = \sum_{k=\max\{0, l^2 - r^2 + m\}}^{\min\{l^2, m\}} \binom{l^2}{k} p^k q^{l^2 - k} \left[\binom{r^2 - l^2}{m - k} p^{m-k} q^{r^2 - l^2 - (m-k)} \right]^2.$$

Routine and crude calculations show that $g(r) \leq \mathbf{E}[X_{r,\rho}]$, while $h(r) \leq \alpha(r) \mathbf{E}[X_{r,\rho}]^2$ where $\alpha((1 - \epsilon) \log n / \kappa(\rho)) \to 0$ as $n \to \infty$, which completes the proof.

Proof 2. Let $X_{r,\rho}^h$, $X_{r,\rho}^l$ be the number of ρ-dense subgraphs induced by only nodes in G_h or G_l, respectively. Let $X_{r,\rho}^b$ be the number of those induced by nodes from both sets. Clearly, $X_{r,\rho} = X_{r,\rho}^h + X_{r,\rho}^l + X_{r,\rho}^b$. The analysis for $G(n, p)$ directly applies for $\mathbf{E}[X_{r,\rho}^h]$ and $\mathbf{E}[X_{r,\rho}^l]$, hence we emphasize on $\mathbf{E}[X_{r,\rho}^b]$. Since $n_h = O(1)$, we have $\mathbf{E}[X_{r,\rho}^b] \leq (1-\rho) r^2 \sum_{k=0}^{n_h} \binom{n_h}{k} \binom{n_l}{r-k} \sum_{l=0}^{2k(r-k)} \binom{2k(r-k)}{l} \binom{(r-k)^2}{\rho r^2 - l} p_b^k q_b^{2k(r-k) - l} p_l^{\rho r^2 - l}$ $q_l^{(r-k)^2 - \rho r^2 + l}$, where $q_b = 1 - p_b$ and $q_l = 1 - p_l$. Then,

$$\mathbf{E}[X_{r,\rho}^b] \leq c(1 - \rho) r^2 n_h \binom{n_l}{r} \sum_{l=0}^{2n_h r} \binom{2n_h r}{l} \binom{r^2}{\rho r^2 - l} p_b^l q_b^{2n_h r - l} p_l^{\rho r^2 - l} q_l^{r^2 - \rho r^2 + l},$$

where c is a constant. Since $l = o(\rho r^2)$, we have $\binom{r^2}{\rho r^2 - l} \leq \binom{r^2}{\rho r^2}$ for $0 \leq l \leq 2n_h r$. Therefore,

$$\mathbf{E}[X_{r,\rho}^b] \leq (1 - \rho) r^2 \binom{n}{r} \binom{r^2}{\rho r^2} p_l^{\rho r^2} q_l^{r^2 - \rho r^2} \sum_{l=0}^{2n_h r} \binom{2n_h r}{l} \left(\frac{p_b q_l}{p_l} \right)^l q_b^{2n_h r - l}.$$

Using $B = \frac{p_b q_l}{p_l} + q_b$ as defined in Theorem 2, we find $P(R_\rho > r) \leq O(2^{f_1(r)})$, where $f_1(r) = -r(r\kappa(\rho) - \log n + \log r - \log e + 2n_h \log B)$. Hence, $P(R_\rho > r_1) \leq O(2^{f_1(r_1)}) \leq O\left(\frac{\log n}{n^{1/\kappa(p_l, \rho)}} \right)$ for large n.

5   Conclusion

In this paper, we attempt on analytically assessing statistical significance of connectivity and conservation in PPI networks. Specifically, we emphasize on the notion of *dense subgraphs*, which is one of the most well-studied pattern structures in extracting biologically novel information from PPI networks. While the analysis based on the $G(n, p)$ model and its extension provides a reasonable means of assessing significance, models that mirror the topological characteristics of PPI networks should also be analyzed. This paper provides a stepping stone for the analysis of such complicated models.

Acknowledgments

This work is supported in part by the NIH Grant R01 GM068959-01, and the NSF Grants CCR-0208709, CCF-0513636, DMS-0202950.

References

1. W. Aiello, F. Chung, and L. Lu. A random graph model for power law graphs. In *Proc. ACM Symp. Theory of Computing*, pages 171–180, 2000.
2. G. D. Bader, I. Donalson, C. Wolting, B. F. Quellette, T. Pawson, and C. W. Hogue. BIND-the Biomolecular Interaction Network Database. *Nuc. Acids Res.*, 29(1):242–245, 2001.
3. G. D. Bader and C. W. V. Hogue. An automated method for finding molecular complexes in large protein interaction networks. *BMC Bioinformatics*, 4(2), 2003.
4. A. Barabási and R. Albert. Emergence of scaling in random networks. *Science*, 286:509–512, 1999.
5. B. Bollobás. *Random Graphs*. Cambridge University Press, Cambridge, UK, 2001.
6. C. Brun, C. Herrmann, and A. Guénoche. Clustering proteins from interaction networks for the prediction of cellular functions. *BMC Bioinformatics*, 5(95), 2004.
7. F. Chung, L. Lu, and V. Vu. Spectra of random graphs with given expected degrees. *PNAS*, 100(11):6313–6318, 2003.
8. A. del Sol, H. Fujihashi, and P. O'Meara. Topology of small-world networks of protein-protein complex structures. *Bioinformatics*, 21(8):1311–1315, 2005.
9. J.-D. J. Han, D. Dupuy, N. Bertin, M. E. Cusick, and M. Vidal. Effect of sampling on topology predictions of protein interaction networks. *Nat. Biotech.*, 23(7):839–844, 2005.
10. E. Hartuv and R. Shamir. A clustering algorithm based on graph connectivity. *Information Processing Letters*, 76:171–181, 2000.
11. H. Hu, X. Yan, Y. Huang, J. Han, and X. J. Zhou. Mining coherent dense subgraphs across massive biological networks for functional discovery. *Bioinformatics*, 21:i213–i221, 2005.
12. S. Itzkovitz, R. Milo, N. KAshtan, G. Ziv, and U. Alon. Subgraphs in random networks. *Physical Review E*, 68(026127), 2003.
13. H. Jeong, S. P. Mason, A. Barabási, and Z. N. Oltvai. Lethality and centrality in protein networks. *Nature*, 411:41–42, 2001.
14. B. P. Kelley, R. Sharan, R. M. Karp, T. Sittler, D. E. Root, B. R. Stockwell, and T. Ideker. Conserved pathways withing bacteria and yeast as revealed by global protein network alignment. *PNAS*, 100(20):11394–11399, 2003.
15. M. Koyutürk, A. Grama, and W. Szpankowski. An efficient algorithm for detecting frequent subgraphs in biological networks. In *Bioinformatics (ISMB'04)*, pages i200–i207, 2004.
16. M. Koyutürk, A. Grama, and W. Szpankowski. Pairwise local alignment of protein interaction networks guided by models of evolution. In *RECOMB'05*, pages 48–65, 2005.
17. M. Koyutürk, Y. Kim, S. Subramaniam, W. Szpankowski, and A. Grama. Detecting conserved interaction patterns in biological networks. submitted.
18. K.-P. Künkele, P. Juin, C. Pompa, F. E. Nargang, J.-P. Henry, W. Neuperr, R. Lill, , and M. Thieffry. The isolated complex of the translocase of the outer membrane of mitochondria. *J Biol Chem*, 273(47):31032–9, 1998.
19. J. B. Pereira-Leal, A. J. Enright, and C. A. Ouzounis. Detection of functional modules from protein interaction networks. *Proteins*, 54(1):49–57, 2004.
20. N. Pržulj. Graph theory analysis of protein-protein interactions. In I. Jurisica and D. Wigle, editors, *Knowledge Discovery in Proteomics*. CRC Press, 2004.

21. N. Pržulj, D. G. Corneil, and I. Jurisica. Modeling interactome: scale-free or geometric?. *Bioinformatics*, 20(18):3508–3515, 2004.
22. N. Pržulj, D. A. Wigle, and I. Jurisica. Functional topology in a network of protein interactions. *Bioinformatics*, 20(3):340–348, 2004.
23. A. W. Rives and T. Galitski. Modular organization of cellular networks. *PNAS*, 100(3):1128–1133, 2003.
24. D. Scholtens, M. Vidal, and R. Gentleman. Local modeling of global interactome networks. *Bioinformatics*, 21(17):3548–57, 2005.
25. P. Shannon, A. Markiel, O. Ozier, N. S. Baliga, J. T. Wang, D. Ramage, N. Amin, B. Schwikowski, and T. Ideker. Cytoscape: a software environment for integrated models of biomolecular interaction networks. *Genome Res.*, 13(11):2498–504, 2003.
26. R. Sharan, T. Ideker, B. P. Kelley, R. Shamir, and R. M. Karp. Identification of protein complexes by comparative analysis of yeast and bacterial protein interaction data. In *RE-COMB'04*, pages 282–289, 2004.
27. R. Sharan, S. Suthram, R. M. Kelley, T. Kuhn, S. McCuine, P. Uetz, T. Sittler, R. M. Karp, and T. Ideker. Conserved patterns of protein interaction in multiple species. *PNAS*, 102(6):1974–1979, 2005.
28. V. Spirin and L. A. Mirny. Protein complexes and functional modules in molecular networks. *PNAS*, 100(21):12123–12128, 2003.
29. M. Stoer and F. Wagner. A simple min-cut algorithm. *J. ACM*, 44(4):585–591, 1997.
30. W. Szpankowski. *Average Case Analysis of Algorithms on Sequences*. John Wiley & Sons, New York, 2001.
31. R. Tatusov, N. Fedorova, J. Jackson, A. Jacobs, B. Kiryutin, and E. Koonin. The cog database: An updated version includes eukaryotes. *BMC Bioinformatics*, 4(41), 2003.
32. A. Thomas, R. Cannings, N. A. Monk, and C. Cannings. On the structure of protein-protein interaction networks. *Biochem Soc Trans.*, 31(6):1491–6, 2003.
33. S. Tornow and H. W. Mewes. Functional modules by relating protein interaction networks and gene expression. *Nuc. Acids Res.*, 31(21):6283–6289, 2003.
34. A. Wagner. The yeast protein interaction network evolves rapidly and contains few redundant duplicate genes. *Mol Bio Evol*, 18(7):1283–92, 2001.
35. A. Wagner. How the global structure of protein interaction networks evolves. *Proc. R. Soc. Lond. Biol. Sci.*, 270(1514):457–466, 2003.
36. M. Waterman. *Introduction to Computational Biology*. Chapman & Hall, London, 1995.
37. M. S. Waterman and M. Vingrons. Rapid and accurate estimates of statistical significance for sequence data base searches. *PNAS*, 91:4625–28, 1994.
38. I. Xenarios, L. Salwinski, X. J. Duan, P. Higney, S. Kim, and D. Eisenberg. DIP: The Database of Interacting Proteins. A research tool for studying cellular networks of protein interactions. *Nuc. Acids Res.*, 30:303–305, 2002.
39. S. H. Yook, Z. N. Oltvai, and A. L. Barabási. Functional and topological characterization of protein interaction networks. *Proteomics*, 4(4):928–942, April 2004.

Clustering Short Gene Expression Profiles

Ling Wang[1], Marco Ramoni[2], and Paola Sebastiani[1]

[1] Department of Biostatistics, Boston University School of Public Health,
Boston, MA, 02118, USA
{wangling, sebas}@bu.edu
http://www.bu.edu/dbin/sph/departments/biostatistics
[2] Children's Hospital Informatics Program, Harvard Medical School,
Boston, MA, 02115, USA
marco_ramoni@harvard.edu

Abstract. The unsupervised clustering analysis of data from temporal
or dose-response experiments is one of the most important and challenging tasks of microarray data anlysis. Here we present an extension
of CAGED (Cluster Analysis of Gene Expression Dynamics, one of the
most commonly used programs) to identify similar gene expression patterns measured in either short time-course or dose-response microarray
experiments. Compared to the initial version of CAGED, in which gene
expression temporal profiles are modeled by autoregressive equations,
this new method uses polynomial models to incorporate time/dosage information into the model, and objective priors to include information
about background noise in gene expression data. In its current formulation, CAGED results may change according to the parametrization. In
this new formulation, we make the results invariant to reparametrization
by using proper prior distributions on the model parameters. We compare the results obtained by our approach with those generated by STEM
to show that our method can identify the correct number of clusters and
allocate gene expression profiles to the correct clusters in simulated data,
and produce more meaningful Gene Ontology enriched clusters in data
from real microarray experiments.

1 Introduction

Since the original development of microarray technology, unsupervised machine
learning methods, clustering methods in particular, have provided a data analytical paradigm and played a central role in the discovery of functionally related
genes. Different unsupervised methods have been used to analyze microarray
data in order to portray various gene functional behaviors. Correlation-based
hierarchical clustering [2] is today one of the most popular analytical methods
to characterize gene expression profiles. In [9], we introduced a Bayesian model-
based clustering method that takes into account the dependency and dynamic
nature of gene expression data measured in temporal experiments. This algorithm, implemented in CAGED (Clustering Analysis of Gene Expression Dynamics), models gene expression temporal profiles by autoregressive equations

A. Apostolico et al. (Eds.): RECOMB 2006, LNBI 3909, pp. 60–68, 2006.

and uses improper prior distributions on the model parameters. As a general framework, CAGED can be used to represent a variety of cross-correlated gene expression data beyond standard temporal experiment, such as dose response data.

It has been recently shown in [3] that the model based formulation implemented in CAGED is more appropriate to cluster long temporal gene expression profiles, possibly measured at regularly spaced time points. There are many scenarios in which experiments are conducted either over a short number of time points, or at a small number of different dosages of drugs. Due to biological considerations, intervals between consecutive time points may not be the same, and the variations in dosages may not be constant. Motivated by these situations, we present an algorithm that uses polynomials models of time or dosage to capture the dynamics of gene expression profiles. The use of polynomial models however requires the specification of proper prior distributions for the regression parameters, so that to ensure the model search algorithm is invariant to reparameterization of time or dosages [7]. A further advantage of the use of proper priors on the model parameters is to include information about background noise of gene expression measured at low intensity, with the effect of making the algorithm more robust to noise and less prone to false positives.

Compared to autoregressive models, polynomial models incorporate information about time/dosage in the design matrix. Therefore, they do not require that the temporal profiles are stationary and appear to be particularly suitable to describe short expression profiles, possibly sampled at irregularly spaced points. By using the same heuristic search strategy in [9], our algorithm can automatically cluster the gene expression data into groups of genes whose profiles are generated by the same process. Furthermore, the Bayesian model-based formulation of the algorithm provides us a principled way to automatically choose the number of clusters with the maximum posterior probability. By properly specifying the prior distribution of the parameters, the clustering model is invariant to linear transformations of time/dosage. In this paper we first describe the Bayesian clustering model in Section 2. In Section 3, we evaluated the accuracy of the results obtained using this method on three simulated datasets and on the immune response data from [4]. We found that compared to STEM, our method is able to reconstruct the generating processes with higher accuracy in simulated data, and produce more Gene Ontology enriched clusters for data from real microarray experiment.

2   Model Formulation

A short time-course/dosage experiment exploring the behavior of J genes usually consists of a set of n microarrays, each measuring the gene expression level x_{jt_i} at a time point/dosage t_i, $i = 1, 2, ..., n$. For each gene, we denote the fold changes of expression levels relative to the first sample (normalized), transformed in natural logarithmic scale, by $S_j = \{x_{jt_1}, x_{jt_2}, ..., x_{jt_n}\}$, $j = 1, 2, ..., J$. These J genes are believed to be generated from an unknown number of processes, and our goal is to group these J genes into clusters by merging genes with similar expression patterns.

The clustering method currently implemented in CAGED is based on a novel concept of similarity for time series from which we derive a model-based description of a set of clusters. We assume that two gene expression profiles are similar when they are generated by the same stochastic process represented by the same parametric model. Under this definition of similarity, the clustering method groups gene expression profiles that are similar into the same cluster. To achieve this objective, CAGED has three components:

1. A model describing the dynamics of gene expression temporal profiles;
2. A probabilistic metric to score different clustering models based on the posterior probability of each clustering model;
3. A heuristics to make the search for the best clustering model feasible. The heuristic was introduced in [8] and adapted to the specific task of clustering gene expression temporal profiles in [9].

In the current implementation, CAGED uses autoregressive models to represent temporal cross-correlation. Here, we replace these models with polynomial models to describe normalized temporal patterns of gene expression data from short temporal/dose-response microarray experiments. The polynomial model describing the temporal pattern of expression for a gene j can be written as

$$x_{jt_i}|\beta_j, \epsilon_{jt} = \mu_j + \beta_{j1}t_i + ... + \beta_{jp}t_i^p + \epsilon_{jt_i}$$

where $\beta_j = (\mu_j, \beta_{j1}, ..., \beta_{jp})^T$ is the vector of regression coefficients that are assumed to be random variables, and ϵ_{jt_i} is random error. Using a matrix notation, we have

$$x_j = F\beta_j + \epsilon_j \tag{1}$$

where $x_j = (x_{jt_1}, x_{jt_2}, ..., x_{jt_n})^T$, F is the $n \times (p+1)$ design matrix with the ith row being $(1, t_i, t_i^2..., t_i^p)$, $\epsilon_j = (\epsilon_{jt_1}, \epsilon_{jt_2}, ..., \epsilon_{jt_n})^T$ is the vector of uncorrelated errors that we assume to be normally distributed, with $E(\epsilon_{jt_i}) = 0$ and $V(\epsilon_{jt_i}) = 1/\tau_j$, and the value p is the polynomial order.

We assume a proper normal-gamma prior density on the parameters β_j and τ_j. Therefore, the marginal distribution of τ_j and the distribution of the regression parameters β_j, conditional on τ_j, are

$$\tau_j \sim \text{Gamma}(\alpha_1, \alpha_2)$$
$$\beta_j|\tau_j \sim N(\beta_0, (\tau_j R_0)^{-1})$$

where R_0 is the identity matrix. The prior hyper-parameters $\alpha_1, \alpha_2, \beta_0$ are identical across genes. One of the advantages offered by this novel parametrization is the possibility to include information about background noise and, in so doing, enables the clustering algorithm to properly handle it. We will show next a method to define the hyper-parameters so that to incorporate information about background noise.

Given the data S_j — a set of observed expression values for gene j — we can then estimate the model parameters β_j and τ_j by updating their prior distribution into the posterior distribution using Bayes' Theorem:

$$f(\beta_j, \tau_j | x_j, p) = \frac{f(x_j | \beta_j, \tau_j, p) f(\beta_j, \tau_j)}{f(x_j | p)}.$$

Standard conjugate analysis leads to compute the marginal likelihood of the data

$$f(x_j | p) = \frac{1}{(2\pi)^{n/2}} \frac{(\det R_0)^{1/2}}{(\det R_{jn})^{1/2}} \frac{\Gamma(\alpha_{j1n})}{\Gamma(\alpha_1)} \frac{\alpha_{j2n}^{\alpha_{j1n}}}{\alpha_2^{\alpha_1}} \tag{2}$$

and hence a closed form solution of the posterior distribution of the model parameters τ_j and β_j [1]

$$\tau_j | x_j \sim \text{Gamma}(\alpha_{j1n}, \alpha_{j2n})$$
$$\beta_j | x_j, \tau_j \sim \text{N}(\beta_{jn}, (\tau_j R_{jn})^{-1})$$

where

$$\alpha_{j1n} = \alpha_1 + \frac{n}{2}$$

$$1/\alpha_{j2n} = \frac{-\beta_{jn}^T R_{jn} \beta_{jn} + x_j^T x_j + \beta_0^T R_0 \beta_0}{2} + \frac{1}{\alpha_2}$$

$$R_{jn} = R_0 + F^T F$$
$$\beta_{jn} = R_{jn}^{-1}(R_0 \beta_0 + F^T x_j)$$

Specification of the hyper-parameters of the prior distribution is an important component of the analysis and we take the approach to define *objective* hierarchical prior distributions on the parameters β_j and τ_j. The main intuition is to use the expression values of genes that are not used in further analysis to model the baseline hyper-variability of gene expression measured with microarrays. Several statistical software for low-level preprocessing of gene expression data score the intensities that represent relative expressions. For example the statistical software implemented in MAS 5.0 and GCOS to process expression data measured with Affymetrix arrays uses a non-parametric statistical method to label gene expression as "absent", "marginally present" or "present". These calls are based on significance tests of differences between intensities of matched probe pairs [10]. Absent calls may denote either technical errors, non-detectable expression or non-expression of the gene in the target, so that investigators are recommended not to use genes that are labelled as absent in the majority of the Affymetrix microarray samples. The more recent Illumina system for microarray data [5] assigns a quality control score to each expression summary and recommends users not to consider genes that have a score lower than 0.99. In both systems, between 25–50% of the total number of genes/probes in the arrays are usually disregarded from further analysis when they are labelled as absent or scored too low. These data however contain information about the variability of non expressed genes and therefore we use them to build our prior distributions.

We assume that disregarded genes do not exhibit any specific patterns, so after normalization and log transformation, they are expected to simply represent noise around zero. Therefore, assuming that $\beta_0 = 0$, then we only need to consider the precision parameters τ_j. We further assume that all absent gene have the same precision. Now let x_{at_i} be the normalized and log-transformed expression of one of these genes at time $t_i, i = 1, ..., n$, then $x_{at_i} | \tau \sim N(0, 1/\tau)$, where τ is the precision parameter whose prior distribution is $\tau \sim \text{Gamma}(\alpha_1, \alpha_2)$. From the properties of conditional mean and conditional variance, it is easy to show that the marginal variance of the data is functionally related to the hyper-parameters:

$$\alpha_2 = \frac{1}{(\alpha_1 - 1)\sigma_a^2}$$

where σ_a^2 is the sample variance of the disregarded expression data. So here, with $\alpha_1 = 2$, we can easily specify the hyper-parameter α_2.

3   Evaluation

We evaluate our algorithm by simulation study and analysis of the data from the microarray experiment on immune response to Helicobacter pylori infection in [4], and compare it to the program STEM recently introduced in [3]. Section 3.1 reports the results from three simulation studies, and section 3.2 presents the analysis of real data from [4]. All the analysis were done with our clustering algorithm and STEM.

3.1   Simulation Study

We simulated three sets of 5,000 gene expression profiles measured over 5 different time points: 0, 1, 2, 3, 4. All the profiles were generated assuming the gene expressions were normalized and transformed into natural logarithmic scale. The first 5,000 profiles were simply noise, and were generated from a normal distribution with mean 0 and a variance representing the average variability of noisy patterns that we inferred from the analysis of previous real microarray experiments. For this dataset, we generated another 1,000 noise profiles to be the data from genes with low intensities and we used these to specify the hyperparameters of the model. The second 5,000 profiles had 4 different baseline patterns and some background noise. Data for each gene expression profile were generated by adding random noise to one of the four baseline patterns (Figure 1 left panel), and the gene expression profiles of the background noise were generated from a normal distribution with mean 0 and a variance inferred from previous analysis of temporal microarray experiments. The number of genes representing each of the four patterns and the background noise was randomly chosen from 11 to 5,000. For this dataset, we simulated another set of 5,000 noise profiles with low intensities to specify the hyperparameters. The third 5,000 expression profiles had 6 different baseline patterns (Figure 1 right panel) that are more difficult to discriminate, plus some background noise. The data were generated using the

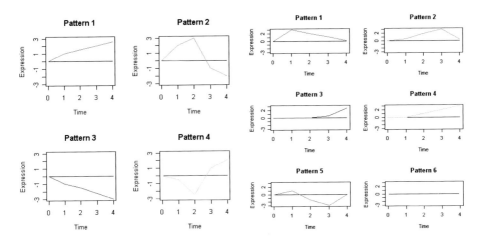

Fig. 1. Left: The 4 distinct baseline patterns of the simulated data. Right: The 6 indistinct baseline patterns of the simulated data.

Table 1. Clustering results of simulated datasets from our program and STEM

Simulated dataset	number of true profiles	# of profiles our program found	# of significant profiles STEM found
noise	0	0	17
4 patterns with noise	4	4	4
6 patterns with noise	6	6	11

same strategy as the second dataset. For the third dataset another 5,000 noise profiles with low intensities were simulated for the specification of hyperparameters. Note that in the last two datasets with planted patterns, the range of variability of the simulated patterns was within the range of variability of the noisy patterns.

Each of the three datasets was analyzed using our clustering algorithm, with polynomial orders 0, 4 and 4 respectively. We also analyzed these three datasets using STEM, with the recommended default settings of $c = 2$, and 50 possible profiles and used Bonferroni correction to control for multiple comparisons. To be consistent, we did not filter out any genes in any of these analysis, but rather used the separately generated noise profiles to specify the hyperparameters. Table 1 reports the clustering results from both our program and STEM, from which we can observe that our program successfully recovered the correct number of patterns, plus the background noise, whereas STEM discovered 17 significant profiles from the noise-only dataset, and 11 significant profiles from the dataset with 6 true patterns.

Figure 2 shows that our program grouped all the gene expression profiles in the noise-only dataset into a single cluster, representing the expected indistinguishability of pure noise. By contrast, STEM found 17 significant profiles in these noise-only data. For the simulated data with 4 different baseline patterns,

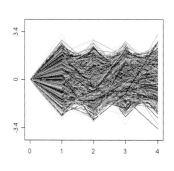 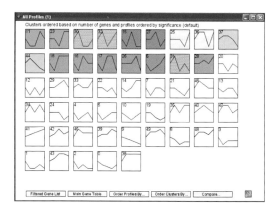

Fig. 2. Left: The only noise cluster found by our program. Right: The results from STEM. The 17 colored profiles are found to be significant by STEM.

our clustering algorithm gave 5 clusters, of which 4 have profiles matching the baseline profiles in Figure 1 left panel, and 1 noise cluster. For these 5,000 genes, 10 were allocated to the wrong cluster, with 3 false negatives (genes with true pattern allocated to noise cluster) and 7 false positives (noise genes allocated to clusters with pattern). The 4 significant profiles that STEM found have 3 profiles that are similar to the baseline patterns, but the up-regulated profile that corresponds to pattern 1 in Figure 1 left panel was not labeled as significant (p value=1). The simulated dataset with 6 different baseline profiles are designed to be harder to discriminate, and our program successfully found 7 clusters, of which 6 had profiles matching the baseline patterns in Figure 1 right panel, and 1 contained only noise. For this set of 5,000 genes, 83 are allocated to the wrong cluster, with 12 false positives and 41 false negatives. STEM analysis found 11 significant profiles for this data.

3.2   Real Data Analysis

We analyzed the data from the microarray experiment on immune response to Helicobacter pylori infection in [4] to further evaluate our clustering algorithm. In this experiment, human cDNA microarrays were used to investigate the temporal behavior of gastric epithelial cells infected with Helicobacter pylori strain G27 and some other mutants. We used the selected 2,243 genes after the data pre-processing in [3] for clustering, and the 17,352 genes that were filtered out were used to specify the hyperparameters. We then normalized and transformed the data into natural log scale, and performed the cluster analysis with polynomial order of 4. The time points we used in the model were the actual time at which the experiments were carried out: 0, 0.5, 3, 6 and 12. Our clustering algorithm returned a total of 11 clusters. Figure 3 shows all the clusters. We then preformed the Gene Ontology enrichment test with EASE [6]. Because there were missing annotations for some genes in each cluster, we carried out the enrichment analysis using only the genes with annotations. Seven out of the 11 clusters

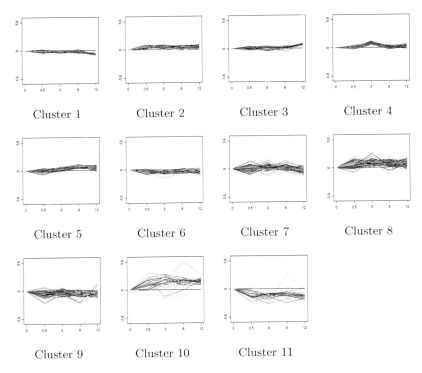

Fig. 3. The 11 clusters our program found in the analysis of data from the microarray experiment on immune response to Helicobacter pylori infection

had EASE scores less than 0.05 and hence 63% of the clusters were significantly enriched for GO categories. Cluster 10, which had 38 genes totally and 11 genes with annotations, represented a stable upregulated pattern over time. This cluster is significantly enriched for the immune response GO category (with EASE score 6.85×10^{-3}). Cluster 1 is significantly enriched for mitotic cell cycle genes (EASE score 4.62×10^{-13}) and cell cycle genes (EASE score 2.05×10^{-10}). The STEM analysis described in [3] identified 10 significant profiles, four of which only were found significantly enriched by the GO analysis. Compared to the 63% significantly GO enriched clusters found by our algorithm, the analysis in STEM therefore produces only 40% significantly GO enriched clusters.

4   Conclusions

We have introduced a model reformulation of CAGED using polynomial models of time/dosage with proper prior distributions. We find this formulation to be well suited for clustering analysis of data from short temporal/dosage microarray experiments. The polynomial models that describe the trend are flexible and do not require the gene expression profile to be stationary. We use proper priors in the model so that we can incorporate the background noise information through specifying the hyperparameters with low-intensity genes, and the clustering algorithm becomes invariant to linear transformation on time/dosage. An empirical comparison on simulated data shows that our clustering algorithm can identify the correct number of generating processes, and allocate genes into clusters with low false positives and false negatives. In the analysis of data from the human cDNA microarray experiment on immune response to Helicobacter pylori infection in [4] we found 11 clusters with our algorithm, 7 out of which are significantly enriched by Gene Ontology analysis. In both the empirical study and the analysis of the immune response data to Helicobacter pylori infection, our algorithm performs better than STEM.

References

1. J. M. Bernardo and A. F. M. Smith: *Bayesian Theory*. Wiley, New York, NY, 1994.
2. M. Eisen, P. Spellman, P. Brown, and D. Botstein. Cluster analysis and display of genome-wide expression patterns. *Proc. Natl. Acad. Sci. USA*, 95:14863–14868, 1998.
3. J. Ernst, G. J. Nau, and Z. Bar-Joseph. Clustering short time series gene expression data. *Bioinformatics*, 21 Suppl. 1:i159-i168, 2005.
4. K. Guillemin, N. Salama, L. Tompkins, and S. Falkow. Cag pathogenicity island-specific response of gastric epithelial cells to helicobacter pylori infection. *Proc. Natl. Acad. Sci. USA*, 99(23):15136–15141, 2002.
5. K. L. Gunderson, S. Kruglyak, M. S. Graigeand F. Garcia, B. G. Kermani, C. Zhao, D. Che, T. Dickinson, E. Wickham, J. Bierle, D. Doucet, M. Milewski, R. Yang, C. Siegmund, J. Haas, L. Zhou, A. Oliphant Ad J. Fan, S. Barnard, and M. S. Chee. Decoding randomly ordered DNA arrays. *Genome Res.*, 14:870–877, 2004.
6. D. A. Hosack, G. Jr. Dennis, B. T. Sherman, H. Clifford Lane, and R. A. Lempicki. Identifying biological themes within lists of genes with EASE. *Genome Biology*, 4(6):4, 2003.
7. R. E. Kass and A. Raftery. Bayes factors. *J. Ameri. Statist. Assoc.*, 90:773–795, 1995.
8. M. Ramoni, P. Sebastiani, and P. R. Cohen. Bayesian clustering by dynamics. *Mach. Learn.*, 47(1):91–121, 2002.
9. M. Ramoni, P. Sebastiani, and I. S. Kohane. Cluster analysis of gene expression dynamics. *Proc. Natl. Acad. Sci. USA*, 99(14):9121-6, 2002.
10. P. Sebastiani, E. Gussoni, I. S. Kohane and M. Ramoni. Statistical challenges in functional genomics (with discussion). *Statist. Sci.*, 18:33–70, 2003.

A Patient-Gene Model for Temporal Expression Profiles in Clinical Studies

Naftali Kaminski[1] and Ziv Bar-Joseph[2,*]

[1] Simmons Center for Interstitial Lung Disease,
University of Pittsburgh Medical School, Pittsburgh, PA 15213, USA
[2] School of Computer Science and Department of Biology,
Carnegie Mellon University 5000 Forbes Ave. Pittsburgh, PA 15213, USA

Abstract. Pharmacogenomics and clinical studies that measure the temporal expression levels of patients can identify important pathways and biomarkers that are activated during disease progression or in response to treatment. However, researchers face a number of challenges when trying to combine expression profiles from these patients. Unlike studies that rely on lab animals or cell lines, individuals vary in their baseline expression and in their response rate. In this paper we present a generative model for such data. Our model represents patient expression data using two levels, a gene level which corresponds to a common response pattern and a patient level which accounts for the patient specific expression patterns and response rate. Using an EM algorithm we infer the parameters of the model. We used our algorithm to analyze multiple sclerosis patient response to Interferon-β. As we show, our algorithm was able to improve upon prior methods for combining patients data. In addition, our algorithm was able to correctly identify patient specific response patterns.

1 Introduction

Time series expression experiments have been used to study many aspects of biological systems in model organisms and cell lines [1, 2]. More recently, these experiments are playing an important role in several pharmacogenomics and clinical studies. For example, the Inflammation and the Host Response to Injury research program [3], a consortium of several leading research hospitals, studies the response of over a hundred trauma and burn patients using time series expression data. Table 1 lists a number of other examples of such studies.

Time series expression experiments present a number of computational problems [4]. These include handling noise, the lack of repeats and the fact that only a small number of points are measured (which is particularly true in clinical experiments since tissue or blood needs to be extracted from the patient for each time point). Clinical experiments, while promising, suffer from all these issues and also raise a number of new computational challenges. In many cases the major goal of these experiments is to combine results from multiple patients to

---

* To whom correspondence should be addressed: `zivbj@cs.cmu.edu`

A. Apostolico et al. (Eds.): RECOMB 2006, LNBI 3909, pp. 69–82, 2006.

Table 1. Examples of time series clinical studies[1]

Reference	Treatment/ condition	Num. patients	Num. time points	Combining method
Inflammation and the Response to Injury[3]	Trauma and burn	over a hundred	varying	not described
Sterrenburg et al[7]	Skeletal my-oblast differen-tiation	3	5	averaging and in-dividual analysis
Weinstock-Guttman et al[6]	IFN-β for multi-ple sclerosis	8	8	averaging
Calvano et al[10]	bacterial endo-toxin	8	6	assumed repeats

[1] Note that in all cases only a few time points are sampled for each patient. In addition, in most cases researchers assume that the different patients represent repeats of the same experiment, even though most of these papers acknowledge that this is not the case (see citation above). Our algorithm, which does not make such assumption and is still able to recover a consensus response pattern is of importance to such studies.

identify common response genes. However, unlike lab animals which are raised under identical conditions, individuals responses may vary greatly. First, individuals may have different baseline expression profiles [5]. These differences result in some genes being expressed very differently from the common response. Second, the *response rate* or patients dynamics varies greatly among individuals [6, 7]. This leads to profiles which, while representing the same response, may appear different.

Previous attempts to address some of these problems have each focused on only one of the two aspects mentioned above. For example, many papers analyzing such data use the average response [6] to overcome individual (baseline) patterns. Such methods ignore the response rate problem, resulting in inaccurate description of the common response. Alignment methods where suggested to overcome response rate problems in time series expression data, especially in yeast [8, 9]. However, these methods rely on *pairwise alignment*, which is not appropriate for large datasets with tens of patients. In addition, it is not clear how to use these methods to remove patient specific response genes.

In this paper we solve the above problems by introducing a model that consists of two levels: The gene level and the patient level. The *gene level* represents the consensus response of genes to the treatment or the disease being studied. The questions we ask at this level are similar to issues that are addressed when analyzing single datasets experiments including overcoming noise, continuous representation of the measured expression values and clustering genes to identify common response patterns [16, 17]. The *patient level* deals with the instances of these genes in specific patients. Here we assume that patient genes follow a mixture model. Some of these genes represent patient specific response, or baseline expression differences. The rest of the genes come from the consensus expression response for the treatment studied. However, even if genes agree with this consensus response, their measured values still depend on the patient unique response rate and starting point. Using the consensus curve from the gene level, we associate with each patient a starting point and speed. These values correspond to difference in the timing of the first sample from that patient (for example, if some patients were admitted later than the others) and the patient dynamics (some patients respond to treatment faster than others [6, 7]).

For the gene level we use a spline based model from [9]. The main focus of this paper is on the patient level and on the relationship between the patient level and the gene level. We describe a detailed generative model for clinical expression experiments and present an EM algorithm for inferring the parameters of this model.

There are many potential uses for our algorithm. For example, researchers comparing two groups of patients with different outcomes can use our algorithm to extract consensus expression curves for each group and use comparison algorithms to identify genes that differ between the two groups. Another use, which we discuss in the results section is in an experiment measuring a single treatment. In such experiment researchers are interested in identifying clusters of genes that respond in a specific way to the treatment. As we show, using our method we obtain results that are superior to other methods for combining patient datasets. Finally, we also show that our algorithm can be used to extract patient specific response genes. These genes may be useful for determining individualized response to treatment and disease course and outcome.

1.1   Related Work

Time series expression experiments account for over a third of all published microarray datasets and has thus received a lot of attention [4]. However, we are not aware of any computational work on the analysis of time series data from clinical studies. Most previous papers describing such data have relied on simple techniques such as averaging. See Table 1 for some examples.

As mentioned above, there have been a number of methods suggested for aligning two time series expression datasets. Aach *et al* [8] used dynamic programming to align two yeast cell cycle expression datasets based on the measured expression values. Such method can be extended to multiple datasets, but the complexity is exponential in the number of datasets combined, making it impractical for clinical studies. Bar-Joseph et al [9] aligned two datasets by minimizing the area between continuous curves representing expression patterns for genes. It is not immediately clear how this methods can be extended to multiple datasets. In addition, the probabilistic nature of our algorithm allows it to distinguish between patient genes that result from a common response pattern and genes with an expression pattern unique to this patient. Again, it is not clear how an area minimization algorithm could have been extended for this goal.

Gaffney *et al* [11] presented an algorithm that extends the splines framework of Bar-Joseph *et al* discussed above to perform joint clustering and alignment. Unlike our goal of combining multiple expression experiments, their goal was to apply alignment to recover similar patterns in a single expression experiment.

In addition, because of the fact that each gene was assumed to have a different response rate, regularization of the translation parameters was required in their approach. In contrast, because we assume one set of translation parameters for all genes in a single patient, such regularization is unnecessary. Finally, their method did not allow for identifying patient specific response patterns.

A number of methods have been suggested to identify differentially expressed genes in time series expression data. These include DiffExp [12] and more recently Edge [13]. It is not clear if and how these could be used to combine large sets of time series data. If we simply treat the different patients as repeats, we falsely identify differentially expressed genes due to differences in patient dynamics.

2   A Generative Model for Expression Profiles in Clinical Experiments

We assume that expression profiles in clinical experiments can be represented using a generative model. Here we discuss the details of this model. In Section 3 we present an algorithm for inferring the parameters of this model. Following the execution of this algorithm we can retrieve consensus expression patterns as well as unique, patient specific, responses.

2.1   Continuous Representation of Time Series Expression Data

In previous work we described a method for representing expression profiles with continuous curves using cubic splines [9]. Here, we extend this model so that we can combine multiple time series expression datasets. We first briefly review the splines based model and then discuss extensions required for combining multiple time series datasets in the next subsection.

Cubic splines are a set of piecewise cubic polynomials, and are frequently used for fitting time-series and other noisy data. Specifically, we use B-splines, which can be described as a linear combination of a set of basis polynomials [18]. By knowing the value of these splines at a set of control points, one can generate the entire set of polynomials from these basis functions. For a single time series experiment, we assume that a gene can be represented by a spline curve and additional noise using the following equation:

$$Y_i = S(t)F_i + \epsilon_i$$

where Y_i is the expression profile for gene i in this experiment, F_i is a vector of spline control points for gene i and S is a matrix of spline coefficients evaluated at the sampling points of the experiment (t). ϵ_i is a vector of the noise terms, which is assumed to be normally distributed with mean 0. Due to noise and missing values, determining the parameters of the above equation (F_i and ϵ_i) for each gene separately may lead to overfitting. Instead, we constrain the control point values of genes in the same class (co-expressed genes) to co-vary, and thus we use other co-expressed genes to overcome noise and missing values in a single gene. In previous work [9], we showed that this method provides a superior fit for time series expression data when compared to previously used methods.

2.2   Extending Continuous Representation to Multiple Experiments

Unlike the above model, which assumes a single measurement for each gene, in clinical experiments we have multiple measurements from different individuals. As mentioned in the introduction, there are two issues that should be addressed. First, we need to allow for patient genes that represent individual response rather than the common response for that gene. Second, we should allow for a patient specific response rate.

To address the first issue, we assume that a patient expression data represents a mixture that includes genes from a patient specific response and genes whose expression follows the common response. This mixture model can be parametrized using a new distribution, w_q, which controls the fraction of genes from patient q that have a unique expression pattern (that is, genes that do not agree with the common response). To constrain the model, all such genes are assumed to have expression values that are sampled from the same Gaussian distribution. Thus, unless a gene instance deviates significantly from its common response it will be assigned to the consensus curve for that gene. To address the second issue, we introduce two new parameters for each patient q, a_q and b_q. These parameters control the stretch (b_q) and shift (a_q) of expression profiles from q w.r.t. the consensus expression curve. In other words, we assume that expression profiles for genes in individual patient lie on different (overlapping) segments of the consensus expression curve. The start and end points of these patient segments are controlled by the time points in which q was sampled and by a_q and b_q. Figure 1 presents the hierarchical graphical model for the generative process we assume in this paper. We denote the vector of values measured for gene g in patient q by $Y_{g,q}$. Below, we summarize the different steps for generating $Y_{g,q}$ according to our model.

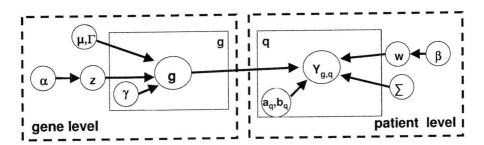

Fig. 1. Graphical representation of our generative model. The left part corresponds to the gene level and the right to the patient level. As can be seen, the two levels are connected. The node annotated with g represents the common response for a gene. An instance of this gene at a certain patients is derived from this curve using the patients response rate parameters (a_q, b_q). Some of the patient genes are not similar to their common response. These genes are assumed to come from separate distribution parameterized by a covariance matrix Σ. See text for complete details on the parameters shown in the figure.

1. The first step corresponds to the gene level (left side of Figure 1) and follows the assumptions discussed in Section 2.1. We assume that in order to generate a common (or consensus) expression curve for a gene g we sample a class z according to the class distribution α. Given a class z we use the class mean (μ_z) and sample a gene specific correction term ($\gamma_{g,z}$) using the class covariance matrix (Γ_z) as discussed in Section 2.1. Summing the vectors μ and γ we obtain the spline control points for the consensus curve.

2. We now turn to the patient level (right side of Figure 1). To generate an instance of gene g in patient q we sample a binary value $w_{g,q}$ according to the *patient specific* individual response distribution, β. If $w_{g,q}$ is 0 then the expression of g in q is unique to q and does not come from the consensus expression pattern for g. We thus sample values for entries in $Y_{g,q}$ according to a normal (Gaussian) distribution with a mean of 0 and diagonal covariance matrix Σ. If $w_{g,q}$ is 1 we continue to step 3.

3. When $w_{g,q}$ is 1, we assume that the expression of g in q lies on the consensus expression curve for g, perhaps with some added noise. Recall that we already generated the control points for this curve (in step 1). Since the *response rate* of patient q determines where on the consensus expression curve for g the values of $Y_{g,q}$ will lie, we use a_q and b_q to construct the basis function matrix, and denote it by $S(at + b)$.

4. Finally, the expression values are generated by adding random noise to the segment of the curve that was extracted in step 3.

3   Inferring the Parameters of Our Model

As described in the previous section, our model is probabilistic. The log likelihood of the data given the model is:

$$L(D|\Theta) = \sum_q \sum_g \delta(w_{g,q} = 0) \log \beta_{q,0} P_0(Y_{g,q}) \tag{1}$$

$$+ \sum_q \sum_g \delta(w_{g,q} = 1) \log \beta_{q,1} \tag{2}$$

$$+ \sum_q \sum_g \sum_i \delta(w_{g,q} = 1)\delta(z_g = i) \log \alpha_i P_{i,q}(Y_{g,q}) \tag{3}$$

$$+ \sum_g \sum_i \log P_i(\gamma_g) \tag{4}$$

δ is the Delta function which is 1 if the condition holds and 0 otherwise. $\beta_{0,q}$ ($\beta_{1,q}$) is the fraction of genes that are patient specific (common) for patient q. α_i is the fractions of genes in class i.

The first row corresponds to the patient specific genes which are not a part of the common response. These are modeled using the P_0 distribution. The second and third rows correspond to patient genes that are expressed as the common response profile. These genes are modeled using the parameters of the class and patient they belong to, $P_{i,q}$ (the distribution differs between patients because of their shift and stretch parameters). The last row is from the gene level and involves the likelihood of the gene specific correction term γ.

To infer the parameters of the model we look for parameter assignment that maximize the likelihood of the data given the model. There are a number of normally distributed parameters in our model, including the noise term ϵ and the gene specific parameters γ. For such a model determining the maximum likelihood estimates is a non convex optimization problem (see [19]). We thus turn the the EM algorithm, which iterates between two steps. In the E step we compute the posterior probabilities (or soft assignments) of the indicator variables. In the M step we use these posteriors to find the parameter assignments that maximize the likelihood of our model. Below we discuss these two steps in detail, focusing mainly on the new parameter that were introduced for the patient level.

E Step: The posterior of our missing data (indicators) is $p(w_{g,q}, z_g|Y)$. As can be seen from the third row of Equation 1 the two indicators are coupled and thus $w_{g,q}$ and z_g are *not* independent given Y. However, we can factorize this posterior if either of these indicators were observed or estimated. Specifically, we can write:

$$p(w_{g,q}, z_g|Y) = p(w_{g,q}|Y, z_g)p(z_g|Y)$$

Which can be further expanded to compute $w_{g,q}$ using Bayes rule by setting:

$$p(w_{g,q}|Y, z_g) = \frac{P_{z_g}(Y_{g,q})\beta_{1,q}}{P_{z_g}(Y_{g,q})\beta_{1,q} + P_0(Y_{g,q})\beta_{0,q}}$$

Alternatively we can factor the posterior by conditioning on $w_{g,q}$ which leads to:

$$p(w_{g,q}, z_g|Y) = p(z_g|Y, w_{g,q})p(w_{g,q}|Y)$$

Again, using Bayes rule we can further derive $p(z_g|Y, w_{g,q})$:

$$p(z_g|Y, w_{g,q}) = \frac{\sum_q w_{g,q}\alpha_{z_g}P_{z_g}(Y_{g,q})}{\sum_q \sum_c w_{g,q}\alpha_i P_i(Y_{g,q})}$$

This observation leads to a message passing algorithm which relies on variational methods to compute a joint posterior [20]. We start with an initial guess for z_g (or the values retained in the previous EM iteration). Next, we compute the posterior for $w_{g,q}$ conditioned on Y and z_g. We now send a 'message' along the edge that connects the gene level and the patient level. This message contains our new estimate of $w_{g,q}$ and is used to compute z_g and so forth. This process is repeated until convergence. In [20] it is shown that for graphical models in which nodes can be separated into disjoint 'clusters' (as is the case for the parameters associated with the gene level and the patient level in our model) such message passing algorithm converges to a lower bound estimate of the joint posterior. The estimates we obtain are used to compute the expected log likelihood which is maximized in the M step.

M Step: In the M step we maximize the parameters of the algorithm. There are two types of parameters. The first are the gene level parameters $\gamma, \mu, \Gamma, \alpha$

and σ^2. These parameters are the same as the parameters in the original model (section 2.1), that assumed a single dataset. The only difference when maximizing these parameters between the single experiment and multiple experiments settings is the weighting by the patient specific posterior value. For example, the mixture fractions in the *single* experiment case are computed by setting:

$$\alpha_i = \frac{\sum_g z_{i,g}}{\sum_g \sum_j z_{j,g}}$$

where $z_{g,i}$ is the posterior computed in the E step for the indicator z_i for gene g. In the multiple experiment setting these sums are weighted by the patient indicator:

$$\alpha_i = \frac{\sum_q \sum_g w_{g,q} z_{i,g}}{\sum_q \sum_g \sum_j w_{g,q} z_{j,g}}$$

where $w_{g,q}$ is the posterior for the patient common response indicator. Due to lack of space we do not include the complete derivation for the rest of the parameters. The interested reader is referred to [9] for these update rules.

The second type of parameters are the patient specific parameters:β, Σ, a_q, b_q. Taking the derivative w.r.t β we arrive at

$$\beta_{p,1} = \frac{\sum_g w_{g,q}}{n}$$

where n is the total number of genes.

Similarly, because we assume a Gaussian distribution with mean 0 for patient specific parameters, Σ can be computed by setting

$$\Sigma = \frac{\sum_q \sum_g w_{g,q} (Y_{p,g})^T (Y_{p,g})}{\sum_q \sum_g w_{g,q}}$$

And zeroing out not diagonal values (Σ is assumed to be diagonal).

Things are more complicated for the stretch and shift parameters. Changing these parameters results in change to the parameterization of the spline basis functions. Fortunately, splines are continuous in the first derivative and so we can compute $\frac{\partial}{\partial a_q} S(a_q + b_q t)$ and similarly for b_q. This leads to a polynomial function in a and b and an optimum can be computed using gradient descent.

In practice, we have found that it is better to maximize these parameters using line search. For example, in an experiment with a total time of one hour we can limit our interest to a scale of 1 minute for a. This leads to a search for a between 0 and 60 with increments of 1. Similarly, we usually can place upper and lower bounds on b and use a search in increments of .05. Denote by $|a|$ the total number of a value we are looking for and similarly for b. Note that the search for a and b is done independently for each patient and so the running time for of the a, b search is $|q||a||b|n$ where $|q|$ is the number of patients. This is linear in $n|q|$ when $|a|$ and $|b|$ are small.

The total running time of the M step is linear in $|q|nTC$ where T is the number of time points measured and C is the number of clusters. Each iteration in the E step involves updates to the $w_{g,q}$ (there are $n|q|$ such parameters) and z_g (n) parameters. These updates take TC time for the $w_{g,q}$ and $TC|q|$ for the z_g parameters. The E step usually converges within a few iteration and thus the total running time of the algorithm is $|q|nTC$ which is linear in n when the number of time points, clusters and patient is small.

4   Results

We have tested our algorithm on simulated data and on two biological datasets. The first biological dataset is from a non clinical experiment in yeast and the second is from a clinical multiple sclerosis (MS) experiment. While the yeast dataset is not the target of this method, since it was studied before using pairwise alignment methods it is an appropriate dataset for comparing the results of the two approaches.

4.1   Simulated Data

We generated three datasets. The first contained 100 rows of sines and 100 rows of cosine measured between 0 and 4π every $\pi/4$. The other two also contained sines and cosines, however, both were measured starting at a different point $(\pi/4)$ and with different rates (one slower by 0.8 and the other faster by 1.2). We added normally distributed random noise to each value in each dataset and used our algorithm to cluster and combine the three dataset.

Our algorithm was able to correctly retrieve both the shift and translation parameters for the second and third datasets. In addition, the algorithm correctly clustered the rows, perfectly separating the (shifted and stretched) sines and cosines. See website [21] for full details and figures.

4.2   Yeast Cell Cycle Data

As mentioned above, a number of previous methods were suggested for aligning the three yeast cell cycle expression experiments from Spellman *et al* [1]. In that paper, yeast cells were arrested using three different methods (cdc15, cdc28 and alpha) and results were combined to identify cell cycle genes. However, as noted in that paper and subsequent papers, each of these methods arrests cells at different points along the cycle. In addition, different arrest methods result in different cell cycle duration (ranging from 60 minutes to almost two hours). This has led to a number of suggestion's for aligning these datasets. Some of the suggested methods use the peak expression time for alignment. Such methods are only appropriate for cell cycle data and are thus not appropriate for the more general setting we consider in this paper. In [9] we presented a method for pairwise alignment of time series data by minimizing the squared integral between expression profiles of genes was presented, and it was shown that this method outperform previously suggested methods.

We have used the algorithm discussed in this paper to combine the the three datasets. Figure 2 (a) presents the average expression value for genes in four

5. Nau, G.J., Richmond, J.F.L., Schlesinger, A., Jennings, E.G., *et al*: Human Macrophage Activation Programs Induced by Bacterial Pathogens. PNAS. **99** (2002) 1503–1508
6. Weinstock-Guttman, B., Badgett, D., Patrick, K., Hartrich, L., *et al*: Genomic effects of IFN-beta in multiple sclerosis patients. J Immunol. **171(5)** (2002) 1503–1508
7. Sterrenburg, E., Turk, R., Peter, A.C., Hoen, P.A., van Deutekom, J.C., *et al*: Large-scale gene expression analysis of human skeletal myoblast differentiation. Neuromuscul Disord. **14(8-9)** (2004) 507–518
8. Aach, J., Church, G. M.: Aligning gene expression time series with time warping algorithms. Bioinformatics. **17** (2001) 495–508
9. Bar-Joseph, Z., Gerber, G., Jaakkola, T.S., Gifford, D.K., Simon, I.: Continuous Representations of Time Series Gene Expression Data. Journal of Computational Biology. **3-4** (2003) 39–48
10. Calvano, S.E., Xiao, W., Richards, D.R., Felciano, R.M., *et al*: A network-based analysis of systemic inflammation in humans. Nature. **437** (2005) 1032–7
11. S. Gaffney, P. Smyth: Joint Probabilistic Curve Clustering and Alignmen. Proceedings of The Eighteenth Annual Conference on Neural Information Processing Systems (NIPS). (2004)
12. Bar-Joseph, Z., Gerber, G., Jaakkola, T.S., Gifford, D.K., Simon, I.: Comparing the Continuous Representation of Time Series Expression Profiles to Identify Differentially Expressed Genes. PNAS. **100(18)** (2003) 10146–51
13. Storey, J.D., Xiao, W., Leek, J.T., Tompkins, R.G., Davis, R.W.: Significance analysis of time course microarray experiments. PNAS. **102(36)** (2005) 12837–42
14. Michalek, R., Tarantello, G.: Subharmonic solutions with prescribed minimal period for nonautonomous Hamiltonian systems. J. Diff. Eq. **72** (1988) 28–55
15. Tarantello, G.: Subharmonic solutions for Hamiltonian systems via a $Z\!\!\!Z_p$ pseudoindex theory. Annali di Matematica Pura (to appear)
16. Troyanskaya, O., Cantor, M., *et al*: Missing value estimation methods for DNA microarrays. Bioinformatics. **17** (2001) 520–525
17. Sharan R, Shamir R.: Algorithmic Approaches to Clustering Gene Expression Data. Current Topics in Computational Biology. (2002) 269–300
18. Piegl, L., Tiller, W.: The NURBS Book. Springer-Verlag. New York (1997)
19. James, G., Hastie, T.: Functional Linear Discriminant Analysis for Irregularly Sampled Curves. Journal of the Royal Statistical Society, Series B. **63** (2001) 533–550
20. Xing, E.P., Jordan, M.I., Russell, S.: A generalized mean field algorithm for variational inference in exponential families. Proceedings of Uncertainty in Artificial Intelligence (UAI). (2003) 583-591
21. Supporting website: URL: www.cs.cmu.edu/~zivbj/comb/combpatient.html
22. Achiron, A., *et al*: Blood transcriptional signatures of multiple sclerosis: unique gene expression of disease activity. Ann Neurol. **55(3)** (2004) 410–17
23. Takeba, Y., *et al*: Txk, a member of nonreceptor tyrosine kinase of Tec family, acts as a Th1 cell-specific transcription factor and regulates IFN-gamma gene transcription. J Immunol. **168(5)** (2002) 2365–70

Global Interaction Networks
Probed by Mass Spectrometry

Anne-Claude Gavin

EMBL, Meyerhofstrasse 1, 69117 Heidelberg, Germany

Abstract. Recent developments in the omics field have provided the community with comprehensive repertoires of RNAs, proteins, metabolites that constitute the cell building blocks. The next challenge resides in the understanding of how the "pieces" of this huge puzzle assemble, combine and contribute to the assembly of a coherent entity: a cell. Biology relies on the concerted action of a number of molecular interactions of gene products and metabolites operationally organized in cellular pathways. Impairment of pathway flow or connections can lead to pathology. The majority of targets of current therapeutics cluster in a limited number of these cellular pathways. However, current appreciation of the "wiring diagram" or "molecular maps" of these pathways is scanty. By applying tandem affinity purification (TAP)/MS approaches to various human pathways that lie beneath major pathologies, we could generate the comprehensive cartography of all proteins functionally involved. For example, the systematic mapping of the protein interaction network around 32 components of the pro-inflammatory TNF-alpha/NF-kappa B signalling cascade led to the identification 221 molecular associations. The analysis of the network and directed functional perturbation studies using RNA interference highlighted 10 new functional modulators that provided significant insight into the logic of the pathway as well as new candidate targets for pharmacological intervention.

    The approach has also been applied to the proteome of a whole organism, *S. cerevisiae*. The proteome-wide analysis provides the first map depicting the organization of a eukaryotic proteome into functional units, or protein complexes. The resulting cellular network goes beyond "classical" protein interaction maps. It depicts how proteins assemble into functional units and contribute their biochemical properties to higher level cellular functions. It also reveals the functional interconnection and circuiting between the various activities of the protein complexes. In collaboration with the structural and bioinformatics program at the EMBL we systematically characterized the structure of protein complexes by electron microscopy and *in silico* approximations. The work does not only bring new insights on how protein complexes operate, but may also open new avenues in the field of systems biology. Recent developments in 3D tomography may soon make it possible to fit such structures into a whole cell tomogram enabling the quantitative and dynamics study of protein complexes in their relevant cellular context; this may contribute a bridge to our understanding of the anatomy of the cell at the molecular level.

A. Apostolico et al. (Eds.): RECOMB 2006, LNBI 3909, p. 83, 2006.
© Springer-Verlag Berlin Heidelberg 2006

Statistical Evaluation of Genome Rearrangement

David Sankoff

Department of Mathematics and Statistics, University of Ottawa,
585 King Edward Avenue, Ottawa, ON, Canada, K1N 6N5
sankoff@uottawa.ca

Abstract. Genomic distances based on the number of rearrangement
steps – inversions, transpositions, reciprocal translocations – necessary
to convert the gene or segment order of one genome to that of another are
potentially meaningful measures of evolutionary divergence. The significance of a comparison between two genomes, however, depends on how it
differs from the case where the order of the n segments constituting one
genome is randomized with respect to the other. In this presentation, we
discuss the comparison of randomized segment orders from a probabilistic
and statistical viewpoint as a basis for evaluating the relationships among
real genomes. The combinatorial structure containing all the information
necessary to calculate genomic distance d is the bicoloured "breakpoint
graph", essentially the union of two bipartite matchings within the set
of $2n$ segment ends, a red matching induced by segment endpoint adjacencies in one genome and black matching similarly determined by the
other genome. The number c of alternating-colour cycles in the breakpoint graph is the key component in formulae for d. Indeed, $d \geq n - c$,
where equality holds for the most inclusive repertory of rearrangement
types postulated to account for evolutionary divergence.

Over a decade ago, it was observed in simulations of random genomes
with hundreds of genes that the distance d seldom differed from n by
more than a few rearrangements, even though it is easy to construct examples where d is as low as $\frac{n}{2}$. Our main result is that in expectation
$c = C + \frac{1}{2}\log n$ for a small constant C, so that $n - d = O(\log n)$, thus
explaining the early observations. We derive this for a relaxed model
where chromosomes need not be totally ordered – they may include circular "plasmids" – since the combinatorics of this case are very simple.
We then present simulations and partial analytical results to show that
the case where all chromosomes are totally linearly ordered (no plasmids)
behaves virtually identically to the relaxed model for large n.

Consider the "reuse" statistic $r = \frac{2d}{n}$. Although r can be as low as 1,
in which case the breakpoint graph contains d cycles of the smallest size
possible, r can also be as high as 2, in which case the cycles become
larger and less numerous. Our results show that the latter is the case for
random gene orders as well. Inference about evolution based on r, then, is
compromised by the fact that a pattern of larger and fewer cycles occurs
both when comparing genomes that have actually diverged through via
high "reuse" rates and in genomes that are purely randomly ordered with
respect to each other.

A. Apostolico et al. (Eds.): RECOMB 2006, LNBI 3909, p. 84, 2006.
© Springer-Verlag Berlin Heidelberg 2006

An Improved Statistic for Detecting Over-Represented Gene Ontology Annotations in Gene Sets

Steffen Grossmann[1], Sebastian Bauer[1,2], Peter N. Robinson[2], and Martin Vingron[1]

[1] Max Planck Institute for Molecular Genetics, Berlin, Germany
{steffen.grossmann, martin.vingron}@molgen.mpg.de
[2] Institute for Medical Genetics, Charité University Hospital,
Humboldt University, Berlin, Germany
{sebastian.bauer, peter.robinson}@charite.de

Abstract. We propose an improved statistic for detecting over-represented Gene Ontology (GO) annotations in gene sets. While the current methods treats each term independently and hence ignores the structure of the GO hierarchy, our approach takes parent-child relationships into account. Over-representation of a term is measured with respect to the presence of its parental terms in the set. This resolves the problem that the standard approach tends to falsely detect an over-representation of more specific terms below terms known to be over-represented. To show this, we have generated gene sets in which single terms are artificially over-represented and compared the receiver operator characteristics of the two approaches on these sets. A comparison on a biological dataset further supports our method. Our approach comes at no additional computational complexity when compared to the standard approach. An implementation is available within the framework of the freely available Ontologizer application.

1 Introduction

The advent of high-throughput technologies such as microarray hybridization has resulted in the need to analyze large sets of genes with respect to their functional properties. One of the most basic approaches to do this is to use the large-scale functional annotation which is provided for several species by several groups in the context of the Gene Ontology (GO) ([1], [2], [3]).

The task is to detect GO terms that are over-represented in a given gene set. The standard statistic for this problem asks for each term whether it appears in the gene set at a significantly higher number than in a randomly drawn gene set of the same size. This approach has been discussed in many papers and has been implemented in numerous software tools ([4], [5], [6], [7], [8], [9], [10]). A p-value for this statistic can easily be calculated using the hypergeometric distribution. Since this approach analyzes each term individually, without respect to any relations to other terms, we refer to it as the *term-for-term* approach.

A. Apostolico et al. (Eds.): RECOMB 2006, LNBI 3909, pp. 85–98, 2006.

The term-for-term approach becomes problematic if one looks at several or all GO terms simultaneously. There are two properties of the GO annotation which result in a complicated dependency structure between the p-values calculated for the individual GO terms. First, the annotation is done in a hierarchical manner such that genes which are annotated to a given GO term are also implicitly annotated to all less specific terms in the hierarchy (the so-called *true path rule*). Second, individual genes can be annotated to multiple GO terms, which reside in very different parts of the GO hierarchy. Both properties have the effect that information about the over-representation of one GO term can carry a substantial amount of information about the over-representation of other GO terms. This effect is especially severe when looking at parent-child pairs. Knowing that a certain term is over-represented in many cases increases the chance that some of its descendant terms also appear to be over-represented. We call this the *inheritance problem* and we consider it to be the main drawback of the term-for-term approach.

In this paper, we propose a different statistic to measure the over-representation of individual GO terms in a gene set of interest. Our method resolves the inheritance problem by explicitly taking into account parent-child relationships between the GO terms. It does this by measuring the over-representation of a GO term *given* the presence of all its parental terms in the gene set. Again, p-values can be calculated using the hypergeometric distribution at no increased computational complexity. We call our approach the *parent-child* approach. A related approach was mentioned as a part of a larger comparative analysis of yeast and bacterial protein interaction data in [11]. However, algorithmic details were not given and a systematic comparison with the term-for-term approach was not carried out.

The rest of the paper is organized as follows. In Section 2 we first review the term-for-term approach and discuss the inheritance problem in more detail. The new parent-child approach is then explained and the rationale behind it is explained. Section 3 is devoted to a comparison of the parent-child approach with the term-for-term approach. We compare the two approaches on gene sets with an artificial over-representation of individual terms. This illustrates that the parent-child approach solves the inheritance problem. We finish the section by comparing the two methods on a biological dataset. The paper is closed by a discussion.

2   Method

Given a set of genes of interest we want to analyze the functional annotation of the genes in the set. A typical example of such an analysis involves a microarray experiment where the gene set would consist of the genes which are differentially expressed under some experimental condition. We will use the name *study set* for such a gene set in the following and denote it by S. We suppose that the study set appears as a subset of the larger set of all the genes which have been considered in the experiment (such as the set of all genes which are represented on the microarray). We will call this set the *population set* and denote it by P.

The functional annotation we want to analyze consists of an assignment of some of the genes in the population set to the terms in the GO. Individual genes can be annotated to multiple GO terms. The relations between the GO terms have the structure of a *directed acyclic graph* (DAG) $G = (T, H)$, where T is the set of GO terms and the relation $H \subset T \times T$ captures the parent-child relationships (i.e. we have $(t_1, t_2) \in H$ whenever t_1 is a direct parent of t_2). In this relationship the children correspond to the *more* specific and the parents to the *less* specific terms. The set of parents of a term t is denoted by pa(t). We also use ρ to denote the unique *root term* of GO which has no parents.

For any GO term t, we denote by P_t the set of genes in the population set that are annotated to this term. The convention is that the annotation of the children of t is also passed to t itself (the so-called *true path rule*). This has the effect that $P_{t'} \subseteq P_t$ whenever $t \in$ pa(t'). When we speak about the *directly assigned* genes of a term t we mean those genes which are assigned to t but not to any of its children. Observe that the population set might also contain genes for which no assignment to any GO term is given. This means that $P \backslash P_\rho$ might be non-empty. As a shorthand notation we will write $m_t := |P_t|$ to denote the size of the set P_t, and the size of the whole population set P will be denoted by m. For the study set we use a corresponding notation by writing S_t and defining $n_t := |S_t|$ and $n := |S|$.

2.1 The Term-for-Term Approach and the Inheritance Problem

The statistic used by the term-for-term approach to measure the over-representation of a GO term t is based on comparing the presence of the term in the study set to its presence in a randomly drawn subset from the population set of the same size as the study set. The over-representation is quantified by calculating the probability of seeing in such a randomly drawn set at least as many term-t genes as in the study set. Formally, let Σ be a set of size n which has been drawn randomly from P. We write $\sigma_t := |\Sigma_t|$ for the number of genes annotated to term t in this random set. The probability of interest can now be easily calculated as the upper tail of a hypergeometric distribution

$$p_t(S) := \mathbb{P}(\sigma_t \geq n_t) = \sum_{k=n_t}^{\min(m_t, n)} \frac{\binom{m_t}{k} \binom{m - m_t}{n - k}}{\binom{m}{n}}.$$

Heuristically formulated, the *inheritance problem* lies in the fact that once it is known that a certain term t is over-represented in the study set there is also an increased chance that descendant terms of t get a small p-value and are also classified as over-represented. The reason for this clearly lies in the fact that the statistical test for each term is carried out in isolation, without taking annotations of other terms into account. The impact of this can be seen by the following thought experiment.

In a typical population set, annotation to GO terms is usually not available for all genes (meaning that $m_\rho < m$). Suppose term-for-term p-values are calculated

3   Comparing the Two Approaches

To compare our new parent-child approach with the term-for-term approach we developed a strategy which allows us to compare the respective false positive rates when over-representation of a certain term is given.

To this end, we generated gene sets in which a given GO term t is artificially over-represented. The most naive way to do this is to take the subset P_t of genes which are annotated to the term in the population set P. More realistic examples of such sets can be obtained by combining a certain proportion of genes from P_t with a certain amount of genes randomly drawn from P. When testing such sets for over-representation of GO terms, the term t itself should be detected along with some other terms which can then be considered as false positives.

The results presented in the next two subsections are based on a population set of 6456 yeast genes for which we downloaded about 32000 annotations to a total of 3870 different GO terms from the *Saccharomyces Genome Database* (http://www.yeastgenome.org/, version as of August 12th, 2005, [14]). We used the yeast annotation for no particular reason, results obtained with other species were comparable.

3.1   All-Subset Minimal p-Values

Suppose we are given a study set S for which we know that a certain term t is over-represented. To detect this, it is necessary that the p-value calculated under the respective method for that term t is small enough to remain significant even after correction for multiple testing.

Since the parent-child method measures over-representation of a term with respect to the presence of its parental terms in the study set, it can happen that there are terms for which it can be already seen from the population set P that any significant over-representation can not occur. This effect can be quantified by looking at what we call the *all-subset minimal p-value* $\hat{p}_t^{\min}$ of a term t. This is the minimal p-value one can obtain when minimizing the $\hat{p}_t(S)$ values over all possible study sets or, formally,

$$\hat{p}_t^{\min} := \min_{S \subseteq P} \hat{p}_t(S) = \hat{p}_t(P_t).$$

The claim of the last equation enables us to calculate the all-subset minimal p-values and can easily be checked using elementary probability theory. The corresponding statement is also true for the term-for-term approach, where we have $p_t^{\min} := \min_{S \subseteq P} p_t(S) = p_t(P_t)$. The behavior of the all-subset minimal p-values differs tremendously between the two approaches.

The histogram in Figure 1 a) shows that for the parent-child approach there is obviously a large number of terms for which the all-subset minimal p-values are *not small*. This can be explained by almost trivial parent-child relations which are already fixed by the annotations of the population set P. More explicitly, denote by $P_t \subseteq P_{\mathrm{pa}(t)} := \bigcup_{t' \in \mathrm{pa}(t)} P_{t'}$ the set of genes annotated to at least one of the parents of t. If there is no sufficiently large (set-)difference between P_t and

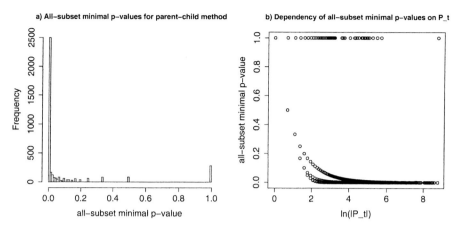

Fig. 1. a) The distribution of all-subset minimal p-values of the parent-child approach. It can be seen that there is a substantial amount of terms for which the all-subset minimal p-value is not small. In contrast, all-subset minimal p-values of the term-for-term approach are always small (below $1.55 \cdot 10^{-4}$ in this dataset). We discuss the reasons for this in the text. b) Scatterplot of the logarithm of the number of genes annotated to a term against the all-subset minimal p-values of the term. It can be seen that high minimal p-values are more likely to appear at terms with only a few genes annotated. The obvious arrangement in curves corresponds to cases where m_t and $m_{\mathrm{pa}(t)}$ differ only by 0, 1, 2,... annotated genes. The maximal value of m_t for which we observe a trivial $\hat{p}_t^{\min}$ value of one goes up to 297 ($\ln(297) \approx 5.7$) (the dot in the upper right corner corresponds to the root term, which is always trivial).

$P_{\mathrm{pa}(t)}$, the value of $\hat{p}_t^{\min}$ cannot be small. In the extreme cases where $\hat{p}_t^{\min} = 1$ we have $P_t = P_{\mathrm{pa}(t)}$.

From Figure 1 b), where we plot $\hat{p}_t^{\min}$ values against the corresponding $\ln(P_t)$ values for all terms t, it can be seen that large values mainly occur for those terms to which only few genes are annotated in the population set (cf. figure legend for more details).

In contrast to the parent-child approach, the term-for-term approach always produces extremely small all-subset minimal p-values. This is not surprising, because since $p_t(P_t)$ is the probability that a set of size $m_t := |P_t|$ drawn randomly from P consists exactly of the genes in P_t it should always be small. This is related to our criticism of the term-for-term approach. We criticize that once we know about the presence of a certain term in the study set, we also have some information about the presence of its descendant terms in the set. This knowledge is reflected by our parent-child approach but neglected by the term-for-term approach.

3.2   False Positive (Descendant) Terms

Our strategy to compare the two approaches with respect to the false positive prediction of over-represented GO terms is the following. We create a large number of (artificial) study sets for each of which a single term is intentionally over-represented. When analyzing such a set with one of the methods, any term found to be over-represented can be counted as a *false positive* classification, unless it is the intentionally over-represented term itself. We compare those false positive counts in terms of *receiver operator characteristics* (ROC) curves to visualize the differences between the two approaches. The technical details of this strategy need a more thorough description which we give now.

We start by selecting those terms which we will intentionally over-represent in the creation of the study sets. According to the results from the last subsection, we restrict ourself to those terms for which a statistically significant over-representation is possible. Therefore, we identified the set

$$T_{\text{good}} := \{t \in T : \hat{p}_t^{\min} < 10^{-7}\}$$

of terms with a small enough all-subset minimal p-value. We chose a cutoff of 10^{-7} because it leaves us enough room to get small p-values even after correction for multiple testing. In our concrete dataset, a total of 1472 out of 3870 terms made it into T_{good}.

For each term in $t \in T_{\text{good}}$ we construct artificial study sets at different levels of over-representation of t and different levels of *noise* as follows. We start with the set P_t from which we keep a certain proportion (called *term proportion*) in the study set by a random selection. To those genes we add another proportion (called *population proportion*) of genes from the whole population set as random noise. We did this for term proportions of 100%, 75% and 50% and population proportions ranging from 0% to 25% at steps of 5% resulting in a total of 18 parameter combinations.

Let S be a study set constructed as just described and let $t_{\text{over}}(S)$ be the term over-represented in the its construction. S is analyzed with both methods and the results are further processed to count the respective false positive and negative predictions. Observe that any analysis of S naturally divides the total set of terms T into two parts. First, there is the set of terms which do not annotate any of the genes in S. We do not consider those terms as true negatives in the calculation of the false positive rate, because both methods will never have a chance to falsely predict any of those terms as over-represented and therefore will agree. Moreover, we restrict ourselves to those terms which reside in the same of the subontologies (defined by the terms *biological process*, *molecular function* and *cellular component*) of GO as the term $t_{\text{over}}(S)$. The reason for this is that there are many biologically meaningful relations between terms in different subontologies which are also respected in the annotation of the genes. The set of terms which is left after this reduction will be considered in the calculation of true/false positives and be denoted by $T_{\text{test}}(S)$. By construction, any term in $T_{\text{test}}(S)$ other than $t_{\text{over}}(S)$ will be treated as a false positive when predicted as over-represented at a certain p-value cutoff by either method. The term $t_{\text{over}}(S)$ itself is counted as a false negative when not detected at that cutoff.

A last distinction has to be explained to understand the two final analyses we present. To better highlight the inheritance problem we first intersect T_{test} with the set of all *strict* descendant terms of $t_{\text{over}}(S)$ and count the false positives only on this set which we denote by $T_{\text{desc}}(S)$. In the second analysis, the counting is done on the whole of T_{test} to compare the general tendency to falsely classify terms as over-represented.

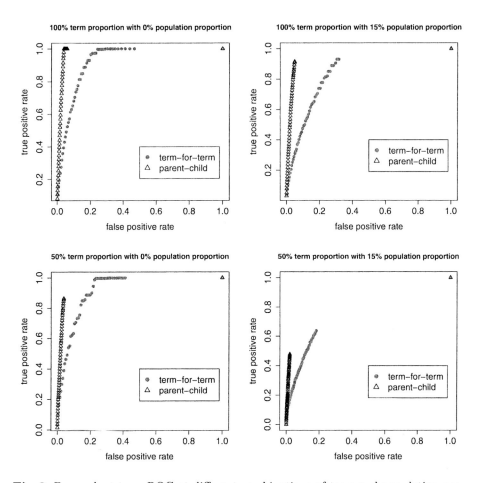

Fig. 2. Descendant-term ROC at different combinations of term and population proportions. Each point corresponds to the false and true positive rate calculated at a certain p-value cutoff π. We did not connect the points with lines, because those would indicate nonexisting combinations of the two rates. The parent-child method drastically reduces the number of descendant terms falsely predicted to be over-represented. Adding noise or reducing the level of over-representation makes it harder for both methods to correctly detect the over-represented term. This is the reason for the breaking off of the curves. ROC analysis of other combinations of term and population proportions always showed a clear advantage of the parent-child approach.

To calculate true/false positive rates, we combine the results from all study sets for a fixed combination of term and population proportions. Let $\mathcal{S}$ be such a collection of study sets.

We begin with the analysis where we count false positives on $T_{\text{desc}}(S)$ only. For a given p-value cutoff π we define the *descendant-term false positive rate* $\text{FPR}_{\text{desc}}(\pi)$ of the term-for-term method over the set $\mathcal{S}$ as

$$\text{FPR}_{\text{desc}}(\pi) := \frac{\sum_{S \in \mathcal{S}} |\{t \in T_{\text{desc}}(S) : p_t(S) < \pi\}|}{\sum_{S \in \mathcal{S}} |T_{\text{desc}}(S)|} \tag{3}$$

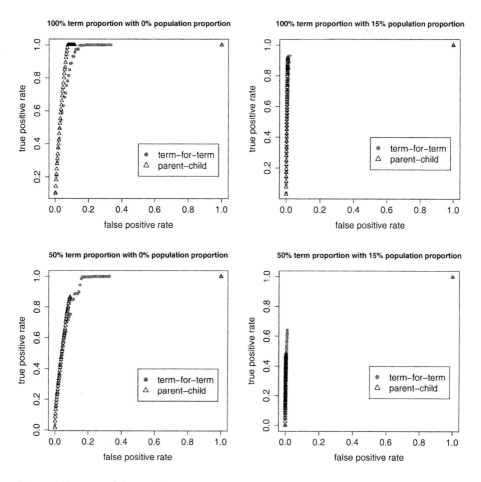

Fig. 3. All-term ROC at different combinations of term and population proportions. It can be seen that the parent-child method performs at least as well as the term-for-term method. Again, adding noise or reducing the level of over-representation has an impact on both method's ability to correctly detect the over-represented term. Additional remarks are in the legend to Figure 2.

and the *descendant-term true positive rate* $\mathrm{TPR}_{\mathrm{desc}}(\pi)$ as

$$\mathrm{TPR}_{\mathrm{desc}}(\pi) := \frac{\left|\{S \in \mathcal{S} : p_{t_{\mathrm{over}}(S)} < \pi\}\right|}{|\mathcal{S}|}. \tag{4}$$

The corresponding descendant-term false and true positive rates for the parent-child method are denoted by $\widehat{\mathrm{FPR}}_{\mathrm{desc}}(\pi)$ and $\widehat{\mathrm{TPR}}_{\mathrm{desc}}(\pi)$ and calculated by replacing p with $\hat{p}$ in (3) and (4).

A *receiver operator characteristics* (ROC) curve is obtained from those values by plotting the false positive rate versus the true positive rate for all p-value cutoffs π between 0 and 1. The results for the descendant-term analysis are shown in Figure 2 for some combinations of term and population proportions. It can be seen that the parent-child method drastically reduces the number of

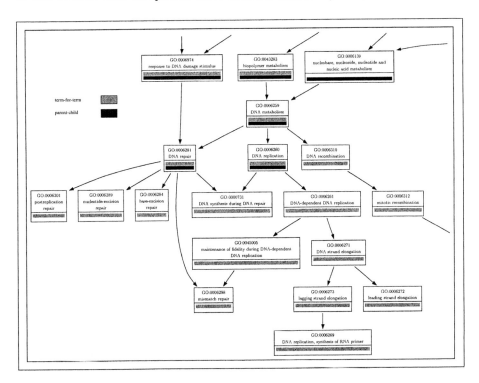

Fig. 4. Excerpt of the graph displaying over-represented terms in a set of 300 yeast genes shown to be specific for cell-cycle phase G1 ([15]). A gray marker below a term means that the term is over-represented in the term-for-term approach while a black marker indicates over-representation in the parent-child method. The inheritance problem of the term-for-term approach can be seen among the descendant terms of the two terms *DNA repair* and *DNA replication*. The range of the specific aspects of *DNA replication* and *DNA repair* found by the term-for-term approach is so wide, that no specific biological information can be gained from this. The figure was generated by post-processing output from the Ontologizer with the *Graphviz* program [16].

descendant terms falsely predicted to be over-represented when compared to the term-for-term approach.

In the second analyses we calculate the false positive rates using all terms in $T_{\text{test}}(S)$. The results in Figure 3 show that the parent-child approach performs comparably to the term-for-term approach with respect to this general counting of false positives.

3.3   A Biological Example

We compare the two approaches on a study set from a set of *Saccharomyces cerevisiae* cell cycle experiments, in which approximately 800 genes were shown to be cell-cycle regulated ([15]). We present an analysis of the 300 G1-specific genes, taking the entire set of genes on the microarray as population set. GO annotations from the *Saccharomyces Genome Database* ([14]) were used.

Although the G1 stage precedes the S, or synthesis, stage when DNA replication occurs, the G1 cluster contains many genes involved in DNA replication and repair, budding, chromatin and the spindle pole body (cf. Fig. 7 of [15]).

In Figure 4 we present a portion of the results of the GO analysis using both the parent-child and the standard term-for-term method. For both methods p-values were corrected by Westfall & Young's step-down resampling correction ([12]). We think that most of the terms which are identified by the term-for-term approach but not by the parent-child method are there because of the inheritance problem. According to the parent-child method, the key terms in this dataset are *DNA repair* and *DNA replication*. The descendant terms which are additionally identified by the term-for-term approach don't show a tendency towards a selection of closely related more specific terms, but rather cover a wide range of different terms. We don't claim that these more specific terms are biologically irrelevant. We only claim that there is no evidence that a certain collection of those terms plays an increased role in the study set.

4   Discussion

With the parent-child approach we have introduced a novel statistic to measure over-representation of GO terms in GO annotated gene sets. The motivation for this was the *inheritance problem* of the term-for-term approach which is the current standard. The inheritance problem refers to the fact that if a certain GO term is over-represented in a gene set, the term-for-term approach has a tendency to incorrectly show an over-representation of some of its descendant terms. We have illustrated this problem by analyzing gene sets in which we artificially introduced different levels of over-representation of individual terms. Analyzing the gene sets with both approaches shows that the parent-child approach drastically reduces the number of descendant terms falsely predicted to be over-represented.

Given this systematic analysis of the advantages of the parent-child approach we think that it should become the future standard. However, it should be clear that, since the two approaches use different statistics, the interpretation of results obtained with the term-for-term approach cannot be carried over to the parent-child approach. The following proper understanding of how the parent-child results have to be interpreted is necessary on the user's side.

One might argue that the inheritance problem of the term-for-term approach is in fact not a problem, but an advantage, since it also detects interesting descendant terms of over-represented terms which the parent-child approach would miss. Still, the parent-child approach does *not* state that those descendant terms are biologically irrelevant. It states that the experiment which resulted in the study set *does not give enough information* to claim that some of those descendant terms are more relevant than others and that therefore all descendant terms might be equally important in further studies. In turn, the additional emergence of descendant terms under the parent-child approach clearly indicates their increased importance. With that interpretation in mind one can claim that the parent-child approach gives more detailed insights into the GO annotation of the study set than the term-for-term approach.

The all-subset minimal p-values which we introduced in Subsection 3.1 are another key quantity which we think is of great importance in the context of the parent-child approach. Knowing about the parent-child combinations for which the all-subset minimal p-values are rather large gives important insights into the nature of the GO annotations of the underlying population set. We therefore plan to incorporate the all-subset minimal p-values into a visualization of the results obtained from the parent-child approach as it is produced by the Ontologizer.

We explicitly did not focus on the problem of multiple testing corrections (MTCs) in the context of finding over-represented GO terms in gene sets. Although some of the standard approaches have meanwhile been implemented and tested in this context, we think that there is still room for improvement and we will broaden our research to that topic. The problem of finding the optimal MTC is hard, because of the complicated dependencies between the GO terms which are caused by the DAG structure and by the annotation of individual genes to multiple terms. The parent-child approach corrects for some of those dependencies, but there remain other non-trivial dependencies between parent-child p-values. The parent-child approach therefore adds a new facet to the topic of MTCs, because it is not clear that the same strategy will turn out to be optimal for both approaches.

References

1. Ashburner, M., Ball, C., Blake, J., Botstein, D., Butler, H., Cherry, J., Davis, A., Dolinski, K., Dwight, S., Eppig, J., Harris, M., Hill, D., Issel-Tarver, L., Kasarskis, A., Lewis, S., Matese, J., Richardson, J., Ringwald, M., Rubin, G., Sherlock, G.: Gene Ontology: tool for the unification of biology. The Gene Ontology Consortium. Nature Genetics **25** (2000) 25–29

2. Harris, M.A., Clark, J., Ireland, A., Lomax, J., Ashburner, M., Foulger, R., Eilbeck, K., Lewis, S., Marshall, B., Mungall, C., Richter, J., Rubin, G.M., Blake, J.A., Bult, C., Dolan, M., Drabkin, H., Eppig, J.T., Hill, D.P., Ni, L., Ringwald, M., Balakrishnan, R., Cherry, J.M., Christie, K.R., Costanzo, M.C., Dwight, S.S., Engel, S., Fisk, D.G., Hirschman, J.E., Hong, E.L., Nash, R.S., Sethuraman, A., Theesfeld, C.L., Botstein, D., Dolinski, K., Feierbach, B., Berardini, T., Mundodi, S., Rhee, S.Y., Apweiler, R., Barrell, D., Camon, E., Dimmer, E., Lee, V., Chisholm, R., Gaudet, P., Kibbe, W., Kishore, R., Schwarz, E.M., Sternberg, P., Gwinn, M., Hannick, L., Wortman, J., Berriman, M., Wood, V., de la Cruz, N., Tonellato, P., Jaiswal, P., Seigfried, T., White, R.: The Gene Ontology (GO) database and informatics resource. Nucleic Acids Res **32** (2004) D258–D261
3. Camon, E., Magrane, M., Barrell, D., Lee, V., Dimmer, E., Maslen, J., Binns, D., Harte, N., Lopez, R., Apweiler, R.: The Gene Ontology Annotation (GOA) Database: sharing knowledge in Uniprot with Gene Ontology. Nucleic Acids Res **32** (2004) D262–D266
4. Castillo-Davis, C.I., Hartl, D.L.: GeneMerge–post-genomic analysis, data mining, and hypothesis testing. Bioinformatics **19** (2003) 891–892
5. Berriz, G.F., King, O.D., Bryant, B., Sander, C., Roth, F.P.: Characterizing gene sets with FuncAssociate. Bioinformatics **19** (2003) 2502–2504
6. Draghici, S., Khatri, P., Bhavsar, P., Shah, A., Krawetz, S.A., Tainsky, M.A.: Onto-Tools, the toolkit of the modern biologist: Onto-Express, Onto-Compare, Onto-Design and Onto-Translate. Nucleic Acids Res **31** (2003) 3775–3781
7. Beissbarth, T., Speed, T.P.: GOstat: find statistically overrepresented Gene Ontologies within a group of genes. Bioinformatics **20** (2004) 1464–1465
8. Martin, D., Brun, C., Remy, E., Mouren, P., Thieffry, D., Jacq, B.: GOToolBox: functional analysis of gene datasets based on Gene Ontology. Genome Biol **5** (2004) R101
9. Robinson, P.N., Wollstein, A., Böhme, U., Beattie, B.: Ontologizing gene-expression microarray data: characterizing clusters with Gene Ontology. Bioinformatics **20** (2004) 979–981
10. Khatri, P., Draghici, S.: Ontological analysis of gene expression data: current tools, limitations, and open problems. Bioinformatics **21** (2005) 3587–3595
11. Sharan, R., Suthram, S., Kelley, R.M., Kuhn, T., McCuine, S., Uetz, P., Sittler, T., Karp, R.M., Ideker, T.: Conserved patterns of protein interaction in multiple species. Proc Natl Acad Sci U S A **102** (2005) 1974–1979
12. Westfall, P.H., Young, S.S.: Resampling-Based Multiple Testing: Examples and Methods for p-Value Adjustment. Wiley-Interscience (1993)
13. Ge, Y., Dudoit, S., Speed, T.: Resampling-based multiple testing for microarray data analysis. TEST **12** (2003) 1–77
14. Dwight, S.S., Harris, M.A., Dolinski, K., Ball, C.A., Binkley, G., Christie, K.R., Fisk, D.G., Issel-Tarver, L., Schroeder, M., Sherlock, G., Sethuraman, A., Weng, S., Botstein, D., Cherry, J.M.: Saccharomyces Genome Database (SGD) provides secondary gene annotation using the Gene Ontology (GO). Nucleic Acids Res **30** (2002) 69–72
15. Spellman, P., Sherlock, G., Zhang, M., Iyer, V., Anders, K., Eisen, M., Brown, P., Botstein, D., Futcher, B.: Comprehensive identification of cell cycle-regulated genes of the yeast Saccharomyces cerevisiae by microarray hybridization. Mol Biol Cell **9** (1998) 3273–3297
16. Gansner, E.R., North, S.C.: An open graph visualization system and its applications to software engineering. Software — Practice and Experience **30** (2000) 1203–1233

Protein Function Annotation Based on Ortholog Clusters Extracted from Incomplete Genomes Using Combinatorial Optimization

Akshay Vashist[1], Casimir Kulikowski[1], and Ilya Muchnik[1,2]

[1] Department of Computer Science
[2] DIMACS Rutgers, The State University of New Jersey,
Piscataway, NJ 08854, USA
vashisht@cs.rutgers.edu, kulikows@cs.rutgers.edu,
muchnik@dimacs.rutgers.edu

Abstract. Reliable automatic protein function annotation requires methods for detecting orthologs with known function from closely related species. While current approaches are restricted to finding ortholog clusters from complete proteomes, most annotation problems arise in the context of partially sequenced genomes. We use a combinatorial optimization method for extracting candidate ortholog clusters robustly from incomplete genomes. The proposed algorithm focuses exclusively on sequence relationships across genomes and finds a subset of sequences from multiple genomes where every sequence is highly similar to other sequences in the subset. We then use an optimization criterion similar to the one for finding ortholog clusters to annotate the target sequences.

We report on a candidate annotation for proteins in the rice genome using ortholog clusters constructed from four partially complete cereal genomes - barley, maize, sorghum, wheat and the complete genome of *Arabidopsis*.

1 Introduction

To reliably annotate a protein sequence, computational methods strive to find a group of proteins, from closely related species, which have evolved from a common ancestral sequence [1]. Such proteins, from different species, related through speciation events are called orthologs [2]. Current methods for protein function annotation can be broadly divided into (complete sequence) homology based and (domain) model based methods, depending on the protein sequence segment they annotate. Homology based methods, such as [1, 3, 4], annotate an entire protein based on the annotation of existing well-studied orthologous proteins whereas model based approaches [5] annotate segment(s) of a protein sequence based on models of protein sequence segments from well studied proteins. Models for protein segments include protein domains (ProDom [6]), functional families (Pfam [7]), structural domains (SCOP, [8]) etc.

A. Apostolico et al. (Eds.): RECOMB 2006, LNBI 3909, pp. 99–113, 2006.
© Springer-Verlag Berlin Heidelberg 2006

Protein annotation by homology hinges on accurately identifying orthologs of a target protein in related species, whereas model based approaches must accurately identify and model conserved protein segments in a set of orthologous proteins. So, a reliable method for extracting orthologs is central to either approach. As the model based approaches build on orthologous sequences, the space of protein functions covered by orthologous sequences is larger than that covered by models of segments of sequences, for instance, many cereal-specific ortholog groups do not have any representation in Pfam, PRODOM and PROSITE. In contrast, ortholog clusters from cereals can be used to annotate proteins from a newly sequenced cereal genome providing accurate and detailed annotations for an entire protein when the members of ortholog clusters are themselves well-studied. An advantage of the model-based approaches is the localization of the functional segments and domain architecture for the target sequence. Recognizing these limitations, current automatic high throughput annotation systems such as TrEMBl [9], Ensembl [10] and PIR [5] are implemented as expert systems using state of the art methods from both categories.

Protein annotation using orthologs is widely recognized [11, 12], but in the absence of reliable automatic tools for identifying such evolutionarily related proteins, functional annotation is carried out by semi-automatic methods requiring expert intervention. A popular method for annotating a target protein is by transferring the annotation of the nearest-neighbors from closely related species [12]. For proteins, the neighbor-relationship is multi-faceted and encompasses similarity in sequence composition, proximity on the chromosome, expression levels in the cell etc. In practice, however, the nearest neighbor relationship is solely guided by sequence similarity, and is often implemented using the closest Blast [13] hit. Annotations based on best-Blast-hit can be error-prone [14], and have resulted in uncontrolled propagation of function annotation errors in protein sequence databases [1]. So, an effective automatic function annotation approach must have (i) a reliable method for ortholog clustering, (ii) a sensitive criterion for identifying ortholog neighbors for a target protein, and (iii) a source of data, with trustworthy annotations, for constructing ortholog clusters. We constructively address these issues by demonstrating how partially complete genomes (whose proteins are experimentally annotated) can be used to build ortholog clusters which then together with a robust criterion allow annotation of target proteins.

Finding candidate orthologs in partially complete genomes is an inadequately addressed problem that is critical to protein function annotation. It allows incorporation of high quality protein functional annotation which has been experimentally determined by biologists. Well before the complete set of proteins in a species is known, biologists experimentally study proteins and proceed to functionally annotate them. During a high throughput annotation for a related proteome, the experimentally studied proteins from the partially completed genomes are a valuable resource.

Current popular approaches for finding orthologs are based on reciprocal best hits [2, 15, 16, 17]. These methods involve finding pairs of mutually most similar proteins in two species as seeds for ortholog clusters, and differ primarily in the method for aggregating the seed clusters to get the final clusters. Finding the reciprocal best hits requires complete genomes, so these approaches cannot work for incomplete genomes. Moreover, it is not robust for detecting orthologs for protein families that have undergone duplication events recently [18]. Such events are frequent in higher species, so ortholog detection based on reciprocal best hits is not robust. Furthermore, even when the reciprocal best hits are reliable orthologs, the aggregation step can be error prone. Consequently, these approaches are best suited for ortholog extraction in a pair of complete genomes and require a careful manual curation by experts [2, 17]. In fact, these approaches are known to produce good results for complete genomes but they have limitations since they were not designed for addressing problems arising in ortholog detection in incomplete genomes.

A solution for detecting orthologs in a large set of incomplete genomes lies in reformulating the problem by finding orthologs, directly, as a subset of highly similar genes from multiple species. Based on this, we use a combinatorial optimization formulation of the ortholog clustering [19]. Here, we use an objective function that is based on sequence similarity viewed in the context of the phylogenetic relationships between the species, and the conserved gene-order such that the resulting method is suitable for ortholog clustering in partially complete genomes. We have also developed a robust criterion for annotating a target sequence with an ortholog cluster. This has been applied to annotate rice protein sequences using ortholog clusters constructed from plants, mostly cereals.

In section 2, we present our solution to ortholog clustering, while section 3 describes the method for extracting ortholog clusters and annotating query proteins with them. Section 4 presents the data used for this study. Section 5 describes the clustering results, while the results of annotating proteins from rice are presented in section 6 with conclusions in Section 7.

2   Addressing Challenges in Ortholog Clustering

A challenge in finding ortholog clusters is to avoid paralogs which result from duplication events. From a functional perspective, correct identification of orthologs is critical since they are usually involved in very similar functions whereas paralogs, although highly similar, functionally diverge to acquire new functions. If a duplication event follows a speciation event, orthology becomes a relationship between a set of paralogs [2]. Due to this complex relationship, paralogs related to ancient duplications are placed in different ortholog clusters whereas recently duplicated paralogs are placed in the same ortholog cluster [2]. This is justified from a functional perspective because recently duplicated genes are most likely to have the same function.

As orthologs cannot be within a species, we only consider similarities between sequences from different species, ignoring similarities between sequences within a species. This allows us to focus exclusively on inter-genome relationships and avoids intra-genome relationships which are paralogs anyway. Accordingly, it is convenient to think of the underlying graph of similarities as a multipartite graph where each of the vertex classes represents a species and protein sequences correspond to vertices in a vertex class. Such a multipartite graph representation provides a powerful framework for many comparative genomics studies as it aptly captures cross-species relationships.

Due to functional constraints, orthologs with the same function are likely to display higher levels of similarity compared to paralogs which tend to diverge in sequence and function. So, we consider only a subset of the most similar matches for any given protein sequence, as in [20]. To be precise, if a sequence i in species $s(i)$ has the sequence j as its best-match in species $s(j)$ with score m_{ij} (considered as a Blast Z-score, the bit-score), we considered all those matches for i from species $s(j)$ which score higher than αm_{ij} $(0 \leq \alpha < 1)$. The idea behind this is to avoid low-scoring spurious matches (most likely paralogs) without filtering out potential orthologs. Although, the precise choice of the value of α seems to be critical, in our experiments we found that clustering results do not vary significantly for $0.5 \leq \alpha \leq 0.8$. So, we chose a conservative value, $\alpha = 0.5$, to avoid ignoring any potential orthologs [1].

A challenge specific to ortholog detection in multiple species is the variation in the observed sequence similarities between orthologs from different pairs of species. Within an ortholog family, orthologs from anciently diverged species are less similar compared to those from the closely related species. So, automatic methods based on quantitative measures of sequence similarity must correct the observed similarities. The issue related to the correction is more complicated due to variations in evolutionary rates across lineages and protein families [21]. Here, we do not wish to correct for the absolute rates of evolution, but only wish to correct the bias in observed similarity values for their use in a computational approach for ortholog clustering. From this perspective, we can rescale the observed similarity values between sequences from a pair of genomes using the distance (time since divergence) between those species. This does not introduce any inconsistency related to diverse rates of evolution because irrespective of protein families, the similarity between sequences in two species is scaled by the same constant (distance between the species); so, the variation in evolutionary rates across protein families is not corrected. We use the species tree for rescaling observed similarities between sequences. Computationally, this requires the ortholog detection method to simultaneously consider two graphs: a multipartite graph between genes from different species and the graph for distances between the species.

One of the indicators of orthology in closely related higher organisms is the synteny (the conservation in the ordering of genetic elements in a set of genomes)

---

[1] Here α is chosen to be a constant for all sequences. However, its value will vary across protein functional families. It can be determined by a separate study of divergence in sequence similarity among paralogs following a duplication event. To be precise, it requires modeling sequence evolution for the duplicates to bring out the critical level of divergence in the sequence similarity that leads to neo-functionalization or sub-functionalization of paralogs.

[21, 22, 23], and is widely used in comparative genomics [24, 25, 26]. This auxiliary information, when available, is useful in correctly identifying orthologs [27] as it can help differentiate orthologs from paralogs. This is particularly useful for discovering orthologs in incomplete genomes of closely related species. If a protein i in species $s(i)$ is similar and syntenous (conserved in gene-order) to protein j in species $s(j)$, we lock them together as orthologs and the optimization procedure treats these genes as a unit when discovering orthologs.

Finally, we address the problems due to the incompleteness of genomes by focusing on all the relationships in a subset of genes from multiple species. Unlike the popular approaches based on reciprocal blast hits, we directly test the subsets of genes from multiple species for candidates of ortholog clusters. Intuitively, by simultaneously analyzing all the relationships between genes in a subset, we extract a subset of genes in which all genes are highly similar to each other.

3   Clustering Method

Let $V = \cup_{k=1}^{n} V_k$ be the set of all proteins from n species where V_k, $k = 1, 2, \ldots, n$ is the set of proteins from species k. An arbitrary subset H ($H \subseteq V$) can be decomposed as $H = \cup_{k=1}^{n} H_k$ where $H_k \subseteq V_k$ is a subset (possibly empty) of sequences from the species k. Let m_{ij} (≥ 0) be the observed similarity between the sequence i from species $s(i)$ and the sequence j from species $s(j)$. Let $p(s(i), s(j))$ be the evolutionary distance (explained later) between $s(i)$ and $s(j)$, then we define the similarity of a protein i ($i \in H$) to the subset H as:

$$\pi(i, H) = \sum_{\substack{t=1 \\ t \neq s(i)}}^{n} p(s(i), t) \left\{ \sum_{j \in H_t} m_{ij} - \sum_{j \in V_t \setminus H_t} m_{ij} \right\} \tag{1}$$

The evolutionary distance $p(s(i), s(j))$ is used to correct the observed sequence similarities by magnifying the sequence similarity between sequences from anciently diverged species. The evolutionary distance can be defined as the estimate of the divergence time between the species $s(i)$ and $s(j)$, but such estimates are usually not available. So, we used the topology of the species tree to define this distance. Using the tree topology, there are various ways to formalize $p(s(i), s(j))$ such as the height, $h_{s(i),s(j)}$, of the subtree rooted at the last common ancestor of $s(i)$ and $s(j)$. The exact definition of $p(s(i), s(j))$ as a function of $h_{s(i),s(j)}$ depends on the species in data for ortholog detection. When the species are closely related (such as the members of the grass family in our data), a function that depends on $h_{s(i),s(j)}$ but grows slower will better model the distance between the species. Choosing an appropriately growing function is critical because a faster growing function will have the undesirable effect of clustering together sequences from distance species but leaving out sequences from closely related species. So, the distance $p(s(i), s(j))$ (≥ 0) between $s(i)$ and $s(j)$ is defined as $(1 + \log_2 h_{s(i),s(j)})$.

The term $\sum_{j \in H_t} m_{ij}$ in (1) aggregates the similarity values between the sequence i from species $s(i)$ and all other sequences in the subset H that do not belong to species $s(i)$, while the second term $\sum_{j \in V_t \setminus H_t} m_{ij}$ estimates how this sequence is related to sequences from species t that are not included in H_t. A large positive difference between these two terms ensures that the gene i is highly similar to genes in H_t but very dissimilar from genes not included in H_t. From a clustering point of view, this ensures large values of intra-cluster homogeneity and inter-cluster separability. Applied to ortholog clustering, this enables separation of the ortholog clusters related to anciently duplicated paralogs.

Using the coefficient of similarity in (1) between a protein and a subset of proteins from other species, any arbitrary subset H of proteins from multiple species is associated with a score, $F(H)$ indicating the strength of the orthologous relationship among proteins in the subset H:

$$F(H) = \min_{i \in H} \pi(i, H), \ \forall i \in H \ \forall H \in V \tag{2}$$

So, $F(H)$ is the $\pi(i, H)$ value of the least similar (outlier) protein in H. Then, a candidate ortholog cluster H^* is defined as the subset that has the maximum score over all possible subsets of proteins from the set of all proteins, V, in the given set of species:

$$H^* = \arg \max_{H \subseteq V} F(H) \tag{3}$$

In other words, H^* contains sequences such that similarity of the least similar sequence in H is maximum, so all sequences in H^* must be highly similar. Equation (3) requires us to solve a combinatorial optimization problem. This problem can be solved efficiently [19] and the details of the algorithm for finding the optimal solution for (3) are given in the Appendix.

The above problem formulation in (1), (2) and (3) yields one ortholog cluster. However, many such clusters are present in a given set of proteins from multiple species. If we assume that these clusters are unrelated (or, weakly related) to one another, then a simple heuristic of iteratively applying the above procedure finds all these clusters. To do this we remove the proteins belonging to the first cluster H^* from V and extract another ortholog cluster in the set of proteins $V \setminus H^*$. This is iteratively applied until no more clusters are left. An advantage of this process is that it automatically determines the number of ortholog clusters. This method has been implemented in C++ and is available upon request.

3.1   Criterion for Annotating Target Proteins

Annotating a query sequence requires finding an ortholog cluster whose proteins are orthologous to the query protein. In an extreme case, ortholog clusters may need to be reconstructed using the target sequences as part of the input data. On the other hand, if the existing ortholog clusters are stable and the annotation criterion is stringent and similar to the one for extracting clusters, then existing clusters will be retained merely as extensions of clusters obtained on extended input data. However, such results depend on the criterion used for annotating the query sequences. We describe such a criterion.

Every ortholog cluster H_t^* is associated with a score value $F(H_t^*)$. A query protein sequence q is annotated with the cluster, $\widehat{H}$ whose proteins are orthologous to it. So, the degree of orthologous membership (1) of q to the ortholog cluster $\widehat{H}$ should be large. Also, since protein q is orthologous to proteins in $\widehat{H}$, we expect the following to hold:

$$\pi(q, \widehat{H}) \geq F(\widehat{H}) \tag{4}$$

In other words, inclusion of q should not adversely affect the score $F(\widehat{H})$, and q should lie close to the center of the cluster. We use (4) to select the candidate ortholog clusters with which a query protein can be potentially annotated. It is possible that no candidate ortholog cluster satisfies (4), and in such a case q cannot be annotated using the ortholog clusters. On the other hand, multiple ortholog clusters, $H_q = \{H^* : \pi(q, H^*) \geq F(H^*)\}$ may satisfy (4). To uniquely annotate q, we select the cluster with highest average similarity to q

$$\widehat{H}_q = \arg \max_{H \in H_q} \pi(q, H) \, / \, |H| \tag{5}$$

The criteria (4) and (5) to annotate proteins are derived from the score function for extracting the clusters. Unlike the nearest-neighbor based criterion for annotation queries with ortholog clusters, our criteria are less sensitive to slight changes in pair-wise similarity values between proteins. By integrating similarity values between the query protein and proteins in an ortholog cluster, our criterion (4) estimates the degree of orthologous relationship between the query and sequences in the ortholog cluster. Additionally, this criterion is very sensitive to the homogeneity of an ortholog cluster, as the condition (4) will not be satisfied unless all members of a cluster are similar. Our criteria are more stringent than the best-Blast-hit criterion (where a query sequence is annotated with the cluster of its best Blast hit). As indicated by (4), the set of query sequences annotated by our criteria is contained in those annotated by the best-Blast-hit criterion. So, our criterion provides smaller but more sensitive annotation coverage.

4   Data

The protein sequences in the partially sequenced genomes from maize (*Zea mays*, 3,318 sequences), sorghum (*Sorghum bicolor*, 468) and wheat (*Triticum aestivum*, 1,693), were downloaded from PlantGDB [28] and the complete *Arabidopsis thaliana* proteome (26,639 sequences) was downloaded from MAtDB at MIPS [29]. These sequences along with the species tree for these plants [30] (shown in Fig. 1) were used for constructing candidate ortholog clusters. The sequences from the rice proteome (61,250), available from TIGR (the size is based on data downloaded from `ftp://ftp.tigr.org/pub/data/Eukaryotic_Projects/o_sativa/annotation_dbs/`), were used as targets for annotation.

To validate the candidate ortholog clusters and the annotations based on them, we used the Pfam family [7] annotations for protein sequences. We

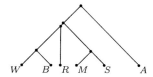

Fig. 1. The species tree for *Arabidopsis* (A), barley (B), maize (M), rice (R), sorghum (S), and wheat (W)

annotated all sequences with Pfam families (rel. 12) using Blast [13] search (with default parameters but filtering out low complexity and coil-coil regions and e-value threshold e^{-10}). The pair-wise sequence similarity between the protein sequences was computed using Blast (as above but using default e-value) and bit-score was used as a measure of similarity. Finally, we used DiagHunter [27] to identify the proteins located in syntenous regions in a pair of genomes and those groups of sequences were treated as single units for clustering.

5   Clustering Results

We applied our method to the 33,227 protein sequences from the five plants. The clustering process took less than 20 minutes (discounting the blast comparison time) on a Pentium 2.8GHz machine, and produced 22,442 sequences as singletons (94 sorghum + 1,038 maize + 185 barley + 286 wheat + 20,839 *Arabidopsis*). The remaining 10,785 sequences were grouped into 1,440 clusters, which we call the ortholog clusters.

The distribution of species in the ortholog clusters is shown in Table 1. As can be seen from this table, most clusters are small and contain only a few sequences from each species. The presence of 66 large clusters (size $\geq$ 25) is intriguing as one would expect ortholog clusters to contain only a few sequences from each genome. We investigated these and found them to be related to various *transposable elements* and *retrotransposons* which are wide spread in plants and, in fact, are not true proteins.

Table 1. Joint distribution of size and species in clusters. S: sorghum M: maize B: barley W: wheat A: *Arabidopsis* cereal: one of S,M,B,W.

Organisms in Clusters	Cluster size						
	2	3	4	5	6-10	11-25	>25
S+M		2		1			1
B+W	8	10	1	3		1	
S+M+B+W			1		1		
A+S+M+B+W				1	5	17	26
A + cereal(s)	251	358	180	132	234	131	39
others	20	10	1	2	5	4	
Total	279	375	183	139	244	155	66

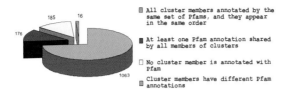

Fig. 2. Consistency of members of ortholog clusters as shown by Pfam annotations

An evaluation of the quality of ortholog clusters requires comparing the clusters with benchmark (manually curated) ortholog clusters, which are not available for this data. So, an assessment of the homogeneity of clusters was performed using Pfam as an independent source of annotation. It must be emphasized that the same Pfam annotation for proteins from different species does not imply orthology, however, when proteins are orthologous, they are likely to be involved in similar functions, so they must have similar Pfam annotations. Moreover, when multiple Pfam families are associated with orthologous proteins, they must appear in the same (domain) order in all members of the ortholog family.

As shown in Fig. 2, almost 75% (1,063) of the clusters contain sequences with the same set of Pfam annotations. Moreover, when multiple Pfam annotations were associated with sequences in a cluster, the order of their appearance (domain architecture) in different sequences was preserved, indicating their orthology. There are 1,239 (86%) clusters in which all sequences share at least one Pfam family annotation, while sequences in 185 (13%) clusters could not be annotated. There are 26 (1%) clusters with sequences from different Pfam families, but the keywords associated with those annotations are very similar. These statistics indicate that our ortholog clusters are homogeneous.

There are 49 clusters with sequences from all the 5 species. Most (33 clusters) of these clusters are large (size ≥ 10) and contain multiple sequences from each species. An inspection of these clusters revealed that their members are consistently annotated and are related to vital processes, such as *transcription* and *RNA synthesis*, and to pathways that are specific to plants such as *disease resistance, chlorophyll, phytochromes, gibberlin, starch branching enzymes* etc . These annotations confirm the obvious: genomic regions known to contain sequences of interest are likely to have been the first sequenced, and so these well-studied proteins are critical for function annotation.

Our clustering results also show some lineage-specific expansion of novel protein families. We analyzed 23 clusters containing sequences exclusively from barley and wheat, and 4 clusters containing maize and sorghum sequences only. The barley-wheat clusters contain genes specific to relatives of wheat, for instance, one cluster (with annotation *Hordoindoline*) containing 19 sequences is related to seed hardness in relatives of wheat [31] while another cluster is annotated as *Sucrose-phosphate synthase*, which has recently diverged significantly from orthologs in other plants [32]. Other barley-wheat and sorghum-maize clusters represent similar cases of family-specific proteins. For instance, a sorghum-maize cluster annotated as *kafirin* and *zein* has 22 zein-proteins from maize and 20 kafirin proteins from sorghum; the zein and kafirin proteins are known to be orthologous proteins found specifically in the sorghum-maize lineage [33].

6   Rice Sequence Annotation

We were able to annotate 25,540 protein sequences with 1,310 of the 1,440 ortholog clusters, and 130 clusters were not used for annotating any sequence. (Computationally, the annotation stage is inexpensive and was completed within 5 minutes on a Pentium 2.8 GHz machine.) Most of the clusters annotated a few sequences (see Fig. 3), for instance 241 clusters annotated 1 sequence each. However, 10,894 were annotated by 35 clusters alone. A manual inspection of annotations of members of these clusters revealed that these are related to families of genetic elements, such as transposable elements (10 clusters), retrotransposons (17 clusters including a polyprotein related cluster that annotates 1728 sequences) and other repeat elements (such as PPR repeats) widely present in plants.

An evaluation of automatic annotation can be performed by comparison with the existing manually curated annotations available for rice sequences. Unfortunately, these annotations do not use standardized terminology and cannot be used for automated validation. So, for a systematic evaluation, we used Pfam annotation for clusters. An ortholog cluster was annotated with the set of Pfam families shared by at least 80% of its member sequences. However, we found that varying this threshold between 50% to 95% has negligible effect on the annotations for clusters as most clusters contain members with the same set of Pfam annotations. The 1,310 candidate ortholog clusters used for annotation can be divided into 1,090 clusters that are annotated with at least one Pfam family and 220 clusters that could not be annotated with any Pfam family.

The annotation results are summarized in Table 2. About 68% (17,356 of 25,540) of the annotated rice sequences exactly match the Pfam annotation of their corresponding clusters (including the organization of different Pfam families when multiple Pfams are present) and about 82% (20,844 of 25,540) share at least one Pfam annotation with their clusters. Among all the annotated sequences, 14% (3,656) rice sequences and the 243 clusters used to annotate them are not associated with any Pfam family. The existing annotations (available as part of the input data) for these clusters and rice sequences show that they are annotated

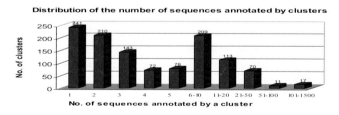

Fig. 3. Distribution of sequences annotated by a cluster

Table 2. Results of annotation and comparison with best hit based annotation approach

Statistics	Annotation using our criterion	Annotation using bBh
Sequences Annotated	25,540	33,144
Exact match between Pfam annotation for rice sequence and corresponding ortholog cluster	17,356	20,224
Partial match between Pfam annotation for rice sequence and corresponding ortholog cluster	3,488	3,301
Clusters used for Annotation	1,310	1,316
Pfam annotated Clusters used for Annotation	1,014	1,030

as *hypothetical* or *putative proteins*. It must be emphasized that these are the cases where ortholog based methods provides more annotations (although not specific) while model based methods fail to provide any annotation at all.

Evaluation of performance of our annotation criterion: We compared our annotation results using the criteria (4) with those obtained using the best-Blast-hit (bBh) criteria (shown in Table 2). The bBh criterion annotated 33,144 rice sequences compared to 25,540 sequences annotated by our criterion. Although more sequences could be annotated using the bBh criterion, it used almost the same (1,310 vs 1,316) number of clusters for annotation. The usage of same clusters for annotation by both criteria further supports our earlier observation that the unused clusters are related to lineage-specific protein families not present in rice. As mentioned earlier, 68% of sequences annotated using our criterion exactly matched the Pfam annotation for their clusters compared to 61% for the bBh criterion. When we require that sequences share at least one Pfam family annotation with the corresponding cluster, results are 82% to 71% in favor of our criterion. Thus, our criterion for annotating sequences with ortholog clusters gives smaller but more sensitive coverage.

The primary reason for smaller coverage is the use of only one available complete plant genome for constructing the ortholog clusters. This limited the space of functions covered by the extracted ortholog clusters. A solution for improving annotation coverage while retaining the advantages (higher sensitivity) of the proposed method is by combining it with the bBh approach - the proposed method should be used whenever it successfully annotates a query sequence but if it fails, we should resort to the bBh approach for annotating the query sequence with sequences that do not belong to any ortholog clusters.

7 Conclusion

Discovering ortholog clusters in partially completed genomes is important for taking advantage of experimentally determined annotations for sequences in those genomes. Existing ortholog finding methods have limitations in solving this problem. We have presented a novel reformulation of the ortholog finding problem by directly finding them as a subset of highly similar genes from multiple species. Computationally, this is formulated as a combinatorial optimization problem whose solution can be found efficiently. This formulation along with the use of an enhanced similarity measure that considers observed pair-wise sequence similarity, distance between the species, and synteny between genomes allowed us to find ortholog clusters in the incomplete genomes. Further, by formulating the problem on a multipartite graph and using gene-specific and genome-specific cut-offs for similarities, we avoided anciently duplicated paralogs (likely to have acquired new functions) but keeping the recently duplicated paralogs (paralogs having similar functions) in ortholog clusters, conforming to the existing expert based ortholog clustering paradigms [2].

We are currently working on further enhancing ortholog detection by incorporating various aspects of similarities for genes that are indicative of their orthology. Such similarities include structural similarity, exon-intron organization, similarity in cellular/tissue location where genes express, and the pathways in which genes are involved. By including these diverse features, we hope to get closer to an expert based ortholog detection. The current software is available upon request and we soon hope to make it available over the web.

Acknowledgments

AV was, in part, supported by the DIMACS graduate student awards. IM was supported by NSF grant CCF - 0325398. The authors thank Dr. Joachim Messing for sharing his expertise in plant genomics. We also thank the anonymous reviewers for their suggestions which helped improve the presentation.

References

1. Abascal, F., Valencia, A.: Automatic annotation of protein function based on family identification. Proteins **53** (2003) 683–692
2. Tatusov, R., Koonin, E., Lipmann, D.: A genomic perspective on protein families. Science **278** (1997) 631–637
3. Enright, A.J., Van Dongen, S., Ouzonis, C.: An efficient algorithm for large-scale detection of protein families. Nucleic Acids Res **30** (2002) 1575–1584
4. Petryszak, R., Kretschmann, E., Wieser, D., Apweiler, R.: The predictive power of the CluSTr database. Bioinformatics **21** (2005) 3604–3609
5. Wu, C.H., Huang, H., Yeh, L.S.L., Barker, W.C.: Protein family classification and functional annotation. Comput Biol Chem. **27** (2003) 37–47
6. Bru, C., Courcelle, E., Carrre, S., Beausse, Y., Dalmar, S., Kahn, D.: The ProDom database of protein domain families: more emphasis on 3D. Nucleic Acids Res **33** (2005) D212–215
7. Bateman, A., Coin, L., Durbin, R., Finn, R.D., Hollich, V., Griffiths-Jones, S., Khanna, A., Marshall, M., Moxon, S., Sonnhammer, E.L.L., Studholme, D.J., Yeats, C., Eddy, S.R.: The Pfam protein families database. Nucleic Acids Res **32** (2004) 138–141

8. Andreeva, A., Howorth, D., Brenner, S.E., Hubbard, T.J.P., Chothia, C., Murzin, A.G.: SCOP database in 2004: refinements integrate structure and sequence family data. Nucleic Acids Res **32** (2004) 226–229

9. Fleishmann, W., Moller, S., Gateau, A., Apweiler, R.: A novel method for automatic functional annotation of proteins. Bioinformatics **15** (1999) 228–233

10. Curwen, V., Wyras, E., Andrews, T.D., Clarke, L., Mongin, E., Searle, S.M., Clamp, M.: The Ensembl automatic gene annotation system. Genome Res **14** (2004) 942–950

11. Eisen, J., Wu, M.: Phylogenetic analysis and gene functional predictions: phylogenomics in action. Theor Popul Biol. **61** (2002) 481–487

12. Galperin, M.Y., Koonin, E.V.: Who's your neighbor? new computational approaches for functional genomics. Nat Biotechnol **18** (2000) 609–613

13. Altschul, S., Madden, T., Schaffer, A., Zhang, J., Zhang, Z., Miller, W., Lipman, D.: Gapped BLAST and PSI-BLAST: a new generation of protein database search programs. Nucleic Acids Res **25** (1997) 3389–3402

14. Koski, L.B., Golding, G.B.: The closest BLAST hit is often not the nearest neighbor. J Mol Biol **52** (2001) 540–542

15. Remm, M., Strom, C.E., Sonnhammer, E.L.: Automatics clustering of orthologs and in-paralogs from pairwise species comparisons. J Mol Biol **314** (2001) 1041–1052

16. Li, L., Stoeckert, C.K., Roos, D.S.: OrthoMCL: Identification of ortholog groups for eukaryotic genomes. Genome Res **13** (2003) 2178–2189

17. Tatusov, R., Fedorova, N., Jackson, J., Jacobs, A., Kiryutin, B., Koonin, E., Krylov, D., R, R.M., Mekhedov, S., Nikolskaya, A., Rao, B., Smirnov, S., Sverdlov, A., Vasudevan, S., Wolf, Y., Yin, J., Natale, D.: The COG database: an updated version includes eukaryotes. BioMed Central Bioinformatics (2003)

18. Abascal, F., Valencia, A.: Clustering of proximal sequence space for identification of protein families. Bioinformatics **18** (2002) 908–921

19. Vashist, A., Kulikowski, C., Muchnik, I.: Ortholog clustering on a multipartite graph. In: Workshop on Algorithms in Bioinformatics. (2005) 328–340

20. Kamvysselis, M., Patterson, N., Birren, B., Berger, B., Lander, E.: Whole-genome comparative annotation and regulatory motif discovery in multiple yeast species. In: RECOMB. (2003) 157–166

21. Huynen, M.A., Bork, P.: Measuring genome evolution. Proc Natl Acad Sci USA **95** (1998) 5849–56

22. Fujibuchi, W., Ogata, H., Matsuda, H., Kanehisa, M.: Automatic detection of conserved gene clusters in multiple genomes by graph comparison and P-quasi grouping. Nucleic Acids Res **28** (2002) 4096–4036

23. Overbeek, R., Fonstein, M., D'Souza, M., Pusch, G.D., Maltsev, N.: The use of gene clusters to infer functional coupling. Proc Natl Acad Sci USA **96** (1999) 2896–2901

24. He, X., Goldwasser, M.H.: Identifying conserved gene clusters in the presence of orthologous groups. In: RECOMB. (2004) 272–280

25. Dandekar, T., Snel, B., Huynen, M., Bork, P.: Conservation of gene order: a fingerprint of proteins that physically interact. Trends Biochem Sci. **23** (1998) 324–328

26. Heber, S., Stoye, J.: Algorithms for finding gene clusters. In: Workshop on Algorithms in Bioinformatics. (2001) 252–263

27. Cannon, S.B., Young, N.D.: OrthoParaMap: distinguishing orthologs from paralogs by integrating comparative genome data and gene phylogenies. BMC Bioinformatics **4** (2003)

28. Dong, Q., Schlueter, D., Brendel, V.: PlantGDB, plant genome database and analysis tools. Nucleic Acids Res **32** (2004) D354–D359
29. Schoof, H., Zaccaria, P., Gundlach, H., Lemcke, K., Rudd, S., Kolesov, G., Mewes, R.A.H., Mayer, K.: MIPS arabidopsis thaliana database (MAtDB): an integrated biological knowledge resource based on the first complete plant genome. Nucleic Acids Res **30** (2002) 91–93
30. Kellogg, E.A.: Relationships of cereal crops and other grasses. Proc Natl Acad Sci USA **95** (1998) 2005–2010
31. Darlingto, H., Rouster, J., Hoffmann, L., Halford, N., Shewry, P., Simpson, D.: Identification and molecular characterisation of hordoindolines from barley grain. Plant Mol Biol. **47** (2001) 785–794
32. Castleden, C.K., Aoki, N., Gillespie, V.J., MacRae, E.A., Quick, W.P., Buchner, P., Foyer, C.H., Furbank, R.T., Lunn1, J.E.: Evolution and function of the sucrose-phosphate synthase gene families in wheat and other grasses. Plant Physiology **135** (2004) 1753–1764
33. Song, R., Llaca, V., Linton, E., Messing, J.: Sequence, regulation, and evolution of the maize 22-kD alpha zein gene family. Genome Res. **11** (2001) 1817–1825
34. Cormen, T., Leiserson, C., Rivest, R., Stein, C.: Introduction to Algorithms, Second Edition. The MIT Press (2001)

Appendix

Algorithm for finding H^*: The objective of (3) is to find the subset H^* for which $F(H)$ is maximum over all possible subsets of V. Observe that $\pi(i, H)$ is an increasing function of the second argument because when an additional element k is included in H, we get $\pi(i, H \cup \{k\}) - \pi(i, H) = 2p(s(i), s(k))m_{ik} \geq 0$. Because of this property, increasing the subset H can only increase the value $\pi(i, H)$ for any $i \in H$. This property is the basis of the algorithm for finding H^*.

Starting from V, the only possibility to get a subset with a larger score than $F(V)$ is to remove the element $i^* = \arg\min_{i \in V} \pi(i, V)$ with the minimum value of the function $\pi(i, V)$. This is because the element i^* is the least similar element in V and defines the score for the set V. By removing any other element, i' ($i' \neq i^*$), the value of the resulting subset $V \setminus \{i'\}$ can only decrease, i.e $F(V \setminus \{i'\}) \leq F(V)$ because $\pi(i^*, V \setminus \{i'\}) \leq \pi(i^*, V)$ due to the property described above. But we are interested in moving toward a subset with the largest score value, and the only way to achieve this is by removing i^* from V. After removing i^*, this argument still holds and we again remove the least similar element, and continue to do so until all elements are exhausted. Pictorially, this approach is described in Fig. 4, and the algorithm is described in Table 3. As result, the subset with the highest score encountered in this shelling-out process of the least similar element is the optimal solution.

The algorithms runs in time $O(|E| + |V|^2)$, where E and V are the set of edges and vertices, respectively, in the sequence similarity graph. For computing the $\pi(i, V)$ values, we must look at all the edges which requires $O(|E|)$ time. At each iteration, the vertex corresponding to the least similar element is found (requiring time $O(|V|)$) and removed. This removal entails deleting the edges (and updating the function $\pi()$ which takes time proportional to the number

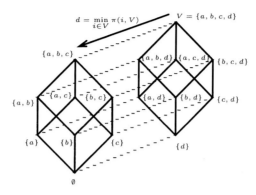

Fig. 4. The motivation for the algorithm explained on a toy example, $V = \{a, b, c, d\}$. All the subsets of V are shown by the Hasse diagram representation of the power set lattice of V, and the objective is to find the subset, H with the largest $F(H)$ value. Assume $d = \arg\min_{i \in V} \pi(i, V)$. Then, $F(V) = \pi(d, V)$ because $F(H) = \min_{i \in H} \pi(i, H)$. As $\pi(i, H)$ is an increasing function of H, for all $S \ni d$, we have $\pi(d, S) \leq \pi(d, V)$. This implies $F(S) \leq F(V) \forall S \ni d$. Then, the only way to obtain a subset with a higher score is to explore the subsets of $V \setminus \{d\}$. After removing d, the same argument holds for the set $V \setminus \{d\}$ and we can continue until all the elements are shelled-out.

Table 3. Pseudocode for extracting H^*

Algorithm 7.1. ALGORITHM FOR FINDING $H^*()$

$t \leftarrow 1; \quad H_t \leftarrow V; \quad H^* \leftarrow V;$
$F(H^*) \leftarrow \min\limits_{i \in V} \pi(i, V)$
while $(H_t \neq \emptyset)$

do $\begin{cases} M_t \leftarrow \{\alpha \in H_t : \pi(\alpha, H_t) = \min\limits_{j \in H_t} \pi(j, H_t)\}; \\ F(H_t) \leftarrow \min\limits_{j \in H_t} \pi(j, H_t); \\ \textbf{if } (H_t \setminus M_t) = \emptyset) \vee (\pi(i, H_t) = 0 \; \forall i \in H_t) \\ \quad \textbf{then } \begin{cases} \text{output } H^* \text{ as the optimal set and} \\ \quad\quad F(H^*) \text{ as the optimal value.} \end{cases} \\ \quad \textbf{else } \begin{cases} H_{t+1} \leftarrow H_t \setminus M_t; \\ t \leftarrow t + 1; \\ \textbf{if } (F(H_t) > F(H^*)) \\ \quad \textbf{then } \{H^* = H_t; \end{cases} \end{cases}$

---

of edges deleted), but each edge is deleted only once, so deleting edges for all iterations together takes $O(|E|)$ time. The maximum number of iterations is bounded by $|V|$, so the algorithm runs in $O(|E| + |V|^2)$ time. However, using the Fibonacci heaps [34] to store the values of $\pi(i, H)$ for elements of V, allows a faster implementation of the algorithm [19] that runs in time $O(|E| + |V| \log |V|)$.

Detecting MicroRNA Targets by Linking Sequence, MicroRNA and Gene Expression Data

Jim C. Huang[1], Quaid D. Morris[2], and Brendan J. Frey[1,2]

[1] Probabilistic and Statistical Inference Group, University of Toronto,
10 King's College Road, Toronto, ON, M5S 3G4, Canada
[2] Banting and Best Department of Medical Research, University of Toronto,
160 College St, Toronto, ON, M5S 3E1, Canada
jim@psi.toronto.edu

Abstract. MicroRNAs (miRNAs) have recently been discovered as an important class of non-coding RNA genes that play a major role in regulating gene expression, providing a means to control the relative amounts of mRNA transcripts and their protein products. Although much work has been done in the genome-wide computational prediction of miRNA genes and their target mRNAs, two open questions are how miRNAs regulate gene expression and how to efficiently detect *bona fide* miRNA targets from a large number of candidate miRNA targets predicted by existing computational algorithms. In this paper, we present evidence that miRNAs function by post-transcriptional degradation of mRNA target transcripts: based on this, we propose a novel probabilistic model that accounts for gene expression using miRNA expression data and a set of candidate miRNA targets. A set of underlying miRNA targets are learned from the data using our algorithm, GenMiR (**Gen**erative model for **miR**NA regulation). Our model scores and detects 601 out of 1,770 targets obtained from TargetScanS in mouse at a false detection rate of 5%. Our high-confidence miRNA targets include several which have been previously validated by experiment: the remainder potentially represent a dramatic increase in the number of known miRNA targets.

1 Introduction

Recent results show that there may not be many more mammalian protein-coding genes left to be discovered [9]. As a result, one of the main goals in genomics is now to discover how these genes are regulated. In the basic model for gene regulation, transcription factors act to enhance or suppress the transcription of a gene into messenger RNA (mRNA) transcripts. Recent evidence points to the existence of an alternative, post-transcriptional mechanism for gene regulation in which the abundances of transcripts and/or their protein products are reduced. In particular, microRNAs (miRNAs), a subclass of so-called non-coding RNA genes [8], have been identified as such a component of the cell's regulatory circuitry. miRNA genes do not go on to produce proteins, but instead produce short, 22-25 nt-long mature miRNA molecules. These then target

A. Apostolico et al. (Eds.): RECOMB 2006, LNBI 3909, pp. 114–129, 2006.

mRNA transcripts through complementary base-pairing to short target sites. miRNAs are believed to either trigger the degradation of their targets [3, 20] or repress translation of the transcript into protein [1]. There is substantial evidence that miRNAs are an important component of the cellular regulatory network, providing a post-transcriptional means to control the amounts of mRNA transcripts and their protein products. Previous work has focused primarily on the genome-wide computational discovery of miRNA genes [5, 19, 23] and their corresponding target sites [13, 16, 17, 24]. Experiments have shown that multiple miRNAs may be required to regulate a targeted transcript [16] and that miR-NAs can regulate the expression of a substantial fraction of protein-coding genes with a diverse range of biological functions [1, 4].

Although many miRNA genes and target sites have been discovered by computational algorithms [5, 6], there remain two open problems in miRNA genomics. One is to determine whether miRNAs regulate their targets through the post-transcriptional degradation mechanism, through the translational repression mechanism, or possibly both. Another problem is the fact that there are relatively few miRNA targets which have experimental support [13, 16]. The computational algorithms used to find targets have limited accuracy [16, 17] due to the short lengths of miRNA target sites and thus empirical methods are needed to tease out true miRNA targets from false ones. Experimental validation of targets is currently done through *in vitro* reporter assays [13, 18] which provide some measure as to whether the miRNA binds to a target site. One concern with this type of assay is that a miRNA-target pair validated *in vitro* might not be biologically relevant inside the cell [1]. In addition, assays performed on a single miRNA-target pair might also erroneously reject the pair given that the combinatorial nature of miRNA regulation isn't taken into account and many miRNAs may be required to observe down-regulation of the targeted transcript. Finally, such assays are relatively expensive and time-consuming to conduct, so that only a handful of targets have been validated using this method. Expression profiling has been proposed as an alternative method for validating miRNA targets [20], but this has the problem of becoming intractable due to the combinatorial nature of miRNA regulation in which the action of many miRNAs must be taken into account.

While computational sequence analysis methods for finding targets and expression profiling methods have their own respective limitations, we can benefit from the advantages of both by combining the two methods [11] to detect miRNA targets. Given the thousands of miRNA targets being output by target-finding programs [13, 16, 17] and given the ability to profile the expression of thousands of mRNAs and miRNAs using microarrays [12, 21], we motivate a high-throughput computational technique for detecting miRNA targets in which both sequence and gene expression data are combined. The pipeline for detecting targets is shown in Fig. 1: a set of candidate miRNA targets is first generated using a target-finding program. Our model uses this set of candidates to account for gene expression using miRNA microarray expression data while taking into account the combinatorial nature of miRNA regulation. In this paper,

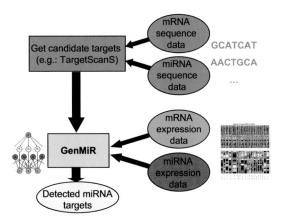

Fig. 1. Pipeline for detecting miRNA targets using GenMiR: a set of candidate targets is generated using a target-finding program (e.g.: TargetScanS). The candidates, along with expression data for mRNAs and miRNAs, are input into the GenMiR probability model. The output of the model consists of a set of miRNA targets which are well-supported by the data.

we first address the question as to how miRNAs regulate gene expression: we will present evidence in favor of the post-transcriptional degradation model for miRNA regulation. From this, we will formulate a probabilistic graphical model in which miRNA targets are learned from expression data. Under this model, the expression of a targeted mRNA transcript can be explained through the regulatory action of multiple miRNAs. Our algorithm, GenMiR (**Gen**erative model for **miR**NA regulation), learns the proposed model to find a set of miRNA targets which are biologically relevant. We will show that our model can accurately identify miRNA targets from expression data and detect a significant number of targets, many of which provide insight into miRNA regulation.

2   Post-transcriptional Degradation (PTD) vs. Translational Repression (TR)

We will begin by addressing the question of whether miRNAs regulate gene expression by post-transcriptional degradation of target mRNAs [3] or by repressing translation of a targeted transcript [1] into proteins. In the first scenario, we expect that both mRNA expression levels and protein abundances will be decreased through the action of a miRNA. In the second scenario, protein abundances would be decreased without any necessary change in the expression of their parent mRNA. To determine which of the two mechanisms of miRNA regulation is most likely given biological data, we will present two simple *Bayesian networks* for the proposed mechanisms, shown in Figure 2a. Each network consists of a directed graph where nodes representing both miRNA and mRNA expression measures as well as protein abundances are linked via directed edges

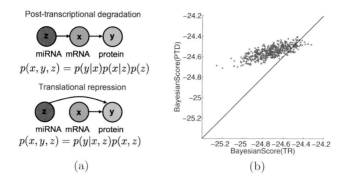

Fig. 2. (a) Bayes nets for degradation and repression regulatory mechanisms: each network presents a particular set of dependencies between miRNA, mRNA and protein measures (b) Scatter plot of scores obtained from both models for each miRNA-mRNA-protein triplet: most of the data is better accounted for by the PTD model than by the TR model

representing dependencies between the 3 variables. Thus, each network encodes a different set of dependencies between miRNA, mRNA and protein measures: our aim here is to see which set of dependencies best describes biological reality.

To do so, we examined data profiling the expression of 78 mouse miRNAs [2] with mRNA expression data [25] paired to protein mass-spectrometry data [15] consisting of the measurements of $3,080$ mRNA-protein pairs across 5 tissues common to the 3 data sets. All the measured values were then ranked across tissues to get discrete rank values. We then used a set of human miRNA targets output from the target-finding program TargetScanS [17, 26]. These consisted of a total of $12,839$ target sites in human genes which were identified based both on miRNA-target site complementarity as well as conservation in 3'-UTR regions across 5 mammalian species (human, mouse, rat, dog and chicken). After mapping these targeted transcripts to the mouse mRNAs and miRNAs in the above data using the Ensembl database and BLAT [7], we were left with 473 candidate miRNA-target interactions involving 211 unique mRNA-protein pairs and 22 unique miRNAs.

With the above data in hand, we gathered statistics over mRNA, protein and miRNA measurements x, y, z across tissues $t = 1, \cdots, 5$ for each putative miRNA targets. We then scored the two regulatory models for each miRNA/mRNA/protein triplet using Bayesian scores [10] computed as

$$\text{BayesianScore}(PTD) = \sum_t \log \Big(p(y_t|x_t)p(x_t|z_t)p(z_t) \Big)$$

$$\text{BayesianScore}(TR) = \sum_t \log \Big(p(y_t|x_t, z_t)p(x_t, z_t) \Big) \tag{1}$$

where each conditional probability term is a multinomial probability averaged over a Dirichlet prior distribution. These scores correspond to the log-likelihood of a miRNA/mRNA/protein data triplet given one of the two models for miRNA regulation. Figure 2b shows a scatter plot of the 2 scores for each mRNA-miRNA-protein triplet: we can see that in the vast majority of cases, the PTD model offers a far better fit to our data than the TR model, providing good evidence in favor of the PTD model within the current debate of whether miRNAs act by degrading targets or by repressing translation [1, 3]. With this result in hand, we will now motivate the use of both mRNA and miRNA expression data to detect miRNA targets.

3   Exploring miRNA Targets Using Microarray Data

To explore putative relationships between mRNAs and miRNAs, we used the above microarray expression data profiling the expression of a total of 41, 699 mRNA transcripts and 78 miRNAs across 17 tissues common to both data sets: expression values consisted of arcsinh-normalized intensity values in the same range, with negative miRNA intensities were thresholded to 0. From the above set of 12, 839 TargetScanS targets, 1, 770 are represented across this set of miR-NAs and mRNAs in the form of 788 unique mRNAs and 22 unique miRNAs: the set of putative miRNA-mRNA pairs are shown in Fig. 3. Given the expression

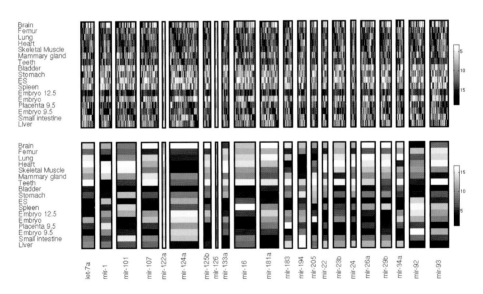

Fig. 3. Rank expression profiles of targeted mRNAs and corresponding miRNAs: each profile measures expression across 17 mouse tissues. For targeted mRNA transcripts (top row), a rank of 17 (black) denotes that the expression in that tissue was the highest amongst all tissues in the profile whereas a rank of 1 (white) denotes that expression in that tissue was lowest amongst all tissues. miRNA intensities are shown using a reverse colormap (bottom row), with a rank of 17 (white) denoting that the expression was highest and a rank of 1 (black) denotes that expression in that tissue was lowest. Each miRNA targets and down-regulates multiple mRNA transcripts and a given mRNA transcript may be targeted by multiple miRNAs.

data and a set of putative targets, we looked for examples of down-regulation in which the expression of a targeted mRNA transcript was low in a given tissue and the targeting miRNA was highly expressed in that same tissue.

Among miRNAs in the data from [2], miR-16 and miR-205 are two that are highly expressed in spleen and embryonic tissue respectively (Fig. 3). The cumulative distribution of expression of their targeted mRNAs in these two tissues is shown in Fig. 4. The plots show that the expression of targeted mRNAs is negatively shifted with respect to the background distribution of expression in these two tissues ($p < 10^{-7}$ and $p < 0.0015$ using a one-tailed Wilcoxon-Mann-Whitney test, Bonferroni-corrected at $\alpha = 0.05/22$). This result suggests that regulatory interactions predicted on the basis of genomic sequence can be observed in microarray data in the form of high miRNA/low targeted transcript expression relationships. While it is feasible to find such relationships for a single miRNA using an expression profiling method [20], to test for the more realistic scenario in which mRNA transcripts are down-regulated by multiple miRNAs, we would require a large number of microarray experiments for a large number of miRNAs. Additional uncertainty would be introduced by miRNAs that are expressed in many tissues. An alternative is to use data which profiles the expression of mRNAs and miRNAs across many tissues and formulate a statistical model which links the two using a set of candidate miRNA targets. A sensible model would account for negative shifts in tissue expression for targeted mRNA transcripts given that the corresponding miRNA was also highly expressed in the same tissue. By accounting for the fact that miRNA regulation is combinatorial in nature [4, 16], we will construct such a model which will hopefully capture the basic mechanism of miRNA regulation. The model takes as inputs a set of candidate miRNA targets and expression data sets profiling both mRNA transcripts and miRNAs: it then accounts for examples of down-regulation in

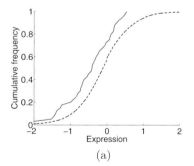
(a)

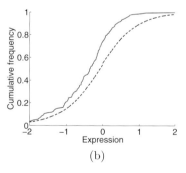
(b)

Fig. 4. Effect of miRNA negative regulation on mRNA transcript expression: shown are cumulative distributions for (a) Expression in embryonic tissue for mRNA transcripts targeted by miR-205 (b) Expression in spleen tissue for mRNA transcripts targeted by miR-16. A shift in the curve corresponds to down-regulation of genes targeted by miRNAs. Targets of miR-205 and miR-16 (solid) show a negative shift in expression with respect to the background distribution (dashed) in tissues where miR-205 and miR-16 are highly-expressed.

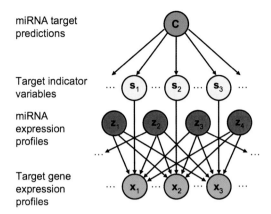

miRNA target
predictions

Target indicator
variables

miRNA
expression
profiles

Target gene
expression
profiles

Fig. 5. Bayesian network used for detecting miRNA targets: each mRNA transcript is assigned a set of indicator variables which select for miRNAs that are likely to regulate it given the data

will be computationally prohibitive for transcripts which are targeted by a large number of miRNAs. Thus, we will turn instead to an approximate method for inference which will make the problem tractable.

5   Variational Learning for Detecting miRNA Targets

For variational learning [14, 22] in a graphical model with latent variables $\mathcal{H}$ and observed variables $\mathcal{E}$, the exact posterior $p(\mathcal{H}|\mathcal{E})$ is approximated by a distribution $q(\mathcal{H}; \phi)$ parameterized by a set of variational parameters ϕ. Variational inference therefore consists of an optimization problem in which we optimize the fit between $q(\mathcal{H}; \phi)$ and $p(\mathcal{H}, \mathcal{E})$ with respect to the variational parameters ϕ. This fit is measured by the Kullback-Leibler (KL) divergence $D(q||p)$ between the q and p distributions, which can be written as

$$D(q||p) = \int_{\mathcal{H}} q(\mathcal{H}; \phi) \log \frac{q(\mathcal{H}; \phi)}{p(\mathcal{H}, \mathcal{E})} \, d\mathcal{H} = \sum_{\mathbf{S}} q(\mathbf{S}|\mathbf{C}) \log \frac{q(\mathbf{S}|\mathbf{C})}{p(\mathbf{X}, \mathbf{S}|\mathbf{C}, \mathbf{Z}, \Theta)}$$

where $\mathbf{X}$, $\mathbf{Z}$, $\mathbf{C}$ and $\mathbf{S}$ have been substituted as the observed and latent variables $\mathcal{E}$ and $\mathcal{H}$ respectively.

The approximating distribution can be further simplified via a mean-field decomposition of the q-distribution in which all the latent s_{ij} variables are assumed to be independent and thus

$$q(\mathbf{S}|\mathbf{C}) = \prod_{i,j} q(s_{ij}|\mathbf{C}) = \prod_{(i,j)\in\mathbf{C}} \beta_{ij}^{s_{ij}} (1 - \beta_{ij})^{1-s_{ij}} \qquad (4)$$

where the variational parameters β_{ij} will be fitted to the observed data $\mathbf{X}$, $\mathbf{Z}$, $\mathbf{C}$. We will therefore approximate the intractable posterior $p(s_{ij}|\mathbf{x}_i, \mathbf{Z}, \mathbf{C}, \Theta)$ with a simpler distribution $q(s_{ij}|\mathbf{C})$ which will make inference tractable. If we write the expected sufficient statistics $\mathbf{u}_i$, $\mathbf{W}$ and $\mathbf{V}$ as

$$\mathbf{u}_i = \sum_{j:(i,j)\in\mathbf{C}} \lambda_j \beta_{ij} \mathbf{z}_j \tag{5}$$

$$\mathbf{W} = \frac{1}{N} \sum_i \left(\mathbf{x}_i - (\boldsymbol{\mu} - \mathbf{u}_i)\right)\left(\mathbf{x}_i - (\boldsymbol{\mu} - \mathbf{u}_i)\right)^T$$

$$\mathbf{V} = \frac{1}{N} \sum_i \sum_{j:(i,j)\in\mathbf{C}} (\beta_{ij} - \beta_{ij}^2)\lambda_j^2 \mathbf{z}_j \mathbf{z}_j^T$$

then the KL divergence $D(q||p)$ can be written simply as

$$D(q||p) = \sum_{(i,j)\in\mathbf{C}} \left(\beta_{ij} \log \frac{\beta_{ij}}{\pi} + (1 - \beta_{ij}) \log \frac{1 - \beta_{ij}}{1 - \pi}\right) + \frac{N}{2} \log |\boldsymbol{\Sigma}|$$
$$+ \frac{N}{2} \mathrm{tr}\left(\boldsymbol{\Sigma}^{-1}(\mathbf{W} + \mathbf{V})\right) + \mathrm{const.} \tag{6}$$

Approximate inference and parameter estimation will be accomplished via the variational EM algorithm [14, 22], which iteratively minimizes $D(q||p)$ with respect to the set of variational parameters (E-step) and the model parameters (M-step) until convergence to a local minimum. Thus, taking derivatives of $D(q||p)$ and setting to zero yields the following updates:

Variational E-step:

$\forall (i,j) \in \mathbf{C}$,

$$\frac{\beta_{ij}}{1 - \beta_{ij}} = \frac{\pi}{1 - \pi} \exp\left[-\lambda_j \mathbf{z}_j^T \boldsymbol{\Sigma}^{-1}\left(\mathbf{x}_i - \left(\boldsymbol{\mu} - \left(\sum_{k\neq j:(i,k)\in\mathbf{C}} \frac{\lambda_k}{2}\beta_{ik}\mathbf{z}_k + \frac{\lambda_j}{2}\mathbf{z}_j\right)\right)\right)\right] \tag{7}$$

Variational M-step:

$$\boldsymbol{\mu} = \frac{1}{N} \sum_i (\mathbf{x}_i + \mathbf{u}_i) \tag{8}$$

$$\boldsymbol{\Sigma} = \mathrm{diag}\left(\mathbf{W} + \mathbf{V}\right)$$

$$\forall j, \quad \lambda_j = \max\left(-\frac{\sum_i \beta_{ij}\mathbf{z}_j^T\boldsymbol{\Sigma}^{-1}\left(\mathbf{x}_i - (\boldsymbol{\mu} - \sum_{k\neq j:(i,k)\in\mathbf{C}} \frac{\lambda_k}{2}\beta_{ik}\mathbf{z}_k)\right)}{\sum_i \beta_{ij}\mathbf{z}_j^T\boldsymbol{\Sigma}^{-1}\mathbf{z}_j}, \epsilon\right)$$

$$\pi = \frac{\sum_{(i,j)\in\mathbf{C}} \beta_{ij}}{\mathrm{card}(\mathbf{C})}$$

$$\tag{9}$$

4. Bartel DP. MicroRNAs: Genomics, biogenesis, mechanism, and function. *Cell* **116**, pp.281-297, 2004.
5. Bentwich I *et al.* Identification of hundreds of conserved and nonconserved human microRNAs. *Nature Genetics* **37**, pp. 766-770, 2005.
6. Berezikov E *et al.* Phylogenetic shadowing and computational identification of human microRNA genes. *Cell* **120(1)**, pp. 21-4, 2005.
7. Kent W.J. BLAT – The BLAST-Like Alignment Tool. *Genome Research* **4**, pp. 656-664, 2002.
8. Eddy S. Non-coding RNA genes and the modern RNA world. *Nature Reviews Genetics* **2**, pp. 919-929, 2001.
9. Frey BJ *et al.* Genome-wide analysis of mouse transcripts using exon-resolution microarrays and factor graphs. *Nature Genetics* **37**, pp. 991-996, 2005.
10. Hartemink A, Gifford D, Jaakkola T, and Young R. Using graphical models and genomic expression data to statistically validate models of genetic regulatory networks. *Proceedings of the Pacific Symposium on Biocomputing 2001, World Scientific: New Jersey*, pp. 422-433.
11. Huang JC, Morris QD, Hughes TR and Frey BJ (2005). GenXHC: A probabilistic generative model for cross-hybridization compensation in high-density, genome-wide microarray data. *Proceedings of the Thirteenth Annual Conference on Intelligent Systems for Molecular Biology*, June 25-29 2005.
12. Hughes TR *et al.* Expression profiling using microarrays fabricated by an ink-jet oligonucleotide synthesizer. *Nature Biotechnol.* **19**, pp. 342-347, 2001.
13. John B *et al.* Human MicroRNA targets. *PLoS Biol* **2(11)**, e363, 2004.
14. Jordan MI, Ghahramani Z, Jaakkola TS and Saul LK. An introduction to variational methods for graphical models. *Learning in Graphical Models*, Cambridge: MIT Press, 1999.
15. Kislinger T *et al.* Global survey of organ and organelle protein expression in mouse: combined proteomic and transcriptomic profiling. Submitted to *Cell*, 2005.
16. Krek A *et al.* Combinatorial microRNA target predictions. *Nature Genetics* **37**, pp. 495-500, 2005.
17. Lewis BP, Burge CB and Bartel DP. Conserved seed pairing, often flanked by adenosines, indicates that thousands of human genes are microRNA targets. *Cell* **120**, pp.15-20, 2005.
18. Lewis BP *et al.* Prediction of mammalian microRNA targets. *Cell* **115**, pp.787-798, 2003.
19. Lim LP *et al.* The microRNAs of Caenorhabditis elegans. *Genes Dev.* **17**, pp. 991-1008, 2003.
20. Lim LP *et al.* Microarray analysis shows that some microRNAs downregulate large numbers of target mRNAs. *Nature* **433**, pp. 769-773, 2005.
21. Lockhart M *et al.* Expression monitoring by hybridization to high-density oligonucleotide arrays. *Nature Biotechnol.* **14**, pp. 1675-1680, 1996.
22. Neal RM and Hinton GE. A view of the EM algorithm that justifies incremental, sparse, and other variants, *Learning in Graphical Models*, Kluwer Academic Publishers, 1998.
23. Xie X *et al.* Systematic discovery of regulatory motifs in human promoters and 3' UTRs by comparison of several mammals. *Nature* **434**, pp. 338-345, 2005.
24. Zilberstein CBZ, Ziv-Ukelson M, Pinter RY and Yakhini Z. A high-throughput approach for associating microRNAs with their activity conditions. *Proceedings of the Ninth Annual Conference on Research in Computational Molecular Biology*, May 14-18 2005.

25. Zhang W, Morris Q *et al.* The functional landscape of mouse gene expression. *J Biol* **3**, pp. 21-43, 2004.
26. Supplemental Data for Lewis *et al.* *Cell* **120**, pp. 15-20. http://web.wi.mit.edu/bartel/pub/Supplemental%20Material/Lewis%20et%20al%202005%20Supp/

Appendix

A supplementary table containing all detected miRNA targets and corresponding human gene identifiers can be found at *http://www.psi.toronto.edu/~GenMiR*

RNA Secondary Structure Prediction Via Energy Density Minimization

Can Alkan[1,*], Emre Karakoc[2,*], S. Cenk Sahinalp[2,*], Peter Unrau[3],
H. Alexander Ebhardt[3], Kaizhong Zhang[4], and Jeremy Buhler[5]

[1] Department of Genome Sciences, University of Washington, USA
[2] School of Computing Science, Simon Fraser University, Canada
[3] Department of Molecular Biology and Biochemistry, Simon Fraser University, Canada
[4] Department of Computer Science, University of Western Ontario, Canada
[5] Department of Computer Science, Washington University in St Louis, USA

Abstract. There is a resurgence of interest in RNA secondary structure prediction problem (a.k.a. the RNA folding problem) due to the discovery of many new
families of non-coding RNAs with a variety of functions. The vast majority of
the computational tools for RNA secondary structure prediction are based on free
energy minimization. Here the goal is to compute a non-conflicting collection of
structural elements such as hairpins, bulges and loops, whose total free energy
is as small as possible. Perhaps the most commonly used tool for structure prediction, `mfold/RNAfold`, is designed to fold a single RNA sequence. More
recent methods, such as `RNAscf` and `alifold` are developed to improve the
prediction quality of this tool by aiming to minimize the free energy of a number of functionally similar RNA sequences simultaneously. Typically, the (stack)
prediction quality of the latter approach improves as the number of sequences
to be folded and/or the similarity between the sequences increase. If the number
of available RNA sequences to be folded is small then the predictive power of
multiple sequence folding methods can deteriorate to that of the single sequence
folding methods or worse.

In this paper we show that delocalizing the thermodynamic cost of forming an RNA substructure by considering the *energy density* of the substructure
can significantly improve on secondary structure prediction via free energy
minimization. We describe a new algorithm and a software tool that we call
`Densityfold`, which aims to predict the secondary structure of an RNA sequence by minimizing the sum of energy densities of individual substructures.
We show that when only one or a small number of input sequences are available,
`Densityfold` can outperform all available alternatives. It is our hope that this
approach will help to better understand the process of nucleation that leads to the
formation of biologically relevant RNA substructures.

1 Introduction

Given an RNA sequence, *RNA secondary structure prediction problem* (sometimes referred to as the *RNA folding problem*) asks to compute all pairs of bases that form

---

* These authors contributed equally to the paper.

A. Apostolico et al. (Eds.): RECOMB 2006, LNBI 3909, pp. 130–142, 2006.
© Springer-Verlag Berlin Heidelberg 2006

hydrogen bonds. Although the problem is one of the earliest in computational biology, it has attracted considerable fresh attention due to the recent discoveries of many non-coding RNA molecules such as siRNAs, riboswitches and catalytic RNAs with a variety of novel functions. (See for example, the September 2, 2005 issue of Science magazine, devoted to "Mapping RNA form and function", which investigates the relationship between RNA structure and functionality [1].)

Much of the literature on RNA secondary structure prediction is devoted to the *free energy minimization* approach. This general methodology (which is sometimes called the *thermodynamic* approach) aims to compute the secondary structure by minimizing the total free energy of its substructures such as *stems*, *loops* and *bulges*.

Free energy minimization can be applied either to a single RNA sequence, or, simultaneously to a number of functionally similar RNA sequences. Free energy minimization of a single RNA sequence has been studied since early 70s [21] and a number of dynamic programming algorithms have been developed [17, 22, 13]. The popular Mfold and its more efficient version RNAfold (from the Vienna package) are implementations of these algorithms.

Despite a 25 year long effort to perfect secondary structure prediction of a single RNA sequence via energy minimization, the end result is still far from perfect. The limitations of this approach are usually attributed to the following factors. The total free energy is affected by tertiary interactions which are currently poorly understood and thus ignored in the energy tables [15] currently used by all structure prediction tools. There are also external, non-RNA related factors that play important roles during the folding process. Furthermore, the secondary structure of an RNA sequence is formed as the molecule is being transcribed. A highly stable substructure, formed only after a short prefix of the RNA sequence is transcribed, can often be preserved after the completion of the transcription, even though it may not conform to a secondary structure with the minimum free energy. [1]

In order to address these issues, much of the recent research on RNA secondary structure is focused on simultaneously predicting the secondary structure of many functionally similar RNA sequences. The intuition underlying this approach is that functional similarity is usually due to structural similarity, which, in many cases, correspond to sequence similarity. Because this approach can utilize the commonly observed co-varying mutations among aligned base pairs in a stem, the accuracy of this approach can outperform single sequence structure prediction approach.

There are two main techniques for simultaneously predicting the secondary structure of multiple sequences via energy minimization.

- The first general technique, used in particular by the alifold program [10] of the Vienna package, assumes that the multiple alignment between the input

---

[1] Another crucial issue that limits the prediction accuracy of many energy minimization based tools is that they do not allow pseudoknots. This is due to the the fact that the energy minimization problem allowing arbitrary pseudoknots is NP-hard [3]. The only software tool we are aware of which allows certain types of pseudoknots (as described by [6]) is Pknots [18], which suffers from efficiency problems. Thus our current implementation does not allow any pseudoknots due to efficiency considerations; however it can easily be extended to allow the class of pseudoknots captured by Pknots.

RNA sequences (in the case of alifold, computed by the Clustal-W program [20]) corresponds to the alignment between their substructures. The structure is then derived by folding the multiple alignment of the sequences. Clearly this method crucially relies on the correctness of the multiple sequence alignment; thus its prediction quality is usually good for highly similar sequences (60% or more) but can be quite poor for more divergent sequences.

- The second general technique aims to compute the sequence alignment and the structure prediction simultaneously [19, 8, 16]. When formulated as a rigorous dynamic programming procedure, the computational complexity of this technique becomes very high; it requires $O(n^6)$ time even for two sequences and is NP-hard for multiple sequences [7]. In order to decrease the computational complexity, it may be possible to restrict the number of substructures from each RNA sequence to be aligned to the substructures from other sequences. In [5], this is done through a preprocessing step which detects all statistically significant potential stems of each RNA sequence by performing a local alignment between the sequence and its reverse complement. When computing the *consensus structure*, only those substructures from each RNA sequence which are enclosed by such stems are considered for being aligned to each other. This strategy is successfully implemented by the RNAscf program recently developed by Bafna et al. [5].

One final approach to multiple sequence structure prediction is the so called *consensus folding* technique. Rather than minimizing free energy, the consensus folding technique first extracts all potential stems of each input RNA sequence. The consensus structure is then computed through determining the largest set of compatible potential stems that are common to a significant majority of the RNA sequences. A good example that uses the consensus folding technique is the comRNA program [11] which, once all stems of length at least ℓ are extracted from individual sequences, computes the maximum number of compatible stems[2] that are common to at least k of the sequences via a graph theoretic approach. As one can expect, the consensus technique also relies on the availability of many sequences that are functionally (and hopefully structurally) similar.

1.1 Our Approach

As described above, the most common objective in secondary structure prediction is total free energy minimization. In the context of multiple sequence structure prediction, this objective can be used in conjunction with additional criteria such as covariation in mutations on predicted stems etc., yet the effectiveness of such criteria very much depends on (1) the availability of sufficient number of RNA sequences with similar functions, and (2) reasonably high sequence similarity between the input sequences. When these two conditions are not met, single sequence energy minimization methods still provide the most accurate prediction. Furthermore, because multiple sequence folding methods generate consensus structures that involve those substructures found in the majority of the sequences, the stems they return get shorter and thus the number of correct base pairs they predict get worse with increasing number of input sequences.

---

[2] The notion of compatibility here allows the types of pseudoknots that are captured by the Pknots program.

The goal of this paper is to show that delocalizing the thermodynamic cost of forming an RNA substructure by considering the *energy density* of the substructure can improve on secondary structure prediction via free energy minimization. We describe a new algorithm and a software tool that we call Densityfold which aims to predict the secondary structure of an RNA sequence by minimizing the sum of energy densities of individual substructures. We believe that our approach may help understand the process of nucleation that is required to form biologically relevant RNA substructures.

Our starting observation is that potential stems that are most commonly realized in the actual secondary structure are those whose *free energy density* (i.e. length normalized free energy) is the lowest. Figure 1(a) depicts the known secondary structure of the *E.coli* 5S rRNA sequence. This sequence is one of the central examples used in [5] for illustrating the advantage of multiple sequence structure prediction approach (i.e. RNAscf) over single sequence structure prediction (i.e. mfold/RNAfold). Indeed, the mfold/RNAfold prediction for this sequence is quite poor as can be seen in figure 1(d). However, although RNAscf prediction using 20 sequences from 5s rRNA family is quite good, as reported in [5], the accuracy of the prediction deteriorates considerably when only 3 sequences, *E.coli*, *asellus aquaticus* and *cyprinus carpio* are used; this is illustrated in figure 1(e).[3] The prediction accuracy of the alifold program is also poor as depicted in figure 1(f). Most importantly, all of the above programs miss the most significant stem (enclosed by the base pair involving nucleotides 79 and 97) depicted in figure 1(b); when normalized by length, the mfold/RNAfold free energy table entry of this base pair is the smallest among all entries. (Compare this to the prediction of our program Densityfold, given in figure 1(c).)

We believe that some of the accuracy loss in structure prediction via total energy minimization can be attributed to "chance stems" which are sometimes chosen over "actual stems" due to problems commonly encountered in *local sequence alignment*. A stem is basically a local alignment between the RNA sequence and its reverse complement. Some of the energy minimization approaches (e.g. RNAscf program [5]) explicitly perform a local alignment search between the input RNA sequence and its reverse complement, in order to extract all potential stems of interest. However not all significant potential stems are realized in the actual secondary structure.

In the context of searching for significant alignments, the problems attributed to Smith-Waterman approach is usually considered to be a result of:

(1) the *shadow effect*, which refers to long alignments with relatively low conservation levels often having a higher score (and thus higher priority) than short alignments with higher conservation levels, and
(2) the *mosaic effect*, which refers to two highly conserved alignments with close proximity being identified as a single alignment, hiding the poorly aligned interval in between.

It is possible that the stem discovery process, which is performed either explicitly (e.g. in RNAscf) or implicitly (e.g. in mfold), may encounter with similar problems.

---

[3] This example is particularly interesting as the independent mfold/RNAfold prediction for some of these sequences are very accurate.

For example, two potential stems, which, by chance, occur in close proximity, can easily be chosen over a conflicting longer stem due to the mosaic effect: the free energy penalty of an internal loop (which will be left in between the two chance stems) is often insignificant compared to the benefit of "merging" two stems.

In the context of local sequence alignment, the impact of these effects could be reduced by the use of *normalized sequence alignment* introduced by Arslan, Egecioglu and Pevzner [4]. The normalized local alignment problem asks to find a pair of substrings with maximum possible alignment score, normalized by their length ($+L$, a user defined parameter to avoid "trivial" alignments of length 1).

Inspired by this approach we propose to apply a *normalized free energy* or *energy density* criteria to compute the secondary structure of one or more RNA sequences. The algorithms we present aim to minimize the sum of *energy densities* of the substructures of an RNA secondary structure.[4] The *energy density of a base pair* is defined as the free energy of the substructure that starts with the base pair, normalized by the length of the underlying sequence. The energy density of an unpaired base is then defined to be the energy density of the closest base pair that encloses it. The overall objective of secondary structure prediction is thus to minimize the total energy density of all bases, paired and unpaired, in the RNA sequence.

The algorithms we describe in this paper also enables one to minimize a linear combination of the total energy density and total free energy of an RNA sequence. Based on these algorithms, we developed the Densityfold program for folding a single sequence and the MDensityfold program for folding multiple sequences. We tested the predictive power of our programs on the RNA sequence families used by Bafna et al. [5] to measure the performance of the RNAscf program. We compare Densityfold and MDensityfold against all major competitors based on energy density minimization criteria - more specifically mfold/RNAfold, the best example of single sequence energy minimization, RNAscf, the best example of multiple sequence energy minimization without an alignment and alifold, the best example of multiple sequence energy minimization with an alignment. We show that when only one or a small number of functionally similar sequences are available, Densityfold can outperform the competitors, establishing the validity of energy density criteria as an alternative to the total energy criteria for RNA secondary structure prediction.

In the remainder of the paper we first describe a dynamic programming approach for predicting the secondary structure of an RNA sequence by minimizing the total free energy density. Then we show how to generalize this approach to minimize a linear combination of the free energy density and total free energy, a criteria that seems to capture the secondary structure of longer sequences. Because the running time of the most general approach is exponential with the maximum number of branches allowed in a multibranch loop we show how to approximate the energy density of such loops through a divide and conquer approach which must be performed iteratively until a satisfactory approximation is achieved. We finally provide some experimental results.

---

[4] Note that, unlike the Arslan, Egecioglu, Pevzner approach we do not need to introduce an additive factor, L, artificially: a base pair in an RNA structure has at least three nucleotides in between.

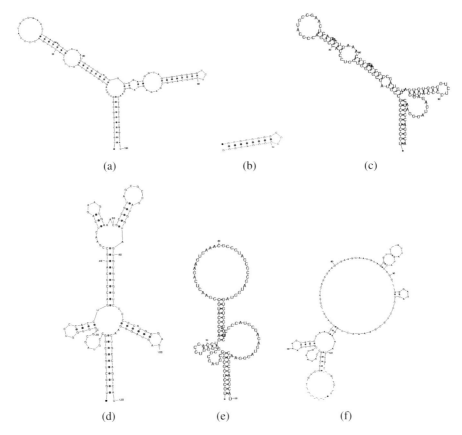

(a)   (b)   (c)

(d)   (e)   (f)

Fig. 1. (a) Known secondary structure of the *E.coli* 5S rRNA sequence. (b) The substructure with minimum energy density (missed by `mfold/RNAfold`, `RNAscf` and `alifold` programs). (c) Structure prediction by our `Densityfold` program. We capture the substructure with minimum energy density and correctly predict 28 of the 37 base pairs in the known structure. (d) Structure prediction by `mfold/RNAfold` program - only 10 of the 37 base pairs correctly predicted (e) Structure prediction by `RNAscf` program (consensus with the the *asellus aquaticus* and *cyprinus carpio* 5S rRNA sequences) - only 10 of the 37 base pairs correctly predicted (f) Structure prediction by `alifold` program (consensus with the *asellus aquaticus* and *cyprinus carpio* 5S rRNA sequences) - only 3 of the 37 base pairs correctly predicted.

2   Energy Density Minimization for a Single RNA Sequence

We start with description of our dynamic programming formulation for minimizing the total free energy density of the secondary structure of an RNA sequence. We denote the input sequence by $S = S[1 : n]$; the ith base of S is denoted by $S[i]$ and $S[i].S[j]$ denotes a base pair. Given input sequence S, its secondary structure $ST(S)$ is a collection of base pairs $S[i].S[j]$. A substructure $ST(S[i,j])$ is always defined for a base pair $S[i].S[j]$ and corresponds to the structure of the substring $S[i,j]$ within $ST(S)$. The base pair $S[i].S[j]$ is said to *enclose* the substructure $ST(S[i,j])$. The free energy

3   Minimizing a Linear Combination of the Energy Density and Energy

The initial tests we performed on the above dynamic programming formulation provided good outcomes for short RNA sequences; however as the sequence length increased, the predictive performance of this formulation deteriorated considerably. We noticed that although the energy density itself can help identify short structural motifs well, it may not provide the right criteria for "stitching them together". Thus, in this section we describe a modified version of the dynamic programming formulation we gave above for energy density minimization. The goal of this modified version is to minimize a linear combination of the energy density and the total free energy. More specifically, for any $x \in \{S, BI, M\}$ let $ELC_x(i, j) = ED_x(i, j) + \sigma \cdot E_x(i, j)$. The function we would like to optimize is thus $ELC(n) = ED(n) + E(n)$.

$$ELC(j) = \min \left\{ \begin{array}{l} ELC(j-1) \\ \min_{1 \le i \le j-1} \{ELC(i-1) + ELC_S(i,j)\} \end{array} \right\}$$

$$ELC_S(i, j) = \min \left\{ \begin{array}{ll} +\infty, & (i) \\ eH(i, j) \cdot (1 + \sigma), & (ii) \\ 2\frac{eS(i,j)+E_S(i+1,j-1)}{j-i+1} + ELC_S(i+1, j-1) + \sigma \cdot eS(i,j), & (iii) \\ ELC_{BI}(i, j), & (iv) \\ ELC_M(i, j) & (v) \end{array} \right\}$$

$$ELC_{BI}(i, j) = \min_{i',j' \mid i < i' < j' < j} \left\{ \begin{array}{l} \frac{eBI(i,j,i',j')+E_S(i',j')}{j-i+1} \cdot [(i'-i) + (j-j')] \\ +ELC_S(i', j') + \sigma \cdot eBI(i, j, i', j') \end{array} \right\}$$

For computing the value of our optimization function for multibranch loops efficiently we have to perform an approximation to the multibranch loop energy density through a divide and conquer approach For this we have to define a new energy table $ELC_M^{[i,j]}(k, \ell) = \overline{ED}_M^{[i,j]}(k, \ell) + \sigma \cdot \overline{E}_M^{[i,j]}(k, \ell)$ where $\overline{E}_M^{[i,j]}(k, \ell)$ and $\overline{ED}_M^{[i,j]}(k, \ell)$ are the free energy and the energy density of the optimal substructures for $S[k, \ell]$ provided that both $S[k]$ and $S[\ell]$ are on a multibranch loop starting with the base pair $S[i].S[j]$.

$$ELC_M(i, j) = \sigma \cdot a + \min_{i < k < j} \left\{ \overline{ELC}_M^{[i,j]}(i, k) + \overline{ELC}_M^{[i,j]}(k+1, j) \right\}$$

Here a is the multibranch loop opening score. Define:

$$\bar{b} = \frac{\widehat{E}_M(i, j)}{(j - i + 1)}$$

where $\widehat{E}_M(i, j)$ is an estimation (a lower bound) for $E_M(i, j)$ of the optimal structure. The initial value of $\widehat{E}_M(i, j)$ is obtained through the following dynamic programming routine.

$$\widehat{E}_M(i, j) = a + \min_{i < k < j} \left\{ \overline{E}_M(i, k) + \overline{E}_M(k+1, j) \right\}$$

$$\overline{E}_M(k, k) = b$$

$$\overline{E}_M(k, \ell) = \min \left\{ \begin{array}{l} E_S(k, \ell) + c + eDA(k-1, k) + eDA(\ell, \ell+1) \\ \min_{k \le h < \ell} \{\overline{E}_M(k, h) + \overline{E}_M(h+1, \ell)\} \end{array} \right\}$$

Here c is the contribution for each base pair on the multibranch loop and b is the unpaired base penalty. Based on this initial estimation $\widehat{E}_M(i, j)$ we have:

$$\overline{ELC}_M^{[i,j]}(k, k) = \bar{b} + \sigma \cdot b$$

$$\overline{ELC}_M^{[i,j]}(k, \ell) = \min \left\{ \begin{array}{l} ELC_S(k, \ell) + \sigma \cdot [c + eDA(k-1, k) + eDA(\ell, \ell+1)] \\ \min_{k \le h < \ell} \{\overline{ELC}_M^{[i,j]}(k, h) + \overline{ELC}_M^{[i,j]}(h+1, \ell)\} \end{array} \right\}$$

The corresponding energies of the substructures are as in the previous section:

$$E_S(i, j) = \left\{ \begin{array}{ll} (i): & +\infty, \\ (ii): & eH(i, j), \\ (iii): & E_S(i+1, j-1) + eS(i, j), \\ (iv): & E_{BI}(i, j), \\ (v): & E_M(i, j) \end{array} \right\}$$

$$E_{BI}(i, j) = eBI(i, j, i', j') + E_S(i', j') \quad for \ i', j' \ computed \ above$$

$$E_M(i, j) = eM(i, j, i_1, j_1, \ldots i_k, j_k) + E_S(i_1, j_1) \ldots + E_S(i_k, j_k)$$
$$for \ i_1, j_1 \ldots i_k, j_k \ computed \ above$$

Note that if $E_M(i, j) \ge \widehat{E}_M(i, j) + \epsilon$ for some user defined (small) value of ϵ we set $\widehat{E}_M(i, j) = \widehat{E}_M(i, j) + \epsilon$ and re-iterate the above procedure for computing $ELC_M(i, j)$. The reader can easily verify that the running time of this dynamic programming algorithm is $O(n^4)$.

3.1   Multiple Sequence Energy Density Minimization

The dynamic programming algorithm for minimizing $ELC(n)$ for a single sequence is generalizable to multiple sequences without difficulty. Here we follow the general approach taken by the `alifold` program: we start with the multiple sequence alignment of the input sequences (obtained by the `Clustal-W` program) and fold the aligned sequences simultaneously, with the objective of minimizing the sum of energy densities of all bases from each sequence. This is somewhat different from the `alifold` and `RNAscf` methods as both of them assigns the *maximum energy* among aligned substructures to the energy of the consensus structure. We assign the *total energy and total energy density* of the aligned substructures to the energy and, respectively, energy density of the consensus structure. The gaps are also included in the calculations as a base.

The reader can verify that for m sequences the running time of this dynamic programming algorithm is $O(m \cdot n^4)$.

4   Experimental Results and Discussion

We implemented and tested the performance of our algorithms for minimizing the linear combination of the energy density and the total free energy of a single sequence as well as of multiple sequences, respectively called Densityfold and MDensityfold. Our test set is comprised of the same 12 RNA families from the Rfam database [9] used by Bafna et al. [5] for testing the performance of RNAscf program. Using this test set, we compared the performance of Densityfold and MDensityfold with varying values of σ (which determines the contribution of the total energy to the optimization function) against mfold/RNAfold, the best single sequence energy minimization program, alifold the best multiple sequence energy minimization program that uses the alignment between the input sequences, and RNAscf the best multiple sequence energy minimization program that computes the alignment and the folding simultaneously. In the context of multiple sequence folding, our goal is to demonstrate the predictive power of MDensityfold when only a limited number of sequences are available; thus we only report on the jointly predicted structures of a pair of sequences, randomly selected from each family.

The most common measure for demonstrating the predictive power of a single sequence secondary structure determination method is the number of correct base pairs (see for example [11]). Unfortunately the Rfam database only provides the consensus structure of a family and not individual sequences; thus it is not possible to reliably count the number of predicted base pairs which appear in the actual structure of an individual sequence and vice versa. To overcome this problem Bafna et al. used an alternative, *stack counting* measure [5] which is defined as the number of actual stacks and predicted stacks that overlap. As mentioned in [5] this measure is intended for comparing methods that explicitly extract stacks - which is not performed by most of the methods we compare.

We thus measure the predictive power of the programs we tested under the *structural edit distance* measure [12, 14]. which considers the differences between two RNA molecules in terms of both sequence/stack composition and structural elements. Given the *tree representation* of two RNA secondary structures, where each branch is labeled with a stack and every node represents a loop, their structural edit distance is defined to be the minimum possible sum of edit distances between the stack compositions of branch pairs and sequences of node pairs that are aligned to each other.

We computed the structural edit distances between the actual (consensus) structure of each of the 12 test families and the structure predictions by each test program via the RNA_align tool, publicly available on the web [2]. A distance of 0 corresponds to an identical sequence and structure, i.e. a perfect prediction. A higher distance value implies a poorer prediction.

The results of our comparative tests are summarized in the table above. (In addition, figure 1 demonstrates the outcome of Densityfold on the *E.coli* 5s_rRNA sequence (from RF00001 family) with that of mfold/RNAfold, alifold and RNAscf.) We used the default parameters in all programs we tested. We list the outcome of Densityfold for $\sigma = 1.5$, 3.0 and 5.0, and list the outcome of MDensityfold for the best possible σ value. As can be seen, Densityfold is at the top or near the top for most of the families. Densityfold with $\sigma = 5.0$ is always better than Densityfold with $\sigma = 3.0$. However Densityfold with $\sigma = 1.5$ outperforms both in a number of examples. Note that as σ approaches to ∞ the outcome

Table 1. Structural edit distances between the actual (consensus) structure of a family and the predicted structures by each one of the programs tested

Name (Rfam_id)	Single sequence methods				Multiple sequence methods		
	mfold/ RNAfold	Densityfold $\sigma = 1.5$	$\sigma = 3$	$\sigma = 5$	MDensity_ fold	RNAscf	alifold
5s_rRNA (RF00001)	149	84	89	89	92	134	122
Rhino_CRE (RF00220)	94	93	93	93	77	88	30
ctRNA_pGA1 (RF00236)	45	83	83	83	48	91	44
glmS (RF00234)	194	288	230	230	189	249	198
Hammerhead_3 (RF00008)	2	2	2	2	74	2	88
Intron_gpII (RF00029)	100	93	103	103	85	113	78
Lysine (RF00168)	182	256	194	186	178	131	173
Purine (RF00167)	64	103	103	103	133	56	141
Sam_riboswitch (RF00162)	124	129	129	99	110	133	121
Thiamine (RF00059)	156	170	179	149	187	179	149
tRNA (RF00005)	31	67	67	67	50	31	32
ykok (RF00380)	158	200	189	189	168	203	157

of Densityfold gets more and more similar to the outcome of mfold/RNAfold.[5] However Densityfold with $\sigma = 5.0$ (the highest value we report) significantly outperforms mfold/RNAfold in a number of examples. Furthermore there is no clear winner between Densityfold and MDensityfold, each one outperforming the other in almost equal number of examples. However, in general, the longer the sequence gets, the better MDensityfold seemed to perform.

In conclusion, Densityfold demonstrates that an energy density minimization objective is a valid alternative to the total energy minimization objective. It can be used both on a single sequence or on multiple sequences. Our goal for the future is to test non-linear combinations of energy density and total energy as well as non-linear normalizations of the free energy as objective functions; we hope that such variations can explain the better performance of MDensityfold over Densityfold on longer sequences.

References

1. Mapping RNA Form & Function. *Science* **309 (5740)**, 2 September 2005.
2. RNA_align tool. http://www.csd.uwo.ca/faculty/kzhang/rna/
3. Akutsu, T., Dynamic programming algorithms for RNA secondary structure prediction with pseudoknots. *Discr. Appl. Math.* **104 (1-3)**, 45-62, 2000.
4. Arslan, A.N., Egecioglu, O., & Pevzner, P.A., A New Approach to Sequence Comparison: Normalized Sequence Alignment. *Proc. RECOMB*, **ACM**, 2-11, 2001.
5. Bafna, V., Tang, H. & Zhang, S., Consensus Folding of Unaligned RNA Sequences Revisited. *Proc. RECOMB*, **LNBI 3500**, 172-187, 2005.
6. Condon, A., Davy, B., Rastegari, B., Zhao, S., Tarrant F., Classifying RNA pseudoknotted structures. *Theor. Comput. Sci.* **320 (1)**, 35-50, 2004.

---

[5] In fact, we observed that for the families tested $\sigma = 100$ gives almost indistinguishable results to that by mfold/RNAfold.

7. Davydov, E.& Batzoglou, S., A Computational Model for RNA Multiple Structural Align-ment. *Proc. Symp. on Combinatorial Pattern Matching*, **LNCS 3103**, 254-269, 2004.
8. Gorodkin, J., Heyer, L. & Stormo, G., Finding the most significant common sequence and structure motifs in a set of RNA sequences. *Nucl. Acids Res.* **25 (18)**, 3724-3732, 1997.
9. Griffiths-Jones, S., Bateman, A., Marshall, M., Khanna, A., Eddy, S., Rfam: an RNA family database. *Nucl. Acids Res.* **31 (1)**, 439-441. 2003.
10. Hofacker, I., Fekete, M., Stadler, P., Secondary structure prediction for aligned RNA se-quences. *J. Mol. Biol.* **319 (5)**, 1059-1066, 2002.
11. Ji, Y., Xu, X., Stormo, G.D., A graph theoretical approach for predicting common RNA secondary structure motifs including pseudoknots in unaligned sequences. *Bioinformatics* **20 (10)**, 1591-1602, 2004.
12. Lin, G., Ma, B., Zhang, K., Edit distance between two RNA structures. *Proc. RECOMB* **ACM**, 211-220, 2001.
13. Lyngso, R.B., Zuker, M. & Pedersen, C.N.S., Fast evaluation of internal loops in RNA sec-ondary structure prediction. *Bioinformatics* **15 (6)**, 440-445, 1999.
14. Ma, B., Wang, L. & Zhang, K., Computing similarity between RNA structures *Theoretical Computer Science* **276 (1-2)**, 111-132, 2002.
15. Mathews, D., Sabina, J., Zuker, M. & Turner, D., Expanded sequence dependence of ther-modynamic parameters improves prediction of RNA secondary structure. *J. Mol. Biol.* **288 (5)**, 911-940, 1999.
16. Mathews, D.& Turner, D., Dynalign: an algorithm for finding the secondary structure com-mon to two RNA sequences. *J. Mol. Biol.* **317 (2)**, 191-203, 2002.
17. Nussinov, R. & Jacobson, A., Fast algorithm for predicting the secondary structure of single stranded RNA. *Proc. Nat. Acad. Sci. USA* **77 (11)**, 6309-6313, 1980.
18. Rivas, E. & Eddy, S.R., A dynamic programming algorithm for RNA structure prediction including pseudoknots. *J. Mol. Biol.* **285 (5)**, 2053-2068, 1999.
19. Sankoff, D., Simultaneous Solution of the RNA Folding, Alignment and Protosequence Prob-lems. *SIAM J. Appl. Math.* **45**, 810-825, 1985
20. Thompson, J., Higgins, D. & Gibson, T., Clustal-W: improving the sensitivity of progressive multiple sequence alignment through sequence weighting, position specific gap penalties and weight matrix choice. *Nucl. Acids Res.* **22**, 4673-4680, 1994.
21. Tinoco, I., Uhlenbeck, O.& Levine, M, Estimation of secondary structure in ribonucleic acids. *Nature* **230 (5293)**, 362-367, 1971.
22. Zuker, M. & Stiegler, P., Optimal computer folding of large RNA sequences using thermo-dynamics and auxiliary information. *Nucleic Acids Res.* **9 (1)**, 133-148, 1981
23. Zuker, M., On finding all suboptimal foldings of an RNA molecule. *Science* **244**, 48-52, 1989.

Structural Alignment of Pseudoknotted RNA

Banu Dost*, Buhm Han*, Shaojie Zhang, and Vineet Bafna

Department of Computer Science and Engineering,
University of California, San Diego,
La Jolla, CA 92093-0404, USA
{bdost, buhan, shzhang, vbafna}@cs.ucsd.edu

Abstract. In this paper, we address the problem of discovering novel non-coding RNA (ncRNA) using primary sequence, and secondary structure conservation, focusing on ncRNA families with pseudo-knotted structures. Our main technical result is an efficient algorithm for computing an optimum structural alignment of an RNA sequence against a genomic substring. This algorithm finds two applications. First, by scanning a genome, we can identify novel (homologous) pseudoknotted ncRNA, and second, we can infer the secondary structure of the target aligned sequence. We test an implementation of our algorithm (PAL), and show that it has near-perfect behavior for predicting the structure of many known pseudoknots. Additionally, it can detect the true homologs with high sensitivity and specificity in controlled tests. We also use PAL to search entire viral genome and mouse genome for novel homologs of some viral, and eukaryotic pseudoknots respectively. In each case, we have found strong support for novel homologs.

1 Introduction

Ribonucleic acid (RNA) is the third, and (until recently) most underrated of the trio of molecules that govern most cellular processes: the other two being proteins and DNA. While much of cellular RNA carries a message encoding an amino-acid sequence, other, 'non-coding' RNA participate directly in performing essential functions. Recent and unanticipated discoveries of novel ncRNA families [1, 2, 3, 4, 5] point to the possibility of a 'Modern RNA world' in which RNA molecules are as abundant, and diverse as protein molecules [6]. The analog of the computational gene-finding problem: "given genomic DNA, identify all substrings that encode ncRNA" is increasingly relevant, and relatively unexplored. While potentially abundant, RNA signals are weaker than proteins making them harder to identify computationally. Possibly, the strongest clue is from secondary structure. Being single-stranded, the base-pairs stabilize by forming hydrogen bonds, leading to a characteristic secondary and tertiary structure. With a few exceptions, the base-pairs are *non-crossing*, and form a tree-like structure. This recursive structure is the basis for efficient algorithms to predict RNA structure [7, 8]. With this extensive work in structure prediction, it is natural to expect that novel non-coding RNA could be discovered simply by looking for genomic sub-strings that fold

---

* These authors contributed equally in the research.

A. Apostolico et al. (Eds.): RECOMB 2006, LNBI 3909, pp. 143–158, 2006.

into low-energy structures. Unfortunately, that idea doesn't work. Rivas and Eddy [9] showed that random DNA (usually with high GC-content) can also 'fold' into low-energy configurations, making it unlikely for a purely *de novo* approach to be successful. Therefore, a comparative approach is employed, often typified by the question: "Given a query RNA with known structure, and a genome, identify all genomic substrings that match the query sequence and structure". The query itself can be either a single molecule or a model (covariance model/stochastic context free grammar) of an RNA structure. This approach has been quite successful and single queries as well as covariance based models are routinely used to annotate genomes with ncRNA [10, 11]. Central to these approaches is an algorithm for computing a local alignment between a query structure and a DNA string. The search itself is simply a scan of the genome to obtain all high scoring local alignments.

Here we pose a related question: *Given a query RNA with known structure, allowing for pseudoknots, and a genome, identify all genomic sub-strings that match the query sequence and structure*. Without being precise, pseudoknots are base-pairs that violate the non-crossing rule (See Figure 1). While not as common as other substructures (bulges,loops), they are often critically important to function. Pseudoknotted RNAs are known to be active as ribozymes [12], self-splicing introns [13], and participate in telomerase activity [14]. They have also been shown to alter gene expression by inducing ribosomal frame-shifting in many viruses [15]. However, understanding the extent and importance of these molecules is partially handicapped by the difficulty of identifying them (computationally). The algorithm presented here will facilitate identification.

In order to compute a local structural alignment, we must start with a formal definition of a pseudoknot in Section 2. Many definitions of pseudoknots have been postulated [16, 17, 18, 19, 20], and recent research investigates the power of these definitions in describing real pseudoknots [21]. We start here with Akutsu's formalism (*simple pseudoknots*) [16], which has a clean recursive structure and encompasses a majority of the known cases [21, 22]. We also present algorithms that extend this class of allowed pseudoknots (standard pseudoknots). Section 3 describes the chaining procedure which is key to the alignment algorithm that follows (Section 4). However, the simple pseudoknots usually do not occur independently, but are embedded in regular RNA structures. In Section 5, we extend the algorithm to handle these cases. Other extensions are considered in Section 7. It has been brought to our attention that a recent publication [23] considers the identical problem using the formation of tree adjoining grammars to model pseudoknots. The pseudoknots considered by them are a restricted version of our simple pseudoknots. Futhermore, our alignment combines sequence and structural similarity. A detailed comparison is deferred to the full version of the paper.

The local alignments can be used in two ways. First, they can be used to infer the structure of the aligned substring that is conserved with the query. We show in Section 8.1 that in a majority of the cases, this leads to a perfect prediction of secondary (pseudoknotted) structure. Next, they can be used to predict novel ncRNA in genomic sequences. While our algorithms are computationally intensive, they can be used in combination with database filtering approaches to search large genomic regions. In Section 8.2, we validate our approach on real sequences embedded in random sequence.

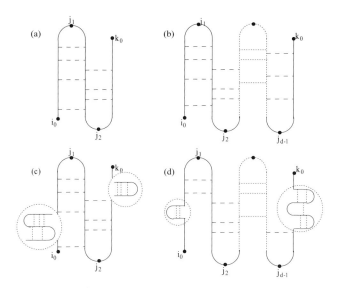

Fig. 1. (a) Simple pseudoknot. (b) Standard pseudoknot of degree d. (c) Recursive simple pseudo-knot. (d) Recursive standard pseudoknot of degree d.

Finally, in Section 9, we identify (putative) novel pseudoknotted ncRNA in a search of viral and eukaryotic genomes.

2 Definitions and Preliminary Information

Let $A = a_1 \ldots a_n$ be an RNA sequence. The secondary structure is represented simply as the set of base-pairs

$$M = \{(i,j) | 1 \leq i < j \leq n, (a_i, a_j) \text{ is a base pair}\}$$

Also, let $M_{i_0,k_0} \subseteq M$ be defined by $M_{i_0,k_0} = \{(i,j) \in M | i_0 \leq i < j \leq k_0\}$. The secondary structure, in the absence of crossing or interweaving base-pairs is called *regular*, and has the following recursive definition.

Definition 1. *An RNA secondary structure M_{i_0,k_0} is regular if and only if $M_{i_0,k_0} = \phi$ or $\exists (i,j) \in M_{i_0,k_0}$ such that*

- $M_{i_0,k_0} = M_{i_0,i-1} \cup M_{i+1,j-1} \cup M_{j+1,k_0} \cup \{(i,j)\}$ *(No base-pairs cross the partitions).*
- *Each of $M_{i_0,i-1}, M_{i+1,j-1}, M_{j+1,k_0}$ is regular.*

Next, we can define the class of allowed pseudoknots ([16]).

Definition 2. *M_{i_0,k_0} is a simple-pseudoknot (see Figure 1(a)) if and only if M_{i_0,k_0} is regular or $\exists j_1, j_2 \in \mathbb{N}$ ($i_0 \leq j_1 < j_2 \leq k_0$) such that the resulting partition, $D_1 = [i_0, j_1 - 1], D_2 = [j_1, j_2 - 1], D_3 = [j_2, k_0]$, satisfies the following:*

- $M_{i_0,k_0} = (S_L \cup S_R)$, where $S_L = \{(i,j) \in M_{i_0,k_0} | i \in D_1, j \in D_2\}$ and $S_R = \{(i,j) \in M_{i_0,k_0} | i \in D_2, j \in D_3, \}$.
- S_L and S_R are regular.

Definition 3. M_{i_0,k_0} is a standard-pseudoknot with degree d ($d \geq 3$, see Figure 1(b)) if and only if M_{i_0,k_0} is regular or $\exists j_1, ..., j_{d-1} \in \mathbb{N}$ ($i_0 \leq j_1 < ... < j_{d-1} \leq k_0$) which divide $[i_0, k_0]$ into d parts, $D_1 = [i_0, j_1 - 1], D_2 = [j_1, j_2 - 1], ..., D_d = [j_{d-1}, k_0]$, and satisfy the following:

- $M_{i_0,k_0} = \bigcup_{l=1}^{d-1} S_l$, where $S_l = \{(i,j) \in M_{i_0,k_0} | i \in D_l, j \in D_{l+1}\}$ for all $1 \leq l < d$.
- S_l is regular for all $1 \leq l < d$,

Note that a simple-pseudoknot is a standard-pseudoknot of degree 3.

Definition 4. M_{i_0,k_0} is recursive-standard-pseudoknot with degree d ($d \geq 3$, see Figure 1(d)) if and only if M_{i_0,k_0} is a standard pseudoknot of degree d or $\exists i_1, k_1, ..., i_t, k_t \in \mathbb{N}$ ($i_0 \leq i_1 < k_1 < i_2 < k_2 < ... < i_t < k_t \leq k_0, t \geq 1$), which satisfy the following:

- ($M_{i_0,k_0} - \bigcup_{l=1}^{t} M_{i_l,k_l}$) is a standard pseudoknot of degree $\leq d$.
- $M_{i_l,k_l} (1 \leq l \leq t)$ is a recursive standard pseudoknot of degree $\leq d$.

A recursive-simple-pseudoknot is a recursive-standard-pseudoknot of degree 3 (Figure 1(c)). While we can devise algorithms to align recursive-standard-pseudoknots, they are computationally expensive, and most known families have a simpler structure. Therefore, we will limit our description and tests to a simpler structure (with a single level of recursion), defined as follows:

Definition 5. M_{i_0,k_0} is embedded-simple-pseudoknot if and only if $\exists i_1, k_1, ..., i_t, k_t \in \mathbb{N}$ ($i_0 \leq i_1 < k_1 < i_2 < k_2 < ... < i_t < k_t \leq k_0, t \geq 1$), which satisfy the following:

- ($M_{i_0,k_0} - \bigcup_{l=1}^{t} M_{i_l,k_l}$) is regular.
- $M_{i_l,k_l} (1 \leq l \leq t)$ is a simple-pseudoknot.

In the full version of the paper, we extend these algorithms to the case of standard-pseudoknots. The full version of the paper will present the algorithm for the most general case (recursive-standard-pseudoknot).

2.1 Structural Alignment Preliminaries

For alignment purposes, we do not distinguish between RNA and DNA, as every substring in the genome might encode an RNA string. Let $q[1 \cdots m]$ and $t[1 \cdots n]$ be two RNA strings over the alphabet $\sum = \{A, C, G, U\}$ where q has a known structure M. An alignment of q and t is defined by a 2-rowd matrix A, in which row 1 (respectively, 2) contains q (respectively, t) interspersed with spaces, and for all columns j, $A[1,j] \neq '-'$ or $A[2,j] \neq '-'$. For $r \in \{1,2\}$, define $\iota_r[i] = i - |\{l < i \text{ s.t. } A[r,l] =' -'\}|$. In other words, if $A[1,i] \neq '-'$, it contains the symbol $q[\iota_1[i]]$. The score of alignment A is given by

$$\sum_j \gamma(A[1,j], A[2,j]) + \sum_{i,j \text{ s.t. } (\iota_1[i],\iota_1[j]) \in M} \delta(\iota_1[i], \iota_1[j], \iota_2[i], \iota_2[j])$$

The function γ scores for sequence similarity, while δ scores for conservation of structure. While this formulation encodes a linear gap penalty, we note here that alignments of RNA molecules may contain large gaps, particularly in the loop regions, and we implement affine penalties for gaps (details omitted). Naturally, we wish to compute alignments with the maximum score.

The key ideas are as follows: First, note that regular and pseudoknotted structures have a recursive formulation. Therefore, the problem of structurally aligning an RNA structure against a subsequence, can be decomposed into the problems of (recursively) aligning its sub-structures against the appropriate sub-sequences, and combining the results. For regular-structures, the structure is tree-like, and the recursion follows the nodes of the tree. For simple-pseudoknots, the structure is more complex, and will be described in Section 4. The structure for embedded-simple-pseudoknots is simply a combination of the two (See Section 7).

However, it is not sufficient to consider structural elements alone, as we wish to score for sequence conservation as well. The recursive structure described only contains a subset of the nucleotides that participate in structure. Therefore, we employ a second trick of introducing spurious structural elements (base-pairs) to M. The augmented structure M' must have the following properties:

- Each nucleotide i appears in M'.
- $|M'| = O(m)$, so that the size of the structure does not increase too much.
- The recursive structure of M is maintained.

Pseudoknots and regular structures have very different recursive structure, and require different augmentation procedures. In Section 3, we present *chaining*, a novel augmentation procedure for simple pseudoknots. An augmentation for regular structures, *binarization* was presented in [24], and is implicit in the covariance models used to align regular RNA [25]. Here, we extend binarization to include chaining for embedded-simple-pseudoknots (Figure 5). These augmentations are used in the alignment algorithms for simple (Section 4), and embedded-simple-pseudoknots (Section 5).

3   Chaining

Before describing the chaining procedure, we revisit the problem of aligning a simple pseudoknot to a genomic sub-string. Unlike regular structures, we cannot partition the genome into contiguous substrings, because of interweaving base pairs. Thus, we need a new substructure for simple pseudoknot structures.

We start by defining a total ordering among the base pairs of a simple pseudoknot. Recall (Definition 3) that a simple-pseudoknot structure M_{i_0,k_0} can be divided into 3 parts: $D_1 = [i_0, j_0 - 1], D_2 = [j_0, j_0' - 1], D_3 = [j_0', k_0]$. (See Figure 2(a)) For each base pair $(i, j) \in M$, exactly one of i and j is in D_2 part. We define an ordering of the base pairs in M by sorting the coordinate in D_2. Formally, define $D_2(i, j)$ for all $(i, j) \in M$ as follows: $D_2(i, j) = i$ if $(i, j) \in S_R$, and $D_2(i, j) = j$ otherwise. For each $(i, j), (i', j') \in M$,

$$(i, j) \geq_p (i', j') \text{ iff } D_2(i, j) \geq D_2(i', j')$$

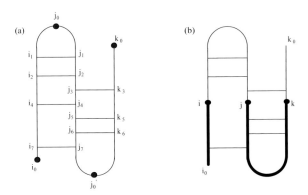

Fig. 2. (a) Base pairs in a simple pseudoknot are ordered according to the index of the endpoint along $[j_0, j_0']$. Therefore, $(i_1, j_1) > (i_2, j_2) > (j_3, k_3) > (i_4, j_4) > (j_5, k_5) > (j_6, k_6) > (i_7, j_7)$. (b) Subpseudoknot structure.

As distinct base-pairs do not share any coordinates, $\geq_p$ defines a total ordering on the actual base-pairs, and can be used to define a partial order on substructures that we can recurse on. Define a *subpseudoknot* $\mathcal{P}(i, j, k)$ as the union of two subintervals $\mathcal{P}(i, j, k) = [i_0, i] \cup [j, k]$ (Figure 2 (b)). Denote the triple (i, j, k) as the *frontier* for $\mathcal{P}(i, j, k)$. Note that i_0 is implicit from the context. Suppose that we are aligning frontier (i', j', k') of the query against frontier (i, j, k) of the target, with the score represented by $B[i, j, k, i', j', k']$. A naive algorithm would need to consider $O(m^3 n^3)$ pairs of frontiers. We improve this as follows: consider the special case of $(i', j') \in M$ where $(i', j') \in S_L$. The following recursion gives the score for B (proof omitted).

Theorem 1

$$B[i, j, k, i', j', k'] = \max\{ MATCH, INSERT, DELETE\} \tag{1}$$

$$MATCH = B[i - 1, j + 1, k, i' - 1, j' + 1, k'] + \delta(q[i'], q[j'], t[i], t[j])$$
$$+ \gamma(q[i'], t[i]) + \gamma(q[j'], t[j]), \tag{2}$$

$$DELETE = \max \begin{cases} B[i - 1, j, k, i' - 1, j' + 1, k'] + \gamma(q[i'], t[i]) + \gamma(q[j'], '-'), \\ B[i, j + 1, k, i' - 1, j' + 1, k'] + \gamma(q[i'], '-') + \gamma(q[j'], t[j]), \\ B[i, j, k, i' - 1, j' + 1, k'] + \gamma(q[i'], '-') + \gamma(q[j'], '-') \end{cases} \tag{3}$$

$$INSERT = \max \begin{cases} B[i - 1, j, k, i', j', k'] + \gamma('-', t[i]), \\ B[i, j + 1, k, i', j', k'] + \gamma('-', t[j]), \\ B[i, j, k - 1, i', j', k'] + \gamma('-', t[k]) \end{cases} \tag{4}$$

Note that in every sub-case of MATCH and DELETE, we move from the query frontier (i', j', k') to the frontier $(i' - 1, j' + 1, k)$, because if either i' or j' is not used, we cannot score for the pair (i', j'). In the INSERT case, we stay at the frontier (i', j', k'). The situation is symmetric when $(j', k') \in S_R \subseteq M$, but is not defined when $(i', j') \notin M \wedge (j', k') \notin M$. The key idea for the chaining procedure is that we can define a unique frontier to move to in all cases, and still ensure that each nucleotide is touched by at least one frontier. By starting with a fixed frontier, and always moving to a fixed child, we only have $O(m)$ frontiers to consider.

From Definition 2, there exist indices j_1, j_2 which divide the simple pseudoknot structure into D_1, D_2 and D_3. We choose $(j_1 - 1, j_1, k_0)$ as the *root* frontier. Note that $\mathcal{P}(j_1 - 1, j_1, k_0)$ represents the entire simple-pseudoknot (See Figure 3(a)). We maintain the invariant that if (i, j, k) is a frontier and j participates in a base-pair, then the base-pair must be 'below' or within the frontier. In other words, if $(i', j) \in S_L$, then $i' \leq i$. Likewise, if $(j, k') \in S_R$, then $k' \leq k$. For a frontier (i, j, k), we have different cases: for example, if $(i', j) \in S_L$, we add spurious base pairs $(i, j), (i - 1, j), \ldots (i', j)$. These base pairs define an ordered set of frontiers $(i, j, k) \geq (i - 1, j, k) \geq \ldots, (i', j, k) \geq (i' - 1, j + 1, k)$. Likewise, if $(j, k') \in S_R$, we add spurious base-pairs $(j, k), (j, k - 1), \ldots, (j, k')$, which define the frontiers $(i, j, k) \geq \ldots \geq (i, j + 1, k' - 1)$. The chaining algorithm, with a complete listing of cases is described in Figure 3. The output of chaining is a directed path of 'frontiers'. The number of nucleotides in a frontier (i, j, k) is given by the expression $((i - i_0 + 1) + (k - j + 1)) \leq m$. Further, this number decreases by at least 1 for each adjacent frontier. Thus the

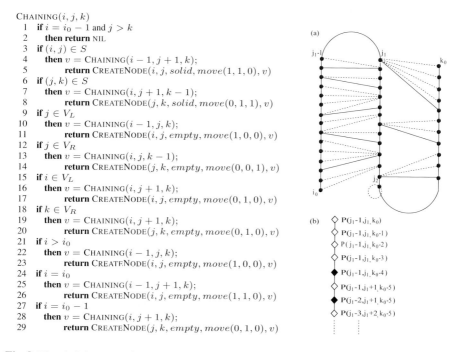

```
CHAINING(i, j, k)
 1   if i = i_0 − 1 and j > k
 2      then return NIL
 3   if (i, j) ∈ S
 4      then v = CHAINING(i − 1, j + 1, k);
 5           return CREATENODE(i, j, solid, move(1, 1, 0), v)
 6   if (j, k) ∈ S
 7      then v = CHAINING(i, j + 1, k − 1);
 8           return CREATENODE(j, k, solid, move(0, 1, 1), v)
 9   if j ∈ V_L
10      then v = CHAINING(i − 1, j, k);
11           return CREATENODE(i, j, empty, move(1, 0, 0), v)
12   if j ∈ V_R
13      then v = CHAINING(i, j, k − 1);
14           return CREATENODE(j, k, empty, move(0, 0, 1), v)
15   if i ∈ V_L
16      then v = CHAINING(i, j + 1, k);
17           return CREATENODE(i, j, empty, move(0, 1, 0), v)
18   if k ∈ V_R
19      then v = CHAINING(i, j + 1, k);
20           return CREATENODE(j, k, empty, move(0, 1, 0), v)
21   if i > i_0
22      then v = CHAINING(i − 1, j, k);
23           return CREATENODE(i, j, empty, move(1, 0, 0), v)
24   if i = i_0
25      then v = CHAINING(i − 1, j + 1, k);
26           return CREATENODE(i, j, empty, move(1, 1, 0), v)
27   if i = i_0 − 1
28      then v = CHAINING(i, j + 1, k);
29           return CREATENODE(j, k, empty, move(0, 1, 0), v)
```

Fig. 3. The chaining procedure on a simple pseudoknot structure M_{i_0, k_0}. (a) Solid base pairs are the actual base pairs, dotted ones are the spurious base pairs. (b) Chain structure representing the simple pseudoknot structure M_{i_0, k_0}. Solid nodes represents a sub-pseudoknot with frontier (i, j, k) where (i, j) or (j, k) is an actual base pair. Empty nodes represents a sub-pseudoknot with frontier (i, j, k) where neither (i, j) nor (j, k) is an actual base pair.

(a)   ALIGN-SP $(M', t[1...n])$
    1   // M' is the chain representing the simple pseudoknot region to be aligned in query q
    2   **for** all intervals (i_0, k_0) in $t[1...n]$
    3   **do for** all (i, j, k), $i_0 \le i < j \le k \le k_0$
    4   **do for** all nodes $v \in M'$
    5   **do if** $v \in M_L$

    6   **then** $B[i, j, k, v] = \max \begin{cases} B[i-1, j+1, k, child(v)] + \delta(q[l_v], q[m_v], t[i], t[j]) \\ \qquad + \gamma(q[l_v], t[i]) + \gamma(q[m_v], t[j]), \\ B[i-1, j, k, child(v)] + \gamma(q[l_v], t[i]) + \gamma(q[m_v], '-'), \\ B[i, j+1, k, child(v)] + \gamma(q[l_v], '-') + \gamma(q[m_v], t[j]), \\ B[i, j, k, child(v)] + \gamma(q[l_v], '-') + \gamma(q[m_v], '-') \end{cases}$

    7   **if** $v \in M_R$

    8   **then** $B[i, j, k, v] = \max \begin{cases} B[i, j+1, k-1, child(v)] + \delta(q[m_v], q[r_v], t[j], t[k]) \\ \qquad + \gamma(q[m_v], t[j]) + \gamma(q[r_v], t[k]), \\ B[i, j, k-1, child(v)] + \gamma(q[m_v], '-') + \gamma(q[r_v], t[k]), \\ B[i, j+1, k, child(v)] + \gamma(q[m_v], t[j]) + \gamma(q[r_v], '-'), \\ B[i, j, k, child(v)] + \gamma(q[m_v], '-') + \gamma(q[r_v], '-') \end{cases}$

    9   **if** $v \in M_S$ and $move(v) = (1, 0, 0)$

    10   **then** $B[i, j, k, v] = \max \begin{cases} B[i-1, j, k, child(v)] + \gamma(q[l_v], t[i]), \\ B[i, j, k, child(v)] + \gamma(q[l_v], '-') \end{cases}$

    11   **if** $v \in M_S$ and $move(v) = (0, 0, 1)$

    12   **then** $B[i, j, k, v] = \max \begin{cases} B[i, j, k-1, child(v)] + \gamma(q[r_v], t[k]), \\ B[i, j, k, child(v)] + \gamma(q[r_v], '-') \end{cases}$

    13   **if** $v \in M_S$ and $move(v) = (0, 1, 0)$

    14   **then** $B[i, j, k, v] = \max \begin{cases} B[i, j+1, k, child(v)] + \gamma(q[m_v], t[k]), \\ B[i, j, k, child(v)] + \gamma(q[m_v], '-') \end{cases}$

    15    $B[i, j, k, v] = \max \begin{cases} B[i, j, k, v] \\ B[i-1, j, k, v] + \gamma('-', t[i]), \\ B[i, j+1, k, v] + \gamma('-', t[j]), \\ B[i, j, k-1, v] + \gamma('-', t[k]) \end{cases}$

    16
    17    $B_{SP}[i_0, k_0, i_{SP}, k_{SP}] = \max_{j=i+1, k=k_0} \{B(i, j, k, \text{ROOT}(M'))\}$

(b)   IMPROVED ALIGN-SP $()$
    1   **for** all $v \in M'$
    2   **do for** $i_0 = 1$ **to** $n - 1$
    3   **do for** $i = i_0 - 1$ **to** $n - 1$
    4   **do for** $j = n + 1$ **downto** $i + 1$
    5   **do for** $k = j - 1$ **to** n
    6   **do** Compute $B[i, j, k, v]$

Fig. 4. (a) Align-SP procedure for alignment of a simple pseudoknot structure to a target sequence $t[1...n]$. (b) Improved Align-SP procedure.

number of nodes in the chain is $O(m)$. We still need to consider $O(n^3)$ target frontiers in aligning, for a complexity of $O(mn^3)$.

4   Alignment Algorithm for Simple-Pseudoknots

Figure 4(a) describes the algorithm ALIGN-SP for aligning a simple-pseudoknot to a DNA substring. Its input is a chain of query sub-pseudoknots, which is aligned to all sub-pseudoknots $\mathcal{P}(i, j, k)$ of the target sequence $t[1 \ldots n]$. Let M_L (respectively M_R) be the set of solid nodes representing subpseudoknots $\mathcal{P}(i, j, k)$ where $(i, j) \in S_L$ (respectively, $(j, k) \in S_R$). Let M_S be set of the nodes representing subpseudoknots $\mathcal{P}(i, j, k)$ where neither $(i, j) \notin S_L$, and $(j, k) \notin S_R$.

As an example, suppose we are aligning sub-pseudoknot $\mathcal{P}(i, j, k)$ in t to the sub-chain rooted at v. Let $B[i, j, k, v]$ be the score of the optimal alignment. First, we have

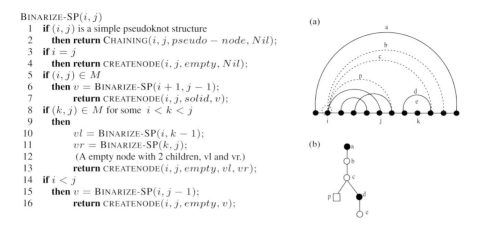

BINARIZE-SP(i, j)
```
 1   if (i, j) is a simple pseudoknot structure
 2     then return CHAINING(i, j, pseudo − node, Nil);
 3   if i = j
 4     then return CREATENODE(i, j, empty, Nil);
 5   if (i, j) ∈ M
 6     then v = BINARIZE-SP(i + 1, j − 1);
 7       return CREATENODE(i, j, solid, v);
 8   if (k, j) ∈ M for some  i < k < j
 9     then
10         vl = BINARIZE-SP(i, k − 1);
11         vr = BINARIZE-SP(k, j);
12         (A empty node with 2 children, vl and vr.)
13         return CREATENODE(i, j, empty, vl, vr);
14   if i < j
15     then v = BINARIZE-SP(i, j − 1);
16       return CREATENODE(i, j, empty, v);
```

Fig. 5. Binarization procedure revised for embedded-simple-pseudoknots and an illustration. (a) An embedded-simple-pseudoknot with spurious base pairs added. (b) Resulting binary tree. Solid nodes correspond to actual base pairs while empty (circular) nodes correspond to spurious base pairs. A '□' represents a pseudonode and subtree rooted at a pseudonode is formed by Chaining procedure.

cases involving insertion of target nucleotides: $t[i], t[j]$, and $t[k]$, as described by the recurrence in Figure 4(a)(Line 15). Next, we have the cases corresponding to match or deletion of v. We consider the case $v \in M_L$ corresponding to the subpseudoknot $\mathcal{P}(l_v, m_v, r_v)$ in q. The following cases can occur

1. $(t[i], t[j])$ is a pair in t corresponding to the pair $(q[l_v], q[m_v])$ in q.
2. $q[l_v]$ is substituted with $t[i]$ and $q[m_v]$ is deleted.
3. $q[m_v]$ is substituted with $t[j]$ and $q[l_v]$ is deleted.
4. $q[l_v]$ and $q[m_v]$ are both deleted.

The corresponding recurrences are shown on Line 6 of the procedure. The other cases are handled in an analogous fashion and are described in Figure 4.

5   Alignment Algorithm for Embedded-Simple-Pseudoknots

We consider now the special case of aligning recursive-simple-pseudoknots in which simple-pseudoknots are embedded in a regular structure. This is by far the most common occurrence of pseudoknots. While it is relatively easy to extend our algorithms to handle the full generality of recursive-pseudoknots, the complexity increase makes the algorithms untractable for real problems. Thus, this special case offers a compromise between generality and practicality.

The first step in the procedure is to *binarize* the query RNA, so that every nucleotide is in a base-pair, and can be represented by a binary tree of size $O(m)$ [24]. The main difference is that we invoke the chaining procedure whenever a simple-pseudoknot is encountered. Thus, in the binary tree, the simple pseudoknot substructure appears as a chain rooted at a *pseudo-node*.

(false negatives), defined as follows: *TP* is the number of base pairs in inferred target structure that are correct: *FP* is the number of base pairs in the inferred structure that are not in the true structure, and *FN* is number of base pairs in the true structure that are not inferred. We define *Specificity* = $TP / (TP + FP)$ and *Sensitivity* = $TP / (TP + FN)$. Good performance is indicated by both being close to 1. Table 2 summarizes the result of testing each pair in the 6 families. As the results show, PAL is a strong predictor of structure, with mean sensitivity and specificity of 0.95. We also investigated the few cases in which the prediction was away from the mean. In most of those cases, the target had stem loops that were longer than the query. As they were not aligned to the query structure, they were not inferred. In practice, we would augment the inferred structure by a local extension of stem loops in both directions. A second source of errors was incorrect annotation in Rfam. Other than these two scenarios, the structure inference was essentially correct.

There is a second caveat in these results which is not apparent. Many (but not all) of the sequences have high sequence similarity, which might be making the alignment task easier. We believe this is because a sequence search tool like Blast is used to fish out candidates, which are then manually aligned, and experimentally validated. We will show in the following sections that our tool can pick out candidates that BLAST cannot find, and also align them structurally. Also, in the cases where there isn't high sequence similarity, the structure inference was just as good.

Table 2. Pairwise tests: Statistics for Specificity and Sensitivity values. Mean is the average of Specificity (Sensitivity) values and median is the mid-point of Specificity (Sensitivity) values over all seed member pairs in an RNA family.

RNA Family	Specificity				Sensitivity			
	Mean	StdDev	Median	Range	Mean	StdDev	Median	Range
UPSK	1.000	0.000	1.000	(1.000-1.000)	1.000	0.000	1.000	(1.000-1.000)
Antizyme	0.991	0.020	1.000	(0.941-1.000)	0.991	0.020	0.941	(0.941-1.000)
Parecho	0.951	0.052	0.976	(0.848-1.000)	0.938	0.053	0.952	(0.844-1.000)
Corona-FSE	0.944	0.100	1.000	(0.737-1.000)	0.937	0.105	1.000	(0.737-1.000)
Corona-pk3	0.971	0.053	1.000	(0.765-1.000)	0.968	0.056	1.000	(0.722-1.000)
IFN-gamma	0.937	0.092	1.000	(0.782-1.000)	0.934	0.093	1.000	(0.782-1.000)

8.2 Searching for Structural Homologs

In this test, we use one of the members of an RNA family as a query, and look for its homolog in a large random sequence, with the other members inserted. Figure 7(a) shows the results for the Corona-FSE family, in which 17 members were embedded in a 19kb random sequence. The windowed scores are shown by solid lines. The actual positions of the remaining 17 members are denoted by '*'. We note that the true hits are easily the highest scoring regions along the sequence, and that all true positives score higher than all the false hits. The lowest scoring TP has a score of 988 and the highest scoring FP has a score of 606. Moreover, the random sequence scores do not

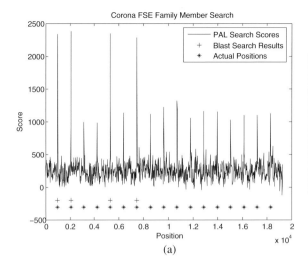

	# Found	
RNA Family	BLAST	PAL
UPSK	3	3
Antizyme	12	12
Parecho CRE	4	4
Corona-FSE	4	17
Corona-pk3	5	13
IFN-gamma	4	4

(a)   (b)

Fig. 7. Use of PAL as a pseudoknot RNA search tool (a) Score plot for Corona-FSE homologue search. '*' denotes actual positions of the members and '+' denotes the members located by Blastn. (b) Comparison against BLAST on other families.

show a large variation. We do not compute P-values on the hits, but in future work, we will use the distribution of scores on random, or genomic sequence (with differing GC-content) to compute the P-value. In general, the distribution is not understood, and we will either use a non-parametric value such as the Chebyshev's inequality [26], or perhaps the Gumbel distribution, which has been shown to be a good approximation to the actual distribution [11]. In contrast, Blastn (E-value 10, Word-size 7) is able to locate only 4 of the members. These results also show the significance of the secondary structure for searching homologue in addition to the primary structure. We repeat the same experiment for RNA families, UPSK, Antizyme, Parecho, Corona-FSE, Corona-pk3 and IFN-gamma. In all cases, PAL locates all members as the topmost hits (See Figure 7(b)). We agree that Blast is not the most appropriate tool for comparison as other tools such as RSEARCH, and our own tool FastR can search for structural homologs of RNA [11, 26]. However, these other tools cannot align psuedoknotted RNA and the search must be followed up with a correct alignment to determine homologs. Also, the complexity of these methods often force a use of Blast to determine initial candidates. In the next section, we show that our tool used in conjunction with RNA filters can efficiently search large genomes.

9   Searching Genomes for Pseudoknots

While PAL is accurate in fishing for structural homologs, it is computationally intensive, making genome scale searches intractable. However, there has been much recent research (including our own work) on *computational filters* for RNA, which quickly eliminate much of the database, while retaining the true homologs [27, 26]. We used PAL in conjunction with sequence based filters [28] to search genomes, for the 3 most interesting families.

```
Query:     Human chromosome 12, minus, 66839786 - 66839618
Subject:   Mouse chromosome 10, plus, 118018890-118019061


             .......AAAAAAA<<<<<<<<           ..<<<<<.   .  <<<<<....<<<<<<<<<..
Query:     CAUUGUUCUGAUCAUCUGAAGA---------UCAGCUAU--U--AGAAGAGAAAGAUCAGUUA
             ||        |+**++ +*+*++++*          ||+***+    |  ** **   |  * **** |
Sbjct:     CA-----GAGAGGUGCAGGCUAUAGCUGCCAUCGGCUGACCUAGAGAAG--ACACAUCAGCU-
             .......AAAAAAA<<<<<<<<...........<<<<<......<<<<<....<<<<<<<<..


             <<<..<<<<....aaaaaaa .......>>>>.>>>>>>>>>>>...>>>>>.>>>>>....>
Query:     AGUCCUUUGGACCUGAUCAG-CUUGAUACAAGAACUACUGAUUUCAACUUCUUUGGCUUAAUU
             ++*||****||  ++**+  ||||| |****||++* +++*  *|  ** ** +***+    *
Sbjct:     GAUCCUUUGGA--CCCUCUGACUUGAGACAGAAGUUCUGGGCUUCUCCUCCUGCGGCC----U
             <<<..<<<<....aaaaaaa........>>>>.>>>>>>>>>>>...>>>>>.>>>>>....>


             >>.>>>>><<<<<<..  <<<<<.....>>>>...<<<<....>>>>...  >>>>>>>
Query:     CUCUCGGAAACGAUGAA--AUAUACAAGUUAUAUCUUGGCUUUUCAGCUCUG---CAUCGUU
             ++|**+**+ *+****||  * +*||   *+ * |||****||  |****|| ***+* *
Sbjct:     AGCUCUGAGACAAUGAACGCUACACA--CUGCAUCUUGGCUUUGCAGCUCUUCCUCAUGGCU
             >>.>>>>><<<<<<....<<<<<.....>>>>...<<<<....>>>>......>>>>>>>
                         Start codon
```

Fig. 8. Structural alignment of the Human Interferon-γ pseudoknot against mouse upstream genomic DNA. The structure of the query is denoted by parenthesis $<, >$", and "A,a" for the pseudoknot. The symbols describe the conservation: (*) sequence and structure is conserved. (+) structure is conserved but not sequence. (|) sequence is conserved, but not structure.

The Corona-FSE family (RF00507) is a conserved pseudoknot in Coronaviruses which can promote ribosomal frameshifting [29]. We searched the entire Viral genome (79 Mb) for homologs of this family in 33.8 CPU hours on 1.6GHz AMD Opteron Grid, and identified 11 novel members of the sub-family. Like other known members, these are found in coronaviruses, murine hepatitis virus, and Avian flu viruses. Only 2 of the 11 were similar enough in sequence to be identified by BLAST. The alignments can be retrieved from (http://www.cse.ucsd.edu/~bdost/RF00507.htm). A similar result was obtained for Corona-pk3. This family has a conserved $\sim$ 55nt pseudoknot structure which has been shown to be necessary for viral genome replication [30]. We identified 20 novel members of this family with significant scores (See http://www.cse.ucsd.edu/~bdost/RF00165.htm). Only 1 of the 20 was similar enough in sequence to be identified by BLAST.

The Interferon-gamma family is an interesting example of a pseudoknot that is found in the 5'UTR of the Interferon-gamma gene. It regulates translation of the downstream gene by binding to the kinase PKR, a known regulator of IFN-gamma translation [31]. After its discovery in 2002, the pseudoknot was found to be conserved in many mammals. Its presence in rodents was speculated, but the homolog was not located. We searched in mouse and rat genomic DNA, and in the complete gene of gerbil. In all 3 species, we clearly identified the homologs as the top-scoring alignment. The alignment of human and mouse pseudoknots are shown in Figure 8. The conserved location in the two species, just upstream of the start codon, and conservation of key elements validates the hit. We are working with collaborators on experimental validation, and to locate more members of this family.

In conclusion, we demonstrate that the algorithm for aligning pseudoknots, implemented as PAL represents a viable tool for searching for novel homologs, and for structural inference. We hope that our tool will help increase the impact and influence of pseudoknotted RNA in cellular function. PAL and supplemental data are available upon request.

Acknowledgement

In this research, Zhang and Bafna are supported by National Science Foundation grant NSF-DBI:0516440.

References

1. Argaman, L., et al.: Novel small RNA-encoding genes in the intergenic regions of *Escherischia coli*. Curr. Biol. **11** (2001) 941–950
2. Novina, C.D., Sharp, P.A.: The RNAi revolution. Nature **430** (2004) 161–164 News.
3. Storz, G.: An expanding universe of noncoding RNAs. Science **296** (2002) 1260–1263
4. Vitreschak, A., Rodionov, D., Mironov, A., Gelfand, M.: Riboswitches: the oldest mechanism for the regulation of gene expression? Trends in Genetics **20** (2003) 44–50
5. Winkler, W.C., Breaker, R.R.: Genetic control by metabolite-binding riboswitches. Chembiochem **4** (2003) 1024–1032
6. Eddy, S.: Non-coding RNA genes and the modern RNA world. Nature Reviews in Genetics **2** (2001) 919–929
7. Jaeger, J., Turner, D., Zuker, M.: Improved prediction of secondary structures for RNA. Proceedings of the National Academy of Sciences **86** (1989) 7706–7710
8. Zuker, M., Sankoff, D.: RNA secondary structures and their prediction. Bull. Math. Biol. **46** (1984) 591–621
9. Rivas, E., Eddy, S.: Secondary structure alone is generally not statistically significant for the detection of noncoding RNAs. Bioinformatics **16** (2000) 583–605
10. Griffiths-Jones, S., Moxon, S., Marshall, M., Khanna, A., Eddy, S.R., Bateman, A.: Rfam: annotating non-coding RNAs in complete genomes. Nucleic Acids Res **33** (2005) 121–124
11. Klein, R., Eddy, S.: Rsearch: Finding homologs of single structured rna sequences. BMC Bioinformatics **4** (2003) 44
12. Rastogi, T., Beattie, T.L., Olive, J.E., Collins, R.A.: A long-range pseudoknot is required for activity of the Neurospora VS ribozyme. EMBO J **15** (1996) 2820–2825
13. Adams, P.L., Stahley, M.R., Kosek, A.B., Wang, J., Strobel, S.A.: Crystal structure of a self-splicing group I intron with both exons. Nature **430** (2004) 45–50
14. Theimer, C.A., Blois, C.A., Feigon, J.: Structure of the human telomerase RNA pseudoknot reveals conserved tertiary interactions essential for function. Mol Cell **17** (2005) 671–682
15. Nixon, P.L., Rangan, A., Kim, Y.G., Rich, A., Hoffman, D.W., Hennig, M., Giedroc, D.P.: Solution structure of a luteoviral P1-P2 frameshifting mRNA pseudoknot. J Mol Biol **322** (2002) 621–633
16. Akutsu, T.: Dynamic programming algorithm for RNA secondary structure prediction with pseudoknots. Disc. Appl. Math. **104** (2000) 45–62
17. Dirks, R.M., Pierce, N.A.: A partition function algorithm for nucleic acid secondary structure including pseudoknots. J Comput Chem **24** (2003) 1664–1677
18. Evans, P.: Algorithms and Complexity for Annotated Sequence Analysis. PhD thesis, University of Victoria, Victoria BC, Canada (1964)
19. Jiang, T., Lin, G., Ma, B., Zhang, K.: A general edit distance between rna structures. Journal of Computational Biology **9** (2002) 371–388

20. Rivas, E., Eddy, S.: A Dynamic Programming Algorithm for RNA Structure Prediction Including Pseudoknots. Journal of Molecular Biology **285** (1999) 2053–2068
21. Condon, A., Davy, B., Rastegari, B., Tarrant, F., Zhao, S.: Classifying RNA Pseudoknotted Structures. Theoretical Computer Science **320** (2004) 35–50
22. Rastegari, B., Condon, A.: Linear time algorithm for parsing rna secondary structure. In: 5th Workshop on Algorithms in Bioinformatics (WABI). (2005)
23. Matsui, H., Sato, K., Sakakibara, Y.: Pair stochastic tree adjoining grammars for aligning and predicting pseudoknot RNA structures. Bioinformatics **21** (2005) 2611–2617
24. Bafna, V., Muthukrishnan, S., Ravi, R.: Computing similarity between RNA strings. Combinatorial Pattern Matching **937** (1995) 1–14
25. Durbin, R., Eddy, S., Krogh, A., Mitchison, G.: 10.3 Covariance models: SCFG-based RNA profiles. In: Biological Sequence Analysis. Cambridge University Press (1998)
26. Zhang, S., Hass, B., Eskin, E., Bafna, V.: Searching genomes for non-coding rna using fastr. IEEE Transactions on Computational Biology and Bioinformatics **2** (200) 366–379
27. Weinberg, Z., Ruzzo, W.L.: Faster genome annotation of non-coding rna families without loss of accuracy. In: Proceedings of the Annual Intl. Conference on Computational Biology (RECOMB). (2004)
28. Zhang, S., Borovok, I., Aharonowitz, Y., Sharan, R., Bafna, V.: A Sequence-Based Filtering Method for ncRNA Identification and its Application to Searching for Riboswitch Elements. Manuscript (2005)
29. Baranov, P.V., Henderson, C.M., Anderson, C.B., Gesteland, R.F., Atkins, J.F., Howard, M.T.: Programmed ribosomal frameshifting in decoding the SARS-CoV genome. Virology **332** (2005) 498–510
30. Williams, G.D., Chang, R.Y., Brian, D.A.: A phylogenetically conserved hairpin-type 3' untranslated region pseudoknot functions in coronavirus RNA replication. J Virol **73** (1999) 8349–8355
31. Ben-Asouli, Y., Banai, Y., Pel-Or, Y., Shir, A., Kaempfer, R.: Human interferon-gamma mRNA autoregulates its translation through a pseudoknot that activates the interferon-inducible protein kinase PKR. Cell **108** (2002) 221–232

Stan Ulam and Computational Biology

Michael S. Waterman

University of Southern California,
Los Angeles, USA

Abstract. In each the 10 years Recomb has existed as an annual conference, there has been an Ulam Lecture, beginning with the 1997 Ulam Lecture delivered in Santa Fe by Eric Lander. While intended by the conference organizers to be a less technical and more casual lecture, it is difficult for a scientist not to speak about their own science and understandably that is just what has happened. In this year's Ulam lecture I will attempt to give a sketch of the life and accomplishments of this fascinating and rather enigmatic person. At the conclusion I will try to make a few comments about the current state of computational biology.

A. Apostolico et al. (Eds.): RECOMB 2006, LNBI 3909, p. 159, 2006.
© Springer-Verlag Berlin Heidelberg 2006

CONTRAlign: Discriminative Training for Protein Sequence Alignment

Chuong B. Do, Samuel S. Gross, and Serafim Batzoglou

Stanford University, Stanford, CA 94305, USA
{chuongdo, ssgross, serafim}@cs.stanford.edu
http://contra.stanford.edu/contralign/

Abstract. In this paper, we present CONTRAlign, an extensible and fully automatic framework for parameter learning and protein pairwise sequence alignment using pair conditional random fields. When learning a substitution matrix and gap penalties from as few as 20 example alignments, CONTRAlign achieves alignment accuracies competitive with available modern tools. As confirmed by rigorous cross-validated testing, CONTRAlign effectively leverages weak biological signals in sequence alignment: using CONTRAlign, we find that hydropathy-based features result in improvements of 5-6% in aligner accuracy for sequences with less than 20% identity, a signal that state-of-the-art hand-tuned aligners are unable to exploit effectively. Furthermore, when known secondary structure and solvent accessibility are available, such external information is naturally incorporated as additional features within the CONTRAlign framework, yielding additional improvements of up to 15-16% in alignment accuracy for low-identity sequences.

1 Introduction

In comparative structural biology studies, analyzing or predicting protein three-dimensional structure often begins with identifying patterns of amino acid substitution via protein sequence alignment. While the evolutionary information obtained from alignments can provide insights into protein structure, constructing accurate alignments may be difficult when proteins share significant structural similarity but little sequence similarity. Indeed, for modern alignment tools, alignment quality drops rapidly when the sequences compared have lower than 25% identity, the "twilight zone" of protein alignment [1].

In recent years, most alignment methods that have claimed improvements in alignment accuracy have done so not by proposing substantially new algorithms for alignment but rather by incorporating additional sources of information. For instance, when structures of some sequences are available, the 3DCoffee program [2] uses pairwise alignments from existing threading-based (FUGUE [3]) and structural (SAP [4] and LSQman [5]) alignment tools to guide sequence alignment construction. When homologous sequences are available and computational expense is of less concern, the PRALINE$_{PSI}$ program [6] uses PSI-BLAST–derived [7] sequence profiles to augment the amount of evolutionary

A. Apostolico et al. (Eds.): RECOMB 2006, LNBI 3909, pp. 160–174, 2006.
© Springer-Verlag Berlin Heidelberg 2006

(a) 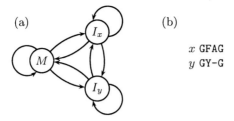 (b)

x GFAG
y GY-G

Fig. 1. Traditional sequence alignment model. (a) A simple three-state HMM for sequence alignment. (b) An example sequence alignment, a.

information available to the aligner. The SPEM program [8] takes the additional step of heuristically incorporating PSIPRED [9] predictions of protein secondary structure, a strategy also adopted in the latest version of PRALINE$_{\text{PSI}}$ [10].

As these programs demonstrate, incorporating additional information can often yield considerable benefits to alignment quality. However, choosing parameters for more complex models can be difficult. In traditional dynamic-programming–based alignment programs, log-odds–based substitution matrices are estimated from large external databases of aligned protein blocks [11], and gap parameters are typically "hand chosen" to maximize performance on benchmark tests [12]. When dealing with more expressive models, however, the high-dimensionality of the parameter space hinders such manual procedures. From the perspective of numerical optimization, the non-convexity of aligner performance as a function of parameters makes hand-tuning difficult for alignment algorithms that rely on complicated ad hoc scoring schemes.

Furthermore, optimizing benchmark performance often leads to *overfitting*, a situation in which the selected parameters are nearly optimal for training benchmark alignments but work poorly on new test data. To combat overfitting, many machine learning studies make use of *cross-validation*, a technique in which an algorithm is trained and tested on independent data sets in order to estimate the ability of the method to generalize to new situations [13].[1]

In this paper, we present CONTRAlign, an extensible and *fully automatic* framework for parameter selection and protein pairwise sequence alignment based on a probabilistic model known as a pair conditional random field (pair-CRF) [15, 16]. In the CONTRAlign methodology, the user first defines an appropriate model topology for pairwise alignment. Unlike for ad hoc algorithms in which model complexity (and hence risk of overfitting) corresponds roughly with the number of free parameters in the model, the effective complexity of a CONTRAlign pair-CRF–based model is controlled by a set of regularization parameters, allowing the user to adjust the trade-off between model expressivity and the risk of overfitting. Given a set of gold standard partially labeled

---

[1] Properly conducted alignment cross-validation studies are extremely rare in the literature. In the past, a typical defense for benchmark tuning was that aligners with few adjustable parameters are less susceptible to overfitting [14]; such reasoning, however, is less applicable to the complicated procedures of some modern aligners.

alignments, CONTRAlign uses gradient-based optimization and holdout cross validation to automatically determine regularization constants and a set of alignment parameters with good expected performance for future alignment problems.

We show that even under stringent cross-validation conditions, CONTRAlign can learn both substitution and gap parameters that generalize well to previously unseen sequences using as few as 20 training alignments. Augmenting the aligner with sequence-based and external features is seamless in the CONTRAlign framework, yielding large accuracy improvements over modern tools for "twilight zone" sequence sets.

2   Methods

In this section, we first review the standard three-state pair hidden Markov model (pair-HMM) formulation of the sequence alignment problem. We also describe the generalization of the standard pair-HMM to a pair conditional random field (pair-CRF), the use of regularization for trading off between the risk of overfitting and expressivity in a pair-CRF, and a standard optimization procedure for learning pair-CRF parameters from data. We then discuss a variety of model topologies and features possible within the CONTRAlign pair-CRF framework.

2.1   Pair-HMMs for Sequence Alignment

Consider the state diagram shown in Figure 1 (a). In the standard model, an alignment corresponds to a sequence of independent events describing a path through the state diagram. First, an initial state s is chosen from $\{M, I_x, I_y\}$ with probability π_s. Then, the alignment process alternates between emitting a pair of aligned residues (c, d) upon entry into some state s with probability $\delta_s^{(c,d)}$ (or a single unaligned residue c with probability $\delta_s^{(c,-)}$ or $\delta_s^{(-,c)}$) and transitioning from some state s to another state t with probability $\tau_{s \to t}$ [17].

Since each event is independent, the probability of the alignment decomposes as a product of several terms. For instance, the joint probability of generating an alignment a and sequences x and y shown in Figure 1 (b) is

$$P(a, x, y) = \pi_M \cdot \delta_M^{(G,G)} \cdot \tau_{M \to M} \cdot \delta_M^{(F,Y)} \cdot \tau_{M \to I_x} \cdot \delta_{I_x}^{(A,-)} \cdot \tau_{I_x \to M} \cdot \delta_M^{(G,G)}. \quad (1)$$

Alternatively, we may rewrite (1) as $P(a, x, y; \mathbf{w}) = \exp(\mathbf{w}^T \mathbf{f}(a, x, y))$ where $\mathbf{w}$ is a parameter vector and $\mathbf{f}(a, x, y)$ is a vector of "feature counts" indicating the number of times each parameter appears in the product on the right-hand side. More explicitly, if $\mathbf{w} = \left[\log \pi_M, \log \delta_M^{(G,G)}, \log \tau_{M \to M}, \cdots \right]^T$, then the corresponding feature count vector is given by

$$\mathbf{f}(a, x, y) = \begin{bmatrix} \# \text{ of times alignment starts in state } M \\ \# \text{ of times alignment generates } (G, G) \text{ in state } M \\ \# \text{ of times alignment follows } M \to M \text{ transition} \\ \vdots \end{bmatrix} = \begin{bmatrix} 1 \\ 2 \\ 1 \\ \vdots \end{bmatrix}. \quad (2)$$

Given two sequences x and y, the Viterbi algorithm computes an alignment a that maximizes $P(a \mid x, y; \mathbf{w})$ in $O(|x| \cdot |y|)$ time. For the model shown in Figure 1, the Viterbi algorithm is equivalent to the Needleman-Wunsch algorithm [18]. In this paper, we use an alternative parsing algorithm for finding alignments with the maximum expected number of correct matches; for details, see [17, 19, 20].

Given a collection of aligned training examples $\mathcal{D} = \left\{ (a^{(i)}, x^{(i)}, y^{(i)}) \right\}_{i=1}^{m}$, the standard parameter estimation procedure (known as *generative* training in the machine learning literature [21]) is to maximize the joint log-likelihood $\ell(\mathbf{w} : \mathcal{D}) := \sum_{i=1}^{m} \log P(a^{(i)}, x^{(i)}, y^{(i)}; \mathbf{w})$ of the data and alignments, subject to constraints ensuring that the original parameters (π_M, $\delta_M^{(\mathtt{G},\mathtt{G})}$, etc.) are nonnegative and normalize. When training with fully-specified alignments, the optimization problem not only is convex but also has a closed-form solution.

In some benchmark alignment databases, such as BAliBASE [22] and PRE-FAB [23], reference alignments are partially ambiguous: certain columns are marked as reliable (known as core blocks) while the alignment of other positions may be left unspecified. In these cases, the training set $\hat{\mathcal{D}} = \left\{ (\hat{a}^{(i)}, x^{(i)}, y^{(i)}) \right\}_{i=1}^{m}$ thus consists of partial alignments $\hat{a}^{(i)}$. Letting $\mathcal{A}^{(i)}$ denote the set of alignments consistent with the known reliable columns of $\hat{a}^{(i)}$, the joint log-likelihood becomes $\ell(\mathbf{w} : \hat{\mathcal{D}}) := \sum_{i=1}^{m} \log \sum_{a \in \mathcal{A}^{(i)}} P(a, x^{(i)}, y^{(i)}; \mathbf{w})$. Despite the nonconvexity of the new optimization problem, most numerical optimization approaches, such as EM or gradient ascent, work well in practice [17].[2]

2.2   From Pair-HMMs to Pair-CRFs

In the pair-HMM formalism, the constraints on the parameters $\mathbf{w}$ to represent initial, transition, or emission log probabilities allowed us to interpret a pair-HMM as defining $P(a, x, y; \mathbf{w})$, the probability of stochastically generating an alignment. Unlike pair-HMMs, pair-CRFs do not define this joint probability but instead directly model the conditional probability,

$$P(a \mid x, y; \mathbf{w}) = \frac{P(a, x, y; \mathbf{w})}{\sum_{a' \in \mathcal{A}} P(a', x, y; \mathbf{w})} = \frac{\exp(\mathbf{w}^T \mathbf{f}(a, x, y))}{\sum_{a' \in \mathcal{A}} \exp(\mathbf{w}^T \mathbf{f}(a', x, y))}, \quad (3)$$

where $\mathcal{A}$ denotes the set of all possible alignments of x and y. As before, the parameter vector $\mathbf{w}$ completely parameterizes the pair-CRF, but this time, we impose no constraints on the entries of $\mathbf{w}$. Here, a parameter entry w_i does not corresponds to the log probability of an event (as in a pair-HMM) but rather is a real-valued feature weight that either raises or lowers the "probability mass" of a relative to other alignments in $\mathcal{A}$. Similar models have been proposed for string edit distance in natural language processing applications [24, 25].

Clearly, pair-CRFs are at least as expressive as their pair-HMM counterparts, as any suitable parameter vector $\mathbf{w}$ for an alignment pair-HMM is a valid parameter vector for its corresponding alignment pair-CRF. Furthermore, while

---

[2] In practice, the only step needed to ensure good convergence was to break symmetries in the model by initializing parameters to small random values.

pair-CRFs assume a particular factorization of the conditional probability distribution $P(a \mid x, y; \mathbf{w})$, they make far weaker independence assumptions regarding feature counts $\mathbf{f}(a, x, y)$. Thus, these models are amenable to using complex feature sets that may be difficult to incorporate within a generative pair-HMM.

Training a pair-CRF involves maximizing the conditional log-likelihood of the data (known as *discriminative* or *conditional* training [21]). Unlike generative training, discriminative training directly optimizes predictive ability while ignoring $P(x, y)$, the model used to generate the input sequences. When a pair-CRF places undue importance on unreliable features (i.e. the magnitude of some parameter w_j is large), overfitting may occur. To prevent this, we place a Gaussian prior, $P(\mathbf{w}) \propto \exp(-\sum_j C_j w_j^2)$, on the parameters $\mathbf{w}$. Thus, we maximize $\ell(\mathbf{w} : \mathcal{D}) := \sum_{i=1}^{m} \log P(a^{(i)} \mid x^{(i)}, y^{(i)}; \mathbf{w}) + \log P(\mathbf{w})$, or equivalently,

$$\sum_{i=1}^{m} \left(\mathbf{w}^T \mathbf{f}(a^{(i)}, x^{(i)}, y^{(i)}) - \log \sum_{a' \in \mathcal{A}} \exp(\mathbf{w}^T \mathbf{f}(a', x^{(i)}, y^{(i)})) \right) - \sum_j C_j w_j^2. \quad (4)$$

The final term in (4) encourages parameters to be "small" unless increased size yields a sufficient increase in likelihood. This technique, known as *regularization*, leads to improved generalization both in theory and in practice [26].

Parameter learning for pair-CRFs using a fixed set of regularization parameters $\mathbf{C} = \{C_j\}$ is straightforward. The objective function in (4) is convex for fully-specified alignments and hence a global maximum of the regularized likelihood can be found using any efficient gradient-based optimization algorithm (such as conjugate gradient, or L-BFGS [27]). The gradient $\nabla_{\mathbf{w}} \ell(\mathbf{w} : \mathcal{D})$ is

$$\sum_{i=1}^{m} \left(\mathbf{f}(a^{(i)}, x^{(i)}, y^{(i)}) - \mathbf{E}_{a \sim P(\mathcal{A}|x^{(i)}, y^{(i)})} \mathbf{f}(a, x^{(i)}, y^{(i)}) \right) - 2\mathbf{C} \circ \mathbf{w}, \quad (5)$$

where $\mathbf{C} \circ \mathbf{w}$ denotes the component-wise product of the vectors $\mathbf{C}$ and $\mathbf{w}$. Disregarding regularization, we see that the partial derivative of the log-likelihood with respect to each parameter w_j is zero precisely when the observed and expected counts for the corresponding feature f_j (taken with respect to the distribution over unobserved alignments) match. For fully-specified alignments $a^{(i)}$, the former term in the parentheses can be directly tabulated from the alignment $a^{(i)}$, and the latter term can be computed using the forward-backward algorithm. The partially-specified alignment case follows similarly [17].

2.3   Pairwise Alignments with CONTRAlign

In the previous subsections, we described the standard pair-HMM model for sequence alignment and its natural extension to pair-CRFs. In this subsection, we present CONTRAlign, a feature-rich alignment framework that leverages the power of pair-CRFs to support large non-independent feature sets while controlling model complexity via regularization.

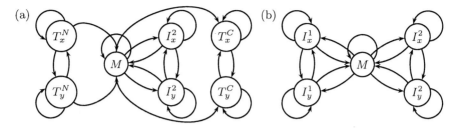

Fig. 2. Model variants. (a) CONTRAlign$_{\text{LOCAL}}$ topology with N/C-terminal flanking inserters, (b) CONTRAlign$_{\text{DOUBLE-AFFINE}}$ topology with two insert state pairs.

Choice of model topology. As a baseline, we used the standard three-state pair-HMM model (CONTRAlign$_{\text{BASIC}}$) shown in Figure 1 (a). We experimented with a variety of other model topologies as well, including:

- CONTRAlign$_{\text{LOCAL}}$: a model with flanking N-terminal and C-terminal insert states to allow for local homology detection (see Figure 2 (a)), and
- CONTRAlign$_{\text{DOUBLE-AFFINE}}$, a model with an extra pair of gap states in order to model both long and short insertions (see Figure 2 (b)).

Hydropathy-based gap context features. The CLUSTALW protein multiple alignment program incorporates a large number of heuristics designed to improve performance on the BAliBASE benchmark reference [28]. One heuristic applicable to pairwise alignment is the reduction of gap penalties in runs of 5 or more hydrophilic residues. Typically, the core regions of globular proteins, where insertions and deletions are less likely, consist of hydrophobic residues. Reducing gap penalties in hydrophilic regions encourages the aligner to place gaps in regions less likely to be part of the hydrophobic core; similar heuristics are incorporated in the MUSCLE [23] alignment program as well.

In CONTRAlign, we tested a variant of this idea (CONTRAlign$_{\text{HYDROPATHY}}$) by incorporating hydropathy-based context features for insertion scoring. Specifically, for each insertion open, insertion continue, or insertion close event in sequence x, we defined the number of hydrophilic residues in a window of length 6 in sequence y to be the *hydrophilic count* context of that event (and vice versa for insertions in sequence y). We added a total of fourteen features to the model, seven indicating whether an insertion open or close occurred with a hydrophilicity context of $0, 1, \ldots,$ or 6, and similarly for insertion continues.

Incorporating external information. To test the ability of CONTRAlign to incorporate external information, we also experimented with giving CONTRAlign information about secondary structure (CONTRAlign$_{\text{DSSP}}$) and solvent accessibility (CONTRAlign$_{\text{ACCESSIBILITY}}$) of the sequences being aligned, as extracted from the PDBFinderII database [29]. In particular, DSSP annotations of sequences from PDBFinderII were converted to a three-letter code using the grouping employed in the EVA automatic structure prediction benchmark server, {{G, H, I}, {E, B}, {T, S, C}} [30]. Similarly, annotations of positional amino acid solvent accessibilities were converted from the PDBFind-erII 0-9 scale using the grouping $\{\{0\}, \{1, 2, 3\}, \{4, 5, 6\}, \{7, 8, 9\}\}$. To assess the value of using predicted external tracks of information, we also tested variants using PSIPRED single ($CONTRAlign_{PSIPRED-SINGLE}$) and multiple ($CONTRAlign_{PSIPRED-MULTI}$) sequence secondary structure predictions.

For each annotation track, we added emission features to the match and insertion states of the basic model that would allow them to simultaneously emit both sequence and annotation. A similar method based on "two-track HMMs" was previously used to improve the quality of fold recognition via predicted local structure [31]. In that work, the authors constructed an HMM that simultaneously emitted two observation signals and relied on the assumed independence of the two character emission tracks during parameter learning. To compensate for the violated independence assumption, the authors added heuristic weights to each emission; thus, the "probability" of a two-track emission was given by $P(o_1|s)^{w_1} P(o_2|s)^{w_2}$, where the weights w_1 and w_2 were selected manually. In contrast, such correction factors are not needed in the pair-CRF model presented here, as pair-CRF learning makes no assumptions regarding the independence of the emission features of each state. Thus, pair-CRFs provide a consistent framework for incorporating multiple sources of evidence without the need for artificial compensation as present in multi-track generalizations of HMMs.

3   Results

In the protein sequence alignment literature, benchmark databases of reference alignments have emerged as the standard metric for evaluating aligner performance. First, the aligner-to-be-tested performs alignments for all sequence sets in the database. Then, accuracy is measured with respect to known reliable columns of a hand-curated reference alignment.

While benchmark tests have been an invaluable asset to the development of alignment algorithms, statistics in the literature often misrepresent the significance of accuracy differences between aligners. Some reference databases, such as BAliBASE and PREFAB, contain multiple copies of a single sequence in several different alignments. Ignoring the non-independence of these test cases artificially lowers p-values when using rank tests to compare the performance of two aligners. Even more dangerous is the common practice of "tuning" parameters to improve performance on individual benchmark datasets. Due to the absence of (or improper use of) cross-validation in most studies in the literature, good benchmark results may not indicate good alignment accuracy for novel proteins.

With this in mind, we designed a series of carefully controlled cross-validation experiments to assess the contribution of the different model topologies/features toward CONTRAlign alignment accuracy, and the ability of the learned alignment model to generalize across different benchmark reference databases.

3.1   Cross-Validation Methodology

We extracted alignments from four standard benchmarking databases:

1. BAliBASE 3.0 [32], a collection of 218 manually refined reference multiple alignments based on 3D structural superpositions;
2. SABmark 1.65 [33], a collection of 236 very low to low identity ("Twilight Zone") and 462 low to intermediate identity ("Superfamilies") sets of all-pairs pairwise consensus structural alignments derived from the SCOP [34] classification;
3. PREFAB 4.0 (beta) [23], a collection of 1932 pairwise structural alignments supplemented by PSI-BLAST homologs from the NCBI nonredundant protein sequence database [35]; and
4. HOMSTRAD (September 1, 2005 release), a curated database of 1032 structure-based multiple alignments for homologous families [36].

We projected the BAliBASE and HOMSTRAD reference multiple alignments into all-pairs pairwise structural alignments. Then, for each multiple sequence set from BAliBASE, HOMSTRAD, and SABmark, we computed percent identity for all pairwise alignments and retained the alignment with median identity.

To construct independent training and testing sets for cross-validation, we relied on the CATH protein structure classification hierarchy [37]; a similar protocol was followed in benchmarking the PSIPRED protein secondary structure prediction program. Specifically, we considered a pair of alignments A and B independent if no two proteins $x \in A$ and $y \in B$ share the same CATH classification at the "homology" level. Using this criterion, we used a greedy procedure to select alignments for training and testing; at each step in the alignment selection process, we selected an alignment, which was independent of all alignments previously selected, from the database with the fewest representatives. The resulting selected pairwise alignments consisted of 38 alignments from BAliBASE, 123 from SABmark, 139 from PREFAB, and 187 from HOMSTRAD.

For parameter learning in CONTRAlign, we considered all matched positions (in core blocks where applicable) to be labeled and treated gapped or unannotated regions as missing data. To select regularization constants in a manner strictly independent of the testing set, we used a staged holdout cross validation procedure on the training data only. Specifically, for a given training collection $\mathcal{D}$, we randomly chose 20% of the alignments for a holdout set and performed training only on the remaining 80%. We manually divided model features into a small number of regularization groups (usually two or three) and constrained the regularization constants for features in each group to be the same. Starting from a model with only transition features, we introduced new features, one group at a time. In each iteration, we used a golden section search and standard L-BFGS optimization to optimize holdout set conditional log-likelihood over possible settings of the regularization parameter for the newly introduced group. Once all features were introduced, we retrained the model on all of the training data using the chosen regularization constants.

We measured alignment accuracy using the Q score [23], the proportion of true alignment character matches correctly predicted. For pairwise alignments, the Q score is equivalent to both the sum-of-pairs (SP) and total column (TC) score commonly used for measuring multiple alignment accuracy [22].

3.2 Comparison of Model Topologies and Feature Sets

In our first set of cross-validation experiments, we selected each of the reference databases in turn as the testing set, and used alignments pooled from the other three databases as the training set.[3] Table 1 compares the various models described in Section 2.3 as evaluated on each of the four databases. As shown in the table, changes in model topology (also possible in pair-HMM aligners) give small improvements in overall accuracy. As expected, the major improvements come with the incorporation of features based on external information, such as DSSP secondary structure or solvent accessibility annotations.

Interestingly, accounting for some sequence features present in the input sequence alone (in particular, hydropathy) gives a larger increase in performance than any change in model topology. We return to this observation in Section 3.3. Also, in contrast to the massive performance gains when using real DSSP secondary structure annotations, our numbers suggest that predicted PSIPRED single sequence secondary structures are not informative for alignment. PSIPRED multiple sequence predictions, however, are substantially more accurate and give strong improvements in aligner performance.

Based on these observations, we constructed the CONTRAlign$_{COMBINED}$ model, which incorporated the four most informative components: double-affine

Table 1. Comparison of CONTRAlign variants. We counted the number of times each variant outperformed or was outperformed by the basic model, and assigned p-values using a simple yet robust statistical sign test to check for deviations from a symmetric distribution in which either aligner is equally likely to do better. Accuracy improvements relative to the basic model are significant in every case with the exceptions of the local and PSIPRED single sequence prediction models.

CONTRAlign variant	BAliBASE (38)	SABmark (123)	PREFAB (139)	HOMSTRAD (187)	Overall (487)	p-value
BASIC	78.93	42.04	74.40	82.61	69.73	n/a
LOCAL	79.10	42.06	74.46	83.34	70.05	7.8×10^{-2}
DOUBLE-AFFINE	78.85	44.50	75.40	84.02	71.17	0.00040
HYDROPATHY	82.07	45.61	76.75	84.78	72.38	1.5×10^{-9}
ACCESSIBILITY	80.80	52.09	79.47	86.84	75.49	3.1×10^{-27}
PSIPRED-SINGLE	77.97	44.94	74.97	82.40	70.47	2.9×10^{-1}
PSIPRED-MULTI	83.13	51.91	79.25	85.35	74.99	2.3×10^{-21}
DSSP	83.01	57.50	81.89	86.88	77.73	1.2×10^{-33}
COMBINED	88.46	61.85	83.66	88.68	80.45	1.2×10^{-44}

---

[3] For most reference databases, with the notable exception of SABmark 1.65, alignment accuracies are roughly consistent. This difference is likely explained by the substantially higher proportion of low-identity alignments in SABmark, though we did not conduct a careful investigation of this phenomenon.

insertion scoring, hydropathy, DSSP secondary structure, and solvent accessibility. To do this, we built an alignment model incorporating the latter two types of features as separate "tracks" of information. A variety of other encodings are possible that allow for more explicit dependencies between secondary structure and solvent accessibility, but we did not explore this further. For the model described, resulting alignments are on average 10% more accurate than those using the basic model alone.

3.3   Comparison to Modern Sequence Alignment Tools

Next, we compared the CONTRAlign$_{\text{HYDROPATHY}}$ model to a variety of modern sequence alignment methods, including MAFFT 5.732 (both L-INS-i and G-INS-i) [38, 39], CLUSTALW 1.83 [28], MUSCLE 3.6 [23], T-Coffee 2.66 [40], and PROBCONS 1.10 [20].[4] In these experiments, we used the existing multiple alignment tools to compute pairwise alignments from the cross-validation setup.

Obtaining a proper cross-validated estimate of an aligner's performance requires tuning the program to multiple training collections, unbiased by testing set performance. For most modern alignment programs, avoiding testing set bias is difficult since parameters are typically tuned by hand. Methods with automatic training procedures, like PROBCONS, permit cross-validation to some extent, with the caveat that the program by default uses BLOSUM62-based amino acid frequencies estimated from data overlapping all testing sets.

In Table 2, the overall accuracies of most modern hand-tuned methods fall within a one percent range (68-69%). The PROBCONS (Bali) method, which uses an automatic unsupervised learning algorithm to infer parameters from all 141 BAliBASE 2 alignments, outperforms most other methods on the BAliBASE dataset except CLUSTALW, which is based on a much more complex model with many internal parameters adjusted to maximize performance on BAliBASE [41]. As previously suggested [41, 42], CLUSTALW's lower relative performance on other databases suggest that it may indeed be overfit to its training set.

To demonstrate the dangers of such overfitting, we trained CONTRAlign on the small set of 38 BAliBASE sequences, with and without regularization. In this situation, omitting regularization leads to tremendous overfitting to BAliBASE, with regularization giving a significant improvement in accuracy. Regularization, however, is not a substitute for proper cross-validation; when overfitting to all four databases, CONTRAlign yields clearly over-optimistic numbers compared to the properly cross-validated test. Similarly, cross-validated PROBCONS (despite using BLOSUM62 amino acid frequencies and thus having an easier learning task than CONTRAlign) performs worse than the non-cross-validated model as expected, confirming that absence of cross-validation can give significantly unrealistic estimates of aligner performance.

As shown, cross-validated CONTRAlign (i.e., CONTRAlign$_{\text{HYDROPATHY}}$) beats current state-of-the-art methods by 3-4% despite (1) estimating all model

---

[4] The Align-m program, which was developed by the creator of the SABmark reference set, could not be tested on pairwise alignments since the current version (2.3) requires at least three input sequences for an alignment.

Table 2. Comparison of modern alignment methods. p-values indicate significance of performance difference between each method and CONTRAlign$_{HYDROPATHY}$ based on a sign test, as in Table 1.

Method	BAliBASE (38)	SABmark (123)	PREFAB (139)	HOMSTRAD (187)	Overall (487)	p-value
MAFFT (G-INS-i)	74.56	41.25	71.37	80.53	67.53	9.8×10^{-22}
MAFFT (L-INS-i)	78.08	39.58	71.95	82.01	68.12	7.1×10^{-17}
T-Coffee	74.73	42.84	72.99	82.40	69.12	1.2×10^{-11}
CLUSTALW	79.43	41.36	73.29	81.62	68.90	1.5×10^{-5}
CLUSTALW (-nohgap)	79.65	40.92	73.51	81.35	68.77	6.2×10^{-7}
MUSCLE	77.42	41.72	72.67	82.63	69.05	2.1×10^{-13}
MUSCLE (-hydrofactor 0.0)	74.78	37.78	69.19	77.83	65.01	7.1×10^{-32}
CONTRAlign (Bali, no reg)	92.57	39.33	68.77	80.45	67.68	5.7×10^{-14}
CONTRAlign (Bali, reg)	84.75	39.08	73.45	82.21	69.01	1.2×10^{-7}
CONTRAlign (All, reg)	82.42	47.39	76.74	85.22	73.03	0.00021
PROBCONS (Bali)	78.62	42.53	73.75	83.64	70.04	4.8×10^{-8}
PROBCONS (cv)	78.48	43.31	71.78	81.36	68.79	9.7×10^{-11}
CONTRAlign$_{HYDROPATHY}$	82.07	45.61	76.75	84.78	72.38	n/a

parameters, including the emission matrix, and (2) following a rigorous cross-validated training procedure. Based on the comparison of the hydropathy and basic models in Table 1, it is clear that these accuracy gains result directly from the use of hydropathy-based gap scoring. Perhaps most striking, however, is that a variety of existing methods, including CLUSTALW and MUSCLE, already incorporate hydropathy-based modifications in their alignment scoring, yet do not manage to achieve above 70% accuracy on our benchmarks. Disabling these modifications in the respective programs gives no substantial change in performance for CLUSTALW and greatly reduces MUSCLE accuracy.[5] Our result confirms that hydropathy is indeed an important signal for protein sequence alignment and that properly accounting for this can yield significantly higher alignment accuracy than the current state-of-the-art.

3.4   Regularization and Generalization Performance

To understand the effects of regularization at low training set sizes, we reserved a set of 200 randomly chosen pairwise alignments pooled from all four reference databases to use as a testing set. We then experimented with learning parameters for the CONTRAlign$_{HYDROPATHY}$ topology using varying training set sizes. For staged regularization, we considered a variant of the basic model in which we introduced amino emission features corresponding to the six-character reduced amino alphabet, {{A, G, P, S, T}, {C}, {D, E, N, Q}, {F, W, Y}, {H, K, R}, {I, L, M, V}}, in addition to the regular twenty-letter amino acid emissions [43]. In the first regularization stage, the program learns a coarse-grained substitution matrix, followed by finer-grained refinements in the second stage.

The results in Figure 3 (a) demonstrate that with intelligent use of regularization, good accuracy can be achieved with only 20 example alignments, far

---

[5] Performing a sign test to compare performance when hydropathy scoring is either enabled or disabled yields p-values of 0.56 and 6.28×10^{-31} for CLUSTALW and MUSCLE, respectively.

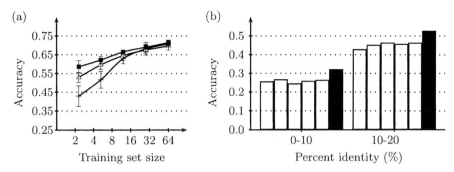

Fig. 3. Alignment accuracy curves. (a) Accuracy as a function of training set size. The three curves give performance when using no (+), simple (△), and staged (●) regularization. All data points are averages over 10 random training/test splits. (b) Accuracy in the "twilight zone." For each conservation range, the uncolored bars (□) give accuracies for MAFFT (L-INS-i), T-Coffee, CLUSTALW, MUSCLE, and PROBCONS (Bali) in that order, and the colored bar (■) indicates the accuracy for CONTRAlign.

fewer than the number of blocks used to estimate traditional alignment substitution matrices such as BLOSUM [11]; nevertheless, the simpler regularization scheme was still quite effective compared to having no regularization at all.

For specific classes of alignments, such as sequences with long insertions or compositional biases, a robust training procedure allows one to tailor the alignment algorithm to the data; when, in addition, training data is sparse, regularization deters overfitting and enables further customization of alignment parameters. Furthermore, as the amount of available training data grows, accuracy will continue to increase as well.

3.5 Alignment Accuracy in the "Twilight Zone"

To understand the situations in which CONTRAlign$_{HYDROPATHY}$ was most effective, we stratified the 487 sequences of our dataset into several percent identity ranges and measured the accuracy of all methods for each range. For alignments with at least 20% identity, all methods obtained similar accuracies, ranging from 87.2% to 88.7%. In the 0-10% and 10-20% identity ranges, however, CONTRAlign accuracy was substantially higher than that of other methods; here, CONTRAlign achieved cross-validated accuracies of 32.2% and 52.8% compared to non-cross-validated accuracy ranges of 25.7-26.8% and 43.0-46.5% for all other methods (see Figure 3 (b)). Incorporating external sequence features such as in the combined model of Section 3.2 yields accuracies of 48.0% and 68.5% (not shown in figure), indicating that external sequence information can significantly increase the reliability of alignments when available.

4 Discussion

Construction of a modern high-performance sequence alignment program involves understanding the variety of biological features available when performing alignment, building a model of interactions demonstrating how those features may be combined in an aligner, and careful cross-validation experiments to ensure good generalization performance of the aligner on future data. In this paper, we presented CONTRAlign, a pair conditional random field for learning alignment parameters effectively even when small amounts of training data are available. Using regularization and holdout cross-validation, our algorithm automatically learns parameters with good generalization performance. Public domain source code for CONTRAlign, datasets used in experiments from this paper, and a web server for submitting sequences are available online at `http://contra.stanford.edu/contralign`.

Since CONTRAlign specifies a conditional probability distribution over pairwise alignments, the PROBCONS methodology provides one straightforward extension of CONTRAlign to multiple alignment. The main limitation of the CONTRAlign framework, however, is training time: L-BFGS gradient-based optimization is expensive, especially in the context of the holdout cross validation procedure used. Typical training runs for the experiments in this paper (including holdout cross-validation to find regularization constants) took approximately an hour on a 40-node Pentium IV cluster. Perceptron learning [44], a recent technique for discriminatively training structured probabilistic models, may provide a scalable alternative to gradient-based optimization.

The primary advantage of CONTRAlign is its ability to free aligner developers to focus on the *biology* of sequence alignment—modelling and feature selection—while transparently taking care of details such as parameter learning and generalization performance. The models described in this paper were only the first steps toward a better understanding of the sequence alignment problem. Combining new CONTRAlign topologies and features with known successful variants should result in even higher performance. A systematic exploration of such possibilities remains to be done.

Acknowledgments

We thank G. Asimenos for his tireless technical assistance in dealing with cluster issues during the development of the program. We also thank R. C. Edgar and A. Sidow for useful discussions regarding aligner features and testing methodology. CBD and SSG were supported by NDSEG fellowships. This work was supported in part by NSF grant 0312459. SB acknowledges support from the NSF CAREER Award and the Alfred P. Sloan Fellowship.

References

1. Rost, B.: Twilight zone of protein sequence alignments. Protein Eng **12** (1999) 85–94
2. O'Sullivan, O., Suhre, K., Abergel, C., Higgins, D.G., Notredame, C.: 3DCoffee: combining protein sequences and structures within multiple sequence alignments. J Mol Biol **340** (2004) 385–395

3. Shi, J., Blundell, T.L., Mizuguchi, K.: FUGUE: sequence-structure homology recognition using environment-specific substitution tables and structure-dependent gap penalties. J Mol Biol **310** (2001) 243–257
4. Taylor, W.R., Orengo, C.A.: Protein structure alignment. J Mol Biol **208** (1989) 1–22
5. Kabsch, W.: A discussion of the solution for the best rotation to relate two sets of vectors. Acta Crystallog Sect A **34** (1978) 827–828
6. Simossis, V.A., Kleinjung, J., Heringa, J.: Homology-extended sequence alignment. Nucleic Acids Res **33** (2005) 816–824
7. Altschul, S.F., Madden, T.L., Schaffer, A.A., Zhang, J., Zhang, Z., Miller, W., Lipman, D.J.: Gapped BLAST and PSI-BLAST: a new generation of protein database search programs. Nucleic Acids Res **25** (1997) 3389–3402
8. Zhou, H., Zhou, Y.: SPEM: improving multiple sequence alignment with sequence profiles and predicted secondary structures. Bioinformatics **21** (2005) 3615–3621
9. Jones, D.T.: Protein secondary structure prediction based on position-specific scoring matrices. J Mol Biol **292** (1999) 195–202
10. Simossis, V.A., Heringa, J.: PRALINE: A multiple alignment toolbox that integrates homology-extended and secondary structure information. Nucleic Acids Res **33 (Web Server issue)** (2005) W289–W294
11. Henikoff, S., Henikoff, J.G.: Amino acid substitution matrices from protein blocks. Proc Nat Acad Sci USA **89** (1992) 10915–10919
12. Vingron, M., Waterman, M.S.: Sequence alignment and penalty choice. Review of concepts, case studies and implications. J Mol Biol **235** (1994) 1–12
13. Kohavi, R.: A study of cross-validation and bootstrap for accuracy estimation and model selection. In: IJCAI. (1995) 1137–1145
14. Raghava, G.P.S., Searle, S.M.J., Audley, P.C., Barber, J.D., Barton, G.J.: OXBench: A benchmark for evaluation of protein multiple sequence alignment accuracy. BMC Bioinformatics **4** (2003)
15. Lafferty, J., McCallum, A., Pereira, F.: Conditional random fields: probabilistic models for segmenting and labeling sequence data. In: Proc. 18th ICML. (2001) 282–289
16. Sha, F., Pereira, F.: Shallow parsing with conditional random fields (2003)
17. Durbin, R., Eddy, S.R., Krogh, A., Mitchison, G.: Biological Sequence Analysis : Probabilistic Models of Proteins and Nucleic Acids. Cambridge University Press (1999)
18. Altschul, S.F.: Amino acid substitution matrices from an information theoretic perspective. J Mol Biol **219** (1991) 555–565
19. Holmes, I., Durbin, R.: Dynamic programming alignment accuracy. J Comp Biol **5** (1998) 493–504
20. Do, C.B., Mahabhashyam, M.S., Brudno, M., Batzoglou, S.: PROBCONS: probabilistic consistency-based multiple sequence alignment. Genome Res **15** (2005) 330–340
21. Ng, A., Jordan, M.: On discriminative vs. generative classifiers: a comparison of logistic regression and naive Bayes. In: NIPS 14. (2002)
22. Thompson, J.D., Plewniak, F., Poch, O.: A comprehensive comparison of multiple sequence alignment programs. Nucleic Acids Res **27** (1999) 2682–2690
23. Edgar, R.C.: MUSCLE: multiple sequence alignment with high accuracy and high throughput. Nucleic Acids Res **32** (2004) 1792–1797
24. McCallum, A., Bellare, K., Pereira, F.: A conditional random field for discriminatively-trained finite-state string edit distance. In: Proc. UAI. (2005)

25. Bilenko, M., Mooney, R.J.: Alignments and string similarity in information integration: A random field approach. In: Proc. Dagstuhl Seminar on Machine Learning for the Semantic Web. (2005)
26. Vapnik, V.N.: Statistical Learning Theory. Wiley-Interscience (1998)
27. Nocedal, J., Wright, S.J.: Numerical Optimization. Springer (1999)
28. Thompson, J.D., Higgins, D.G., Gibson, T.J.: CLUSTAL W: Improving the sensitivity of progressive multiple sequence alignment through sequence weighting, position-specific gap penalties, and weight matrix choice. Nucleic Acids Res **22** (1994) 4673–4680
29. Krieger, E., Hooft, R.W.W., Nabuurs, S., Vriend, G.: PDBFinderII—a database for protein structure analysis and prediction. Submitted (2004)
30. Eyrich, V.A., Mart'i-Renom, M.A., Przybylski, D., Madhusudhan, M.S., Fiser, A., Pazos, F., Valencia, A., Sali, A., Rost, B.: EVA: continuous automatic evaluation of protein structure prediction servers. Bioinformatics **17** (2001) 1242–1243
31. Karchin, R., Cline, M., Mandel-Guttfreund, Y., Karplus, K.: Hidden markov models that use predicted local structure for fold recognition: alphabets of backbone geometry. Proteins: Structure, Function, and Genetics **51** (2003) 504–514
32. Thompson, J.D., Koehl, P., Ripp, R., Poch, O.: BAliBASE 3.0: latest developments of the multiple sequence alignment benchmark. Proteins **61** (2005) 127–136
33. Walle, I.V., Lasters, I., Wyns, L.: SABmark—a benchmark for sequence alignment that covers the entire known fold space. Bioinformatics **21** (2005) 1267–1268
34. Murzin, A.G., Brenner, S.E., T., H., C., C.: SCOP: a structural classification of proteins database for the investigation of sequences and structures. J Mol Biol **247** (1995) 536–540
35. Pruitt, K.D., Tatusova, T., Maglott, D.R.: NCBI Reference Sequence project: update and current status. Nucleic Acids Res **31** (2003) 34–37
36. Mizuguchi, K., Deane, C.M., Blundell, T.L., Overington, J.P.: HOMSTRAD: a database of protein structure alignments for homologous familes. Protein Sci **7** (1998) 2469–2471
37. Orengo, C.A., Michie, A.D., Jones, S., Jones, D.T., Swindells, M.B., Thornton, J.M.: CATH—a hierarchic classification of protein domain structures. Structure **5** (1997) 1093–1108
38. Katoh, K., Misawa, K., Kuma, K., Miyata, T.: MAFFT: a novel method for rapid multiple sequence alignment based on fast fourier transform. Nucleic Acids Res **30** (2002) 3059–3066
39. Katoh, K., Kuma, K., Toh, H., Miyata, T.: MAFFT version 5: improvement in accuracy of multiple sequence alignment. Nucleic Acids Res **33** (2005) 511–518
40. Notredame, C., Higgins, D., Heringa, J.: T-Coffee: a novel method for multiple sequence alignments. J Mol Biol **302** (2000) 205–217
41. Heringa, J.: Local weighting schemes for protein multiple sequence alignment. Computers and Chemistry **26** (2002) 459–477
42. Edgar, R.C.: MUSCLE: low-complexity multiple sequence alignment with T-Coffee accuracy. In: ISMB/ECCB. (2004)
43. Edgar, R.C.: Local homology recognition and distance measures in linear time using compressed amino acid alphabets. Nucleic Acids Res **32** (2004) 380–385
44. Collins, M.: Discriminative training methods for hidden markov models: theory and experiments with perceptron algorithms. In: EMNLP. (2002)

Clustering Near-Identical Sequences for Fast Homology Search

Michael Cameron[1], Yaniv Bernstein[1], and Hugh E. Williams[2]

[1] School of Computer Science and Information Technology,
RMIT University, GPO Box 2476V, Melbourne 3001, Australia
{mcam, ybernste}@cs.rmit.edu.au
[2] Microsoft Corporation, One Microsoft Way,
Redmond, Washington 98052, USA
hughw@microsoft.com

Abstract. We present a new approach to managing redundancy in sequence databanks such as GenBank. We store clusters of near-identical sequences as a representative *union-sequence* and a set of corresponding edits to that sequence. During search, the query is compared to only the union-sequences representing each cluster; cluster members are then only reconstructed and aligned if the union-sequence achieves a sufficiently high score. Using this approach in BLAST results in a 27% reduction is collection size and a corresponding 22% decrease in search time with no significant change in accuracy. We also describe our method for clustering that uses *fingerprinting*, an approach that has been successfully applied to collections of text and web documents in Information Retrieval. Our clustering approach is ten times faster on the GenBank nonredundant protein database than the fastest existing approach, CD-HIT. We have integrated our approach into FSA-BLAST, our new Open Source version of BLAST, available from http://www.fsa-blast.org/. As a result, FSA-BLAST is twice as fast as NCBI-BLAST with no significant change in accuracy.

1 Introduction

Comprehensive genomic databases such as the GenBank non-redundant protein database contain a large amount of internal redundancy. Although exact duplicates are removed from the collection, there remain large numbers of near-identical sequences. Such near-duplicate sequences can appear in protein databases for several reasons, including the existence of closely-related homologues or partial sequences, sequences with expression tags, fusion proteins, post translational modifications, and sequencing errors. These minor sequence variations lead to the over-representation in databases of certain protein domains, particularly those that are under intensive research. For example, the GenBank database contains several thousand near-identical protein sequences from the human immunodeficiency virus.

Database redundancy has several pernicious effects. First, a larger database takes longer to query; as sequencing efforts continue to outpace improvements in

A. Apostolico et al. (Eds.): RECOMB 2006, LNBI 3909, pp. 175–189, 2006.

computer hardware, this is a problem that will continue to worsen. Second, redundancy can lead to highly repetitive search results for any query that matches
closely with an over-represented sequence. Third, large-scale redundancy has
the effect of skewing the statistics used for determining alignment significance,
ultimately leading to decreased search effectiveness. Fourth, the PSI-BLAST algorithm (1) can be misled by redundant matches during iteration, causing it
to bias the profile towards over-represented domains; this can result in a less
sensitive search or even profile corruption (2; 3).

Redundancy has been managed in the past by the creation of representative-
sequence databases (RSDBs), culled collections in which no two sequences share
more than a given level of identity. Such databases have been shown to significantly improve profile training in iterative search tools such as PSI-BLAST by
reducing the amount of over-representation of certain protein domains and consequently reducing profile corruption. However, they are less suitable for regular
search algorithms such as BLAST (4; 1) and FASTA (5; 6) because, by definition, RSDBs are not comprehensive. This leads to search results that are both
less accurate—the representative sequence for a cluster may not be the one that
aligns best with a given query—and less authoritative because the user is only
shown one representative sequence from a family of similar sequences.

In this paper, we describe a sequence clustering methodology that efficiently
and effectively identifies and manages redundancy. Importantly, it lacks the
drawbacks of previous representative-sequence databases. Previous approaches
choose one sequence from each near-duplicate cluster as a representative to the
database and delete the other sequences. In contrast, we generate for each cluster
a special union-sequence that—through use of wildcard characters—represents
all of the sequences in the cluster simultaneously. Through careful choice of wildcards, we are able to achieve near-optimal alignments while still substantially
reducing the number of sequences against which queries need to be matched.
Further, we store all sequences in a cluster as a set of edits against the union-
sequence. This achieves a form of compression and allows us to retrieve cluster
members for more precise alignment against a query should the union-sequence
achieve a good alignment score. Thus, both space and time are saved with no
significant loss in accuracy or sensitivity.

Our method supports two modes of operation: users can choose to see all
alignments or only the best alignment from each cluster. In the former mode,
the clustering is transparent and the results comparable to searches on an unclustered collection. In the latter mode, the search output is similar to the result
of searching a culled representative database, except that our approach is guaranteed to display the best alignment from each cluster and is also able to report
the number of similar alignments that have been suppressed.

Our work also improves on previous approaches by reducing the time and
resources required to create clusters. The most successful existing algorithms use
a form of all-against-all comparison that is quadratic in the number of sequences
in the database. Our innovative clustering approach uses a technique known as
fingerprinting that leads to significantly faster clustering; we are able to process
the entire GenBank collection in one hour on a commodity workstation. By

contrast, the fastest previously available system, CD-HIT (7), takes almost ten hours on the same machine.

To investigate the effectiveness of our clustering approach we have integrated it with our freely available open-source software package, FSA-BLAST. When applied to the GenBank non-redundant (NR) database, our method reduces the size of sequence data in the NR database by 27% and improves search times by 22% with no significant effect on accuracy.

2   Existing Approaches

Reducing redundancy in a sequence database is essentially a two-stage process: first, redundancy within the database must be identified by grouping similar sequences into clusters; then, the clusters must be managed in some way. In this section we describe past approaches to these two stages.

The first stage of most clustering algorithms involves identifying pairs of similar sequences. An obvious approach to this is to align each sequence with every other sequence in the collection using a pairwise alignment scheme such as Smith-Waterman local alignment (8). This is the approach taken by several existing clustering algorithms, including d2_cluster (9), OWL (10), and KIND (11). However, this approach is impractical for any collection of significant size; each pairwise comparison is computationally intensive and the number of pairs is quadratic in the number of sequences.

Several schemes, including CLEANUP (12), NRDB90 (13), RSDB (3), CD-HI (14) and CD-HIT (7), use fast clustering approaches based on greedy incremental algorithms. In general, each proceeds as follows. To begin, the collection sequences are sorted by decreasing order of length. Then, each sequence is extracted in turn and used as a query to search an initially-empty representative database for high-scoring matches. If a similar sequence is found, the query sequence is discarded; otherwise, it is added to the database as the representative of a new cluster. When the algorithm terminates, the database consists of the representative (longest) sequence of each cluster. This greedy approach reduces the number of pairwise comparisons but has three drawbacks: first, a match is only identified when one sequence is a substring of another; second, cases where the prefix of one sequence matches the suffix of another are neglected; and, third, clusters form around longer sequences instead of natural centroids, potentially leading to a suboptimal set of clusters.

Existing greedy incremental algorithms also use a range of BLAST-like heuristics to quickly identify high-scoring pairwise matches. The CLEANUP algorithm (12) builds a rich inverted index of short substrings or *words* in the collection and uses this structure to score similarity between sequence pairs. NRDB90 (13) and RSDB (3) use in-memory hashtables of decapeptides and pentapeptides for fast identification of possible high-scoring sequence pairs before proceeding with an alignment. CD-HI (14) and CD-HIT (7) use lookup arrays of very short subsequences to more efficiently identify similar sequences. However, despite each scheme having fast methods for comparing sequence pairs, the algorithms still operate on a pairwise basis and remain $O(n^2)$ in the size of the database. Indeed, we show in Section 7 that CD-HIT — the fastest of the greedy

incremental algorithms mentioned and the most successful existing approach —
scales poorly, with superlinear complexity in the size of the collection.

One way to avoid an all-against-all comparison is to pre-process the collection
using an index that can efficiently identify high-scoring candidate pairs. Malde
et al. 2003 (15) and Gracey et al. 1998 (16) investigated the use of suffix structures such as suffix trees (17) and suffix arrays (18) to identify groupings of
similar sequences in linear time. However, traditional suffix structures consume
large amounts of memory and are not suitable for processing large sequence
collections such as GenBank on desktop workstations. Malde et al. 2003 (15)
report results for only a few thousand EST sequences. The algorithm described
by Gracey et al. 1998 (16) requires several days to process a collection of around
60,000 sequences. External suffix structures, which record information on disk,
are also unsuitable; they use a large amount of disk space, are extremely slow
for searching, or have slow construction times (19). Nonetheless, we believe that
investigating data structures for identifying all pairs of similar sequences in a
fixed number of passes is the correct approach.

Once a set of clusters have been identified, most existing approaches retain a
single representative sequence from each cluster and delete the rest (13; 3; 14; 7).
The result is a representative database with fewer sequences and less redundancy.
However, purging near-duplicate sequences can significantly reduce the quality
of results returned by search tools such as BLAST. There is no guarantee that
the representative sequence from a cluster is the sequence that best aligns with a
given query. Therefore, some queries will fail to return matches against a cluster
that contains sequences of interest, which reduces sensitivity. Further, results
of a search lack authority because they do not show the best alignment from
each cluster. Also, the existence of highly-similar alignments, even if strongly
mutually redundant, may be of interest to a researcher.

3   Clustering Using Wildcards

In this section we describe our approach to representing and searching clusters
of highly-similar sequences using union-sequences and special-purpose wildcard
characters to represent clusters.

Let us define $E = \{e_1, ..., e_n\}$ as the set of sequences in a collection where each
sequence is a string of residues $e_i = r_1...r_n \mid r \in R$. Our approach represents the
collection as a set of clusters C, where each cluster contains a union-sequence
U and edit information for each member of the cluster. The union-sequence
is a string of residues and wildcards $U = u_1...u_n | u_i \in R \cup W$ where $W =
\{w_1, ..., w_n \mid w_i \subseteq R\}$ is the set of available wildcards. Each wildcard represents
a set of residues and is able to act as a substitute for any of these residues. By
convention, w_n is assumed to be the *default wildcard* w_d that can represent any
residue; that is, $w_n = R$.

Figure 1 shows an example cluster constructed using our approach. The union-
sequence is shown at the top and cluster members are aligned below. Columns
where the member sequences differ from each another and a wildcard has been
inserted are shown in bold face. In this example, $W = \{w_d\}$ — that is, only the
default wildcard is used and it is represented by an asterisk.

```
KNQVAMN * QNTVFDAKRLIGRKFDEPTVQADMKHWPFKV * QAEVDV * RFRSNT * ER   (union-seq)
        P QNTVFDAKRLIGRKFDEPTVQADMKHWPFKV I QAEV                   (gi 156103)
KNQVAMN P QNTVFDAKRLIGRKFDEPTVQADMKHWPFKV I QAEV                   (gi 156105)
          QNTVFDAKRLIGRKFDEPTVQADMKHWPFKV V QAEVDV L RFRSNT K ER   (gi 156121)
KNQVAMN P QNTVFDAKRLIGRKFDEPTVQADMKHWPFKV V QAEVDV L RFRSNT K      (gi 552059)
KNQVAMN P QNTVFDAKRLIGRKFDEPTVQADMKHWPFKV I QAEVDV Q RFRSNT R      (gi 552055)
KNQVAMN P QNTVFDAKRLIGRKFDEPTVQADMKHWPFKV I QAEVDV Q RFRSNT R E    (gi 552057)
        P QNTVFDAKRLIGRKFDEPTVQADMKHWPFKV V QAEVDV L RFRSNT K ER   (gi 156098)
          QNTVFDAKRLIGRKFDEPTVQADMKHWPFKV V QAEVDV L RFRS          (gi 156100)
     VFDAKRLIGRKFDEPTVQADMKHWPFKV I QAEVDV Q RFRSNT R E            (gi 156111)
    N QNTVFDAKRLIGRKFDEPTVQADMKHWPFKV I QAEVDV Q RFRSNT R          (gi 552056)
```

Fig. 1. Example cluster of heat shock proteins from the GenBank NR database. The union-sequence is shown at the top, followed by the ten member sequences.

When a cluster is written to disk, the union-sequence — shown at the top of the figure — is stored in its complete form, and each member of the cluster is recorded using edit information. The edit information for each member sequence includes start and end offsets that specify a range within the union-sequence, and a set of residues that replace the wildcards in that range. For example, the first member of the cluster with GI accession 156103 would be represented by the tuple (8,44,PI); the member sequence can be reconstructed by copying the substring between positions 8 and 44 of the union-sequence and replacing the wildcards at union-sequence positions 8 and 40 with characters P and I respectively. Note that we do not permit gaps; insertions and deletions are heavily penalised during alignment and any scheme that allows gaps in representative sequences is likely to reduce search accuracy. A more complex cluster representation such as a partial-order graph (20) could tolerate gaps; while the increased complexity of such a representation leads inevitably to larger on-disk footprint and longer alignment times, the potential increase in cluster size that gapping would allow means that such a technique merits future investigation.

Our clustering method is designed so that each union-sequence aligns to the query with a score that is—with high probability—equal to or higher than the best score for aligning the query to members of the cluster (see Section 5). During search, the query is compared to the union-sequence of each cluster; if the union-sequence produces a statistically significant alignment, then the members of the cluster are restored from their compressed representations and aligned to the query. Our approach supports two modes of operation: users can choose to see all high-scoring alignments, or only the best alignment from each cluster. The latter mode reduces redundancy in the results.

4 Clustering Algorithm

In this section we briefly describe our approach to efficiently clustering large sequence collections. A more detailed description of the algorithm is given in Bernstein and Cameron 2006 (21).

In our approach, we use a largely linear-time algorithm that has low main-memory overheads for identifying candidate pairs. Document fingerprinting (22; 23; 24; 25; 26) has been used for grouping highly similar documents in extremely large collections and has been successfully applied to text and web data for several applications including plagiarism detection, copyright protection, and search-engine optimisation. Fingerprinting operates by selecting fixed-length subsequences — known as *chunks* — from each document. This set of chunks is known as the document *fingerprint* and acts as a compact surrogate for the document. As highly similar documents are expected to share a large number of chunks, fingerprints are used to efficiently detect similar documents in a collection.

The basic process of fingerprinting can be applied to biological sequence data by substituting sequences for documents, although some alterations in approach are necessary because genomic sequences do not contain natural word-delimiters such as punctuation and whitespace. We have modified our DECO fingerprinting package (26; 27) for use with sequence data and it is used as the first stage of our clustering algorithm.

The fingerprinting process identifies chunks that occur in the collection more than once. In the context of sequence data we use subsequences or *words* of length W as our chunks. For each word, DECO outputs a *postings list* of sequences that contain the word and the offset into each sequence where the word occurs. Our clustering algorithm uses these lists to calculate the number of identical words shared by each pair of sequences in the collection. The number of matching words is normalised by the length of the overlapping region between the two sequences; this provides a good quality estimate of the degree of mutual redundancy between the sequences. If this measure exceeds a threshold then the two sequences are aligned using the similarity score measure we describe next. Highly similar candidate pairs with a score below threshold T are then recorded.

Given the list of candidate pairs, we use a variation on single-linkage hierarchical clustering (28) to identify clusters. Each sequence is initially considered as a cluster with one member. Candidate pairs are processed in increasing order of similarity score, from most- to least-similar, and clusters are merged. To merge a pair of candidate clusters C_X and C_Y with union-sequences X and Y respectively, the overlapping regions of X and Y are aligned. A new union-sequence U is then created by replacing each mismatched residue in the overlap region with a suitable wildcard w. The clusters will only be merged if the mean alignment score increase $\bar{Q}$ in the overlap region is below a specified threshold T — this prevents union-sequences from containing too many wildcards and reducing search performance.

If the clusters are merged, a new cluster C_U is created consisting of all members of C_X and C_Y. When inserting wildcards into the union-sequence, if more than one wildcard is suitable then the one with the lowest expected match score $e(w) = \sum_R s(w,r)p(r)$ is selected, where $p(r)$ is the background probability of residue r (29) and $s(w,r)$ is the alignment score for matching wildcard w to residue r. We discuss how alignment vectors $s(w, \cdot)$ are constructed in Section 5 and how wildcards are chosen in Section 6.

The alignment score increase Q for a wildcard w is calculated as

$$Q(w) = \sum_R s(w,r)p(r) - \sum_{R \times R} s(r_1, r_2)p(r_1)p(r_2)$$

where $s(r_1, r_2)$ is the score for matching a pair of residues as defined by a scoring matrix such as BLOSUM62 (30). This value estimates the increase in alignment score one can expect against arbitrary query residues by aligning against w instead of against the actual residue at that position.

The above approach has a quadratic complexity in the length of each postings list. While most lists remain quite short even in large databases, a small proportion of words can appear a large number of times. As the database being processed grows, common words can come to dominate overall processing time. For example, a postings lists with 500 entries produces 124,750 potential sequence pairs, which will take a very long time to process. We therefore process frequently occurring words —those with more than M occurrences in the collection, where we use $M = 100$ by default—in a different, top-down manner before proceeding to the standard hierarchical clustering approach described above.

Given a list of sequences l containing a particular frequently occurring word, the top-down approach extracts all sequences in l and selects an exemplar; this is the sequence with the highest percentage identity to the other sequences in the list. The exemplar is then aligned against each sequence in l and used to create a cluster as defined above. All sequences in the new cluster are removed from l and the process is repeated until $|l| < M$. The shortened list is then processed using the hierarchical clustering method. This process still has an $O(n^2)$ worst-case in the length of the list, but is significantly quicker than the hierarchical approach when processing long lists in practice.

5   Scoring Wildcards

We have modified BLAST to work with our clustering algorithm. Instead of comparing the query sequence to each member of the database, our approach compares the query only to the union-sequence representing each cluster, where the union-sequence may contain wildcard characters. If a high-scoring alignment between the union-sequence and query is identified, the members of the cluster are reconstructed and aligned to the query. In this section we discuss how, given a set of wildcards W, we determine the scoring vectors $s(w_i, \cdot)$ for each $w_i \in W$.

Ideally, we would like the score between a query sequence Q and a union-sequence U to be precisely the highest score that would result from aligning Q against any of the sequences in cluster C_U. This would result in no loss in sensitivity as well as no false positives. Unfortunately, such a scoring scheme is not likely to be achievable without aligning against each sequence in every cluster, defeating much of the purpose of clustering in the first place.

To maintain the speed of our approach, scoring of wildcards against residues must be on the basis of a standard scoring vector $s(w, \cdot)$ and cannot take into consideration any data about the sequences represented by the cluster. Thus, scoring will involve a compromise between sensitivity (few false negatives) and speed (few false positives). We describe two such compromises below, and finally show how to combine them to achieve a good balance of sensitivity and speed.

During clustering, wildcards are inserted into the union-sequence to denote residue positions where the cluster members differ. Let us define $S = s_1 ... s_x \mid s_i \in W$ as the ordered sequence of x wildcards substituted into union-sequences during clustering. Each occurrence of a wildcard is used to represent a set of residues that appear in its position in the members of the cluster. We define $o \subseteq R$ as the set of residues represented by an occurrence of a wildcard in the collection and $O = o_1...o_x \mid o_i \subseteq R$ as the ordered sequence of substituted residue sets. The kth wildcard s_k that is used to represent the set of residues o_k must be chosen such that $o_k \subseteq s_k$.

Our first scoring scheme, s_{exp}, builds the scoring vector by considering the actual occurrence pattern of residues represented by the wildcard in the collection. Formally, we calculate the expected best score s_{exp} as:

$$s_{exp}(w, r) = \frac{\sum\limits_{k \in P_i} \max\limits_{f \in o_k} s(r, f)}{|P_i|}$$

where P_i is the set of ordinal numbers of all substitutions using the wildcard w_i:

$$P_i = \{j \mid j \in \mathbb{N}, \ j \leq x, \ s_j = w_i\}.$$

This score can be interpreted as the mean score we would get by aligning residue r against the actual residues represented by the wildcard w. This score has the potential to reduce search accuracy; however, it distributes the scores well, and provides an excellent tradeoff between sensitivity and speed.

The second scoring scheme, s_{opt}, calculates the optimistic alignment score of the wildcard w against each residue. The optimistic score is the highest score for aligning residue q to any of the residues represented by wildcard w. This is calculated as follows:

$$s_{opt}(w, r) = \max\limits_{f \in w} s(r, f)$$

The optimistic score guarantees no loss in sensitivity: the score for aligning against a union-sequence U using this scoring scheme is at least as high as the score for any of the sequences represented by U. The problem is that in many cases the score for U is significantly higher, leading to false-positives where the union-sequence is flagged as a match despite none of the cluster members being sufficiently close to the query. The result is substantially slower search.

The expected and optimistic scoring schemes represent two different compromises between sensitivity and speed. We can adjust this balance by combining the two approaches using a mixture model. We define a mixture parameter, λ, such that $0 \leq \lambda \leq 1$. The mixture-model score for aligning wildcard w to residue r is defined as:

$$s_\lambda(w, r) = \lambda s_{opt}(w, r) + (1 - \lambda) s_{exp}(w, r)$$

The score $s_\lambda(w, r)$ for each w, r pair is calculated when the collection is being clustered and then recorded on disk. During a BLAST search, the wildcard scores are loaded from disk and used to perform the search. We report experiments with varying values of λ in Section 7.

6   Selecting Wildcards

Having defined a system for assigning a scoring vector to an arbitrary wildcard, we now describe a method for selecting a set of wildcards to be used during the clustering process. Each wildcard w represents a set of residues $w \subseteq R$ and can be used in a union-sequence to substitute for any set of residues of which it is a superset. A set of wildcards, $W = \{w_1, ..., w_n\}$ is used during clustering. We assume that one of these wildcards w_n is the default wildcard that can be used to represent any of the 24 residue and ambiguous codes, that is $w_n = R$. The remaining wildcards must be selected carefully; large residue sets can be used more frequently but provide poor discrimination with higher average alignment scores and more false positives. Small residue sets can be used less frequently, increasing the use of larger residue sets such as the default wildcard.

The first aspect of choosing a set of wildcards to use for substitution is to decide on the size of this set. It would be ideal to use as many wildcards as necessary, so that each substitution $s_i = o_i$. However, each wildcard must be encoded as a different character leading to an extremely large alphabet. An enlarged alphabet would in turn lead to inefficiencies in BLAST due to larger lookup and scoring data structures. Thus, a compromise is required. BLAST uses a set of 20 character codes to represent residues, as well as 4 IUPAC-IUBMB ambiguous residue codes and an end-of-sequence code, for a total of 25 distinct codes. Each code is represented using 5 bits, permitting a total of 32 codes. This leaves 7 unused character codes. We have therefore chosen to use $|W| = 7$ wildcards.

We treat the task of selecting a good set of wildcards as an optimisation problem. To do this, we first cluster the collection as described in Section 4 using only the default wildcard, $ie.$ $W = \{w_d\}$. We use the residue-substitution sequence O from this clustering to create a set W^* of candidate wildcards. Our goal can then be defined as follows: we wish to select the set of wildcards $W \subseteq W^*$ such that the total average alignment score $A = \sum_{w \in S} \sum_{r \in R} s(w, r)p(r)$ for all substitutions S is minimised. A lower A implies a reduction in the number of high-scoring matches between a typical query sequence and union-sequences in the collection, thereby reducing the number of false-positive situations in which cluster members are fruitlessly recreated and aligned to the query.

In selecting the wildcard set W that minimises A we use the following greedy approach: first, we initialize W to contain only the default wildcard w_d. We then scan through W^* and select the wildcard that leads to the greatest overall reduction in A. This process is repeated until the set W is filled, at each iteration considering the wildcards already in W in the calculation of A. Once W is full we employ a hill-climbing strategy where we consider replacing each wildcard with a set of residues from W^* with the aim of further reducing A.

A set of wildcards was chosen by applying this strategy to the GenBank NR database described in Section 7. The following wildcards were identified and are used for all reported experiments: LVIFM, GEKRQH, AVTIX, SETKDN, LVTPRFYMHCW, AGSDPH, LAGSVETKDPIRNQFYMHCWBZXU.

We also considered defining wildcards based on groups of amino acids with similar physico-chemical properties by using the amino acid classifications described in Taylor 1986 (31). However, a preliminary investigation of this approach resulted in 3% slower search times and reduced search accuracy compared to the approach we have described.

7 Results

The Structural Classification of Proteins (SCOP) database (32; 33) is widely used to evaluate the accuracy of sequence search tools (34; 35). For our own assessments, we used version 1.65 of the ASTRAL Compendium (36) that uses information from the SCOP database to classify sequences with fold, superfamily, and family information. The database contains a total of 67,210 sequences classified into 1,538 superfamilies.

A set of 8,759 test queries were extracted from the ASTRAL database such that no two of the queries shared more than 90% identity. To measure search accuracy, each query was searched against the ASTRAL database and the Receiver Operating Characteristic (ROC) score (37) was calculated. A match between two sequences was considered positive if they came from the same superfamily, otherwise it was considered negative. The ROC_{50} score provides a measure between 0 and 1, where a higher score represents better sensitivity (detection of true positives) and selectivity (ranking true positives ahead of false positives).

The SCOP database is too small to provide an accurate measure of search time, so we use the GenBank non-redundant (NR) protein database to measure search times. The GenBank collection was downloaded August 18, 2005 and contains 2,739,666 sequences in around 900 megabytes of sequence data. Performance was measured using 50 queries randomly selected from GenBank NR. Each query was searched against the entire collection three times with the best runtime recorded and the results averaged. Experiments were conducted on a Pentium 4 2.8GHz machine with two gigabytes of main memory.

We used FSA-BLAST[1]—our open-source version of BLAST—with default parameters as a baseline. To assess the clustering scheme, the GenBank and ASTRAL databases were clustered and FSA-BLAST was configured to report all high-scoring alignments, rather than only the best alignment from each cluster. All reported collection sizes include sequence data and edit information but exclude sequence descriptions. CD-HIT version 2.0.4 beta was used for experiments with 95% clustering threshold and maximum memory set to 1.5 Gb. We also report results for NCBI-BLAST version 2.2.11 and our own implementation of Smith-Waterman that uses the exact same scoring functions and statistics as BLAST (38). No sequence filtering was performed.

The overall results of our clustering method are shown in Table 1. When used with default settings of $\lambda = 0.2$ and $T = 0.25$, our clustering approach reduces the overall size of the NR database by 27% and improves search times by 22%. Importantly, the ROC score indicates that there is no significant effect on search accuracy, with the highly redundant SCOP database reducing in size by 80% when clustered. If users are willing to accept a small loss in accuracy, then the parameters $\lambda = 0$ and $T = 0.3$ improve search times by 27% and reduce the size of the sequence collection by 28% with a decrease of 0.001 in ROC score when compared to our baseline. Since we are interested in improving performance with no loss in accuracy we do not consider these non-default settings further. Overall, our clustering approach with default parameters combined with improvements to the gapped alignment (39) and hit detection (40) stages of BLAST more than

---

[1] Available from: http://www.fsa-blast.org

Table 1. Average runtime for 50 queries searched against the GenBank NR database, and SCOP ROC$_{50}$ scores for the ASTRAL collection

Scheme	GenBank NR		ASTRAL
	Time	Sequence data	
	secs (% baseline)	Mb (% baseline)	ROC$_{50}$
FSA-BLAST			
No clustering (baseline)	28.75 (100%)	900 (100%)	0.398
Cluster $\lambda = 0.2, T = 0.25$	22.54 (78%)	655 (73%)	0.398
Cluster $\lambda = 0, T = 0.3$	20.97 (73%)	650 (72%)	0.397
NCBI-BLAST	45.75 (159%)	898 (100%)	0.398
Smith-Waterman	—	—	0.415

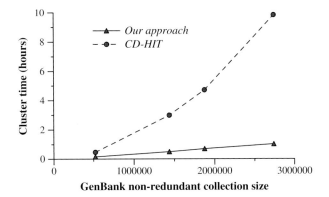

Fig. 2. Clustering performance for GenBank NR databases of varying sizes

double the speed of FSA-BLAST compared to NCBI-BLAST with no significant effect on accuracy. Both versions of BLAST produce ROC scores 0.017 below the optimal Smith-Waterman algorithm.

Figure 2 shows a comparison of clustering times between CD-HIT and our clustering approach for four different releases of the GenBank NR database; details of the collections used are given in Table 2. The results show that the clustering time of our approach is linear with the collection size and the CD-HIT approach is superlinear (Figure 2). On the recent GenBank non-redundant collection, CD-HIT is almost 10 times slower than our approach; we expect this ratio to further increase with collection size.

Table 2. Redundancy in GenBank NR database over time

Release date	Number of sequences	Collection Size (Mb)	Overall size reduction (Mb)	Percentage of collection
16 July 2000	521,662	157	45	28.9%
22 May 2003	1,436,591	443	124	28.1%
30 June 2004	1,873,745	597	165	27.4%
18 August 2005	2,739,666	900	245	27.3%

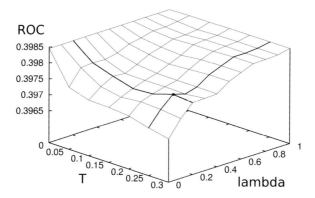

Fig. 3. Search accuracy for collections clustered with varying values of λ and T. Default values of $\lambda = 0.2, T = 0.25$ are highlighted.

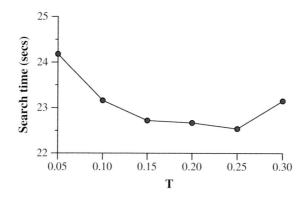

Fig. 4. Average BLAST search time using $\lambda = 0.2$ and varying values of T

Table 2 shows the amount of redundancy in the GenBank NR database as it has grown over time, measured using our clustering approach. We observe that the degree of redundancy is not changing significantly with the percentage reduction through clustering remaining between 27% and 29% across versions of the collection tested.

Figure 3 shows the effect on accuracy for varying values of λ and T. We have chosen $\lambda = 0.2$ as a default value because smaller values of λ result in a larger decrease in search accuracy, and larger values reduce search speed. We observe that for $\lambda = 0.2$ there is little variation in search accuracy for values of T between 0.05 and 0.3.

Figure 4 shows the effect on search times for varying values of T where $\lambda = 0.2$. As T increases the clustered collection becomes smaller, leading to faster search times. However, if T is too large then union-sequences with a high percentage of wildcards are permitted, leading to an increase in the number of cluster members that are recreated and a corresponding reduction in search speed. We have chosen the value $T = 0.25$ that maximises search speed.

8   Conclusion

Sequence databanks such as GenBank contain a large number of redundant sequences. Such redundancy has several negative effects including larger collection size, slower search, and difficult-to-interpret results. Redundancy within a collection can lead to over-representation of alignments within particular protein domains, distracting the user from other potentially important hits.

We have proposed a new scheme for managing redundancy. Instead of discarding near-duplicate sequences, our approach identifies clusters of redundant sequences and constructs a special union-sequence that represents all members of the cluster through the careful use of wildcard characters. We present a new approach for searching clusters that, when combined with a well-chosen set of wildcards and a system for scoring matches between wildcards and query residues, leads to faster search times without a significant loss in accuracy. Moreover, by recording the differences between the union-sequence and each cluster member using edit information our approach compresses the collection. Our scheme is general and can be adapted to most homology search tools.

We have integrated our algorithm into FSA-BLAST, a new version of BLAST that is substantially faster than NCBI-BLAST and freely available for download at http://www.fsa-blast.org/. Our results show that our clustering scheme reduces BLAST search times against the GenBank non-redundant database by 22% and compresses sequence data by 27% with no significant effect on accuracy. We have also described a new system for identifying clusters that uses fingerprinting, a technique that has been successfully applied to duplicate-document detection in information retrieval. Our implementation can cluster the entire GenBank NR protein database in one hour on a standard workstation and scales linearly in the size of the collection. We propose that pre-clustered copies of the GenBank collection be made publicly available for download.

We have confined our experimental work to protein sequences and plan to investigate the effect of our clustering scheme on nucleotide data as future work. We also plan to investigate the effect of our approach on iterative search algorithms such as PSI-BLAST, and how our scheme can be used to improve the current measure of the statistical significance of BLAST alignments.

Acknowledgements

The authors thank Peter Smooker and Michelle Chow for valuable suggestions. This work was supported by the Australian Research Council.

References

[1] Altschul, S., Madden, T., Schaffer, A., Zhang, J., Zhang, Z., Miller, W., Lipman, D.: Gapped BLAST and PSI–BLAST: A new generation of protein database search programs. Nucleic Acids Research **25** (1997) 3389–3402

[2] Li, W., Jaroszewski, L., Godzik, A.: Sequence clustering strategies improve remote homology recognitions while reducing search times. Protein Engineering **15** (2002) 643–649

[3] Park, J., Holm, L., Heger, A., Chothia, C.: RSDB: representative sequence databases have high information content. Bioinformatics **16** (2000) 458–464

[4] Altschul, S., Gish, W., Miller, W., Myers, E., Lipman, D.: Basic local alignment search tool. Journal of Molecular Biology **215** (1990) 403–410

[5] Pearson, W., Lipman, D.: Improved tools for biological sequence comparison. Proceedings of the National Academy of Sciences USA **85** (1988) 2444–2448

[6] Pearson, W., Lipman, D.: Rapid and sensitive protein similarity searches. Science **227** (1985) 1435–1441

[7] Li, W., Jaroszewski, L., Godzik, A.: Tolerating some redundancy significantly speeds up clustering of large protein databases. Bioinformatics **18** (2001) 77–82

[8] Smith, T., Waterman, M.: Identification of common molecular subsequences. Journal of Molecular Biology **147** (1981) 195–197

[9] Burke, J., Davison, D., Hide, W.: d2_cluster: A validated method for clustering EST and full-length DNA sequences. Genome Research **9** (1999) 1135–1142

[10] Bleasby, A.J., Wootton, J.C.: Construction of validated, non-redundant composite protein sequence databases. Protein Engineering **3** (1990) 153–159

[11] Kallberg, Y., Persson, B.: KIND — a non-redundant protein database. Bioinformatics **15** (1999) 260–261

[12] Grillo, G., Attimonelli, M., Liuni, S., Pesole, G.: CLEANUP: a fast computer program for removing redundancies from nucleotide sequence databases. Computer Applications in the Biosciences **12** (1996) 1–8

[13] Holm, L., Sander, C.: Removing near-neighbour redundancy from large protein sequence collections. Bioinformatics **14** (1998) 423–429

[14] Li, W., Jaroszewski, L., Godzik, A.: Clustering of highly homologous sequences to reduce the size of large protein databases. Bioinformatics **17** (2001) 282–283

[15] Malde, K., Coward, E., Jonassen, I.: Fast sequence clustering using a suffix array algorithm. Bioinformatics **19** (2003) 1221–1226

[16] Gracy, J., Argos, P.: Automated protein sequence database classification. i. integration of compositional similarity search, local similarity search, and multiple sequence alignment. Bioinformatics **14** (1998) 164–173

[17] Gusfield, D.: Algorithms on Strings, Trees, and Sequences. Cambridge University Press (1997)

[18] Manber, U., Myers, G.: Suffix arrays: a new method for on-line string searches. SIAM Journal on Computing **22** (1993) 935–948

[19] Cheung, C.F., Yu, J.X., Lu, H.: Constructing suffix tree for gigabyte sequences with megabyte memory. IEEE Transactions on Knowledge and Data Engineering **17** (2005) 90–105

[20] Lee, C., Grasso, C., Sharlow, M.F.: Multiple sequence alignment using partial order graphs. Bioinformatics **18** (2002) 452–464

[21] Bernstein, Y., Cameron, M.: Fast discovery of similar sequences in large genomic collections. In: Proc. European Conference on Information Retrieval. (2006) To appear.

[22] Manber, U.: Finding similar files in a large file system. In: Proceedings of the USENIX Winter 1994 Technical Conference, San Fransisco, CA, USA (1994) 1–10

[23] Heintze, N.: Scalable document fingerprinting. In: 1996 USENIX Workshop on Electronic Commerce, Oakland, California, USA (1996) 191–200

[24] Brin, S., Davis, J., García-Molina, H.: Copy detection mechanisms for digital documents. In Carey, M., Schneider, D., eds.: Proceedings of the ACM SIGMOD Annual Conference, San Jose, California, United States, ACM Press (1995) 398–409

[25] Broder, A.Z., Glassman, S.C., Manasse, M.S., Zweig, G.: Syntactic clustering of the web. Computer Networks and ISDN Systems **29** (1997) 1157–1166

[26] Bernstein, Y., Zobel, J.: A scalable system for identifying co-derivative documents. In Apostolico, A., Melucci, M., eds.: Proc. String Processing and Information Retrieval Symposium (SPIRE), Padova, Italy, Springer (2004) 55–67

[27] Bernstein, Y., Zobel, J.: Redundant documents and search effectiveness. In: CIKM '05: Proceedings of the 14th ACM international conference on Information and knowledge management, New York, NY, USA, ACM Press (2005) 736–743

[28] Johnson, S.: Hierarchical clustering schemes. Psychometrika **32** (1967) 241–254

[29] Robinson, A., Robinson, L.: Distribution of glutamine and asparagine residues and their near neighbors in peptides and proteins. Proceedings of the National Academy of Sciences USA **88** (1991) 8880–8884

[30] Henikoff, S., Henikoff, J.: Amino acid substitution matrices from protein blocks. Proceedings of the National Academy of Sciences USA **89** (1992) 10915–10919

[31] Taylor, W.: The classification of amino-acid conservation. Journal of Theoretical Biology **119** (1986) 205–218

[32] Murzin, A., Brenner, S., Hubbard, T., Chothia, C.: SCOP: a structural classification of proteins database for the investigation of sequences and structures. Journal of Molecular Biology **247** (1995) 536–540

[33] Andreeva, A., Howorth, D., Brenner, S., Hubbard, T., Chothia, C., Murzin, A.: SCOP database in 2004: refinements integrate structure and sequence family data. Nucleic Acids Research **32** (2004) D226–D229

[34] Brenner, S., Chothia, C., Hubbard, T.: Assessing sequence comparison methods with reliable structurally identified distant evolutionary relationships. Proceedings of the National Academy of Sciences USA **95** (1998) 6073–6078

[35] Park, J., Karplus, K., Barrett, C., Hughey, R., Haussler, D., Hubbard, T., Chothia, C.: Sequence comparisons using multiple sequences detect three times as many remote homologues as pairwise methods. Journal of Molecular Biology **284** (1998) 1201–1210

[36] Chandonia, J., Hon, G., Walker, N., Conte, L.L., Koehl, P., Levitt, M., Brenner, S.: The ASTRAL compendium in 2004. Nucleic Acids Research **32** (2004) D189–D192

[37] Gribskov, M., Robinson, N.: Use of receiver operating characteristic (ROC) analysis to evaluate sequence matching. Computers & Chemistry **20** (1996) 25–33

[38] Karlin, S., Altschul, S.: Methods for assessing the statistical significance of molecular sequence features by using general scoring schemes. Proceedings of the National Academy of Sciences USA **87** (1990) 2264–2268

[39] Cameron, M., Williams, H.E., Cannane, A.: Improved gapped alignment in BLAST. IEEE Transactions on Computational Biology and Bioinformatics **1** (2004) 116–129

[40] Cameron, M., Williams, H.E., Cannane, A.: A deterministic finite automaton for faster protein hit detection in BLAST. Journal of Computational Biology (2005) To appear.

New Methods for Detecting Lineage-Specific Selection*

Adam Siepel[1],[**], Katherine S. Pollard[1],[***], and David Haussler[1],[2]

[1] Center for Biomolecular Science and Engineering, U.C. Santa Cruz,
Santa Cruz, CA 95064, USA
[2] Howard Hughes Medical Institute, U.C. Santa Cruz,
Santa Cruz, CA 95064, USA

Abstract. So far, most methods for identifying sequences under selection based on comparative sequence data have either assumed selectional pressures are the same across all branches of a phylogeny, or have focused on changes in specific lineages of interest. Here, we introduce a more general method that detects sequences that have either come under selection, or begun to drift, on any lineage. The method is based on a phylogenetic hidden Markov model (phylo-HMM), and does not require element boundaries to be determined *a priori*, making it particularly useful for identifying noncoding sequences. Insertions and deletions (indels) are incorporated into the phylo-HMM by a simple strategy that uses a separately reconstructed "indel history." To evaluate the statistical significance of predictions, we introduce a novel method for computing P-values based on prior and posterior distributions of the number of substitutions that have occurred in the evolution of predicted elements. We derive efficient dynamic-programming algorithms for obtaining these distributions, given a model of neutral evolution. Our methods have been implemented as computer programs called DLESS (Detection of LinEage-Specific Selection) and phyloP (phylogenetic P-values). We discuss results obtained with these programs on both real and simulated data sets.

1 Introduction

In recent years, abundant sequence data has led to widespread interest in methods for detecting genomic sequences that are evolving faster, slower, or by different patterns of substitution than would be expected under neutral drift. While some such sequences could result from non-uniformities in mutational and repair processes, the majority are thought to be subject to pressure by natural selection, and to have evolutionarily important biological functions. The genomes of most species of interest are too vast, and laboratory assays are still too labor-intensive,

---

* This paper is presented here in abbreviated form; the complete version is available from http://www.bscb.cornell.edu/Homepages/Adam_Siepel/dless.pdf
** Current address: Dept. of Biol. Stats. & Comput. Biol., Cornell Univ., Ithaca, NY.
*** Current address: U.C. Davis Genome Center & Dept. of Stats., Davis, CA.

A. Apostolico et al. (Eds.): RECOMB 2006, LNBI 3909, pp. 190–205, 2006.
© Springer-Verlag Berlin Heidelberg 2006

to permit exhaustive wet-laboratory searches for functional elements. Computational screens based on comparative sequence data allow whole genomes to be reduced to much smaller sets of candidate functional elements, which can more feasibly be tested in the lab (e.g., [1, 2]).

In the comparative genomics community, much attention has focused on two problems in particular: (1) identifying (especially noncoding) sequences that are unusually conserved across species, and thus are likely to be subject to negative selection (e.g., [3, 4, 5, 6]); and (2) identifying protein-coding genes that show unusually high d_N/d_S ratios, and thus might be subject to positive selection (e.g., [7, 8, 9, 10, 11, 12]). Methods focused on problem (1) generally have made the assumption (explicitly or implicitly) that selectional pressures are the same across all branches of a phylogeny—i.e., that each candidate sequence is under selection in all species or not under selection in any species. This assumption is sometimes relaxed in methods focused on problem (2) (e.g., [8, 10, 11]), but these methods generally can be used only with protein-coding sequences, whose boundaries are predetermined by annotations of known genes. In addition, most methods that allow for lineage-specific selection have required *a priori* specification of the branches of the tree on which the mode of selection may change [8, 10].

In recent work, we have developed methods for identifying sequences (coding or noncoding) that are significantly changed in the human lineage (K. Pollard, S. Salama, B. King, et al., submitted). These methods are efficient enough to be applied at the scale of complete vertebrate genomes, given the locations of candidate sequences. Our aim here is to develop more general methods capable of detecting sequences that have been subject to lineage-specific selection on any (unspecified) branch of a phylogeny, and that do not require predefined element boundaries. These methods must remain highly efficient and suitable for use with noncoding sequences. We focus on the case of negative selection, although our methods can be extended to positive selection (see Discussion). We describe two programs, called DLESS (Detection of LinEage Specific Selection) and phyloP (phylogenetic P-values), that address the problem of detecting lineage-specific selection, and show good power and low false positive rates in simulation experiments. These programs are fast enough to run in a few minutes on multiple alignments for the ENCODE regions [13] (which span ~1% of the human genome), using a small compute cluster. We describe our methods in detail, and discuss results for both real and simulated data.

2   Methods

2.1   The Model

HMM and Phylogenetic Models. DLESS is based on a phylogenetic hidden Markov model (phylo-HMM), an HMM that emits columns of a multiple alignment according to probability distributions that are defined by phylogenetic models associated with its states [14, 15] (reviewed in [16]). DLESS's model is a generalization of the two-state phylo-HMM used by the phastCons program [6].

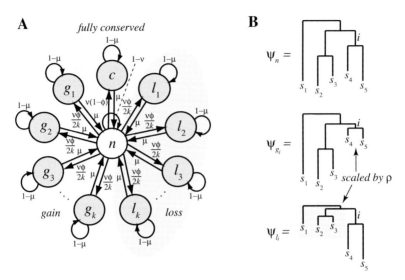

Fig. 1. (A) State-transition diagram for DLESS. The probability of beginning with each state (not shown) is taken to be that state's probability at stationarity. (B) Neutral phylogenetic model (ψ_n), with a branch i indicated, and derived phylogenetic models for a "gain" (ψ_{g_i}) and "loss" (ψ_{l_i}) of a conserved element on branch i.

PhastCons has a state c for conserved sequences and a state n for nonconserved sequences; these states are associated with two phylogenetic models, ψ_c and ψ_n, respectively, which are identical except that the branch lengths of ψ_c are scaled by a factor $\rho \in (0,1)$. Based on this two-state model, phastCons parses an alignment into likely "conserved" and "nonconserved" segments. DLESS works by the same principle, but also allows for conserved elements that have been "gained" or "lost" on any branch of the phylogeny. The new model has $2k + 2$ states, labeled c (the "fully conserved" state), n ("nonconserved"), $g_1, \ldots, g_k$ ("gain"), and $l_1, \ldots, l_k$ ("loss"), where k is the number of branches in the tree in question (Fig. 1A). (For a phylogeny of N present-day species, $k = 2N - 3$, assuming a reversible model and an unrooted tree.).

To limit the number of parameters, the states are arranged in a "hub and spokes" configuration (Fig. 1A). As a result, predicted conserved elements are required to be separated from one another by at least one base of nonconserved sequence. In practice, this is not a severe limitation, because, conserved elements in vertebrates are relatively sparse. In addition, conserved elements of all classes are assumed to have the same (geometric) length distribution, and all lineage-specific elements are assumed to occur with the same (prior) probability. Three parameters—μ, ν, and ϕ—define all transition probabilities in the HMM (Fig. 1A). For interpretability, it is useful to reparameterize μ and ν as $\omega = \frac{1}{\mu}$, the expected length of conserved elements, and $\gamma = \frac{\nu}{\mu + \nu}$, the expected fraction of bases in conserved elements [6]. The third free parameter, ϕ, is the probability that an element is lineage-specific given that it is conserved. Note that this model fails to allow for scenarios in which single elements undergo multiple "gain" and "loss" events over evolutionary time, even if these events occur on separate lineages (see Discussion).

As with phastCons, the phylogenetic models associated with the states are identical, except that certain branches are scaled by the parameter $\rho \in (0,1)$. The neutral model, ψ_n, is assumed to be given (it can be estimated, e.g., from fourfold degenerate sites in coding regions), and all other models are derived from it. The model for a gain event on branch i, ψ_{g_i}, is equal to ψ_n, except that branch i and all branches in the subtree beneath it are scaled by the factor ρ. Similarly, the model for a loss event on branch i, ψ_{l_i} is equal to ψ_n, except that all branches outside the subtree beneath (and including) branch i are scaled by the factor ρ (Fig. 1B). The model parameters can be estimated by maximum likelihood or treated as "tuning" parameters to be set according to some other principle (see below).

Model for Indels. Most efforts to use phylogenetic models in the identification of functional elements have finessed the issue of indels (which have long been a thorny problem in phylogenetic analysis), by treating alignment gaps as missing data (e.g., [6]), treating indels like substitutions (e.g., [17,18]), or using other heuristics (e.g., [19,5]). Previous approaches, however, are inadequate for the problem of identifying elements under lineage-specific selection. In some cases, alignment gaps are the strongest indication that an element has been "lost" or "gained" (consider an element that was completely deleted on some branch of the tree), so they cannot be treated as missing data. On the other hand, methods that assume site-independence of gaps tend to be too sensitive to occasional indels of moderate length. Other methods (e.g., [19]) cannot be readily applied.

Ideally, one would sample over indel scenarios conditional on an alignment, and possibly sample over alignments as well (e.g., [20,21]), but we take a short-cut here, which is simpler, faster, and adequate for our purposes. Briefly, we reconstruct an "indel history" (a history of insertion and deletion events on all branches of the tree) by parsimony, using a slightly modified version of the inferAncestors program [22]. We then compute emission probabilities of indels for a phylo-HMM conditional on this history. Given an alignment and indel history, probabilities of indels can be computed using well-known pair-HMM methods, and indel parameters can easily be estimated by maximum likelihood. In addition, it turns out to be straightforward to integrate this indel model into a phylo-HMM (see full paper at http://www.bscb.cornell.edu/Homepages/Adam_Siepel/dless.pdf). This approach, of course, is only as good as the accuracy of the alignment and the indel history, but simulation experiments suggest that their accuracy is quite good, at least for mammalian genomes at modest evolutionary distances [22].

2.2   Assessing Significance

The significance of predicted conserved elements is summarized by a P-value, indicating how surprising the aligned sequences (within the region of the prediction) would be under neutral evolution. We introduce a novel method for computing P-values that is based on counts of substitutions. It appears to have nearly as much power as the likelihood ratio tests (LRTs) more commonly used in statistical phylogenetics (e.g., [7, 8, 11]), and it has certain advantages over LRTs (see Discussion).

The Distribution for One Branch. We first derive a solution to the problem of finding the distribution of the number of substitutions along a single branch of a phylogenetic tree, under a general continuous-time Markov model of substitution. To our knowledge, a general solution for this problem has not been published, although methods for computing the mean and variance of such a distribution have been developed [23].

Consider a branch of length t in a phylogenetic tree, connecting a "child" sequence and a "parent" sequence. Assume that substitutions occur by a continuous-time Markov model, defined by a rate matrix $\mathbf{Q} = \{q_{a,b}\}$, where $q_{a,b}$ is the instantaneous rate at which base a changes to base b. Thus, the probability of a base b in the child species given an orthologous base a in the parent species, denoted $P(b|a, t)$, is given by element (a, b) of the matrix $\mathbf{P}(t) = \exp(\mathbf{Q}t) = \sum_{i=0}^{\infty} \frac{(\mathbf{Q}t)^i}{i!}$. We assume that $\mathbf{Q}$ is scaled such that t has units of expected substitutions per site.

The probability mass function of interest is $P(n|t)$, where n is the number of substitutions per site, allowing for so-called "multiple hits"—i.e., substitutions that obscure other substitutions. The expected value of this distribution, $E[n|t]$, is equal to t, but the entire distribution is not known; it depends not only on t but on the particular process by which sequences of substitutions occur, as defined by the matrix $\mathbf{Q}$. This distribution is sometimes assumed to be Poisson with rate t, and indeed, under certain conditions (e.g., when all substitutions occur at the same rate, as in the Jukes-Cantor model [24]) this assumption is correct. In general, however, the Poisson postulates are violated by the dependency of substitution rates on the starting base in the continuous-time Markov chain. For example, this state-dependency causes the numbers of events in disjoint time intervals to be dependent. It is possible to come up with matrices $\mathbf{Q}$ that cause $P(n|t)$ to be quite unlike a Poisson distribution (Fig. 2A). The mean and variance of $P(n|t)$, for general $\mathbf{Q}$, are of interest in computing the widely used index of dispersion [25, 23].

Solving directly for $P(n|t)$ appears to be difficult (for general $\mathbf{Q}$), but the distribution can be obtained fairly easily by working with the embedded discrete Markov process associated with $\mathbf{Q}$. We decompose the substitution process into a "jump process," which does obey the Poisson postulates, and a substitution process conditional on jumps. The construction is such that every substitution follows a jump, but not every jump is followed by a substitution. Let λt be the rate of the jump process, and let $\mathbf{R} = \{r_{a,b}\}$ be a matrix of conditional probabilities of substitution given a single jump; i.e., $r_{a,b} = P(b|a, 1 \text{ jump})$. Both λ and $\mathbf{R}$ can be derived from $\mathbf{Q}$ (see full paper). The desired distribution can now be written as:

$$P(n|t) = \sum_{j=0}^{\infty} P(n|j)\mathrm{Pois}(j|\lambda t), \qquad (1)$$

where $P(n|j)$ is the probability of n substitutions given j jumps and $\mathrm{Pois}(j|\lambda t)$ is the probability of j jumps in time t. $P(n|j)$ is a function of $\mathbf{R}$ only, and can be precomputed for all n and j less than some adequately large j_{max} and stored in a table. This computation can be done efficiently by dynamic programming (see full paper). Subsequently, $P(n|t)$ can be approximated arbitrarily closely, for any n and t of interest, by taking the sum of the first j_{max} terms of the RHS of equation 1. Using similar methods, it is also possible to obtain the posterior distribution, $P(n|a, b, t)$, and the distribution in the presence of rate variation (see full paper).

To compute P-values of conservation for conserved elements, we need the distribution of the number of substitutions in an interval consisting of m sites. As long as m is not too large, this distribution can be obtained by taking a convolution of the individual-site distributions, assuming site independence. In the case of the prior distribution, these individual-site distributions are identical, but in the case of the posterior distribution, they differ according to the bases observed at each site.

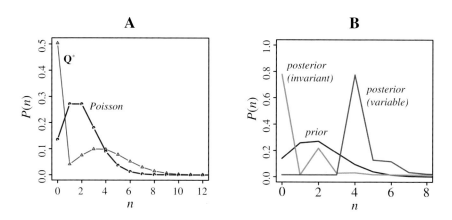

Fig. 2. (A) Distribution of the number of substitutions per site (n) for a rate matrix $\mathbf{Q}^*$ and a single branch of length $t = 2$, obtained with phyloP, and the Poisson distribution of the same mean. $\mathbf{Q}^*$ has very high rates of substitution between bases A and G, but much lower rates ($1000\times$) between other pairs of bases. The stationary distribution is uniform, and the model is reversible. The left peak in the bimodal distribution reflects a starting base of C or T and the right peak reflects a starting base of A or G. (B) Prior and posterior distributions for a phylogeny of 10 vertebrates and for two alignment columns, one highly variable and one invariant (see full paper for details). The prior distribution is indistinguishable from a Poisson distribution. The posterior distributions have modes corresponding to maximum parsimony solutions but also give considerable weight to nonparsimonious scenarios. Note that some numbers of substitutions have zero probability in the posterior.

The Distribution for a Full Phylogeny. The distribution of the number of substitutions per site for a general phylogeny can be computed by an algorithm that takes a convolution of the distributions for each branch, using the recursive structure of the tree.

Let X be a (possibly observed) column in an alignment, let u be a node in the tree, let t_u be the length of the branch above node u, let X_u be a random variable representing the base at node u, and let $x_{\underline{u}}$ indicate any observed data at the leaves beneath node u. In addition, let v and w be the children of node u. The key recurrence relation is:

$$P(n, x_{\underline{u}} | X_u = a) = \sum_{i=0}^{n} \left[\sum_b \sum_{j=0}^{i} P(j, x_{\underline{v}} | X_v = b) P(i - j, b | a, t_v) \right] \times$$
$$\left[\sum_c \sum_{k=0}^{n-i} P(k, x_{\underline{w}} | X_w = c) P(n - i - k, c | a, t_w) \right] \tag{2}$$

where $P(n, x_{\underline{u}} | X_u = a)$ is the probability of n substitutions beneath node u and the data beneath node u, given that the base at node u is a. The terms $P(i - j, b | a, t_v)$ and $P(n - i - k, c | a, t_w)$ represent branch-specific distributions related to those of the previous section (see full paper).

The algorithm for computing the distribution resembles Felsenstein's pruning algorithm [26]. It differs only slightly in the cases of the prior distribution and the posterior distribution. Details are given in the full paper. As for a single branch, the distribution for m sites can be obtained by taking a convolution of the individual-site distributions.

The Joint Distribution for a Subtree and Supertree. In the case of lineage-specific selection, the tree is partitioned at some branch B of interest into a subtree and its complementary "supertree." What is of interest in this case is the joint distribution of n_{sub}, the number of substitutions in the subtree beneath B, and n_{sup}, the number of substitutions in the supertree. If the substitution model is reversible, then the tree can be rerooted at the node above branch B, so that the original subtree becomes one subtree of the root, and the original supertree becomes the other subtree of the root. The joint distribution of interest can then be computed by a slight modification of the algorithm described in the previous section. Only the termination step of the algorithm, which is applied at the root of the tree, needs to be altered. Details are given in the full paper.

As above, the distribution for m sites can be computed by taking a convolution of individual-site distributions. In this case, however, these distributions are bivariate.

The Computation of P-Values. The methods above allow a prior distribution $P(n | m, \boldsymbol{\psi}_n)$ and a posterior distribution $P(n | \mathbf{X}, \boldsymbol{\psi}_n)$ to be computed for any alignment fragment $\mathbf{X}$ of length m and any neutral model $\boldsymbol{\psi}_n$. To compute a P-value, we interpret the prior distribution as a null distribution, reflecting the hypothesis of neutral evolution, and we take the mean of the posterior distribution as a proxy for an "observed" number of substitutions. With ample data and branches of modest length, the variance of the posterior distribution is fairly small (Fig. 2B), and it is reasonable to summarize the distribution by its mean. In computing the posterior distribution, the neutral model can be used as a prior, but this influences the posterior mean toward the prior mean, making the P-values conservative. To avoid this problem, we use an empirical Bayes approach: based on the alignment fragment of interest, we estimate a scale factor for the neutral model by maximum likelihood (using a numerical optimization algorithm), then use the scaled neutral model as a prior when computing the posterior distribution. A P-value for a posterior mean $E[n|\mathbf{X}, \hat{\rho}\psi_n]$ is computed as:

$$P = \sum_{0 \leq i \leq E[n|\mathbf{X}, \hat{\rho}\psi_n]} P(i|m, \psi_n) \tag{3}$$

where $\hat{\rho}$ is the estimated scale factor and $\hat{\rho}\psi_n$ denotes the scaled neutral model.

In the case of lineage-specific selection, separate scale factors are estimated for the subtree and supertree, and four P-values are computed. The first two P-values are as described above, except that they are based on marginal distributions (prior and posterior, derived from the corresponding joint distributions) for the subtree and supertree in question. These P-values indicate whether, considered separately, the numbers of substitutions in the subtree and supertree are surprising in comparison to the null model. The other two P-values are conditional P-values, indicating how surprising are the numbers of substitutions in the subtree and supertree, given the total number of substitutions in the whole tree. These P-values allow for the possibility that the substitution rate across the whole tree does not fit the neutral model well, and focus attention more directly on differences between the subtree and supertree. Note that all four P-values are computed independently and do not account for correlation between tests. Adjustments for multiple hypothesis tests are needed when jointly interpreting the marginal P-values for a collection of elements.

2.3   Implementation and Experimental Design

The DLESS (Detection of LinEage Specific Selection) and phyloP (phylogenetic P-value) programs were implemented in C, as new modules in the PHAST (Phylogenetic Analysis with Space/Time models) package [6]. Simulation experiments were conducted to test the false positive rates and power of both programs. All experiments were based on a set of about 100,000 fourfold degenerate sites extracted from alignments of up to 19 species for the 44 ENCODE regions [13], and on a model of neutral evolution estimated from these sites using the REV substitution model (E. Margulies, pers. comm.). We looked at both the full 19-species set and a subset of 10 species (Fig. 3). We simulated neutral alignments using both a "parametric" method (generating sites from the estimated neutral model) and a "nonparametric" method (randomly drawing sites from the original alignment, with replacement). The phyloBoot program in PHAST was used. False positive rates were estimated by running the two programs on these neutral alignments.

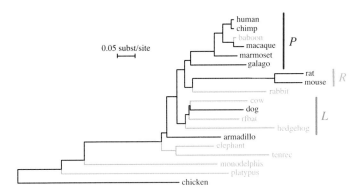

Fig. 3. Phylogenetic tree for the 19-species considered, with neutral branch lengths estimated from fourfold degenerate sites in the ENCODE regions. The 10-species subset is highlighted in black, and three subtrees of interest are indicated: the primates (P), rodents (R), and laurasiatherians (L).

To estimate power, we measured the ability of the programs to correctly identify simulated conserved elements, after controlling for false positive rates. Conserved alignment columns were generated parametrically using versions of the neutral model in which either all branches, or some subset of branches (in the case of lineage-specific conservation), were scaled by a factor $\rho \in (0,1)$. In tests of the power to detect fully conserved elements, we also used a nonparametric method in which columns were drawn randomly from protein-coding sites extracted from the ENCODE alignments. (Note that some of these sites are in reality not conserved.) Conserved elements of 15–200bp were generated. In tests of DLESS, conserved elements were embedded within neutral alignments of 300bp. Predictions of the correct types overlapping the embedded elements were counted as correct.

The programs were also run on a full set of ENCODE alignments, consisting of 19 species and about 35 million sites (including gaps in the human reference sequence), and produced by the TBA program [27]. The DLESS predictions are available as a track in the UCSC Genome Browser (http://genome.ucsc.edu/encode). Clicking on individual predictions causes the statistics computed by phyloP to be displayed.

3   Results

3.1   Simulation Results

Power of DLESS. Based on simulated neutral data sets of 1 million columns, we adjusted the tuning parameters of DLESS to permit a false positive rate of approximately one base per thousand, as estimated by the parametric method. We found that γ and ϕ needed to be increased substantially from their maximum likelihood estimates (MLEs; based on the ENCODE data)—from $\gamma = 0.06$ to $\gamma = 0.35$, and from $\phi = 0.18$ to $\phi = 0.8$—to achieve a reasonable tradeoff between false positive and false negative rates. The MLEs led to high specificity but relatively weak sensitivity, especially for lineage-specific elements having short lengths, weak conservation, or small subtrees. This tendency to under-predict lineage-specific elements presumably results from these elements effectively being supported by less data than are fully conserved elements—i.e., it is primarily only the sequences in the subtree (in the case of a gain) or supertree (in the case of a loss) of interest that support the hypothesis of lineage-specific conservation, while all sequences support the hypothesis of full conservation. The parameter ω was set to 20 and the parameter ρ was set to 0.3, based on our experience with the phastCons program.

The power of DLESS to detect conserved elements depends on many factors, including the lengths of the elements, the sizes of the whole phylogeny and of the subtree and supertree in question (numbers of species and total branch length), and the degree of conservation (Fig. 4). The power is generally quite good when elements are of length 50bp or greater and the scaling parameter $\rho \leq 0.3$. The power for detecting fully conserved elements is excellent, even when element lengths are as small as 15bp. The power is also reasonably good for detecting elements gained or lost in subtrees with relatively large numbers of species and large total branch length (e.g., the laurasiatherian subtree; see Fig. 3), but it is significantly reduced when the number of species in a subtree is small (e.g., the rodent subtree), or when the total branch length is small (e.g., the primate subtree). Still, in these cases, longer and more conserved elements can be detected fairly reliably. Interestingly, the method has considerably more power to detect lineage specific "losses" than lineage-specific "gains," particularly for smaller subtrees (see full paper). Apparently, there is more to be gained by switching

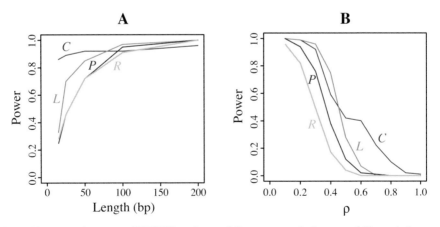

Fig. 4. Estimated power of DLESS to detect fully conserved elements (C), and elements gained or lost in the primate (P), rodent (R), or laurasiatherian (L) clades. Plots show the power to detect (A) lineage-specific losses as a function of element length (for $\rho = 0.3$), and (B) lineage-specific gains as a function of the scale parameter ρ (for elements of length 100). Results are for the 19-species phylogeny and 100 simulated data sets.

states in the HMM when the supertree has short branches and the subtree has long ones (as in a loss) than when the supertree has long branches and the subtree has small ones (as in a gain). This effect might be compensated for by alternative parameterizations of the transition probabilities of the model.

False Positive Rates and Power of PhyloP. Because of the estimation of the scale factors and the use of the posterior mean as a proxy for an observed number of substitutions, the P-values reported by phyloP for data sets drawn from the null model are not guaranteed to be uniformly distributed. In practice, the reported P-values are nearly uniform but usually slightly conservative—i.e., the fraction of reported P-values below some p_0 is generally less than p_0, implying a false positive rate below the target value. The P-values are somewhat more conservative for short elements than for longer elements, and somewhat more conservative in the case of lineage-specific selection than in the case of fully conserved elements (see full paper).

Despite the conservative P-values, the method has good power. The power to detect fully conserved elements is excellent with $\rho \leq 0.5$, very good with $\rho = 0.7$, and respectable even for $\rho = 0.9$ at lengths of ≥ 100bp (Fig. 5A). With smaller values of ρ, elements as short as 15bp can reliably be detected. The nonparametric results, based on protein-coding sites ("CDS" curve in Fig. 5A), suggest that the method's performance in detecting these more conserved elements may be a reasonable indication of its ability to detect real functional elements. For lineage-specific elements (Fig. 5B), the power is reduced but still quite good as long as ρ is not too large, elements are not too short, and subtrees have adequate phylogenetic information. As with DLESS, the power to detect losses

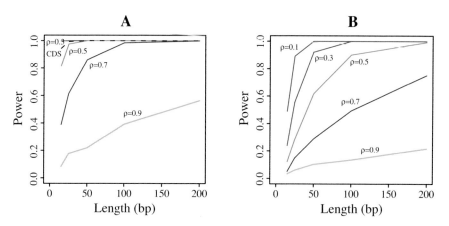

Fig. 5. Power of phyloP in simulation experiments to detect (A) fully conserved elements and (B) elements gained in primates. Power is estimated as the fraction of 1000 simulated data sets in which the null hypothesis was correctly rejected, with a P-value threshold of 0.05. Conditional P-values were used for lineage-specific elements. These experiments were based on the 10-species set. The results of the nonparametric test, based on sites from coding regions, are shown as the dashed "CDS" line in (A).

is greater than the power to detect gains, but the difference between losses and gains is less pronounced with phyloP than with DLESS. The primate (Fig. 5B) and rodent (not shown) subtrees had similar power curves, and power increased in the laurasiatherian subtree. The method has low power to detect elements in very small subtrees, such as the one consisting of just human (which is of obvious interest). Nevertheless, across a wide range of parameter values, the power of the method is nearly as good as that of an empirically calibrated likelihood ratio test (LRT), which is asymptotically most powerful, and is expected to be close to optimal in practice. The gap between phyloP and the LRT is greatest for short elements, and is more evident with lineage-specific elements than with fully conserved elements (see full paper).

3.2 Results for ENCODE Data

DLESS predicted 24,011 elements covering 5.7% of the 29.9 million human bases in ENCODE regions. Retaining only fully conserved predictions with $P < 0.05$ (as reported by phyloP) and lineage-specific predictions with conditional $P < 0.05$ reduced these numbers to 20,959 and 4.8%, respectively. The resulting set of predictions is conservative because maximum-likelihood estimates were used for γ and ϕ, which, as noted above, produced high specificity (and relatively low sensitivity) in simulation experiments, especially for short elements, small subtrees, and weak conservation. In addition, phyloP does not consider indels in computing its P-values, so some of the discarded predictions (with $P \geq 0.05$) might be strongly supported by indel evidence.

About 76% of the predictions (covering 52% of bases) were of fully conserved elements, 14% (36%) were of lineage-specific losses, and the remaining 9% (12%) were of lineage-specific gains. These numbers are undoubtedly affected by differences in power in detecting lineage-specific versus conserved elements, and gains versus losses (Fig. 4). Still, because the method favors fully conserved elements, these results suggest that the number of elements under negative selection in any species is at least 30% higher than the number conserved in all species, and nearly twice as many bases are conserved in any species as are conserved in all species. The predictions covered 70% of bases in coding regions, of which 74% were fully conserved, 21% were losses, and only 5% were gains. In contrast, predictions in introns and intergenic regions covered about 3% of bases, and in these regions we saw more gains (18–20% of predicted bases) and fewer losses (29–30%) than average, while the fully conserved fraction was about average (52%). The most common type of lineage-specific prediction, by far, was a gain on the branch above the last common ancestor of the eutherian mammals, suggesting extensive gain-of-function evolution on this branch.

4 Discussion

In this paper, we have introduced DLESS, a phylo-HMM-based program for identifying sequences that are subject to lineage-specific selection, and phyloP, a new method for computing P-values of conservation (or acceleration) based on prior and posterior distributions of numbers of substitutions. These methods have performed quite well in simulation experiments and yielded promising results with real data. Nevertheless, much work remains to be done in this area.

DLESS currently allows only for neutral evolution and negative selection, and it allows conserved sequence to undergo at most a single "gain" or "loss" event on the branches of the tree. A straightforward extension to positive selection would be possible if the assumption of at most one change in selective "mode" per element were maintained. It is likely, however, that some sequences experiencing adaptive evolution undergo multiple changes in their selective mode. The codon model of Guindon et al. [11] allows for any number of such changes in continuous time, but assumes no correlation between codons. This approach may be reasonable for very deep alignments of protein coding sequences, but a similar model applied to current alignments of noncoding DNA would likely have weak power. A more promising approach for our purposes may be to use a generalization of a phylo-HMM with a separate state-transition Markov chain per node of the tree, rather than one shared chain for all nodes. Unfortunately, models of this type (like models for context-dependent substitution [16]) have "loops" of dependency and do not permit exact probabilistic inference; Markov chain Monte Carlo (MCMC) methods or variational methods would be needed for likelihood evaluation, parameter estimation, and prediction of lineage-specific elements.

A more standard way to compute P-values of conservation would be to use a likelihood ratio test (LRT) (e.g., [28]). LRTs have many appealing statistical properties, and have some advantages over our method. For example, they allow for the fact that some substitutions (e.g., transitions) are generally less surprising than others (e.g., transversions), while our test statistic (a count of substitutions) does not. Also, an LRT would avoid the problem that the statistic in question is not actually observed, which leads to some loss of power in our method. On the other hand, in our method, the exact null distribution of the test statistic can be computed from a model of neutral evolution, without needing to assume asymptotic behavior (e.g., [28]) or to conduct extensive simulation experiments (e.g., [12]), specific to each new set of model parameters. (Among other things, this means that very small P-values can be accurately computed—something which is important when ranking extreme cases, as in genome-wide screens for sequences of interest.) In addition, the test statistic—a count of substitutions—has a clear, intuitive meaning, unlike a likelihood ratio. The descriptions of the prior and posterior distributions produced by phyloP (mean, variance, 95% confidence interval, etc.) are easy to interpret and informative to the user. Moreover, the cost in power of using phyloP appears to be minimal. Interestingly, phyloP seems to have better power, in comparison with an LRT, than methods for the identification of positively selected amino acid sites that were also based on substitution counts [29]. This may be because the method is more similar than these methods to an LRT, the main difference being in the choice of the test statistic (R. Nielsen, pers. comm.). Given that relatively little power is sacrificed, interpretability and convenience of application may make our method an attractive alternative to an LRT for a variety of purposes.

While the number of substitutions is not Poisson-distributed for general $\mathbf{Q}$ matrices (Fig. 2A), it appears to have *essentially* a Poisson distribution for most realistic $\mathbf{Q}$ matrices. (Here we restrict ourselves to models of DNA substitution; the situation may be quite different with amino acid models [J. Felsenstein, pers. comm.].) We generated $\mathbf{Q}$ matrices under the HKY model [30] for a wide range of base compositions and transition/transversion ratios, then computed the distribution $P(n|t)$ for a range of values of t and compared each one to a Poisson distribution of the same mean. The symmetric KL divergence of the two distributions was never more than 0.053 bits, suggesting that in many cases it may be quite reasonable to assume a Poisson distribution (see related observations by Zheng [23]). We have not, however, considered rate variation in these experiments (either among sites, among lineages, or along individual branches), which is known to alter the distribution of the number of substitutions. Certain kinds of rate variation can be accommodated with our methods (see full paper). With or without rate variation, of course, the posterior distribution of the number of substitutions is decidedly non-Poisson (Fig. 2B).

Related to detecting lineage-specific selection is the issue of the rate of "turnover" of functional elements. The rate of turnover is a critical factor in the relationship between the fraction of sites in a genome that are conserved (e.g., between human and mouse) and the fraction that are functional [31]. The methods described here may lead to improved estimates of the rate of turnover, but certain hurdles remain to be cleared. In particular, the strong dependency of the power of the method on element length, degree of conservation, properties of the subtree and supertree in question, and whether an event is a "gain" or a "loss," make it difficult to estimate the rate of turnover accurately. Obtaining good estimates of turnover rates remains an exciting challenge.

Acknowledgments

This project was inspired by questions raised by Bob Harris and Webb Miller about rates of turnover of functional elements, and by independent interest in the ENCODE working groups in detecting lineage-specific selection. The decomposition into a jump process and conditional substitution process was suggested by Rick Durrett. We thank Elliott Margulies for preparing the multiple alignments for the ENCODE regions and for providing us with phylogenetic models estimated from neutral sites; Brian Raney for his work distinguishing indels from missing data; Mathieu Blanchette for providing us with the inferAncestors program; and Greg Cooper, Arend Sidow, George Asimenos, Joe Felsenstein, and Rasmus Nielsen for helpful discussions.

References

1. Nobrega, M.A., Ovcharenko, I., Afzal, V., Rubin, E.M.: Scanning human gene deserts for long-range enhancers. Science **302** (2003) 413
2. Woolfe, A., Goodson, M., Goode, D., Snell, P., McEwen, G., Vavouri, T., Smith, S., North, P., Callaway, H., Kelly, K., et al.: Highly conserved non-coding sequences are associated with vertebrate development. PLoS Biol **3** (2005) e7
3. Boffelli, D., McAuliffe, J., Ovcharenko, D., Lewis, K.D., Ovcharenko, I., Pachter, L., Rubin, E.M.: Phylogenetic shadowing of primate sequences to find functional regions of the human genome. Science **299** (2003) 1391–1394
4. Margulies, E.H., Blanchette, M., NISC Comparative Sequencing Program, Haussler, D., Green, E.D.: Identification and characterization of multi-species conserved sequences. Genome Res **13** (2003) 2507–2518
5. Cooper, G.M., Stone, E.A., Asimenos, G., Green, E.D., Batzoglou, S., Sidow, A.: Distribution and intensity of constraint in mammalian genomic sequence. Genome Res **15** (2005) 901–913
6. Siepel, A., Bejerano, G., Pedersen, J.S., Hinrichs, A.S., Hou, M., Rosenbloom, K., Clawson, H., Spieth, J., Hillier, L.W., Richards, S., et al.: Evolutionarily conserved elements in vertebrate, insect, worm, and yeast genomes. Genome Res **15** (2005) 1034–1050
7. Nielsen, R., Yang, Z.: Likelihood models for detecting positively selected amino acid sites and applications to the HIV-1 envelope gene. Genetics **148** (1998) 929–936
8. Yang, Z., Nielsen, R.: Codon-substitution models for detecting molecular adaptation at individual sites along specific lineages. Mol Biol Evol **19** (2002) 908–917
9. Clark, A.G., Glanowski, S., Nielsen, R., Thomas, P.D., Kejariwal, A., Todd, M.A., Tanenbaum, D.M., Civello, D., Lu, F., Murphy, B., et al.: Inferring nonneutral evolution from human-chimp-mouse orthologous gene trios. Science **302** (2003) 1960–1963
10. Forsberg, R., Christiansen, F.B.: A codon-based model of host-specific selection in parasites, with an application to the influenza A virus. Mol Biol Evol **20** (2003) 1252–1259
11. Guindon, S., Rodrigo, A.G., Dyer, K.A., Huelsenbeck, J.P.: Modeling the site-specific variation of selection patterns along lineages. Proc Natl Acad Sci U S A **101** (2004) 12957–12962
12. Nielsen, R., Bustamante, C., Clark, A.G., Glanowski, S., Sackton, T.B., Hubisz, M.J., Fledel-Alon, A., Tanenbaum, D.M., Civello, D., White, T.J., et al.: A scan for positively selected genes in the genomes of humans and chimpanzees. PLoS Biol **3** (2005) e170
13. ENCODE Project Consortium: The ENCODE (ENCyclopedia Of DNA Elements) Project. Science **306** (2004) 636–640
14. Felsenstein, J., Churchill, G.A.: A hidden Markov model approach to variation among sites in rate of evolution. Mol Biol Evol **13** (1996) 93–104
15. Yang, Z.: A space-time process model for the evolution of DNA sequences. Genetics **139** (1995) 993–1005
16. Siepel, A., Haussler, D.: Phylogenetic hidden Markov models. In Nielsen, R., ed.: Statistical Methods in Molecular Evolution. Springer, New York (2005) 325–351
17. Cooper, G.M., Brudno, M., Stone, E.A., Dubchak, I., Batzoglou, S., Sidow, A.: Characterization of evolutionary rates and constraints in three mammalian genomes. Genome Res **14** (2004) 539–548

18. McAuliffe, J.D., Pachter, L., Jordan, M.I.: Multiple-sequence functional annotation and the generalized hidden Markov phylogeny. Bioinformatics **20** (2004) 1850–1860
19. Siepel, A., Haussler, D.: Computational identification of evolutionarily conserved exons. In: Proc. 8th Int'l Conf. on Research in Computational Molecular Biology. (2004) 177–186
20. Holmes, I., Bruno, W.J.: Evolutionary HMMs: a Bayesian approach to multiple alignment. Bioinformatics **17** (2001) 803–820
21. Lunter, G., Miklos, I., Drummond, A., Jensen, J.L., Hein, J.: Bayesian coestimation of phylogeny and sequence alignment. BMC Bioinformatics **6** (2005) 83
22. Blanchette, M., Green, E.D., Miller, W., Haussler, D.: Reconstructing large regions of an ancestral mammalian genome in silico. Genome Res **14** (2004) 2412–2423
23. Zheng, Q.: On the dispersion index of a Markovian molecular clock. Math Biosci **172** (2001) 115–128
24. Jukes, T.H., Cantor, C.R.: Evolution of protein molecules. In Munro, H., ed.: Mammalian Protein Metabolism. Academic Press, New York (1969) 21–132
25. Gillespie, J.: Lineage effects and the index of dispersion of molecular evolution. Mol Biol Evol **6** (1989) 636–647
26. Felsenstein, J.: Evolutionary trees from DNA sequences. J Mol Evol **17** (1981) 368–376
27. Blanchette, M., Kent, W.J., Riemer, C., Elnitski, L., Smit, A.F.A., Roskin, K.M., Baertsch, R., Rosenbloom, K., Clawson, H., Green, E.D., et al.: Aligning multiple genomic sequences with the threaded blockset aligner. Genome Res **14** (2004) 708–715
28. Felsenstein, J.: Inferring Phylogenies. Sinauer Associates, Inc., Sunderland, Massachusetts (2004)
29. Nielsen, R., Huelsenbeck, J.P.: Detecting positively selected amino acid sites using posterior predictive P-values. Pac Symp Biocomput (2002) 576–588
30. Hasegawa, M., Kishino, H., Yano, T.: Dating the human-ape splitting by a molecular clock of mitochondrial DNA. J Mol Evol **22** (1985) 160–174
31. Smith, N.G.C., Brandstrom, M., Ellegren, H.: Evidence for turnover of functional noncoding DNA in mammalian genome evolution. Genomics **84** (2004) 806–813

A Probabilistic Model for Gene Content Evolution with Duplication, Loss, and Horizontal Transfer

Miklós Csűrös[1] and István Miklós[2]

[1] Department of Computer Science and Operations Research, Université de Montréal,
C.P. 6128, succ. Centre-Ville, Montréal, Québec, H3C 3J7, Canada
csuros@iro.umontreal.ca
[2] Department of Plant Taxonomy and Ecology, Eötvös Lóránd University,
1117 Budapest, Pázmány Péter Sétány 1/c, Hungary
miklosi@ramet.elte.hu

Abstract. We introduce a Markov model for the evolution of a gene family along a phylogeny. The model includes parameters for the rates of horizontal gene transfer, gene duplication, and gene loss, in addition to branch lengths in the phylogeny. The likelihood for the changes in the size of a gene family across different organisms can be calculated in $O(N + hM^2)$ time and $O(N + M^2)$ space, where N is the number of organisms, h is the height of the phylogeny, and M is the sum of family sizes. We apply the model to the evolution of gene content in Proteobacteria using the gene families in the COG (Clusters of Orthologous Groups) database.

1 Introduction

At this time, 294 microbial genomes have been sequenced, and that figure is expected to soon double (this in addition to 19 complete eukaryotic genomes, see http://www.ncbi.nlm.nih.gov/Genomes/). These numbers continue to grow exponentially with advances in technology and expertise [1]. The wealth of genome sequence data has already caused a revolution in molecular evolution methods [2,3]. A few years ago, scientific studies had to focus on nucleotide-level differences between orthologous genes, mainly because of the technical and financial limitations on DNA sequence collection. With increasing amounts of whole genome information, however, it becomes possible to analyze genome-scale differences between organisms, and to identify the evolutionary forces responsible for these changes. In particular, sizes of gene families can be compared, allowing us to better understand adaptive evolutionary mechanisms and organismal phylogeny. Several studies suggest that gene content may carry sufficient phylogenetic signal for the construction of evolutionary trees [4,5,6,7,8,9,10,11,12,13]. Comparative analyses of genome-wide protein domain content [7,14,15] have also provided important insights into evolution. Gene content and similar features have been used to construct viral [16,17], microbial [4,5,12], and universal

A. Apostolico et al. (Eds.): RECOMB 2006, LNBI 3909, pp. 206–220, 2006.

trees [6,14,18]. Comparative gene content analysis is also used to estimate ancestral genome composition [19,20]. The presence-absence pattern of homologs in different organisms, the so-called phyletic pattern [21,22], provides clues about gene function [23] and the evolution of metabolic pathways [20].

A number of processes shape the gene content of an organism. New genes may be created by duplication of an existing gene, horizontal transfer from a different lineage, and rarer events such as gene fusion and fission [19]. It has been widely debated how the extent of horizontal gene transfer (HGT) compares to vertical inheritance [18,19,24,25,26,27,28]. It is clear that horizontal gene transfer plays a major role in microbial evolution [29], but there is still need for adequate mathematical models in which that role can be measured.

We introduce a probabilistic model for the evolution of gene content along a phylogeny. Our model accounts for gene duplication, gene loss and horizontal transfer. We consider the evolution of the size of a gene family, where the different processes add new genes to the family or erase members of it, and arrive at the family sizes observed at the terminal taxa. We describe an algorithm that calculates the likelihood of gene family sizes in different organisms, given an evolutionary tree. The algorithm computes the likelihood of family sizes in $O(N + M^2 h)$ time where M is the total number of genes in the family, N is the number of genomes, and h is the height of the tree. Note that the tree height is at most linear in N, and on average, it is $O(\sqrt{N})$ or $O(\log N)$ for uniform or Yule-Harding distribution of random trees.

To our knowledge, no tractable stochastic model has yet been introduced that simultaneously accounts for horizontal transfer, gene loss, and duplication. These processes cannot be modeled by using only two parameters: whereas the intensity of gene loss and duplication depend on the size of a gene family, the rate of horizontal transfer has a constant component. Among other applications, a model that accounts for duplication and transfer is useful for analyzing the evolution of metabolic networks [30]: do new paths evolve by gene duplication and adaptive selection, or by accommodating genes with new functions via horizontal gene transfer?

A few probabilistic models were proposed for gene content evolution, which are less general than ours. Most studies use stochastic models with two parameters. Huson and Steel [11] analyzed a two-parameter model that accounts for gene loss and horizontal transfer but not for gene duplication. They derived a distance measure based on gene family sizes using likelihood maximization arguments. They further showed that traditional scores for shared gene content [5] are not as suitable for phylogeny reconstruction as either Dollo parsimony or their own distance function. Gu and Zhang [12] relied on a model that includes gene loss and gene duplication but no other modes of gene genesis, and assumes identical rates across different branches. They showed how gene family sizes can be used to define additive distances in such a model. Interestingly enough, the data can be reduced to a three-letter alphabet for the purposes of distance calculations: only 0, 1 or "many" homologs per family need to be counted. The distance metric relies on estimates of the rate parameters, which are obtained through likelihood optimization. Hahn et al. [31] developed an alternative likelihood-based approach for the same two-parameter model with constant rates across lineages. Karev et al. developed a rich probabilistic model of gene content evolution in a series of papers [32,33,34]. The model explains the distribution of gene family sizes found in different organisms. It is, however, too general for exact detailed calculations, and for likelihood computations in particular. Our likelihood algorithm is also notable for its computational efficiency. For instance, the likelihood calculations of [31] in a two-parameter model take cubic time in M, and involve the evaluation of infinite sums that are truncated heuristically.

Not all comparative studies of gene content rely on gene family sizes. A frequently employed approach is to measure shared gene content [5,6,8,9,10] by identifying orthologs between each pair of genomes. Pairwise scores of shared gene content can be analyzed using distance-based methods of phylogeny construction or other clustering techniques. Lake and Rivera [13] proposed an improved technique of assessing shared gene content: for each genome, the presence and absence of homologs are marked with respect to genes of a reference genome. The presence-absence marks are encoded in a binary sequence for every genome. The sequences are used to compute a pairwise distance matrix using standard methods of phylogeny construction. Finally, a number of studies rely on families of homologous genes across many organisms, and record the absence or presence of each family in the genomes [4,7,24,35]. The resulting absence-presence data are further analyzed with traditional parsimony or distance-based methods. Some specialized parsimony methods were purposely devised to analyze absence-presence data [20,36] for gene families. Our work is concerned with the actual numbers of paralogs within the gene families, which give an even richer signal for evolutionary analyses [11,19,31].

The paper is organized in the following manner. Section 2 introduces our stochastic model of gene content evolution, and describes formulas for computing various associated probabilities, including likelihood. The formulas are used in an algorithm described in Section 3. Section 4 describes our initial experiments in modeling gene content evolution in 51 proteobacteria and 3555 gene families from the database of Clusters of Orthologous Groups (COGs) [22]. Section 5 concludes the paper.

2   Mathematical Model

Let T be a phylogenetic tree over a set of organisms S. The tree T is a rooted tree with node set $V(T)$ and edge set $E(T)$, in which leaves are bijectively labeled with elements of S. Non-leaf nodes have at least two children. Every edge e has a length $t_e > 0$. We are interested in modeling the evolution of a gene family. The family size changes along the edges: genes may be duplicated, lost, or gained from an unknown source. We model the evolution of *gene counts* (family size) at the tree nodes: the gene count at every node $u \in V(T)$ is a random variable $\chi(u)$ that can take non-negative integer values. In addition to its length, each edge is equipped with a *duplication rate* λ, a *loss rate* μ, and a *transfer rate* κ. The loss rate accounts for all possible mechanisms of gene loss, including deletion and pseudogenization. The transfer rate accounts for processes of gene genesis, including HGT from another lineage in the same tree, or HGT from an unknown organism. The tree topology, the edge lengths and rates determine the joint distribution of the gene counts.

In our model, the evolution of the gene counts on a branch follows a linear birth-and-death process [37] parametrized by λ, κ, and μ. Let $\{X(t) : t \geq 0\}$ denote the continuous-time Markov process formed by the gene counts along an edge uv: $\chi(u) = X(0)$ and $\chi(v) = X(t_{uv})$. The transition probabilities of the process are the following:

$$\mathbb{P}\Big\{X(t+\epsilon) = n+1 \,\Big|\, X(t) = n\Big\} = \Big(\kappa + n\lambda\Big)\epsilon + o(\epsilon)$$

$$\mathbb{P}\Big\{X(t+\epsilon) = n-1 \,\Big|\, X(t) = n\Big\} = n\mu\epsilon + o(\epsilon)$$

$$\mathbb{P}\Big\{|X(t+\epsilon) - n| > 1 \,\Big|\, X(t) = n\Big\} = o(\epsilon).$$

In other words, every existing gene produces an offspring through duplication with an intensity of λ, or disappears with an intensity of μ, and new genes are acquired with an intensity of κ, independently from the number of existing genes.

REMARK. For simplicity of notation, we impose the same rates across all edges throughout the paper. Nevertheless, the presented method accommodates branch-dependent rates in a straightforward manner.

The histories of individual genes on an edge form a *Galton-Watson* forest, see Figure 1. The figure illustrates a scenario where the gene count changes from three to five. The gene count at the child node is the result of many duplication, transfer and loss events. The change involves three horizontally transferred genes, from among which one survives, another one does not, and the third one produces two surviving paralogs.

While it is not too difficult to calculate the probabilities for any particular gene count on a branch (see §2.1), the likelihood L of observed gene counts at the leaves involves an infinite number of possible gene counts at intermediate nodes:

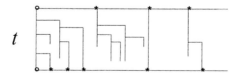

Fig. 1. Galton-Watson forest showing the evolution of genes in the same family along a tree edge. The top line represents the ancestral genome with three genes; the bottom line represents the descendant genome, in which there are five family members. Symbol o represents the source from which genes might be transferred horizontally, symbols ⋆ represent paralogous genes in the genome at the beginning and the end of the investigated time span t. Each o or ⋆ in the ancestral genome is the root of a Galton-Watson tree. Note that the physical order of genes is immaterial: here they are simply drawn next to each other for clarity.

$$L = \sum_{\langle m_x : \; x \in V(T) \rangle} \gamma(m_{\text{root}}) \prod_{xy \in E(T)} \mathbb{P}\Big\{ \chi(y) = m_y \; \Big| \; \chi(x) = m_x \Big\}, \tag{1}$$

where $\gamma(\cdot)$ defines the gene count distribution at the root, and the summation over the $\langle m_x \rangle$ vectors takes all values in agreement with the gene counts at the leaves in the input data. Our main technique for computing the likelihood is to restrict the computation to genes that have at least one surviving descendant at the leaves. In what follows we develop the formulas to compute the likelihood.

2.1   Basic Transition Probabilities

First we analyze the *blocks* of homologs at a node comprising genes of common origin. A *xenolog* block consists of the genes that trace back to a horizontal transfer event on the branch from the parent. For every gene at the parent, its descendants form an *inparalog* block. (Our terminology follows [38].) The homologs in Figure 1 belong to four blocks: a xenolog block of size three, an inparalog block of size zero for the deceased parental gene, and two inparalog blocks of size one. The independent birth-and-death processes associated with the blocks have been analyzed in the statistical literature.

Definition 1. *Define the following basic transition probabilities for gene count evolution on a branch. Let $h_t(n)$ denote the probability that there are n genes of foreign origin after time t. Let $g_t(n)$ denote the probability that a single gene has n copies after time t.*

In other words, $h_t(n)$ is the probability mass function for the number of xenologs at time t, and $g_t(n)$ defines the size distribution of an inparalog block at time t.

Theorem 1. *The basic transition probabilities can be written as follows.*

$$h_t(n) = \binom{\frac{\kappa}{\lambda} + n - 1}{n} \left(1 - \lambda\beta(t) \right)^{\frac{\kappa}{\lambda}} \left(\lambda\beta(t) \right)^n \tag{2}$$

where $\beta(t) = \frac{1 - e^{-(\mu - \lambda)t}}{\mu - \lambda e^{-(\mu - \lambda)t}}$, and

$$\binom{\frac{\kappa}{\lambda} + n - 1}{n} = \begin{cases} 1 & \text{if } n = 0; \\ \dfrac{\left(\frac{\kappa}{\lambda}\right)\left(\frac{\kappa}{\lambda}+1\right)\cdots\left(\frac{\kappa}{\lambda}+n-1\right)}{n!} & \text{if } n > 0. \end{cases}$$

Furthermore,

$$g_t(n) = \begin{cases} \mu\beta(t) & \text{if } n = 0; \\ \left(1 - \mu\beta(t) \right)\left(1 - \lambda\beta(t) \right)\left(\lambda\beta(t) \right)^{n-1} & \text{if } n > 0. \end{cases} \tag{3}$$

Proof. The size of the xenolog block follows a birth-and-death process with a constant immigration rate κ and no emigration. The transition probabilities of (2) for such a process were analyzed by Karlin and McGregor [39]. An inparalog block evolves by a simple birth-and-death process: the transition probabilities of (3) are derived in, e.g., [37]. □

2.2 Gene Extinction and Survival

Definition 2. *A surviving gene at a node x is such that it has at least one modern descendant at the leaves below x.*

Let D_x denote the probability that a gene present at node x is not surviving, i.e., that it has no modern descendants.

Lemma 1. *The extinction probability D_x can be calculated as follows. If x is a leaf, then $D_x = 0$. Otherwise, let x be the parent of $x_1, x_2, \ldots, x_d$.*

$$D_x = \prod_{j=1}^{d} \left(\mu\beta(t_j) + \left(1 - \mu\beta(t_j)\right)\left(1 - \lambda\beta(t_j)\right)\frac{D_{x_j}}{1 - \lambda\beta(t_j)D_{x_j}} \right) \qquad (4)$$

where t_j is the length of the branch leading from x to x_j.

Proof. For leaves, the statement is trivial. When x is not a leaf, condition on the gene counts at the children:

$$D_x = \prod_{j=1}^{d} \sum_{m=0}^{\infty} g_{t_j}(m)\left(D_{x_j}\right)^m.$$

Plugging in $g_t(m)$ from Eq. (3) and replacing the infinite series with a closed form gives (4). $\qquad \square$

2.3 Effective Transition Probabilities

We introduce two new probabilities, denoted by $H_x(n)$ and $G_x(n)$, for having n surviving genes in a block at node x. The effective transition probabilities are related to $h_t(n)$, and $g_t(n)$, but take into consideration eventual extinction below node x. A formal definition follows.

Definition 3. *Let y be a non-root node. Define the following effective transition probabilities. Let $H_y(n)$ denote the probability that the xenolog block at node y contains n surviving genes. Let $G_y(n)$ denote the probability that an inparalog block at node y contains n surviving genes.*

Lemma 2. *Let y be a non-root node, let x be its ancestor, and let t be the length of the edge xy. The effective transition probabilities can be written as follows.*

$$H_y(n) = \binom{\frac{\kappa}{\lambda} + n - 1}{n} \left(\frac{1 - \lambda\beta(t)}{1 - D_y\lambda\beta(t)} \right)^{\frac{\kappa}{\lambda}} \left(\frac{(1 - D_y)\lambda\beta(t)}{1 - D_y\lambda\beta(t)} \right)^n \qquad (5)$$

$$G_y(0) = 1 - \frac{\left(1 - \mu\beta(t)\right)\left(1 - D_y\right)}{1 - D_y\lambda\beta(t)}; \qquad (6a)$$

$$G_y(n) = \frac{\left(1 - \mu\beta(t)\right)\left(1 - \lambda\beta(t)\right)}{\left(\lambda\beta(t)\right)\left(1 - D_y\lambda\beta(t)\right)} \left(\frac{(1 - D_y)\lambda\beta(t)}{1 - D_y\lambda\beta(t)} \right)^n, \qquad n > 0. \qquad (6b)$$

Proof. We condition on the number of xenologs at y (whether or not they survive).

$$H_y(n) = \sum_{i=0}^{\infty} \binom{n+i}{i} h_t(n+i)(D_y)^i(1-D_y)^n.$$

Using Eq. (2) leads to an infinite series that can be simplified to get (5). Similarly, write

$$G_y(n) = \sum_{i=0}^{\infty} \binom{n+i}{i} g_t(n+i)(D_y)^i(1-D_y)^n.$$

Taking the values of $g_t(n+i)$ from Eq. (3) and simplifying the resulting infinite series yields (6). □

2.4   Number of Surviving Genes on a Branch

Definition 4. *Let y be a non-root node, and let x be its ancestor. Let $p_y(m|n)$ denote the* survival probability *defined as the probability of the event that there are m surviving genes at node y under the condition that there are n genes at node x (not necessarily surviving).*

Lemma 3. *The survival probabilities can be computed as follows.*

$$p_y(m|0) = H_y(m) \tag{7a}$$

$$p_y(0|n) = H_y(0)\big(G_y(0)\big)^n \qquad\qquad 0 < n \tag{7b}$$

$$p_y(1|n) = G_y(0)p_y(1|n-1) + G_y(1)p_y(0|n-1) \qquad\qquad 0 < n \tag{7c}$$

$$\begin{aligned} p_y(m|n) = {}& \alpha p_y(m-1|n) \qquad\qquad 0 < n, 1 < m \\ &+\big(G_y(1) - \alpha G_y(0)\big)p_y(m-1|n-1) \\ &+G_y(0)p_y(m|n-1) \end{aligned} \tag{7d}$$

where

$$\alpha = \frac{(1-D_y)\lambda\beta(t)}{1-D_y\lambda\beta(t)}. \tag{8}$$

Proof. For $p_y(m|0)$ and $p_y(0|n)$, the equations are straightforward. Otherwise, we condition on the surviving copies of a single gene at y:

$$p_y(m|n) = \sum_{i=0}^{m} G_y(i)p_y(m-i|n-1). \tag{9}$$

Now, using that $G_y(i+1) = \alpha G_y(i)$ whenever $i > 0$, and comparing (9) for $p_y(m|n)$ and $p_y(m-1|n)$, we can write $p_y(m|n)$ in a recursive form as shown. □

2.5   Conditional Likelihoods

Definition 5. *Let x be a node in the tree. Define the* conditional likelihood *$L_x(n)$ for all n as the probability of having the observed gene counts at the leaves in the subtree rooted at x, under the condition that there are n surviving genes at x.*

Theorem 2. *The conditional likelihoods can be calculated as follows. In the case when x is a leaf, $L_x(n) = 1$ if n is the observed gene count at x, otherwise the likelihood is 0. If x is not a leaf, and has children $x_1, x_2, \ldots, x_d$, then the following recursions hold.*

$$L_x(0) = \prod_{j=1}^{d} \sum_{m=0}^{M_j} p_{x_j}(m|0) L_{x_j}(m); \tag{10a}$$

$$L_x(n) = (1 - D_x)^{-n} \left(\prod_{j=1}^{d} \sum_{m=0}^{M_j} p_{x_j}(m|n) L_{x_j}(m) \right.$$

$$\left. - \sum_{i=0}^{n-1} \binom{n}{i} (D_x)^{n-i}(1 - D_x)^i L_x(i) \right); \qquad 0 < n \leq \sum_{j=1}^{d} M_j, \tag{10b}$$

where M_j is the sum of gene counts at the leaves in the subtree rooted at x_j. If $n > \sum_{j=1}^{d} M_j$, then $L_x(n) = 0$.

Proof. For a leaf node, or for $n > \sum_{j=1}^{d} M_j$, the theorem is trivial. Otherwise, consider the likelihood $\ell_x(n)$ of the observed gene counts at the leaves in the subtree rooted at x, conditioned on the event that there are n genes present at x, which may or may not survive. We write the likelihood in two ways. First, by conditioning on the number of surviving genes at the children,

$$\ell_x(n) = \prod_{j=1}^{d} \sum_{m=0}^{M_j} p_{x_j}(m|n) L_{x_j}(m). \tag{11}$$

Secondly, by conditioning on the number of surviving genes at x,

$$\ell_x(n) = \sum_{i=0}^{n} \binom{n}{i} (D_x)^{n-i}(1 - D_x)^i L_x(i). \tag{12}$$

Now, rearranging the equality of the two right-hand sides gives the desired result. $\qquad \square$

REMARK. Clearly, the gene counts M_x of Theorem 2 are easily computed for all x. If $m(x)$ is the gene count for every leaf x then

$$M_x = \begin{cases} m(x) & \text{if } x \text{ is a leaf;} \\ \sum_{j=1}^{d} M_{x_j} & \text{if } x_1, \ldots, x_k \text{ are the children of } x. \end{cases} \tag{13}$$

2.6   Likelihood

It is assumed that the family size at the root is distributed according to the equilibrium probabilities:

$$\gamma(n) = h_\infty(n) = \binom{\frac{\kappa}{\lambda} + n - 1}{n} \left(1 - \frac{\lambda}{\mu}\right)^{\frac{\kappa}{\lambda}} \left(\frac{\lambda}{\mu}\right)^n. \tag{14}$$

Theorem 3. *Let M be the total number of genes at the leaves. The likelihood of the observed gene counts equals*

$$L = \sum_{n=0}^{M} L_{\text{root}}(n) \frac{\binom{\frac{\kappa}{\lambda}+n-1}{n}\left(1-\frac{\lambda}{\mu}\right)^{\frac{\kappa}{\lambda}}\left((1-D_{\text{root}})\frac{\lambda}{\mu}\right)^{n}}{\left(1-\frac{\lambda}{\mu}D_{\text{root}}\right)^{\frac{\kappa}{\lambda}+n}}. \qquad (15)$$

Proof. By summing the likelihoods conditioned on the surviving genes at the root,

$$L = \sum_{n=0}^{M} L_{\text{root}}(n) \sum_{i=0}^{\infty} \gamma(n+i) \binom{n+i}{i} (D_{\text{root}})^{i}(1-D_{\text{root}})^{n}. \qquad (16)$$

Now, plugging in the values of $\gamma(\cdot)$ from Eq. (14) and replacing the infinite series by a closed form gives the theorem's formula. $\qquad \square$

REMARK. In place of the equilibrium probabilities of (14), many other prior distributions can be accommodated by the summation in (16).

3   Algorithm

This section employs the formulas of Section 2 in a dynamic programming algorithm to compute the likelihood exactly. More precisely, the algorithm computes the likelihood of gene counts at the tree leaves, given the duplication rate λ, the transfer rate κ, and the loss rate μ. Algorithm COMPUTELIKELIHOOD below proceeds by a depth-first traversal; the necessary variables are calculated from the leaves towards the root. Let $m(u)$ denote the gene count at every leaf u.

COMPUTELIKELIHOOD
Input λ, κ, μ, T, gene counts $m(u)$: u is a leaf of T
Output likelihood of the $m(\cdot)$ values
1 **for** each node $x \in V(t)$ in a depth-first traversal
2   Compute D_x using Eq. (4).
3   Compute the sum of gene counts M_x by Eq. (13).
4   **if** x is not the root **then**
5   Let y be the parent of x.
6   **for** $n = 0, \ldots, M_y$ **do**
7   **for** $m = 0, \ldots, M_x$ **do** compute $p_x(m|n)$ by Eq. (7).
8   **for** $n = 0, \ldots, M_x$ **do** compute $L_x(n)$ by Eq. (10).
9 Compute the likelihood L at the root using Eq. (15).
10 **return** L.

Theorem 4 below analyzes the algorithm's complexity in terms of the topology of T. In particular, it uses the notions of *height of a node* x, defined as the number of edges on the path leading from the root to x, *levels* of nodes, which are sets of nodes with the same height, and *height of the tree*, which is the maximum of the leaf heights.

Theorem 4. *Let h be the height of T in Algorithm* COMPUTELIKELIHOOD, *let N be the number of its leaves, and let $M = M_{\text{root}}$ be the sum of gene counts. The algorithm can be implemented in such a way that it uses $O(N + M^2)$ space and runs in $O(N + hM^2)$ time.*

Proof. Computing D_x and M_x takes $O(1)$ time when x is a leaf, or $O(d)$ for an inner node with d children. There are $O(N)$ nodes in the tree and, thus, computing D_x and M_x for all x is done in $O(N)$ time. The computed values are stored in $O(N)$ space.

In order to analyze the computations in Lines 4–8, we consider nodes at the same level. Line 8 computes $L_x(n)$ for all $n = 0, \ldots, M_x$ in $O((M_x + 1)(M_x + d_x))$ total time where d_x is the number of children of node x. Lines 5–7 compute $p_x(m|n)$ for $(M_x + 1)(M_y + 1)$ pairs of n, m values. (Notice that $H_y(m)$ can be computed in $O(1)$ time for each m in the iteration over m using that $H_y(m) = \alpha \frac{m + \kappa/\lambda - 1}{m} H_y(m-1)$ with the α of Eq. (8).) For the children $x_1, \ldots, x_{d_y}$ of the same node y, the total time spent in Lines 5–7 is $O((M_y + 1)(M_y + d_y))$. Terms of the type $O(d_x)$ sum up to $O(N)$ in the tree. Considering all nodes at the same level k, other terms' contribution to the running time is

$$O\left(\sum_{\text{all } y \text{ at level } k - 1} (M_y^2 + dM_y) + \sum_{\text{all } x \text{ at level } k} (M_x^2 + dM_x) \right),$$

where d is the maximum number of children. Clearly, $\sum_x M_x \leq M$ if the summation goes over x for which their subtrees do not overlap, such as nodes at the same level. Now, $\sum_x M_x^2 \leq (\sum_x M_x)^2 \leq M^2$, and, thus, $O(M^2 + Md)$ time is spent on each level. Therefore, the total time spent in the loop of Line 4 is $O(N + h(M^2 + Md))$. Line 9 takes $O(M)$ time. Ignoring degenerate cases with $M \ll d$, the theorem's claim follows.

In order to obtain the space complexity result, notice that at the end of the loop in Line 8 the computed variables for the children of x are not needed anymore. Therefore, the nodes for which $p_x(\cdot|\cdot)$ is needed are such that their subtrees do not overlap. By the same type of argument as with time spent on a level, the number of variables that need to be kept in memory is $O(M^2)$.   □

4   Gene Content Evolution in Proteobacteria

Proteobacteria form one of the most diverse groups of prokaryotes. Proteobacteria provide an excellent case study for gene content evolution: they include pathogens, endosymbionts, and free-living organisms. Genome sizes vary tenfold within this group, and horizontal transfer is abundant [25]. Their phylogeny is still not resolved to satisfaction [40,41,42,43]. We used 51 proteobacteria in the first application of our likelihood method. Gene counts were based on the newer version [22] of the COG database. Each COG is a manually curated protein family of homologs. The COGs are classified into 23 functional categories. (For each of the 51 proteobacteria, the number of genes in each COG family was established by Pál et al. [30]. There are 3555 COG families that have at

Acknowledgments

We would like to thank Eugene Koonin, Hervé Philippe and Yuri Wolf for useful discussions concerning gene content evolution, as well as Csaba Pál and Martin Lercher for providing us with pre-publication data. This work was supported in part by the e-Science Regional Knowledge Center at Eötvös Lóránd University, Budapest, sponsored by the Hungarian National Office for Research and Technology (NKTH). M.Cs. is supported by grants from the Natural Sciences and Engineering Research Council of Canada and the *Fonds québecois de la recherche sur la nature et les technologies*. I.M. is supported by a Békésy György postdoctoral fellowship.

References

1. Green, E.D.: Strategies for the systematic sequencing of complex genomes. Nature Reviews Genetics **2** (2001) 573–583
2. Wolfe, K.H., Li, W.H.: Molecular evolution meets the genomic revolution. Nature Genetics **33** (2003) 255–265
3. Delsuc, F., Brinkmann, H., Philippe, H.: Phylogenomics and the reconstruction of the tree of life. Nature Reviews Genetics **6** (2005) 361–375
4. Fitz-Gibbon, S.T., House, C.H.: Whole genome-based phylogenetic analysis of free-living microorganisms. Nucleic Acids Research **27** (1999) 4218–4222
5. Snel, B., Bork, P., Huynen, M.A.: Genome phylogeny based on gene content. Nature Genetics **21** (1999) 108–110
6. Tekaia, F., Lazcano, A., Dujon, B.: The genomic tree as revealed from whole proteome comparisons. Genome Research **9** (1999) 550–557
7. Lin, J., Gerstein, M.: Whole-genome trees based on the occurrence of folds and orthologs: implications for comparing genomes on different levels. Genome Research **10** (2000) 808–818
8. Clarke, G.D.P., Beiko, R.G., Ragan, M.A., Charlebois, R.L.: Inferring genome trees by using a filter to eliminate phylogenetically discordant sequences and a distance matrix based on mean normalized BLASTP scores. Journal of Bacteriology **184** (2002) 2072–2080
9. Korbel, J.O., Snel, B., Huynen, M.A., Bork, P.: SHOT: a web server for the construction of genome phylogenies. Trends in Genetics **18** (2002) 158–162
10. Dutilh, B.E., Huynen, M.A., Bruno, W.J., Snel, B.: The consistent phylogenetic signal in genome trees revealed by reducing the impact of noise. Journal of Molecular Evolution **58** (2004) 527–539
11. Huson, D.H., Steel, M.: Phylogenetic trees based on gene content. Bioinformatics **20** (2004) 2044–2049
12. Gu, X., Zhang, H.: Genome phylogenetic analysis based on extended gene contents. Molecular Biology and Evolution **21** (2004) 1401–1408
13. Lake, J.A., Rivera, M.C.: Deriving the genomic tree of life in the presence of horizontal gene transfer: conditioned reconstruction. Molecular Biology and Evolution **21** (2004) 681–690
14. Yang, S., Doolittle, R.F., Bourne, P.E.: Phylogeny determined by protein domain content. Proceedings of the National Academy of Sciences of the USA **102** (2005) 373–378

15. Deeds, E.J., Hennessey, H., Shakhnovich, E.I.: Prokaryotic phylogenies inferred from protein structural domains. Genome Research **15** (2005) 393–402
16. Montague, M.G., Hutchison III, C.A.: Gene content phylogeny of herpesviruses. Proceedings of the National Academy of Sciences of the USA **97** (2000) 5334–5339
17. Herniou, E.A., Luque, T., Chen, X., Vlak, J.M., Winstanley, D., Cory, J.S., O'Reilly, D.R.: Use of whole genome sequence data to infer baculovirus phylogeny. Journal of Virology **75** (2001) 8117–8126
18. Simonson, A.B., Servin, J.A., Skophammer, R.G., Herbold, C.W., Rivera, M.C., Lake, J.A.: Decoding the genomic tree of life. Proceedings of the National Academy of Sciences of the USA **102** (2005) 6608–6613
19. Snel, B., Bork, P., Huynen, M.A.: Genomes in flux: the evolution of archaeal and proteobacterial gene content. Genome Research **12** (2002) 17–25
20. Mirkin, B.G., Fenner, T.I., Galperin, M.Y., Koonin, E.V.: Algorithms for computing evolutionary scenarios for genome evolution, the last universal common ancestor and dominance of horizontal gene transfer in the evolution of prokaryotes. BMC Evolutionary Biology **3** (2003) 2
21. Koonin, E.V., Galperin, M.Y.: Sequence-Evolution-Function: Computational Approaches in Comparative Genomics. Kluwer Academic Publishers, New York (2002)
22. Tatusov, R.L., Fedorova, N.D., Jackson, J.D., Jacobs, A.R., Kiryutin, B., Koonin, E.V., Krylov, D.M., Mazumder, R., Mekhedov, S.L., Nikolskaya, A.N., Rao, B.S., Smirnov, S., Sverdlov, A.V., Vasudevan, S., Wolf, Y.I., Yin, J.J., Natale, D.A.: The COG database: an updated version includes eukaryotes. BMC Bioinformatics **4** (2003) 441
23. Pellegrini, M., Marcotte, E.M., Thompson, M.J., Eisenberg, D., Yeates, T.O.: Assigning protein functions by comparative genome analysis: protein phylogenetic profiles. Proceedings of the National Academy of Sciences of the USA **96** (1999) 4285–4288
24. Jordan, I.K., Makarova, K.S., Spouge, J.L., Wolf, Y.I., Koonin, E.V.: Lineage-specific gene expansions in bacterial and archaeal genomes. Genome Research **11** (2001) 555–565
25. Gogarten, J.P., Doolittle, W.F., Lawrence, J.G.: Prokaryotic evolution in light of gene transfer. Molecular Biology and Evolution **19** (2002) 2226–2238
26. Kurland, C.G., Canback, B., Berg, O.G.: Horizontal gene transfer: a critical view. Proceedings of the National Academy of Sciences of the USA **100** (2003) 9658–9662
27. Kunin, V., Goldovsky, L., Darzentas, N., Ouzounis, C.A.: The net of life: reconstructing the microbial phylogenetic network. Genome Research **15** (2005) 954–959
28. Ge, F., Wang, L.S., Kim, J.: The cobweb of life revealed by genome-scale estimates of horizontal gene transfer. PLoS Biology **3** (2005) e316
29. Boucher, Y., Douady, C.J., Papke, R.T., Walsh, D.A., Boudreau, M.E.R., Nesbø, C.L., Case, R.J., Doolittle, W.F.: Lateral gene transfer and the origin of prokaryotic groups. Annual Review of Genetics **37** (2003) 283–328
30. Pál, C., Papp, B., Lercher, M.: Adaptive evolution of bacterial metabolic networks by horizontal gene transfer. Nature Genetics **37** (2005) 1372–1375
31. Hahn, M.W., De Bie, T., Stajich, J.E., Nguyen, C., Cristianini, N.: Estimating the tempo and mode of gene family evolution from comparative genomic data. Genome Research **15** (2005) 1153–1160
32. Karev, G.P., Wolf, Y.I., Rzhetsky, A.Y., Berezovskaya, F.S., Koonin, E.V.: Birth and death of protein domains: a simple model of evolution explains power law behavior. BMC Evolutionary Biology **2** (2002) 18

33. Karev, G.P., Wolf, Y.I., Koonin, E.V.: Simple stochastic birth and death models of genome evolution: was there enough time for us to evolve? Bioinformatics **19** (2003) 1889–1900
34. Karev, G.P., Wolf, Y.I., Berezovskaya, F.S., Koonin, E.V.: Gene family evolution: an in-depth theoretical and simulation analysis of non-linear birth-death-innovation models. BMC Evolutionary Biology **4** (2004) 32
35. Wolf, Y.I., Rogozin, I.B., Grishin, N.V., Tatusov, R.L., Koonin, E.V.: Genome trees constructed by five different approaches suggest new major bacterial clades. BMC Evolutionary Biology **1** (2001) 8
36. Kunin, V., Ouzounis, C.A.: GeneTRACE-reconstruction of gene content of ancestral species. Bioinformatics **19** (2003) 1412–1416
37. Feller, W.: An Introduction to Probability Theory and Its Applications. Wiley & Sons (1950)
38. Sonnhammer, E.L.L., Koonin, E.V.: Orthology, paralogy and proposed classification for paralog subtypes. Trends in Genetics **18** (2002) 619–620
39. Karlin, S., McGregor, J.: Linear growth, birth, and death processes. Journal of Mathematics and Mechanics **7** (1958) 643–662
40. Lerat, E., Daubin, V., Moran, N.A.: From gene trees to organismal phylogeny in Prokaryotes: the case of the γ-Proteobacteria. PLoS Biology **1** (2003) E19
41. Boussau, B., Karlberg, E.O., Frank, A.C., Legault, B.A., Andersson, S.G.E.: Computational inference of scenarios for α-proteobacterial genome evolution. Proceedings of the National Academy of Sciences of the USA **101** (2004) 9722–9727
42. Herbeck, J.T., Degnan, P.H., Wernegren, J.J.: Nonhomogeneous model of sequence evolution indicates independent origins of endosymbionts within the Enterobacteriales (γ-Proteobacteria). Molecular Biology and Evolution **22** (2005) 520–532
43. Belda, E., Moya, A., Silva, F.J.: Genome rearrangement distances and gene order phylogeny in γ-Proteobacteria. Molecular Biology and Evolution **22** (2005) 1456–1467
44. Reed, W.J., Hughes, B.D.: A model explaining the size distribution of gene families. Mathematical Biosciences **189** (2004) 97–102
45. Pupko, T., Pe'er, I., Shamir, R., Graur, D.: A fast algorithm for joint reconstruction of ancestral amino acid sequences. Molecular Biology and Evolution **17** (2000) 890–896
46. Csűrös, M.: Likely scenarios of intron evolution. In McLysaght, A., Huson, D.H., eds.: Comparative Genomics. Volume 3678 of LNBI., Heidelberg, Springer-Verlag (2005) 47–60

A Sublinear-Time Randomized Approximation Scheme for the Robinson-Foulds Metric

Nicholas D. Pattengale and Bernard M.E. Moret

Department of Computer Science,
University of New Mexico,
Albuquerque, NM 87131, USA
{nickp, moret}@cs.unm.edu

Abstract. The Robinson-Foulds (RF) metric is the measure most widely used in comparing phylogenetic trees; it can be computed in linear time using Day's algorithm. When faced with the need to compare large numbers of large trees, however, even linear time becomes prohibitive. We present a randomized approximation scheme that provides, with high probability, a $(1 + \varepsilon)$ approximation of the true RF metric for all pairs of trees in a given collection. Our approach is to use a sublinear-space embedding of the trees, combined with an application of the Johnson-Lindenstrauss lemma to approximate vector norms very rapidly. We discuss the consequences of various parameter choices (in the embedding and in the approximation requirements). We also implemented our algorithm as a Java class that can easily be combined with popular packages such as Mesquite; in consequence, we present experimental results illustrating the precision and running-time tradeoffs as well as demonstrating the speed of our approach.

1 Introduction

The need to compare phylogenetic trees is common. Many reconstruction methods (particularly maximum parsimony and Bayesian methods) produce a large number of possible trees. Trees are also built for the same collection of organisms from different types of data (e.g., nucleotide or codon sequences for one or more genes, gene-order data, protein folds, but also metabolic and morphological data). Phylogenetic trees can be compared and the result summarized in many ways; for instance, consensus methods [1] return a single tree that best summarizes the information present in the entire collection, while supertree methods (typically used when the trees are built on different, overlapping subsets of organisms) [2] combine the individual trees into a single larger one. A more elementary step is to produce estimates of how much the trees differ from each other, by computing pairwise similarity or distance measures. Here again, many approaches have been used, such as computing pairwise edit distances based on tree rearrangement operators [3, 4]; the most common distance measure between two trees, however, is the Robinson-Foulds (RF) metric [5]. This measure is in widespread use because it can be computed in linear time [6], is based directly on the edge structure of the trees and their induced bipartitions, and is a lower bound on the more expensive edit distances. Yet, as the size of datasets used by researchers grows ever larger, even a linear-time computation of pairwise distances becomes onerous.

A. Apostolico et al. (Eds.): RECOMB 2006, LNBI 3909, pp. 221–230, 2006.
© Springer-Verlag Berlin Heidelberg 2006

In this paper, we present the first sublinear-time algorithm to compute pairwise RF distances among a collection of trees. Our algorithm is a randomized approximation scheme: it returns, with high probability, an approximation that is guaranteed to be within $(1 + \varepsilon)$ of the true distance, where $\varepsilon > 0$ can be chosen arbitrarily small. Our approach uses a sublinear-space embedding of the trees, combined with an application of the Johnson-Lindenstrauss lemma [7] to approximate vector norms rapidly. We discuss the consequences of various parameter choices (in the embedding and in the approximation requirements). We also implemented our algorithm as a Java class that can easily be combined with popular packages such as Mesquite [8]; in consequence, we present experimental results illustrating the precision and running-time tradeoffs as well as demonstrating the speed of our approach.

2   Terminology and Definitions

A *phylogenetic tree* is an undirected, connected, acyclic graph; its leaves (also called tips) correspond to the *taxa* about which data was collected, while its internal nodes all have degree at least 3. If every internal node of a phylogenetic tree has degree equal to 3, the tree is said to be *binary* or *fully resolved*. We will use $\mathcal{T}_n$ to denote a set of phylogenetic trees on n taxa.

Removing an edge (a, b) from a tree T disconnects the tree, creating two smaller trees, T_a (containing a) and T_b (containing b). Note that a (resp., b) might now have only degree 2 in T_a (resp., T_b), in which case we remove it (connecting its two neighbors directly to each other) in order to preserve the constraint that each internal node have degree at least 3. Cutting T into T_a and T_b induces a *bipartition* (or *split*) of the set S of taxa of T into the set A of taxa of T_a and the set B of taxa of T_b, a bipartition that we denote $A|B$. Thus there exists a one-to-one correspondence between the bipartitions of S and the edges of T, so that each tree is uniquely characterized by the set of bipartitions it induces. If S has n taxa, then any (unrooted) phylogenetic tree for S has at most $2n - 3$ edges and so induces at most $2n - 3$ bipartitions, only a small subset of the

$$NB = \sum_{i=1}^{\lfloor \frac{n}{2} \rfloor} \binom{n}{i} \approx 2^{n-1}$$

possible bipartitions of the set S. Moreover, n of these bipartitions are *trivial bipartitions* that split S into a one-element set against the remaining $n - 1$ elements—trivial, because these n bipartitions are common to all phylogenetic trees on S and thus need not be explicitly recorded. We shall denote by $B(T)$ the set of (at most $n - 3$) nontrivial bipartitions of S induced by T.

The *Robinson-Foulds distance* [5] between two trees on the same set S of taxa is simply a normalized count of the bipartitions induced by one tree, but not by the other.

Definition 1. *Given a set S of taxa and two phylogenetic trees, T_1 and T_2, on S, the Robinson-Foulds distance between T_1 and T_2 is*

$$RF(T_1, T_2) = \frac{1}{2} \left(|B(T_1) - B(T_2)| + |B(T_2) - B(T_1)| \right)$$

This measure of dissimilarity is easily seen to be a metric [5] and can be computed in linear time [6].

We mentioned that one could also define edit distances between trees, in the context of one or more operators that alter the structure of a tree. Two commonly used operators are the *Nearest Neighbor Interchange (NNI)* and the more powerful *Tree Bisection and Reconnection (TBR)*—see [4, 9] for definitions and discussions of these operators. Applying the NNI operator to T_1 can change $RF(T_1, T_2)$ by at most 1, while applying the TBR operator to T_1 can change $RF(T_1, T_2)$ almost arbitrarily. We use these operators in generating test sets for our RF approximation routine, as discussed in Section 5.

3   Theoretical Basis for the Algorithm

The key concept in our approach is *representation*. Our approximation algorithm is a reduction to the computation of vector norms in a suitable vector space and the sublinear running time results from our ability to represent the necessary characteristics of phylogenetic trees in sublinear space. More specifically, we represent phylogenetic trees as vectors in such a way that RF distances become simply the $\| \cdot \|_1$-norm of the difference vector, then generalize the result to arbitrary $\| \cdot \|_p$-norms for $p \geq 1$.[1] We then borrow a technique from high-dimensional geometry to reduce the dimensionality of tree vectors while maintaining pairwise $\| \cdot \|_2$-norms. Finally we combine these techniques to obtain a fast approximation algorithm for computing RF distances.

3.1   Bit-Vector Representation

Consider a (bijective) function $f \colon \bigcup_{T \in \mathcal{T}_n} B(T) \to \mathbb{N}$ that assigns a unique integer in the interval $[1, NB]$ (recall that NB is the number of possible bipartitions of the set) to each bipartition.

Definition 2. *The* bit-vector representation *of a phylogenetic tree T is $v_T \in \mathbb{R}^b$ where we have*

$$v_T[i] = \begin{cases} 1 & f^{-1}(i) \in T \\ 0 & \textit{otherwise} \end{cases}$$

Obviously, this representation is quite space-consuming and proportionally time-consuming to produce; fortunately, we need only consider the bits set to 1, as discussed in Section 4.2. By construction the $\|.\|_1$-norm between tree vectors is the (non-normalized) RF distance.

Theorem 1. $\forall T_1, T_2 \in \mathcal{T}_n,\ RF(T_1, T_2) = \frac{1}{2} \| v_{T_1} - v_{T_2} \|_1$.

Proof. For all $s \in B(T_1) - B(T_2)$ (resp., $B(T_2) - B(T_1)$), we have $v_{T_1}[f(s)] = 1$ (resp., $v_{T_2}[f(s)] = 1$) and $v_{T_2}[f(s)] = 0$ (resp., $v_{T_2}[f(s)] = 0$). For all $s \in B(T_1) \cap B(T_2)$, we also have $v_{T_1} = v_{T_2} = 1$ and, for all $s \in \bigcup_{T \in \mathcal{T}_n} B(T) - (B(T_1) \cup B(T_2))$, we have $v_{T_1} = v_{T_2} = 0$. Thus we can conclude

$$\| v_{T_1} - v_{T_2} \|_1 = |B(T_1) - B(T_2)| + |B(T_2) - B(T_1)| = 2 \cdot RF(T_1, T_2) \qquad \square$$

---

[1] The $\| \cdot \|_p$-norm of a vector $v = (v_1 v_2 \ldots v_k)$ is $\|v\|_p = \left(\sum_{i=1}^{k} |v_i|^p \right)^{\frac{1}{p}}$.

3.2   Properties of $\| \cdot \|_p$-Norms of Bit-Vectors

Theorem 2. *For an arbitrary vector $v \in \mathbb{R}^b$ where every element is chosen from the set $\{-k, 0, k\}$ (for arbitrary $k > 0$), we have $\|v\|_1 = k^{1-p} \cdot (\|v\|_p)^p$.*

Proof. Assume that v has c entries of value $\pm k$; we can write

$$\|v\|_p = \left(\sum_{i=1}^{b} (|v_i|)^p \right)^{\frac{1}{p}} = (ck^p)^{\frac{1}{p}} = c^{\frac{1}{p}} k$$

$$\|v\|_1 = \sum_{i=1}^{b} |v_i| = ck = c^{\frac{p-1}{p}} (c^{\frac{1}{p}} k) = c^{\frac{p-1}{p}} \|v\|_p$$

Raising the first result to the power $(p-1)$ and solving for $c^{\frac{p-1}{p}}$ yields

$$c^{\frac{p-1}{p}} = k^{1-p} \cdot (\|v\|_p)^{p-1}$$

and substituting into the second result finally yields

$$\|v\|_1 = k^{1-p} \cdot (\|v\|_p)^p \qquad \qquad \square$$

Corollary 1. *For bit-vectors $(k = 1)$ we have $\|v\|_1 = (\|v\|_p)^p$; in particular, we have $\|v\|_1 = (\|v\|_2)^2$.*

3.3   Reducing Dimensionality

We briefly outline a result of Johnson and Lindenstrauss [7] for norm-preserving embeddings; see [10, 11, 12] for a more detailed treatment and proofs.

Consider an $m \times NB$ matrix V in which we want to compute the $\| \cdot \|_2$-norm between pairs of row vectors. Naïvely calculating a pairwise norm costs $O(NB)$ time. The Johnson-Lindenstrauss lemma states that, if we first multiply V by another matrix F of size $NB \times \frac{4\ln m}{\varepsilon^2}$, filled with random numbers from the normal distribution $(0, 1)$, we can then use the pairwise norms between rows of $V \cdot F$ as good approximations of the pairwise norms between corresponding rows of V. Specifically, for given ε and F, we have, with probability at least $1 - m^{-2}$,

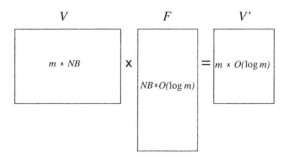

Fig. 1. A sketch of randomized embedding. Each tree is a row in V; F is a random matrix; each row of V' is the embedded representation of the corresponding row vector in V.

$$\forall u, v \in V, \ (1-\varepsilon)\|u-v\|_2 \leq \|(u-v)F\|_2 \leq (1+\varepsilon)\|u-v\|_2$$

The dimensionality of $(u-v)F$ is now $\frac{4\ln m}{\varepsilon^2}$ and thus independent of NB. Other probability distributions can also be used for populating the elements of F, as discussed in Section 4.3. Figure 1 illustrates the basic embedding technique.

3.4 Assembling the Pieces

By combining Theorem 1, the Johnson-Lindenstrauss Lemma, and Corollary 1, we can produce our algorithm. Given a set of m phylogenetic trees:

1. stack their bit-vector representations (recall that each has dimensionality NB) to form an $m \times NB$ matrix;
2. perform the embedding of Section 3.3 (see also Section 4.3 for implementation considerations) thereby compacting the row dimensionality of the matrix while preserving pointwise $\|\cdot\|_2$-norms between row vectors; and
3. for any pair of row vectors v_{T1}, v_{T2} (i.e., embedded trees), compute the approximate RF distance by computing $(\|v_{T1} - v_{T2}\|_2)^2$.

However, this is the theoretical form of the algorithm. In practice, we do not compute the large matrix, but use a compact representation from the beginning, one whose size is determined by the number of bipartitions present in a tree, which is just $n-3$. We also use a sampling matrix with entries in $\{-1, 0, +1\}$ (see Section 4.3) rather than arbitrary reals in $[0,1]$. Since the dimensionality of the embedded row vectors is $O(\log m)$, the time complexity of for our approximate RF distance between two trees is also $O(\log m)$, so that our technique is asymptotically faster whenever we have $\log m = o(n)$.

4 Implementation

We have implemented our algorithm as a Java class that can be used as one of many distance functions from within popular packages such as TreeViz [14, 15] and Mesquite [8]. Our source code can be obtained from http://compbio.unm.edu.

Implementation raises a number of nontrivial issues, which we now address.

4.1 Dimensionality Is Not Prohibitive

Although the bit-vector representation is presented as having dimensionality $NB \approx 2^{n-1}$, we need not produce, store, or manipulate exponentially large vectors. Since the number of bipartitions present in a single tree is at most $n-3$, tree vectors are very sparse; by representing them as lists of indices (corresponding to the bits that are set), we avoid the issue of exponentially large vectors.

4.2 Indexing Bipartitions

In order to embed a tree we must read the entire tree, so that the embedding step, for each tree, must run in $\Omega(n)$ time. Day's algorithm [6] computes the RF distance in $O(n)$ time. Thus our algorithm (with inclusion of the embedding step) cannot asymptotically outperform Day's algorithm for a single distance computation. We can expect an asymptotic speedup, however, if we compute an all-pairs distance matrix for a set of trees: for such a computation the standard technique costs $O(n)$ per pair on $\binom{m}{2} = O(m^2)$ pairs, yielding a total complexity of $O(m^2 n)$. If embedding costs some function $g(n)$, our technique will cost $O(m \cdot g(n))$ for the embedding step and $O(m^2 \log m)$ for the distance computations on pairs of embedded trees, for a total of $O(m \cdot (g(n) + m \log m))$.

One of the notable attributes of Day's algorithm, that we have not yet achieved in our embedding routine, is the ability to determine in constant time whether a bipartition found in one tree exists in the other tree. As currently implemented, our cluster-matching routine takes $O(n)$ time, inflating the cost of embedding a single tree to $O(n^2)$ and thus causing the overall all-pairs algorithm to run in $O(m(n^2 + m \log m))$ time. We are pursuing the design of a subquadratic-time embedding algorithm, but note that the current implementation already performs well in practice, especially in the common situation where the number of trees to be compared far exceeds the number of taxa in these trees.

4.3   Filling the F Matrix

Generating a large number of Gaussian random numbers and performing floating-point arithmetic (for matrix-vector multiplications) on them is costly. Fortunately Achlioptas [13] has shown that simpler distributions can populate the embedding matrix. The best distribution in implementation terms is as follows:

$$p(X = -\sqrt{3}) = 1/6$$
$$p(X = 0) = 2/3$$
$$p(X = \sqrt{3}) = 1/6$$

The $\sqrt{3}$ is just a normalizing factor and can be omitted until the very end of the computation, so we use values in $\{-1, 0, 1\}$, with two major advantages: (i) it is an easy distribution to sample with a uniform random number generator; and (ii) multiplying by elements in $\{0, \pm 1\}$ can be done through additions embedded in a three-way conditional. Using this distribution requires a slightly different row dimensionality for F: for some $\beta > 0$, F must have size $NB \times k_0$ with

$$k_0 \geq \frac{(4 + 2\beta) \cdot \log m}{\frac{\varepsilon^2}{2} - \frac{\varepsilon^3}{3}}$$

and the embedded matrix is normalized by $\sqrt{k_0}$. The dimensionality is $O(\log m)$ and the error bound of $(1 + \varepsilon)$ is obeyed with probability at least $1 - m^{-\beta}$.

5   Experiments

Two major factors influence the usefulness of our technique in practice: the effect of ε and the overhead of embedding. We ran a series of experiments to assess both factors. The experiments were run on the CIPRES cluster at the San Diego Supercomputing Center, a 16-node Western Scientific Fusion A8 running Linux, in which each node is an 8-way Opteron 850 system with 32GB of memory.

5.1 The Effect of ε on Clustering Quality

Choosing a large ε cuts down the dimensionality of embedded trees and thus speeds up the computation. To assess the effect on clustering quality of a big ε, we generated a large set of test data using the following procedure:

1. Generate a phylogenetic tree T_{seed} uniformly at random from $\mathcal{T}_{numTaxa}$
2. do numClusters times
   (a) create a new tree $T_{clusterSeed}$ by doing a random number ($0 \leq k < $ maxTBR) of TBR operations to T_{seed}.
   (b) write $T_{clusterSeed}$ to file
   (c) do treesPerCluster times
       i. create a new tree T' by doing a random number ($0 \leq j < $ maxNNI) of NNI operations to $T_{clusterSeed}$.
       ii. write T' to file

with the following parameters:

- $numTaxa = 100$
- numClusters $= 2, 3, 4, 5, 6, 7, 8, 9, 10$
- treesPerCluster $= 50, 100$
- maxTBR $= 5, 8, 11, 14$
- maxNNI $= 10, 20, 30, 40, 50, 60, 70, 80, 90, 100$

This procedure creates the classic "islands" of trees [16] by providing pairwise distant trees as seeds and generating a cluster of new trees around each seed tree. We created 10 files for each combination of parameters, yielding a total of 7,200 data files, varying from easy to very hard to cluster correctly.

For each data set we performed hierarchical agglomerative clustering with a range of ε values. We then compared the results with the known intended clustering by using the Rand index [17]. Figure 2 shows the results. These results indicate that, even with ε = 1.0, we identify the correct clustering quite often, even on very challenging datasets.

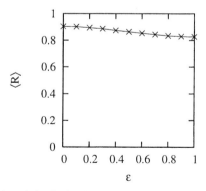

Fig. 2. The mean Rand index, $\langle R \rangle$, obtained by running the entire battery of 7,200 datasets for values of ε ranging from 0.1 to 1.0 by increments of 0.1. The datapoint at ε = 0 was obtained by the standard RF algorithm.

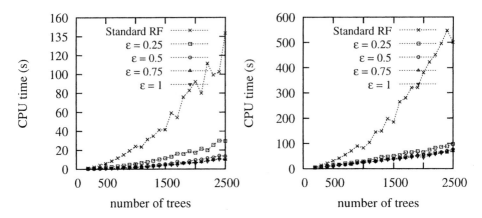

Fig. 3. The running time for computing an all-pairs distance matrix as a function of the number of trees with 250 taxa (left) and 1,000 taxa (right)

Given that biological datasets tend to yield well resolved clusters of trees (due to the nature of the algorithms that produce them), we expect that our algorithm will perform well in analyses of biological data.

5.2 The Effect of Embedding on Running Time

Figure 3 demonstrates the speedup afforded by our technique when we hold constant the number of taxa and plot the running time (of an all-pairs distance matrix computation) as a function of the number of trees. We include results form experiments in which the number of taxa is held fixed at 250 (typical of trees being used today) as well as 1,000 (typical of trees to be used in the near future). Times shown are somewhat pessimistic as a "cold" run (with memory management overhead) was averaged into each of the trials.

If we fix the number of trees and vary the number of taxa—admittedly not a realistic scenario—, we are limited by the $O(mn^2)$ embedding cost. In this case we do not perceive a *significant* advantage to using our technique, but our algorithm will easily outperform the standard technique even in this setting if we can design an embedding technique that runs in subquadratic time.

6 Conclusion

We used an embedding in high-dimensional space and techniques for computing vector norms from high-dimensional geometry to design the first sublinear-time approximation scheme to compute Robinson-Foulds distances between pairs of trees. We implemented our algorithm and provided experimental support for its computational advantages. As computational biologists everywhere increasingly turn to phylogenetic computations to further their understanding of genomic, proteomic, and metabolomic data, and do so on larger and larger datasets, a fast computational method to compare large collections of trees will enable interactive analyses (in the type of setting provided by Mesquite).

While our algorithm easily outperforms repeated applications of Day's algorithm for large collections of trees, its relatively expensive embedding step prevents it from achieving similarly spectacular speedups for smaller collections of very large trees (although even there it runs nearly as fast as Day's algorithm). A natural question is whether the embedding can be run in subquadratic time. Given the close connection between RF distances and the strict consensus tree, another natural question is whether similar randomized techniques could be used to speed up the computation of consensus trees.

Acknowledgments

This work is supported by the US National Science Foundation under grants EF 03-31654 (the CIPRES project), IIS 01-13095, IIS 01-21377, and DEB 01-20709. N. Pattengale wishes to thank T. Berger-Wolf for introducing him to norm-preserving embeddings.

References

1. Bryant, D.: A classification of consensus methods for phylogenetics. In: Bioconsensus. Volume 61 of DIMACS Series in Discrete Mathematics and Theoretical Computer Science. American Math. Soc. (2002) 163–184
2. Bininda-Edmonds, O., ed.: Phylogenetic Supertrees: Combining information to reveal the Tree of Life. Kluwer Publ. (2004)
3. DasGupta, B., He, X., Jiang, T., Li, M., Tromp, J., Zhang, L.: On computing the nearest neighbor interchange distance. In: Proc. DIMACS Workshop on Discrete Problems with Medical Applications. Volume 55 of DIMACS Series in Discrete Mathematics and Theoretical Computer Science. American Math. Soc. (2000) 125–143
4. Allen, B., Steel, M.: Subtree transfer operations and their induced metrics on evolutionary trees. Annals of Combinatorics **5** (2001) 1–15
5. Robinson, D., Foulds, L.: Comparison of phylogenetic trees. Math. Biosciences **53** (1981) 131–147
6. Day, W.: Optimal algorithms for comparing trees with labeled leaves. J. of Classification **2** (1985) 7–28
7. Johnson, W., Lindenstrauss, J.: Extensions of Lipschitz mappings into a Hilbert space. Cont. Math. **26** (1984) 189–206
8. Maddison, W., Maddison, D.: Mesquite: A modular system for evolutionary analysis. Version 1.06 http://mesquiteproject.org (2005)
9. Bryant, D.: The splits in the neighborhood of a tree. Annals of Combinatorics **8** (2004) 1–11
10. Indyk, P.: Algorithmic applications of low-distortion geometric embeddings. In: Proc. 42nd IEEE Symp. on Foundations of Computer Science FOCS'01, IEEE Computer Society (2001) 10–33
11. Linial, N., London, E., Rabinovich, Y.: The geometry of graphs and some of its algorithmic applications. Combinatorica **15** (1995) 215–245
12. Indyk, P., Motwani, R.: Approximate nearest neighbors: Towards removing the curse of dimensionality. In: Proc. 13th ACM Symp. on Theory of Computing STOC'98. (1998) 604–613
13. Achlioptas, D.: Database-friendly random projections: Johnson-Lindenstrauss with binary coins. J. Comput. Syst. Sci. **66** (2003) 671–687

14. Hillis, D., Heath, T., St John, K.: Analysis and visualization of tree space. Syst. Bio. **54** (1995) 471–482
15. Amenta, N., Klingner, J.: Case study: Visualizing sets of evolutionary trees. In: Proc. IEEE Symp. on Information Visualization INFOVIS'02, IEEE Computer Society (2002) 71–73
16. Maddison, D.: The discovery and importance of multiple islands of most-parsimonious trees. Syst. Zoology **40** (1991) 315–328
17. Rand, W.: Objective criteria for the evaluation of clustering methods. J. American Stat. Assoc. **66** (1971) 846–850

Algorithms to Distinguish the Role of Gene-Conversion from Single-Crossover Recombination in the Derivation of SNP Sequences in Populations

Yun S. Song[1,*], Zhihong Ding[1], Dan Gusfield[1],
Charles H. Langley[2], and Yufeng Wu[1]

[1] Department of Computer Science, University of California, Davis, CA 95616, USA
[2] Section of Evolution and Ecology, University of California, Davis, CA 95616, USA
yssong@cs.ucdavis.edu

Abstract. Meiotic recombination is a fundamental biological event and one of the principal evolutionary forces responsible for shaping genetic variation within species. In addition to its fundamental role, recombination is central to several critical applied problems. The most important example is "association mapping" in populations, which is widely hoped to help find genes that influence genetic diseases [3, 4]. Hence, a great deal of recent attention has focused on problems of inferring the historical derivation of sequences in populations when both mutations and recombinations have occurred. In the algorithms literature, most of that recent work has been directed to single-crossover recombination. However, *gene-conversion* is an important, and more common, form of (two-crossover) recombination which has been much less investigated in the algorithms literature.

In this paper we explicitly incorporate gene-conversion into discrete methods to study historical recombination. We are concerned with algorithms for identifying and locating the extent of historical crossing-over and gene-conversion (along with single-nucleotide mutation), and problems of constructing full putative histories of those events. The novel technical issues concern the incorporation of gene-conversion into recently developed discrete methods [20, 26] that compute *lower* and *upper-bound* information on the amount of needed recombination without gene-conversion. We first examine the most natural extension of the lower bound methods from [20], showing that the extension can be computed efficiently, but that this extension can only yield weak lower bounds. We then develop additional ideas that lead to higher lower bounds, and show how to solve, via integer-linear programming, a more biologically realistic version of the lower bound problem. We also show how to compute effective upper bounds on the number of needed single-crossovers and gene-conversions, along with explicit networks showing a putative history of mutations, single-crossovers and gene-conversions.

We validate the significance of these methods by showing that they can be effectively used to distinguish simulation-derived sequences gen-

---

* To whom correspondence should be addressed.

A. Apostolico et al. (Eds.): RECOMB 2006, LNBI 3909, pp. 231–245, 2006.

erated without gene-conversion from sequences that were generated with gene-conversion. We apply the methods to recently studied sequences of *Arabidopsis thaliana*, identifying many more regions in the sequences than were previously identified [22], where gene-conversion may have played a significant role. Demonstration software is available at www.cs. ucdavis.edu/~gusfield.

1   Introduction

Sequence variations in populations (in a single species) are caused in part by mutations at single nucleotide sites, and in part by recombination during meiosis, which creates a chimeric genome in an individual from the genomes of the individual's two parents. Sites where two alleles (states) occur in a population with a frequency above some threshold are called Single Nucleotide Polymorphism (SNP) sites. Much recent attention has focused on problems of inferring the historical derivation of SNP sequences in populations when both mutations and recombinations have occurred. In the algorithms literature, most of that work has been directed to single-crossover recombination. (Previous methods for single-crossover recombination appear in [1, 9, 10, 11, 12, 13, 14, 20, 24, 25, 26].) However, *gene-conversion* is a form of two-crossover recombination that has large biological significance, and there has been much less algorithmic work devoted to the study of models that incorporate gene-conversion as well as mutation and single-crossover recombination. Some exceptions are the papers [6, 9, 19], and statistical methods have also been developed [5, 23, 27, 30] to address gene-conversion.

Tools to study gene-conversion are important because gene-conversion is a fundamental biological process [18] that is not fully understood (partly because fine-scale data is needed which is only now becoming available, and partly because of the lack of algorithmic tools); because gene-conversion is a cause of genomic sequence variation in populations [8, 21]; and because gene-conversion has the potential to cause problems in association-mapping [15, 29]. Association mapping depends on understanding the structure of linkage disequilibrium (LD) in population data: "Standard population genetics models of recombination generally ignore gene conversion, even though crossovers and gene conversions have different effects on the structure of LD" [29].

In this paper, we extend recently developed tools for the study of historical (single-crossover) recombination and mutation, to explicitly incorporate gene-conversion events. We validate the biological significance of these methods by showing that the methods can be effectively used to distinguish sequences that were generated without gene-conversion from sequences that were generated with gene-conversions, and we apply these methods to identify regions in *Arabidopsis thaliana* sequences where gene-conversions may have played a significant role, in comparison to single-crossover recombination, in the derivation of the sequences.

In contrast to our methods, existing statistical methods (for example [23]), do not provide information on the necessary amount of recombination in the history of the sequences, or produce an explicit derivation of those sequences using mutation and recombination. Those methods also do not assess the relative importance of single-crossover recombination compared to gene-conversion. Those methods are based on patterns in the sequences rather than on how well a full history can be obtained to explain the derivation of the sequences, or on how much recombination is needed.

2   Recombination: Crossing-Over and Gene-Conversion

There are two major forms of recombination that occur during meiosis: single-crossover recombination (called "crossing-over" in the genetics literature), and gene-conversion. We will use "crossing-over" and "single-crossover recombination" interchangeably, and use "recombination" to refer to either crossing-over or gene-conversion.

Meiotic Crossing-Over: The best studied form of recombination is *crossing-over*, where during meiosis two equal length sequences produce a third sequence of the same length consisting of some prefix of one of the sequences, followed (at the "breakpoint") by a suffix of the other sequence.

Gene-Conversion: The other major form of meiotic recombination, called "gene-conversion", involves two crossovers at two breakpoints. In gene-conversion, a new sequence is formed from a prefix of one sequence, followed by an internal segment of a second sequence, followed by a suffix of the first sequence. All three sequences are of the same length. The endpoints of the internal segment are the "breakpoints" of the gene-conversion. Gene-Conversion is a small-scale meiotic event; the internal segment (called a "conversion-tract", or "tract") is short, around 50 to 2000 base pairs. Gene-conversion has been hard to study in populations because of the lack of analytical tools and the lack of fine-scale data. For example, little is known about the distribution of tract lengths. However, genomic data produced over the next several years should allow quantification of the fundamental parameters of gene conversion, and the contribution of gene conversion to the overall patterns of sequence variations in a population.

3   Minimizing the Total Number of Recombination Events

Given a set M of binary (SNP) sequences, we would like to determine the true history of mutations, crossing-over events and gene-conversions that derived the sequences from some ancestral sequence. This is of course impossible and instead previous research has focused on computing or estimating the *minimum* number, denoted $R_{\min}(M)$, of crossing-over events needed to derive the sequences from some known or unknown ancestral sequence, when only one mutation is allowed per site in the entire history of the sequences. Although the true history of the sequences may have involved more than $R_{\min}(M)$ recombinations, $R_{\min}(M)$ and particular *lower-bounds* on $R_{\min}(M)$, have proven to be useful reflections of the true historical number, for example allowing or contributing to the identification of recombination hotspots in genomic sequences [2, 7, 28].

In this paper we move from a focus on $R_{\min}(M)$, to incorporate gene-conversions. We define $T_{\min}(M)$ as the minimum total number of recombination events needed to derive M from an ancestral sequence (either known or unknown in different versions of the problem) where a recombination event is *either* a crossing-over or a gene-conversion event. Because gene-conversion tract length is typically small, we will often bound its permitted length, and define $T_{\min}(M,t)$ as the minimum number of recombination events needed to derive M, where each gene-conversion has tract length at most t (nucleotides). In the next section we discuss a practical method to compute *lower bounds* on $T_{\min}(M,t)$, and in Section 5 we discuss a practical method to compute an actual sequence of events that derives M. Since $T_{\min}(M) = T_{\min}(M,t)$ when t is sufficiently large (for example, the physical distance between the first and the last sites in M), these methods can be used to compute bounds on $T_{\min}(M)$.

4   Lower Bounds on Crossing-Over and Gene-Conversion

Since the effect of one gene-conversion can be obtained by two crossing-over events, $LBCO(M)/2$ is a valid lower-bound on $T_{\min}(M,t)$, where $LBCO(M)$ is any lower bound $R_{\min}(M)$. Several such lower bounds on $R_{\min}(M)$ have been developed and extensively studied [1, 12, 17, 20]. We will prove that when t is unconstrained, the most natural extensions of these methods to include gene-conversions, yield only weak lower bounds. However, we introduce additional ideas to increase these lower bounds, and show how to obtain higher lower bounds when t is bounded.

Our methods to compute lower bounds on $T_{\min}(M,t)$ are based on an a general approach developed by Myers and Griffiths [20] to compute lower bounds on $R_{\min}(M)$. Their approach has two essential parts: methods to compute *local* lower bounds for intervals of sites, and a simple, polynomial-time method to combine those local bounds into a composite *global* lower bound. All of the known methods (HK [17], Haplotype and History [20], and Connected Component [1, 12]) to compute local lower bounds on crossing-over extend immediately to the case that gene-conversions are allowed, but the issue of how to combine those local bounds into a composite global bound on $T_{\min}(M,t)$ is more complex.

The Haplotype Local Bound: Due to its centrality, we discuss the local Haplotype bound of Myers and Griffiths [20] in detail. Consider the set of sequences M arrayed in a matrix, and an interval I of sites. Let $M(I)$ be the sequences M restricted to the sites of I. Then $h(M(I))$ is defined as the number of *distinct* rows of M, minus the number of *distinct* columns of M, minus one. $h(M(I))$ is a valid lower bound on $R_{\min}(M(I))$, and in fact a lower bound on the number of breakpoints that must be located *inside* interval I. To see this, assume first that all the columns of M are distinct, since removal of (duplicate) columns cannot increase the number of crossing-over events needed, so any lower-bound computed for the reduced matrix will be a lower-bound for the original M. Next, consider the derivation of $M(I)$ using $R_{\min}(I)$ crossing-over events and one mutation per site in I, and consider an actual specification of the relative order that the events and mutations occur (this is called a "history"). Any mutation of a site in I can only occur once in the history, and it can result in only one new sequence (not yet derived in the history). Similarly, each crossing-over event with breakpoint in I can only create one new sequence. A crossing-over event with breakpoint outside of I, or a mutation outside of I cannot create a new sequence in the set $M(I)$. The history must create all the distinct sequences in $M(I)$ starting from some ancestral sequence which might itself be in $M(I)$. It follows that there must be at least $h(M(I))$ crossing-over events whose breakpoint is (strictly) in I, so $h(M(I)) \leq R_{\min}(M(I))$. Clearly, $h(M(I))$ is also a lower bound on $T_{\min}(M(I), t)$ for any t, because a single gene-conversion can also create at most one new sequence in $M(I)$.

In [20], local haplotype lower bounds for an interval I were raised by considering subsets of sites in the interval. That approach was further explored in [2] and optimized in [26]. In our software, we use the latter approach to obtain the highest possible local haplotype bounds for each interval I.

The Composite Global Lower Bound: We are interested in a lower bound on $T_{\min}(M, t)$, not just a bound on $T_{\min}(M(I), t)$ for a single interval I. Of course, $h(M)$ (the haplotype bound applied to the interval consisting of all the sites) is a lower bound on $T_{\min}(M, t)$, but in the computations we have done, it is a very poor lower bound, often a negative number. The same is true of $R_{\min}(M)$, but a much better composite global lower bound on $R_{\min}(M)$ can be obtained from the local lower bounds. We say a point p "covers" an interval I if p is contained in I with at least one SNP site on each side of p. Then we obtain a composite global bound on $R_{\min}(M)$ by solving the **Crossover Coverage Problem**: Find the smallest set of points B so that each interval I is covered by at least $h(M(I))$ points of B. $|B|$ is a valid lower bound on $R_{\min}(M)$, and B can be found using a simple polynomial-time algorithm[20]. However, $|B|$ is not necessarily a lower bound on $T_{\min}(M, t)$. To use the local bounds to obtain a composite bound on $T_{\min}(M, t)$, we next formulate a natural generalization of the Crossover Coverage Problem.

We say a line-segment "covers" an interval I if at least one end p of the line-segment covers I. Note that a line-segment that strictly contains I does not cover it, and that a line-segment covers I only once even if both of its endpoints are in I. The intuitive meaning is that a line-segment represents a gene-conversion, and a line-segment covers an interval I only if the action of the gene-conversion it represents could create a new sequence in $M(I)$. Then we obtain a composite bound on $T_{\min}(M, t)$ by solving the **Gene-Conversion Coverage Problem**: Find the smallest set consisting of points P, and line segments S with length at most t, so that each interval I is covered by at least $h(M(I))$ elements of $P \cup S$. Each point in P represents a crossing-over and each line-segment in S represents a gene-conversion. Clearly, a derivation of M using exactly $T_{\min}(M, t)$ recombinations, where the conversion tract of any gene-conversion is of length at most t, defines a set of points and line-segments that cover each interval I at least $h(M(I))$ times. Therefore, a solution to the Gene-Conversion Coverage Problem is a valid lower bound on $T_{\min}(M, t)$.

4.1   A Special Case of the Gene-Conversion Coverage Problem

We first show that when t is unbounded, the Gene-Conversion Coverage Problem has a simple, yet disappointing solution. For the following discussion, let B be a minimum-sized set of breakpoints that solves the Crossover Coverage Problem, and let $|B| = n$. Number the breakpoints in B left to right choosing an arbitrary ordering of breakpoints that lie on the same point. For any $k \leq \lfloor \frac{n}{2} \rfloor$, let $P(B, k)$ be a pairing of the leftmost k breakpoints to the rightmost k breakpoints of B under the mapping $i \rightarrow n - k + i$, and create a line-segment between the two endpoints of each pair in $P(B, k)$. Let S be the set of these k line-segments, and let P be the set $n - 2k$ unpaired points in B. We will show that there is a solution of the Gene-Conversion Coverage Problem that has this form for some k, and show that the best k can be easily obtained.

Note that in $P(B, k)$, if $i < j \leq k$, then i maps to a breakpoint to the left of the breakpoint that j maps to, a property that we call "monotonicity". Define $L(I)$ as the number of breakpoints in B to the left of I, and $R(I)$ as the number of breakpoints in B to the right of I. We say a line-segment is "contained in" I if both of its ends are contained in I. Define the "coverage" of interval I as the number of elements of $P \cup S$ that cover I.

Lemma 1. *Let I be any interval where some line-segment in S is contained in I. Then exactly $k - (L(I) + R(I))$ line-segments in S are contained in I.*

Proof. First, if a line-segment (a, b) in S is contained in I, then $k \geq L(I) + 1$ so breakpoint $L(I) + 1$ (the leftmost breakpoint in I) must be the left end of some line-segment in S. Moreover, by monotonicity, the right endpoint of that segment (which is at breakpoint $n - k + L(I) + 1$) must be at or to the left of b, and hence in I. Since the pairing $P(B, k)$ involves the rightmost k breakpoints in B, all the breakpoints in I to the right of $n - k + L(I) + 1$ must be right endpoints of some line-segment in S, and again by monotonicity, their paired left endpoints must be to the right of $L(I) + 1$, and hence must be in I. The rightmost breakpoint in I is $n - R(I)$, so there are exactly $n - R(I) - [n - k + L(I) + 1] + 1 = k - (L(I) + R(I))$ line-segments in S that are contained in I. $\qquad\square$

Lemma 2. *For $k \leq n - \max_I(h(M(I)))$, the coverage of any interval I is at least $h(M(I)))$.*

Proof. Let $B(I)$ be the number of breakpoints in B that are contained in I. The coverage of I is exactly $B(I)$ minus the number of line-segments in S contained in I. Since, $B(I) \geq h(M(I)))$ for all I, we only need to examine intervals where some line-segment in S is contained in the interval. Let I be such an interval. By assumption, $k \leq n - \max_I h(M(I))) \leq n - h(M((I)))$, so

$$k - (L(I) + R(I)) \leq n - h(M((I))) - (L(I) + R(I)) = B(I) - h(M((I))).$$

Therefore $h(M((I))) \leq B(I) - [k - (L(I) + R(I))]$, and by the Lemma 1, the coverage of I is at least $h(M(I)))$. $\qquad\square$

Corollary 1. *If $k = \min(\lfloor \frac{n}{2} \rfloor, n - \max_I h(M(I)))$, then the coverage of I is at least $h(M(I))$, for each interval I.*

Theorem 1. *If B is a minimum sized set of breakpoints (of size n) solving the Crossover Coverage Problem, then the optimal solution to the Gene-Conversion Coverage Problem has size exactly $\max(\lceil \frac{n}{2} \rceil, \max_I h(M(I)))$.*

Proof. By the Corollary, if we set k to $\min(\lfloor \frac{n}{2} \rfloor, n - \max_I h(M(I)))$, then every interval I has coverage at least $h(M(I))$, and $|S \cup P|$ is exactly $n - k = \max(\lceil \frac{n}{2} \rceil, \max_I h(M(I)))$. But both of those terms are trivial lower bounds on the number of needed line-segments and single breakpoints in any solution to the Gene-Conversion Coverage Problem, and hence that choice of k gives the optimal solution. □

So when t is unbounded, we have a simple, efficient algorithm for the Gene-Conversion Coverage Problem: solve the Crossover Coverage Problem, yielding set B, and then apply Theorem 1. Note that Theorem 1 holds regardless of which (optimal) solution B is used, and provides a lower bound on $T_{\min}(M, t)$ for any t, as well as for $T_{\min}(M)$. It can also be shown that Theorem 1 holds even if we use HK or history or connected component local lower bounds, instead of Haplotype local lower bounds.

4.2 Improving the Bounds

Theorem 1 establishes trivial lower bounds on $T_{\min}(M, t)$ and $T_{\min}(M)$. It's importance is that it proves that the natural extension of the way that good lower bounds on $R_{\min}(M)$ were obtained, will not yield non-trivial lower bounds when gene-conversion is included. To get higher bounds we have to use additional constraints. The first such constraint is to bound the permitted tract length to t in any solution to the Gene-Conversion Coverage Problem. We do not have a polynomial-time algorithm for this version of the problem, but next show how to effectively solve it using integer linear programming.

An ILP Formulation for Bounded t: We define $\phi(i)$ as the physical position in the chromosome of site i. Given an input matrix M with m sites, and a bound t, we define an integer-valued variable $K_{i,j}$ for each pair of integers i, j where $0 < i \leq m - 1$, $0 < j \leq m - 1$, $i \leq j$, and either $i = j$ or $\phi(j) - \phi(i + 1) < t$. The value that variable $K_{i,j}$ takes on in the ILP solution specifies the number of line-segments $[i, j]$ (whose two endpoints are between sites $i, i + 1$ and sites $j, j + 1$) that will be used in the solution. For an interval $I = [a, b]$, we define the set $A(I) = \{K_{i,j} : a \leq i < b \ or \ a \leq j < b\}$. Set $A(I)$ is the set of the variables that can specify a line-segment that covers I. We allow $i = j$ to indicate a single point. Then the following ILP solves the Gene-Conversion Coverage Problem when t is bounded:

Minimize $\sum_{(i,j)} K_{i,j}$

Subject to

$\sum_{K_{i,j} \in A(I)} K_{i,j} \geq h(M(I))$, for each interval I.

Further Improvements: We can often increase the composite global lower bound on $T_{\min}(M, t)$ with the following observation. We say sites p, q are *incompatible* when all four binary combinations $00, 01, 10, 11$ appear at those two sites. If p, q are incompatible, then there must be at least one breakpoint in the interval $[p, q]$, and this can introduce additional constraints on possible gene conversions. For example, consider the following four sequences: $000, 011, 110, 101$. All three sites are pairwise incompatible. Let a, b, c denote the first, second, and third sites, respectively. Intervals $[a, b], [b, c], [a, c]$ all have a local bound of 1. We can cover those intervals using a single line-segment with one endpoint between a and b, and the other endpoint between b and c. That single segment covers all three of the intervals. However, it is easy to see that a gene conversion corresponding to that single segment cannot make sites a and c incompatible. Thus, there must be at least *two* gene-conversion or crossing-over events for this example. Those additional constraints can increase the resulting global lower bound, and can be incorporated into the ILP with the following constraints:

$$\sum_{p \leq a < q, b \geq q} K_{a,b} + \sum_{a < p, p \leq b < q} K_{a,b} + \sum_{p \leq a < q} K_{a,a} \geq 1, \text{ for each pair of}$$

incompatible sites p, q.

Note that $K_{a,a}$ defines a crossing-over event rather than gene-conversion.

The above ILP formulation can be solved reasonably fast for data of the size of current biological interest. Some timing details will be presented in Section 6. This approach has been implemented in the program HapBound-GC. A demonstration version of HapBound-GC uses the free GNU GLPK package to solve the ILP.

Another way to raise the composite lower bound involves the interaction of local bounds and global bounds that use those local bounds. Consider a subset of sites K that span an interval I, and let $M(K)$ be the sequences M restricted to the sites in K. The action of a gene-conversion can create a sequence in $M(K)$ differing from both of the parent sequences only if there are sites in K to the left and right of one of the two breakpoints of the gene-conversion. Moreover, if the two ends of a gene-conversion are in the interval spanned by K, then the gene-conversion can create a new sequence in $M(K)$ only if there is a site in K between those two breakpoints. Those observations constrain where we must place points and line-segments in a solution to the Gene-Conversion Coverage Problem. In fact, such constraints are used in the ILP for the subsets described in Section 4 that yield the highest local haplotype bounds. However, we can further raise the composite global bound by enumerating each subset of sites K up to a certain size, and computing the haplotype lower bound on the sequences $M(K)$. Then, we generate constraints for the ILP requiring that the number of points and line-segments covering K must be at least the computed haplotype bound for $M(K)$, and requiring that the selection of covering points and line-segments be constrained as described above. These additional ideas result in larger lower-bounds at the cost of increasing the size of the ILP and the time needed to solve it. Our experience shows that when CPLEX is used, the ILP formulation can still be solved reasonably fast when enumerating all size-3 (or even 4) subsets of sites. Unfortunately, the free GNU GLPK runs much slower than CPLEX when all size-3 subsets are used for data containing more than 30 sites.

5   Computing Upper Bounds on $T_{\min}(M, t)$: A Practical Algorithm to Derive M Explicitly

In this section we describe our algorithm that produces an explicit sequence of recombination and mutation events to generate an input set of sequences M. This of course gives an upper bound on $T_{\min}(M, t)$.

The overall framework of our upper bound on $T_{\min}(M, t)$ is similar to that of the upper bound described in [26] where only mutations and crossing-over events are allowed. The main distinction lies in that we here use the following more general *derivation cost*: Given a row r in a binary matrix A and a specified maximum tract length t, $c(r|A - r, t)$ is defined as the minimum total number of crossing-over events and gene-conversions with tract length at most t that are needed to derive row r from some other rows in A. In [26], only single-crossovers were allowed in defining the derivation cost, denoted $w(r|A - r)$.

The following procedure, called **Procedure History**, is the key component of our method to compute an upper bound on $T_{\min}(M, t)$.

Step 0. Set $A = M$.
Step 1. Set $W=0$.
Step 2. Repeat Steps 2a and 2b until neither operation is possible.
    Step 2a. Collapse two identical rows of A into one row.
    Step 2b. Remove any column k of A containing less than two 0s or less than two 1s.
Step 3. If A is empty, then stop. Otherwise, remove a row from A, say row r, set $W \leftarrow W + c(r|A - r, t)$, and go to step 2.

The final upper bound $\tau(M, t)$ on $T_{\min}(M, t)$ is defined as the *minimum* final value of W over all possible executions of Procedure History. (Two inequivalent executions have different sequences of row removals in Step 3.)

Of course, if we explicitly explore all possible executions, then the method would only be practical for very small problem instances. Instead, we use branch and bound ideas to find $\tau(M, t)$ without explicitly exploring all possible executions of Procedure History. The details of the branch and bound method are similar to those in [26], and their use results in dramatic speedups allowing practical computation of $\tau(M, t)$ for moderate size data. Some timing details are presented in Section 6. We implemented this method into a program called SHRUB-GC which is available on the web (www.cs.ucdavis.edu/~gusfield).

The fact that $\tau(M, t)$ is an upper bound on $T_{\min}(M, t)$ follows from the observation that every execution of Procedure History, with final cost denoted W^*, specifies backwards in time a series of W^* recombination events (along with one mutation per site) that derive M. These events can be represented in an directed acyclic graph (DAG), and our program can produce this DAG. More

Figure 1 illustrates our upper bound results on datasets generated with $n = 20, s = 30, k = 5000, \lambda = 500$ and $\rho = 2$ or 5. An important thing to note is that, for $f = 0$ (i.e., when no gene-conversions were used in the generation of M), both $\alpha(t)$ and $\beta(t)$ remain close to zero as t changes. That is, when the data were generated without gene-conversions, the allowance of gene-conversions in the bounding methods does not reduce the bounds very much. In effect, the methods cannot "make up" gene-conversions that did not actually occur in the generation of the sequences. Similarly, for fixed f, both $\alpha(t)$ and $\beta(t)$ first increase rapidly as t increases from zero, but increasing t beyond 1500 does not seem to influence them very much. This behavior is consistent with the fact that, for $\lambda = 500$, the probability of the tract length being less than or equal to 1500 is about 95%. These characteristics are a very strong validation of the biological relevancy of our methods.

As f increases, both $\alpha(t)$ and $\beta(t)$ grow, and this growth is more pronounced for larger t. In general $\alpha(t)$ is more sensitive to changes in f and t than is $\beta(t)$. For fixed f and t, both $\alpha(t)$ and $\beta(t)$ tend to increase as the recombination rate ρ increases, with $\alpha(t)$ more so than $\beta(t)$. The general behavior of our lower bound is quite similar to that of our upper bound.

In general, as t, f or ρ increases, the running times of our programs increase. The average running time per dataset of SHRUB-GC ranged from a fraction of a second to a bit over a minute on a 2 GHz pentium PC. HapBound-GC is faster than SHRUB-GC, with the average running time per dataset being less than a second.

6.2   Gene-Conversion Presence (GCP) Test

Based on the simulation results of our methods, one can devise various tests for determining whether gene-conversion was used to generate a given dataset. We here suggest a simple test involving $\gamma(M, t)$: For a given maximum tract length t, we say that $\gamma(M, t) > 0$ indicates the presence of gene-conversion. Percentages of simulated datasets with $\gamma(M, t) > 0$ are summarized in Table 1, for mean tract length $\lambda = 500$. Percentages for $f = 0$ can be regarded as false positive rates, whereas percentages for $f > 0$ can be regarded as sensitivity. Results for three different methods are shown in the table: (U) requiring $\gamma(M, t) > 0$ in the upper bound method only, (L) requiring $\gamma(M, t) > 0$ in the lower bound method only, and (U&L) requiring $\gamma(M, t) > 0$ in both upper and lower bound methods.

The outcome of these tests depends on the value of t used, but our results indicate that increasing t beyond a certain point does not change the percentages by a considerable amount; that is, in our simulations using $\lambda = 500$, the difference between $t = 1$ and $t = 1000$ is much more significant than that between $t = 1000$ and $t = 2000$. In practice, the user is advised to decide on an appropriate t based on what is believed to be the mean tract length for the species being studied.

Results in Table 1 suggest that for small ρ (say, $\rho \leq 2$), false positive rates are low and method U seems to work better than method L or method U&L. For $\rho = 5$, however, both methods U and L lead to somewhat high false positive rates. Since using the combined method U&L reduces the false positive rate, a

Table 1. Percentage of datasets with $\gamma(M,t) > 0$ for $n = 20, s = 30, k = 5000$ and $\lambda = 500$. "U" denotes having $\gamma(M,t) > 0$ in the upper bound method, "L" having $\gamma(M,t) > 0$ in the lower bound method, and "U&L" having $\gamma(M,t) > 0$ in both upper and lower bound methods.

ρ	f	$t=1$			$t=500$			$t=1000$			$t=1500$			$t=2000$		
		U	L	U&L	U	L	U&L	U	L	U&L	U	L	U&L	U	L	U&L
2	0.0	0.8	0.6	0.6	0.8	1.0	0.8	1.2	1.6	1.2	1.6	3.0	1.6	1.8	3.0	1.8
	2.5	20.6	15.0	12.4	32.6	30.2	25.2	36.6	35.6	29.8	37.4	38.4	31.0	38.6	40.4	32.6
	5.0	38.6	31.4	24.8	56.8	55.0	46.8	61.2	61.2	53.4	63.8	63.0	56.0	64.6	65.2	58.0
	7.5	46.8	37.0	29.8	62.6	61.0	52.0	65.8	67.0	57.2	68.2	70.4	60.6	69.6	72.0	62.2
	10.0	57.0	45.4	36.0	74.4	72.6	64.4	78.8	78.2	71.0	79.8	80.4	73.0	81.8	81.2	74.8
5	0.0	4.6	4.6	2.8	8.4	9.4	6.0	10.2	12.6	7.4	11.8	16.2	8.8	12.6	18.4	9.6
	2.5	40.6	37.2	25.0	63.6	64.6	54.4	68.0	71.8	61.2	69.6	72.6	62.8	71.8	73.8	64.6
	5.0	64.4	50.6	42.2	81.4	82.6	74.2	87.4	88.6	81.6	89.8	90.4	85.2	91.0	90.6	86.2
	7.5	72.6	63.2	51.4	92.8	92.4	88.0	96.0	94.4	92.2	96.8	95.4	93.8	97.0	95.4	94.0
	10.0	81.6	69.4	61.8	95.8	93.6	91.4	97.2	96.8	94.8	97.4	97.0	95.2	98.0	97.0	95.8

conservative strategy would be to use that method if $\rho > 2$ or if ρ is unknown. All three methods perform significantly better with increasing f (i.e., high sensitivity can be achieved for $f \geq 5$). Further, we remark that, although not shown here, our GCP test performs better with increasing number of segregating sites, given that all other parameters remain fixed.

6.3   GCP Test on *Arabidopsis thaliana* Data

We applied the above GCP Test to the *Arabidopsis thaliana* data of Plagnol et al.[22] To be conservative, we used method U&L. The data consist of 96 samples broken up into 1338 short fragments, each of length between 500 and 600 bps.

Most fragments contain a significant fraction of missing data, which were handled in [22] as follows. Given a fragment, they first found the set of all pairs of columns containing $00, 01, 10$ and 11. Then, if there were less than or equal to ten missing data restricted to that set, they tried all possible assignments of values to those missing data and declared that the fragment contains a clear gene-conversion event if their test produced an affirmative answer for any assignment. In our approach, instead of assigning values to missing data, we removed certain columns and rows so that the remaining dataset was free of missing data.

Typically $\rho < 1$ for each fragment, as estimated in [22], so the above simulation results imply that our GCP test should have a low false positive rate, provided that the actual evolution of *Arabidopsis thaliana* has been consistent with the model used for simulation. Since we ignored homoplasy events (recurrent or back mutations) in our simulations, we decided to account for their possibility as follows. First, we ran our GCP test using $t = 1$, thus allowing for at most one SNP in the conversion tract. Note that a gene-conversion event in such a case has similar effects as does a homoplasy event. Second, when we performed GCP tests for $t > 1$, we ignored those datasets that had affirmative GCP test results for $t = 1$.

Plagnol et al. [22] identified four fragments as containing clear signals for gene-conversion, with potential tracts being $55, 190, 200$ and 400 bps long. In contrast, 22 fragments passed our test when the maximum tract length was set to 200. (Increasing t beyond 200 did not change our results.) Of these 22 fragments, three coincided with those found in [22]. We believe that the fact that we handled missing data differently is responsible for our not detecting any signal for gene-conversion in the remaining one fragment (whose potential tract length is 400 bps) identified in [22]. All in all, our detection methods are more general than the method used in [22], and we believe that that led us to identify many more fragments than they did. Effectively, the method of Plagnol et al. [22] can only detect fragments with $\tau(M, t) = 1$ and $\gamma(M, t) = 1$. All 19 additional fragments we identified had $\tau(M, t) > 1$ and $\gamma(M, t) \geq 1$.

Acknowledgement

We thank Vincent Plagnol for providing us with the *Arabidopsis thaliana* data and for communicating with us regarding the data. This research is supported by the National Science Foundation grants EIA-0220154 and IIS-0513910.

References

1. V. Bafna and V. Bansal. The number of recombination events in a sample history: conflict graph and lower bounds. *IEEE/ACM Transactions on Computational Biology and Bioinformatics*, 1:78–90, 2004.
2. V. Bafna and V. Bansal. Improved recombination lower bounds for haplotype data. In *Proceedings of RECOMB 2005*, pages 569–584, 2005.
3. C. Carlson, M. Eberle, L. Kruglyak, and D. Nickerson. Mapping complex disease loci in whole-genome association studies. *Nature*, 429:446–452, 2004.
4. A. G. Clark. Finding genes underlying risk of complex disease by linkage disequilibrium mapping. *Curr. Opin. Genet. Dev.*, 13:296–302, 2003.
5. G. Drouin, F. Prat, M. Ell, and G.D. Clarke. Detecting and characterizing gene conversion between multigene family members. *Mol. Bio. Evol.*, 16:1369–1390, 1999.
6. N. El-Mabrouk. Deriving haplotypes through recombination and gene conversion pathways. *J. Bioinformatics and Computational Biology*, 2(2):241–256, 2004.
7. P. Fearnhead, R.M. Harding, J.A. Schneider, S. Myers, and P. Donnelly. Application of coalescent methods to reveal fine scale rate variation and recombination hotspots. *Genetics*, 167:2067–2081, 2004.
8. L. Frisse, R.R. Hudson, A. Bartoszewicz, J.D. Wall, J. Donfack, and A. Di Rienzo. Gene conversion and different population histories may explain the contrast between polymorphism and linkage disequilibrium levels. *Am. J. Hum. Genet.*, 69:831–843, 2001.
9. D. Gusfield. Optimal, efficient reconstruction of Root-Unknown phylogenetic networks with constrained and structured recombination. *JCSS*, 70:381–398, 2005.
10. D. Gusfield, S. Eddhu, and C. Langley. The fine structure of galls in phylogenetic networks. *INFORMS J. on Computing, special issue on Computational Biology*, 16:459–469, 2004.

11. D. Gusfield, S. Eddhu, and C. Langley. Optimal, efficient reconstruction of phylogenetic networks with constrained recombination. *J. Bioinformatics and Computational Biology*, 2(1):173–213, 2004.
12. D. Gusfield, D. Hickerson, and S. Eddhu. An efficiently-computed lower bound on the number of recombinations in phylogenetic networks: Theory and empirical study. To appear in *Discrete Applied Math, special issue on Computational Biology*.
13. J. Hein. Reconstructing evolution of sequences subject to recombination using parsimony. *Math. Biosci*, 98:185–200, 1990.
14. J. Hein. A heuristic method to reconstruct the history of sequences subject to recombination. *J. Mol. Evol.*, 36:396–405, 1993.
15. J. Hein, M. Schierup, and C. Wiuf. *Gene Genealogies, Variation and Evolution: A primer in coalescent theory*. Oxford University Press, UK, 2004.
16. R. Hudson. Generating samples under the Wright-Fisher neutral model of genetic variation. *Bioinformatics*, 18(2):337–338, 2002.
17. R. Hudson and N. Kaplan. Statistical properties of the number of recombination events in the history of a sample of DNA sequences. *Genetics*, 111:147–164, 1985.
18. A. J. Jeffreys and C. A. May. Intense and highly localized gene conversion activity in human meiotic crossover hot spots. *Nature Genetics*, 36:151–156, 2004.
19. M. Lajoie and N. El-Mabrouk. Recovering haplotype structure through recombination and gene conversion. *Bioinformatics*, 21(Suppl 2):ii173–ii179, 2005.
20. S. R. Myers and R. C. Griffiths. Bounds on the minimum number of recombination events in a sample history. *Genetics*, 163:375–394, 2003.
21. B. Padhukasahasram, P. Marjoram, and M. Nordborg. Estimating the rate of gene conversion on human chromosome 21. *Am. J. Hum. Genet.*, 75:386–97, 2004.
22. V. Plagnol, B. Padhukasahasram, J. D. Wall, P. Marjoram, and M. Nordborg. Relative influences of crossing-over and gene conversion on the pattern of linkage disequilibrium in Arabidopsis thaliana. *Genetics*, in press. Ahead of Print: 10.1534/genetics.104.040311.
23. S. Sawyer. Statistical tests for detecting gene conversion. *Mol. Biol. Evol.*, 6:526–538, 1989.
24. Y.S. Song and J. Hein. Parsimonious reconstruction of sequence evolution and haplotype blocks: Finding the minimum number of recombination events. In *Proc. of 2003 Workshop on Algorithms in Bioinformatics*, pages 287–302, 2003.
25. Y.S. Song and J. Hein. On the minimum number of recombination events in the evolutionary history of DNA sequences. *J. Math. Biol.*, 48:160–186, 2004.
26. Y.S. Song, Y. Wu, and D. Gusfield. Efficient computation of close lower and upper bounds on the minimum number of needed recombinations in the evolution of biological sequences. *Proc. of ISMB 2005, Bioinformatics*, 21:i413–i422, 2005.
27. J. C. Stephens. Statistical methods of DNA sequence analysis: Detection of intragenic recombination or gene conversion. *Mol. Bio. Evol.*, 2:539–556, 1985.
28. The International HapMap Consortium. A haplotype map of the human genome. *Nature*, 437:1299–1320, 2005.
29. J. D. Wall. Close look at gene conversion hot spots. *Nat. Genet.*, 36:114–115, 2004.
30. T. Wiehe, J. Mountain, P. Parham, and M. Slatkin. Distinguishing recombination and intragenic gene conversion by linkage disequilibrium patterns. *Genet. Res., Camb.*, 75:61–73, 2000.

Inferring Common Origins from mtDNA

Ajay K. Royyuru[1], Gabriela Alexe[1], Daniel Platt[1], Ravi Vijaya-Satya[2],
Laxmi Parida[1], Saharon Rosset[1], and Gyan Bhanot[1,3]

[1] Computational Biology Center, IBM Thomas J. Watson Research Center, Yorktown
Heights, NY 10598
[2] Department of Computer Science, University of Central Florida,
Orlando, FL 32816
[3] Center for Systems Biology, Institute for Advanced Study,
Princeton, NJ 08540
ajayr@us.ibm.com, galexe@us.ibm.com, watplatt@us.ibm.com,
rvijaya@cs.ucf.edu, parida@us.ibm.com,
srosset@us.ibm.com, gyan@us.ibm.com

Abstract. The history of human migratory events can be inferred from
observed variations in DNA sequences. Such studies on non-recombinant
mtDNA and Y-chromosome show that present day humans outside Africa
originated from one or more migrations of small groups of individuals between 30K-70K YBP. Coalescence theory reveals that, any collection of
non-recombinant DNA sequences can be traced back to a common ancestor. Mutations fixed by genetic drift act as markers on the timeline
from the common ancestor to the present and can be used to infer migration and founder events that occurred in ancestral populations. However,
most mutations seen in the data today are relatively recent and do not
carry useful information about deep ancestry. The only ones that can
be used reliably are those that can be shown to robustly distinguish
large clusters of individuals and thus qualify as true representatives of
population events in the past.

In this talk, we present results from the analysis of 1737 complete
mtDNA sequences from public databases to infer such a robust set of
mutations that reveal the haplogroup phylogeny. Using principal component analysis we identify the samples in L, M and N clades and with
unsupervised consensus ensemble clustering we infer the substructure in
these clades. Traditional methods are inadequate to handle data of this
size and complexity.

The substructure is inferred using a new algorithm that mitigates
the usual problems of sample size bias within haplogroups as well as
the sampling bias across haplogroups. First, we cluster the data in each
of the M, N, L clades separately into $k = 2, 3, 4, \ldots k_{max}$ groups using
an agreement matrix derived from multiple clustering techniques and
bootstrap sampling. Repeated training/test splits of the samples identify
robust clusters and patterns of SNPs which can assign haplogroup labels
with a reliability greater than 90%. Even though the clustering at each
k is done independently, the clusters split in a way that suggests that
the data is revealing population events; a cluster at level k has $k - 2$
clusters which are identical with those at level $k - 1$ plus two more that

A. Apostolico et al. (Eds.): RECOMB 2006, LNBI 3909, pp. 246–247, 2006.
© Springer-Verlag Berlin Heidelberg 2006

obtain from a split of one of the clusters at level $k - 1$. The clustering is repeated with equal number of samples from the first level clusters. The sequence in which the clusters now split defines a binary network which reveals population events unbiased by sample size. We root the network using an out-group and, assuming a molecular clock, identify an internal node in the bifurcation process which is equidistant from the leaves. This rooting removes the bias across haplogroups which would otherwise influence the order in which the clusters emerge.

Our analysis shows that the African clades L0/L1, L2 and L3 have the greatest heterogeneity of SNPs, in agreement with their ancient ancestry. It also suggests that the M, N clades originated from a common ancestor of L3 in two separate migrations. The first migration gave rise to the M haplogroup, whose descendents currently populate South-East Asia and Australia. The second migration resulted in the N haplogroup, accounting for the current populations in China, Japan, Europe, Central Asia and North and South America. We reveal and robustly label many branches of the mtDNA tree, improving current results significantly. We find that for our choice of robust SNPs, the genetic distances between the NA and NRB haplogroups is smaller compared to that between B and J/T/H/V/U. The detailed N migratory sub-tree is rooted so that the T, J and U haplogroups are on one side of the root and the F, V/H, I, X, R5, B, N9, A and W are on the other. We also find a detailed structure for the M tree consistent with prior literature and we infer additional branches for the MD haplogroup. Finally we provide detailed SNP patterns for each haplogroup identified by our clustering. Our patterns can be used to infer a haplogroup assignment with reliability greater than 90%.

Efficient Enumeration of Phylogenetically Informative Substrings

Stanislav Angelov[1], Boulos Harb[1], Sampath Kannan[1],
Sanjeev Khanna[1], and Junhyong Kim[2]

[1] Department of Computer and Information Sciences, University of Pennsylvania
{angelov, boulos, kannan, sanjeev}@cis.upenn.edu
[2] Department of Biology, University of Pennsylvania
junhyong@sas.upenn.edu

Abstract. We study the problem of enumerating substrings that are common amongst genomes that share evolutionary descent. For example, one might want to enumerate all identical (therefore conserved) substrings that are shared between all mammals and not found in non-mammals. Such collection of substrings may be used to identify conserved subsequences or to construct sets of identifying substrings for branches of a phylogenetic tree. For two disjoint sets of genomes on a phylogenetic tree, a substring is called a *discriminating substring* or a *tag* if it is found in all of the genomes of one set and none of the genomes of the other set. Given a phylogeny for a set of m species, each with a genome of length at most n, we develop a suffix-tree based algorithm to find all tags in $O(nm \log^2 m)$ time. We also develop a sublinear space algorithm (at the expense of running time) that is more suited for very large data sets. We next consider a stochastic model of evolution to understand how tags arise. We show that in this setting, a simple process of tag generation essentially captures all possible ways of generating tags. We use this insight to develop a faster tag discovery algorithm with a small chance of error. However, tags are not guaranteed to exist in a given data set. We thus generalize the notion of a tag from a single substring to a set of substrings whereby each species in one set contains a large fraction of the substrings while each species in the other set contains only a small fraction of the substrings. We study the complexity of this problem and give a simple linear programming based approach for finding approximate generalized tag sets. Finally, we use our tag enumeration algorithm to analyze a phylogeny containing 57 whole microbial genomes. We find tags for all nodes in the phylogeny except the root for which we find generalized tag sets.

1 Introduction

Genomes are related to each other by evolutionary descent. Thus, two genomes share sequence identities in regions that have not experienced mutational changes; i.e., the genomes share common subsequences. While common subsequences can also arise by chance, sufficiently long common sequences are homologous (identity by descent) with high probability. The pattern of common

A. Apostolico et al. (Eds.): RECOMB 2006, LNBI 3909, pp. 248–264, 2006.
© Springer-Verlag Berlin Heidelberg 2006

subsequences in a set of genomes can be informative for reconstructing the evolutionary history of the genomes. Furthermore, since stabilizing selection for important functions can suppress fixed mutational differences between genomes, long common subsequences can be indicative of important biological function. This hypothesis has been extensively used in comparative genomics to scan genomes for novel putatively functional sequences [1, 2, 3].

Typical approaches for obtaining such subsequences involve extensive pairwise comparison of sequences using BLAST-like approaches along with additional modifications. Alternatively, one can use a dictionary-based approach, scanning the genomes for presence of common k-mers (which is also the base approach for BLAST heuristics). The presence and absence of such k-mers can be also used to identify unlabeled genomes or reconstruct the evolutionary history. Detection of particular k-mers can be experimentally implemented using oligonucleotide microarrays leading to a laboratory genome identification device.

For detecting functionally important common subsequences or for identifying unlabeled genomes, it is important that k is sufficiently large to ensure homologous presence with high probability. However, the required address space increases exponentially with k. Furthermore, not all patterns of k-mer presence are informative for detecting common subsequences. If the phylogenetic relationship of the genomes is known, then the phylogeny can become a guide to delineating the most informative common subsequences. For any given branch of the tree, there will be substrings common to all genomes on one side of the branch and not present on the other side. For example, there will be a collection of common subsequences unique to the mammalian lineage of the Vertebrates. Such substrings will be parts of larger subsequences that are conserved in the mammalian genomes; and, such substrings will be indicators of mammalian genomes. If we had an enumeration of all such informative common substrings, we can apply the information to efficiently detect conserved subsequences or to an experimental detection protocol to identify unlabeled genomes. In this paper, we describe a procedure to efficiently enumerate all such informative common substrings (which we call "sequence tags") with respect to a guide phylogeny. In particular, here we explore the application to the construction of an identification oligonucleotide detection array, which can be applied to high-throughput genome identification and reconstructing the tree of life.

More specifically, given complete genomes for a set of organisms S and the binary phylogenetic tree that describes their evolution, we would like to be able to detect all *discriminating* oligo tags. We will say that a substring t is discriminating at some node u of the phylogeny if all genomes under one branch of u contain t while none of the genomes under any other branch of u contain t. Thus, a set of discriminating tags, or simply tags, for all the nodes of a phylogeny allows us to place a genome *that is not necessarily sequenced* in the phylogeny by a series of binary decisions starting from the root. This procedure can be implemented experimentally as a microarray hybridization assay, enabling a rapid determination of the position of an unidentified organism in a predetermined phylogeny or classification. It is noteworthy that heuristic construction of short sequence tags has been used before for identification and classification (reviewed in [4]), but no algorithm has been presented for data driven tag design.

Our Results

- We first present an efficient algorithm for enumerating substrings common to the extant sequences under every node of a given phylogeny. The algorithm runs in linear time and space.
- We use our common-substrings algorithm to develop a near-linear time algorithm for generating the discriminating substrings for every node of the phylogeny. Specifically, if S is the set of given genomes, the discriminating-substrings algorithm runs in time $O(n|S|\log^2|S|)$ where n is the length of each genome. This improves an earlier bound of $O(n|S|^2)$ given in [5].
- Even though all the above algorithms require linear space, due to physical memory limitations they may be impractical for analyzing very large genomic data sets. We therefore give a *sublinear* space algorithm for finding all discriminating substrings (or all common substrings). The space complexity of the algorithm is $O(n/k)$ and its running time is $O(kn|S|^2)$. The tradeoff between the time and space complexities is controlled by the parameter k.
- We demonstrate the existence of tags in the prokaryotes data set of [6]. The genomes represented in the data set span one of the three recognized domains of life. We find that either left or right tags exist for all nodes of the phylogeny except the root.
- Motivated by our results on the microbial genomes, we study the potential application to arbitrary scale phylogenetic problems using a stochastic model of molecular evolution. We assume that the given species set S is generated according to this model. We first analyze the case where the phylogeny is a balanced binary tree with a uniform probability of change on all its edges. *We show that in this setting, if t is a tag that discriminates a set S' of species from set $\bar{S}'$, then w.h.p. that increases with the number of species— probability $\geq 1/(1+O(\ln(n)/|S|))$, t is present in the common ancestor of S' (occurs early in the evolution) and is absent from the common ancestor of $\bar{S}'$ (is absent from the beginning).* Our study of the stochastic model allows us to design faster algorithms for tag generation with small error. Even when we allow arbitrary binary trees, we show that this probability is $\geq 1/2$.
- As observed in our experiments and subsequent analysis, tags are not guaranteed to exist in a given data set. We consider a relaxed notion of tags to deal with such a scenario. Given a partition $(S', \bar{S}')$ of species, we say that a set T of tags is an (α, β)-*generalized tag set* for some $\alpha > \beta$, if every species in S' contains at least an α fraction of the strings in T and every species in $\bar{S}'$ contains at most a β fraction of them. Clearly, such a tag set can still be used to decide whether a genome belongs to S' or to $\bar{S}'$. We show that the problem of computing generalized tag sets may be viewed as a set cover problem with certain "coverage" constraints. We also show that this generalization of tags is both NP-hard and LOGSNP–hard when $(\alpha, \beta) = (\frac{2}{3}, \frac{1}{3})$. However, if $|T| = \Omega(\log m)$, a simple linear programming based approach

can be used to compute approximate generalized tag sets. As an example, we find $(\frac{2}{3}, \frac{1}{3})$-generalized tag sets for the root of the prokaryotes phylogeny (where we did not find tags).

Due to lack of space, proofs are omitted from this version of the paper.

2 Preliminaries

Formally, the problems we consider are the following:

Hierarchical Common Substring Problem (HCS)
Input: A set of strings $S = \{s_1, \cdots, s_m\}$ drawn from a bounded-size alphabet with $|s_i| \leq n$ for all i; and an m-leaf binary tree P whose leaves are labeled $s_1, \cdots, s_m$ sequentially from left to right.

Goal: For all $u \in P$, find the set of *right-maximal* substrings common to the strings in S_u, where S_u is the set of all the input strings in the subtree rooted at u. A substring t common to a set of strings is right-maximal if for any non-empty string α, $t\alpha$ is no longer a common substring. (Right-maximal substrings efficiently encode all common substrings.)

Discriminating Substring. A substring t is said to be a *discriminating substring* or a *tag* for a node u in a phylogeny if all strings under one branch of u contain t while none of the strings under the other branch contain t. The input to the discriminating substring problem is the same as that for the first problem:

Discriminating Substring Problem
Input: A set of strings S and a binary tree P.
Goal: Find sets D_u for all nodes $u \in P$, such that D_u contains all *discriminating* substrings for u.

Suffix trees, introduced in [7], play a central role in our algorithms. A suffix tree T of a string s is a trie-like data structure that represents all suffixes of s. We adopt the following definitions from [8]. The *path-label* of a node v in T is the string formed by following the path from the root to v. The path-labels of the $|s|$ leaves of T spell out the suffixes of s, and the path-labels of internal nodes spell out substrings of s. Furthermore, the suffix tree ensures that there is a unique path from the root, not necessarily ending at an internal node, that represents each substring of s. We also say that the path-label of node v is *the string corresponding* to v in the tree.

The algorithms we present are based on *generalized suffix trees* [8, p. 116]. A generalized suffix tree extends the idea of a suffix tree for a string to a suffix tree for a set of strings. Conceptually, it can be built by appending a unique terminating marker to each string, then concatenating all the strings and building a suffix tree for the resultant string. The tree is post-processed so that each path-label of a leaf in the tree spells a suffix of one of the strings in the set and, hence, is terminated with that string's unique marker.

We will also need the notion of a generalized tag set.

(α, β)**-Generalized Tag Set.** Given a partition $(S', \bar{S}')$ of species, we say that a set T of tags is an (α, β)-*generalized tag set* for some $\alpha > \beta$, if every species in S' contains at least an α fraction of the strings in T and every species in $\bar{S}'$ contains at most a β fraction of them.

3   The Hierarchical Common Substring Problem

Long common substrings among genomes can be indicative of important biological functions. In this section we give a linear time/linear space algorithm that enumerates substrings common to all sequences under every node of a given binary phylogeny. This is a significant improvement over naively running the linear time common substrings algorithm of [9] for every node of the phylogeny. By carefully merging sets of common substrings along the nodes of the phylogeny and eliminating redundancies we are able to achieve the desired running time. Note that for a given node, there may be $O(n^2)$ substrings common to its child sequences. The algorithm will therefore list all *right-maximal* common substrings. Such substrings efficiently encode all common substrings. The formal problem description is given in Section 2. We start with two definitions.

Definition 1. *Let C be a collection of nodes of a suffix tree. A node $p \in C$ is said to be* redundant *if its path-label is empty or it is the prefix of some other node in C.*

Definition 2. *For a tree T, let $o(v)$ be the postorder index of node $v \in T$.*

Algorithm. We preprocess the input as follows: (a) Build a generalized suffix tree T for the strings in S by using two copies of every $s_i \in S$, each with a unique terminating marker: $s_1 \#_{1_a} s_1 \#_{1_b} \cdots s_m \#_{m_a} s_m \#_{m_b}$; (b) Process T so that lowest common ancestor (lca) queries can be answered in constant time; and, (c) Label the nodes of T with their postorder index.

1. For each node $u \in P$, build a list C_u of nodes in T with the properties:
    (P0) A substring t is common to the strings in S_u iff t is a prefix of the path-label of a node in C_u.
    (P1) $|C_u| \leq n$.
    (P2) The elements of C_u are sorted based on their postorder index.
    (P3) $\nexists p \in C_u : p$ is redundant.
    The lists are built bottom-up starting with the leaves of P:
    (a) For a leaf $u \in P$, since $|S_u| = 1$, compute C_u by removing the redundant suffixes of $s \in S_u$.
    (b) For each internal node $u \in P$, let $l(u)$ and $r(u)$ be the left and right children of u respectively. We compute $C_u = C_{l(u)} \sqcap C_{r(u)}$, where $A \sqcap B = \{p = lca(a, b) : a \in A, \ b \in B, \ p \text{ not redundant}\}$.
2. For each $u \in P$, output C_u.

Analysis. The time and space complexities of the preprocessing phase is $O(nm)$ [10, 11, 12, 13]. We now analyze Step 1. The lists C_u for the leaves of P are first simultaneously built in Step 1(a) by performing a postorder walk on T. Assuming $S_u = \{s_i\}$, node $p \neq root(T)$ is appended to list C_u if it has an outgoing edge labeled "$\#_{i_a}$". Suffix tree properties guarantee that C_u will consist of all suffixes of s_i. Since an input string may only have n suffixes, the lists constructed in this fashion will posses P1 and P2. Property P3 is obtained by scanning each list from left to right and removing redundant nodes. Observe that if p is an ancestor of q, then p is an ancestor of all q' satisfying $o(q) \leq o(q') \leq o(p)$. Hence, we can remove redundancies from each C_u in time linear in $|C_u| = O(n)$ by examining only adjacent entries in the list.

Lemma 1. C_u *possesses properties P0, P1, P2 and P3 for each leaf* $u \in P$.

Proof. Let $S_u = \{s_i\}$ where u is a leaf of P. Since every substring of s_i is a prefix of some suffix of s_i, and we removed only redundant suffixes, C_u possesses P0.

We now show how to compute the lists C_u for the internal nodes of P. We first show that the operation $\sqcap$ as defined in Step 1(b) preserves P0.

Lemma 2. *Let* $u \in P$ *be the parent of* $l(u)$ *and* $r(u)$. *If* $C_{l(u)}$ *and* $C_{r(u)}$ *possess P0, then* $C_u = C_{l(u)} \sqcap C_{r(u)}$ *also possesses P0.*

Proof. The string t is a common substring to the strings in S_u iff t is common to the strings in $S_{l(u)}$ and $S_{r(u)}$. This is equivalent to the existence of $p \in C_{l(u)}$ and $q \in C_{r(u)}$ such that t is a prefix of the path-labels of both p and q. That is, t is a prefix of the path-label of $lca(p, q)$ as required.

For each internal node $u \in P$, we construct a merged list Y_u containing all the elements of $C_{l(u)}$ and $C_{r(u)}$ with repetitions. Let $src(a)$ be the source list of node $a \in Y_u$. When computing $C_{l(u)} \sqcap C_{r(u)}$, the following lemma allows us to only consider the lca of consecutive nodes in Y_u whose sources are different.

Lemma 3. *If* $a, a', b, b' \in T$ *satisfy* $o(a') \leq o(a) \leq o(b) \leq o(b')$, *then* $lca(a', b')$ *is an ancestor of* $lca(a, b)$.

Proof. By postorder properties, $lca(a', b')$ is an ancestor of both a and b so it is an ancestor of $lca(a, b)$.

Let $a, a', b, b' \in Y_u$, where $o(a') \leq o(a) \leq o(b) \leq o(b')$. If $src(a) \neq src(b)$ and $src(a') \neq src(b')$, then, since $lca(a', b')$ is an ancestor of $lca(a, b)$, the former is redundant. This suggests the following procedure for computing C_u's. Suppose at step i, $a = Y_u[i]$ and $b = Y_u[i+1]$: If $src(a) = src(b)$, proceed to next step; else, let $p' = lca(a, b)$. If $p' = root(T)$ then we discard it and proceed to the next step. In order to avoid redundancies before adding p' to C_u, we compute $lca(p, p')$ where p is the last node appended to C_u. If $lca(p, p') = p'$ we discard p', and if $lca(p, p') = p$ we replace p with p'.

Each step of the above procedure requires constant time. Hence, since $|Y_u| \leq 2n$, the procedure runs in $O(n)$ time. The next lemma shows the correctness of the procedure, and Theorem 1 follows.

Lemma 4. *For an internal node $u \in P$, the above procedure correctly computes $C_u = C_{l(u)} \sqcap C_{r(u)}$. Furthermore, the list C_u is sorted and $|C_u| \leq n$.*

Theorem 1. *The Hierarchical Common Substring Problem can be solved in $O(nm)$ time and $O(nm)$ space.*

4 The Discriminating Substring Problem

In this section we will use the phylogeny for extracting the most informative substrings common to the child sequences of every node in the phylogeny. Suppose we know that a sequence belongs to a certain subtree of the phylogeny that is rooted at u. We wish to know whether the sequence belongs to the left or right branch of u. If we knew the substrings common to the left subtree of u but not present in the right subtree (or vice versa), then we would know to which of the two subtrees the sequence belongs. Hence for a given node, the substrings that are common to its children but not present in the children of its sibling are more informative than only the common ones. Below we show two methods with certain tradeoffs for finding such discriminating substrings or tags, for every node in the phylogeny. It is easy to see that the set of tags obtained by selecting a tag from each node on a root-leaf path *uniquely distinguishes the sequence at the leaf from all other sequences in the phylogeny.*

4.1   A Near-Linear Time Algorithm

The HCS algorithm finds all the common substrings for each node in P. The common substrings are encoded as the prefixes of the path-labels of the nodes in C_u for each $u \in P$. However, these substrings may not be discriminating. That is, the prefix of the path-label of a node $p \in C_{l(u)}$ (symmetrically $C_{r(u)}$) may also be a substring of one of the strings in $S_{r(u)}$ ($S_{l(u)}$). The following algorithm finds for each node in $C_{l(u)}$ its longest path-label prefix that is not discriminating.

Algorithm. Let C_u for all $u \in P$ be the output of the HCS algorithm and let T be the computed suffix tree.

1. For each $u \in P$, build a list A_u of nodes in T with the following properties:
   - (P4) A string t is a substring of a string in S_u iff t is a prefix of the path-label of some node in A_u.
   - (P5) The elements of A_u are sorted based on their postorder index.
     - (a) For the leaves of P, $A_u = C_u$.
     - (b) For each internal node $u \in P$, compute, $A_u = A_{l(u)} \cup A_{r(u)}$ in a bottom up fashion.
2. For each internal node $u \in P$, compute the set of discriminating substrings encoded with D_u, where,

$$D_u = \{(p, w) : p \in C_{l(u)}, \ o(p) < o(w), \ w = lca(p, \arg\min_{q \in A_{r(u)}}[o(lca(p, q))]) \}.$$

In the above expression for D_u, w is the node in the suffix tree whose path-label is the longest proper prefix of the path-label of p that is present in some string in the right subtree of u. For all $q \in A_{r(u)}$, $\arg\min_{q \in A_{r(u)}}[o(lca(p,q))]$ finds the q that has the deepest lowest common ancestor with p, i.e. the q whose path-label shares the longest prefix with that of p. The condition $o(p) < o(w)$ guarantees that the least common ancestor found is a proper ancestor to p.

Analysis. We first show how each D_u encodes the discriminating substrings for $u \in P$.

Lemma 5. *A string t is discriminating for an internal node $u \in P$ iff $\exists (p,w) \in D_u$ such that the path-label of w is a proper prefix of t and t is a prefix of the path-label of p.*

Proof. (if) Let $(p,w) \in D_u$, and suppose t is a string s.t. the path-label of w is a proper prefix of t and t is a prefix of the path-label of p. By P0, t is a common substring of the strings in $S_{l(u)}$. Assume for contradiction that t is a substring of some string in $S_{r(u)}$. Then, by P4, $\exists q \in A_{r(u)}$ s.t. t is a prefix of the path-label of q. But then, since w is a proper prefix of t, it is a proper prefix of the path-labels of both p and q. Hence, $o(w) > o(lca(p,q))$; a contradiction.

(only-if) Suppose t is a discriminating string for u. Then, $\exists p \in C_{l(u)}$ s.t. t is a prefix of the path-label of p, and, by P4, $\nexists q \in A_r(u)$ s.t. t is a prefix of the path-label of q. Hence, the path-label of $w = lca(p, \arg\min_{q \in A_{r(u)}}[o(lca(p,q))])$ is a proper prefix of t.

The following corollary will allow us to efficiently compute w as defined in D_u for a given $p \in C_{l(u)}$.

Corollary 1. *Given $p \in C_{l(u)}$. Let $q', q'' \in A_{r(u)}$ be such that*

$$q' = \operatorname*{arg\,max}_{q \in A_{r(u)}:o(q) \leq o(p)} [o(q)] \,, \qquad q'' = \operatorname*{arg\,min}_{q \in A_{r(u)}:o(q) > o(p)} [o(q)].$$

If q' and q'' exist and $lca(q',p) \neq p$, then $w = \operatorname*{arg\,min}_{q \in \{lca(q',p),lca(q'',p)\}} [o(q)].$

It remains to show how to compute the lists D_u for the internal nodes of P. The computation is bounded by the union operations needed to construct the A_u lists in Step 2(b). Note that for a leaf $u \in P$, since $|S_u| = 1$ and C_u is sorted, $A_u = C_u$ trivially possesses both P4 and P5. Furthermore, for an internal node $u \in P$, the union operation maintains P4. Now merging two sorted lists of sizes N and M, with $M \leq N$, requires at least $\lceil \log \binom{N+M}{N} \rceil = \Theta(M \log \frac{N}{M})$ comparisons to distinguish among the $\binom{N+M}{N}$ possible placements of the elements of the larger list in the output. We can use the results of [14], for example, to match this lower bound. The analysis assumes that the A_u lists are represented as linked-level 2-3 trees [14]. Conversion of these lists for the leaves is direct since they are sorted.

Lemma 6. *The lists A_u, for all $u \in P$, can be computed in $O(nm \log^2 m)$ time.*

Now, we can compute D_u for an internal node $u \in P$ by finding the position of each $p \in C_{l(u)}$ in the sorted $A_{r(u)}$, determining its immediate neighbors q' and q'', and computing w as in Corollary 1. If we consider the elements of $C_{l(u)}$ in their sorted order, then by [14], and since $|A_{r(u)}| \leq nm$, finding the positions of all the elements of $C_{l(u)}$ in $A_{r(u)}$ takes $O(|C_{l(u)}| \log(nm/|C_{l(u)}|))$ time. Moreover, finding the neighbors of each one of these elements takes constant time. This leads to the following lemma.

Lemma 7. *The lists D_u, for all internal nodes $u \in P$, can be computed in $O(nm \log m)$ time.*

Note that we can *simultaneously* compute the lists C_u, A_u, and D_u, for each internal node $u \in P$, in a bottom-up fashion discarding $A_{l(u)}$ and $A_{r(u)}$ at the end of the computation for each u. Hence, the total size of the A_u lists we store at any point is no more than $\sum_{\text{leaf } v} A_v = O(nm)$. Finally, since, by definition, $|D_u| \leq |C_u|$, the space required to store C_u and D_u for all $u \in P$ is $O(nm)$.

Theorem 2. *The Discriminating Substring Problem can be solved in $O(nm \log^2 m)$ time and $O(nm)$ space.*

4.2 A Sublinear Space Algorithm

The above algorithms are optimal or near optimal in terms of their running times and they require only linear space. For very large genomes, even linear space might not fit in primary memory. It is important to further reduce the algorithms' space requirements for such situations to avoid expensive access to secondary storage. Intuitively, it seems that we should be able to run the algorithms on chunks of the data at a time in order to reduce the space complexity. Below we describe such a *sublinear* space algorithm, with a time-space tradeoff, for finding all discriminating substrings (or all common substrings). The precise tradeoff is stated in Theorem 3.

For a node $u \in P$, we can find the set of discriminating substrings D_u by using the matching statistics algorithm introduced in [15]. Given strings s and s', the algorithm computes the length $m(s, j, s')$ of the longest substring of s starting at position j in s and matching some substring of s'. This is done by first constructing the suffix tree for s', and then walking the tree using s. The algorithm requires $O(|s'|)$ time and space for the construction of the suffix tree, and $O(|s|)$ additional time and space to compute and store $m(s, j, s')$ for all j.

Let S_u be the union of two disjoint sets L_u and $R_u = S_u \backslash L_u$ where L_u and R_u are the sets of strings under the two branches of $u \in P$. A substring starting at position j of $s \in L_u$ is discriminating for u iff

$$\min_{s_i \in L_u} m(s, j, s_i) - \max_{s_i \in R_u} m(s, j, s_i) > 0 \ . \tag{1}$$

That is, there exists a substring of s starting at j that is common to all strings in L_u and is sufficiently long so that it does not occur in any string in R_u. If a position j satisfies (1), then a substring t starting at j such that $\max_{s_i \in R_u} m(s, j, s_i) < |t| \leq \min_{s_i \in L_u} m(s, j, s_i)$ is discriminating. Clearly, all tags are substrings of s and thus the outlined procedure computes D_u. The running time of the algorithm is $\sum_{s_i \in S_u} O(|s| + |s_i|) = O(nm)$ and requires $O(\max_{s_i \in S_u} |s_i|) = O(n)$ space. Computing the set of tags for all nodes of P with height h requires $O(nmh)$ time and $O(n)$ space matching the running time of [5]. By limiting the maximum allowed length of a tag, we can obtain a tradeoff between the running time and memory required by the algorithm as stated in the following theorem.

Theorem 3. *The Discriminating Substring Problem for tags of length $O(n/k)$, for some threshold $k \leq n$, can be solved in time $O(knmh)$ and $O(n/k)$ space, where h is the height of P.*

5   Experimental Results

The existence of tags in small (relative to the full genomes) homologous data sets was demonstrated on the CFTR data set [16] and the RDP-II data set [17] in [5]. However, we are interested in finding tags in whole genomes. Further, the existence of tags in subsequences of a given set of strings does not necessarily imply the existence of tags in the strings. Here the existence of tags in data sets containing whole genomes is confirmed on the prokaryotes phylogeny obtained from [6][1]. The genomes represented in the data set span a broad evolutionary distance, at the level of one of the three recognized domains of life. But these genomes are also some of the smallest (1000-fold smaller than the human genome), allowing less sampling space for tags. Thus, they represent cases on the hard extremes of potential applications. We find left and right tags for all nodes of the phylogeny except for the root and the lowest common ancestor of Cac (Clostridium acetobutylicum) and jHp (Helicobacter pylori), where for the latter node only left tags are found. Relaxing the definition of a tag set as in Section 2, we show two $(\frac{2}{3}, \frac{1}{3})$-generalized tag sets for the root as examples.

The prokaryotes phylogeny consists of 57 genomes where the average genome length is roughly 2.75 Mbp and the total length is about 157 Mbp. There are 11 sequences in the root's left subtree (Archaea) and 43 sequences in its right subtree (Bacteria). We implemented the sublinear space algorithm described in Section 4.2 to find all tags for every node in the tree if they exist. Figure 1(a) shows both left and right tags for the lowest common ancestor of Ape (Aeropyrum pernix) and Pho (Pyrococcus horikoshii), and Fig. 1(b) shows the tags for lowest common ancestor of Nos (Nostoc sp. PCC 7120) and Cac (Clostridium acetobutylicum). As mentioned earlier, we did not find tags for the root of the phylogeny. Hence we generated two $(\frac{2}{3}, \frac{1}{3})$-generalized tag sets for the root. Table 1 displays those two sets. We also enumerated the common substrings for this phylogeny as shown in Fig. 1. As expected, longer common substrings are

---

[1] Our experimental results can be found at `http://www.cis.upenn.edu/~angelov/phylogeny`

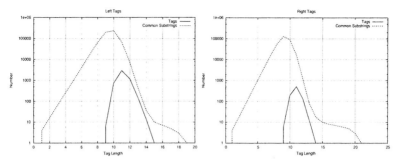

(a) Tags and common substrings for the lowest common ancestor of Ape (Aeropyrum pernix) and Pho (Pyrococcus horikoshii).

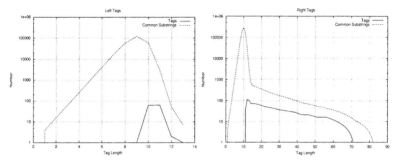

(b) Tags and common substrings for the lowest common ancestor of Nos (Nostoc sp. PCC 7120) and Cac (Clostridium acetobutylicum).

Fig. 1. Length (log-scale) distribution of tags and common substrings for two nodes of the prokaryotes phylogeny of [6]. The left (right) panel displays discriminating tags present in the left (right) subtree of the corresponding node.

Table 1. Left and Right $(\frac{2}{3}, \frac{1}{3})$-generalized tag sets for the root of the prokaryotes tree shown in [6]. Tags in the left (resp. right) tag set have length 14 (resp. 12). *Left Tag Set*: Nine genomes in the left clade contain all 3 left tags and 2 genomes contain 2 tags, while 3 genomes in the right clade contain 1 tag from the set and the remaining 43 contain no left tags. *Right Tag Set*: 21 genomes in the right clade contain all right tags and 25 genomes contain 2 of the tags, while 5 genomes in the left clade contain 1 tag and the remaining 6 genomes contain no tags.

Left Tag Set	Right Tag Set
CCGGGATTTGAACC	CCAACTGAGCTA
GTTCAAATCCCGGC	GTACGAGAGGAC
GGGATTTGAACCCG	TGCTTCTAAGCC

also discriminating tags; i.e., the longer the shared substrings, the more likely they are shared by evolutionary descent (what we call *type–I* tags in the analysis below). The experimental data suggests that, at least for this range of diversity, our approach will be successful at recovering informative substrings.

6   Discriminating Tags Under a Stochastic Model of Evolution

Motivated by our results on the microbial genomes, we next study the potential application to arbitrary scale phylogenetic problems using a simplified assumption of molecular evolution. We analyze statistical properties of tags under this model and we make the first steps toward generalizing the capability of tags to placing new sequences in the given phylogeny. We show that there is a *primary mechanism* for generating tags which suggests that tags are indicative of shared evolutionary history.

We use the *Jukes-Cantor* model [18, pp. 21-132] for our analysis. In this model each position in the genome evolves independently according to an identical stochastic process where the probability of a mutation occurring per unit time is given by a parameter λ. Further, it is assumed that the probability λ of change is distributed equally between the 3 possible changes at a site. Thus if a site currently has the nucleotide A, then it has probability $\lambda/3$ of changing to C, for example, in unit time. When branching occurs at a node in a phylogeny, then the two branches start with identical sequences but evolve independently according to the stochastic process. Finally, we assume that the sequence at the root of the phylogeny is a random sequence of length n. Since we only allow substitutions all genomes will have the same length. Given the actual time durations between evolutionary events, it is possible to represent the Jukes-Cantor model by specifying the probabilities of change along each edge in the phylogeny where these probabilities depend on the time duration represented by the edge (e.g. if an edge is infinitely long, the probability of change is 3/4).

Even with this simple model, obtaining a closed-form representation of tag length distribution as a function of the probabilities of change along each edge is a complex task. We therefore start with a simplifying assumption—the phylogeny is a complete binary tree and the probability of change along each edge is p. We let h be the height of our tree and label the sequences at its leaves with $s_1 \cdots s_{2^h}$. We label the sequence at the root with r. We will focus on tags present in the left subtree of the root, which we call *left tags*. Similar analysis holds for right tags and for other nodes in the tree. In Section 6.3, we generalize the analysis to arbitrary binary tree topologies and probabilities of change along the edges.

6.1   The Primary Mechanism for Generating Tags

Given the stochastic model of evolution we show that there is a dominant process by which tags are generated. We first prove that if the probability of change p along an edge is more than $\ln(n)/(2^{h-2}k)$, we do not expect tags to be generated. Using this bound on p, we show in Theorem 4 that the primary mechanism by which a tag t that discriminates a set S' of species from set $\bar{S}'$ arises is one where t is present in the common ancestor of the species in S' and is absent from the common ancestor of those in $\bar{S}'$. In particular, if we let T denote the set of all tags and T' the set of tags generated by the primary mechanism, then we show that $|T| \leq |T'|(1 + O(\frac{\ln n}{|S' \cup \hat{S}'|}))$. Thus the error term decays inversely in the number of species. We start with the following two lemmas bounding the minimum tag length and the maximum probability of change p.

Lemma 8. *Tags have length greater than* $(1 - \epsilon) \log_4 n$ *w.h.p. where* $0 < \epsilon < 1$.

Lemma 9. *If* $p > \frac{3 \ln n}{k(2^h - 2)}$, *the expected number of tags of length k is* < 1.

Henceforth, we will assume that $p \leq 3 \ln(n)/(k(2^h - 2))$. Let A_i for $1 \leq i \leq n - k + 1$ be the event that position i in the root sequence, r, is *good*. Position i is said to be good if the k-mer starting at i in the left child of r differs from that in the right child. Therefore, $P(A_i = 1) = 1 - (p^2/3 + (1 - p)^2)^k$. If the event A_i results in a tag being generated, we will say that this tag is a *type–I* tag. The following theorem shows that type–I tags are dominant.

Theorem 4. *Let t be a sequence that either does not occur at the left child of the root or occurs at the right child of the root. Then the probability that any such t is a left tag is negligible compared to the probability of type–I tags.*

Expected number of length k tags. Define B_i for $1 \leq i \leq n - k + 1$ to be the event that the ith k-mer at each leaf of the left subtree agrees with that at the root of the left subtree. A lower bound on $P(B_i = 1)$ is obtained when there are no changes in the left subtree. That is,

$$P(B_i = 1) \geq (1 - p)^{\#\{\text{edges in the left subtree of } r\} \cdot k} = (1 - p)^{(2^h - 2)k}$$

One way a type–I tag is generated is if A_i occurs, the k-mer does not change anywhere in the left subtree and a position that changed due to the occurrence of A_i remained unchanged in the right subtree. Let the random variable X_i indicate if a type–I tag of length k occurs at position i. Then, $E[X_i] \geq P(A_i = 1)P(B_i = 1)(1 - p)^{(2^h - 2)}$. Finally, let the random variable X equal the number of tags of length k. Then, $X \geq \sum_{i=1}^{n-k+1} X_i$, implying that $E[X] \geq (n - k + 1)E[X_i]$.

6.2   A Sampling Based Approach

Consider the phylogeny described above, and suppose event B_i occurred. That is, suppose that the k-mer starting at position i is common to all the sequences in the left branch of the root. Call this k-mer t_i. Let $R = \{s_{2^{h-1}+1}, \cdots, s_{2^h}\}$ be the set of sequences at the leaves of the right subtree of r. For t_i to be discriminating, it should not occur in any of the sequences in R. Instead of testing the occurrence of t_i in every one of those sequences, we will only test a sample of those sequences. Let $\mathcal{M}$ be the sample we pick. We will consider t_i to be a tag if it does not occur in any of the sequences in $\mathcal{M}$. If t_i is a tag, then our test will succeed. However, we need to bound the probability that we err. Specifically, we bound the ratio of the expected number of false positive tags to the expected number of tags our algorithm produces.

Algorithm. We use the sampling idea to speed up our tag detection algorithm:

1. Run the HCS algorithm to compute C_u for all u in our phylogeny P.
2. For each $u \in P$,
   - Pick a set $\mathcal{M}_u$ of sequences from the right subtree of u.
   - For each $s \in \mathcal{M}_u$, trim $C_{l(u)}$ as in Step 2 of the algorithm in Section 4.1.

Assuming $\mathcal{M}$ is the sample of maximum size, then the running time of the above algorithm is $O(nm|\mathcal{M}|)$.

Sampling Error. How well does the sampling based approach work? Even with a sample of constant size, the probability that we err decreases with the tag size k. Theorem 4 shows that if t occurs at the right child of the root, then t is not a left tag *w.h.p.* Hence, assuming that the k-mer t is a left tag at position i, we need only consider the case when the right child of the root contains a k-mer $t' \neq t$ at position i. We do not err when a differentiating bit in t' is preserved in the right subtree which is at most $(1-p)^{2^h - 2}$ implying the following theorem.

Theorem 5. *The sampling algorithm errs with probability $< 1/2$ for $k = \Omega(\ln n)$.*

6.3   General Tree Topologies

We generalize our stochastic analysis to arbitrary binary topologies and probabilities of change along edges of the phylogeny. Given a phylogeny P with root r, let L (resp. R) be the total length of the edges in the left (resp. right) subtree of r, and let E be the total length of the two edges incident on r. Recall that at a given site a nucleotide changes to one of the three remaining nucleotides with a rate of λ per year. Hence, the position i will experience x number of mutations on a branch of length ℓ with probability $e^{-\lambda \ell}(\lambda \ell)^x / x!$. Again, we will focus on left tags occurring at homologous sites. The following is the analog of Lemma 9.

Lemma 10. *Let $k > \zeta \log_4 n$ where $\zeta < 2$ is a constant. If $\lambda L > \frac{\zeta \log_4 n}{k}$, the expected number of left tags of length k is less than 1.*

Assuming that both left and right tags occur in the given phylogeny, we show that type–I tags constitute the majority of tags if $\lambda E = \Omega(1/k)$.

Theorem 6. *The probability that a tag t is of type–I is $> 1/2$ if $\lambda E = \Omega(1/k)$.*

7   Generalized Tag Sets

The stochastic analysis in Section 6 shows that tags may not always exist even in data sets generated by stochastic evolutionary processes. When tags are not present, we can relax the definition of discriminating substrings and still be able to distinguish if a genome comes from a node's left or right subtree. Recall that given a partition $(S', \bar{S}')$ of species, we say that a set T of tags is an (α, β)-generalized tag set for some $\alpha > \beta$, if every species in S' contains at least an α fraction of the strings in T and every species in $\bar{S}'$ contains at most a β fraction of them. Clearly, such a tag set can still be used to decide whether a species belongs to S' or to $\bar{S}'$. The problem of computing generalized tag sets may be viewed as a set cover problem with certain "coverage" constraints as we next show. W.l.o.g. assume we are computing generalized tag sets at the root.

(α, β)–**Set Cover.** Given a universe $U = U' \cup U''$ of m elements, and a collection of subsets of U, $\mathcal{S} = \{S_1, \ldots, S_N\}$, find a minimum size subcollection $\mathcal{C}$ of $\mathcal{S}$ such that each element of U' is contained in at least $\alpha|\mathcal{C}|$ sets in $\mathcal{C}$, and each element of U'' is contained in at most $\beta|\mathcal{C}|$ sets in $\mathcal{C}$.

The set U corresponds to the m input strings each of length n with U' and U'' being the strings in the left and right subtrees of the root of the given phylogeny. Each $S_i \in \mathcal{S}$ represents the set of strings that share a substring t_i drawn from a suitable collection of substrings with cardinality $N = O(n^2 m)$. In [5] it was shown how to efficiently compute and represent the corresponding sets of all substrings in $O(nm^2)$ time and space with the help of a generalized suffix tree. A biologically motivated pruning sub-step may be applied to reduce their number [19]. We note that the Discriminating Substring Problem corresponds to the $(1,0)$–Set Cover problem when the objective is to maximize the size of $\mathcal{C}$ since we find all tags. The subcollection $\mathcal{C}$ is a tag set when $\alpha > \beta$.

The next theorem follows via a reduction from Set Cover. In the main reduction, the size of all feasible subcollections $\mathcal{C}$ is the same and therefore the results hold even for the existence version of the problem.

Theorem 7. $(\frac{2}{3}, \frac{1}{3})$-*Set Cover is NP-hard. Furthermore,* $(\frac{2}{3}, \frac{1}{3})$-*Set Cover is LOGSNP-hard.*

The reduction relies on the construction of a collection $\mathcal{Q}$ of subsets of U'' such that for each proper subcollection of $\mathcal{Q}$, there is an element that appears in more than β/α-fraction of the sets while each element occurs in exactly β/α-fraction of the sets in $\mathcal{Q}$. By suitably padding $\mathcal{Q}$ with elements of U' we are able to bound the solution size. We can therefore extend the analysis for rational α and β s.t. $\alpha = 1 - \beta$ and $\beta = 1/c$ for a fixed integer $c > 2$.

The (α, β)–Set Cover problem can be formulated as an ILP in a straightforward manner. When there exists an optimal solution of size value $\Omega(\log m)$, standard randomized rounding of the fractional solution can be used to derive from it an (α', β')-cover where $\alpha' \geq (1 - \epsilon)\alpha$ and $\beta' \leq (1 + \epsilon)\beta$ for some small ϵ.

8   Conclusion

The data-driven approach to choosing discriminating oligonucleotide sequences appears to be novel. In this paper we have described how such sequences can be chosen given a "complete" data set consisting of a phylogeny where all the input sequences are present at the leaves. In this situation when our algorithms produce tags we can use them for high-throughput identification of an unlabeled sequence which is known to be one of the sequences in the input. Each tag found (at any node in the phylogeny) identifies an exactly conserved sequence shared by a clade. Such conserved segments can be used as seeds (in a BLAST-like fashion) to identify longer segments with high-similarity multiple alignments. When our algorithm fails to find tags, or even sufficiently long, shared sequences this is also informative. We learn that there is no strong conservation of segments within the clade. A natural extension of the problem considered here is to the situation where our knowledge is less complete. For example, how can one generalize to the case when the phylogeny is not fully known? If we attempt to place a new sequence in the phylogeny using the tags to guide us, how good is the placement as a function of the position of the new sequence in the phylogeny *vis a vis* the sequences from which the tag set was built? These are some of the directions that we plan to explore.

References

1. Bejerano, G., Siepel, A., Kent, W., Haussler, D.: Computational screening of conserved genomic DNA in search of functional noncoding elements. Nature Methods **2**(7) (2005) 535–45
2. Siepel, A., Bejerano, G., Pedersen, J., Hinrichs, A., Hou, M., Rosenbloom, K., Clawson, H., Spieth, J., Hillier, L., Richards, S., Weinstock, G., Wilson, R.K., Gibbs, R., Kent, W., Miller, W., Haussler, D.: Evolutionarily conserved elements in vertebrate, insect, worm, and yeast genomes. Genome Research **15**(8) (2005) 1034–1050
3. Bejerano, G., Pheasant, M., Makunin, I., Stephen, S., Kent, W., Mattick, J., Haussler, D.: Ultraconserved elements in the human genome. Science **304**(5675) (2004) 1321–1325
4. Amann, R., Ludwig, W.: Ribosomal RNA-targeted nucleic acid probes for studies in microbial ecology. FEMS Microbiology Reviews **24**(5) (2000) 555–565
5. Angelov, S., Harb, B., Kannan, S., Khanna, S., Kim, J., Wang, L.S.: Genome identification and classification by short oligo arrays. In: Proceedings of the Fourth Annual Workshop on Algorithms in Bioinformatics. (2004)
6. Wolf, Y.I., Rogozin, I.B., Grishin, N.V., Koonin, E.V.: Genome trees and the tree of life. Trends in Genetics **18**(9) (2002) 472–479
7. Weiner, P.: Linear pattern matching algorithms. In: Proc. of the 14th IEEE Symposium on Switching and Automata Theory. (1973) 1–11
8. Gusfield, D.: Algorithms on Strings, Trees, and Sequences. Cambridge University Press, New York (1997)
9. Hui, L.: Color set size problem with applications to string matching. In: 3rd Symposium on Combinatorial Pattern Matching. Volume 644 of Lecture Notes in Computer Science., Springer (1992) 227–240
10. McCreight, E.M.: A space-economical suffix tree construction algorithm. Journal of the ACM (JACM) **23**(2) (1976) 262–272
11. Ukkonen, E.: On-line construction of suffix-trees. Algorithmica **14** (1995) 249–260
12. Harel, D., Tarjan, R.E.: Fast algorithms for finding nearest common ancestors. SIAM Journal of Computing **13**(2) (1984) 338–355
13. Schieber, B., Vishkin, U.: On finding lowest common ancestors: Simplifications and parallelization. SIAM Journal of Computing **17** (1988) 1253–1262
14. Brown, M.R., Tarjan, R.E.: Design and analysis of data structures for representing sorted lists. SIAM Journal of Computing **9**(3) (1980) 594–614

15. Chang, W.I., Lawler, E.L.: Sublinear approximate string matching and biological applications. Algorithmica **12** (1994) 327–343
16. Thomas, J., et al.: Comparative analyses of multi-species sequences from targeted genomic regions. Nature **424**(6950) (2003) 788–793
17. Maidak, B.L., Cole, J.R., Lilburn, T.G., Parker, Charles T., J., Sax man, P.R., Farris, R.J., Garrity, G.M., Olsen, G.J., Schmidt, T.M., Tie dje, J.M.: The RDP-II (ribosomal database project). Nucl. Acids. Res. **29**(1) (2001) 173–174
18. Jukes, T.H., Cantor, C.: Mammalian Protein Metabolism, chapter Evolution of protein molecules. Academic Press, New York (1969)
19. Matveeva, O.V., Shabalina, S.A., Nemtsov, V.A., Tsodikov, A.D., Gesteland, R.F., Atkins, J.F.: Thermodynamic calculations and statistical correlations for oligo-probes design. Nucl. Acids. Res. **31**(14) (2003) 4211–4217

Phylogenetic Profiling of Insertions and Deletions in Vertebrate Genomes

Sagi Snir and Lior Pachter

Department of Mathematics,
University of California, Berkeley, CA
{ssagi, lpachter}@math.berkeley.edu

Abstract. Micro-indels are small insertion or deletion events (indels) that occur during genome evolution. The study of micro-indels is important, both in order to better understand the underlying biological mechanisms, and also for improving the evolutionary models used in sequence alignment and phylogenetic analysis. The inference of micro-indels from multiple sequence alignments of related genomes poses a difficult computational problem, and is far more complicated than the related task of inferring the history of point mutations. We introduce a tree alignment based approach that is suitable for working with multiple genomes and that emphasizes the concept of *indel history*. By working with an appropriately restricted alignment model, we are able to propose an algorithm for inferring the optimal indel history of homologous sequences that is efficient for practical problems. Using data from the ENCODE project as well as related sequences from multiple primates, we are able to compare and contrast indel events in both coding and non-coding regions. The ability to work with multiple sequences allows us to refute a previous claim that indel rates are approximately fixed even when the mutation rate changes, and allows us to show that indel events are not neutral. In particular, we identify indel hotspots in the human genome.

1   Introduction

Sequence insertion and deletion events (indels) play a major role in shaping the evolution of genomes. Such events range in scale from transposable element replication within genomes, to single nucleotide events. Despite the importance of indels in modifying the function of genes and genomes [5, 24, 26], the underlying biological mechanisms are not well understood [12]. This is particularly true of small indels, also called micro-indels [25]. Analysis of micro-indels has also been limited by the availability of tractable models of indel evolution. Examples of statistical models of micro-indels include the TKF model [27], and others [17, 18, 19], however, in contrast to the large literature on evolutionary models of point mutations [11], there has been far less work on micro-indels.

The difficulty in inferring the history of insertions and deletions from a multiple sequence alignment is illustrated by a simple example. Consider a tree on three taxa (Figure 1, where the top leaf is human, the middle leaf mouse and the

A. Apostolico et al. (Eds.): RECOMB 2006, LNBI 3909, pp. 265–280, 2006.

bottom rat) and four events: two speciation events and two micro-indel events, but no point substitution event. Suppose that the primates-rodent ancestor consists of three bases. Upon the primates-rodents speciation, both ancestors keep the same three bases. Next comes the rat-mouse speciation which is followed by two parallel events: a deletion of all the three bases in the rat and a deletion of the middle base in the mouse. There is no indel event along the branch leading to the human. The true alignment of this section in the three species human, rat and mouse consists of the three human bases aligned with three gaps in the rat and a base-gap-base sequence at the mouse. In order to trace the optimal history, one may consider a site-by-site approach, however the resulting optimal sequence at the ancestral rodent is base-gap-base, yielding a history of three indel events: two deletions at sites 1 and 3 along the rat lineage and one deletion along the rodents ancestor lineage (or alternatively an insertion along the human lineage) at site 2. Obviously, this is not the true history, nor the most parsimonious one.

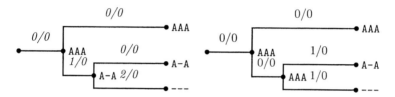

Fig. 1. An example of an alignment and two histories

The parsimony model for indel analysis has been avoided in large scale analyses, in part because the naive algorithm for reconstructing an indel history requires time that is exponential in the length of the alignment. One of our main contributions is implementing an algorithm whose running time is exponential in the number of sequences, but not in their length. This observation has already been utilized in the simplest cases. For example, as part of a broad analysis of micro-evolutionary features across the genomes of the human, mouse and rat, micro-indels and their variability was studied in [4]. It was found that there is a constant ratio between the rates of indel events and point substitutions along the mouse and rat lineages. One of the issues in such a study is the relevance of alignment quality to the results of the indel analysis, an issue which is discussed at length in [16] and which we return to when discussing indels among primates. The paper [25] restricts analysis to human and rodent coding sequences, in particular to 8148 orthologous genes among the three genomes. Only codon indel events were examined. Among the main findings was that slippage like indel events [13] are substantially more frequent than expected.

Micro-indels have also been considered in the context of reconstructing large portions of the ancestral mammalian genome [1]. Although the main goal was not the study of indels, this work was the first to deal with a non-trivial set of species and large datasets. For their purpose, a heuristic was devised in order to infer a plausible indel history and subsequently reconstruct the ancestral sequences at gap-less positions. Although this heuristic is accurate in general, it can fail to reconstruct the true history.

In this paper we introduce the notion of an *indel history*. We assert that such a history can explain the sequence of events that occurred during the evolution of a set of species, via inference from a multiple alignment of the respective genome sequences. Our model of evolution is a restricted tree alignment model, where gap extensions receive no penalty. We argue that this specialization is biologically interesting, and computationally appealing. In particular, we develop, via a series of simplifications, an algorithm for inferring an indel history that is linear in the number of events. We also discuss the possibilities and limitations of approximation algorithms for our problem.

We applied the algorithm to coding data from the ENCODE project as well as non-coding sequences from multiple alignment of primates. Working with primates is important, as it improves the reliability of alignments [2] which are crucial for obtaining meaningful results in analysis of indels. Our findings extend the results of [4, 25] and we compare and contrast indel events in both coding and non-coding regions. The ability to work with multiple sequences allows us to refute an assumption made in [22] that indel rates are approximately fixed even when the mutation rate changes (also observed by [1]), and allows us to show that indel events are not neutral.

2   Notations and Definitions

Let us denote by $\Sigma_S = \{A, C, G, T\}$ and $\Sigma_A = \{*, -\}$. A *multiple alignment* $\mathbf{a} = a^1, \ldots, a^m$ consists of a set of m sequences with $a^i \in (\Sigma_A)^n$. We use the notation a^i_j to denote the j-th element of a^i and $[a]_j$, the j-*th column*, to denote the set a^i_j $1 \le i \le m$. We say that $\mathbf{a}$ has size m and length n. A multiple alignment $\mathbf{a}$ describes homology between a set of sequences $\mathbf{s} = s^1, \ldots, s^m$, $s^i \in (\Sigma_S)^{|\{j : a^i_j = *\}|}$, where each sequence element is associated with a $*$ in $\mathbf{a}$. $a^i_j = a^{i'}_{j'} = *$ means that two elements, from sequences i and i' respectively, are 'matched' in the multiple alignment. Let $X = \{1, \ldots, m\}$ be our set of taxa and T a phylogenetic tree with leaves X. Let $\mathbf{a}$ be a multiple alignment of size m and length n An *insertion-deletion history* h (or *indel history*) consists of a labeling of vertices of T with sequences, such that each internal vertex v is labeled by a sequence $a^v \in (\Sigma_A)^n$ and each leaf i is labeled by a^i. We consider T to be a directed graph with edges directed away from the root.

Indel histories are therefore records of insertion and deletion events. Every time a $*$ switches to a $-$ there has been a deletion, and similarly every switch of a $-$ to a $*$ corresponds to an insertion. An *insertion event* corresponds to a sequence of consecutive (along one sequence) changes from $-$ to $*$, whereas a *deletion event* corresponds to consecutive (along one sequence) changes from $*$ to $-$. A history explanation, to be defined formally in the sequel, associates indel events to the given history. Let P_n denote the path of length n. Observe that we can view a history h as a function from the graph product $T \times P_n$ to Σ_A where for $v \in V(T)$ and $1 \le j \le n$, $h((v, j)) = a^v_j$. Let $G = T \times P_n$ and $slice_{j,G}$ (or just

$slice_j$ for short) be the graph induced by the set $\{(v, j) \in V(G)\}$. We extend the notion of parenthood of trees to G as follows: we say that $x = (v, j) \in G$ is the *parent* of $x' = (v', j') \in G$, or $x = p(x')$, if $j = j'$ and $(v \rightarrow v') \in E(T)$ (or, by the definition of $E(G)$, $(x \rightarrow x') \in E(G)$). In the sequel, we will interchangeably refer to a node either as a node in the graph, or as a combination of a tree node and an index in the path.

A leaf in G is defined analogously as in trees: a node with out-degree zero. Let $I(G)$ ($L(G)$) be the internal (leaf) nodes of G. Observe that for $j \in P_n$, $x = (v, j)$ is an internal (leaf) node in G if and only if x is an internal (leaf) node in T. Let r be the root of T and set $R = \{(r, j) \in V(G) : 1 \leq j \leq n\}$ to be the roots in G.

A convex coloring C of a graph G is a mapping of vertices of G to colors, i.e., $C : V(G) \rightarrow \{1, \ldots, k\}$ such that for each color c, the subgraph of G induced by the vertices $\{v : C(v) = c\}$ is connected [6]. We will use the notation $|C| = k$ for the number of colors in the coloring.

Given a history h, an explanation to h assigns different colors to indel events under the following rules: Two neighboring nodes in G, $x = (v, j)$ and $x' = (v', j')$ can have the same color if either $v = v' =$ the root $r \in V(T)$, or $x = p(x')$ and $h(x) = h(x')$, or $v = v'$, $j' = j - 1$, $h(x) = h(x')$ and $h(p(x')) \neq h(x')$. In addition, we require the coloring induced by the explanation to be convex on G. It is easy to see that even the naive explanation where every vertex has a different color is legal. However, we are interested in the explanation(s) with minimal number of colors. The following algorithm produces a coloring from a given history h:

1. Begin by coloring the path $r \times P_n$ monochromatically, i.e., let $C((r, j)) = 1$ for all $j \in P_n$.
2. Given a vertex $v_1 \in T$ for which all the vertices $(v_1, j), j \in P_n$ have been colored, and a child v_2 of v_1, we color the vertices $(v_2, j'), j' \in P_n$ as follows: First partition P_n into three sets $S_1 = \{j' : h(v_1, j') = h(v_2, j')\}$, $S_2 = \{j' : h(v_1, j') = * \wedge h(v_2, j') = -\}$, $S_3 = \{j' : h(v_1, j') = - \wedge h(v_2, j') = *\}$. Now set $C(v_2, j') \leftarrow C(v_1, j')$ if $j' \in S_1$. Then color each connected component of $v_2 \times S_2$ or $v_2 \times S_3$ with a unique new color (so that components get different colors from each other and from previously assigned colors). Thus, the number of new colors in C after assigning colors to $v_2 \times P_n$ is equal to the number of connected components in S_2 plus the number of connected components in S_3.

Observation 1. *The coloring obtained by the above algorithm is optimal and unique (up to the choice of colors). The number of colors corresponds to the number of indels required to explain the given history.*

By the observation above, we identify every history h with its optimal coloring, C_h. Our problem is to find the indel history h_{opt} and associated indel coloring $C_{h_{opt}}$ for which $|C_{h_{opt}}|$ is minimized.

3   Algorithm

In this section we first describe an algorithm that runs in time exponential in the number of species in the alignment and linear in the alignment length. Specifically, for m the number of species and n the length of the alignment, our algorithm runs in time $O(2^{2m-2}n)$. We then explain an improvement of the algorithm that reduces the linear factor significantly.

Let h be a history. For the purpose of the algorithm, h can be viewed as an assignment of $\{0,1\}$ to the nodes of G where 0 corresponds to a gap and 1 corresponds to an existing character state. Therefore, from now on we identify a history with its corresponding assignment. We denote by $U \subseteq V(G)$, $h|_U$ the restriction of h to the vertices of U. Recall that we index the species (the tree leaves/alignment sequences) with i and the columns of the alignment/history with j. We call $h_j = h|_{slice(j)}$ a *history slice*. We say that history slice s is *valid for j* if for every $i \in L(T)$, $s(i) = 1$ if $a_j^i = *$ and $s(i) = 0$ otherwise (that is, the slice s is consistent with the alignment at the leaves). A history h is *valid* if for every j, the history slice h_j is valid for j. Henceforth, we will restrict ourselves to valid histories and slices only. We denote by $pref(G, j)$ (or $pref(j)$ for short), the subgraph of G induced by slices $1 \ldots j$.

Definition 1. *For $1 \le j \le n$, and a history slice s which is valid for j, let:*

$$opt(G, j, s) = \min_{h' \,:\, h'_j = s} |C_{h'}|_{pref(j)}|.$$

That is, $opt(G, j, s)$ is the value of an optimal history (with the least number of colors) over $pref(j)$ among all histories h' such that the j-th slice equals to s. In the sequel, we will remove G from the notation as it is clear by the context. Let $opt(j)$ be the optimal history for $pref(j)$. Since $opt(j) = \min_{s'} opt(j, s')$, the answer to the optimal indel history problem $|C_{h_{opt}}|$, is $opt(n)$. For a vertex $x \in V(G) \setminus R$ and a history h, the *sign of x under h*, $sign(h, x)$, is defined by $h(x) - h(p(x))$. In the context of slices, sign is defined for vertices of T (by omitting the index of the path).

For two history slices s and s', we have

Definition 2. $dist(s, s') = \sum_{v \in (T) \setminus r} |sign(s', v)| \delta_{sign(s,v), sign(s',v)}.$

where δ_{x_1, x_2} is the complement of the Kronecker delta (i.e. δ_{x_1, x_2} is one if $x_1 \ne x_2$ and zero otherwise). The distance between two assigned slices $dist(s, s')$ is just the sum over all vertices $v \in V(T)$, where a vertex contributes to the distance if (1) it has a different assignment than its father under s' (i.e. $sign(s', v) \ne 0$) and (2) it has a different sign under s' than under s (see Figure 2(b)) . Note that this distance function is *not* symmetric and therefore is not a metric. This leads us to the following observation:

Observation 2. *For $1 \le j \le n$, and a valid history slice for j and s, $opt(j, s) = \min_{s'} (opt(j - 1, s') + dist(s', s))$ where $opt(0, s) = \sum_{v \in V(T) \setminus \{r\}} |sign(s, v)|$.*

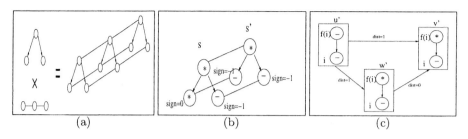

Fig. 2. (a) The graph product of a cherry and P_3. (b) The distance between s and s' is computed by summing over all vertices except the root in s. If a vertex has sign $\neq 0$ and it has a different sign than its left brother, it contributes to the distance. (c) In order for i not to attain a new color under $s^{v'}$, it needs to have the same sign under $s^{v'}$ as under $s^{w'}$. This implies that i attains a new color on the edge from u' to w'.

We now define a new graph G' over the set of binary characters over T. Let $G' = (V', E', w)$ be a complete weighted directed graph where V' is the set of all binary characters (i.e. slices) over T. For an edge $e' = (u' \to v') \in E'$ such that $u' = s$ and $v' = s'$, $w(e') = dist(s, s')$. Practically, we can restrict ourselves to the sub graph of G' induced by vertices corresponding to valid slices, i.e., vertices $v' \in V'$ such that v' is valid for some j. Our problem can now be formulated slightly differently:

Problem 1. Given a tree T and an alignment $a = [a]_1 \ldots [a]_n$, find a minimum weight path $P' = v_{j_1}, \ldots, v_{j_n}$ in $G'(T)$ such that for every $v_{j_k} = s \in P'$, $s|_{L(T)} = [a]_k$.

Lemma 1. *w satisfies the triangle inequality.*

Proof. Consider $u', v', w' \in V'$ with corresponding history slices $s^{u'}$, $s^{v'}$ and $s^{w'}$. Recall, by the definition of $dist$, $w(u' \to v') = dist(s^{u'}, s^{v'})$ which is the number of vertices attaining new colors by moving from $s^{u'}$ to $s^{v'}$. Now, a vertex $v \in V(T)$ attains a new color upon moving from a slice s to s' when it has $sign(v) \neq 0$ under s' (i.e. different assignment than its father $p(v)$) and $sign(v)$ under s is not equal to $sign(v)$ under s' (see Figure 2(b)). Consider now the path (u', w', v') in G' (see Figure 2(c)). Let $i \in V(T)$ be a vertex such that i changes its color by moving from u' to v'. Observe that i has a different assignment than $p(i)$ (i.e. $sign(i) \neq 0$) under $s^{v'}$. In order for i to use an existing color upon moving from w' to v' on the path (u', w', v'), there must be that $sign(i)$ under $s^{w'}$ is equal to $sign(i)$ under $s^{v'}$. This implies that $sign(i)$ under $s^{w'}$ is not equal to $sign(i)$ under $s^{u'}$ and also $sign(i) \neq 0$ under $s^{w'}$, implying i obtains a new color upon moving from u' to w'.

Let $\mathbf{a} = ([a]_1, \ldots, [a]_n)$ be an alignment. Then $\mathbf{a}'$ is a *subalignment* if it contains a subset of the columns $\{[a]_1 \ldots, [a]_n\}$ in the same order as in $\mathbf{a}$. The following corollary follows from the lemma above:

Corollary 1. *Let X be a set of species, and $\mathbf{a}$ and T are an alignment and a phylogenetic tree (resp.) over X with a' a subalignment of $\mathbf{a}$. Then, $opt(\mathbf{a}, T) \geq opt(\mathbf{a}', T)$.*

Claim. Let $\mathbf{a}^*$ be a subalignment of $\mathbf{a}$ obtained by removing every column $[a]_i$ such that $[a]_i = [a]_{i-1}$. Then $opt(\mathbf{a}, T) = opt(\mathbf{a}', T)$.

Proof. $\geq$: By Corollary 1.
$\leq$: Let H^* be the history attaining the minimal cost on $\mathbf{a}'$. It can be noted that, for every removed column, assigning the same history slice as the remaining column in $\mathbf{a}'$ under H^* (note that of every block of identical columns, exactly one remains in $\mathbf{a}'$), achieves the same cost.

Observation 2 and Claim 3 give rise to a straightforward dynamic programming algorithm which runs in time $O(2^{2m-2}n^*)$.

IndelHistory(a,T=(V,E))

  1. remove identical adjacent columns and let n^* be the new length.
  2. for every slice s valid for column 0, $opt(0, s) \leftarrow \sum_{v \in V(T) \setminus \{r\}} |sign(s, v)|$.
  3. for i from 1 to n^*, $opt(i, s) \leftarrow \min_{s'}(opt(i - 1, s') + dist(s', s))$.
  4. return $\min_s(opt(n^*, s))$.

A more careful analysis allows us to give a better asymptotic bound.

Claim. Let $(\mathbf{a}, T)$ be an input to the problem and let $\mathbf{a}^*$ be its subalignment as in Claim 3 and with length n^*. Then $opt(\mathbf{a}, T) \geq \frac{n^*}{2}$.

Proof. Let h^* be an optimal history for (a^*, T). The proof is based on the observation that at every column (site) in h^*, at least one event is either starting or ending. We exclude the case of a whole gapped column as that does not occur in an alignment. We look at sites j and $j + 1$. We divide into two cases:

1. One sequence does not change:
   Let a^i be a sequence s.t. $a^i_j \neq a^i_{j+1}$ and let $a^{i'}$ be a sequence s.t. $a^{i'}_j = a^{i'}_{j+1}$. Again we look at the cherry T_C induced by leaves i and i'. Then any history for the graph $G_C = T_C \times P_2$ with the above sequences at the leaves, must have one event.
2. The case when all sequences change is proved similarly.

Since it takes $O(nm)$ to process the alignment and by the above claim $opt(\mathbf{a}, T) \geq \frac{n^*}{2}$, we can bound the linear component in the running time by the size of the optimal solution. This implies that the time complexity of the IndelHistory algorithm is $O(mn + 2^{2m-2}|C_{h_{opt}}|)$.

4 Implementation

Although our algorithm has running time linear in the length of the alignment and the number of events, a major drawback is the exponential factor in the number of species. Our model is a special case of tree alignment (see, e.g., [23]) which has been extensively studied, and was shown to be NP-hard [28] (including a recent generalization by [9]). More recently, it has been show that there is a

3 to align with the human reading frame. We believe this in general resulted in a more accurate alignment.

5.1   Comparisons with Previous Studies

The excess of deletion over insertion has already been highlighted in previous studies of both coding [25] and non coding regions [26] . Our results are consistent with those studies but also reveal differences between the two types of regions. Since Taylor et al. considered only codon insertion and deletion events (i.e., indels of length divisible by 3), we filtered out all events of other lengths. We ran our software on these alignments and obtained the number of events along all branches not emanating from the root. The distribution of events obtained along the tree branches is shown in Figure 4 (right tree).

The first value we examined is the ratio between insertions and deletions along each branch of the tree. The del/ins ratio at the mouse lineage is 1.05 (versus 1.1 obtained by Taylor et. al.) and 1.50 at the rat (versus 1.7 obtained by Taylor et. al.). The second value we measured is the frequency of events along a sequence. This measure should not be confused with the rate of events along a branch in the tree. The latter indeed measures the number of events with respect to the edge length, while the frequency of events ignores this factor. In our data, there were 108,000 codons (twice the alignments length, 156,000 for both rat and mouse divided by 3) for the rat and mouse sequences and total of 73 events (sum of events for rat and mouse), yielding a frequency of one event per 1,479 codons (versus 1,736 obtained by Taylor et. al.). The agreement is striking considering that our trees contain more than twice the number of species and four-fold more branches (eight vs. two, not counting branches emanating from the root), and events could have been attributed to other branches of the tree. We now elaborate on the above argument. While in Taylor et. al. a gap in the mouse and human is automatically inferred as an insertion in the rat, in our method, based on the whole set of sequences, this scenario can be interpreted as a multi event site (see exact definition in the sequel) in which both mouse and human exhibit two different deletion events at that site.

Cooper et. al [4] found a constant ratio between the rate of indel events (insertions and deletions) measured as the number of events per site, to the rate of point substitutions per site (expressed as the length of the tree branch). This ratio, calculated only in the two mouse and rat branches and along the whole genome, was found to be 0.05. Since our non-coding data was comprised of closer species, we cannot make an exact comparison. However, the value we obtained at the rodents in coding regions was 0.0073 (obtained by summing the number of events for rat and mouse, normalized by the alignment length and divide by the length of the rat-mouse path. See values at Figure 4). Considering a ratio of 10 between coding to non-coding regions (see values at Table 1), we obtain approximately a ratio of 0.07 for non-coding regions. Taking into account the distance between human and the rodents which may lead to alignment inaccuracies, we believe the agreement is satisfactory.

5.2 Events in Coding Regions

Table 2 illustrates our finding regarding events in both non-coding and coding regions. Information regarding coding regions is the left number at every column. Both insertions and deletions decay exponentially in length. An exception to this exponential decay in the length is events of 7 codons (21 bases) that stand out in both insertions and deletions (not shown in table). We do not have an explanation for this. It can also be seen that deletions are slightly longer on average than insertions.

We also wanted to measure if, and how, the rate of indels changes along the branches of the tree. We normalized the number of events on each tree edge, by the length of the edge. This measurement enables us to estimate the correlation between the length of an edge (the expected number of substitutions on the edge) and the number of indel events accumulated on it. Another question we examined is whether the indel process is homogeneous over time, or changes along different lineages of the tree. Our coding data is composed of many genes and different

Table 2. Length distribution of indel events in coding/non-coding regions

Events in Coding/Non-Coding Regions						
	total events distribution		insertions distribution		deletions distribution	
event length	#events	total length	#events	total length	#events	total length
1	-/1895	-/1895	-/174	-/174	-/1721	-/1721
2	-/606	-/1212	-/33	-/66	-/573	-/1146
3	578/388	1734/1164	132/35	396/105	446/353	1338/1059
4	-/379	-/1516	-/20	-/80	-/359	-/1436
5	-/175	-/875	-/12	-/60	-/163	-/815
6	177/123	1062/738	48/11	288/66	129/112	774/672
7-8	-/211	-/1584	-/18	-/138	-/193	-/1446
9	55/66	495/594	11/3	99/27	44/63	396/567
10-11	-/151	-/1577	-/12	-/126	-/139	-/1451
12	56/72	672/864	7/1	84/12	49/71	588/852
13-18	34/224	543/3403	3/20	45/302	31/204	498/3101
19-30	42/175	945/4112	13/23	297/541	29/152	648/3571
total	942/4465	5451/19534	214/362	1209/1697	728/4103	4242/17837
average event length		5.786/4.37		5.649/4.68		5.826/4.34

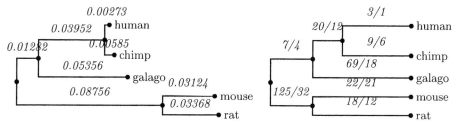

Fig. 4. Point mutation (left) and indel (right) statistics along tree edges for coding regions. The indels were computed over 156,000 sites.

sets of species. In order to obtain enough information we concatenated many genes over a common set of species. Figure 4 at the right shows the number of del/ins along each edge of the tree. We used the same data in order to obtain the edge lengths corresponding to the tree (by ML estimate according to the HKY [14] model). As the dog served as an outgroup to the rest of the species it was excluded from the figure. The tree on the left of the figure shows the edge lengths inferred for the tree (measured by the expected number of substitutions along an edge). The correlation between the edge length and the total number of events (insertions plus deletions) is notable. Specifically, for every edge e, we computed the ratio $\frac{id_e}{l_e}$ where id_e is the expected number of indel events (total number of events divided by sequence length) per site along e and l_e is the length of e (the expected number of substitution per site along it). We found that $\frac{id_e}{l_e}$ is centered around mean 0.009 (std. dev. 0.0038) with a ratio of 3.2 between the lowest value (0.005 for the ancestor of human chimp) and the highest (0.0165 for the pendant edge of the chimp). This should be contrasted to a ratio of 40 between the number of events along the pendant edge of the human (4) and 157 along the edge leading to the rodent ancestral vertex.

In [22] it was postulated that the indel process obeys the rule of molecular clock. This means that if we measure the length of the path along the tree, from any internal vertex to any of its descendants, this length will be the same. It is well known [31] that with respect to point substitutions, this hypothesis does

Table 3. Amino acid indel events

AA	#ins.	#del.	percent in insertions	percent in deletions	percent in population	relative insertion	relative deletion
A	45	133	10.56	9.38	7.41	1.41	1.26
C	10	14	2.34	0.98	1.73	1.34	0.56
D	10	48	2.34	3.38	4.59	0.5	0.73
E	28	117	6.57	8.25	7.16	0.91	1.15
F	5	20	1.17	1.41	3.52	0.33	0.4
G	49	129	11.50	9.1	6.56	1.73	1.38
H	15	41	3.52	2.89	2.47	1.41	1.16
I	5	23	1.17	1.62	4.05	0.28	0.39
K	19	47	4.46	3.31	5.37	0.82	0.61
L	30	143	7.04	10.09	9.89	0.7	1.02
M	5	23	1.17	1.62	2.47	0.47	0.65
N	16	55	3.75	3.88	3.24	1.14	1.19
P	39	116	9.15	8.18	6.78	1.33	1.2
Q	24	108	5.63	7.62	4.82	1.15	1.58
R	15	52	3.52	3.66	5.91	0.59	0.62
S	49	166	11.50	11.71	8.48	1.34	1.38
T	24	79	5.63	5.57	5.4	1.03	1.03
V	33	73	7.74	5.15	6.28	1.22	0.81
W	3	10	0.70	0.7	1.13	0.61	0.62
Y	2	20	0.46	1.41	2.46	0.18	0.57

Indel Events for Amino Acids

not apply to the set of species we investigated here. There was acceleration in the rate of mutations in the rodents' lineage after the speciation event from the primates. This causes a substitution rate twice as much bigger in the rodents, than as in primates. Our findings refute this hypothesis. It can easily be seen that the number of events on the path from the root to the mouse is exactly 200 while to the human it is only 47. We comment here that although there are deviations in the $\frac{id_e}{l_e}$ ratio that might explain small differences, the difference here in the number of events is statistically significant.

At the amino acid level, we examined whether there was a preference for certain kinds of insertions or deletions. The composition of amino acids in insertion and deletion events is depicted in Tables 3. We inferred amino acid insertion and deletion events in both extant species (i.e. in the aligned sequences) and the ancestral nodes. An event is determined to be an insertion/deletion by the optimal explanation. It is notable that some amino acids maintain the same ratio in both processes (e.g. Arginine, Serine, Threonine) although this deviates from a neutral rate of relative value of one (e.g Arginine, Serine, Phenylalanine). Another characteristic is that most of the amino acids are either overrepresented or underrepresented in both insertions and deletions. Exceptions include Cysteine, Valine, and Glutamic acid that are over represented in one process but underrepresented in the other.

5.3 Events in Non-coding Regions

Our non-coding data was taken from homologous sequences of primates surrounding various genes (see [21]). Here the emphasis was to examine the deleted and inserted sequences and their properties. Values are shown in Table 2 (right number of every column). There are 4465 events with total length of 19534 bases, which yields an average event length of 4.37 bases per event. Events of a single base comprise 42.4 percent of the total number of events and of length two, 13.5 percent. Of the total number of events, there are 362 insertion events with total length of 1697, yielding an average insertion size of 4.68 bases. In turn, there are 4103 deletion events with total size of 17837, yielding average deletion size of 4.34 bases.

Table 4 shows the base composition of indels in non-coding regions. We used the same method here for the inference of the content of the indel events as we did for coding regions, except for the fact that we considered indels of all length. We found that the percentage of Gs and Cs in indel events was even lower than the population GC content. In insertion events C is substantially underrepresented (0.81 of its background frequency) while T is similarly overrepresented (1.13). In deletion events, both A and T are similarly overrepresented (around 1.05 of their ground frequency) while C and G are similarly underrepresented (0.92). C and T exhibit the largest variation between insertion and deletions.

Similarly to coding regions, we wanted to measure the correlation between rate of indel events to the rate of point substitutions along the tree branches. Figure 5 depicts our findings in the CFTR region (ENCODE region number 1). The right tree depicts the distribution along the edges. The edge lengths of

Table 4. Distribution of bases in insertions and deletion events in non-coding regions

	Bases Distribution in Non-Coding Regions				
base	% in population	% in insertions	% in ins relative to % in population	% in deletions	% in ins del relative to % in population
A	28.3	30.6	1.08	30.3	1.06
T	29.5	33.3	1.13	30.8	1.04
C	20.8	17.0	0.81	19.1	0.92
G	21.2	18.9	0.89	19.6	0.92

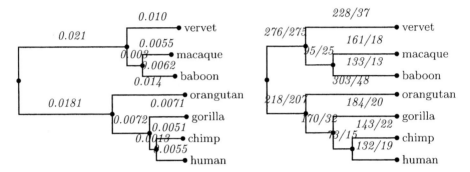

Fig. 5. Point mutation (left) and indel (right) statistics along tree edges for the CFTR region. The indels were computed over 209,000 sites.

the tree in the left correspond to the point substitution probabilities. Here the value $\frac{id_e}{l_e}$ (see definition in the coding region section) is centered around a mean of 0.146 (std. dev. 0.061) with a single big exception for the ancestral edge of human and chimp which is double that value.

5.4   Indel Hotspots

Multi event sites (MES) are sites where an indel event has occurred on more than one branch of the tree. Indel events at MES sites are called parallel events.

Our findings show that in both datasets, the frequency of parallel events was more than two fold above its expected value. Specifically, for coding regions, the number of sites containing a single event was 9,553 yielding an expected value of 0.0295 (recall the total number of sites was 323,673) and a probability of 0.000871 of finding a parallel event at a site. The actual number of parallel events was 1093, yielding a frequency of 0.00337 parallel events per site, 3.876 times its expected value. For non-coding regions, we found 30,616 sites containing an event, yielding a frequency of 0.0714 sites containing events, and a probability of 0.0051 for a parallel event at a site. The actual frequency of parallel events was 0.0111, which is 2.178 times its expected value.

These findings are consistent with the findings in [25] about the effect of slippage at indel events. [25] found that the frequency of indel events is about

50% higher than expected in proximity to small regions where the same amino acids is duplicated multiple times. As we have shown, such indel "hotspots", are also evident in non-coding sites. Although some indel hotspots may be due to alignment artifacts as suggested in [16], we believe that our results confirm that indel hotspots exist.

Acknowledgments

SS and LP were partially funded by a grant from the NIH (R01: HG2362-3). We thank Yifat Felder, Benny Chor, and the anonymous referees for comments that improved the manuscript. We also thank Colin Dewey for his generous help.

References

1. BLANCHETTE, M., GREEN, E. D., MILLER, W., HAUSSLER, D. (2004). Reconstructing large regions of an ancestral mammalian genome in silico. Genome Res. 14, p 2412-2423.
2. BOFFELLI, D., MCAULIFFE, J., OVCHARENKO, D., LEWIS, K.D., OVCHARENKO, I., PACHTER, L., RUBIN, E.M. (2003). Phylogenetic shadowing of primate sequences to find functional regions of the human genome, Science, Volume 299, Number 5611 (2003), p 1391–1394.
3. BRAY, N. AND PACHTER, L.. (2004) MAVID: Constrained ancestral alignment of multiple sequences Genome Res. 14, p 693-699.
4. COOPER, G.M., BRUDNO, M., STONE, E.A., DUBCHAK, I., BATZOGLOU, S., AND SIDOW, A. (2004). Characterization of evolutionary rates and constraints in three mammalian genomes. Genome Res. 14, p 539-548.
5. CHUZHANOVA N.A., ANASSIS E.J., BALL E.V., KRAWCZAK M., COOPER D.N. (2003), Meta-analysis of indels causing human genetic disease: mechanisms of mutagenesis and the role of local DNA sequence complexity. Human Mutation 21(1), p 28-44.
6. DRESS, A., STEEL, M.A. (1992). Convex tree realizations of partitions. Applied Mathematics Letters 5(3), p 3–6,
7. THE ENCODE PROJECT CONSORTIUM (2004). The ENCODE (ENCyclopedia of DNA Elements) Project. Science, 306(5696), p 636–640.
8. THE BERKELEY ENCODE WEBSITE. http://bio.math.berkeley.edu/encode/
9. ELIAS, I. (2003). Settling the Intractability of Multiple Alignment. Int. Symp. on Algorithms and Computation (ISAAC) p 352-363.
10. FITCH, W.M. (1981). A non-sequential method for constructing trees and hierarchical classifications. J. Mol. Evol. 18(1), p 30–37.
11. FELSENSTEIN, J. (2004). Inferring Phylogenies. Sinauer Associates Inc., Mass.
12. FRAZER K.A., CHEN X., HINDS D.A., PANT P.V., PATIL N., COX D.R. (2003). Genomic DNA insertions and deletions occur frequently between humans and non-human primates. Genome Res. 13(3), p 341-6.
13. HANCOCK J.M., VOGLER A.P. (2000). How slippage-derived sequences are incorporated into rRNA variable-region secondary structure: Implications for phylogeny reconstruction. Mol. Phylogenet. Evol. 14, p 366-374.
14. HASEGAWA, M., KISHINO H., AND YANO, T. (1985). Dating the human-ape splitting by a molecular clock of mitochondrial DNA. J. Mol. Evol. 22, p 160–174

15. LAI, Y. AND SUN, F. (2003). The relationship between microsatellite slippage mutation rate and the number of repeat units. Mol. Biol. Evol. 20, p 2123-2131

16. LÖYTYNOJA, A. AND GOLDMAN, N. (2005). An algorithm for progressive multiple alignment of sequences with insertions. Proc. Natl. Acad. Sci. 102,p 10557-10562.

17. McGUIRE, G., DENHAM, M.C., AND BALDING, D.J. (2001). Models of sequence evolution for DNA sequences containing gaps. Mol. Biol. Evol. 18, p 481-490.

18. MITCHISON, G. J. (1999). A probabilistic treatment of phylogeny and sequence alignment. J. Mol. Evol. 49, p 11-22.

19. MITCHISON, G. J., AND DURBIN R. M. (1995). Tree-based maximal likelihood substitution matrices and hidden Markov models. J. Mol. Evol. 41, p 1139-1151.

20. PETROV, D.A., SANGSTER, T.A., JOHNSTON, J.S., HARTL, D.L., AND SHAW, K.L. (2000). Evidence for DNA loss as a determinant of genome size. Science 287, p 1060-1062.

21. BERKELEY PGA. http://pga.lbl.gov/

22. SAITOU N. AND UEDA S. (1994). Evolutionary rates of insertion and deletion in noncoding nucleotide sequences of primates. Mol. Biol. Evol. 11(3), p 504-12.

23. SANKOFF, D., AND CEDERGREN, R. (1983). Simultaneous comparisons of three or more sequences related by a tree, in D. Sankoff and J. Kruskal (eds), Time Warp, String Edits, and Macromolecules: the Theory and Practice of Sequence Comparison, p 253-264, Addison Wesley, Reading Mass.

24. SODING J. AND LUPAS A.N. (2003). More than the sum of their parts: on the evolution of proteins from peptides, Bioessays, Sep 25(9), p 837-46.

25. TAYLOR, M.S., PONTING, C.P., COPLEY, R.R. (2004). Occurrence and consequences of coding sequence insertions and deletions in mammalian genomes Genome Res. 14, p 555–566.

26. THOMAS, J.W., TOUCHMAN, J.W., BLAKESLEY, R.W., BOUFFARD, G.G., BECKSTROM-STERNBERG, S.M., MARGULIES, E.H., BLANCHETTE, M., SIEPEL, A.C., THOMAS, P.J., McDOWELL, J.C., ET AL. (2003). Comparative analyses of multi-species sequences from targeted genomic regions. Nature 424, p 788-793.

27. THORNE, J. L., KISHINO H, AND FELSENSTEIN J. (1991). An evolutionary model for maximum likelihood alignment of DNA sequences. J. Mol. Evol. 33, p 114-124.

28. WANG, L., AND JIANG, T. (1994). On the complexity of multiple sequence alignment. Journal of Computational Biology 1(4), p 337-348.

29. WANG, L., JIANG, T., AND LAWLER, E.L. (1996). Approximation algorithms for tree alignment with a given phylogeny. Algorithmica 16(3), p 302-315.

30. WANG, L., AND GUSFIELD, D. (1997). Improved approximation algorithms for tree alignment. J. Algorithms 25(2), p 255-273.

31. WU, C. AND LI W.H. (1985). evidence for higher rates of nucleotide substitution in rodents than in man. Proc. Natl. Acad. Sci. 82, p 1741-1745.

Maximal Accurate Forests from Distance Matrices[*]

Constantinos Daskalakis, Cameron Hill, Alexandar Jaffe, Radu Mihaescu,
Elchanan Mossel, and Satish Rao

University of California, Berkeley

Abstract. We present a fast converging method for distance-based phylogenetic inference, which is novel in two respects. First, it is the only method (to our knowledge) to guarantee accuracy when knowledge about the model tree, i.e bounds on the edge lengths, is *not* assumed. Second, our algorithm guarantees that, with high probability, no false assertions are made. The algorithm produces a maximal forest of the model tree, in time $\tilde{O}\left(n^3\right)$ in the typical case. Empirical testing has been promising, comparing favorably to Neighbor Joining, with the advantage of making few or no false assertions about the topology of the model tree; guarantees against false positives can be controlled as a parameter by the user.

1 Introduction

The shortcomings of "naive" distance methods in phylogenetic reconstruction, such as Neighbor Joining (NJ) [12], are well-known, and reconstructing trees from small subtrees is evidently both desirable and increasingly popular. All quartet-based methods are examples of this paradigm. However, this divide-and-conquer approach presents at least two serious difficulties: (1) identifying those subsets of taxa on which a tree topology can be accurately inferred; and (2) retaining accuracy when some subtree topologies cannot be correctly determined. In particular, quartet methods, such as the Dyadic Closure Method of [4] and the series of Disk-Covering Methods (DCM) [8, 13] are confined to considering only quartets of small diameter, so-called short quartets, in the hope that these provide enough information for a complete reconstruction. These methods, moreover, are compelled to reconstruct the entire tree; consequently, errors are incurred when attempting to combine subtrees when the given distance matrix simply does not justify the attempt.

The first DCM method, DCM1, is a good illustration of these difficulties. That method iterates over thresholds $\widehat{D}(i,j)$ where $\widehat{D}$ is the given distance matrix–estimated from sequences, for example. At threshold w, a graph G_w is constructed, where the vertices of G_w are the taxa, with an edge between i, j whenever $\widehat{D}(i,j) \leq w$. Trees are built on maximal cliques of a triangulation G_w^* of G_w using a base method such as NJ and merged according to a perfect

---

[*] Research supported by CIPRES (NSF ITR grant # NSF EF 03-31494).

A. Apostolico et al. (Eds.): RECOMB 2006, LNBI 3909, pp. 281–295, 2006.

elimination order of G_w^*. In some cases, there may be no accuracy guarantees for the trees built on maximal cliques of G_w^*, and the merging procedure–using strict consensus merger–is provable only when $\widehat{D}$ is nearly additive (so that G_w itself is chordal).

Much recent work in the study of distance-based methods has focused on the notion of *fast convergence*. Indeed, the work of [4,5] can be considered a breakthrough in this vein; there, the authors delineate an algorithm which accurately infers almost all trees on n leaves when provided sequences of length $O(poly(log(n)))$, and all trees with $O(poly(n))$ length sequences. By way of comparison, the venerable NJ requires exponentially long sequences. A notable drawback of the Dyadic Closure method of [4], however, is the dearth of useful performance guarantees when sequence lengths are small. In this paper, we will present an algorithm which achieves fast convergence, to the same extent and with similar time complexity as in [5], and further, is guaranteed to return accurate subtrees even when sequences are too short to infer the whole tree correctly.

To this end, we adapt the work of [9], a method which reconstructs a collection of subtrees of the model tree from which only a constant fraction of edges is omitted, when given $O(\log n)$ characters. We have improved on the framework of [9], for we do away with the need for parameters f and g, the lower and upper bounds on the lengths of edges of the model tree. Specifically, we prove a *local quartet reliability criterion*, which is blissfully ignorant of f and g. This permits our algorithm to produce an accurate subforest which is as large as possible from the data provided–it builds everything that can be built. Subsequently, such a forest can be used to boost other reconstruction methods by, for example, inferring sequences at ancestral nodes.

In the following subsection, we will present a number of definitions towards formulating the optimization problem for which our algorithm is a solution, namely, the Maximal Forest (MF) problem. In Section 2 we delineate the subtree reconstruction and forest construction algorithms and analyze their performance. This section also constitutes a significant simplification of the arguments in [9], and the efficiency of our methods is such that we have been able to implement them. Experimental results are examined in Section 4. In Section 3, we prove that our method reconstructs almost all n-leaf trees accurately given sequences of length $O(poly(log(n)))$; our method achieves this guarantee with marked improvements in efficiency.

1.1   Definitions and Notation

Let T be an edge-weighted, unrooted binary tree. (In the sequel, all trees are assumed to be unrooted.) Then, we define $\mathcal{L}(T)$ to be the set of leaves of T. For any subset X of $\mathcal{L}(T)$, $T|X$ denotes the restriction of T to X. We assume that T is leaf-labelled by a set of taxa, S, of size n and that S is equipped with a distance matrix $\widehat{D}$. For each taxon $v \in S$, let $L(v)$ denote a subset of S such that if $\widehat{D}(v,y) < \widehat{D}(v,x)$ and $x \in L(v)$, then $y \in L(v)$. For $x, y \in S$, let $P(x,y)$ denote the set of edges of the path from x to y in T. We say that $L(u)$ and $L(v)$ are *edge-sharing* if there exist $x, y \in L(u)$ and $x', y' \in L(v)$ such that $P(x, y) \cap P(x', y')$ is nonempty; otherwise, $L(u)$ and $L(v)$ are *edge-disjoint*. For $U \subseteq S$, $\mathcal{E}(U)$ is the graph with vertex set $\{L(x) | x \in U\}$ and edges determined by the edge-sharing relation. Naturally, $\mathcal{E}(U)$ is called an *edge-sharing graph* on U. For convenience, we will freely identify a node $L(x)$ of $\mathcal{E}(S)$ with x itself. Let $N(v)$ denote the set of neighbors of v in $\mathcal{E}(S)$. Then, we define $SL(v) = L(v) \cup \bigcup_{u \in N(v)} L(u)$.

We will make use of the *strict consensus merger* [3] method for constructing supertrees. The strict consensus merger of two unrooted leaf-labelled trees is defined as follows. Let t and t' be trees. Let $L = \mathcal{L}(t) \cap \mathcal{L}(t')$ and let $z = t|L$ and $z' = t'|L$; let Z be the maximally resolved tree that is a contraction of both z and z'. We call Z the *backbone* of t and t'. Finally, reattach the remaining pieces of t and t' to Z appropriately (ambiguities and conflicts induce nodes of degree higher than three). Note that the strict consensus merger of two trees is unique.

Generally, each taxon $s \in S$ is identified with a sequence over some alphabet Σ–for example, $\Sigma = \{A, C, G, T\}$. S is equipped with a distance matrix $\widehat{D}$, which is, by definition, symmetric, zero along the diagonal, and positive off the diagonal. The following several definitions and Theorem 1 motivate the algorithms of this paper.

Definition 1. *Let T be an edge-weighted binary tree, leaf-labelled by S, and let D be the associated additive matrix. Suppose $0 < \epsilon < M$. We say that $\widehat{D} : S \times S \to \mathbb{R}^+$ is a local (ϵ, M) distortion for $S' \subseteq S$ if*

1. *$\widehat{D}$ is a distance matrix.*
2. *$\widehat{D}(x, y) = \infty$ implies $D(x, y) > M$, for all $x, y \in S'$*
3. *$\widehat{D}(x, y) < M$ implies $|\widehat{D}(x, y) - D(x, y)| < \epsilon$, for all $x, y \in S'$*

Definition 2. *Let T be an edge-weighted binary tree, leaf-labelled by S, and let D be the associated additive matrix. Suppose $S = C_1 \sqcup ... \sqcup C_\alpha$ such that $T|C_i$ and $T|C_j$ are edge-disjoint for each $1 \leq i < j \leq \alpha$. For each $i \leq \alpha$, let $0 < \epsilon_i < M_i$ be given. Suppose $\widehat{D} : S \times S \to \mathbb{R}^+$. We say that $\mathcal{C} = \{(C_i, \epsilon_i, M_i) : 0 \leq i \leq \alpha\}$ is a local distortion decomposition of $\widehat{D}$ if $\widehat{D}$ is a local (ϵ_i, M_i) distortion for C_i, for each $i = 1, ..., \alpha$.*

Furthermore, let f_i be the weight of the smallest edge in $T|C_i$, and let $\epsilon_i < \frac{f_i}{2}$; and let $r_i \leq \frac{M_i - 7\epsilon_i}{6}$, and assume $M_i > 7\epsilon_i$. For each $v \in C_i$, let $L(v)$ be the ball of radius r_i about v. If $\mathcal{E}(C_i)$ are the connected components of $\mathcal{E}(S)$, then we say that $\mathcal{C}$ is constructive.

The component reconstruction procedure presented below justifies the use of the word "constructive"; in the case described, we can accurately reconstruct $T|C_i$ in polynomial time.

Theorem 1 ([9]). *Let T be an edge-weighted binary tree, leaf-labelled by S, and let D be the associated additive matrix. Suppose $\widehat{D}$ is an (ϵ, M) distortion for S with $\epsilon < f/2$ and $M > 7\epsilon$, where f is the weight of the smallest edge in T. Let g be the weight of the largest edge in T. Let $\mathcal{E}(S)$ be the edge-sharing graph of*

Since there are at most n^2 iterations in Algorithm 1 there are at most n^2 executions of Algorithm 3. Therefore, the total time spent in executions of Algorithm 3 is $O(n^3\kappa^4)$, typically $\widetilde{O}(n^3)$. On the other hand, each time Algorithm 4 is called the number of trees in the forest decreases by one. And since we start off with n trees, Algorithm 4 is called at most n times, hence $O(n^3)$ time is spent in executions of this algorithm overall. Thus, the total running time is typically $\widetilde{O}(n^3)$.

Finally, we note that, for clarity of exposition, the described algorithms are not optimized. Using hash tables to store the results of Algorithm 2 and the partial $T|SL(v)$ trees, each quartet is evaluated once along the course of the algorithm, and $T|SL(v)$ trees are built at each step on top of partially reconstructed topologies.

3   Log-Length Sequences

In this section, we will prove that our method reconstructs almost all n-leaf trees provided that the sequence length k is $O(poly(log(n)))$ under the Cavender-Farris-Neyman 2-state model of evolution [2, 6, 11]. More specifically, we argue that our method achieves the same performance guarantees as does the Dyadic Closure Method of [4]. A key notion in the analysis is the *depth* of a tree T, defined as follows: for an edge e of T, let T_1 and T_2 be the rooted subtrees obtained by deleting e, and let $d_i(e)$ denote the topological distance from the root of T_i to its nearest leaf in T_i; subsequently, we define

$$depth(T) = \max_e \{\max(d_1(e), d_2(e))\}$$

letting e range over the set of internal edges of T. A quartet $\{i, j, k, l\}$ is called *short* if $T|\{i, j, k, l\}$ consists of a single edge connected to four disjoint paths of topological length no more than $depth(T) + 1$. Let Q_{short} denote the set of short quartets of T. Given a set of quartets Q, we let Q^* denote the set of quartet topologies induced by T.

Given sequences x, y of length k, let $h_{xy} = H(x, y)/k$ where $H(x, y)$ is the Hamming distance of the sequences. Let $E_{xy} = \mathbb{E}[h_{xy}]$.

Let Q_w denote the set of quartet topologies q such that $h_{ij} \leq w$ for all $i, j \in q$. In [4], it is proved that if $Q^*_{short} \subseteq Q_w$ and Q_w is consistent, then $cl(Q_w) = Q(T)$ where $cl(Q)$ is the dyadic closure of a set of quartet topologies. But observe that by lemma 4 if $Q^*_{short} \subseteq Q_w \subseteq Q_{6w} \subseteq Q(T)$ for some w, then Algorithm 1 correctly reconstructs T. Let E denote this event, and further, define the following events: A for $Q^*_{short} \subseteq Q_w$; B for $Q_{6w} \subseteq Q(T)$; and C for "Q_w contains all quartets containing pairs i, j such that $E_{ij} < b$, and Q_{6w} does not contain any pairs i, j such that $E_{ij} > 13b$." If i, j lie in a short quartet, then $E_{ij} \leq \frac{1-e^{-2g(2depth(T)+3)}}{2} = b$. We take $w = 2b$.

It's easy to see that

$$\mathbb{P}[E] = \mathbb{P}[A \cap B] \geq \mathbb{P}[A \cap B \cap C] =$$

$$= \mathbb{P}[C] \cdot \mathbb{P}[A|C] \cdot \mathbb{P}[B|A, C] = \mathbb{P}[C] \cdot \mathbb{P}[B|C]$$

We will bound probability $\mathbb{P}[\overline{B}|C]$ first. Suppose $q = \{u, v, w, z\} \in \binom{n}{4}$ s.t. $\forall i, j \in q : E_{ij} \leq 13b$. Then, the quartet split of q is found with probability at least $1 - \delta_1$ if:

(I) $(1 - 26b)\left(1 + \frac{2\epsilon}{1-\epsilon}\right) < \left(1 - \frac{2\epsilon}{1-\epsilon}\right) \Leftrightarrow \epsilon < \frac{13b}{2-13b}$

(II) $\frac{1}{\epsilon} = min_{i,j \in \{u,v,w,z\}}\left\{\frac{c(i,j)}{\alpha(k,\delta_1)}\right\} > 1$

If $k > \frac{8 \ln \frac{12}{\delta_1}(2-13b)^2}{(1-26b)^2(13b)^2}$, by the Azuma-Hoeffding inequality it follows that the probability that event $I \cap II$ does not hold is at most $6\exp\left\{-\frac{(1-26b)^2k}{8}\right\}$ so $\mathbb{P}[I \cap II] \geq 1 - \exp\left\{-\frac{(1-26b)^2k}{8}\right\}$. Now, we can lower bound the probability of estimating quartet q correctly as follows:

$$\mathbb{P}[q \text{ is estimated correctly}] \geq 1 - \delta_1 - \mathbb{P}[\overline{II \cap I}] \geq 1 - \delta_1 - \exp\left\{-\frac{(1-26b)^2k}{8}\right\}$$

Since the quartets are at most $\binom{n}{4}$ we can bound the probability of $\mathbb{P}[B|C]$ roughly as follows:

$$\mathbb{P}[B|C] \geq 1 - \binom{n}{4}\delta_1 - \binom{n}{4}\exp\left\{-\frac{(1-26b)^2k}{8}\right\}$$

It remains to bound $\mathbb{P}[C]$. Define $S_r = \{\{i,j\} \mid h_{ij} < \frac{1}{2} - r\}$. Then, if i, j are such that $E_{ij} \geq \frac{1}{2} - 13b$, then

$$\mathbb{P}[\{i,j\} \in S_{12b}] = \mathbb{P}[h_{ij} < \frac{1}{2} - 12b] \leq$$

$$\leq \mathbb{P}[h_{ij} - E_{ij} < \frac{1}{2} - 12b - E_{ij}] \leq \mathbb{P}[h_{ij} - E_{ij} \leq -b] \leq e^{-b^2k/2}$$

by the Azuma-Hoeffding inequality. A similar analysis shows that if $E_{ij} < \frac{1}{2} - 3b$, then $\mathbb{P}[\{i,j\} \notin S_{2b}] \leq e^{-b^2k/2}$. Thus, $\mathbb{P}[C] \geq 1 - \binom{n}{2}e^{-b^2k/2}$, and $\mathbb{P}[E]$ is not less than

$$1 - \binom{n}{4}\delta_1 - \binom{n}{4}\exp\left(-\frac{(1-26b)^2}{8}k\right) - \binom{n}{2}e^{-b^2k/2}$$

We have, therefore, proved

Lemma 5. *Suppose k sites evolve on binary tree T according to the Cavender-Farris-Neyman model, such that $f \leq D(e) \leq g$ for each edge e of T. Then Algorithm 1 reconstructs T with probability $1 - o(1)$ whenever*

$$k > \frac{c \cdot \ln \delta_1}{(1-26b)^2b^2} = \frac{c' \cdot \log n}{(1-26b)^2b^2}$$

and δ_1 is chosen $\delta_1 < n^{-5}$

where $b = \frac{1-e^{-2g(2depth(T)+3)}}{2}$.

In [4], it is also proven that a random n-leaf binary tree T has

$$depth(T) \leq (2 + o(1)) \log \log 2n$$

with probability $1 - o(1)$. Thus,

Theorem 4. *Under the Cavender-Farris-Neyman model, Algorithm 2 correctly reconstructs almost all trees on n leaves with sequences of length $k = O(poly (\log n))$.*

4   Experiments

In all of our experiments, we used the CFN 2-state model of evolution. Empirical distances were computed as described in Section 2. Random trees were obtained via the r8s package, with mutation probabilities scaled into the range $[0.1, 0.3]$ by affine transformation.

If M is a forest reconstruction method and D is a distance matrix, then $M[D]$ denotes the set of trees returned by M applied to D. If T is a binary edge-weighed tree and k is a positive integer, then D_T^k is a distance matrix on the leaves of T obtained by generating binary sequences of length k to the leaves of T according to the CFN model of evolution and computing empirical distances as discussed previously.

4.1   Experiment 1: Comparisons of Variations on the Theme

In this experiment, we examine the practicality of the quartet reliability criterion. The Global Radius (GR) method is a strict implementation of [9], recovering a global accuracy threshold as in that result via binary search on the list of pairwise distances between leaves. The Local Radii (LR) method is implementation of our algorithm *without the quartet reliability criterion*–that is, of some heuristics underlying the algorithm. In LR, the accuracy threshold is not read from the model tree *a priori*; rather, balls around leaves are grown dynamically during the run of the algorithm. Finally, $LR + Q_\delta$ denotes the method described in previous sections of this paper, wherein balls around leaves grow dynamically and only statistically reliable quartets (with error tolerance δ, see theorems 2 and 3) are permitted in construction.

Method: For each method, we examined both the number of subtrees of a model tree the method returned and the aggregated accuracy of the subtrees.

Our measure of accuracy is as follows. For a pair of trees T and T' with a common leaf set S, $RF(T, T')$ denotes the Robinson-Foulds distance between them. In our case, it is impossible to compare a forest $\mathcal{F}$ and a tree using the Robinson-Foulds distance directly, so we will apply the distance measure only to subtrees of the model tree induced by the leaf sets of trees in $\mathcal{F}$. Let T be a model tree, and suppose $\mathcal{F} = \{t_1, ..., t_k\}$ is the forest returned by one the reconstruction methods from a distance matrix generated on T. Then we may assess the accuracy of the forest $\mathcal{F}$ with respect to T by $A(\mathcal{F}, T) = \sum_{i=1}^{k} RF(T|\mathcal{L}(t_i), t_i)$. We refer to this measure as IRF (Induced Robinson-Foulds) distance.

We compared the three methods–GR, LR, and $LR + Q_\delta$–on randomly generated n-leaf model trees, for $n = 16, 32, 64$, and 128, and on each model tree we generated sequences of length k^2, k^3, k^4, k^5, k^6, and k^7 and $k = 4$. That is to say, for each n, we generated s trees, say $T_1, ..., T_s$, and for $i = 1, ..., s$, we generated binary sequences of length k^t, for $t = 2, ..., 7$. For $M = GR, LR$, we recorded $IRF(n, k^t)$, the mean IRF distance of M on n-leaf trees with sequences of length k^t, and $Dis_M(n, k^t)$, the average number of disjoint subtrees. For $M = LR + Q_\delta$, we need to consider the error tolerance submitted to the quartet test (i.e. δ in Algorithm 2); therefore, we recorded $IRF_{M,\delta}(n, k^t)$ and $Dis_{M,\delta}(n, k^t)$ for several values of δ.

As expected the IRF distance of GR and LR is similar while LR produces forests with fewer subtrees than does GR. As δ increases, we expect that $Dis_{M,\delta}(n, k^t)$ will decrease while $IRF_{M,\delta}(n, k^t)$ increases.

4.2 Experiment 2: Local Accuracy Comparison with Existing Methods

We compare $LR + Q$ to an industry-standard implementation of the Neighbor-Joining (NJ) method, examining the latter for local accuracy in two different ways. That is, we wish to compare the accuracy of NJ on the disjoint leaf sets induced by our method. Suppose $LR + Q$ returns a forest $\mathcal{F} = \{t_1, ..., t_\alpha\}$ when given a distance matrix D generated on a model tree T. Define

$$pre_{NJ}(\mathcal{F}, T) = \sum_{i=1}^{\alpha} RF(T|L(t_i), NJ[D|L(t_i)])$$

measuring the accuracy of NJ when applied to subsets of $L(T)$ independently, and

$$post_{NJ}(\mathcal{F}, T) = \sum_{i=1}^{\alpha} RF(T|L(t_i), NJ[D]|L(t_i))$$

measuring the accuracy of NJ applied to D and subsequently restricted to disjoint subsets of $L(T)$. Then, following Experiment 1, we define $pre_{NJ}(n, k^t)$ to be the mean over pre_{NJ}'s and $post_{NJ}(n, k^t)$, the mean over $post_{NJ}$'s. It is then reasonable to compare pre_{NJ} and $post_{NJ}$ with IRF_{LR+Q_δ}. We expect $LR + Q$ to outperform NJ under both of these measures.

4.3 Results and Discussion

Detailed results are available on the web at the following URL:

http://www.cs.berkeley.edu/~satishr/recomb2006

Herein, we present a brief summary. As anticipated, LR outperforms GR significantly in terms of the number of subtrees, producing smaller forests for each

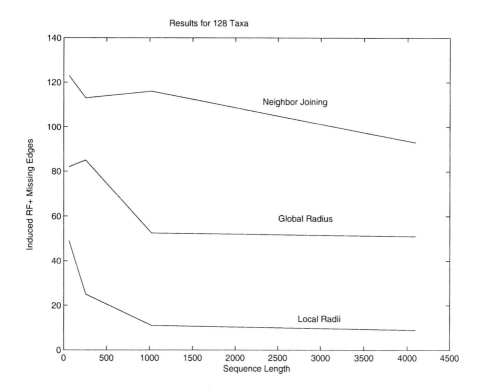

Fig. 1. Comparison of Neighbor-Joining, Global-Radius and Local-Radii methods on 128 taxa for various sequence lengths

sequence length. For example, for 64 taxa, GR returns 38, 29, 13, 9, and 5 trees with sequences of length 64, 256, 1024, 4096, and 16384, respectively, and by comparison, LR returns 13, 7, 5, 2, and 1 trees, respectively. Simultaneously, LR turns out to be more accurate for long sequences, attaining Induced Robinson-Foulds distance of 13.5, 5.5, 3, 1 and 0.5 at the corresponding sequence lengths; GR obtained IRF distance 3.5, 6.0, 5.5, 4.5, and 2.5. Moreover, the advantages of our method seems to be amplified for larger sets of taxa. This advantage also holds in comparison to NJ applied to the distance matrix naively. For example, for 128 taxa with sequences of length 4096, LR returns 6 trees with IRF 3, whereas GR returns 40 trees with IRF 11 and NJ achieves RF distance 93 (while returning one tree). A graphical illustration can be found at figure 1.

We did not measure running-times carefully; however, they appear comparable to popular algorithms.

Due to optimization issues and the delicacy of the probabilistic bounds, we must still look forward to detailed testing of $LR + Q$, and detailed analyses will also appear at the URL above. Results of experiment 2 are also to be found there, and are similarly promising.

Acknowledgements

Radu Mihaescu was supported by the Fannie and John Hertz Foundation Graduate Fellowship.

References

1. Buneman, P. 1971 . The recovery of trees from measures of dissimilarity, 387395 . In *Mathematics in the Archaeological and Historical Sciences*, Edinburgh University Press, Edinburgh .
2. Cavender, J. 1978. Taxonomy with confidence. *Mathematical Biosciences*, 40:271-280.
3. Day, W. 1995. Optimal algorithms for comparing trees with labelled leaves. *J. Class.* 2, 7-28.
4. Erdos, P., Steel, M., Szekely, L., Warnow, T. 1999. A few logs suffice to build (almost) all trees (part 1). *Random Structures and Algorithms*, 14(2):153-184.
5. Erdos, P., Steel, M., Szekely, L., Warnow, T. 1999. A few logs suffice to build (almost) all trees (part 2). *Theoretical Computer Science*, 221:77-118.
6. Farris, J. 1973. A probability model for inferring evolutionary trees. *Systematic Zoology*, 22:250-256.
7. Golumbic , M. 1980. *Algorithmic Graph Theory and Perfect Graphs*. Academic Press, New York.
8. Huson, D., Nettles, S., Warnow, T. 1999. Disk-Covering, A fast converging method for phylogenetic tree reconstruction. *Journal of Computational Biology*, 6:369-386.
9. Mossel, E. Distorted metrics on trees and phylogenetic forests. 2004, to appear in *IEEE Comp. Biol. and Bioinformatics*. Availible at: http://arxiv.org/abs/math.CO/0403508.
10. Mossel, E. Phase Transitions in Phylogeny. 2004. *Trans. Amer. Math. Soc.* 356 no.6 2379–2404. (electronic)
11. Neyman, J. 1971. Molecular studies of evolution: a source of novel statistical problems. In *Statistical Decision Theory and Related Topics*, S. Gupta and J Yackel (ed.) Academic Press, New York.
12. Saitou, N., Nei, M. 1987. The neighbor-joing method: A new method for reconstructing phylogenetic trees. *Mol. Biol. Evol.* 4: 406-425.
13. Usman, R., Moret, B., Warnow, T., Williams, T. 2004. Rec-I-DCM3: A fast algorithmic technique for reconstructing large phylogenetic trecs. Proc. IEEE Computer Society Bioinformatics Conference CSB 2004, Stanford Univ.

Leveraging Information Across HLA Alleles/Supertypes Improves Epitope Prediction

David Heckerman, Carl Kadie, and Jennifer Listgarten

Microsoft Research, Redmond, WA 98052, USA
heckerma@microsoft.com

Abstract. We present a model for predicting HLA class I restricted CTL epitopes. In contrast to almost all other work in this area, we train a single model on epitopes from all HLA alleles and supertypes, yet retain the ability to make epitope predictions for specific HLA alleles. We are therefore able to leverage data across all HLA alleles and/or their supertypes, automatically learning what information should be shared and also how to combine allele-specific, supertype-specific, and global information in a principled way. We show that this leveraging can improve prediction of epitopes having HLA alleles with known supertypes, and dramatically increases our ability to predict epitopes having alleles which do not fall into any of the known supertypes. Our model, which is based on logistic regression, is simple to implement and understand, is solved by finding a single global maximum, and is more accurate (to our knowledge) than any other model.

1 Introduction

The human adaptive immune response is composed of two core elements: antibody-mediated response (sometimes called humoral response), and T-cell-mediated response (sometimes called cellular response). To date, essentially all successful human vaccines have been made by exploiting the underlying mechanisms of the antibody-mediated response, for example with diseases such as polio and measles. However, for these diseases, it was known that people could recover upon acquisition of humoral immunity. In contrast, for certain viruses—for example, HIV—there are no known documented cases of a person recovering from the infection, and it is highly unlikely that the same principles of vaccine design could be successfully applied in these cases. In particular, it is thought that vaccines for diseases such as HIV must prime the cellular immune response rather than or in addition to the humoral response in order to be successful [15, 12].

At the core of cellular response is the ability of certain antigen-presenting cells to ingest and digest viral proteins into smaller peptides, and then to present these peptides, known as *epitopes*, at the surface of the cell. This process is mediated by HLA (Human Leukocyte Antigen) molecules which form a complex with the epitope before it is presented. The epitope/HLA complexes can then be recognized by a T-cell, thereby activating the T-cell to subsequently recognize and kill virally infected cells. Several types of T-cells exist, each playing its own role.

A. Apostolico et al. (Eds.): RECOMB 2006, LNBI 3909, pp. 296–308, 2006.
© Springer-Verlag Berlin Heidelberg 2006

In ongoing HIV vaccine research, the elicitation of a CD8+ T-cell response has shown promise. Since CD8+ T-cells recognize only HLA class I bound epitopes (which range in length from eight to eleven amino acids), our application focuses on such epitopes. Furthermore, we concentrate on the prediction of 9mer epitopes, as this length is the most common.

Due to specificity in a number of sequential mechanisms, only certain epitopes are both presented at the surface of antigen-presenting cells and then subsequently recognized by T-cells. This specificity is determined in part by the sequence and properties of the presented epitope and by the genetic background (i.e. allelelic diversity) of the host (humans have up to six HLA class I alleles arising from the A,B and C loci). A crucial task in vaccine development is the identification of epitopes and the alleles that present them, since it is thought that a good vaccine will include a robust set of epitopes (robust in the sense of broad coverage and of covering regions that are essential for viral fitness in a given population characterized by a particular distribution of HLA alleles). Because experiments required to prove that a peptide is an epitope for a particular HLA allele [e.g., Goulder et al., 2001] are time-consuming and expensive, epitope prediction can be of tremendous help in identifying new potential epitopes whose identity can then be confirmed experimentally. Beyond vaccine design, epitope prediction may have important applications such as predicting infectious disease susceptibility and transplantation success.

In this work, we present a logistic regression (LR) model for epitope prediction which is more accurate than the most accurate model that we can find in the literature—DistBoost [Yanover and Hertz, 2005], and also has several practical advantages: (1) it is a well known model with many readily-available implementations, (2) its output is easy to interpret, (3) training requires $O(N)$ memory whereas DistBoost requires $O(N^2)$ memory, where N is the sample size of the data, (4) the parameters of LR given data have a single, globally optimal value that is easily learned (in contrast to DistBoost and artificial-neural-network based predictors such as NetMHC [4] which have many hidden units), and (5) it produces probabilities that tend to be well calibrated [e.g., Platt, 1999] and hence useful for making decisions about (e.g.) whether to confirm a prediction in the lab.

Another important contribution of this paper is that we show how to leverage information across multiple HLA alleles to improve predictive accuracy for a specific allele. An epitope is defined with respect to one or more HLA alleles. That is, a peptide which is an epitope for HLA-allele X may not also be an epitope for HLA-allele Y. Thus, epitope prediction takes as input both a peptide and an HLA allele, and returns the probability (or some score) reflecting how likely that pair is to be an epitope. Note that HLA alleles are encoded in a hierarchy, where extra digits are used to refer to more specific forms of the allele. For example, moving up the hierarchy from more specific to less specific, we have, A*020101, A*0201, and A02. In addition, many 4-digit alleles belong to a "supertype"—for example, A*0201 belongs to the A2 supertype.

Typically, a single classifier is trained and tested *for each HLA allele* (where the allele is defined with respect to one specific level of the hierarchy) [e.g., Buus et al.,2003] or for each HLA supertype [e.g., Larsen et al., 2005]. These approaches have several shortcomings. One can build classifiers only for alleles with a large number of known epitopes or for alleles which fall in to one of the currently defined supertypes—a fairly strong restriction. Also, if one builds allele-specific or supertype-specific classifiers, then any information which could have been shared across somewhat similarly behaving alleles or supertypes is lost. Because sample sizes are usually extremely small, this shortcoming could be huge in some cases. With supertype classifiers, one is dependent upon the current definitions of supertypes, which has not been rigorously tested in a quantitative way. It may also be the case that some information contained in epitopes is very general, not specific to either alleles or supertypes. Thus, it would be desirable to *simultaneously leverage* epitope information from a number of sources when making epitope predictions:

1. within specific HLA alleles (as available and appropriate),
2. within specific HLA supertypes (as available and appropriate),
3. across all epitopes, regardless of supertype or allele (as appropriate).

That is, in predicting whether a peptide is an epitope for a given HLA allele, we would like to use all information available to us, not just information about epitopes for this allele, but from information about epitopes for other alleles within this allele's supertype (if it has one), and from information about other epitopes of any HLA type. Also, we would like to learn automatically when each type of information is appropriate, and to what degree, allowing us to combine them in a principled way for prediction.

The essence of how we achieve this goal is in the features we use, and is also related to the fact that we train on all HLA alleles and supertypes simultaneously with these features even though our model makes predictions on whether a peptide is an epitope for a specific HLA allele. In the simplest application to epitope prediction, a separate model would either be built for each HLA-allele, or for each supertype, and the features (inputs to the model) would be the amino acid sequence of the peptide, or some encoding of these, such as those discussed for example in [14]. Standard elaborations to this simple approach, in any domain, include using higher order moments of the data (*e.g.*, pairwise statistics of neighboring amino acids) as features in addition to the features of single amino acids. While such higher-order statistics may improve epitope prediction, such experimentation is not the focus of our work. Instead, as mentioned above, we seek to leverage information across HLA alleles and supertypes, and do so by learning a single model for all HLA alleles using features of the form (1) position i has a particular amino acid or chemical property and the epitope's HLA allele is Y, which when used alone would be roughly equivalent to simultaneously building separate models for each HLA allele, as well as (2) position i has a particular amino acid or chemical property and the epitope's HLA has supertype Y, which helps leverage information across HLA alleles for a given supertype, and (3) position i has a particular amino acid or position i has an amino acid with a particular chemical property, which helps leverage information across all HLA alleles and supertypes (see Table 2). This leveraging approach can be applied to various classification models including logistic regression, support vector machines, and artificial neural networks. In our experiments, we show that our leveraging approach applied to logistic regression yields more accurate predictions than those generated from models learned on each supertype individually.

2   Related Work

The general idea of leveraging has been described previously under the names "multitask learning" and "transfer learning" (e.g., [5]). To our knowledge, the only published epitope prediction algorithm that might leverage information across alleles or supertypes is DistBoost [19], which could do so indirectly by learning a distance function across the entire space of epitopes (*i.e.*, for all alleles or supertypes). However, they did not explicitly seek to leverage information in the way we have described, and therefore did not explicitly show that their algorithm does in fact leverage this type of information.

Other approaches to the problem of epitope prediction (or the slightly different problem of binding affinity prediction) include the use of weight matrices (sometimes called PSSMs—position-specific scoring matrices), whereby a probability distribution or score over amino acids at each position is used to make a prediction [18, 1, 6], artificial-neural-network approaches which are said to model amino acid position correlations in a fruitful way [1, 4, 13, 20], support vector machine (SVM) approaches [1, 2, 20, 7] and decision trees [20]. In addition, there is the mostly hand-crafted SYFPEITHI classifier [17]. The approach of Nielsen *et al.* also uses a Hidden Markov Model (HMM) whose output is used as feature for their neural network [14]. In the recent approach of Larsen *et al.* in [11], they demonstrate that their binding affinity neural network approach combined with TAP transport efficiency predictors and proteasomal cleavage predictors does better than a non-integrated approach where the latter two pieces of information are not used.

Among the aforementioned papers, [18, 4, 2, 6, 20, 7] build classifiers for individual HLA alleles (or just a single HLA allele) using only data from each respective HLA class for training. [11] build classifiers for individual supertypes using only data from each respective supertype for training, while [14] use some combination of the two, but never train on data outside of a the respective allele or supertype. Furthermore, perhaps with the exception of PSSM-based approaches, our method is simpler to understand and to implement, yet outperforms PSSM-based methods, and also achieves better results than the most sophisticated methods.

3   Logistic Regression

Let y denote the binary variable (or class label) to be predicted and $\mathbf{x} = x_1, \ldots, x_k$ denote the binary (0/1) or continuous features to be used for prediction. In our case, y corresponds to whether or not a peptide–HLA pair is an epitope and the features correspond to 0/1 encodings of properties of the peptide–HLA pair. In this notation, the logistic regression model is

$$\log \frac{p(y|\mathbf{x})}{1 - p(y|\mathbf{x})} = w_0 + \sum_{i=1}^{k} w_i \cdot x_i \tag{1}$$

where $\mathbf{w} = (w_0, \ldots, w_k)$ are the model parameters or weights. Given a data set of cases $(y^1, \mathbf{x}^1), \ldots, (y^n, \mathbf{x}^m)$ that are independent and identically distributed given the model parameters, we learn the weights by assuming that they are mutually independent, each having a Gaussian prior $p(w_i|\sigma^2) = N(0, \sigma^2)$, and determining the weights that have the maximum a posteriori (MAP) probability. That is, we find the weights that maximize the quantity

$$\sum_{j=1}^{n} \log p(y^i|\mathbf{x}^i, \mathbf{w}) + \sum_{i=0}^{k} \log p(w_i|\sigma^2) \tag{2}$$

This optimization problem has a global maximum which can be found by a variety of techniques including gradient descent. We use the method (and code) of Goodman [2002], which he calls sequential conditional generalized iterative scaling. We tune σ^2 using ten-fold cross validation on the training data.

4   Data and Methods

We used two data sets to evaluate our approach. The first, called MHCBN, contains selected 9mer–HLA and 9mer–supertype pairs from the MHCBN data repository. In this repository, both epitopes and non-epitopes are experimentally confirmed. See [19] for details.

The second, called SYFPEITHI+LANL, includes all unique 9mer–HLA epitopes from the SYFPEITHI database (www.syfpeithi.de) in March 2004 and the Los Alamos HIV Database (www.hiv.lanl.gov) in December 2004. Examples not classified as human MHC class I (HLA-A, HLA-B, or HLA-C) were excluded, yielding 1287 and 339 positive examples of epitopes from SYFPEITHI and LANL, respectively. Neither SYFPEITHI nor LANL contains experimentally confirmed negatives, so we generated examples of non-epitope HLA–9mer pairs by randomly drawing from the distributions of HLAs and amino acids in the positive examples. The amino acid at each position in a 9mer was generated independently.[1] For each positive example, we generated 100 negative examples.

---

[1] In preliminary experiments, we found that, in contrast to the findings of [19] on MHCBN, the use of real negatives from a proprietary data source and the use of randomly generated negatives produced essentially the same results. Here, we report results for the randomly generated negatives, so that we may publish the data on which these results are based.

Table 1. Mapping from HLA to supertype (available at www.hiv.lanl.gov/content/immunology/motif_scan/supertype.html)

Supertype	HLAs
A1	A01, A25, A26, A32, A36, A43, A80
A2	A02, A6802, A69
A3	A03, A11, A31, A33, A6801
A24	A23, A24, A30
B7	B07, B1508, B35, B51, B53, B54, B55, B56, B67, B78
B27	B14, B1503, B1509, B1510, B1518, B27, B38, B39, B48, B73
B44	B18, B37, B40, B41, B44, B45, B49, B50
B58	B1516, B1517, B57, B58
B62	B13, B13, B1501, B1502, B1506, B1512, B1513, B1514, B1519, B1521, B46, B52

As the research in our lab focuses primarily on the prediction of HIV epitopes, we trained our models on both SYFPEITHI and LANL data, but then tested only on LANL data with appropriate cross validation. In particular, we used ten-fold cross validation where the training data of a given fold consisted of all SYFPEITHI data and nine-tenths of the LANL data, and the test data consisted of one-tenth of the LANL data. If an epitope appeared in both SYF-PEITHI and LANL, we treated it as if it were in LANL only. As mentioned, HLA alleles are encoded in a hierarchy. Because many examples in the SYF-PEITHI and LANL databases have HLA alleles encoded only to two digits, we encoded all our examples with two-digit HLA alleles, except for the allele classes B15xx and A68xx, which have elements that belong to different supertypes. There are several supertype classifications; we used the one available from LANL shown in Table 1. The train–test splits of each fold are available at ftp://ftp.research.microsoft.com/users/heckerma/recomb06.

As discussed, we introduced a variety of feature types in an effort to leverage information across HLA alleles and supertypes. The types of features that we used are described in Table 2. In addition to features representing the presence or absence of amino acids at positions along the epitope, we included features representing the chemical properties of the amino acids in our LR models. We used the chemical properties available (e.g.) at www.geneinfinity.org/rastop/manual/aatable.htm: cyclic, aliphatic, aromatic, hydrophobic, buried, large, medium,

Table 2. Feature types used for prediction. Examples are shown for the peptide SLYNTVATL which is an epitope for HLA allele A*0201, which in turn belongs to the A2 supertype.

Feature type	Description
HLA	The HLA allele with 2 or 4 digit encoding; HLA=A02
Supertype (S)	The supertype of the HLA allele; S=A2
HLA ∧ amino acid (AA)	Conjunction of HLA and AA; HLA=A02 and AA1=Ser
HLA ∧ chemical property (CP)	Conjunction of HLA and CP; HLA=A02 and polar(AA1)
S ∧ AA	Conjunction of S and AA; S=A2 and AA1=Ser
S ∧ CP	Conjunction of S and CP; S=A2 and polar(AA1=Ser)
AA	Amino acid at a given position in the peptide; AA1=Ser
CP	Chemical property of amino acid at given position; polar(AA1)

small, negative, positive, charged, and polar. We note, however, that in a separate comparison using LR, the addition of the chemical-property features did not improve predictive accuracy.

Using a large number of features in LR can lead to poor prediction unless some method for feature selection is used [e.g., Kohavi, 1995]. In our experiments, we set the Z weights with the smallest magnitudes to zero, where Z was determined by optimizing the average log probability of prediction on a ten-fold cross validation of the training set. (We used these same cross-validation runs to tune σ^2.) In our largest model, which used all feature types and was trained on all of the data, this feature selection method chose 3,180 out of 23,852 features.[2]

Finally, to evaluate prediction accuracy, we used ROC curves—in particular, plots of the false-positive rate (% non-epitopes identified as epitopes) versus the false-negative rate (% epitopes missed). We summarized the prediction accuracy for a given method using the area under the curve (AUC) of the ROC. To determine whether two methods are significantly different, for each distinct false-negative value, we determined corresponding false-positive values for the two methods, and applied the resulting pairs to a two-sided Wilcoxon matched-pairs signed-ranks test. We deemed a difference to be significant if it's p-value (corrected for multiple tests when appropriate) was less than 0.05.

5   Results

First, we examined whether LR with our features can leverage information about epitopes associated with a variety of supertypes and/or HLA alleles to help predict epitopes associated with different supertypes and/or alleles. To do so, for each supertype (including "none"), we compared the predictive accuracy of a leveraged model that was learned from all training examples with a non-leveraged or individual model that was trained only on epitopes (and non-epitopes) associated with that supertype. Our comparison used ten-fold cross validation, stratified by class label. We pooled the results across the ten folds before generating the ROC curves. In this case, pooling was justified because LR models produce calibrated probabilities. Figure 1 shows ROC curves for leveraged and individual models for each supertype. Leveraging helps significantly for two of the supertypes (A24 and B7)[3], and helps dramatically when predicting epitopes whose HLA alleles have no supertype. In two cases (B27 and B62), the AUC for predictions of the leveraged model is greater than that for non-leveraged model, but the differences are not significant.

Second, we compared the predictions of our (leveraged) LR model with those of DistBoost. In their paper, Yanover and Hertz [19] compared their approach to

---

[2] Many more than 23,852 features were possible, but only this many were warranted based on the training data (e.g., if amino acid Arg was never found in position 3, then no corresponding feature was created).

[3] The p-value of 0.0267 for A2 is not significant after Bonferroni correction.

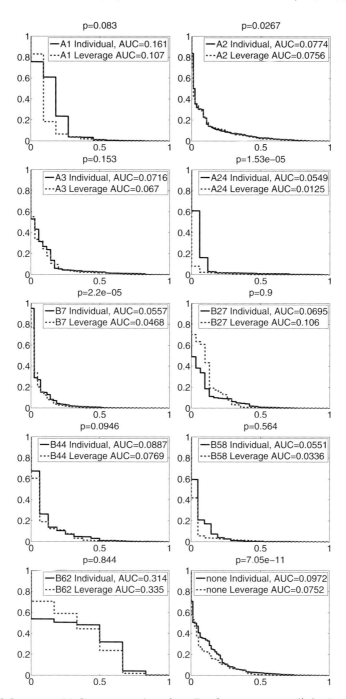

Fig. 1. ROC curves, AUCs, and p-values (not Bonferroni corrected) for leveraged and non-leveraged (individual) predictions of epitopes having alleles in each supertype (including "none", *i.e.*, those not belonging to any supertype). ROC curves plot false-positive rate versus false-negative rate.

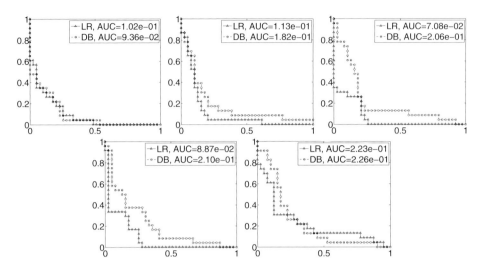

Fig. 2. ROC curves for LR and DistBoost applied to five-fold cross validation 9mer data from MHCBN. The two-sided p-value from false-positive rates pooled across the five folds is 1.8210e-08.

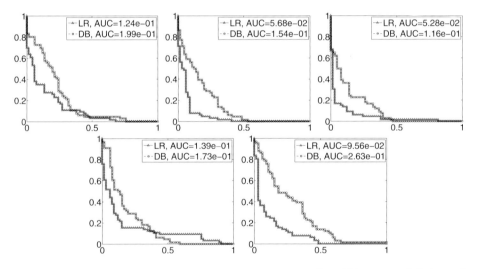

Fig. 3. ROC curves for LR and DistBoost applied to five-fold cross validation of the SYFPEITHI+LANL data. The two-sided p-value from false-positive rates pooled across the five folds is 5.1581e-29.

RANKPEP (PSSM), NetMHC (artificial neural network), and SVMHC (support-vector machine). Their comparison used a 70/30 train–test split of the MHCBN data set, and evaluated performance on A2 supertype epitopes. Yanover and Hertz found that DistBoost predicted significantly better than the other methods. Here,

Table 3. A portion of a model learned from the full SYFPEITHI+LANL data set. The forty features with the largest magnitude weights are shown. Positive weights increase the probability of being an epitope. Feature names are described in Table 2.

weight	feature
-3.87821	large(AA1)
-3.01895	S=A1
2.8267	S=B27 and AA2=Arg
-2.61487	polar(AA1)
-2.48691	large(AA2)
-2.09559	HLA=A01
-1.83075	polar(AA2)
1.73488	S=A1 and polar(AA1)
1.71218	S=A1 and charged(AA1)
1.66352	S=B27 and positive(AA2)
-1.62407	charged(AA1)
1.47669	S=A24 and AA2=Tyr
-1.4628	aliphatic(AA3)
1.45694	negative(AA2)
1.44531	S=A1 and large(AA1)
-1.39833	AA1=Pro
1.35753	S=B44 and large(AA2)
-1.32388	buried(AA4)
1.31555	HLA=B27 and large(AA2)
1.29462	AA4=Trp
1.28076	HLA=B27 and AA2=Arg
1.27827	S=B44 and AA2=Glu
1.26313	HLA=A02 and AA2=Leu
-1.26253	medium(AA1)
1.24698	S=A1 and hydrophobic(AA3)
1.24487	S=B62 and AA2=Gln
1.22292	S=A24 and charged(AA1)
1.19599	S=A24 and positive(AA1)
1.18911	S=A1 and aliphatic(AA3)
-1.17646	charged(AA2)
1.16866	S=A3 and positive(AA1)
1.09196	S=B27 and large(AA2)
1.08261	HLA=A02 and large(AA1)
1.07628	S=B7 and AA2=Pro
-1.07365	S=B44 and hydrophobic(AA2)
1.04742	AA4=Pro
1.04397	S=none and large(AA1)
-1.0417	S=B27 and hydrophobic(AA2)
1.03173	AA9=Leu
1.02222	HLA=A02 and polar(AA1)

we compared DistBoost with LR on this same data and on the SYFPEITHI+ LANL data[4], in both cases using five-fold cross validation stratified by class label and HLA. We ran DistBoost for 30 boosting iterations on the MHCBN data set, and for 50 iterations on the larger SYFPEITHI+LANL data set. (We also tried 100 boosting iterations for each data set, with no substantial change in results.) The results, illustrated in Figures 2 and 3, indicate that the predictive accuracy of LR is better than that of DistBoost. Two-sided p-values computed from false-positive rates pooled across the five folds of the MHCBN and SYFPEITHI+LANL data are 1.8210e-08 and 5.1581e-29, respectively.

Finally, it is interesting to look at the learned features and their weights to see where leveraging is taking place. Table 3 contains a portion of a model trained on the full SYFPEITHI+LANL data set. The forty features with the largest magnitude weights are shown. Many of these strong features are general (e.g., large(AA1) and polar(AA1)) or contain conjunctions with supertypes (e.g., Supertype=A1 and polar(AA1)) and thereby facilitate leveraging.

6   Discussion

We have presented a model for predicting HLA class I restricted CTL epitopes. Our model, which is based on logistic regression, is simple to implement and understand, is solved by finding a single global maximum, and is more accurate (to our knowledge) than the best published results. In addition, we have shown how to leverage information about epitopes having one allele or supertype to predict epitopes having different alleles or supertypes. We have shown that this leveraging can improve prediction of epitopes having HLA alleles with known supertypes, and dramatically increases our ability to predict epitopes having alleles which do not fall into any of the known supertypes.

Our next steps will be to build and evaluate LR predictors for HLA class I epitopes of lengths eight, ten, and eleven amino acids. In addition, rather than use a predefined set of supertypes, we plan to learn a set of (overlapping) supertypes that lead to accurate prediction. In particular, we plan to extend the LR model to include hidden variables that represent these new supertypes. Finally, we are looking at whether the inclusion of additional features such as distances between amino acids in the epitope and in the HLA molecule when the epitope and HLA molecule are in their minimum-energy configuration can improve prediction accuracy.

---

[4] After the publication of [19], the entry AYAKAAAAF–A02 was deleted from the MHCBN repository. We similarly deleted this entry from the MHCBN data set. The SYFPEITHI+LANL data set contained nine entries with unique HLA types. We deleted these entries as they could not be processed by DistBoost. In addition, we used only one negative example for every positive example to accommodate DistBoost's computational requirements, and used the feature encoding of [19] when training and testing with DistBoost. The train–test splits of each fold for both comparisons are available at ftp://ftp.research.microsoft.com/users/ heckerma/recomb06.

Acknowledgments

We thank Vladimir Jojic for extracting and parsing the SYFPEITHI data set as well as Chen Yanover and Tomer Hertz for providing us with their data and software. Work done by JL was supported by an internship at Microsoft Research while away from her studies in the Department of Computer Science, University of Toronto.

References

1. Bhasin, M. and Raghava, G. (2004a). Prediction of CTL epitopes using QM, SVM and ANN techniques. *Vaccine*, 22:3195–204.
2. Bhasin, M. and Raghava, G. P. S. (2004b). SVM based method for predicting HLA-DRB1*0401 binding peptides in an antigen sequence. *Bioinformatics*, 20(3):421–423.
3. Bhasin, M., Singh, H., and Raghava, G. (2003). MHCBN: A comprehensive database of MHC binding and non-binding peptides. *Bioinformatics*, 19:665–666.
4. Buus, S., Lauemoller, S., Worning, P., Kesmir, C., Frimurer, T., Corbet, S., Fomsgaard, A., Hilden, J., Holm, A., and Brunak, S. (2003). Sensitive quantitative predictions of peptide-MHC binding by a 'query by committee' artificial neural network approach. *Tissue Antigens*, 62:378–84.
5. Caruana, R. (1997). *Multitask Learning*. PhD thesis, School of Computer Science, Carnegie Mellon University, Pittsburgh, PA.
6. Dong, H.-L. and Suie, Y.-F. (2005). Prediction of HLA-A2-restricted CTL epitope specific to HCC by SYFPEITHI combined with polynomial method. *World Journal of Gastroenterology*, 2:208–211.
7. Donnes, P. and Elofsson, A. (2002). Prediction of MHC class I binding. *BMC Bioinformatics*, 3.
8. Goodman, J. (2002). Sequential conditional generalized iterative scaling. In *ACL*.
9. Goulder, P., Addo, M., Altfeld, M., Rosenberg, E., Tang, Y., Govender, U., Mngqundaniso, N., Annamalai, K., Vogel, T., Hammond, M., Bunce, M., Coovadia, H., and Walker, B. (2001). Rapid definition of five novel HLA-A*3002-restricted human immunodeficiency virus-specific cytotoxic T-lymphocyte epitopes by Elispot and intracellular cytokine staining assays. *J Virol*, 75:1339–1347.
10. Kohavi, R. (1995). A study of cross-validation and bootstrap for accuracy estimation and model selection. In *Proceedings of the Fourteenth International Joint Conference on Artificial Intelligence*, Montreal, QU. Morgan Kaufmann, San Mateo, CA.
11. Larsen, M., Lundegaard, C., Lamberth, K., Buus, S., Brunak, S., Lund, O., and Nielsen, M. (2005). An integrative approach to CTL epitope prediction: a combined algorithm integrating MHC class i binding, TAP transport efficiency, and proteasomal cleavage predictions. *European Journal of Immunology*, 35:2295–303.
12. McMichael, A. and Hanke, T. (2002). The quest for an aids vaccine: Is the CD8+ T-cell approach feasible? *Nature Reviews*, 2:283–291.
13. Milik, M., Sauer, D., Brunmark, A., Yuan, L., Vitiello, A., Jackson, M., Peterson, P., Skolnick, J., and Glass, C. (1998). Application of an artificial neural network to predict specific class I MHC binding peptide sequences. *Nature Biotechnology*, 16:753–756.

14. Nielsen, M., Lundegaard, C., Worning, P., SL, S. L., Lamberth, K., Buus, S., Brunak, S., and Lund, O. (2003). Reliable prediction of T-cell epitopes using neural networks with novel sequence representations. *Protein Science*, 5:1007–17.
15. Parham, P. (2004). *The Immune System*. Garland Science Publishing.
16. Platt, J. (1999). Probabilities for support vector machines. In *Advances in Large Margin Classifiers*, pages 61–74. MIT Press.
17. Rammensee, H., Bachmann, J., Emmerich, N., O.A. Bachor, O., and Stevanovic, S. (1999). SYFPEITHI: database for MHC ligands and peptide motifs. *Immunogenetics*, 50:213–219.
18. Reche, P., Glutting, J., Zhang, H., and Reinher, E. (2004). Enhancement to the Rankpep resource for the prediction of peptide binding to MHC molecules using profiles. *Immunogenetics*, 26:405–419.
19. Yanover, C. and Hertz, T. (2005). Predicting protein-peptide binding affinity by learning peptide-peptide distance functions. In *RECOMB*. Springer.
20. Zhao, Y., Pinilla, C., Valmori, D., Martin, R., and Simon, R. (2003). Application of support vector machines for T-cell epitopes prediction. *Bioinformatics*, 19(15):1978–1984.

Improving Prediction of Zinc Binding Sites by Modeling the Linkage Between Residues Close in Sequence

Sauro Menchetti[1], Andrea Passerini[1], Paolo Frasconi[1],
Claudia Andreini[2], and Antonio Rosato[2]

[1] Machine Learning and Neural Networks Group,
Dipartimento di Sistemi e Informatica,
Università degli Studi di Firenze, Italy
{menchett, passerini, p-f}@dsi.unifi.it
http://www.dsi.unifi.it/neural/
[2] Magnetic Resonance Center (CERM),
Dipartimento di Chimica,
Università degli Studi di Firenze, Italy
{andreini, rosato}@cerm.unifi.it
http://www.cerm.unifi.it/

Abstract. We describe and empirically evaluate machine learning methods for the prediction of zinc binding sites from protein sequences. We start by observing that a data set consisting of single residues as examples is affected by autocorrelation and we propose an ad-hoc remedy in which sequentially close pairs of candidate residues are classified as being jointly involved in the coordination of a zinc ion. We develop a kernel for this particular type of data that can handle variable length gaps between candidate coordinating residues. Our empirical evaluation on a data set of non redundant protein chains shows that explicit modeling the correlation between residues close in sequence allows us to gain a significant improvement in the prediction performance.

1 Introduction

Automatic discovery of structural and functional sites from protein sequences can help towards understanding of protein folding and completing functional annotations of genomes. Machine learning approaches have been applied to several prediction tasks of this kind including the prediction of phosphorylation sites [1], signal peptides [2, 3], bonding state of cysteines [4, 5] and disulfide bridges [6, 7]. Here we are interested in the prediction of metal binding sites from sequence information alone, a problem that has received relatively little attention so far. Proteins that must bind metal ions for their function (metalloproteins) constitute a significant share of the proteome of any organism. A metal ion (or metal-containing cofactor) may be needed because it is involved in the catalytic mechanism and/or because it stabilizes/determines the protein tertiary or quaternary structure. The genomic scale study of metalloproteins could significantly

A. Apostolico et al. (Eds.): RECOMB 2006, LNBI 3909, pp. 309–320, 2006.
© Springer-Verlag Berlin Heidelberg 2006

benefit from machine learning methods applied to prediction of metal binding sites. In fact, the problem of whether a protein needs a metal ion for its function is a major challenge, even from the experimental point of view. Expression and purification of a protein may not solve this problem as a metalloprotein can be prepared in the demetallated form and a non-metalloprotein can be prepared as associated to a spurious metal ion. In this paper, we focus on an important class of structural and functional sites that involves the binding with zinc ions. Zinc is essential for Life and is the second most abundant transition metal ion in living organisms after iron. In contrast to other transition metal ions, such as copper and iron, zinc(II) does not undergo redox reactions thanks to its filled D-shell. In Nature, it has essentially two possible roles: catalytic or structural, but can also participate in signalling events in quite specific cellular processes. A major role of zinc in humans is in the stabilization of the structure of a huge number of transcription factors, with a profound impact on the regulation of gene expression. Zinc ions can be coordinated by a subset of amino acids (see Table 2) and binding sites are locally constrained by the side chain geometry. For this reason, several sites can be identified with high precision just mining regular expression patterns along the protein sequence. The method presented in [8] mines patterns from metalloproteins having known structure to search gene banks for new metalloproteins. Regular expression patterns are often very specific but may give a low coverage (many false negatives). In addition, the amino acid conservation near the site is a potentially useful source of information that is difficult to take into account by using simple pattern matching approaches. Results in [9] corroborate these observations showing that a support vector machine (SVM) predictor based on multiple alignments significantly outperforms a predictor based on PROSITE patterns in discriminating between cysteines bound to prosthetic groups and cysteines involved in disulfide bridges. The method used in [9] is conceptually very similar to the traditional 1D prediction approach originally developed for secondary structure prediction [10], where each example consists of a window of multiple alignment profiles centered around the target residue.

Although effective, the above approaches are less than perfect and their predictive performance can be further improved. In this paper we identify a specific problem in their formulation and propose an ad-hoc solution. Most supervised learning algorithms (including SVM) build upon the assumption that examples are sampled *independently*. Unfortunately, this assumption can be badly violated when formulating prediction of metal binding sites as a traditional 1D prediction problem. The autocorrelation between the metal bonding state is strong in this domain because of the linkage between residues that coordinate the same ion. The linkage relation is not observed on future data but we show in Section 2.3 that a strong autocorrelation is also induced by simply modeling the close-in-sequence relation. This is not surprising since most binding sites contain at least two coordinating residues with short sequence separation. Autocorrelation problems have been recently identified in the context of relational learning [11] and *collective classification* solutions have been proposed based on probabilistic learners [12, 13]. Similar solutions do not yet exist for extending in the same direction other statistical learning algorithms such as SVM. Our solution is based on a reformulation of the learning problem where examples formed by pairs of sequentially close residues are considered. We test our method on a representative non redundant set of zinc proteins in order to assess the generalization power of the method on new chains. Our results show a significant improvement over the traditional 1D prediction approach.

2   Data Set Description and Statistics

2.1   Data Preparation

We generated a data set of high quality annotated sequences extracted from the Protein Data Bank (PDB). A set of 305 unique zinc binding proteins was selected among all the structures deposited in the PDB at June 2005 and containing at least one zinc ion in the coordinate file. Metal bindings were detected using a threshold of 3Å and excluding carbon atoms and atoms in the backbone. In order to provide negative examples of non zinc binding proteins, an additional set was generated by running UniqueProt [14] with zero HSSP distance on PDB entries that are not metalloproteins. We thus obtained a second data set of 2,369 chains. Zinc binding proteins whose structure was solved in the apo (i.e. without metal) form, were removed from the ensemble of non-metalloproteins.

2.2   A Taxonomy of Zinc Binding Sites and Sequences

Zinc binding sites of zinc metalloenzymes are traditionally divided into two main groups [15]: catalytic (if the ions bind a molecule directly involved in a reaction) and structural (stabilizing the folding of the protein but not involved in any reaction). In addition, zinc may influence quaternary structure; we consider these cases as belonging to a third site type (interface site), which also lacks a catalytic role. Site types can be heuristically correlated to the number of coordinating residues in the same chain. The distribution of site types obtained in this way is reported in Table 1.

Table 2 reports the observed binding frequencies grouped by amino acid type and site type. As expected, cysteines, histidines, aspartic acid and glutamic acid are the only residues that bind zinc with a high enough frequency. It is interesting to note that such residues show different binding attitudes with respect to the site type. While cysteines are mainly involved in structural sites and histidines participate to both Zn4 and Zn3 sites with similar frequency, aspartic and glutamic acids are much more common in catalytic sites than in any other site type.

2.3   Bonding State Autocorrelation

Jensen & Neville [11] define relational autocorrelation as a measure of linkage between examples in a data set due to the presence of binary relations that link examples to other objects in the domain (e.g. in a domain where movies are the examples, linkage might be due to the fact that two movies were made by the

Table 1. Top: Distribution of site types (according to the number of coordinating residues in the same chain) in the 305 zinc-protein data set. The second column is the number of sites for each site type; the third column is the number of chains having at least one site of the type specified in the row. Bottom: Number of chains containing multiple site types. The second row gives the number of chains that contain at least one site for each of the types belonging to the set specified in the first row.

Number of Coordinating Residues	Site Number	Chain Number
1 (Zn1)	37	20
2 (interface - Zn2)	65	53
3 (catalytic - Zn3)	123	106
4 (structural - Zn4)	239	175
Any	464	305

Site types	{1, 2}	{1, 3}	{1, 4}	{2, 3}	{2, 4}	{3, 4}	{1, 2, 3}	{1, 2, 4}	{1, 3, 4}	{2, 3, 4}	{1, 2, 3, 4}
# Chains	14	9	3	21	4	8	7	1	0	2	0

Table 2. Statistics over the 305 zinc proteins (464 binding sites) divided by amino acid and site type. N_a is the amino acid occurrence number in corresponding site type; f_a is the observed percentage of each amino acid in a given site type; f_s is the observed percentage of each site type for a given amino acid. All is the total number of times a given amino acid binds zinc in general.

Site type	Zn4			Zn3			Zn2			Zn1			All
Amino acid	N_a	f_a	f_s	N_a	f_a	f_s	N_a	f_a	f_s	N_a	f_a	f_s	N_a
C	663	69.3	91.8	45	12.2	6.2	10	7.7	1.4	4	10.8	0.6	722
H	220	23.0	45.7	194	52.6	40.3	59	45.4	12.3	8	21.6	1.7	481
D	48	5.0	27.6	83	22.5	47.7	30	23.1	17.2	13	35.1	7.5	174
E	18	1.9	17.5	46	12.5	44.7	28	21.5	27.2	11	29.7	10.7	103
N	5	0.5	83.3	0	0.0	0.0	1	0.8	16.7	0	0.0	0.0	6
Q	2	0.2	33.3	1	0.3	16.7	2	1.5	33.3	1	2.7	16.7	6
Total	956	100	-	369	100	-	130	100	-	37	100	-	1492

same studio). Here we expect the bonding state of candidate residues be affected by autocorrelation because of the presence of at least two relations causing linkage: `coordinates(r,z)`, linking a residue r to a zinc ion z, and `member(r,c)`, linking a residue r to a protein chain c. Unfortunately the first kind of linkage cannot be directly exploited by a classifier as the relation `coordinates` is hidden on new data. However, we may hope to capture some information about this relation by looking at the sequence separation between two candidate residues. In particular, there should be some positive correlation between the bonding state of pairs of residues within the same chain, and it should also depend on the sequence separation between them.

Correlation was empirically measured on the data set described in Section 2. In Figure 1(a) the prior probability of zinc binding for a residue is compared to the same probability conditioned on the presence of another zinc binding residue within a certain separation, for different values of the separation threshold.

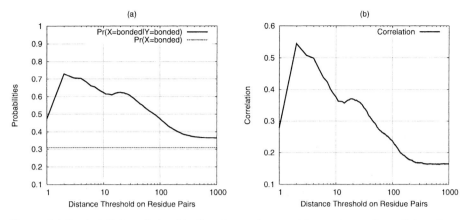

Fig. 1. (a) Probabilities of zinc binding for a given residue: prior and conditioned on the presence of another zinc binding residue within a certain separation. (b) Correlation between the targets of pairs of residues within a given distance.

Table 3. Binding site patterns ordered by frequency of occurrence in the 464 sites. Square brackets denote alternative binding residues, $x(\cdot)$ denotes a sequence of residues of an arbitrary length, $x(n-m)$ denotes a sequence of between n and m residues, $x(> n)$ denotes a sequence of more than n residues. The type column highlights some common binding site patterns: S refers to $x(0\text{-}7)$, L refers to $x(> 7)$.

Binding Site Patterns	N	Type
[CHDE] x($\cdot$) [CHDE] x($\cdot$) [CHDE] x($\cdot$) [CHDE]	232	
[CH] x($\cdot$) [CH] x($\cdot$) [CH] x($\cdot$) [CH]	196	
[CHDE] x(0-7) [CHDE] x($\cdot$) [CHDE] x(0-7) [CHDE]	161	
[CHDE] x(0-7) [CHDE] x(> 7) [CHDE] x(0-7) [CHDE]	141	SLS
[CHDE] x($\cdot$) [CHDE] x($\cdot$) [CHDE]	122	
[C] x($\cdot$) [C] x($\cdot$) [C] x($\cdot$) [C]	85	
[CHDE] x($\cdot$) [CHDE]	62	
[CHDE] x(0-7) [CHDE] x(> 7) [CHDE]	55	SL
[CH] x($\cdot$) [CH] x($\cdot$) [CH]	37	
[CHDE] x(> 7) [CHDE] x(0-7) [CHDE]	24	LS
[CH] x($\cdot$) [CH]	21	
[CHDE] x(0-7) [CHDE] x(> 7) [CHDE] x(> 7) [CHDE]	17	SLL
[CHDE] x(> 7) [CHDE] x(0-7) [CHDE] x(0-7) [CHDE]	16	LSS
[DE] x($\cdot$) [DE]	15	
[DE] x($\cdot$) [DE] x($\cdot$) [DE]	10	
[CHDE] x(> 7) [CHDE] x(> 7) [CHDE] x(0-7) [CHDE]	10	LLS
[CHDE] x(0-7) [CHDE] x(0-7) [CHDE] x(> 7) [CHDE]	8	SSL
[DE] x($\cdot$) [DE] x($\cdot$) [DE] x($\cdot$) [DE]	1	

Figure 1(b) reports the correlation coefficient between the bonding state of pairs of residues, again varying the separation threshold between them. Both curves show a very similar behavior, with the highest peak for a distance of less then three residues, and a small one for a distance of around twenty residues. It can be

Table 4. Site and chain coverage for the [CHDE] x(0-7) [CHDE] semi-pattern. N is absolute, while f is the percentage over the total number of chains/sites of that type.

Site Type	Chain Coverage		Site Coverage	
	N	f	N	f
All	261	85.5	338	72.8
Zn4	168	96.0	227	94.9
Zn3	85	80.1	86	69.9
Zn2	35	66.0	25	38.4
Zn1	13	65.0	0	0.0

noted that correlation tends to a non zero residual asymptotic value as distance grows. This effect is due to the relation `member`, by which two residues are linked by the fact of belonging to the same chain.

Patterns of Binding Sites. Metal binding sites can be described by patterns characterized by the type of residues coordinating the same ion and their sequence separation. Table 3 reports the most commonly occurring zinc binding patterns together to their frequencies within our data set. Many of these sites, especially the structural ones, contain pairs of coordinating residues whose sequence separation is less than seven residues. In the following, a pattern formed by a single pair of nearby coordinating residues is called a *semi-pattern*. Most structural sites consist of two semi-patterns whose distance ranges between 8 and 29. Catalytic sites typically contain a semi-pattern and a single residue. Finally, interface sites are observed as a single semi-pattern in each chain. Table 4 shows the fraction of sites and zinc proteins containing at least once the semi-pattern [CHDE] x(0-7) [CHDE]. These observations suggest a partial solution to the relational auto-correlation problem based on binary classification of semi-patterns to predict binding sites.

3   Methods

3.1   Standard Window Based Local Predictor

Many applications of machine learning to 1D prediction tasks use a simple vector representation obtained by forming a window of flanking residues centered around the site of interest. Following the seminal work of Rost & Sander [10], evolutionary information is incorporated in these representations by computing multiple alignment profiles. In this approach, each example is represented as a vector of size $d = (2k + 1)p$, where k is the size of the window and p the size of the position specific descriptor.

We enriched multiple alignment profiles by two indicators of profile quality, namely the entropy and the relative weight of gapless real matches to pseudo-counts. An additional flag was included to mark positions ranging out of the sequence limits, resulting in an all-zero profile. We thus obtained a position specific descriptor of size $p = 23$. A baseline classifier was constructed using this representation in conjunction with an SVM classifier trained to predict the zinc bonding state of individual residues (cysteine, histidine, aspartic acid and glutamic acid).

Support Vector Machines. Support vector machines [16] are a well established machine learning algorithm capable of effectively handling extremely large and sparse feature spaces. Given a training set $D_m = \{(x_i, y_i)\}_{i=1}^{m}$, where $y_i \in \{-1, 1\}$ is the class label of example x_i, a new instance x is classified as

$$f(x) = \sum_{i=1}^{m} \alpha_i y_i K(x, x_i) \tag{1}$$

where the sign of $f(x)$ gives the predicted class, and the actual value is a measure of the confidence of such prediction. K is a real valued positive semidefinite kernel function measuring the similarity between pairs of examples, and the weights α_i are learned by a convex optimization function trading off between training errors and complexity of the learned hypothesis. Details on kernel machines can be found in several textbooks [17,18]. We employed the dot product between example vectors as a baseline linear kernel, to be combined with more complex kernels as described in the experimental section.

3.2   Semi–pattern Based Predictor

A standard window based local predictor such as the one described in the previous section does not explicitly model the correlation analyzed in Section 2.3, missing a strong potential source of information. Thus, we developed an ad-hoc semi-pattern predictor for pairs of residues in nearby positions within the sequence. A candidate semi-pattern is a pair of residues (cysteine, histidine, aspartic acid or glutamic acid) separated by a gap of δ residues, with δ ranging from zero to seven. The task is to predict whether the semi-pattern is part of a zinc binding site. Each example is represented by a window of local descriptors (based on multiple alignment profiles) centered around the semi-pattern, including the gap between the candidate residues. A semi-pattern containing a gap of length δ is thus encoded into a vector of size $d = (2k + 2 + \delta)p$, where k is the window size and p is the size of the position specific descriptor as described in Section 3.1. In order to address this task, the predictor must be able to compare pairs of semi-patterns having gaps of different lengths. We thus developed an ad-hoc *semi-pattern kernel* in the following way. Given two vectors x and z, of size d_x and d_z, representing semi-patterns with gap length δ_x and δ_z respectively,

$$
\begin{aligned}
K_{semi-pattern}(x, z) = &\langle x[1 : w], z[1 : w] \rangle \\
&+ \langle x[d_x - w : d_x], y[d_z - w : d_z] \rangle \\
&+ K_{gap}(x[w + 1 : \delta_x p + w], z[w + 1 : \delta_z p + w]) \tag{2}
\end{aligned}
$$

where $v[i : j]$ is the sub-vector of v that extends from i to j, and $w = (k + 1)p$. The first two contributions compute the dot products between the left and right windows around the semi-patterns, included the two candidate residues, whose sizes do not vary regardless of the gap lengths. K_{gap} is the kernel between the gaps separating the candidate residues, and is computed as:

$$K_{gap}(u, v) = \begin{cases} K_{\mu gap}(u, v) + \langle u, v \rangle & \text{if } |u| = |v| \\ K_{\mu gap}(u, v) & \text{otherwise} \end{cases}$$

with

$$K_{\mu gap}(u, v) = \left\langle \sum_{i=1}^{|u|} u[(i-1)p + 1 : ip], \sum_{i=1}^{|v|} v[(i-1)p + 1 : ip] \right\rangle$$

$K_{\mu gap}$ computes the dot product between the position specific descriptors within each gap, and if the two gaps have same length, the full dot product between the descriptors in the gaps is added.

3.3 Gating Network

The coverage of the [CHDE] x(0-7) [CHDE] semi-pattern (see Table 4) makes it a good indicator of zinc binding, but a number of binding sites remain uncovered. Moreover, the semi-pattern can match a subsequence which, while not being part of a binding site as a whole, still binds zinc with just one of the two candidate residues. Semi-patterns having a single coordinating residue are considered to be negative examples. This implies that one of the two residues would by construction receive an incorrect label. However, in these cases we can still rely on the local predictor (see Section 3.1) to predict its bonding state. For any given residue we combine the single output from the local predictor, and the (possibly empty) set of outputs from the semi-pattern based predictor, as we get one prediction for each subsequence matching the semi-pattern and containing the residue as one of the two binding candidates. The functional margin calculated by a single SVM (see Eq. (1)) cannot be directly interpreted as a degree of confidence of the predictor, as its magnitude depends on artifacts such as the number of support vectors and the dimension of the feature space. For this reason, in order to combine two predictors, it is preferable to first convert their margins into conditional probabilities using e.g. the sigmoid function approach suggested in [19]:

$$P(Y = 1|x) = \frac{1}{1 + \exp(-Af(x) - B)}$$

where $f(x)$ is the SVM output for example x, and sigmoid slope (A) and offset (B) are parameters to be learned from data. The probability $P(Y_b = 1|x)$ that a single residue binds zinc can now be computed by the following *gating network*:

$$P(Y_b = 1|x) = P(Y_s = 1|x) + (1 - P(Y_s = 1|x))P(Y_l = 1|x) \tag{3}$$

where $P(Y_l = 1|x)$ is the probability of zinc binding from the local predictor, while $P(Y_s = 1|x)$ is the probability of x being involved in a positive semi-pattern, approximated as the maximum between the probabilities for each semi-pattern x is actually involved in.

4   Experimental Results

We run a series of experiments aimed at comparing the predictive power of the local predictor alone to that of the full gating network. While aspartic and glutamic acids coordinate zinc ions less frequently than cysteines and histidines (see Table 2), they are far more abundant in protein chains. This yields a highly unbalanced data set (the ratios of positive to negative examples were found to be 1:59 and 1:145 for the local and the semi-pattern predictor, respectively). We thus initially focused on cysteines and histidines, bringing the unbalancing down to 1:16 and 1:11 at the residue and semi-pattern level respectively. Moreover, we labelled a [CH] x(0-7) [CH] semi-pattern as positive if both candidate residues bound a zinc ion, even if they were not actually binding the same ion. Preliminary experiments showed this to be a better choice than considering such a case as a negative example, allowing to recover a few positive examples, especially for semi-pattern matches with longer gaps.

Multiple alignment profiles were computed using PSI-Blast [20] on the non-redundant (nr) NCBI protein database. In order to reduce noise in the training data we discarded examples whose profile had a relative weight less than 0.015, indicating that too few sequences had aligned at that position. This also allowed to discard poly-histidine tags which are attached at either the N- or C-terminus of some chains in the PDB, as a result of protein engineering aimed at making protein purification easier. We employed a Gaussian kernel on top of both the linear kernel of the local predictor and the semi-pattern kernel (Eq. (2)). A stratified 4-fold cross validation procedure was used to tune Gaussian width, C regularization parameter, window size and parameters of the sigmoids of the gating network. Due to the strong unbalancing of the data set, accuracy is not a reliable measure of performance. We used the area under the recall-precision curve (AURPC) for both model selection and final evaluation, as it is especially suitable for extremely unbalanced data sets. We also computed the area under the ROC curve (AUC) to further assess the significance of the results.

The best models for the local predictor and the gating network were tested on an additional stratified 5-fold cross validation procedure, and obtained an AURPC equal to 0.554 and 0.611 respectively. Figure 2 reports full recall precision curves, showing that the gating network consistently outperforms the local

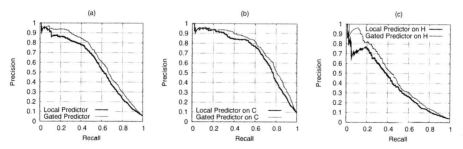

Fig. 2. Residue level recall-precision curves for the best [CH] local and gated predictors. (a) cysteines and histidines together, (b) cysteines only, (c) histidines only.

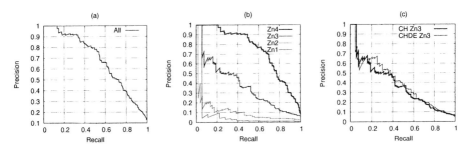

Fig. 3. Protein level recall-precision curves for the best [CH] gated predictor. (a) all proteins together, (b) proteins divided by zinc site type, (c) proteins with Zn3 sites, comparison with the best [CHDE] gated predictor.

predictor. While cysteines are far better predicted with respect to histidines, both predictions are improved by the use of the gating network. AUC values were 0.889 ± 0.006 and 0.911 ± 0.006 for local predictor and gating network respectively, where the method for obtaining the confidence intervals is only available for the AUC computing the standard error of the Wilcoxon-Mann-Whitney statistic, confirming that the gating network attains a significant improvement over the local predictor.

Protein level predictions were obtained by choosing the maximum prediction between those of the residues contained in the chain. Figure 3(a) reports the recall precision curve obtained at a protein level for the best gated predictor, while Figure 3(b) shows the results separately for proteins containing different binding site types. As expected, Zn4 sites were the easiest to predict, being the ones showing the strongest regularities and most commonly containing the [CH] x(0-7) [CH] semi-pattern.

Finally, we investigated the viability of training a predictor for all the four amino acids involved in zinc binding, trying to overcome the disproportion issue. On the rationale that binding residues should be well conserved because of their important functional role, we put a threshold on the residue conservation in the multiple alignment profile in order to consider it as a candidate target. By requiring that $\Pr(D) + \Pr(E) \geq 0.8$, we reduced the unbalancing in the data set for the local predictor to 1:24. At the level of semi-patterns, we realized that such a threshold produced a reasonable unbalancing only for gap lengths between one and three, and thus decided to ignore semi-patterns containing aspartic or glutamic acid with gaps of different lengths. While global performances were almost unchanged, aspartic acid and glutamic acid alone obtained a value of the AURPC of 0.203 and 0.130 respectively. Due to the still high unbalancing, AURPC values for a random predictor are as low as 0.007 for aspartic acid and 0.015 for glutamic acid. AUC values of 0.78 ± 0.03 and 0.70 ± 0.04, respectively (with respect to the 0.5 baseline) confirm that results are significantly better than random. However, results on these two residues are still preliminary and further work has to be done in order to provide a prediction quality comparable to that obtained for cysteines and histidines. It is interesting to note that at the level of protein classification, the only difference that can be noted by using [CHDE] instead of [CH] is a slight improvement in the performances for the Zn3 binding sites, as shown in Figure 3(a). This is perhaps not surprising given that half of [DE] residues binding zinc are contained in Zn3 sites, as reported in Table 2.

5   Conclusions

We have enlightened the autocorrelation problem in the prediction of metal binding sites from sequence information, and presented an improved approach based on semi-pattern classification as a simple linkage modeling strategy. Our results, focused on the prediction of zinc binding proteins, appear to be very promising, especially if we consider that they have been obtained on a non redundant set of chains. Sites mainly coordinated by cysteines and histidines are easier to predict thanks to the availability of a larger number of examples. Linkage modeling allows us to gain a significant improvement in the prediction of the bonding state of these residues. Sites coordinated by aspartic acid and glutamic acid are more difficult to predict because of data sparsity, but our results are significantly better than chance.

The method has been also evaluated on the task of predicting whether a given protein is a zinc protein. Good results were obtained in the case of chains where zinc plays a structural role (Zn4). In the case of chains with catalytic sites (Zn3) the inclusion of D and E targets does allow us to obtain slightly improved predictions. In future work, we plan to test the effectiveness of this method at the level of entire genomes.

References

1. Blom, N., Gammeltoft, S., Brunak, S.: Sequence and structure-based prediction of eukaryotic protein phosphorylation sites. J Mol Biol **294** (1999) 1351–1362
2. Nielsen, H., Brunak, S., von Heijne, G.: Machine learning approaches for the prediction of signal peptides and other protein sorting signals. Protein Eng **12** (1999) 3–9
3. Nielsen, H., Engelbrecht, J., Brunak, S., von Heijne, G.: Identification of prokaryotic and eukaryotic signal peptides and prediction of their cleavage sites. Protein Eng **10** (1997) 1–6
4. Martelli, P.L., Fariselli, P., Casadio, R.: Prediction of disulfide-bonded cysteines in proteomes with a hidden neural network. Proteomics **4** (2004) 1665–1671
5. Fiser, A., Simon, I.: Predicting the oxidation state of cysteines by multiple sequence alignment. Bioinformatics **16** (2000) 251–256
6. Fariselli, P., Casadio, R.: Prediction of disulfide connectivity in proteins. Bioinformatics **17** (2001) 957–964
7. Vullo, A., Frasconi, P.: Disulfide connectivity prediction using recursive neural networks and evolutionary information. Bioinformatics **20** (2004) 653–659
8. Andreini, C., Bertini, I., Rosato, A.: A hint to search for metalloproteins in gene banks. Bioinformatics **20** (2004) 1373–1380

9. Passerini, A., Frasconi, P.: Learning to discriminate between ligand-bound and disulfide-bound cysteines. Protein Eng Des Sel **17** (2004) 367–373

10. Rost, B., Sander, C.: Improved prediction of protein secondary structure by use of sequence profiles and neural networks. Proc Natl Acad Sci U.S.A. **90** (1993) 7558–7562

11. Jensen, D., Neville, J.: Linkage and autocorrelation cause feature selection bias in relational learning. In: Proceedings of the Nineteenth International Conference on Machine Learning (ICML2002). (2002)

12. Taskar, B., Abbeel, P., Koller, D.: Discriminative probabilistic models for relational data. In: Proceedings of the Eighteenth Conference on Uncertainty in Artificial Intelligence, Morgan Kaufmann (2002)

13. Jensen, D., Neville, J., Gallagher, B.: Why collective inference improves relational classification. In: Proceedings of the 10th ACM SIGKDD International Conference on Knowledge Discovery and Data Mining. (2004)

14. Mika, S., Rost, B.: Uniqueprot: creating sequence-unique protein data sets. Nucleic Acids Res. **31** (2003) 3789–3791

15. Vallee, B.L., Auld, D.S.: Functional zinc-binding motifs in enzymes and DNA-binding proteins. Faraday Discuss (1992) 47–65

16. Cortes, C., Vapnik, V.: Support vector networks. Machine Learning **20** (1995) 1–25

17. Schölkopf, B., Smola, A.: Learning with Kernels. The MIT Press, Cambridge, MA (2002)

18. Shawe-Taylor, J., Cristianini, N.: Kernel methods for pattern analysis. Cambridge Univ. Press (2004)

19. Platt, J.: Probabilistic outputs for support vector machines and comparisons to regularized likelihood methods. In Smola, A., Bartlett, P., Schölkopf, B., Schuurmans, D., eds.: Advances in Large Margin Classifiers. MIT Press (2000)

20. Altschul, S., Madden, T., Schaffer, A., Zhang, J., Zhang, Z., Miller, W., Lipman, D.: Gapped blast and psi-blast: a new generation of protein database search programs. Nucleic Acids Res **25** (1997) 3389–3402

An Important Connection Between Network Motifs and Parsimony Models

Teresa M. Przytycka

National Center for Biotechnology Information,
US National Library of Medicine, National Institutes of Health,
Bethesda, MD 20894
przytyck@mail.nih.gov

Abstract. We demonstrate an important connection between network motifs in certain biological networks and validity of evolutionary trees constructed using parsimony methods. Parsimony methods assume that taxa are described by a set of characters and infer phylogenetic trees by minimizing number of character changes required to explain observed character states. From the perspective of applicability of parsimony methods, it is important to assess whether the characters used to infer phylogeny are likely to provide a correct tree. We introduce a graph theoretical characterization that helps to select correct characters. Given a set of characters and a set of taxa, we construct a network called character overlap graph. We show that the character overlap graph for characters that are appropriate to use in parsimony methods is characterized by significant under-representation of subnetworks known as holes, and provide a mathematical validation for this observation. This characterization explains success in constructing evolutionary trees using parsimony method for some characters (e.g. protein domains) and lack of such success for other characters (e.g. introns). In the latter case, the understanding of mathematical obstacles to applying parsimony methods in a direct way has lead us to a new approach for dealing with inconsistent and/or noisy data. Namely, we introduce the concept of persistent characters which is similar but less restrictive than the well known concept of pairwise compatible characters. Application of this approach to introns produces the evolutionary tree consistent with the Coelomata hypothesis. In contrast, the direct application of a parsimony method, using introns as characters, produces a tree which is inconsistent with any of the two competing evolutionary hypotheses. Similarly, replacing persistence with pairwise compatibility does not lead to a correct tree. This indicates that the concept of persistence provides an important addition to the parsimony metohds.

1 Introduction

The term *biological network* is used in connection to any network where nodes correspond to biological entities (like proteins, genes, metabolites, etc.) and edges are defined by a particular relation between these biological units. Can such biological networks help us to understand evolutionary processes? A number of

A. Apostolico et al. (Eds.): RECOMB 2006, LNBI 3909, pp. 321–335, 2006.

studies have focused on the scale free property – a characteristic power-law like distribution of node degrees observed in various biological networks [4]. However, it has been demonstrated [26, 21] that different evolutionary mechanisms can lead to non-distinguishable scale free-like characteristics. Thus, analysis of degree distribution alone does not bring sufficient insight into the evolution of a network. Recently, small size subgraphs, termed *network motifs*, attracted significant attention [23,34,22,24]. The idea is to consider, exhaustively, all possible subnetworks up to a certain size and identify network motifs which are present more frequently than expected by chance.

In this work, we introduce the concept of a character overlap graph and relate the frequency of occurrences of certain network motifs in these graphs to the evolution of the corresponding character traits. Consider a set of taxa, where each taxon is described by a vector of attributes, the so called *characters*. Assume that each character can assume binary values: one – if the taxon has the property described by the character (we will simply say that the taxon contains the character) and zero – otherwise. We further assume that during the evolution characters are gained and/or lost. This acquisition and loss of character traits is the basis for inferring evolutionary trees using parsimony methods. Maximum parsimony methods search for the evolutionary tree with the topology that can explain the observed characters with the minimum number of character changes (here insertions and deletions). The problem of finding most parsimonious tree, under most parsimony models, is NP-complete [9] and thus the corresponding algorithms are computationally intense. However, a more significant drawback comes from the observation that evolutionary trees constructed with these methods are sometimes incorrect. In this work, we focus on the second problem.

The correctness of the evolutionary tree obtained using a parsimony method depends strongly on the characters used to infer the tree. Intuitively, characters that are easy to gain and easy to lose are not appropriate to use with maximum parsimony methods. Extensive independent acquisition and/or loss of characters in several lineages can make it difficult, if not impossible, to recover the correct evolutionary relationships. At the same time, any realistic approach has to tolerate some events of this type. Therefore, it is important to be able to distinguish characters that provide a consistent evolutionary signal from those which do not. We propose a graph theoretical approach to address this problem.

As mentioned above, we use a particular type of network - a character overlap graph. The vertices of a character overlap graph are characters, and there is an edge between two such characters if and only if there exists a taxon that contains both characters. First, we focus on characters that we call *persistent*. A character is persistent if it is gained exactly once and lost at most once. Thus, the assumption of persistence is weaker than what is required in *perfect parsimony* (where a character can change state only once) but stronger than in Dollo parsimony (where there is no restriction on the number deletions of any given character). We show that a character overlap graph for persistent characters cannot contain network motifs known as *holes*. The simplest hole is a cycle of four nodes with no diagonal edges (*chords*) and is also referred to as a *square*. In general, a *hole* is a chordless cycle of length at least four (Figure 2).

The requirement that all characters be persistent, although weaker than the assumption of perfect parsimony, is still very restrictive. However, the criterion for recognizing persistent characters suggests a heuristic for evaluating whether a given set of characters is hard-to-gain and hard-to-lose in a less restrictive sense. Our simple measure relies on counting squares in the character overlap graph constructed for a given set of characters, and comparing the count to the number of squares expected by chance. (This approach can easily be extended to counting also larger holes, e.g. of size 5, but identifying all holes in a large graph is computationally infeasible.) Furthermore, nodes involved in a large number of squares can be used to identify characters whose removal is likely to improve the results of a parsimony method.

We applied our technique to two types of characters: protein domains and introns. In eukaryotic organisms, most of the proteins are made up of several domains. Domains are conserved evolutionary units that are assumed to fold independently, and are observed in different proteins with different neighboring domains. Introns are non-coding DNA sequences that interrupt the flow of a gene coding sequence in eukaryotic genes. It has been widely accepted that the probability of gaining an intron independently at the same position in two different organisms is relatively low [11]. In terms of introns persisting through the evolution, the picture is mixed. They are remarkably conserved between some lineages (e.g. between Arabidopsis and Human), but they are lost at a significant rate in other organisms (e.g. worm) [27].

We tested a large set of domain overlap graphs and found that squares are significantly under-represented as compared to what is expected by chance. This is in line with the results of Deeds *et al.* [10] and Winstanley *et al.* [29]. They report a successful reconstruction of evolutionary trees using the Dollo parsimony where (structural) domains are taken as characters. In contrast, the intron overlap graph has nearly as many squares as is expected by chance, indicating a very noisy signal. This explains the observation that the tree constructed from intron data using Dollo parsimony method is incorrect [27].

Examining the distribution of squares in each network provides additional insight into the properties of corresponding characters. For both character types, we find that the distribution of squares is non-uniform. For example, in the domain overlap graph, a small number of domains is involved in a large number of holes (see Figure 2). Removal of about 3% of the domains leaves the domain overlap graph square-free. Characteristically, the group of removed domains contains known promiscuous domains (domains known to appear in a large number of diverse proteins). It is indeed appropriate not to include them on equal footing with other characters in parsimony methods.

As mentioned before, the number of squares in the intron overlap graph is very large and it was not clear if removal of the inconsistencies represented by these squares would lead to a meaningful result. We devised a heuristic algorithm to remove squares from the intron overlap graph. Interestingly, we obtained the evolutionary tree consistent with the Coelomata hypothesis [1, 2, 5, 30]. One can think of squares removal as a process that selects a set of characters that are likely to yield a correct tree. This is very much like choosing pairwise compatible characters and building the tree based on these characters alone. However, it is important to point out that, since the concept of persistence is less stringent than pairwise compatibility, this method can be successful when the compatibility method fails. In particular, as shown later in the paper, replacing persistence by pairwise compatibility in the context of intron data does not lead to a correct tree.

2   Characters, Character Overlap Graphs and Parsimony Methods

Characters and parsimony methods. Assume that we are given a set of taxa such that each taxon is characterized by a vector of characters. Intuitively, a character can be anything that describes the properties of a taxa, e.g. external characteristics (like wings, legs, etc.) or a molecular information (like genes, protein domains, etc.). In this work, we assume binary characters, that is, characters that take either value one or value zero (interpreted respectively as the presence/absence of the given characteristics in the taxon). Assume that, during the evolution, characters can be gained and/or lost. Under this assumption, the evolution of a given set of taxa is often reconstructed using parsimony methods. The underlying assumption of parsimony methods is that the characters evolve in a way that minimizes character changes. The maximum parsimony tree is a tree whose leaves are labeled with the character vectors associated with the input taxa, and internal nodes are labeled with the inferred character vectors of ancestral taxa such that the total number of character changes along the tree branches is minimized. Additional restrictions on the type, number, and direction of changes lead to a variety of specific parsimony models [11]. For example, in Dollo parsimony, a character may be inserted (change state from zero to one) only once, but it can be lost multiple times [15]. In Camin-Sokal parsimony, no reversal of character changes is allowed [8]. The problem of computing the maximum parsimony tree is NP-complete for most of parsimony models, including Dollo parsimony and Camin-Sokal parsimony mentioned above [9].

A major problem with parsimony methods is (in addition to their computational cost) that they sometimes produce an obviously incorrect tree. This elucidates the importance of being able to decide if a given character set is likely to be misleading when used in conjunction with a parsimony method. Intuitively, we are interested in characters that are not very easy to gain (thus the number of independent insertions of the same character is limited) and which persist through evolution, i.e. they are not too easy to lose. We propose a graph-theoretical measure that can be used to test whether a given selection of characters is likely to produce the correct evolutionary tree.

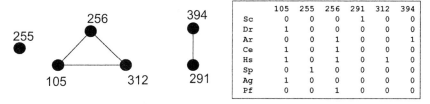

	105	255	256	291	312	394
Sc	0	0	0	1	0	0
Dr	1	0	0	0	0	0
Ar	0	0	1	0	0	1
Ce	1	0	1	0	0	0
Hs	1	0	1	0	1	0
Sp	0	1	0	0	0	0
Ag	1	0	0	0	0	0
Pf	0	0	1	0	0	0

Fig. 1. The intron overlap graph for KOG0009 [28]. The introns are identified by the position in the multiple alignment of the corresponding genes. In the matrix on the right side, of the figure rows correspond to the species included in the KOG *Arabidopsis thaliana* (At), *Homo sapiens* (Hs), *C.elegans* (Ce), *Drosophila melanogaster* (Dm), *Anopheles gambaie* (Ag), *Saccharomyces cerevisiae* (Sc), *Schizosaccharomyces pombe* (Sp), and *Plasmodium falciparum* (Pf), and colums correspond to introns identified by Rogozin *et al.* [27] where 1 correspond to the presence and 0 to the absence of the intron at a given position in the multiple alignment.

Character overlap graph. To answer the question whether a given set of characters is hard-to-gain and hard-to-lose, we introduce the concept of a character overlap graph. A character overlap graph is a graph $G = (V, E)$, where V is a set of characters, and $(u, v) \in E$ if there exists a taxon T in the set such that both u and v are present T.

In this paper, we consider two examples of character overlap graphs: a domain overlap graphs and intron overlap graphs. The first family of graphs, also known as domain co-occurrence graphs or domain graphs, has been studied before [31,3,25,32]. A set of taxa used to construct a domain overlap is a family of multidomain proteins. The vertices of the domain overlap graph correspond to protein domains and two domains are connected by an edge if and only if there is a protein that contains both domains. In turn, a set of taxa used in the construction of an intron overlap is a set of completely sequenced genomes. The nodes of an intron overlap graph correspond to the introns and there is an edge between two introns if and only if there is a genome that contains both introns (see Figure 2). No construction equivalent to intron overlap graph has been considered before.

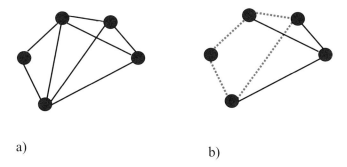

a)   b)

Fig. 2. a) An example of a chordal graph. b) A graph that is not chordal. The red (dotted) circle forms a hole of size four – a square.

Holes and chordal graphs. Chordal graphs constitute a well studied family of graphs [14]. A *chord* in a graph is an edge that connects two non-consecutive vertices of a cycle. A *chordal graph* is a graph which does not have chordless cycles of length greater than three. Chordless cycles of length more than three are called *holes*. Figure 2 (a) shows an example of a chordal graph and Figure 2 (b) a graph which contains a hole of size four – a *square*.

There is a powerful connection between chordal graphs and trees [12,7,19,18, 25], which has been exploited before in the context of phylogenetic trees. We do not use this connection in the paper explicitly, but it is a key result in chordal graph theory and our approach is motivated by this relation.

3   Holes and Parsimony Methods

Graph chordality and persistent characters. We start with an extreme case in which we assume that each character can be gained exactly once and lost at most once. We call such characters *persistent*. Thus a persistent character can undergo at most two changes and these changes are required to respect the order: $0 \to 1 \to 0$. Note that the persistence property is independent of the way a tree is rooted. We show the following simple theorem about persistent characters.

Theorem [characterization of persistent characters]. *If all characters are persistent then the corresponding character overlap graph is chordal.*

Proof. By induction on the size k of the hole.[1] For $k = 4$, assume that there exists a square spanning nodes (characters) A, B, C, and D. This implies that there are four taxa containing respectively pairs of characters AB, BC, CD, and DA, but there does not exist a taxon containing diagonal pairs AC or BD. In fact, no taxon can contain three or more of A, B, C, D simultaneously. Ignoring all other characters, there are, (up to symmetry), two possible binary topologies for the parsimony tree for the four taxa (Figure 3). Since there can be only one insertion per character, all taxa (ancestral or not) containing a specific character must form a connected subtree in the parsimony tree. For example, all nodes on the path from the taxon with characters AB to the taxon with characters BC must contain character B (see Figure 3). Repeating this argument for all pairs of taxa, we infer that the labeling of the two internal nodes in Figure 3 a must contain, respectively, characters A, B, D and B, C, D. By examining all the possibilities it can be seen that this labeling cannot be achieved without deleting at least one character twice. The argument for the case represented in Figure 3 (b) is similar.

Assume now that the graph has a hole $A_0, A_1, \ldots A_{k-1}$ of size k, where $k > 4$. Then for any i there exists a taxon containing the pair of characters $A_i A_{i+1}$ (index additions/subtractions are mod k) but not containing any other A_j where $j \neq i, i+1$. Assume that there exists a parsimony tree T that allows for at most

---

[1] A shorter proof can be made based on the relation between chordal graphs and trees mentioned in the previous section, but in the interest of keeping the paper self-contained we present here a direct argument.

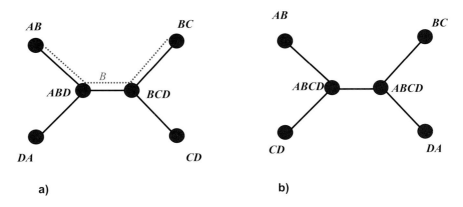

Fig. 3. The two possible (up to symmetry) topologies for an evolutionary tree for four taxa containing respectively characters: AB, BC, CD and DA. Under the assumption that only one insertion per character is allowed, in each case there must exist a character which is deleted twice.

one insertion and one deletion of each character. Consider the subtree of T spanning the taxa involved in the cycle. Let X be an internal node of this tree which is adjacent to two leaves (such node must exist). First, we argue that the two leaves must correspond to two consecutive taxa on the circle. Assume otherwise and let the two leaves be described by character pairs $A_i A_{i+1}$ and $A_j A_{j+1}$ where $j \neq i + 1$ and $i \neq j + 1$. Then X must contain characters $A_i A_{i+1} A_j A_{j+1}$. (This observation follows from the fact that each of the four characters also belongs to a taxon other than the two taxa corresponding to the leaves $A_i A_{i+1}$ and $A_j A_{j+1}$ and that no double insertions are allowed.) Consider now the subtree spanned by the leaves $A_i A_{i+1}$, $A_j A_{j+1}$, the internal node X, and the leaves containing characters $A_i, A_{i+1}, A_j, A_{j+1}$ other than the leaves corresponding to the pairs $A_i A_{i+1}$ and $A_j A_{j+1}$. By a case analysis similar to the one for the base case if find that the topology of this tree contradicts the assumption of single insertion/deletion. Thus the two leaves must correspond to two consecutive taxa in the circle, that is without loss of generality, $A_{i+1} = A_j$. Now we are ready to use the inductive hypothesis. Replace the pair of taxa with characters $A_i A_{i+1}$ and $A_{i+1} A_{i+2}$ with one taxon with characters $A_i A_{i+1} A_{i+2}$ and consider the tree T' obtained from the tree T by removing leaves corresponding to $A_i A_{i+1}$ and $A_{i+1} A_{i+2}$. If T is a tree that does not require more than one insertion and more than one deletion per character so is T' with respect to the modified set of taxa. By the inductive hypothesis, this is impossible since the character overlap graph for the reduced set of taxa contains a cycle of size $k - 1$. QED.

Persistence versus Compatibility. The persistence criterion provided above is similar to the well known compatibility criterion [11] at the basic level. Namely, they both seek to identify characters that are in some sense inconsistent. Then, one can look for a set of characters whose removal leaves a set of consistent characters and construct the tree based on these consistent characters. There are, however, important differences. In the case of persistent characters, a character

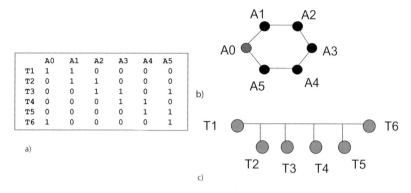

a)

	A0	A1	A2	A3	A4	A5
T1	1	1	0	0	0	0
T2	0	1	1	0	0	0
T3	0	0	1	1	0	1
T4	0	0	0	1	1	0
T5	0	0	0	0	1	1
T6	1	0	0	0	0	1

Fig. 4. a) An artificial example of 6 taxa and 6 characters; (b) corresponding character overlap graph; c)the most parsimonious tree after removing character A_0. Note that in this case half of the characters would have to be removed to obtain a pairwise compatible set.

can change the state at most twice (one insertion and at most one deletion) while for pairwise compatible characters each character changes state at most once. Thus the assumption of persistence is a weaker assumption than that of compatibility. In particular it is easy to see that every edge in a hole identifies a pair of non-compatible characters. Consider for example a set of n taxa each described by two characters $A_0 A_1, A_1 A_2, \ldots A_n A_0$. Then characters $A_i A_{i+1}$ are incompatible. Removing just one character will ensure persistence (and later in this paper we propose a method to decide which one) while one has to remove half of the characters to obtain pairwise compatible set (see figure 4). This weaker consistency requirement is particularly useful when one cannot assume that there exist a sufficiently large set of characters which once inserted are never lost. An example of such situation occurs in the case of intron evolution.

Finally, we shall point out that, unlike the compatibility criterion, the theorem shown provides a necessary but not sufficient condition for persistence of characters.

Graph motifs and persistent characters. The requirement that each character must be persistent is very restrictive. For example, the fact that bats and birds gained wings independently (that is the character wings is gained twice) does not lead to an incorrect evolutionary tree as long as other characters are used to complement this information. So, even if characters are occasionally gained/lost more than once, we may still be able to apply parsimony methods successfully. However, if characters are gained and/or lost independently on massive scale, then there is not much hope of recovering the correct tree. How can one distinguish between these two cases?

One solution would be to measure how far the corresponding character overlap graph is from being a chordal graph. This can be measured, for example, by counting the minimal number of edges whose addition makes the graph chordal. Unfortunately, the problem of finding such a minimal set is NP-complete [17].

We propose a simple heuristic based on the network motifs approach. Rather than considering all holes, we consider only holes of size four (squares). All squares can be easily enumerated. The number of squares in an attribute overlap graph can then be compared to the number of squares in a null model, where the characters are gained/lost randomly. The ratio of these two counts can be used to measure how easily the characters are gained/lost.

Our null model assumes the same number of taxa as the real data and the same set of characters. Furthermore, there is a one-to-one correspondence between the real taxa and the taxa in the null model. In the null model, the characters of each taxon are selected randomly in a way that for each taxon the expected number of characters equals the number of characters of the corresponding real taxon.

In general, each square in a character overlap graph indicates existence of a non-persistent character. While a small number of squares is clearly an indication of a persistent nature of most of the characters, the large number of squares does not necessarily indicate that the number of non-persistent characters is equally large. Squares can overlap and a small number of non-persistent characters can result in a relatively large number of squares. Thus, characters involved in a large number of squares introduce significant noise to the data. One can address this problem by assigning a smaller weight to these characters, or simply by removing them from the data.

4   Applications and Experimental Results

Construction of intron overlap graph and domain overlap graphs. To construct intron overlap graph, we used the data from a study by Rogozin *et al.* [27]. This data contains information about introns found in conserved (and orthologous) genes of eight fully sequenced organisms: *Arabidopsis thaliana* (At), *Homo sapiens* (Hs), *C.elegans* (Ce), *Drosophila melanogaster* (Dm), *Anopheles gambaie* (Ag), *Saccharomyces cerevisiae* (Sc), *Schizosaccharomyces pombe* (Sp), and *Plasmodium falciparum* (Pf). Introns are identified by their starting position with respect to the coding sequence. The data contains information about 7236 introns, however most of these introns are observed in one organism only. After eliminating these single-organism entries, we were left with 1790 introns.

To construct domain overlap graphs, we used the data from a study by Przytycka *et al.* [25] containing 479 multi-domain superfamilies. This data was built using the non-redundant multidomain proteins in Swiss-prot [6], where the domains were recognized using CDART [13]. Proteins in this set are grouped into overlapping superfamilies. Each superfamily is defined to be the maximal set of proteins that have a specific domain in common. For example, all proteins containing the kinase domain form one superfamily, proteins containing the SH2 domain form another superfamily and these two superfamilies intersect. Each such superfamily is considered to have its own evolutionary history, therefore, each superfamily is treated separately. For each such superfamily there is a separate domain overlap graph. Domain overlap graphs with less than four nodes

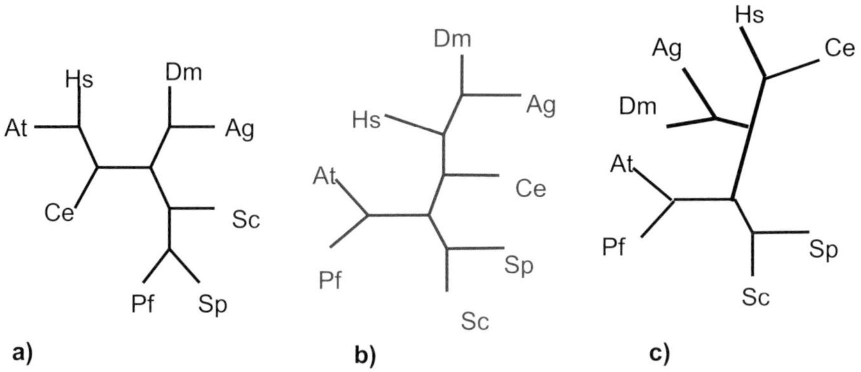

Fig. 5. Three tree topologies for organisms: *Arabidopsis thaliana* (At), *Homo sapiens* (Hs), *C.elegans* (Ce), *Drosophila melanogaster* (Dm), *Anopheles gambaie* (Ag), *Saccharomyces cerevisiae* (Sc), *Schizosaccharomyces pombe* (Sp), and *Plasmodium falciparum* (Pf) a) The incorrect Dollo parsimony tree computed from intron data b) The tree consistent with Coelomata hypothesis. This is also exactly the tree obtained after applying our squares removal procedure. c) The tree consistent with Ecdysozoa hypothesis.

Table 1. The frequencies of occurrences of squares in intron overlap graph and domain overlap graph relative to the corresponding null random model. Observe significant under-representation of squares in the domain overlap graph.

Character type	# squares in character overlap graph (s)	# squares expected by chance
Introns	954 667 368	1 389 751 510
Domains	251	3 822

were ignored. Similarly, networks in which the number of edges was smaller than the number of nodes were disregarded.

Counting squares. The relative numbers of squares for both types of overlap graphs are summarized in Table 1. In the domain overlap graphs, the total number of squares is relatively small. This indicates that domains tend to be persistent and thus provide a good set of characters to be used by parsimony methods. In contrast, the intron overlap graph contains nearly as many squares as it is expected by chance. This suggests that applying parsimony methods to this data is likely to give an incorrect result. Indeed, Rogozin *et al.* constructed such tree (using Dollo parsimony) and found that it is completely wrong. Figure 5 shows the result of this construction, Figure 5 (b) the tree consistent with the Coelomata hypothesis, and Figure 5 (c) the tree consistent with the Ecdysozoa hypothesis. Interestingly, the incorrect Dollo tree is supported by high bootstrap values [27] suggesting that the incorrectness of the tree is due to a systematic bias rather than a random noise.

Eliminating squares in domain overlap graphs. Figure 6 shows the distribution of number squares summed up over all domain overlap graphs. We observe that a few domains are involved in a large number of squares. Since the

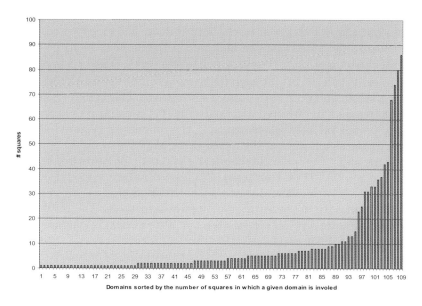

Fig. 6. The distribution of squares. The domains (x axis) are sorted in the increasing order of the number of squares they belong to.

problem of removing the smallest number of nodes to obtain a hole-free graph is NP-complete ([33]), we iteratively removed the node involved in the highest number of squares (re-computing the number of squares each time). The first two domains removed by our greedy approach are two functionally uncharacterized domains (smart00267 and smart00052 [16]). Subsequently, the algorithm identifies for removal two known promiscuous domains (domains that are known to appear in many diverse multidomain proteins): SH2, ABC-ATPase. Removal of these four domains reduces the already small number of squares by nearly 80%. After this step, there are still a few domains involved in squares including: PDZ, PH, SH3, EGF, IG-like. However, none of these domains are involved in more than 11 squares.

Eliminating squares in the intron overlap graph. In the case of intron overlap graph, the number of squares is not much smaller than what is expected by chance. The most frequently occurring squares are of type: (At, Hs, Ce, X), where X is Dm or Ag. Note that each intron is represented by a binary pattern of length eight (the number of genomes in the data) where one corresponds to the intron being present in the given genome and zero to its absence. Introns with the same pattern are indistinguishable from the perspective of parsimony methods and are involved in the same number of squares. Note further that with eight species there are $2^8 - 9$ intron patterns (the subtraction corresponds to the assumption that each intron must be in at least two species) out of which 90 patterns were populated. Thus, some patterns are represented multiple times. The patterns that appear significantly more often than it is expected by chance are considered to be more informative (more significant). Let n_i be the number of times pattern i is observed in the intron data, and r_i expected number of occurrences of the pattern in the null model. Define $p_i = \frac{n_i}{r_i}$ to be the significance the intron pattern i. (Using $p_i = \max(\log \frac{n_i}{r_i}, \epsilon)$, where ϵ is a real number closed to zero (here $= 10^{-10}$) gave the same results.) Let S_i be the number of squares in which an intron with pattern i is involved. Our greedy square removal algorithm removes iteratively intron patterns that maximize the value $\frac{S_i}{p_i}$. This provides a trade off between maximizing the number of removed squares and minimizing the significance of the removed intron patterns. After all squares are removed, we apply the Dollo parsimony to the remaining introns. The procedure removed intron 52 (57 %) patterns. We also introduce a modification to the Dollo parsimony with enforces that the contribution of each intron is weighted with the significance of the corresponding intron pattern. The resulting tree is presented on Figure 5 (b). Thus we obtained a tree which is consistent with the Coelomata hypothesis.

We also applied the same greedy approach with the persistence criterion replaced with the compatibility criterion. The procedure removed 86 (95 %) of intron patterns and produced 15 incorrect trees.

5   Discussion, Conclusions, and Further Work

We demonstrated that the character overlap graph for persistent characters is chordal. This suggests that the character overlap graph for characters that are hard-to-gain and hard-to-lose is expected to contain relatively few holes as compared to a null model. In particular, the number of holes of size four (squares) is also expected to be relatively small. The last property is easily testable, and provides a fast method for checking whether a set of characters can be used to produce a correct evolutionary tree. In practical applications, we found that the number of squares in the domain overlap graph is very small, supporting the findings that domains can be used as characters in a parsimony approach. In contrast, the number of squares in the intron overlap graph is not much smaller than it is expected by chance. This explains why the Dollo tree built based on intron data is incorrect.

A large number of squares does not necessarily indicate that all characters are non-persistent. For example, we demonstrated that in the domain overlap graph, the majority of squares come from the existence of a handful of promiscuous domains. Consequently, removing a small number of domains from this character set leaves the domain overlap graph square free.

A similar approach applied to the intron overlap graph also produced an interesting result. While it is known that introns can be remarkably conserved in some lineages, they are not so conserved in others. This leads to a large number of squares. However, we found that the distribution of these squares is non-uniform. Thinking of squares as inconsistencies in the data, we applied a greedy algorithm to remove introns that are involved in square formation, choosing introns of low significance and high involvement in square motifs. We used this truncated intron data to construct a weighted Dollo parsimony tree. That is, we weighted the contribution of each intron according to the significance of the corresponding intron pattern. With these two changes, we obtained a parsimony tree which is consistent with the tree constructed using other methods [30]. This is in contrast to the previous applications of parsimony methods, which have been unable to recover a tree consistent with any of the proposed evolutionary hypotheses.

The results of this work strongly suggest that removal of non-persistent characters involved in a large number of squares may significantly improve the applicability of parsimony methods.

Acknowledgments. All algorithms have been implemented with the help of LEDA tools [20]. The author thanks David Lipman (NCBI), Igor Rogozin (NCBI), Uri Wolf (NCBI) for stimulating discussions, and Raja Jothi (NCBI) and Elena Zotenko (NCBI/UMD) for discussions and critical comments on the manuscript. This research was supported by the intramural research program of the National Institutes of Health.

References

[1] A. Adoutte, G. Balavoine, N. Lartillot, O. Lespinet, Benjamin Prud'homme, and Renaud de Rosa. Special Feature: The new animal phylogeny: Reliability and implications. *PNAS*, 97(9):4453–4456, 2000.

[2] A.M. Aguinaldo, J.M. Turbeville, L.S. Linford, M.C. Rivera, J.R. Garey, R.A. Raff, and J.A. Lake. Evidence for a clade of nematodes, arthropods and other moulting animals. *Nature*, 387:489–93, 1997.

[3] G. Apic, W. Huber, and S.A. Teichmann. Multi-domain protein families and domain pairs: Comparison with known structures and a random model of domain recombination. *J. Struc. Func. Genomics*, 4:67–78, 2003.

[4] A.-L. Barabasi and R. Albert. Emergence of scaling in random networks. *Science*, 286:509–512, 1999.

[5] Jaime Blair, Kazuho Ikeo, Takashi Gojobori, and S Blair Hedges. The evolutionary position of nematodes. *BMC Evolutionary Biology*, 2(1):7, 2002.

[6] B. Boeckmann, A. Bairoch, R. Apweiler, M.-C. Blatter, A. Estreicher, E. Gasteiger, M.J. Martin, K. Michoud, C. O'Donovan, I. Phan, S. Pilbout, and M. Schneider. The SWISS-PROT protein knowledgebase and its supplement TrEMBL in 2003. *Nucleic Acids Res.*, 31:365–370, 2003.

[7] P. Buneman. A characterisation of rigid circuit graphs. *Discrete Math.*, 9:205–212, 1974.

[8] J. H. Camin and R.R. Sokal. A method for deducting branching sequences in phylogeny. *Evolution*, 19:311–326, 1965.

[9] W.H.E. Day, D. Johnson, and D. Sankoff. The computational complexity of inferring rooted phylogenies by parsimony. *Mathematical Biosciences*, 81:33–42, 1986.

[10] E.J. Deeds, Hooman Hennessey, and Eugene I. Shakhnovich. Prokaryotic phylogenies inferred from protein structural domains. *Genome Res.*, 15(3):393–402, 2005.

[11] J. Felsenstein. *Inferring Phylogenies*. Sinauer Associates, 2004.

[12] F. Gavril. The intersection graphs of subtrees in trees are exactly the chordal graphs. *J. Comb. Theory (B)*, 16:47–56, 1974.

[13] L.Y. Geer, M. Domrachev, D.J. Lipman, and S.H. Bryant. CDART: protein homology by domain architecture. *Genome Res.*, 12(10):1619–23, 2002.

[14] M. Golumbic. *Algorithmic Graph Theory and Perfect Graphs*. Academic Press, New York, 1980.

[15] J.S.Farris. Phylogenetic analysis under Dollo's law. *Systematic Zoology*, 26(1):77–88, 1977.

[16] I. Letunic, L. Goodstadt, N.J. Dickens, T. Doerks, J. Schultz, R. Mott, F. Ciccarelli, R.R. Copley, C.P. Ponting, and P. Bork P. Recent improvements to the SMART domain-based sequence annotation resource. *Nucleic Acids Res.*, 31(1):242–244, 2002.

[17] J. M. Lewis and M. Yannakakis. The node-deletion problem for hereditary properties is NP- complete. *J. Comput. Syst. Sci.*, 20(2):219–230, 1980.

[18] T.A. McKee and F.R. McMorris. *Topics in intersection graph theory*. SIAM Monographs on Discrete Mathematics and Applications, 1999.

[19] F.R. McMorris, T. Warnow, and T. Wimer. Triangulating vertex colored graphs. *SIAM J. on Discrete Mathematics*, 7(2):296–306, 1994.

[20] K. Mehlhorn and S. Naher. *The LEDA Platform of Combinatorial and Geometric Computing*. Cambridge University Press, 1999.

[21] M. Middendorf, E. Ziv, and C. H. Wiggins. From The Cover: Inferring network mechanisms: The Drosophila melanogaster protein interaction network. *PNAS*, 102(9):3192–3197, 2005.

[22] R. Milo, S. Itzkovitz, N. Kashtan, R. Levitt, S. Shen-Orr, I. Ayzenshtat, M. Sheffer, and U. Alon. Superfamilies of Evolved and Designed Networks. *Science*, 303(5663):1538–1542, 2004.

[23] R. Milo, S. Shen-Orr, S. Itzkovitz, N. Kashtan, D. Chklovskii, and U. Alon. Network Motifs: Simple Building Blocks of Complex Networks. *Science*, 298(5594):824–827, 2002.

[24] N. Przulj, D. G. Corneil, and I. Jurisica. Modeling interactome: scale-free or geometric? *Bioinformatics*, 20(18):3508–3515, 2004.

[25] T.M. Przytycka, G. Davis, N. Song, and D. Durand. Graph theoretical insight into evolution of multidomain proteins. *Lecture Notes in Computational Biology (RECOMB 2005)*, 3500:311321, 2005.

[26] T.M. Przytycka and Y.K. Yu. Scale-free networks versus evolutionary drift. *Computational Biology and Chemistry*, 28:257–264, 2004.

[27] I.B. Rogozin, I.Y Wolf, A.V. Sorokin, B.G. Mirkin, , and V Koonin, E. Remarkable interkingdom conservation of intron positions and massive, lineage-specific intron loss and gain in eukaryotic evolution. *Current Biology*, 13:1512–1517, 2003.

[28] R. Tatusov, N. Fedorova, J. Jackson, A. Jacobs, B. Kiryutin, E. Koonin, D. Krylov, R. Mazumder, S. Mekhedov, A. Nikolskaya, B.S. Rao, S. Smirnov, A. Sverdlov, S. Vasudevan, Y. Wolf, J. Yin, and D. Natale. The cog database: an updated version includes eukaryotes. *BMC Bioinformatics*, 4(1):41, 2003.

[29] Henry F. Winstanley, Sanne Abeln, and Charlotte M. Deane. How old is your fold? *Bioinformatics*, 21(suppl1):i449–458, 2005.

[30] Y.I. Wolf, I.B. Rogozin, and E.V. Koonin. Coelomata and Not Ecdysozoa: Evidence From Genome-Wide Phylogenetic Analysis. *Genome Res.*, 14(1):29–36, 2004.

[31] S. Wuchty. Scale-free behavior in protein domain networks. *Mol. Biol. Evol.*, 18:1694–1702, 2001.

[32] S. Wuchty and E. Almaas. Evolutionary cores of domain co-occurrence networks. *BMC Evolutionary Biology*, 5(1):24, 2005.

[33] M. Yannakakis. Computing the minimum fill-in is NP- complete. *SIAM J. Alg and Discrete Math*, 2:77–79, 1981.

[34] E. Yeger-Lotem, S. Sattath, N. Kashtan, S. Itzkovitz, R. Milo, R. Y. Pinter, U. Alon, and H. Margalit. Network motifs in integrated cellular networks of transcription-regulation and protein-protein interaction. *PNAS*, 101(16):5934–5939, 2004.

Ultraconserved Elements, Living Fossil Transposons, and Rapid Bursts of Change: Reconstructing the Uneven Evolutionary History of the Human Genome

David Haussler

Howard Hughes Medical Institute,
UC Santa Cruz, USA

Abstract. Comparison of the human genome with the genomes of the mouse, rat, dog, cow, and other mammals reveals that at least 5% of the human genome is under negative selection. Negative selection occurs in important functional segments of the genome where random (mostly deleterious) mutations are rejected by natural selection, leaving the orthologous segments in different species more similar than would be expected under a neutral substitution model. Protein coding regions account for at most 1/3 of the segments that are under negative selection. In fact, the most conserved segments of the human genome do not appear to code for protein. These "ultraconserved" elements, of length from 200-800bp, are totally unchanged between human mouse and rat, and are on average 96% identical in chicken. The function of most is currently unknown, but we have evidence that many may be distal enhancers controlling the expression of genes involved in embryonic development. Other ultraconserved elements appear to be involved in the regulation of alternative splicing. Evolutionary analysis indicates that many of these elements date from a period very early in the evolution of vertebrates, as they have no orthologous counterparts in sea squirts, flies or worms. At least one group, involving a conserved enhancer of one gene and an ultraconserved altspliced exon of another, evolved from a novel retrotransposon family that was active in lobe-finned fishes, and is still active today in the "living fossil" coelacanth, the ancient link between marine and land vertebrates.

In contrast with the slowly changing ultraconserved regions, in other areas of the genome recent genetic innovations that are specific to primates or specific to humans have caused relatively rapid bursts of localized changes, possibly through positive selection. Via simulation, we estimate that most of the DNA sequence of the common ancestor of all placental mammals, which lived in the last part Cretaceous period about 80-100 million years ago, can be predicted with 98% accuracy. We recently reconstructed and entire chromosome arm from the genome of this ancient species, and are currently working on a full genome reconstruction. Given this as a basis, and enough well-placed primate genomes to reconstruct intermediate states, we should eventually be able to document most of the genomic changes that occurred in the evolution of

A. Apostolico et al. (Eds.): RECOMB 2006, LNBI 3909, pp. 336–337, 2006.

the human lineage from the placental ancestor over the last 100 million years, including innovations that arose by positive selection.

Credits

UCSC Genome Bioinformatics Group and Genome Laboratory; enhancer work is collaboration with Eddy Rubin lab at Berkeley, reconstruction project is collaboration with Webb Miller group at Penn State, Mathieu Blanchette at McGill, and Eric Lander.

References

see http://genome.ucsc.edu/goldenPath/pubs.html

Permutation Filtering: A Novel Concept for Significance Analysis of Large-Scale Genomic Data

Stefanie Scheid and Rainer Spang

Max Planck Institute for Molecular Genetics,
Computational Diagnostics Group,
Ihnestrasse 63-73, D-14195, Berlin, Germany
firstname.lastname@molgen.mpg.de

Abstract. Permutation of class labels is a common approach to build null distributions for significance analyis of microarray data. It is assumed to produce random score distributions, which are not affected by biological differences between samples. We argue that this assumption is questionable and show that basic requirements for null distributions are not met.

We propose a novel approach to the significance analysis of microarray data, called permutation filtering. We show that it leads to a more accurate screening, and to more precise estimates of false discovery rates. The method is implemented in the Bioconductor package *twilight* available on http://www.bioconductor.org.

1   Introduction

Screening thousands of candidate genes using some scoring function is a widely applied strategy in the analysis of microarrays. A typical scenario is the search for differentially expressed genes, where the sores can be fold changes or t-statistics. Screening inherently leads to a multiple testing problem, which requires the definition of a null distribution of scores. It is common practice to use simulated distributions obtained from randomizations of the original data [1]. With a set of samples (arrays) and corresponding class labels for the samples, one calculates scores for the original class labels, and compares them to the distribution of scores obtained from random shuffling of the class labels. Permutation approaches are popular because the correlation structure of gene expression levels is unknown, which makes the definition of a theoretical joint null distribution difficult. By randomly assigning the class labels to the samples and recomputing scores one circumvents this difficulty and generates a set of random scores, which serves as a null distribution for statistical inference. One transforms the scores obtained from the original class labels to empirical p-values by using the distribution of simulated scores from the permutation null model. Under the assumption that not a single gene is differentially expressed, one expects that this set of p-values is uniformly distributed. Several methods for estimating global or

A. Apostolico et al. (Eds.): RECOMB 2006, LNBI 3909, pp. 338–347, 2006.

local false discovery rates rely on the assumption that the p-value distribution for a set of non-differentially expressed genes is uniform [2, 3, 4, 5, 6, 7].

To borrow information across genes, empirical p-values are computed using a pooled set of scores from all genes on the array [8]. The combined use of class label permutations and score pooling leads to a conceptual problem. In real applications, one typically has both differentially and non-differentially expressed genes. While permutations produce a justifiable null distribution of scores for the non-differentially expressed genes, one expects that they produce wider score distributions for the differentially expressed genes. Wide score distributions are not only expected for genes that are differentially expressed between the class distinction of interest, but also for genes that are differentially expressed regarding some hidden non-random structure in the data, such as the gender of patients or experimental artefacts. As a consequence, the pooled set of scores is contaminated by signals resulting from differentially expressed genes and does not yield a pure null distribution.

In the next section we recall the notation for permutation approaches to multiple testing in microarray studies. In Section 3 we use a clinical data set to show that random permutations produce distributions, which do not meet basic requirements for a null distribution. As a way out of this dilemma, we describe in Section 4 the details of a novel approach to permutation tests termed *permutation filtering*. In Section 5 we show that permutation filtering produces valid null distributions, increases the accuracy of the screening, and leads to more precise estimates of false discovery rates.

2   Notation

Let matrix $\mathbf{X}$ be an $m \times n$ gene-expression matrix with genes in rows and samples in columns. Entry x_{ij} is the value of the ith gene observed for the jth sample with genes $i = 1, \ldots, m$ and samples $j = 1, \ldots, n$. In addition, we have a vector $\mathbf{c}_0 = (c_1, \ldots, c_n)$ with c_j being the class label of the jth sample. For simplicity of presentation we only consider binary class labels here. As a real world example, we shall later discuss a breast-cancer data set, where the class label is either one of two clinically defined risk groups.

Let $\mathbf{s}_0$ denote the vector of scores with entries $(s_{i0})_{i=1,\ldots,m}$. Let $\mathbf{c}$ be a random permutation of the entries of vector $\mathbf{c}_0$. Note that we shuffle only the class labels to preserve the correlation structure between the genes. We recompute the score of each gene based on $\mathbf{c}$ and derive a set of scores $\mathbf{s}$. Say, we do B permutations $\mathbf{c}_1, \ldots, \mathbf{c}_B$ in total. This yields B random score vectors $\mathbf{s}_1, \ldots, \mathbf{s}_B$. We join the original and the random scores into the $m \times (B + 1)$ score matrix $\mathbf{S}$ defined as:

$$\mathbf{S} := (\mathbf{s}_0 \, \mathbf{s}_1 \ldots \mathbf{s}_B) = (s_{ij}) \text{ with } i = 1, \ldots, m \text{ and } j = 0, \ldots, B.$$

To compute the empirical p-value for score s_{i0}, we count how often a random score exceeds the observed score of the gene of interest:

$$p_{i0} = \frac{1}{m(B+1)} \sum_{k=1}^{m} \sum_{l=0}^{B} I\{|s_{kl}| \geq |s_{i0}|\} \tag{1}$$

with $I\{x\}$ being an indicator function that returns 1 if x is true and 0 otherwise. For simplicity of notation, we summarize the whole process in function $U_{\mathcal{C}}$, which maps a fixed vector of class labels c_0 to the vector $p_0 = (p_{i0})_{i=1,...,m}$ of associate p-values

$$U_{\mathcal{C}}(c_0) = p_0 \qquad (2)$$

where $\mathcal{C} = \{c_0, c_1, \ldots, c_B\}$ is the set of permutations on which we assess the significance of scores.

3   Random Permutations Can Produce Invalid Null Distributions

In this section we show that random permutations can produce score distributions, which do not meet basic requirements of a null distribution. We use a clinical microarray study comprising a total of 89 samples from breast-cancer patients measured on Affymetrix GeneChip® HGU95Av2 arrays, which code for $m = 12625$ transcripts/genes [9]. We applied the following preprocessing steps. The background was calculated similar as in the Affymetrix® software Microarray Suite 5.0 [10]. The only difference was that we did not use a correction to avoid negative values. After background correction, we normalized on probe level using a variance-stabilizing procedure [11]. Perfect match probes within a probe set were summarized by the median-polish method [12]. For each probe set, an additive model with probe set, chip and overall effect was fitted using a robust median-polish procedure. Mismatch probes were not taken into account at all.

We compare two risk groups, that is 18 patients with high risk of relapse to 19 low-risk patients ($n = 37$). Again, c_0 is a binary vector of length n of class labels where "1" corresponds to the high-risk and "0" to the low-risk class. We score each gene i by computing absolute z-scores as described in [13]. The z-scores are defined as regularized t-statistics with a positive fugde factor added to their denominators. The fudge factor prevents genes with small variances from having high scores. We set the fudge factor to the median value of the pooled standard deviations across genes.

We draw $B = 1000$ random permutations of the original labeling c_0, compute the matrix $\mathbf{S}$ of z-scores and empirical p-values $p_0 = U_{\mathcal{C}}(c_0)$. Each permutation is assumed to destroy all biological signals in the data, such that the resulting set of scores consists of random scores, which are not driven by biological signals at all. Deviations in the scores obtained from the original (not permuted) class labels give evidence for differentially expressed genes.

Next we introduce a key requirement for a valid null distribution. We let each permutation of class labels c_b in turn play the role of the original class labels and calculate $p_b := U_{\mathcal{C}}(c_b)$. Hence, we use the function $U_{\mathcal{C}}$ not only for assigning a vector of p-values to the original class labels, but also to each permuted vector of class labels. If the permutation process truly has destroyed all biological signal one would expect to observe uniform distributions of p-values. In panel A of

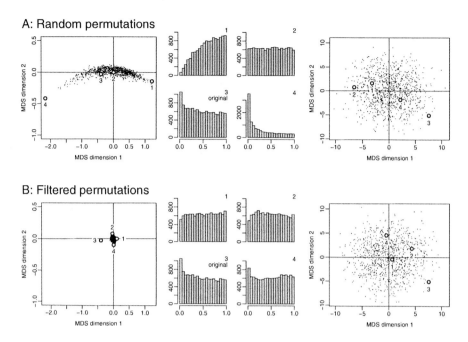

Fig. 1. A: Random permutation does not always produce valid null distributions. The multi-dimensional scaling plot on the left-hand side shows distributional distances between 1000 sets of p-values resulting from random permutations. Euclidean distances between the CDFs of the p-value sets were used. The four numbered examples show that permutations on the right side in the MDS plot have increasing densities, permutations on the left side have decreasing densities, and only permutations close to the origin produce uniform densities. No. 3 represents the original class labels c_0. The scatterplot on the right-hand side shows a second MDS mapping of the permutations, now based directly on the Hamming distances of permuted class labels. The permutations do not cluster but scatter randomly around the origin. **B: Filtering of permutations leaves uniform p-value distributions.** The filtering algorithm returns 1000 permutations that produce uniform p-value distributions, which cluster around the origin in the MDS plot on the left-hand side. Again, no. 3 represents the original labeling c_0 while the other three permutations were chosen from the extremes of the filtered set to show that these are still admissible. The MDS plot based on Hamming distances between permutations is similar to the one in A. Filtered permutations still spread evenly in the permutation space. Note that both pairs of MDS plots were derived from joint sets of filtered and unfiltered permutations.

Fig. 1 one can see that this is not the case. The top left plot shows a multi-dimensional scaling (MDS) representation of the p-value distributions obtained by fixing single permutations. We derived the mapping into two dimensions from the Euclidean distances between the empirical cumulative distribution functions (CDFs) of the associated sets of p-values. Close points represent permutations c_b, which produce similarly distributed p-values $U_\mathcal{C}(c_b)$.

We annotated four exemplary permutations by numbers including the original labels, whose p-value distributions are shown in the top middle plot. Only permutations close to the MDS origin produce uniform p-value distributions. The majority of permutations, however, deviates substantially from uniformity, and often produces distributions, which deviate stronger from uniformity than that of the original class labels.

These results show that random permutations do not produce valid null distributions. Many permutations produce more differential gene expression than the original labels. The scores are not random and the randomization process has not destroyed all biological signal in the data.

4   Permutation Filtering

We now present the permutation filtering procedure. The key idea is to apply the function $U_\mathcal{C}$ not only to the original class labels, but also to the permuted ones, as was already done in the previous section. We argue that a valid permutation-based null distribution has to be derived from a set $\mathcal{C}$ of permutations, satisfying the requirement that $U_\mathcal{C}(c)$ is uniformly distributed for all $c \in \mathcal{C}$.

Assume we have identified a set of permutations $\mathcal{C}_0$, which consists only of permutations that represent valid null hypotheses across all genes. We expect that $U_{\mathcal{C}_0}(c)$ is uniform for all $c \in \mathcal{C}_0$. If however, we observe strong deviations from uniformity, either c_0 or large parts of the remaining permutations in $\mathcal{C}_0$ correlate with some non-random structure in the data.

We propose the following filtering procedure to derive a set of permutations $\mathcal{C}_0$, which consistently produces uniform p-value distributions when calculating p-values for a fixed permutation using the remaining permutations in $\mathcal{C}_0$:

1. Let $\mathcal{C} = \{c_1, \ldots, c_B\}$ be a set of unique random permutations of the original class labels c_0. Apply function $U_\mathcal{C}$ to all $c_b \in \mathcal{C}$, which yields the p-value vectors $p_1, \ldots, p_B$. Choose a stepsize k and set $v = 1$.

2. Let F_b be the empirical CDF of the p-values in p_b. Test each permutation for uniformity of its p-value CDF by computing the Kolmogoroff-Smirnoff statistic

$$\mathrm{KS}_b = \max_{i=1,\ldots,m} |F_b(p_{ib}) - p_{ib}|.$$

   Keep the $v \cdot k$ permutations with the smallest KS statistic in the set $\mathcal{C}_0$. Increase $v = v + 1$.

3. Generate a new set of unique random permutations $\mathcal{C}$, join it with $\mathcal{C}_0$ and apply $U_{\mathcal{C}_0 \cup \mathcal{C}}$ to all $c_b \in \mathcal{C}_0 \cup \mathcal{C}$.

4. Iterate steps 2 and 3 until $|\mathcal{C}_0|$ reaches a predefined number of permutations.

5. Compute the final vector of empirical p-values $p_0 = U_{\mathcal{C}_0 \cup c_0}(c_0)$ for the original class labels.

We chose an iterative design to reduce computational time and save memory. Only a subset of unique random permutations is drawn and tested for uniformity in each step. We keep the admissible permutations in $\mathcal{C}_0$. We do not have to recompute the corresponding scores for the kept permutations. Only when we join $\mathcal{C}_0$ with a new set of permutations, we need to recompute the p-values since we then use an altered set of permutations.

The proposed algorithm is flexible and adaptable to various types of screening studies. We provide an implemention of the procedure in the statistical software language R. We included the algorithm in the Bioconductor package *twilight* for estimating global and local false discovery rates [7, 14, 15, 16, 17].

5   Results

We apply permutation filtering to the breast-cancer data set described in Section 3. Again we use the z-scores for testing. As default parameters in the filtering process we set the stepsize to $k = 50$, the number of permutations per iteration to 1000 and the stopping criterion to $|\mathcal{C}_0| \geq 1000$.

5.1   Permutation Filtering Produces Valid Null Distributions

The effect of permutation filtering is shown in panel B of Fig. 1. Both pairs of plots were derived from joint sets of filtered and unfiltered permutations. Hence the axes of the MDS plots equal those in panel A. As expected, the filtered permutations lie closer to the origin, and even permutations from the margins of the cloud produce acceptable uniform p-value distributions (middle plot).

We removed identical permutations within the iterative filtering. One might suspect that filtering introduces a selection bias in that the filtered permutations cluster strongly and do not spread over the entire permutation space. To show that this is not the case, we display a two-dimensional MDS mapping of the permutations that we derived from the Hamming distances between the binary

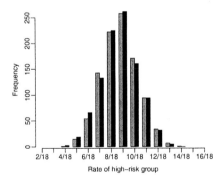

Fig. 2. Permutation filtering does not introduce biased permutations. Distribution of the percentage of the 18 high-risk patients re-assigned to the high-risk group based on random (grey bars) and filtered permutations (black bars). The two distributions do not differ substantially indicating that the filtering does not introduce biased permutations.

vectors of permuted class labels before (panel A) and after (panel B) filtering. Filtered permutations do not form clusters but spread evenly over the permutation space in the MDS representation. There is no visible difference to the corresponding plot for random permutations. We further examine the distribution of the number of samples being randomly re-assigned to their original group. To this end, we count the occurrences of the 18 high-risk patients in the high-risk group for both random and filtered permutations. The result is shown in Fig. 2. We do not observe substantial difference between the two distributions and hence conclude that filtering does not lead to a biased selection of permutations.

5.2   Permutation Filtering Leads to More Significant Genes

A widely used approach to account for multiplicity in microarray studies is to estimate the false discovery rate (FDR) of a list of genes with scores above some prespecified cutoff [18, 19]. The FDR is the expected rate of false positives in this list of genes. Filtering has the effect that one identifies more genes on the same FDR level than without filtering. Hence it increases the sensitivity of the screening for differentially expressed genes.

To show this, we compute p-values of the original labeling c_0 based on the random as well as on the filtered set of permutations. For both sets, we estimate false discovery rates as defined in [8]. In Fig. 3, we display FDRs versus the corresponding number of significant genes. As an example, we marked the FDR cutoff of 0.2 with the dashed line. With filtering, this leads to a list of 103 significant genes, which more than doubles the size of a list without filtering (45 genes).

The increase of significant genes is due to the removal of permutations with p-value distributions similar to that of the original labeling, that is with more small p-values than expected. These distributions correspond to score distributions with heavy tails. The removal of these distributions increases the empirical p-values of genes with high scores.

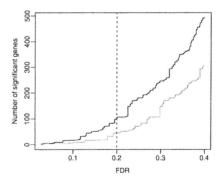

Fig. 3. Permutation filtering leads to more significant genes. FDR cutoffs are plotted versus corresponding numbers of significant genes. The same FDR cutoff leads to more significant findings with filtering (black line) than without (grey line).

5.3 Permutation Filtering Leads to a Higher Accuracy of the Screening

On real data, we can only show an increase of sensitivity since we do not know a priori whether a significant gene is truly induced or not. If the higher sensitivity came for the price of a reduced specificity nothing would be won. To show that this is not the case we use a simulation experiment where the true positives genes are known by design of the simulation.

To this end, we generate random data for 2500 genes and 10 samples per condition. We draw a vector of 2500 random values from a lognormal distribution with location parameter 2 and scale parameter 0.3, and, taking these as mean values, generate 20 random samples from a normal distribution with variance 1. To induce the first 500 genes, we add a value of 2 to the samples of one condition. By adding a value of 4 to five samples of each condition, we introduce hidden non-random structure affecting the following 1000 genes. Note that only the first 500 genes are differentially expressed between populations.

We proceed with the analysis as before and compute p-values based on 1000 filtered and 1000 unfiltered permutations. We rank the genes by p-values and for every rank we estimate the FDR as in [8]. We repeat the data generating procedure 100 times, each time calculating the number of truly induced genes within the list of genes with estimated FDR $\leq 5\%$. Filtering increases the number of correctly identified genes. Without filtering, the list of significant genes includes an average of 457 true positive findings out of 500. Filtering improves the accuracy to 482 correctly identified genes on average. This difference is highly significant in a t-test ($p < 0.0001$).

Hence the filtering increases the sensivity, that is the number of true positives among 500 induced genes, from 0.9134 to 0.9639 on average. The specificity, that is the number of true negatives among 2000 non-induced genes, decreased slightly from 0.9957 to 0.9892. We argue that this loss is negligible regarding the improved sensitivity.

5.4 Permutation Filtering Produces More Precise Estimates of the False Discovery Rate

We use the simulation data from the previous section. The thick black line in Fig. 4 shows the true fraction of induced genes among the top ranking genes. To calculate this line, one has to know a priori which genes are differentially expressed between populations. Hence we can only calculate it in a simulation. The false discovery rate estimates this quantity without knowing the truly differentially expressed genes. Again, one can use both random permutations and filtered permutations to estimate the FDR. The two thin lines in Fig. 4 are the estimated FDR based on filtered (black line) and unfiltered permutations (grey line). While random permutations yield conservative estimates of the false discovery rate, they substantially overestimate it. In contrast, filtered permutation based estimates match the gold standard well. Hence, permutation filtering improves the accuracy of estimated false discovery rates.

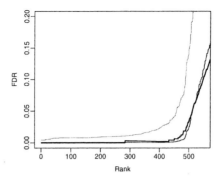

Fig. 4. Permutation filtering leads to more precise FDR estimates. Ranks of high-scoring genes versus the true and estimated FDRs. The FDRs based on filtered permutations (thin black line) estimate the true FDR (thick black line) with high accuracy for the first 500 ranks. FDRs computed without filtering (grey line) lead to conservative but inaccurate estimates.

6   Conclusion

We propose a filtering algorithm that searches for a set of class label permutations where each permutation produces a uniform distribution of p-values. The filtered permutations are then used for calculating empirical p-values and for estimating false discovery rates. The benefits of filtering are valid null distributions, increased numbers of significant genes, a higher accuracy of the screening and more precise estimates of false discovery rates.

We have implemented permutation filtering in the Bioconductor package *twilight* where it is used for calculating both local and global false discovery rates. Permutation filtering is a general concept applicable in many screening studies. It is a novel approach for building valid null distributions. We expect that it will improve the accuracy of high-throughput screenings in various applications in bioinformatics.

Acknowledgments

This work was done within the context of the Berlin Center for Genome Based Bioinformatics (BCB), part of the German National Genome Network (NGFN), and supported by BMBF grants 031U209 and 01GR0455. We thank Anja von Heydebreck and all members of the Computational Diagnostics Group for inspiring discussions and helpful comments.

References

1. Dudoit, S., Yang, Y.H., Callow, M.J., Speed, T.P.: Statistical methods for identifying differentially expressed genes in replicated cDNA microarray experiments. Statistica Sinica **12** (2002) 111–139

2. Broberg, P.: A new estimate of the proportion unchanged genes in a microarray experiment. Genome Biology **5** (2004) P10
3. Dalmasso, C., Broët, P., Moreau, T.: A simple procedure for estimating the false discovery rate. Bioinformatics **21** (2005) 660–668
4. Liao, J., Lin, Y., Selvanayagam, Z.E., Shih, W.J.: A mixture model for estimating the local false discovery rate in DNA microarray analysis. Bioinformatics **20** (2004) 2694–2701
5. Nettleton, D., Hwang, J.G.: Estimating the number of false null hypothesis when conducting many tests. Technical Report 9, Department of Statistics & Statistical Laboratory, Iowa State University (2003)
6. Pounds, S., Morris, S.W.: Estimating the occurrence of false positives and false negatives in microarray studies by approximating and partitioning the empirical distribution of p-values. Bioinformatics **19** (2003) 1236–1242
7. Scheid, S., Spang, R.: A stochastic downhill search algorithm for estimating the local false discovery rate. IEEE Transactions on Computational Biology and Bioinformatics **1** (2004) 98–108
8. Storey, J.D., Tibshirani, R.: Statistical significance for genomewide studies. Proceedings of the National Academy of Sciences **100** (2003) 9440–9445
9. Huang, E., Cheng, S., Dressman, H., Pittman, J., Tsou, M., Horng, C., Bild, A., Iversen, E., Liao, M., Chen, C., West, M., Nevins, J., Huang, A.: Gene expression predictors of breast cancer outcomes. Lancet **361** (2003) 1590–1596
10. Affymetrix: Microarray Suite User Guide, Version 5.0. Affymetrix, Santa Clara, CA, USA. (2001)
11. Huber, W., von Heydebreck, A., Sültmann, H., Poustka, A., Vingron, M.: Variance stabilization applied to microarray data calibration and to the quantification of differential expression. Bioinformatics **18** (2002) 96–104
12. Irizarry, R., Bolstad, B., Collin, F., Cope, L., Hobbs, B., Speed, T.: Summaries of Affymetrix GeneChip probe level data. Nucleic Acids Research **31** (2003) e15
13. Efron, B., Tibshirani, R., Storey, J.D., Tusher, V.: Empirical Bayes analysis of a microarray experiment. Journal of the American Statistical Society **96** (2001) 1151–1160
14. R Development Core Team: R: A language and environment for statistical computing. R Foundation for Statistical Computing, Vienna, Austria. (2005) ISBN 3-900051-07-0.
15. Gentleman, R., Carey, V., Bates, D., Bolstad, B., Dettling, M., Dudoit, S., Ellis, B., Gautier, L., Ge, Y., Gentry, J., Hornik, K., Hothorn, T., Huber, W., Iacus, S., Irizarry, R., Leisch, F., Li, C., Maechler, M., Rossini, A., Sawitzki, G., Smith, C., Smyth, G., Tierney, L., Yang, J., Zhang, J.: Bioconductor: Open software development for computational biology and bioinformatics. Genome Biology **5** (2004) R80
16. Scheid, S., Spang, R.: twilight; a Bioconductor package for estimating the local false discovery rate. Bioinformatics **21** (2005) 2921–2922
17. Scheid, S., Spang, R.: Estimation of local false discovery rate - User's guide to the Bioconductor package twilight. CompDiag Technical Report 1, Computational Diagnostics Group, Max Planck Institute for Molecular Genetics, Berlin, Germany (2004)
18. Benjamini, Y., Hochberg, Y.: Controlling the false discovery rate: A practical and powerful approach to multiple testing. Journal of the Royal Statistical Society: Series B **57** (1995) 289–300
19. Storey, J.D.: The positive false discovery rate: A Bayesian interpretation and the q-value. Annals of Statistics **31** (2003) 2013–2035

Genome-Wide Discovery of Modulators of Transcriptional Interactions in Human B Lymphocytes

Kai Wang[1,2], Ilya Nemenman[2], Nilanjana Banerjee[2],
Adam A. Margolin[1,2], and Andrea Califano[1,2,3,*]

[1] Department of Biomedical Informatics, Columbia University,
622 West 168th Street, Vanderbilt Clinic 5th Floor, New York 10032, New York
[2] Joint Centers for Systems Biology, Columbia University,
1130 St. Nicholas Ave, Rm 801A, New York 10032, New York
[3] Institute of Cancer Genetics, Columbia University,
Russ Berrie Pavilion, 1150 St. Nicholas Ave, New York 10032

Abstract. Transcriptional interactions in a cell are modulated by a variety of mechanisms that prevent their representation as pure pairwise interactions between a transcription factor and its target(s). These include, among others, transcription factor activation by phosphorylation and acetylation, formation of active complexes with one or more cofactors, and mRNA/protein degradation and stabilization processes.

This paper presents a first step towards the systematic, genome-wide computational inference of genes that modulate the interactions of specific transcription factors at the post-transcriptional level. The method uses a statistical test based on changes in the mutual information between a transcription factor and each of its candidate targets, conditional on the expression of a third gene. The approach was first validated on a synthetic network model, and then tested in the context of a mammalian cellular system. By analyzing 254 microarray expression profiles of normal and tumor related human B lymphocytes, we investigated the post transcriptional modulators of the MYC proto-oncogene, an important transcription factor involved in tumorigenesis. Our method discovered a set of 100 putative modulator genes, responsible for modulating 205 regulatory relationships between MYC and its targets. The set is significantly enriched in molecules with function consistent with their activities as modulators of cellular interactions, recapitulates established MYC regulation pathways, and provides a notable repertoire of novel regulators of MYC function. The approach has broad applicability and can be used to discover modulators of any other transcription factor, provided that adequate expression profile data are available.

1 Introduction

The reverse engineering of cellular networks in prokaryotes and lower eukaryotes [1, 2], as well as in more complex organisms, including mammals [3, 4, 5], is

---

* Correspondence should be addressed to `califano@c2b2.columbia.edu`

A. Apostolico et al. (Eds.): RECOMB 2006, LNBI 3909, pp. 348–362, 2006.

unraveling the remarkable complexity of cellular interaction networks. In particular, the analysis of targets of specific transcription factors (TF) reveals that target regulation can change substantially as a function of key modulator genes, including transcription co-factors and molecules capable of post-transcriptional modifications, such as phosphorylation, acetylation, and degradation. The yeast transcription factor STE12 is an obvious example, as it binds to distinct target genes depending on the co-binding of a second transcription factor, TEC1, as well as on the differential regulation by MAP kinases FUS3 and KSS1 [6]. Although the conditional, dynamic nature of cellular interactions was recently studied in yeast [7, 8, 9, 10], methods to identify a genome-wide repertoire of the modulators of a specific transcription factor are still lacking.

In this paper, we explore a particular type of "transistor like" logic, shown in Fig. 1a, where the ability of a transcription factor g_{TF} (emitter) to regulate a target gene g_t (collector) is modulated by a third gene g_m (base), which we shall call a modulator. Pairwise analysis of mRNA expression profiles will generally fail to reveal this complex picture because g_m and g_{TF} (e.g., a kinase and a transcription factor it activates) are generally statistically independent and because the correlation between the expression of g_{TF} and g_t is averaged over an

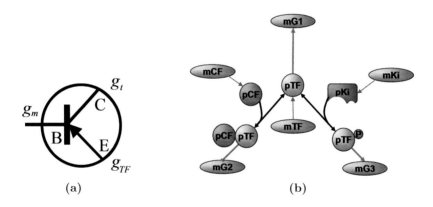

(a)   (b)

TF Interaction	I	I^-_{Ki}	I^+_{Ki}	ΔI_{Ki}	I^-_{CF}	I^+_{CF}	ΔI_{CF}
Ki	−		−	−	0.007	0.016	0.009
CF	−	0.001	0.018	0.008	−	−	−
G_1	0.732	0.542	0.579	0.037	0.552	0.548	0.004
G_2	0.064	0.079	0.093	0.014	0.007	0.378	0.371
G_3	0.097	0.007	0.351	0.344	0.071	0.042	0.029

(c)

Fig. 1. Synthetic network model of transistor-like regulatory logic. (a) Transistor model. (b) Schematic representation of the synthetic network. (c) Unconditional MI, conditional MI and conditional MI difference for the TF interactions conditioning on the expression level of the Ki and the CF; entries colored in red are determined to be statistically significant.

entire range of values of g_m and thus significantly reduced. However, we show that by conditioning on the expression of the modulator gene (e.g., an activating kinase), a statistically significant change in the $g_{TF} \leftrightarrow g_t$ correlation can be measured, thus directly identifying key post-transcriptional regulation mechanisms, including modifications by signaling molecules, co-factor binding, chromatin accessibility, modulation of protein degradation, etc. An important element of this analysis is that, while signaling proteins are conventionally viewed as constitutively expressed, rather than transcriptionally modulated, in practice their abundance within a cell population is subject to fluctuations (either functional or stochastic). Depending on the number of available microarray expression profiles and on the range of fluctuation, this may be sufficient to establish a $g_{TF} \leftrightarrow g_t$ statistical dependency, conditional on the availability of one or more signaling molecules.

We validated the approach on a simple synthetic network and then applied it to the identification of key modulators of MYC, an important TF involved in tumorigenesis of a variety of lymphomas. We identify a set of 100 putative modulators, which is significantly enriched in genes that play an obvious post-transcriptional or post-translational modulation role, including kinases, acyltransferases, transcription factors, ubiquitination and mRNA editing enzymes, etc. Overall, this paper introduces the first genome-wide computational approach to identify genes that modulate the interaction between a TF and its targets. We find that the method recapitulates a variety of known mechanisms of modulation of the selected TF and identifies new interesting targets for further biochemical validation.

2 Method

As discussed in [11, 12], the probability distribution of the expression state of an interaction network can be written as a product of functions of the individual genes, their pairs, and higher order combinations. Most reverse engineering techniques are either based on pairwise statistics [5, 11, 13], thus failing to reveal third and higher order interactions, or attempt to address the full dependency model [14], making the problem computationally untractable and under-sampled. Given these limitations, in this paper we address a much more modest task of identifying the "transistor-like" modulation of specific regulatory interaction, a specific type of third order interactions that is biologically important and computationally tractable in a mammalian context. Furthermore, given the relatively high availability of microarray expression profile data, we restrict our analysis to only genes that modulate transcriptional interactions, i.e., a TF regulating the expression of its target gene(s).

In our model, just like in an analog transistor where the voltage on the base modulates the current between the other terminals, the expression state of the modulator, g_m, controls the statistical dependence between g_{TF} and g_t, which may range from statistically independent to strongly correlated. If one chooses mutual information (MI) to measure the interaction strength (see [11] for the rationale), then the monotonic dependence of $I(g_{TF}, g_t | g_m)$ on g_m, or lack thereof, can reveal respectively the presence or the absence of such a transistor-like interaction.

Analysis along the lines of [11] indicates that currently available expression profile sets are too small to reliably estimate $I(g_{TF}, g_t | g_m)$ as a function of g_m. To reduce the data requirements, one can discretize g_m into well sampled ranges g_m^i. Then, $|I(g_{TF}, g_t | g_m^{i_1}) - I(g_{TF}, g_t | g_m^{i_2})| > 0$ (at the desired statistical significance level) for any range pair (i_1, i_2) is a sufficient condition for the existence of the transistor logic, either direct (i.e., g_m is causally associated with the modulation of the TF targets) or indirect (i.e., g_m is co expressed with a true modulator gene). Below we present details of an algorithm that, given a TF, explores all other gene pairs (g_m, g_t) in the expression profile to identify the presence of the transistor logic between the three genes.

2.1   Selection of Candidate Modulator Genes

Given a expression profile dataset with N genes and an a-priori selected TF gene g_{TF}, an initial pool of candidate modulators g_m, $\{m\} \in 1, 2, \ldots, M$, is selected from the N genes according to two criteria: (a) each g_m must have sufficient expression range to determine statistical dependencies, (b) genes that are not statistically independent of g_{TF} (based on MI analysis) are excluded. The latter avoids reducing the dynamic range of g_{TF} due to conditioning on g_m, which would unnecessarily complicate the analysis of significance of the conditional MI change. It also removes genes that transcriptionally interact with g_{TF}, which can be easily detected by pair-wise co-expression analysis, and thus are not the focus of this work. We don't expect this condition to substantially increase the false negative rate. In fact, it is reasonable to expect that the expression of a post-transcriptional modulator of a TF function should be statistically independent of the TF's expression. For instance, this holds true for many known modulators of MYC function (including MAX, JNK, GSK, and NFκB).

Each candidate modulator g_m is then used to partition the expression profiles into two equal-sized, non-overlapping subsets, L_m^+ and L_m^-, in which g_m is respectively expressed at its highest (g_m^+) and lowest (g_m^-) levels. The conditional MI, $I^\pm = I(g_{TF}, g_t | g_m^\pm)$, is then measured as $I(g_{TF}, g_t)$ on the subset $L^\pm$. Note that this partition is not intended to identify the over or under expression of the modulator, but rather to estimate g_m^i. Then, $|I(g_{TF}, g_t | g_m^+) - I(g_{TF}, g_t | g_m^-)| > 0$ for target genes using the two tails of the modulator's expression range. The size of $L_m^\pm$ is constrained by the minimal number of samples required to accurately measure MI, as is discussed in [11]. Mutual information is estimated using an efficient Gaussian kernel method on rank-transformed data, and the accuracy of the measurement is known [11].

2.2   Conditional Mutual Information Statistics

Given a triplet (g_m, g_{TF}, g_t), we define the conditional MI difference as:

$$\Delta I(g_{TF}, g_t | g_m) = |I^+ - I^-| = |I(g_{TF}, g_t | g_m^+) - I(g_{TF}, g_t | g_m^-)| \tag{1}$$

For simplicity, hereafter we use I for the unconditional MI (i.e., the MI across all samples) and ΔI for conditional MI difference. To assess the statistical significance of a ΔI value, we generate a null hypothesis by measuring its distribution across 10^4 distinct (g_{TF}, g_t) pairs with random conditions. That is, for each gene pair, the non-overlapping subsets $L_m^{\pm}$ used to measure $I^{\pm}$ and ΔI are generated at random rather than based on the expression of a candidate modulator gene (1000 ΔI from random sub-samples are generated for each gene pair). Since the statistics of ΔI should depend on I, we binned I into 100 equiprobable bins, resulting in 100 gene pairs and $10^5 \Delta I$ measurements per bin. Within each bin, we model the distribution of ΔI as an extended exponential, $p(\Delta I) = exp(-\alpha \Delta I^n + \beta)$, which allows us to extrapolate the probability of a given ΔI under this null hypothesis model. As shown in Fig.2, both the mean and the standard deviation of ΔI increase monotonically with I (as expected) and the extended exponentials produce an excellent fit for all bins. Specifically, for small I, the exponent of the fitted exponential distribution is $n \approx 1$. This is

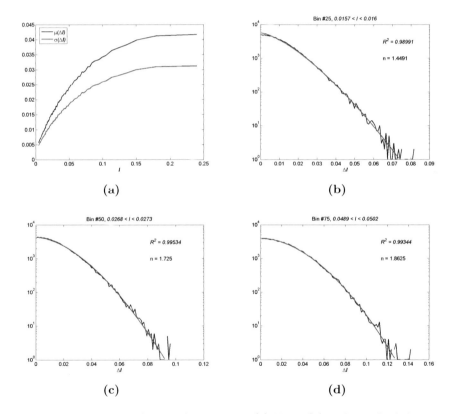

Fig. 2. Null distribution for the ΔI statistics. (a) Mean (μ) and standard deviation (σ) of the ΔI statistics in each bin as a function of I. (b) - (d), distribution of the ΔI statistics (blue curves) and the extended exponential function, $p(\Delta I) = exp(-\alpha \Delta I^n + \beta)$ (red curves), obtained by least square fitting in bin 25, 50 and 75; a goodness-of-fit measure, R^2, and the value of n are also shown for each bin.

because in this case both I^+ and I^- are close to zero and ΔI is dominated by the estimation error, which falls off exponentially [11]. For large I, the estimation error becomes smaller than the true mutual information difference between the two random sub-samples, hence $n \approx 2$ from the central limit theorem.

2.3 Interaction-Specific Modulator Discovery

Given a TF, g_{TF}, and a set of candidate modulators g_m selected as previously discussed, we compute $I(g_{TF}, g_t)$ and $\Delta I(g_{TF}, g_t | g_m)$ for all genes g_t in the expression profile such that $g_t \neq g_m$ and $g_t \neq g_{TF}$. Significance of each ΔI is then evaluated as a function of I, using the extended exponentials from our null hypothesis model. Gene pairs with a statistically significant p-value ($p < 0.05$), after Bonferroni correction for multiple hypothesis testing, are retained for further analysis.

Significant pairs are further pruned if the interaction between g_{TF} and g_t is inferred as an indirect one in both conditions $g_m^\pm$, based on the ARACNE [5, 11] analysis on the two subsets $L_m^\pm$. This is accomplished by using the Data Processing Inequality (DPI), a well-know property of MI introduced in [5, 11], which states that the interaction between g_{TF} and g_t is likely indirect (i.e. mediated

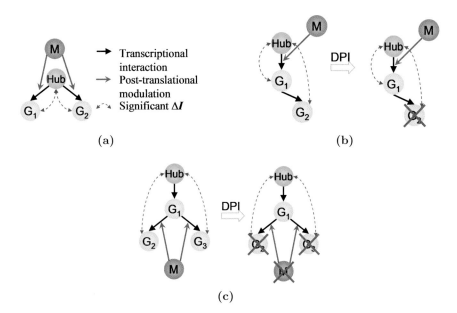

(a)

(b)

(c)

Fig. 3. Schematic diagram of the effect of DPI on eliminating indirect regulatory relationships. (a) Correct modulation model where the modulator (M) significantly changes the regulatory relationship between the TF (Hub) and its direct targets (G_1 and G_2). (b) Removal of indirect connections to the hub eliminates the detection of a significant ΔI on indirect targets. (c) Modulators that affect the downstream target of the TF hub, thus causing significant ΔI between the TF and its indirect neighbors, will be removed by applying DPI.

through a third gene g_x), if $I(g_{TF}, g_t) < min[I(g_{TF}, g_x), I(g_t, g_x)]$. This step eliminates some specific cases, illustrated in Fig. 3, where g_m can produce a significant ΔI even though it does not directly affect the $g_{TF} \leftrightarrow g_t$ interaction. Briefly, two cases will be addressed by the use of the DPI: (a) g_m affects the $g_{TF} \leftrightarrow g_x$ interaction instead of $g_{TF} \leftrightarrow g_t$ (Fig. 3b); and (b) g_m modulates g_x, therefore affecting the $g_x \leftrightarrow g_t$ interaction instead of the $g_{TF} \leftrightarrow g_t$. Thus g_m is not a modulator of the g_{TF} gene and should be removed (Fig. 3c). As discussed in [5, 11], the DPI was applied with a 15% tolerance to minimize the impact of potential MI estimation errors.

3   Results

3.1   Synthetic Model

We first tested our approach on a simple synthetic network (Fig. 1b) that explicitly models two post-translational modifications (activation by phosphorylation and by co-factor binding) that modulate the ability of a TF to affect its targets. The synthetic network includes a TF, a protein kinase (Ki) that phosphorylates the TF, a co-factor (CF) that can bind to TF forming a transcriptionally active complex, and three downstream targets of the TF's isoforms. The transcription activation/inhibition was modeled using Hill kinetics with exponential decay of mRNA molecules. Phosphorylation and cofactor binding were modeled using Michaelis Menten and mass-action kinetics respectively (see Supplementary Table 1 for kinetic equations).

A set of 250 synthetic expression profiles was generated from this model using Gepasi (ver 3.30) [15] by (a) randomly sampling the independent variables (concentration of mRNA for the TF, Ki, and CF) from a uniform distribution, so that they were statistical independent (b) simulating network dynamics until a steady state was reached, and (c) measuring the concentration of all mRNA species that were explicitly represented in the network (using a Gaussian experimental noise model with mean 0 and standard deviation equal to 10% of the mean concentration for each variable). Note that only the mRNA concentrations were used as inputs to our algorithm, even though all molecular species (including all proteins isoforms) were explicitly represented in the model. By conditioning on the expression of the Ki and CF, using the 40% of expression profiles with their most and least expressed values, our approach correctly identified the two and only two significant ΔI, associated with the pairs (CF, g_2) and (Ki, g_3), as shown in Fig. 1c.

3.2   Analysis of Human B Lymphocyte Data

We then used our method to identify a genome-wide repertoire of post-transcriptional modulators of the MYC proto-oncogene – a TF which represents a major hub in the B cell transcriptional network [5]. The analysis was performed on a collection of 254 gene expression profiles, representing 27 distinct cellular phenotypes derived from populations of normal and neoplastic human B lymphocytes. The gene expression profiles were collected using the Affymetrix HG-U95A GeneChip®System (approximately 12,600 probes). Probes with absolute expression mean $\mu < 50$ and standard deviation $\sigma < 0.3\mu$, were considered non-informative and were excluded a-priori from the analysis, leaving 7907 genes.

We further selected 1117 candidate modulators with sufficient expression range ($\mu > 200$ and $\sigma > 0.5\mu$) that were statistically independent of MYC based on MI (significance was established as in [11]). The top 40% and bottom 40% of the expression profiles in which a candidate modulator g_m is expressed at its highest and lowest levels, respectively, were used to define the two conditional subsets $L_m^{\pm}$. The choice of the 40% threshold was specific to this dataset. It ensured that ≥ 100 samples were available within each conditional subset for estimating MI with a reasonable accuracy [11], while keeping the modulators' expression range within the two subsets as separated as possible.

The analysis inferred a repertoire of 100 genes, at a 5% statistical significance level (Bonferroni corrected), which are responsible for modulating 205 regulatory relationships between MYC and its 130 inferred targets in an interaction-specific fashion. See Supplementary Fig. 1 for a map of the modulators and the affected interactions. A complete list is available in the Supplementary Table 2.

3.3   Gene Ontology Enrichment Analysis

To analyze the biological significance of these putative modulator genes, we studied the enrichment of the Gene Ontology [16] Molecular Function categories among the 100 modulators compared to the initial list of 1117 candidate modulators. As shown in Table 1, the top enriched categories represent functions consistent with their activities as modulators of cellular interactions. In particular, putative modulators were enriched in kinases (PKN2, MAP4K4, BRD2, CSNK1D, HCK, LCK, TRIB2, BRD2 and MARCKS), acyltransferase (GGT1, SAT and TGM2) and transcriptional regulators (CUTL1, SSBP2, MEF2B, ID3, AF4, BHLHB2, CREM, E2F5, MAX, NR4A1, CBFA2T3, REL, FOS and NFKB2). This is in agreement with the established evidence that MYC is modulated

Table 1. Most enriched Gene Ontology Molecular Function categories for the inferred MYC modulators. False discovery rate (FDR) are calculated from Fisher's exact test and adjusted for multiple hypothesis testing. Only categories with at least 5 genes from the initial 1117 candidate modulators were used.

Gene Ontology Molecular Function Categories	Enrichment FDR
DNA binding	0.007
Transferase activity	0.010
Acyltransferase activity	0.010
Antioxidant activity	0.018
Phosphoric monoester hydrolase activity	0.026
Adenyl nucleotide binding	0.028
Transcription regulator activity	0.052
Protein serine/threonine kinase activity	0.066

through phosphorylation and acetylation events, affecting its protein stability [17, 18], and that MYC requires broadly distributed effector proteins to influence its genomic targets [19]. We also found that 4 of the 6 modulators with the largest number of affected targets (e.g. UBE2G1, HCK, USP6 and IFNGR1), are associated with non-target-specific functions (e.g. protein degradation, upstream signaling pathway components and receptor signaling molecules, etc). On the other hand, the 14 modulators that are transcription factors (and may thus be MYC co-factors) tend to be highly interaction-specific, affecting only 1-4 target genes (see Supplementary Fig. 1).

3.4   Literature Validation of Known MYC Modulators

Closer scrutiny, through literature review reveals that a number of the inferred modulators play a role in the post-transcriptional and post-translational modulation of MYC, either by direct physical interaction, or by modulating well-characterized pathways that are known to affect MYC function.

Among the list of putative modulators, we found two well known co-factors of MYC: MAX and MIZ-1. Numerous studies, [20] among many others, have shown that transcriptional activation by MYC occurs via dimerization with its partner MAX. Similarly, MIZ-1 has been shown to specifically interact with MYC through binding to its helix-loop-helix domain, which may be involved in gene repression by MYC [21]. Several protein kinases identified by our method are also notable: CSNK1D, a member of the Casein Kinase I gene family, is a reasonable MYC modulator since one of its related family member, Casein Kinase II, has been demonstrated to phosphorylate MYC and MAX, thus affecting the DNA-binding kinetics of their heterodimer [22, 23]. MYC is also know to be phosphorylated by JNK [24] and GSK [25], which affect the stability of its protein. Although both kinases were excluded from our initial candidate modulator set due to their insufficient expression range, our approach was able to identify some of their upstream signaling molecules, such as MAP4K4 and HCK. Both MAP4K4 and HCK are members of the BCR singling pathway that is known to control MYC activation and degradation [26]. In particular, MAP4K4 has been previously reported to specifically activate JNK [27].

MYC stability is also known to be regulated through ubiquitin-mediated proteolysis [28]. Two enzymes in this process, USP6 and UBE2G1, were identified as putative modulators of MYC. Although there is no biochemical evidence implicating these two proteins specifically, they serve as a reasonable starting point for biochemical validation. We also identified putative modulators that could potentially influence the MYC mRNA stability. One of them, APOBEC3B, is closely related to APOBEC1, which has been well characterized as a RNA-editing enzyme capable of binding MYC mRNA in the 3' untranslated region, thus increasing its stability [29]. While APOBEC1 was excluded from our analysis due to its insufficient expression range, the identification of its closely related family member, APOBEC3B, may suggest a similar mechanism. Another protein from this category, HNRPDL, encodes a heterogeneous nuclear ribonucleoprotein which interacts with mRNA and may have a role in mRNA nuclear export. MYC stimulates gene expression in part at the level of chromatin, through its association with co-factors that affect the histone acetylation and DNA methylation. DNMT1, which encodes a DNA methyltransferase, was found in our putative modulator list. Current literature suggests that MYC may repress transcription through the recruitment of DNA methyltransferase as corepressor, which may in turn lead to hypoacetylated histones that are often associated with transcriptional silencing [30, 31].

Many other genes in our list of putative modulators of MYC also present relevant biological functionality, such as transcription factors FOS, CREM, REL and NFKB2, anti-apoptosis regulator BCL-2, to name but a few. Those for which functional relevance can not be established from the current literature likely belong to two groups: (a) novel bona fide MYC modulators requiring further biochemical validation and (b) genes that are co-expressed with a bona fide modulator, such as gene from the same biological pathway. A likely example of the latter case is NFKB2 and its inhibitor NFKBIA, which are both identified as modulators of MYC, while having substantially correlated expression profiles (Pearson correlation 0.55).

Table 2. Promoter analysis of the MYC target genes affected by TF-modulators. Binding signatures of 8 of the 14 TFs in the putative modulator list were obtained through TRANSFAC. Promoter sequences of the target genes (2Kb upstream and 2Kb downstream of the transcription initiation site) were retrieved from the UCSC Golden Path database [33] and masked for repetitive elements. Statistical significance is assessed by considering a null score distribution computed from random sequences using an order-2 Markov model learned from the actual promoter sequences, where P_{BS} is calculated as the probability of finding at least one binding site per 1Kb sequence under the null hypothesis. We used a significance threshold of 0.05; findings below this threshold are shown in red.

Modulator	Binding Signature	MYC targets	P_{BS}
CUTL1		NP	0.032
CREM		PRKDC	0.041
BHLHB2		TLE4	0.030
		MLL	0.039
		IL4R	0.140
MEF2B		KLF12	0.391
		CR2	0.326
		CYBB	0.045
FOS		ZNF259	0.049
REL		KEL	0.125
E2F5		IL4R	0.870
NFKB2		FOXK2	0.041

3.5   Transcription Factors Co-binding Analysis

For the putative modulators annotated as TF, one potential mechanism of modulation is as MYC co-factors. We thus searched for the binding signatures of both MYC and the modulator-TF within the promoter region of the genes whose interaction with MYC appeared to be modulated by the TF. Of the 14 TFs in our putative modulator list, 8 have credibly identified DNA binding signatures from TRANSFAC [32] (represented as position-specific scoring matrix). These TFs affect 12 MYC interactions with 11 target genes. Additionally, 4 of the 5 target genes whose expressions are positively correlated with MYC present at least one E-Box in their promoter region ($p < 4.03 \times 10^{-4}$)[1] . As is shown in Table 2, of the 12 instances of statistically significant modulator target pairs, 7 target genes harbor at least one high specificity TF binding signature ($P_{BS} < 0.05$) in their promoter region. The overall p-value associated with this set of events is $p < 0.0025$ (from the binomial background model). This strongly supports the hypothesis that these TFs are target specific co-factors of MYC.

4   Conclusion and Discussion

Cellular interactions can be neither represented as a static relationship nor modeled as pure pairwise processes. The two issues are deeply interlinked as higher order interactions are responsible for the rewiring of the cellular network in a context dependent fashion. For transcriptional interactions, one can imagine a transistor-like model, in which the ability of a TF to activate or repress the expression of its target genes is modulated, possibly in a target-specific way, by one or more signaling proteins or co-factors. Such post-transcriptional and post-translational conditional interactions are necessary to create complex rather than purely reactive cellular behavior and should be abundant in biological systems. Unfortunately, most post-translational interactions (e.g. phosphorylation or complex formation) do not affect the mRNA concentration of the associated proteins. As a result, they are invisible to naïve co-expression analysis methods. However, proteins that are involved in post-translational regulation may be themselves transcriptionally regulated. At steady state, the concentration of such post-translationally modified proteins and complexes can then be expressed as a function of some mRNA expressions, albeit in a non-obvious, conditional fashion. With this in mind, we show that conditional analysis of pairwise statistical dependencies between mRNAs can effectively reveal a variety of transient interactions, as well as their post-transcriptional and post-translational regulations.

In this paper, we restrict our search to genes that affect the ability of a given TF to transcriptionally activate or repress its target(s). While the identification of the targets of a transcription factor is a rather transited area, the identification

---

[1] MYC is known to transcriptional activate its targets through binding to E-box elements. Repression by MYC occurs via a distinct mechanism, not involving E-boxes, which is not yet well characterized.

of upstream modulators is essentially unaddressed at the computational level, especially in a TF-target interaction specific way. Experimentally, it constitutes an extremely complex endeavor that is not yet amenable to high-throughput approaches. For instance, while hundreds of MYC targets are known, only a handful of genes have been identified that are capable of modifying MYC's ability to activate or repress its targets. Even fewer of these are target specific.

We show that such modulator genes can be accurately and efficiently identified from a set of candidates using a conditional MI difference metric. One novelty of our approach is that it requires no a-priori selection of the modulator genes based on certain functional criteria: the candidate modulators include all genes on the microarray that have sufficient dynamic range and no significant MI with the TF gene. The first requirement can be actually lifted without affecting the method other than making it more computationally intensive and requiring more stringent statistical tests (as the number of tested hypotheses would obviously increases). This is because conditioning on a gene with an expression range comparable to the noise is equivalent to random sub-sampling of the expression profiles, an event that will be filtered out by our statistical test.

Another critical element of our method is the phenotypic heterogeneity of the expression profiles. This ensures that no sufficiently large subset of the expression profiles can be obtained without sampling from a large number of distinct phenotypes, including both normal and malignant cells. In fact, the average number of distinct cellular phenotypes in any subset of gene expression profiles used in the analysis is about 20, with no subset containing fewer than 13. Thus, the modulators identified in this paper are not associated with a specific cellular phenotype.

To derive a null model for estimating the significance of individual conditional mutual information differences, ΔI, we investigated the statistics of ΔI as a function of the unconditional mutual information I. Other models, such as dependence of ΔI on I^+ or I^-, were also investigated, but they proved to be less informative. It is possible that a more accurate null model may be learned by studying the variation of ΔI as a function of both I and either I^+ or I^-. For example, this may answer questions such as: given measured values of I and I^-, what is the probability of seeing a specific difference in ΔI? While this may provide finer-grained estimates of the statistical significance, this also would dramatically increases the number of Monte Carlo samples necessary for achieving a reasonable numerical precision, which would prohibit the actual deployment of the strategy.

Method limitations follow broadly into three categories: (a) The computational deconvolution of molecular interaction is still manifestly inaccurate. This has obvious effects on the discovery of interaction-specific modulators, i.e. we may identify modulators of "functional" rather than physical interactions. (b) The method cannot dissect modulators that are constitutively expressed (housekeeping genes) and activated only at the post-translational level (e.g., the p53 tumor suppressor gene), nor modulators that are expressed at very low concentrations. However, in both cases a gene upstream of the most direct modulator may be identified in its place. For instance, JNK is a known modulator of MYC activity, which is weakly expressed in human B cells and, therefore it is not even included in the initial candidate modulator list. However, MAP4K4, which is upstream of JNK in the signaling cascade, is identified as a MYC modulator in its place. (c) The method cannot disambiguate true modulators from those co-expressed with them.

Techniques to deal with all these drawbacks are currently being investigated. However, we believe that, even in its current state, our approach presents a substantial advancement in the field of reverse engineering of complex cellular networks.

5   Supplementary Material

Supplementary Materials are available at:
http://www.dbmi.columbia.edu/~kaw7002/recomb06/supplement.html.

Acknowledgment

This work was supported by the NCI (1R01CA109755-01A1) and the NIAID (1R01AI066116-01). AAM is supported by the NLM Medical Informatics Research Training Program (5 T15 LM007079-13).

We thank R. Dalla-Favera, K. Basso and U. Klein for sharing the B Cell gene expression profile dataset and helpful discussions.

References

1. Friedman, N. Inferring cellular networks using probabilistic graphical models. Science **303** (2004) 799–805
2. Gardner, T. S. and di Bernardo, D. and Lorenz, D., Collins, J. J.: Inferring genetic networks and identifying compound mode of action via expression profiling. Science **301** (2003) 102–105
3. Elkon, R., Linhart, C., Sharan R., Shamir, R., Shiloh, Y.: Genome-Wide In Silico Identification of Transcriptional Regulators Controlling the Cell Cycle in Human Cells. Genome Res. **13** (2003) 773–780
4. Stuart, J. M., Segal, E., Koller, D., Kim, S. K.: A gene-coexpression network for global discovery of conserved genetic modules. Science **302** (2003) 249–55
5. Basso, K., Margolin, A. A., Stolovitzky, G., Klein, U., Dalla-Favera, R., Califano, A: Reverse engineering of regulatory networks in human B cells. Nature Genetics **37** (2005) 382–390
6. Zeitlinger, J., Simon, I., Harbison, C. T., Hannett, N. M., Volkert, T. L., Fink, G. R., Young, R. A.: Program-Specific Distribution of a Transcription Factor Dependent on Partner Transcription Factor and MAPK Signaling. Cell **113** (2003) 395–404
7. Luscombe, N. M., Babu, M. M., Yu, H., Snyder, M., Teichmann, S. A., Gerstein, M.: Genomic analysis of regulatory network dynamics reveals large topological changes. Nature **431** (2004) 308–12

8. Segal, E., Shapira, M., Regev, A., Pe'er, D., Botstein, D., Koller, D., Friedman, N.: Module networks: identifying regulatory modules and their condition-specific regulators from expression data. Nature Genetics **34** (2003) 166–176
9. de Lichtenberg, U., Jensen, L. J., Brunak, S., Bork, P.: Dynamic Complex Formation During the Yeast Cell Cycle. Science **307** (2005) 724–727
10. Pe'er, D., Regev, A., Tanay, A.: Minreg: Inferring an active regulator set. Bioinformatics **18** (2002) S258–S267
11. Margolin, A., Nemenman, I., Basso, K., Klein, U., Wiggins, C., Stolovitzky, G., Dalla-Favera, R., Califano, A.: ARACNE: An algorithm for reconstruction of genetic networks in a mammalian cellular context. BMC Bioinformatics (2005) In press (manuscript available online at http://arxiv.org/abs/q-bio.MN/0410037)
12. Nemenman, I.: Information theory, multivariate dependence, and genetic network inference KITP, UCSB, NSF-KITP-04-54, Santa Barbara, CA (2004) (manuscript available online at http://arxiv.org/abs/q-bio/0406015)
13. Butte, A.J., Kohane, I. S.: Mutual information relevance networks: functional genomic clustering using pairwise entropy measurements. Pac. Symp. Biocomput. (2000) 418–29
14. Friedman, N., Linial, M., Nachman, I., Pe'er, D.: Using Bayesian networks to analyze expression data Journal of Computational Biology **7** (2000) 601–620
15. Mendes, P.: Biochemistry by numbers: simulation of biochemical pathways with Gepasi 3. Trends Biochem Sci. **22** (1997) 361–363
16. Ashburner, M. et al.: Gene Ontology: tool for the unification of biology. Nature Genetics **25** (2000) 1061–4036
17. Sears, R., Nuckolls, F., Haura, E., Taya, Y., Tamai, K., Nevins, J. R.: Multiple Ras-dependent phosphorylation pathways regulate Myc protein stability Genes Dev. **14** (2000) 2501–2514
18. Patel, J. H. et al.: The c-MYC Oncoprotein Is a Substrate of the Acetyltransferases hGCN5/PCAF and TIP60. Mol. Cell. Biol. **24** (2004) 10826–10834
19. Levens, D. L.: Reconstructing MYC. Genes Dev **17** (2003) 1071–1077
20. Amati, B., Brooks, M. W., Levy, N., Littlewood, T. D., Evan, G. I., Land, H: Oncogenic activity of the c-Myc protein requires dimerization with Max. Cell **72** (1993) 233–245
21. Peukert, K. et al.: An alternative pathway for gene regulation by Myc. EMBO J. **16** (1977) 5672-5686
22. Luscher, B., Kuenzel, E. A., Krebs, E. G., Eisenman, R. N.: Myc oncoproteins are phosphorylated by casein kinase II. EMBO J. **8** (1989) 1111–1119
23. Bousset, K., Henriksson, M., Luscher-Firzlaff, J. M., Litchfield, D. W., Luscher, B: Identification of casein kinase II phosphorylation sites in Max: effects on DNA-binding kinetics of Max homo- and Myc/Max heterodimers. Oncogene **8** (1993) 3211–3220
24. Noguchi, K. et al.: Regulation of c-Myc through Phosphorylation at Ser-62 and Ser-71 by c-Jun N-Terminal Kinase. J. Biol. Chem. **274** (1999) 32580–32587
25. Gregory, M. A., Qi, Y., Hann, S. R.: Phosphorylation by glycogen synthase kinase-3 controls c-myc proteolysis and subnuclear localization. J. Biol. Chem. **278** (2003) 51606–51612
26. Niiro, H., Clark, E. A.: Regulation of B-cell fate by antigen-receptor signals. Nature Reviews Immunology **2** (2002) 945–956
27. Machida, N. et al.: Mitogen-activated Protein Kinase Kinase Kinase Kinase 4 as a Putative Effector of Rap2 to Activate the c-Jun N-terminal Kinase. J. Biol. Chem. **279** (2004) 15711–15714

28. Salghetti, S. E., Kim, S. Y., Tansey, W. P.: Destruction of Myc by ubiquitin-mediated proteolysis: cancer-associated and transforming mutations stabilize Myc. EMBO J. **18** (1999) 717–726

29. Anant, S., Davidson, N. O.: An AU-Rich Sequence Element (UUUN[A/U]U) Downstream of the Edited C in Apolipoprotein B mRNA Is a High-Affinity Binding Site for Apobec-1: Binding of Apobec-1 to This Motif in the 3' Untranslated Region of c-myc Increases mRNA Stability. Mol. Cell. Biol. **20** (2000) 1982–1992

30. Brenner, C. et al.: Myc represses transcription through recruitment of DNA methyltransferase corepressor. EMBO J. **24** (2005) 336–346

31. Robertson, K. D. et al. DNMT1 forms a complex with Rb, E2F1 and HDAC1 and represses transcription from E2F-responsive promoters. Nature Genetics **25** (2000) 338–342

32. Wingender, E. et al.: The TRANSFAC system on gene expression regulation Nucl. Acids Res. **29** (2001) 281–283

33. Karolchik, D. et al.: The UCSC Genome Browser Database. Nucl. Acids Res. **31** (2003) 51–54

A New Approach to Protein Identification

Nuno Bandeira, Dekel Tsur, Ari Frank, and Pavel Pevzner

University of California, San Diego, Dept. of Computer Science and Engineering,
9500 Gilman Drive, La Jolla, CA 92093, USA
{nbandeir, dtsur, amfrank, ppevzner}@cs.ucsd.edu

Abstract. Advances in tandem mass-spectrometry (MS/MS) steadily increase the rate of generation of MS/MS spectra and make it more computationally challenging to analyze such huge datasets. As a result, the existing approaches that compare spectra against databases are already facing a bottleneck, particularly when interpreting spectra of post-translationally modified peptides. In this paper we introduce a new idea that allows one to perform MS/MS database search ... without ever comparing a spectrum against a database. The idea has two components: experimental and computational. Our experimental idea is counter-intuitive: we propose to intentionally introduce chemical damage to the sample. Although it does not appear to make any sense from the experimental perspective, it creates a large number of "spectral pairs" that, as we show below, open up computational avenues that were never explored before. Having a spectrum of a modified peptide paired with a spectrum of an unmodified peptide, allows one to separate the prefix and suffix ladders, to greatly reduce the number of noise peaks, and to generate a small number of peptide reconstructions that are very likely to contain the correct one. The MS/MS database search is thus reduced to extremely fast pattern matching (rather than time-consuming matching of spectra against databases). In addition to speed, our approach provides a new paradigm for identifying post-translational modifications.

1   Introduction

Most protein identifications today are performed by matching spectra against databases using programs like SEQUEST [1] or MASCOT [2]. While these tools are invaluable, they are already too slow for matching large MS/MS datasets against large protein databases. Moreover, the recent progress in mass spectrometry instrumentation (a single LTQ-FT mass-spectrometer can generate 100,000 spectra per day) may soon make them obsolete. Since SEQUEST compares every spectrum against every database peptide, it will take a cluster of about 60 processors to analyze the spectra produced by a single such instrument in real time (if searching through the Swiss-Prot database). If one attempts to perform a time-consuming search for post-translational modificatoions, the running time may further increase by orders of magnitude. We argue that new solutions are needed to deal with the stream of data produced by shotgun proteomics projects. Beavis et al, 2004 [3] and Tanner et al., 2005 [4] recently developed X!Tandem and InsPecT algorithms that prune (X!Tandem) and filter (InsPecT) databases to speed-up the search. However, these tools still have to compare every spectrum against a (smaller) database.

A. Apostolico et al. (Eds.): RECOMB 2006, LNBI 3909, pp. 363–378, 2006.

In this paper we explore a new idea that allows one to perform MS/MS database search without ever comparing a spectrum against a database. The idea has two components: experimental and computational. Our experimental idea, while counter-intuitive, is trivial to implement. We propose to slightly change the experimental protocol by intentionally introducing chemical damage to the sample and generating many modified peptides. The current protocols try to achieve the opposite goal of minimizing the chemical damage since (i) modified peptides are difficult to interpret and (ii) chemical adducts do not provide any useful information. Nevertheless, the existing experimental protocols unintentionally generate many chemical modifications (sodium, potassium, Fe(III), etc.)[1] and below we show that existing MS/MS datasets often contain modified versions for many peptides. In addition, even in a sample without chemical damage, exopeptidases routinely create a variety of peptides that differ from each other by a deletion of terminal amino acids.

From the experimental perspective, subjecting a sample to chemical damage does not make any sense.[2] However, from the computational perspective, it creates a large number of "spectral pairs" that, as we show below, open up computational avenues that were never explored before. Having a pair of spectra (one of a modified and another of an unmodified peptide) allows one to separate the prefix and suffix mass ladders, to greatly reduce the number of noise peaks, and to generate a small number of peptide reconstructions that are very likely to contain the correct one. In difference from our recent approach to generating *covering sets* of short 3–4 amino acid *tags* (Frank et al., 2005 [7], Tanner et al., 2005 [4]), this approach generates a small covering set of *peptides* 7–9 amino acids long. This set typically has a single perfect hit in the database that can be instantly found by hashing and thus eliminates the need to ever compare the spectrum against the database.[3]

Let $S(P)$ and $S(P^*)$ be spectra of an unmodified peptide P and its modified version P^* (spectral pair). The crux of our computational idea is a simple observation that a "database" consisting of a single peptide P is everything one needs to interpret the spectrum $S(P^*)$. If one knows P then there is no need to scan $S(P^*)$ against the database of all proteins! Of course, in reality one does not know P since only $S(P)$ is available. Below we show that the spectrum $S(P)$ is nearly as good as the peptide P for interpreting $S(P^*)$ thus eliminating the need for database search. This observation opens the possibility of substituting MS/MS database search with finding spectral pairs and further interpreting the peptides that produced them. Below we show that these

---

[1] Hunyadi-Goulyas and Medzihradszky, 2004 [5] give a table of over 30 common chemical adducts that are currently viewed as annoyances.

[2] Probably the easiest way to chemically damage the sample is to warm it up in urea solution or to simply bring it into mildly acidic pH and add a hefty concentration of hydrogen peroxide. See Levine et al., 1996 [6] for an example of a slightly more involved protocol that generates samples with desired extent of oxidation in a controlled fashion. Also, to create a mixture of modified and unmodified peptides, one can split the sample in half, chemically damage one half, and combine both halves together again.

[3] We remark that the Peptide Sequence Tag approach reduces the number of considered peptides but does not eliminates the need to match spectra against the *filtered* database. For example, Tanner et al., 2005 [4] describe a dynamic programming approach for matching spectra against a filtered database.

problems can be solved using a variation of the spectral alignment approach [8]. We further show how to transform the spectral pair (S_1, S_2) into virtual spectra $S_{1,2}$ and $S_{2,1}$ of extremely high quality; with nearly perfect b and y ion separation and the number of noisy peaks reduced twelvefold, these spectra (albeit virtual) are arguably the highest quality spectra mass-spectrometrists ever saw.

2   Dataset

We describe our algorithm and illustrate the results using a sample of MS/MS spectra from IKKb protein. The IKKb dataset consists of 45,500 spectra acquired from a digestion of the inhibitor of nuclear factor kappa B kinase beta subunit (IKKb protein) by multiple proteases, thereby producing overlapping peptides (spectra were acquired on a Thermo Finigan LTQ mass spectrometer). The activation of the inhibitor kappaB kinase (IKK) complex and its relationships to insulin resistance were the subject of recent intensive studies. The IKK complex represents an ideal test case for algorithms that search for post-translationally modified (PTM) peptides. Until recently, phosphorylations were the only known PTMs in IKK, which does not explain mechanisms of signaling and activation/inactivation of IKK by over 200 different stimuli, including cytokines, chemicals, ionization and UV radiation, oxidative stress, etc. It is likely that different stimuli use different mechanisms of signaling involving different PTM sites. Revealing the combinatorial code responsible for PTM-controlled signalling in IKK remains an open problem.

The IKKb dataset was extensively studied in Tanner et al., 2005 [4] and Tsur et al., 2005 [9] resulting in 11760 identified spectra and 1154 annotated peptides (p-value ≤ 0.05). This IKKb sample presents an excellent test case for our protocol since 77% of all peptides in this sample have spectral pairs even without intentionally subjecting the sample to chemical damage. 639 out of 1154 annotated peptides are modified. 448 out of 639 modified peptides have an unmodified variant. 208 out of 515 unmodified peptides have a modified version, and 413 out of 515 unmodified peptides have either a modified version or a prefix/suffix peptide in the sample. The sample contains 571 peptides with 3 or more spectra (345 unmodified and 226 modified), 191 peptides with 2 spectra (71 unmodified and 120 modified) and 392 peptides with a single spectrum (99 unmodified and 293 modified). The dataset has not been manually validated and the unusually high proportion of modified peptides with a single spectrum as compared to peptides annotated by multiple spectra may be an indication that some annotations of peptides explained by a single spectrum may be incorrect.

3   Detecting Spectral Pairs

3.1   Clustering Spectra

Clustering multiple spectra of the same peptide achieves a twofold goal: (i) the consensus spectrum of a cluster contains much fewer noise peaks than the individual spectra, and (ii) clustering speeds up and simplifies the search for spectral pairs. The clustering step capitalizes on the fact that true peaks consistently occur in multiple spectra from the

Table 1. Statistics of single spectra, consensus spectra, spectral pairs, and star spectra. Satellite peaks include fragment ions correlated with b and y peaks ($b-H_2O$, $b-NH_3$, a, b^2, etc.). Signal-to-noise ratio (SNR) is defined as $\frac{\#b-ions}{\#unexplained\ peaks}$. Spectral pairs separate prefix and suffix ladders and make interpretations of resulting spectra $S^b_{i,j}$ straightforward. Spectral stars further increase the number of b and y peaks in the resulting spectra. Note that b peaks are responsible for about 90% of the score in both paired and star spectra. The results are given only for the $S^b_{i,j}$ spectra since the $S^y_{i,j}$ spectra have the same statistics.

Type of spectra		#Explained			#Unexplained	#Total	SNR
		b	y	Satellite			
Single spectra	# peaks:	9.48	9.26	20.07	35.25	74.05	**0.27**
(11760 spectra)	% peaks:	13%	13%	26%	48%		
	% score:	28%	28%	19%	25%		
Consensus spectra	# peaks:	9.47	9.39	10.42	13.74	43.06	**0.69**
(567 spectra)	% peaks:	22%	22%	24%	32%		
	% score:	37%	36%	13%	14%		
Spectral pairs $S^b_{i,j}$	# peaks:	6.47	0.2	0.38	1.69	8.64	**3.83**
(1569 pairs)	% peaks:	75%	2%	4%	19%		
	% score:	87%	2%	4%	7%		
Star spectra	# peaks:	8.38	0.52	0.92	2.90	12.72	**2.89**
(745 stars)	% peaks:	66%	4%	7%	23%		
	% score:	88%	3%	2%	7%		

same peptide, while noise peaks do not. Our clustering approach follows Bandeira et al., 2004 [10] with some improvements outlined below. We first transform every spectrum into its scored version that substitutes peak intensities with log likelihood scores. Any scoring used in *de novo* peptide sequencing algorithms can be used for such transformation (we have chosen to use scoring from Frank and Pevzner [11]). We also transform every spectrum into a PRM spectrum (see [10]).

Bandeira et al. [10] use a spectral similarity measure to decide whether two spectra come from the same peptide. While spectral similarity largely succeeds in identifying related spectra, it may in some cases pair non-related spectra. Although such false pairings are rare, they may cause problems if they connect two unrelated clusters. To remove false pairs we use a heuristic approach from Ben-Dor et al. [12]. This clustering procedure resulted in 567 clusters representing 98% of all unmodified and 96% of all modified peptides with three or more spectra in the original sample.

Each cluster of spectra is then collapsed into a single *consensus spectrum* that contains peaks present in at least k spectra in the cluster. The parameter k is chosen in such a way that the probability of seeing a peak in k spectra by chance is below 0.01.[4] We further sum up the scores of matching peaks to score the peaks in the consensus spectrum. As shown in Table 1, the resulting consensus spectra have unusually high

---

[4] We model the noise peak generation as a Bernoulli trial and the occurrence of k matching peaks in a cluster of n spectra as random variable with a Binomial distribution.

signal-to-noise ratio (the number of unexplained peaks in the consensus spectra is reduced by a factor of 2.5). We also observed some consistently co-occurring unexplained peaks possibly due to co-eluting peptides or unexplained fragment ions (e.g., internal ions). After clustring we end up with 567 consensus spectra (that cover 93% of all individual spectra) and 862 unclustered spectra.

3.2   Spectral Pairs

Peptides P_1 and P_2 form a *peptide pair* if either (i) P_1 differs from P_2 by a single modification/mutation, or (ii) P_1 is either a prefix or suffix of P_2[5]. Two spectra form a *spectral pair* if their corresponding peptides are paired. Although the peptides that give rise to a spectral pair are not known in advance, we show below that spectral pairs can be detected with high confidence using uninterpreted spectra.

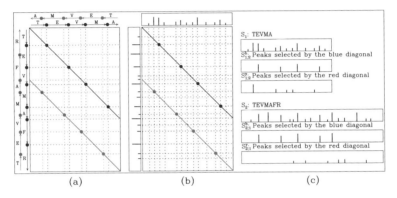

Fig. 1. Spectral product for the spectra of the peptides TEVMA and TEVMAFR. Figure (a) shows the spectral product for the theoretical spectra of these peptides (all points at the intersections between the vertical and horizontal lines). The blue (resp., red) circles correspond to matching b ions (resp., y ions) in the two spectra. The blue and red circles are located on the blue and red diagonals. Figure (b) shows the spectral product for uninterpreted spectra of the peptides TEVMA and TEVMAFR. The two diagonals in the spectral product matrix still reveal the points where peaks from the spectrum at the top match peaks from the spectrum on the left. Figure (c) illustrates how the blue and red diagonals define the spectra $S_{1,2}^b$ and $S_{1,2}^y$.

For two spectra S_1 and S_2, the *spectral product* [8] of S_1 and S_2 is the set of points $(x, y) \in \mathbb{R}^2$ for every $x \in S_1$ and $y \in S_2$ (S_1 and S_2 are represented as sets of masses). Figure 1a shows the spectral product for the theoretical spectra of two peptides. The

---

[5] Condition (ii) can be viewed as a variation of (i) if one considers a pair of peptides differing by a few prefix/suffix residues as a single mutation (such variations are common in MS/MS samples). More generally, peptides P_1 and P_2 form a peptide pair if either (i) P_1 is a modified/mutated version of P_2, or (ii) P_1 and P_2 overlap. While our techniques also work for this generalization, we decided to limit our analysis to simple peptide pairs described above. We found that such simple pairs alone allow one to interpret most spectra. Adding pairs of spectra with more subtle similarities further increases the number of spectral pairs but slows down the algorithm.

the first segment is on the blue diagonal and the second segment is on the red diagonal (one of the segments is empty when $a = b$ or $b = c$). We say that the *endpoints* of the path are the leftmost and rightmost points on the path.

The spectral alignment algorithm [8] finds the path from $(0,0)$ to $(M(S_1), M(S_1) + \Delta)$ that contains the maximum number of points from $\mathcal{M}(S_1, S_2)$. For the optimal path P, the projection of P onto S_i (i.e. the set $\{x_1 : (x_i, x_{i-1}) \in P\}$) gives a subset of S_i which usually contains many b-ion peaks. However, this set can also contain many peaks corresponding to y and neutral loss ion peaks. In order to obtain better b/y separation, we change the spectral alignment problem by selecting only a subset of the points in P: (1) Since the minimum mass of an amino acid is 57 Da, we will choose peaks with distance at least 57 between every two peaks, and (2) We will not select two points that are generated by a peak and its complement peak in S_1 or S_2.

Formally, we say that two peaks x and x' in a spectrum S are *complements* if $x + x' = M(S) + 18$. A subset A of a spectrum S is called *anti-symmetric* if it does not contain a complement pair. A set A is called *sparse* if $|x - x'| \geq 57$ for every $x, x' \in A$. Given a path P, a set $A \subseteq P$ is called sparse if the projection of A onto S_1 is sparse, and it is called anti-symmetric if the projections of A onto S_1 and S_2 are anti-symmetric (w.r.t. S_1 and S_2, respectively). Our goal is to find the largest sparse anti-symmetric subset of $\mathcal{M}(S_1, S_2)$ that is contained in some path from $(0,0)$ to $(M(S_1), M(S_1) + \Delta)$, and contains the points $(0,0)$ and $(M(S_1), M(S_1) + \Delta)$.

Our algorithm for solving the problem above is similar to the algorithm of Chen et al. [14] for de-novo peptide sequencing. But unlike de-novo peptide sequencing, our problem is two-dimensional, and this adds additional complication to the algorithm. We use dynamic programing to compute optimal sets of points that are contained in two paths, one path starting at $(0,0)$ and the other path starting at $(M(S_1), M(S_1) + \Delta)$. By keeping two paths, we make sure that for each set of points we build, its projection on S_1 is anti-symmetric. In order to keep the projection on S_2 anti-symmetric, we need additional information which is kept in a third dimension of the dynamic programming table. The full details of the algorithm are given in the appendix.

3.5   Spectral Stars

A set of spectra incident to a spectrum S_1 in the spectral pairs graph is called a *spectral star*. For example, the spectral star for the spectrum derived from peptide 3 in Figure 3 consists of multiple spectra from 5 different peptides. Even for a single spectral pair (S_1, S_2), the spectra $S_{1,2}^b$ and $S_{1,2}^y$ already have high signal-to-noise ratio and rich prefix and suffix ladders. Below we show that spectral stars allow one to further enrich the prefix and suffix ladders (see Table 1). A spectral star consisting of spectral pairs $(S_1, S_2), (S_1, S_3), \ldots, (S_1, S_n)$ allows one to increase the signal-to-noise ratio by considering $2(n-1)$ spectra $S_{1,i}^b$ and $S_{1,i}^y$ for $2 \leq i \leq n$. We combine all these spectra into a *star spectrum* S_1^* using our clustering approach. This needs to be done with caution since spectra $S_{1,i}^b$ and $S_{1,i}^y$ represent separate prefix and suffix ladders. Therefore, one of these ladders needs to be reversed to avoid mixing prefix and suffix ladders in the star spectrum. The difficulty is that the assignments of upper indexes to spectra $S_{1,i}^b$ and $S_{1,i}^y$ are arbitrary and it is not known in advance which of these spectra represents b ions and which represents y ions (i.e., it may be that $S_{1,i}^b$ represents the suffix ladder while $S_{1,i}^y$ represents the prefix ladder).

A similar problem of reversing DNA maps arises in *optical mapping* (Karp and Shamir, 2000 [15], Lee et al., 1998 [16]). It was formalized as *Binary Flip-Cut* (BFC) Problem [17] where the input is a set of n 0-1 strings (each string represents a snapshot of a DNA molecule with 1s corresponding to restriction sites). The problem is to assign a *flip* or *no-flip* state to each string so that the number of consensus sites is maximized. We found that for the case of spectral stars, a simple greedy approach to the BFC problem works well. In this approach, we arbitrarily select one of the spectra $S_{1,i}^b$ and $S_{1,i}^y$ and denote it $S_{1,i}$. We select $S_{1,2}$ as an initial consensus spectrum. For every other spectrum $S_{1,i}$ ($2 \leq i \leq n$), we find whether $S_{1,i}$ or its reversed copy $S_{1,i}^{rev}$ better fits the consensus spectrum. In the former case we add $S_{1,i}$ to the growing consensus, in the latter case we do it with $S_{1,i}^{rev}$.

After the greedy solution of the BFC problem we know the orientations of all spectra in the spectral star. The final step in constructing *star spectrum* S^* from the resulting collection of $S_{1,i}$ spectra using the consensus spectrum approach from Section 3.1. Table 1 illustrates the power of spectral stars in further enriching the prefix/suffix ladders.

4   Interpretation of Spectral Pairs/Stars

The high quality of the spectra derived from spectral pairs ($S_{i,j}$) and spectral stars (S_i^*) makes *de novo* interpretation of these spectra straightforward (Figure 4). Since these spectra feature excellent separation of prefix and suffix ladders and a small number of noise peaks, de novo reconstructions of these spectra produce reliable (gapped) sequences that usually contain long correct tags.[7] On average, de novo reconstructions of our consensus spectra correctly identify 72% of all possible "cuts" in a peptide (i.e., on average, $0.72 \cdot (n - 1)$ b-ions (y-ions) in a peptide of length n are explained). This is a very high number since the first (e.g., b_1) and the last (e.g., b_{n-1}) b-ions are rarely present in the MS/MS spectra thus making it nearly impossible to explain more than 80% of "cuts" in the IKKb sample. Moreover, on average, the explained b-peaks account for 95% of the total score of the de novo reconstruction implying that unexplained peaks usually have very low scores.[8] In addition to the optimal de novo reconstruction, we also generate suboptimal reconstructions and long peptide tags.

Benchmarking in mass-spectrometry is inherently difficult due to shortage of manually validated large MS/MS samples that represent "golden standards". While the ISB dataset [18] represents such a golden standard for unmodified peptides, large validated samples of spectra from modified peptides are not currently available. As a compromise, we benchmarked our algorithm using a set of 11760 spectra from the IKKb dataset

---

[7] We use the standard longest path algorithm to find the highest scoring path (and a set of suboptimal paths) in the spectrum graph of spectra $S_{i,j}$ and S_i^*. In difference from the standard de novo algorithms we do not insist on reconstructing the entire peptide and often shorten the found path by removing its prefix/suffix if it does not explain any peaks. As a result, the found path does not necessarily start/end at the beginning/end of the peptide. We also do not invoke the antisymmetric path restriction [14] since the spectra $S_{i,j}$ and S_i^* already separate prefix and suffix ladders.

[8] We realize that our terminology may be confusing since, in reality, it is not known whether a spectrum $S_{i,j}^b$ describes b- or y-ions. Therefore, in reality we average between prefix and suffix ladders while referring to b-ions.

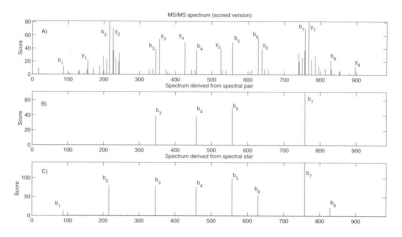

Fig. 4. Improvements in signal-to-noise. The scored MS/MS spectrum for peptide SEELVAEAH has both prefix and suffix peaks along with several noise peaks (A). Using the spectral product of a pair of spectra, many of the noise and suffix peaks that do not reside on the selected diagonal are eliminated. Though paired spectra provide very good separation of prefix/suffix ladders they may sometimes be too selective (e.g. causing the loss of the b_1, b_2, b_6, b_8 peaks) (B). By incorporating more paired spectra to form a spectral star, all noise peaks are removed and all missing prefix peaks are adequately recovered (C).

that were annotated by InsPecT (with p-values ≤ 0.05) and extensively studied in recent publications [4, 9] (including comparisons with Sequest, Mascot and X!Tandem). The entire analysis (starting from clustering and ending with interpretations) of the IKKb dataset took 32 minutes on a regular desktop machine, well below the expected running time of searching the same dataset against even a medium sized database. Below we give results for both spectral pairs and spectral stars.

InsPecT identified 515 unmodified peptides[9] in the IKKb sample, 413 of which have some other prefix/suffix or modified variant in the sample and are thus amenable to pairing. We were able to find spectral pairs for 386 out of these 413 peptides. Moreover, 339 out of these 386 peptides had spectral pairs coming from two (or more) different peptides, i.e., pairs (S_1, S_2) and (S_1, S_3) such that spectra S_2 and S_3 come from different peptides.

The average number of (gapped) de novo reconstruction (explaining at least 85% of optimal score) for spectral stars was 10.4. While the spectral stars generate a small number of gapped reconstructions, these gapped sequences are not well suited for fast membership queries in the database. We therefore transform every gapped de-novo reconstruction into an ungapped reconstruction by substituting every gap with all possible combinations of amino acids.[10] On average, it results in 165 sequences of length 9.5 per spectrum. It turned out that for 86% of peptides, one of these tags is correct.

---

[9] We remark that 99 of them are represented by a single spectrum and thus are more likely to be interpretation artifacts.

[10] In rare cases the number of continuous sequences becomes too large. In such cases we limit the number of reconstructions to 500.

While checking the membership queries for 165 sequences[11] can be done very quickly with database indexing (at most one of these sequences is expected to be present in the database), there is no particular advantages in using such super-long tags (9.5 amino acids on average) for standard database search: a tag of length 6-7 will also typically have an unique hit in the database. However, the long 9-10 amino acid tags have distinct advantages in difficult non-standard database searches, e.g., discovery of new alternatively spliced variants or fusion genes via MS/MS analysis. Moreover, for standard search one can generate the smaller set of shorter (6-7 amino acids) tags based on the original gapped reconstruction and use them for membership queries. We used the obtained gapped reconstruction to generate such short 6 aa tags (each such tag was allowed to have at most one missing peak) and enumerated all possible continuous l-mers by substituting every gap with all possible combinations of amino acids.[12] On average, each consensus spectrum generates about 50 6-mer tags. It turned out that 82% of spectra derived from spectral stars contain at least one correct 6-mer tag.

5 Using Spectral Pairs to Identify Post-translational Modifications

Our approach, for the first time, allows one to detect modifications without any reference to a database. The difference in parent masses within a spectral pair either

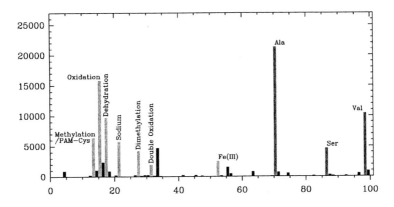

Fig. 5. Histogram of absolute parent mass differences for all detected spectral pairs; the y-axis represents the number of spectral pairs with a given difference in parent mass. For clarity, we only show mass range 1–100. The peaks at masses 71, 87, and 99 correspond to amino acid masses, and the peaks at masses 14, 16, 18, 22, 28, 32, and 53 correspond to known modifications which were also found by Tsur et al. [9]. The peak at mass 34 corresponds to a modification that remains unexplained to date.

---

[11] The actual number of queries is twice as large since we have to check every "reversed" sequence as well. However, this doubling in the number of database queries can be avoided by accounting for reverse variants during the database indexing step.

[12] In rare cases the number of continuous sequences becomes too large. In such cases we limit the number of reconstructions to 100.

correspond to a modification offset (case (i) above) or to a sum of amino acid masses (case (ii)). Therefore, the modification offsets present in the sample can be revealed by the parent mass differences within spectral pairs while their positions and specificities can be determined from *de-novo* reconstructions. While not every difference in parent mass corresponds to a PTM offset (some spectral pairs may be artifacts), the histogram of parent mass differences (Fig. 5) reveals the PTMs present in the IKKb sample. Indeed, 7 out of 8 most frequent parent mass differences in Fig. 5 are listed among 8 most common PTMs in IKKb sample in Tsur et al., 2005 [9]. We emphasize that Fig. 5 was obtained without any reference to database while Tsur et al., 2005 [9] found these PTMs via database search. The only modification from [9] not represented in Fig. 5 is deamidation with (small) offset 1 that is difficult to distinguish from parent mass errors and isotopic peaks artifacts. Interestingly enough, our approach reveals an offset +34 (present in thousands of spectral pairs) that was missed in [9].

6   Conclusions

We have demonstrated the utility of using spectral pairs and stars for protein identification. The key idea of our approach is that correlations between MS/MS spectra of modified and unmodified peptides allow one to greatly reduce noise in individual MS/MS spectra and, for the first time, make de novo interpretations so reliable that they can substitute the time-consuming matching of spectra against databases.

Tandem mass-spectra are inherently noisy and mass-spectrometrists have long been trying to reduce the noise by advancing both instrumentation and experimental protocols. In particular, Zubarev and colleagues [19, 20] recently demonstrated the power of using both CAD and ECD spectra. We emphasize that, in difference from our approach, this technique requires highly accurate Fourier transform mass-spectrometry. Another approach to reduce the complexity of spectra involves stable isotope labeling [21]. However, the impact of this approach (for peptide interpretation) has been restricted, in part by the cost of the isotope and the high mass resolution required. Alternative end-labeling chemical modification approaches have disadvantages such as low yield, complicated reaction conditions, and unpredictable changes in ionization and fragmentation. As a result, the impact of these important techniques is mainly in protein quantification rather than interpretation [21]. The key difference between our approach and labeling techniques is that, instead of trying to introduce a specific modification in a controlled fashion, we take advantage of multiple modifications naturally present in the sample. Our clustering and spectral alignment approaches allow one to decode these multiple modifications (without knowing in advance what they are) and thus provide a computational (rather than instrumentation-based or experiment-based) solution to the problem of increasing signal-to-noise ratio in MS/MS spectra.

Acknowledgements

The authors would like to thank Ebrahim Zandi for providing the MS/MS dataset used to benchmark our algorithm and Vineet Bafna and Stephen Tanner for insightful discussions and help in annotating the data using InsPecT. This project was supported by NIH grant NIGMS 1-R01-RR16522.

References

1. Eng, J., McCormack, A., Yates, J.: An approach to correlate tandem mass-spectral data of peptides with amino acid sequences in a protein database. Journal Of The American Society For Mass Spectrometry **5** (1994) 976–989

2. Perkins, D., Pappin, D., Creasy, D., Cottrell, J.: Probability-based protein identification by searching sequence databases using mass spectrometry data. Electrophoresis **20** (1999) 3551–3567

3. Craig, R., Beavis, R.: TANDEM: matching proteins with tandem mass spectra. Bioinformatics **20** (2004) 1466 – 1467

4. Tanner, S., Shu, H., Frank, A., Wang, L., Zandi, E., Mumby, M., Pevzner, P., Bafna, V.: InsPecT: Fast and accurate identification of post-translationally modified peptides from tandem mass spectra. Anal. Chem. **77** (2005) 4626–4639

5. Hunyadi-Gulyas, E., Medzihradszky, K.: Factors that contribute to the complexity of protein digests. Drug Discovey Today: Targets - mass spectrometry in proteomics supplement **3** (2004) 3–10

6. Levine, R., Mosoni, L., Berlett, B., Stadtman, E.: Methionine residues as endogenous antioxidants in proteins. Proc Natl Acad Sci U S A. **93** (1996) 15036–40

7. Frank, A., Tanner, S., Bafna, V., Pevzner, P.: Peptide sequence tags for fast database search in mass-spectrometry. J. of Proteome Research **4** (2005) 1287–95

8. Pevzner, P., Dancík, V., Tang, C.: Mutation-tolerant protein identification by mass spectrometry. J. Comput. Biol. **7** (2000) 777–787

9. Tsur, D., Tanner, S., Zandi, E., Bafna, V., Pevzner, P.A.: Identification of post-translational modifications by blind search of mass spectra. Nat Biotechnol **23** (2005) 1562–1567

10. Bandeira, N., Tang, H., Bafna, V., Pevzner, P.: Shotgun protein sequencing by tandem mass spectra assembly. Analytical Chemistry **76** (2004) 7221–7233

11. Frank, A., Pevzner, P.: PepNovo: De novo peptide sequencing via probabilistic network modeling. Analytical Chemistry **77** (2005) 964–973

12. Ben-Dor, A., Shamir, R., Yakhini, Z.: Clustering gene expression patterns. J Comput Biol **6** (1999) 281–297

13. Bern, M.W., Goldberg, D.: Eigenms: de novo analysis of peptide tandem mass spectra by spectral graph partitioning. Proceedings of the 9-th annual International Conference on Research in Computational Molecular Biology (RECOMB 2005) (2005) 357–372

14. Chen, T., Kao, M., Tepel, M., Rush, J., Church, G.: A dynamic programming approach to de novo peptide sequencing via tandem mass spectrometry. J Comput Biol **8** (2001) 325–337

15. Karp, R., Shamir, R.: Algorithms for optical mapping. Journal of Computational Biology **7** (2000) 303–316

16. Lee, J.K., Dancík, V., Waterman, M.S.: Estimation for restriction sites observed by optical mapping using reversible-jump Markov Chain Monte Carlo. J. Comput. Biol. **5** (1998) 505–515

17. Dancík, V., Hannenhalli, S., Muthukrishnan, S.: Hardness of flip-cut problems from optical mapping. Journal of Computational Biology **4** (1997) 119–126

18. Keller, A., Purvine, S., Nesvizhskii, A., Stolyar, S., Goodlett, D., Kolker, E.: Experimental protein mixture for validating tandem mass spectral analysis. OMICS **6** (2002) 207–212

19. Savitski, M.M., Nielsen, M.L., Zubarev, R.A.: New data base-independent, sequence tag-based scoring of peptide ms/ms data validates mowse scores, recovers below threshold data, singles out modified peptides, and assesses the quality of ms/ms techniques. Mol Cell Proteomics **4** (2005) 1180–1188
20. Savitski, M.M., Nielsen, M.L., Kjeldsen, F., Zubarev, R.A.: Proteomics-grade de novo sequencing approach. J Proteome Res **4** (2005) 2348–2354
21. Shevchenko, A., Chernushevich, I., Ens, W., Standing, K., Thomson, B., Wilm, M., Mann, M.: Rapid 'de novo' peptide sequencing by a combination of nanoelectrospray, isotopic labeling and a quadrupole/time-of-flight mass spectrometer. Rapid Commun Mass Spectrom **11** (1997) 1015–1024

A   The Anti-symmetric Spectral Alignment Algorithm

In this section we describe the algorithm for solving the maximum sparse anti-symmetric subset problem that was presented in Section 3.4. We use the notations and definitions from that section. For simplicity of the presentation, we will first give a simple algorithm, and then describe several enhancements to the algorithm.

Recall that the input to the problem are two PRM spectra S_1 and S_2 and the goal is to find the largest sparse anti-symmetric subset of $\mathcal{M}(S_1, S_2)$ that is contained in some path from $(0,0)$ to $(M(S_1), M(S_1) + \Delta)$, and contains the points $(0,0)$ and $(M(S_1), M(S_1) + \Delta)$.

In a preprocessing stage, we remove every element x of S_1 if $x \notin S_2$ and $x + \Delta \notin S_2$. Denote $S_1 = \{x_1, \ldots, x_n\}$ and $S_2 = \{y_1, \ldots, y_m\}$, where $x_1 < x_2 < \cdots < x_n$ and $y_1 < y_2 < \cdots < y_m$. Let N be the largest index such that $x_N \le (M(S) + 18)/2$.

A peak x_i in S_1 will be called *left-critical* (resp., *right-critical*) if $x_i + \Delta \in S_1$ (resp., $x_i - \Delta \in S_1$). Denote by S_1^L and S_1^R the left-critical and right-critical peaks in S_1, respectively.

For $i \le n$, let $\mathrm{Left}(i)$ be the set of all sparse anti-symmetric subsets of $S_1^L \cap [x_i - \Delta, x_i - 57]$, and let $\mathrm{Right}(i)$ be the set of all sparse anti-symmetric subsets of $S_1^R \cap [x_i + 57, x_i + \Delta]$. Note that if $\Delta < 57$ then $\mathrm{Left}(i) = \mathrm{Right}(i) = \phi$ for all i, which simplifies the algorithm. In the following, we shall assume that $\Delta \ge 57$.

For $i \le N$ and $j > N$, define $D_1(i, j)$ to be the maximum size of a sparse anti-symmetric set $A \subseteq \mathcal{M}(S_1, S_2)$ such that

1. A is contained in the union of a path from $(0,0)$ to (x_i, x_i) and a path from $(x_j, x_j + \Delta)$ to $(M(S_1), M(S_1) + \Delta)$.
2. A contains the points $(0,0)$, $(M(S_1), M(S_1) + \Delta)$, (x_i, x_i), and $(x_j, x_j + \Delta)$.

If there is no set that satisfies the requirements above, $D_1(i, j) = 0$.

We define tables D_2 and D_3 in a similar way: For $i \le N < j$ and $S \in \mathrm{Left}(i)$, $D_2(i, j, S)$ is the maximum size of a sparse anti-symmetric set $A \subseteq \mathcal{M}(S_1, S_2)$ such that

1. A is contained in the union of a path from $(0,0)$ to $(x_i, x_i + \Delta)$ and a path from $(x_j, x_j + \Delta)$ to $(M(S_1), M(S_1) + \Delta)$.
2. A contains the points $(0,0)$, $(M(S_1), M(S_1) + \Delta)$, and $(x_j, x_j + \Delta)$. Moreover, if $i > 1$ then A contains the point $(x_i, x_i + \Delta)$.
3. $\{x \in S_1^L : x_i - \Delta \le x \le x_i - 57 \text{ and } (x, x + \Delta) \in A\} = S$.

For $i \leq N < j$ and $S \in \text{Right}(j)$, $D_3(i, j, S)$ is the maximum size of a sparse anti-symmetric set $A \subseteq \mathcal{M}(S_1, S_2)$ such that

1. A is contained in the union of a path from $(0, 0)$ to (x_i, x_i) and a path from (x_j, x_j) to $(M(S_1), M(S_1) + \Delta)$.
2. A contains the points $(0, 0)$, $(M(S_1), M(S_1) + \Delta)$, and (x_i, x_i). If $j < n$ then A also contains the point (x_j, x_j).
3. $\{x \in S_1^R : x_j + 57 \leq x \leq x_j + \Delta \text{ and } (x, x) \in A\} = S$.

We also need the following definitions: For $i \leq n$, $\text{prev}(i) = i'$, where i' is the maximum index such that $x_{i'} \leq x_i - 57$. If no such index exists then $\text{prev}(i) = 1$. Similarly, $\text{next}(i) = i'$, where i' is the minimum index such that $x_{i'} \geq x_i + 57$. If no such index exists then $\text{next}(i) = n$. Define

$$M_1^L(i, j) = \max_{i' \leq i} D_1(i', j)$$

$$M_1^R(i, j) = \max_{j' \geq j} D_1(i, j')$$

$$M_2^R(i, j, S) = \max_{j' \geq j} D_2(i, j', S)$$

and

$$M_3^L(i, j, S) = \max_{i' \leq i} D_3(i', j, S).$$

We also define

$$M_2^L(i, j, S) = \max_{i' \leq i} \max_{S'} D_2(i', j, S'),$$

where the second maximum is taken over all sets $S' \in \text{Left}(i')$ that are consistent with S, namely $S' \cap [x_i - \Delta, x_i - 57] = S$. Similarly,

$$M_3^L(i, j, S) = \max_{j' \geq j} \max_{S'} D_3(i, j', S'),$$

where the second maximum is taken over all sets $S' \in \text{Right}(j')$ such that $S' \cap [x_j + 57, x_j + \Delta] = S$. We now show how to efficiently compute $D_1(i, j)$, $D_2(i, j, S)$, and $D_3(i, j, S)$ for all i, j, and S.

Computing $D_1(i, j)$. If either $x_i \notin S_2$ or $x_j + \Delta \notin S_2$, then by definition, $D_1(i, j) = 0$. We also have $D_1(i, j) = 0$ when x_i and x_j are complements or when $x_j - x_i < 57$. Furthermore, if $i = 1$ and $j = n$ then $D_1(i, j) = 2$. Now, suppose that none of the cases above occurs. Then,

$$D_1(i, j) = \begin{cases} M_1^L(\text{prev}(i), j) + 1 & \text{if } x_i > M(S_1) + 18 - x_j \\ M_1^R(i, \text{next}(j)) + 1 & \text{otherwise} \end{cases}.$$

Computing $D_2(i, j, S)$. Suppose that $x_i + \Delta, x_j + \Delta \in S_2$, x_i and x_j are not complements, and $x_j - x_i \geq 57$. If $x_{i'} + \Delta$ is complement of $x_{j'} + \Delta$ (w.r.t. S_2) for some $i' \in \{i, j\}$ and $j' \in S \cup \{j\}$, then $D_2(i, j, S) = 0$. Otherwise,

$$D_2(i, j, S) = \begin{cases} M_2^L(\text{prev}(i), j, S) + 1 & \text{if } x_i > M(S_1) + 18 - x_j \\ M_2^R(i, \text{next}(j), S) + 1 & \text{otherwise} \end{cases}.$$

Computing $D_3(i, j, S)$. Suppose that $x_i, x_j \in S_2$, x_i and x_j are not complements, and $x_j - x_i \geq 57$. If $x_{i'}$ is complement of $x_{j'}$ (w.r.t. S_2) for some $i' \in \{i, j\}$ and $j' \in S \cup \{j\}$, then $D_3(i, j, S) = 0$. Otherwise,

$$D_3(i, j, S) = \begin{cases} M_3^L(\text{prev}(i), j, S) + 1 & \text{if } x_i > M(S_1) + 18 - x_j \\ M_3^R(i, \text{next}(j), S) + 1 & \text{otherwise} \end{cases}.$$

Computing $M_1^L(i, j)$. The recurrence formula for M_1^L is straightforward: For $i = 1$, $M_1^L(i, j) = D_1(i, j)$, and for $i > 1$,

$$M_1^L(i, j) = \max\left\{ D_1(i, j), M_1^L(i - 1, j) \right\}.$$

The recurrence formulae of M_1^R, M_2^R, and M_3^L are similar.

Computing $M_2^L(i, j, S)$. For $i > 1$,

$$M_2^L(i, j, S) = \max\left\{ D_2(i, j, S), \max_{S'} M_2^L(i - 1, j, S') \right\},$$

where the second maximum is taken over all sets $S' \in \text{Left}(i - 1)$ that are consistent with S. The computation of $M_3^R(i, j, S)$ is similar.

Finding the Optimal Solution. After filling the tables D_1, D_2, and D_3, we can find the size of the optimal set of points by taking the maximum value in these tables. The corresponding optimal set can be found by traversing the dynamic programming tables starting from the cell containing the maximum value.

Time Complexity. Using additional data structures, each cell of D_1, D_2, and D_3 can be computed in constant time (we omit the details). Thus, the time complexity of the algorithm is $O(kn^2)$, where

$$k = \max\{|\text{Left}(1)|, \ldots, |\text{Left}(N)|, |\text{Right}(N + 1)|, \ldots, |\text{Right}(n)|\}.$$

Although k can be exponential in n, in practice, k has small values.

Markov Methods for Hierarchical Coarse-Graining of Large Protein Dynamics

Chakra Chennubhotla and Ivet Bahar

Department of Computational Biology, School of Medicine,
University of Pittsburgh, Pittsburgh, PA 15261
{chakra, bahar}@ccbb.pitt.edu

Abstract. Elastic network models (ENMs), and in particular the Gaussian Network Model (GNM), have been widely used in recent years to gain insights into the machinery of proteins. The extension of ENMs to supramolecular assemblies/complexes presents computational challenges, however, due to the difficulty of retaining atomic details in mode decomposition of large systems dynamics. Here, we present a novel approach to address this problem. Based on a Markovian description of communication/interaction stochastics, we map the full-atom GNM representation into a hierarchy of lower resolution networks, perform the analysis in the reduced space(s) and reconstruct the detailed models dynamics with minimal loss of data. The approach (hGNM) applied to chaperonin GroEL-GroES demonstrates that the shape and frequency dispersion of the dominant 25 modes of motion predicted by a full-residue (8015 nodes) GNM analysis are almost identically reproduced by reducing the complex into a network of 35 soft nodes.

1   Introduction

With advances in sequence and structure genomics, an emerging view is that to understand and control the mechanisms of biomolecular function, knowledge of sequence and structure is insufficient. Additional knowledge in the form of *dynamics* is needed. In fact, proteins do not function as static entities or in isolation; they are engaged in functional motions, and interactions, both within and between molecules. The resulting motions can range from single amino acid side chain reorientations (*local*) to concerted domain-domain motions (*global*). The motions on a local scale can be explored to a good approximation by conventional molecular dynamics (MD) simulations, but the motions at a global scale are usually beyond the range of such simulations. Elastic network models (ENM), based on polymer mechanics, succeed in providing access to global motions [1, 2, 3].

A prime example of an EN is the Gaussian Network Model (GNM) [4, 5]. In graph-theoretic terms, each protein is modeled by an undirected graph $\mathcal{G}$, given by $\mathcal{G} = (\mathcal{V}, \mathcal{E})$, with residues $\mathcal{V} = \{v_i | i = 1, \ldots, n\}$ defining the nodes of the network, and edges $\mathcal{E} = \{e_{ij}\}$ representing interactions between residues v_i and v_j. The set of all pairwise interactions is described by a non-negative, symmetric

A. Apostolico et al. (Eds.): RECOMB 2006, LNBI 3909, pp. 379–393, 2006.
© Springer-Verlag Berlin Heidelberg 2006

affinity matrix $\boldsymbol{A} = \{a_{ij}\}$, with elements $a_{ij} = a_{ji}$. GNM chooses a simple interaction model, which is to set the affinity $a_{ij} = a_{ji} = 1$, for a pair of residues v_i and v_j whose C^α atoms are within a cut-off distance of r_c. The interactions represent both bonded and non-bonded contacts in the native configuration of the protein. The cutoff distance represents the radius of the first coordination shell around residues observed in Protein Data Bank (PDB) [6] structures and is set to be 7 Å [7, 8].

The motions accessible under native state conditions are obtained from the Kirchhoff matrix $\boldsymbol{\Gamma}$, defined in terms of the affinity and degree matrices as $\boldsymbol{\Gamma} = \boldsymbol{D} - \boldsymbol{A}$. Here $\boldsymbol{D}$ is a diagonal matrix: $\boldsymbol{D} = \mathrm{diag}(d_1, \ldots, d_n)$ and d_j represents the degree of a vertex v_j: $d_j = \sum_{i=1}^{n} a_{ij} = \sum_{j=1}^{n} a_{ji}$. $\boldsymbol{\Gamma}$ is referred to as the *combinatorial Laplacian* in graph theory [9]. The Kirchhoff matrix multiplied by a force constant γ that is uniform over all springs defines the *stiffness* matrix of an equivalent mass-spring system. The eigenvalue decomposition of $\boldsymbol{\Gamma}$ yields the shape and frequency dispersion of equilibrium fluctuations. In most applications it is of interest to extract the contribution of the most cooperative modes, i.e. the low frequency modes that have been shown in several systems to be involved in functional mechanisms [1, 2]. Also, of interest is the inverse of $\boldsymbol{\Gamma}$, which specifies the covariance matrix for the Boltzmann distribution over equilibrium fluctuations.

GNM is a linear model, and as such it cannot describe the transition between configurations separated by an energy barrier (or any other non-linear effect), so it only applies to fluctuations in the neighborhood of a single energy minimum. The energy well is approximated by a harmonic potential, which limits the magnitude of the predicted motion. The topology of inter-residue contacts in the equilibrium structure is captured by the Kirchhoff matrix $\boldsymbol{\Gamma}$. Also, there is no information on the 'directions' of motions in different vibrational modes, but on their sizes only. The fluctuations are assumed to be isotropic and Gaussian, but for anisotropic extension of GNM called ANM see [10, 11] or equivalent EN-based normal mode analyses (NMA) [12, 13]. Despite this simplicity, many studies now demonstrate the utility of GNM and other EN models in deducing the machinery and conformational dynamics of large structures and assemblies (for a recent review see [2]).

The application and extension of residue-based ENMs to more complex processes, or larger systems, is computationally expensive, both in terms of memory and time, as the eigen decomposition scales on the order of $O(n^3)$, where n is the number of nodes in the graph. Given that the Kirchhoff matrix is sparse, there are a plethora of efficient sparse eigensolvers that one can use [14, 15, 16, 17], including eigensolvers designed specifically for decomposing graph Laplacians [18].

Another way to reduce complexity is to adopt coarser-grained models. For example, in the hierarchical coarse-graining (HCG) approach, sequences of m consecutive amino acids are condensed into unified nodes - which reduces the computing time and memory by factors of m^3 and m^2, respectively [19]; or a mixed coarse-graining has been proposed in which the substructures of interest are modeled at single-residue-per-node level and the surrounding structural units at a lower resolution of m-residues-per node [20]; another common representation of the structure is to adopt rigidly translating and rotating blocks (RTB) [21, 22], or the so-called block normal mode analysis (BNM) [23].

While these methods have been useful in tackling larger systems, the choice and implementation of optimal model parameters to retain physically significant interactions at the residue-, or even atomic level, has been a challenge. The level of HCG has been arbitrarily chosen in the former group of studies, requiring *ad-hoc* readjustments to spring constants or cutoff distances of interaction. In the case of RTB or BNM approaches, all atomic, or residue level information is lost, and substructures that may contain internal degrees of freedom – some of which being functional – are assumed to move as a rigid block. Overall, information is lost on local interactions as structures are coarse-grained. Clearly, the challenge is to map a high resolution model to a low resolution, with a minimal loss of information. In this paper, we present a novel approach to address this problem.

Our approach is to model structures as networks of interacting residues and study the Markov propagation of "information" across the network. We rely on the premise that, the components (residues) of a protein machinery (network) communicate with each other and operate in a coordinated manner to perform their function successfully. Using the Markov chain perspective, we map the full atom network representation into a hierarchy of intermediate ENMs, while retaining the Markovian stochastic charactersitcs, i.e. transition probabilities and stationary distribution, of the original network. The communication properties at different levels of the hierarchy are intrinsically defined by the network topology. This new representation has several features, including: soft clustering of the protein structure into stochastically coherent regions thus providing a useful assessment of elements serving as hubs and/or transmitters in propagating information/interaction; automatic computation of the contact matrices for ENMs at each level of the hierarchy to facilitate computation of both Gaussian and anisotropic fluctuation dynamics; and a fast eigensolver for NMA. We illustrate the utility of the hierarchical decomposition by presenting its application to the bacterial chaperonin GroEL–GroES.

2   A Markov Model for Network Communication

We model each protein as a weighted, undirected graph $\mathcal{G}$ given by $\mathcal{G} = (\mathcal{V}, \mathcal{E})$, with residues $\mathcal{V} = \{v_i | i = 1, \dots, n\}$ defining the nodes of the network, and edges $\mathcal{E} = \{e_{ij}\}$ representing interactions between residues v_i and v_j. The set of all pairwise interactions is described by a non-negative, symmetric *affinity* matrix $A = \{a_{ij}\}$, with elements $a_{ij} = a_{ji}$ and where a_{ij} is the total number of *atom–atom* contacts made within a cutoff distance of $r_c = 4.5$ Å between residues v_i and v_j. The self-contact a_{ii} is similarly defined, but all bonded pairs are excluded. This representation takes into account the difference in the size of amino acids, and captures to a first approximation the strong (weak) interactions expected to arise between residue pairs with large (small) number of atom-atom contacts.

The degree of a vertex v_j is defined as $d_j = \sum_{i=1}^{n} a_{ij} = \sum_{j=1}^{n} a_{ji}$, which are organized in a diagonal matrix of the form $\boldsymbol{D} = \text{diag}(d_1, \ldots, d_n)$.

A *discrete-time, discrete-state Markov* process of network communication is defined by setting the communication (or signalling) probability m_{ij} from residue v_j to residue v_i in *one time-step* to be proportional to the affinity between nodes, $a_{i,j}$. In matrix notation, this conditional probability matrix $\boldsymbol{M} = \{m_{ij}\}$, also called the Markov transition matrix, given by

$$\boldsymbol{M} = \boldsymbol{A}\boldsymbol{D}^{-1}. \tag{1}$$

defines the stochastics of a *random walk* on the protein graph $\mathcal{G}$. Note, $m_{ij} = d_j^{-1} a_{ij}$ where d_j gives a measure of local packing density near residue v_j and serves as a normalizing factor to ensure $\sum_{i=1}^{n} m_{ij} = 1$. Alternatively, m_{ij} can be viewed as the conditional probability of interacting with residue v_i, that is transmitting information to residue v_i, given that the signal (or perturbation) is initially positioned, or originates from, v_j. Suppose this initial probability is p_j^0. Then, the probability of reaching residue v_i using link e_{ij} is $m_{ij}p_j^0$. In matrix notation, the probability of ending up on any of the residues $\boldsymbol{v} = [v_1, v_2, \cdots, v_n]$ after one time step is given by the distribution $\boldsymbol{p}^1 = \boldsymbol{M}\boldsymbol{p}^0$, where $\boldsymbol{p}^k = [p_1^k, \ldots, p_n^k]$. Clearly this process can be iterated, so that after β steps we have

$$\boldsymbol{p}^\beta = \boldsymbol{M}^\beta \boldsymbol{p}^0. \tag{2}$$

Assume the graph is connected, i.e. there is a path connecting every pair of residues in the graph. Then, as $\beta \to \infty$ the Markov chain $\boldsymbol{p}^\beta$ approaches a unique *stationary* distribution $\boldsymbol{\pi}$, the elements of which are given by: $\pi_i = d_i / \sum_{k=1}^{n} d_k$. While the evolution of the random walk is a function of the starting distribution,

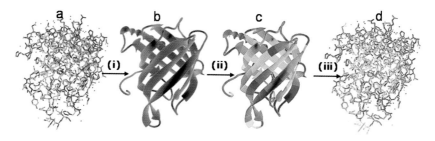

Fig. 1. Hierarchical Network Decomposition Overview: step (i) map the structure (a) to its optimal reduced level representation (illustrated here for retinol-binding protein mapped from full atomic scale to intermediate-chain representation). This step may involve several intermediate levels of resolution (b) (e.g. see Fig. 2); step (ii) perform structural analysis (e.g. GNM) at a coarse-grained scale (c); and step (iii) reconstruct the detailed structure-dynamics (d). The communication/coupling of residues at a given level are assumed to obey a Markov process controlled by atom-atom contact topology. The steps (i) and (iii) are achieved by two operators, $\boldsymbol{R}$ for model reduction, and $\boldsymbol{K}$ for model reconstruction. $\boldsymbol{R}$ and $\boldsymbol{K}$ ensure that similar stochastic characteristics (transition probabilities and stationary distributions) are retained between successive levels of the hierarchy.

the stationary distribution is invariant to the precise details of how the random walk is initiated.

The main goal in undertaking random walks is to reveal the communication patterns *inherent* to the network because of its architecture. However, a naive random walk on a large protein, as will be presented below for the GroEL-GroES complex, is computationally challenging. We address this problem by building a hierarchy of intermediate resolution network models, performing the analysis in the reduced space and mapping the results back to the high resolution representation as illustrated in Fig. 1.

3   Network Hierarchy to Reduce Communication Complexity

The objective in designing a network hierarchy is to map the Markov process operating at the highest resolution onto successively lower resolution network models, while maintaining its stochastic characteristics [24]. In particular, using the stationary distribution π and the Markov transition matrix M, we build a coarse-scale Markov propagation matrix $\widetilde{M}$ (size: $m \times m$, where $m \ll n$) and its stationary distribution δ. The random walk initiated on the coarse-scale network $\widetilde{G}(m)$, and reaching distribution δ, is equivalent to the random walk on the full resolution network $G(n)$ with stationary distribution π. To build a hierarchy of intermediate resolution networks we devise two sets of new operators at each level of the hierarchy: R for model reduction, and K for model expansion/reconstruction.

3.1   Deriving Stationary Distribution in the Reduced Model

We begin by expressing the stationary distribution $\pi = [\pi_1, \pi_2, \cdots, \pi_n]$ as a probabilistic mixture of *latent* distributions,

$$\pi = K\delta, \tag{3}$$

where $\delta = [\delta_1, \delta_2, \cdots, \delta_m]$ is an unknown stationary distribution in a reduced ($m-$dimensional) representation of the structure; $K = \{K_{ij}\}$ is an $n \times m$ non-negative *kernel* matrix with elements K_{ij} and columns K_j being latent probability distributions that each sum to 1, and $m \ll n$. The kernel matrix acts as an *expansion* operator, mapping the low-dimensional distribution δ to a high-dimensional distribution π.

We derive a maximum likelihood approximation for δ using an expectation-maximization (EM) type algorithm [25]. To this aim we minimize the *Kullback-Liebler* distance measure [26, 27] between the two probability distributions π and $K\delta$, subject to the constraint that $\sum_{j=1}^m \delta_j = 1$ and ensured by the Lagrange multiplier λ in the equation below:

$$E = -\sum_{i=1}^n \pi_i \ln \sum_{j=1}^m K_{ij}\delta_j + \lambda\left(\sum_{j=1}^m \delta_j - 1\right). \tag{4}$$

Setting the derivative of E with respect to δ_j to be zero we obtain

$$\sum_{i=1}^{n} \frac{\pi_i K_{ij} \delta_j}{\sum_{k=1}^{m} K_{ik} \delta_k} = \lambda \delta_j. \tag{5}$$

The contribution made by kernel j to a node i (or its stationary probability π_i) is given by K_{ij} (or the product $K_{ij}\delta_j$), and hence we can define an *ownership* of node i in the high resolution representation by a node j in the low resolution representation as

$$R_{ij} = \frac{K_{ij} \delta_j}{\sum_{k=1}^{m} K_{ik} \delta_k}. \tag{6}$$

R_{ij} is also referred to as the responsibility of node j in the low resolution representation, for node i in the high resolution. We note that the mapping between the two resolutions is not deterministic, but probabilistic in the sense that $\sum_{j=1}^{m} R_{ij} = 1$.

Using this relation, and the equalities $\sum_{j=1}^{m} \delta_j = 1$ and $\sum_{i=1}^{n} \pi_i = 1$, summing over j in Eq. 5 gives $\lambda = 1$. This further leads to the stationary distribution δ at the coarse scale

$$\delta_j = \sum_{i=1}^{n} \pi_i R_{ij}. \tag{7}$$

The matrix R therefore maps the high dimensional distribution π to its low-dimensional counterpart δ and hence the name *reduction* operator. Following Bayes theorem, K_{ij} can be related to the *updated* δ values as

$$K_{ij} = \frac{R_{ij} \pi_i}{\delta_j}. \tag{8}$$

In summary, the operators K and R and stationary distribution δ are computed using the following EM type procedure: (1) select an initial estimate for K and δ (see § 3.2); (2) *E-step*: compute ownership maps R using Eq. 6; (3) *M-step*: estimate δ and update K using Eqs. 7 and 8 respectively; and finally, (4) repeat *E-* and *M-* steps until convergence.

3.2   Kernel Selection Details

As an initial estimate for δ, a uniform distribution is adopted. The kernel matrix K is conveniently constructed by diffusing M to a small number of iterations β to give M^β and selecting a small number of columns. In picking the columns of M^β, a greedy decision is made. In particular, column i in M^β corresponds to information diffusion from residue v_i. The first kernel K_i that is picked corresponds to the residue v_i with the highest stationary probability π_i. Following the selection of K_i, all other residues j (and the corresponding columns K_j in M^β) that fall within the half-height of the peak value of the probability distribution in K_i are eliminated from further consideration. This approach generates kernels that are spatially disjoint. The selection of kernels continues until every residue in the protein is within a half-height of the peak value of at least one kernel. While other kernel selection procedures are conceivable, we chose the greedy method for computational speed. In practice, we observed the EM algorithm generates results of biological interest that are insensitive to the initial estimates of K and δ.

3.3   Transition and Affinity Matrices in the Reduced Model

The Markov chain propagation at the reduced representation obeys the equation $q^{k+1} = \widetilde{M} q^k$, where q^k is the coarse scale m-dimensional probability distribution after k steps of the random walk. We *expand* q^k into the fine scale using $p^k = K q^k$, and *reduce* p^k back to the coarse scale by using the ownership value $R_{i,j}$ as in $q_j^{k+1} = \sum_{i=1}^{n} p_i^k R_{i,j}$. Substituting Eq. 6 for ownerships, followed by the expression for p^k, in the equation for q_j^{k+1}, we obtain

$$\widetilde{M} = \operatorname{diag}(\delta)\, K^{\mathsf{T}} \operatorname{diag}(K\delta)^{-1} K. \tag{9}$$

Using the definition of $\widetilde{M}$, and the corresponding stationary distribution δ, we generate a *symmetric* affinity matrix $\widetilde{A}$ that describes the node-node interaction strength in the low resolution network

$$\widetilde{A} = \widetilde{M} \operatorname{diag}(\delta). \tag{10}$$

To summarize, we use the stationary distribution π and Markov transition matrix M at the fine-scale to derive the operator K and associated reduced stationary distribution δ, using the EM algorithm described in the previous section. K and δ are then used in Eq. 9 and 10 to derive the respective transition $\widetilde{M}$ and affinity $\widetilde{A}$ matrices in the coarse-grained representation. Clearly, this procedure can be repeated recursively to build a hierarchy of lower resolution network models.

4   Hierarchical Decomposition of the Chaperonin GroEL-GroES

We examine the structure and dynamics of the bacterial chaperonin complex GroEL-GroES-$(ADP)_7$ [28], from the perspective of a Markov propagation of information/interactions. GroEL is a cylindrical structure, 150 Å long and 140 Å wide, consisting of 14 identical chains organized in two back-to-back stacked rings (*cis* and *trans*) of seven subunits each. The GroES co-chaperonin, also heptameric, binds to the apical domain of GroEL and closes off one end of the cylinder. During the allosteric cycle that mediates protein folding, the *cis* and *trans* rings alternate between open (upon binding of ATP and GroES) and closed (unliganded) forms, providing access to, or release from, the central cylindrical cavity, where the folding of an encapsulated (partially folded or misfolded) protein/peptide is assisted.

First, the inter-residue affinity matrix A based on all atom-atom contacts is constructed (Fig. 2a), from which the fine-scale Markov transition matrix M is derived using Eq. 1. The kernel selection algorithm applied to M^β ($\beta = 4$) yields 1316 (reduced level 1) kernels. Using these kernels as an initialization, a recursive application of the EM procedure derives stationary distributions δ (Eq. 7), updated expansion matrices K (Eq. 8), reduced level probability transition matrices $\widetilde{M}$ (Eq. 9) and the corresponding residue interaction matrices $\widetilde{A}$

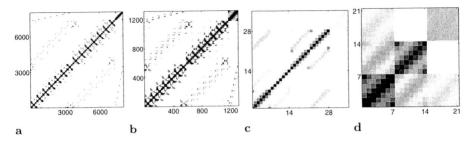

Fig. 2. Affinity matrix hierarchy for the protein GroEL/GroES (PDB: **1AON**). The respective sizes of the reduced models, and the associated affinity matrices, across the hierarchy are $n = 8015$ (fine-scale, panel **a**) and $m = 1316$ (coarse-scale 1, panel **b**), 483 (coarse-scale 2), 133 (coarse-scale 3), 35 (coarse-scale 4, panel **c**) and 21 (coarse-scale 5, panel **d**). The affinity matrices are real-valued but are shown here as *dot* plots (panels **a-b**), to highlight the similarity in the matrix structure across the hierarchy. The affinity matrices for the two lowest resolution models (panels **c-d**) are shown as images, where the affinity value is *inversely* proportional to the brightness of a pixel.

(Eq. 10). The respective dimensions of $\widetilde{A}$ turn out to be 483 (reduced level 2), 133 (reduced level 3), 35 (reduced level 4, Fig. 2c) and 21 (reduced level 5, Fig. 2d). We note that the individual subunits of the GroEL/GroES are distinguished by their strong intra-subunit interactions, and a number of inter-subunit contacts are maintained at all levels, which presumably establish the communication across the protein at all levels. The dimension m of the reduced model is automatically defined during the kernel selection at each level of the hierarchy. The method thus avoids the arbitrary choices of sampling density and interaction cutoff distances at different hierarchical levels.

In contrast to the *deterministic* assignment of one-node-per-residue in the original ENM, the Markov-chain-based representation adopts a *stochastic* description in the sense that each node *probabilistically* 'owns', or 'is responsible for' a subset of residues. To see this, consider the ownership matrix $R^{(l,l+1)} = \{R_{ij}^{(l,l+1)}\}$ that relates information between two adjacent levels l and $l+1$ of the hierarchy. Likewise, the matrix $R^{(0,L)} = \prod_{l=0}^{L-1} R^{(l,l+1)}$ ensures the passage from the original high resolution representation 0 to the top level L of the hierarchy. In particular, the ij^{th} element $R_{ij}^{(0,L)}$ describes the probabilistic participation of residue v_i (at level 0) in the cluster j (at level L), and $\sum_j R_{ij}^{(0,L)} = 1$. Hence, the nodes at level L perform a *soft partitioning* of the structure. *This type of soft distribution of residues among the m nodes, or their partial/probabilistic participation in neighboring clusters, establishes the communication between the clusters, and is one of the key outcomes of the present analysis.* Of interest is to examine the ownership of clusters at a reduced representation. We select the coarse-scale 4, for example, which maps the structure into a graph of 35 clusters (Fig. 2c). Fig. 3 demonstrates the ownership of the individual clusters at this level. Essentially there are five sets of seven clusters each, centered near the apical and equatorial domains of the *cis* and *trans* rings, and at the individual

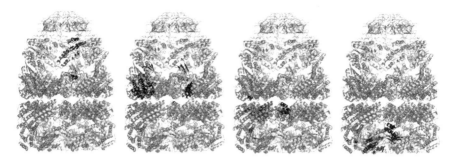

Fig. 3. Four different *soft* clusters located on GroEL

GroES chains. The intermediate domains are being shared between the clusters at the apical and equatorial domains. As such, they play a key role in establishing intra-subunit communication. The color-coded ribbon diagrams in Fig. 3 display the loci of representative clusters from each of these four distinct types (excluding the GroES clusters). The color code from red-to-blue refers to the higher-to-lower involvement (or responsibility) of the individual residues in the indicated clusters. Evidently, the regions colored red serve as hubs for broadcasting the information within clusters, and those colored blue play the key role of establishing the communication, or transferring information between clusters. Detailed examination of the ownership of these clusters reveal several interesting features, correlating with the experiments and summarized in §6.

Next, we benchmark the utility and robustness of the presently introduced methodology in so far as the equilibrium dynamics of the examined structure is concerned. Mainly, we compare the collective modes of motion predicted for the GroEL-GroES complex using a full-residue (8015 nodes) ENM [29], with those captured by the hierarchy of reduced models. The newly introduced representation hierarchy will be shown below to successfully map structure-dynamics information between successive levels with minimal loss in accuracy[1].

5   Hierarchical Gaussian Network Model (hGNM)

Here we present a methodology for generating GNM modes at different levels of coarse-graining the information on contact topology inherent in $\mathcal{G}$, and reconstructing the detailed mode behavior by projecting the eigenvectors and eigenvalues generated at low levels of resolution back to their fine scale counterparts using the Markov chain propagation formalism, a method shortly referred to as hierarchical GNM (hGNM).

For hGNM, assume that the dimensions of the Kirchhoff matrices at the coarse, intermediate and fine scales are e, m and n respectively, where $e \leq m \ll n$.

---

[1] The ownership matrix can also be used to propagate the location information of the residues from one level of the hierarchy to another. This in turns help perform anisotropic fluctuation modeling, but for lack of space this procedure will not be elaborated any further.

The affinity and Kirchhoff matrices at the coarsest level are not likely to be sparse, however a full eigen decomposition of the coarsest Kirchhoff matrix (size: $e \times e$) will be computationally the least expensive step.

To reconstruct the eigen information at the fine-scale, assume we have access to the leading eigenvectors $\widehat{U}$ (size: $m \times e$) for $\widehat{\Gamma}$ (size: $m \times m$). Using this we generate the leading eigenvectors $\widetilde{U}$ (size: $n \times e$), and the leading eigenvalues $\widetilde{\Lambda} = [\lambda_1, \lambda_2 \cdots \lambda_e]$ (size: $e \times 1$) of the fine-scale Kirchhoff matrix Γ (size: $n \times n$). Let $\{U, \Lambda\}$ denote the eigenvectors and eigenvalues obtained from a direct decomposition of Γ. There are several steps to the eigen reconstruction process. (i) The coarse-scale eigenvectors $\widehat{U}$ can be transformed using the kernel matrix K as $\widetilde{U} = K\widehat{U}$ to generate $\widetilde{U}$ as an approximation to U. (ii) This transformation alone is unlikely to set the directions of $\widetilde{U}$ exactly aligned with U. So, we update the directions in $\widetilde{U}$ by repeated application of the following iteration (called *power* iterations [30]): $\widetilde{U} \Leftarrow \Gamma_g \widetilde{U}$ Note, here instead of using Γ we use an adjusted matrix Γ_g given by $\Gamma_g = \nu I - \Gamma$, where ν is a constant and I is an identity matrix. The power iterations will direct the eigenvectors to directions with large eigenvalues. But for fluctuation dynamics, we are interested in the *slow* eigen modes with *small* eigenvalues, hence the adjustment Γ_g is made. In particular, because of Gerschgorin disk theorem [30] the eigenvalues of Γ are bound to lie in a disk centered around the origin with a radius ν that is no more than twice the largest element on the diagonal of Γ. (iii) Steps i and ii need not preserve orthogonality of the eigenvectors in U. We fix this by a Gram-Schmidt orthogonalization procedure [30]. Finally, the eigenvalues are obtained from $\widetilde{\Lambda} = \text{diag}(\widetilde{U}^{\mathsf{T}} \Gamma \widetilde{U})$. In [24] we present more details of this coarse to fine eigen mapping procedure, including a discussion on the number of power iterations to use; setting the thresholds for convergence and a comparison of the speed ups obtained over a standard sparse eigensolver for large matrices.

5.1 Collective Dynamics in the Reduced Space: Benchmarking Against GNM

As discussed earlier, the eigenvalue decomposition of Γ yields the shape and frequency dispersion of equilibrium fluctuations. The shape of mode k refers to the normalized distribution of residue displacements along the principal axis k, given by the elements $u_i^{(k)}$ $(1 \leq i \leq n)$ of the kth eigenvector $u^{(k)}$, and the associated eigenvalue λ_k scales with the frequency of the kth mode. In most applications, it is of interest to extract the contribution of the most cooperative modes, i.e. the low frequency modes that have been shown in several systems to be involved in functional mechanisms. To this end, we used the Markov-chain based hierarchy to build reduced Kirchhoff matrices $\widetilde{\Gamma}$ at increasingly lower levels of resolution. We then performed their mode decompositions and propagated the information back over successive levels of the hierarchy, so as to generate the eigenvectors and eigenvalues for the fine-scale Kirchhoff matrix Γ. We now show that hGNM maps the structure-dynamics information between successive levels of the hierarchy with minimal loss in accuracy.

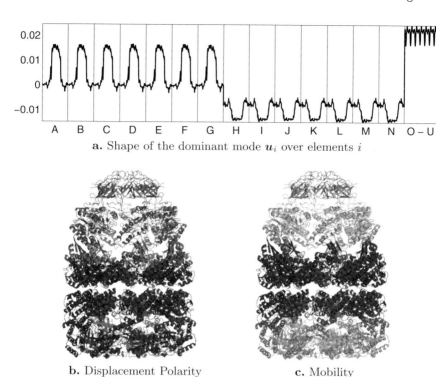

a. Shape of the dominant mode $\boldsymbol{u}_i$ over elements i

b. Displacement Polarity   **c.** Mobility

Fig. 4. Dominant mode shape and mobility **a.** The labels on the abscissa indicate the chain identities, A-G belong to the *cis* ring, H-N come from the *trans* ring and O-U are from the GroES cap. The black curve gives the shape of the *slowest* eigen mode. The ordinate value is the normalized distribution of residue displacements along the dominant mode coordinate. **b.** Ribbon diagram illustrating the polarity of the displacement, color coded to be red for positive and blue for negative, indicating the anticorrelated motions of the two halves of the complex. **c.** Ribbon diagram color-coded after residue mobilities in mode 1. The mobility of residue v_i given by the squared displacement: $\left(u_i^{(1)}\right)^2$, with a color code that is red for high and blue for low.

First, our previous study identified ten slowest modes of interest, including the counter-rotation of the two rings around the cylindrical axis (non-zero mode 1) and other collective deformations proposed to be involved in chaperonin function [29]. Results presented in Fig. 4 show the mechanism of the dominant mode, mainly a global twisting of the structure where the *cis* and *trans* undergo counter rotation about the cylindrical axis (mode 1). The most important point is that these results corroborate previous findings [29, 1] and are reproduced here by adopting a reduced representation down to $m = 21$ nodes and mapped back to full-residue level.

Second, Figure 5a compares the frequencies obtained by the full-residue-level representation, with those obtained by hGNM, upon propagation of the topology information from reduced level 4 (Fig. 2c). An excellent agreement is observed between the reconstructed eigenvalues $\widetilde{\lambda}$ (red curve) and their original values λ

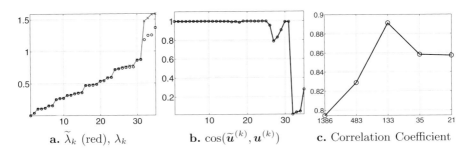

a. $\widetilde{\lambda}_k$ (red), λ_k **b.** $\cos(\widetilde{u}^{(k)}, u^{(k)})$ **c.** Correlation Coefficient

Fig. 5. hGNM results (a) comparing eigenvalues λ (circles) from a direct decomposition of the Γ with multi-scale eigensolver spectrum $\widetilde{\lambda}$ (red line). For the direct eigen decomposition, we use the Matlab program `svds.m` which invokes the compiled ARPACKC routine [14], with a default convergence tolerance of `1e-10`. (b) Mode shape correlation: diag $\left(|\widetilde{U}^{\mathsf{T}} U|\right)$, between the matrix of eigenvectors $\widetilde{U}$ derived by hGNM and U from direct decomposition. (c) Correlation coefficient between the theoretical B-factors (derived at each level of the hierarchy) vs experiment. The abscissa labels indicate the size m of the network at successive levels of the hierarchy.

(open circles). In Fig. 5**b**, we display the correlation cosine between the eigenvectors $u^{(k)}$ and $\widetilde{u^{(k)}}$ obtained by the full-residue representation and the reconstruction from reduced level 4 respectively. Notably, the reduced representation contains only 35 nodes. Yet, the correlation cosine with the detailed representation containing 8015 nodes is almost unity throughout all the leading 25 modes, and above 0.8 for all modes, except the terminal four modes. The contribution of the latter to the overall dynamics is negligibly small compared to the large group of slow modes.

Finally, in order to assess the effect of coarse-graining on fluctuation dynamics, we compared in Fig. 5**c** the mean-square fluctuations obtained from different levels of the hierarchy with the experimental B-factor values. The theoretical B-factor for each residue v_i is computed using [31]

$$
B_i = \frac{8\pi^2 k_B T}{\gamma} \sum_{k=2}^{n} \lambda_k^{-1} \left(u_i^{(k)}\right)^2 , \tag{11}
$$

where the summation is performed over all $n-1$ modes in the GNM, or over all the $m-1$ reduced eigenvectors and eigenvalues reconstructed from different levels of the hierarchy in hGNM. Because experimental B-factors correspond to each atom and our representation at the fine-scale is a summary of atom-atom contact information for each residue, we average the experimental B-factors over all atoms for each residue. As shown in Fig. 5**c**, a correlation coefficient value of 0.86 is achieved between the experimental and theoretical B-factors after mapping the structure of 8015 residues into a representative network of 21 nodes. Thus, the fluctuation behavior of individual residues is accurately maintained despite a drastic reduction in the complexity of the examined network. Interestingly, a maximum in correlation coefficient is obtained at an intermediate level of resolution, $m = 133$, which may be attributed to an optimal elimination of noise in line with the level of accuracy of experimental data at this level of representation.

6   Conclusions

A new method is introduced in the present study, which permits us to use structural information at atomic level in building network representations of different complexity, which lend themselves to efficient analysis of collective dynamics and information propagation stochastics. The approach is particularly useful for analyzing large structures and assemblies, or cooperative/allosteric processes that are usually beyond the range of conventional molecular simulations.

We illustrated the utility of the methodology by way of application to the chaperonin GroEL-GroES, a widely studied structure composed of $n = 8015$ residues. Notably, we start with the full-atomic representation of the complex, which involves a total of $\approx 10^6$ atom-atom contacts (based on an interaction range of $4.5\,\text{Å}$). Interatomic contacts define the affinities of pairs of residues, which are, in turn, used to define the weights of the connectors between residues (nodes) in the graph/network representation of the structure. The affinities also define the conditional probabilities of information transfer across residues following a Markovian process. The original network of n nodes is mapped into lower dimensional representations, down to $m = 21$ nodes, by an EM algorithm that maintains two basic properties of the original stochastic process: its Markovian conditional probabilities and stationary distribution (i.e. communication probability/potential) of individual residues. Two sets of operators, ensuring model reduction and reconstruction at different hierarchical levels permit us to perform the analysis at reduced scales.

Acknowledgments. Partial support by the NSF-ITR grant # EIA-0225636 and the NIH grant #1 R01 LM007994-01A1 is gratefully acknowledged.

References

1. Ma J.: Usefulness and limitations of normal mode analysis in modeling dynamics of biomolecular systems, *Structure*, **13**, 373–380, (2005).
2. Bahar, I. and Rader, A. J.: Coarse-grained normal mode analysis in structural biology, *Curr. Opi. Struct. Bio.* **15**, 1–7, 2005.
3. Rader, A. J., Chennubhotla, C., Yang, L.-W. & Bahar, I: The Gaussian Network Model: Theory and Applications, in *Normal Mode Analysis: Theory and Applications to Biological and Chemical Systems*, eds. Cui, Q. & Bahar, I., CRC Press, (2005).
4. Bahar, I., Atilgan, A. R. & Erman, B.: Direct evaluation of thermal fluctuations in protein using a single parameter harmonic potential, *Folding & Design* **2**, 173–181, 1997.
5. Haliloglu, T., Bahar, I. & Erman, B.: Gaussian dynamics of folded proteins, *Phys. Rev. Lett.* **79**, 3090, 1997.

6. Berman, H. M., Westbrook, J., Feng, Z., Gilliland, G., Bhat, T. N., Weissig, H., Shindyalov, I. N. and Bourne, P. E.: The Protein Data Bank, *Nucleic Acids Research*, **28**, 235–242, 2000.

7. Miyazawa, S. & Jernigan, R. L.: Estimation of effective inter-residue contact energies from protein crystal structures: quasi-chemical approximation, *Macromolecules*, **18**, 534, (1985).

8. Bahar, I. & Jernigan, R. L.: Inter-residue potentials in globular proteins and the dominance of highly specific hydrophilic interactions at close separation, *J. Mol. Biol.*, **266**, 195, (1997).

9. Chung, F. R. K. *Spectral Graph Theory* CBMS Lectures, AMS, (1997).

10. Doruker, P., Atilgan, A. R. & Bahar, I.: Dynamics of proteins predicted by molecular dynamics simulations and analytical approaches: Application to α-amylase inhibitor, *Proteins* **40**, 512–524, (2000).

11. Atilgan, A. R., Durell, S. R., Jernigan, R. L., Demirel, M. C., Keskin, O. & Bahar, I.: Anisotropy of fluctuation dynamics of proteins with an elastic network model, *Biophys. J.* **80**, 505, (2001).

12. Hinsen K.: Analysis of domain motions by approximate normal mode calculations, *Proteins* **33**, 417, (1998).

13. Tama, F. & Sanejouand, Y. H.: Conformational change of proteins arising from normal mode calculations, *Protein Eng.* **14**, 1–6, (2001).

14. Lehoucq, R. B., Sorensen, D. C. & Yang, C. *ARPACK User Guide: Solution of Large Scale Eigenvalue Problems by Implicitly Restarted Arnoldi Methods*, TR, Dept. of CAM, Rice University, (1996).

15. Simon, H. and Zha, H.: Low-rank matrix approximation using the Lanczos bidiagonalization process with applications, *SIAM J. of Sci. Comp.* **21**, 2257-2274, (2000).

16. Barnard, S. and Simon H.: A fast multi-level implementation of recursive spectral bisection for partitioning usntructured grid, *Concurrency: Practice and Experience*, **6**, 101–117, (1994).

17. Fowlkes, C., Belongie, S., Chung, F. & Malik, J.: Spectral Grouping Using the Nystrm Method, *IEEE PAMI*, **26**, 2, (2004).

18. Koren, Y., Carmel, L. & Harel, D.: Drawing Huge Graphs by Algebraic Multigrid Optimization, Multiscale Modeling and Simulation, 1:4, 645–673, SIAM, (2003).

19. Doruker, P., Jernigan, R.L. & Bahar, I.: Dynamics of large proteins through hierarchical levels of coarse-grained structures, *J. Comp. Chem.*, **23**, 119, (2002).

20. Kurkcuoglu, O., Jernigan, R. L. & Doruker, P.: Mixed levels of coarse-graining of large proteins using elastic network model methods in extracting the slowest motions, *Polymers*, **45**, 649–657, (2004).

21. Marques O. *BLZPACK: Description and User's Guide*, TR/PA/95/30, CERFACS, Toulouse, France, (1995).

22. Tama, F., Gadea, F. X., Marques, O. & Sanejouand, Y. H.: Building-block approach for determining low-frequency normal modes of macromolecules, *Proteins*, **41**, 1–7, (2000).

23. Li, G. H. & Cui, Q.: A coarse-grained normal mode approach for macromolecules: an efficient implementation and application to Ca^{2+}-ATPase, *Bipohys. J.*, **83**, 2457, (2002).

24. Chennubhotla, C. & Jepson, A.: Hierarchical Eigensolver for Transition Matrices in Spectral Methods, *NIPS* **17**, 273–280, (2005).

25. McLachlan, G. J. & Basford, K. E. (1988) *Mixture Models: Inference and Applications to Clustering*, Marcel Dekker, N.Y.

26. Kullback, S.: *Information Theory and Statistics* Dover Publications, New York, (1959).
27. Kullback, S. & Leibler, R. A.: On Information and Sufficiency, *Ann. of Math. Stat.* **22**, 79–86, (1951).
28. Xu, Z. H., Horwich, A. L. & Sigler, P. B.: The crystal structure of the asymmetric GroEL-GroES(ADP)$_7$ chaperonin complex, *Nature* **388**, 741–750, (1997).
29. Keskin, O., Bahar, I., Flatow, D. Covell, D. G. & Jernigan, R. L.: Molecular Mechanisms of Chaperonin GroEL-GroES Function, *Biochemistry* **41**, 491–501, (2002).
30. Watkins, D. S. *Fundamentals of Matrix Computations*, Wiley-Interscience, (2002).
31. Kundu, S., Melton, J. S., Sorensen, D. C. & Phillips, G. N.: Dynamics of proteins in crystals: comparison of experiment with imple models, *Biophys. J.*, **83**, 723–732, (2002).
32. Landry, S. J., Zeilstra-Ryalls, J., Fayet, O., Georgopoulos, C., and Gierasch, L. M.: Characterization of a functionally important mobile domain of GroES, *Nature* **364**, 255–258, (1993).
33. Hohfeld, J., and Hartl, F. U.: Role of the chaperonin cofactor Hsp10 in protein folding and sorting in yeast mitochondria, *J. Cell Biol.* **126**, 305–315, (1994) .
34. Kovalenko, O., Yifrach, O., and Horovitz, A.: Residue lysine-34 in GroES modulates allosteric transitions in GroEL, *Biochemistry* **33**, 14974–14978, (1994).
35. Richardson, A., van der Vies, S. M., Keppel, F., Taher, A., Landry, S. J., and Georgopoulos, C.: Compensatory changes in GroEL/Gp31 affinity as a mechanism for allele-specific genetic interaction, *J. Biol. Chem.* **274**, 52–58, (1999).
36. Richardson, A., and Georgopoulos, C.: Genetic analysis of the bacteriophage T4-encoded cochaperonin Gp31, *Genetics* **152**, 1449–1457, (1999).
37. Richardson, A., Schwager, F., Landry, S. J., and Georgopoulos, C.: The importance of a mobile loop in regulating chaperonin/ co-chaperonin interaction: humans versus Escherichia coli, *J. Biol. Chem.* **276**, 4981–4987, (2001).
38. Shewmaker, F., Maskos, K., Simmerling, C., and Landry, S.J.: A mobile loop order-disorder transition modulates the speed of chaperoin cycling, *J. Biol. Chem.* **276**: 31257–31264, (2001).
39. Ma, J., Sigler, P. B., Xu, Z. H. & Karplus, M.: A Dynamic Model for the Allosteric Mechanism of GroEL, *J. Mol. Biol.* **302**, 303–313, (2000).
40. Braig, K., Otwinowski, Z., Hegde, R., Boisvert, D.C., Joachimiak, A., Horwich, A.L., and Sigler, P.B.: The crystal structure of the bacterial chaperonin GroEL at 2.8 Å, *Nature* **371**, 578–586, (1994).
41. Yifrach, O. and Horovitz, A.: Nested cooperativity in the ATPase activity of the oligomeric chaperonin GroEL, *Biochemistry* **34**, 5303–5308, (1995).
42. Saibil, H. R., and Ranson, N. R.: The chaperonin folding machine, *Trends in Biochem. Sci.* **27**, 627–632, (2002).

Simulating Protein Motions
with Rigidity Analysis[*]

Shawna Thomas, Xinyu Tang, Lydia Tapia, and Nancy M. Amato

Parasol Lab, Dept. of Comp. Sci., Texas A&M University,
College Station, TX 77843

Abstract. Protein motions, ranging from molecular flexibility to large-scale conformational change, play an essential role in many biochemical processes. Despite the explosion in our knowledge of structural and functional data, our understanding of protein movement is still very limited. In previous work, we developed and validated a motion planning based method for mapping protein folding pathways from unstructured conformations to the native state. In this paper, we propose a novel method based on rigidity theory to sample conformation space more effectively, and we describe extensions of our framework to automate the process and to map transitions between specified conformations. Our results show that these additions both improve the accuracy of our maps and enable us to study a broader range of motions for larger proteins. For example, we show that rigidity-based sampling results in maps that capture subtle folding differences between protein G and its mutations, NuG1 and NuG2, and we illustrate how our technique can be used to study large-scale conformational changes in calmodulin, a 148 residue signaling protein known to undergo conformational changes when binding to Ca^{2+}. Finally, we announce our web-based protein folding server which includes a publically available archive of protein motions: http://parasol.tamu.edu/foldingserver/

1   Introduction

Protein motions, ranging from molecular flexibility to large-scale conformational change, play an essential role in many biochemical processes. For example, conformational change often occurs in binding. While no consensus has been reached regarding models for protein binding, the importance of protein flexibility in the process is well established by the ample evidence that the same protein can exist in multiple conformations and can bind to structurally different molecules.

Our understanding of molecular movement is still very limited and has not kept pace with the explosion of knowledge regarding protein structure and function. There are several reasons for this. First, the structural data in repositories

---

[*] Supported in part by NSF Grants EIA-0103742, ACR-0081510, ACR-0113971, CCR-0113974, ACI-0326350, and by the DOE. Thomas supported in part by an NSF Graduate Research Fellowship.

A. Apostolico et al. (Eds.): RECOMB 2006, LNBI 3909, pp. 394–409, 2006.

like the Protein Data Bank (PDB) [8] consists of the spatial coordinates of each atom. Unfortunately, the experimental methods used to collect this data cannot operate at the time scales necessary to record detailed large-scale protein motions. Second, traditional simulation methods such as molecular dynamics and Monte Carlo methods are computationally too expensive to simulate long enough time periods for anything other than small peptide fragments.

There has been some attention focused on methods for modeling protein flexibility and motion. One notable effort is the Database of Macromolecular Movements [15, 14]. They generate and archive protein 'morphs' that interpolate between two different protein conformations. While the method used is more chemically realistic than straight-line interpolation (as described in Section 2), it was selected over other more accurate methods for computational efficiency and is known to have problems for some kinds of large deformations.

In previous work [3, 2, 42, 41], we developed a new computational technique for studying protein folding that builds an approximate map of a protein's potential energy landscape. This map contains thousands of feasible folding pathways to the known native state enabling the study of global landscape properties. We obtained promising results for several small proteins (60–100 amino acids) and validated our pathways by comparing secondary structure formation order with known experimental results [3].

Our Contribution. We augment our framework with three powerful new concepts that enable us to study a broader range of motions for larger proteins:

– We propose a new method based on rigidity theory to sample conformations.
– We generalize our PRM framework to map specified transitions.
– We present a new framework to automate the map building process.

Our new rigidity-based sampling allows us to study larger proteins by more efficiently characterizing the protein's energy landscape with fewer, more realistic conformations. We exploit rigidity information by focusing sampling on (currently) flexible regions. This results in smaller, better maps. In one dramatic case study, we show that rigidity-based sampling and analysis reveals the folding differences between protein G and its mutants, NuG1 and NuG2, which is an important 'benchmark' set that has been developed by the Baker Lab [36].

Extending our framework to focus on particular conformations enables us to investigate questions related to the transition between particular conformations, e.g., when studying folding intermediates, allostery, or misfolding. We provide evidence that the transitions mapped by our approach are more realistic than those provided by the computationally less expensive Morph Server [14], especially for transitions requiring large conformational changes.

The accuracy of our approach heavily depends on how densely we sample the conformation space. Previously, this was user specified and fixed. Here, we use an extension of our basic technique which incrementally samples the conformation space at increasingly denser resolution until our map of the landscape stabilizes.

Finally, we announce our protein folding server which uses our technique to generate protein transitions to the native state or between selected

Table 1. Comparison of protein motion models

Approach	Landscape	# Paths	Path Quality	Computation	Native Required
Molecular Dynamics	No	1	Good	Long	No
Monte Carlo	No	1	Good	Long	No
Statistical Model	Yes	0	N/A	Fast	Yes
PRM-Based (Our Approach)	Yes	Many	Approx	Fast	Yes
Lattice Model	Not used on real proteins				

conformations. We invite the community to help enrich our publicly available database by submitting to our server: http://parasol.tamu.edu/foldingserver/

2   Related Work

Protein Motion Models. Several computational approaches have been used to study protein motions and folding, see Table 1. These include lattice models [10], energy minimization [30, 44], molecular dynamics [29, 16], and Monte Carlo methods [13, 26]. Molecular dynamics and Monte Carlo methods provide a single, high quality transition pathway, but each run is computationally intensive. Statistical mechanical models [35, 1, 6], while computationally efficient, are limited to studying global averages of the energy landscape and kinetics and are unable to produce individual pathways.

Computing Macromolecular Motions. Gerstein et al. have developed the Database of Macromolecular Movements [15, 14] to classify protein motions. Their server produces a 'morph' movie between two target conformations in just a few minutes on a desktop PC. Their database currently includes more than 240 distinct motions.

To 'morph' between two target conformations, they first perform an alignment. Then, an iterative 'sieve-fit' procedure produces a superposition of the target conformations. The superimposed conformations are 'morphed' by interpolating the $C\alpha$ atom positions. Each intermediate conformation is energy minimized. This interpolation method, called adiabatic mapping, was selected because it has modest computational requirements yet produces chemically reasonable 'morphs.' Adiabatic mapping, however, is not guaranteed to produce accurate trajectories and in fact cannot model many large deformations.

Motion Planning and Molecular Motions. The motion planning problem is to find a valid path for a movable object from a start to a goal. The probabilistic roadmap method (PRM) [23] has been highly successful in solving high degree of freedom (dof) problems.

PRMs first sample random points in the movable object's conformation space (C-space). C-space is the set of all possible positions and orientations of the movable object, valid or not. Only those samples that meet feasibility requirements (e.g., collision free or low potential energy) are retained. Neighboring samples are connected to form a graph (or roadmap) using some simple local planner (e.g., a straight line). This roadmap can then be used to find the motion between different start and goal pairs by connecting them to the roadmap and extracting a path, if one exists. PRMs are simple to apply, even for high dof problems, only requiring the ability to generate random samples in C-space and test them feasibility.

PRMs have been applied to model molecular motions by modeling the molecule as an articulated linkage and replacing the typical collision detection validity check with some measure of physical viability (e.g., potential energy). Singh, Latombe and Brutlag first applied PRMs to protein/ligand binding [40]. In subsequent work, our group used another PRM variant on this problem [7]. Our group was the first to adapt PRMs to model protein folding pathways [3, 2, 42, 41]. Apaydin et. al. [5, 4] also applied PRMs to proteins, however their work differs from ours in several aspects. First, they model the protein at a much coarser level, considering all secondary structure elements in the native state to already be formed and rigid. Second, while our focus is on studying the transition process, their focus has been to compare the PRM approach with other computational methods such as Monte Carlo simulation. Cortes and Simeon used a PRM-based approach to model long loops in proteins [12]. Recently, we adapted the PRM framework to study RNA folding kinetics [45].

Rigidity Theory and Protein Flexibility. Several computational approaches study protein rigidity and flexibility. One approach infers rigidity and flexibility by comparing different known conformations [37, 9]. Molecular dynamics has been used to extract flexibility information from simulated motion [32, 11, 24]. A third method studies rigidity/flexibility of a single conformation [21, 22, 33]. Here, we use a rigidity analysis technique belonging to the third class of approaches called the pebble game [19, 18] to better simulate motion. It is fast and efficient; we can apply it to every conformation we sample.

The pebble game is a constraint counting algorithm which determines the dof in a two-dimensional graph, along with its rigid/flexible regions. In 2D, the pebble game assigns each vertex two pebbles, representing its two dof, see Figure 1a. Each edge/constraint is examined to determine if it is independent or redundant. If two free pebbles can be placed on both endpoints of the edge, then it is marked independent and covered by a pebble from one of its incident vertices. Once an edge is covered by a pebble, it remains covered, although which vertex the pebble comes from may change. Pebbles may be rearranged as shown

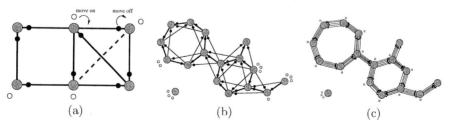

Fig. 1. (a) The result of the pebble game on a 2D graph. Pebbles may be free (white) or covering (black). Constraints are marked as independent (solid) or redundant (dashed). Pebbles may be rearranged as shown. Rigidity models for a sample molecule: (b) bar-joint and (c) body-bar.

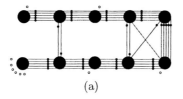 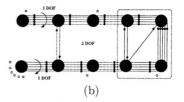

(a)   (b)

Fig. 2. (a) Model of a 5 residue protein. Each residue has two rigid bodies. We model peptide bonds and disulphide bonds with 5 bars, hydrogen bonds with 2 bars, and hydrophobic interactions with 1 bar. Redundant constraints (dashed lines) identified by the pebble game. (b) Pebble game results: a rigid cluster (dotted box), a 2 dof dependent hinge set (dashed lines), and independently flexible bonds (arcs).

the rest of the bonds in the system. Dependently flexible bonds form a set of bonds such that perturbing any one of these bonds results in a corresponding perturbation in the rest of the set.

If the bond is independently flexible, we perturb with a high probability, P_{flex}. If the bond is rigid, we perturb with a low probability, P_{rigid}. For each dependently flexible set, we randomly select d bonds to perturb with probability P_{flex} and perturb the remaining bonds with probability P_{rigid}, where d is the internal dof in the set. Perturbing rigid dof ensures good coverage of the space.

Rigidity Model. We employ the body-bar model to analyze a conformation's rigidity. With the body-bar model, we can represent the protein at a residue level, a closer match to our phi-psi model for sampling than the bar-joint model with a more detailed all-atoms view.

We model the protein simply as a chain of rigid bodies, each representing one torsional dof, see Figure 2a. We model each peptide bond and disulphide bond with 5 bars, each hydrogen bond with 2 bars, and each hydrophobic contact with 1 bar. On all conformations tested, this yields the same rigid and flexible regions as the equivalent bar-joint model on an all-atoms representation of the protein.

Rigidity Map. We can also use rigidity analysis to define a new residue mapping and distance metric. A rigidity map, r, is similar to a contact map. Rigid body pairs (i, j) from the rigidity model are marked if they have the same rigidity relationship: 2 if they are in the same rigid cluster, 1 if they are in the same dependent hinge set, and 0 otherwise. (Recall that there are two rigid bodies for each residue representing the two torsional dof.) Figure 3a shows the rigidity map of the native state for protein G with rigid clusters (black) and dependent hinge sets (green/shaded). Rigidity maps provide a convenient way to define a rigidity distance metric, $r_{\text{dist}}(q_1, q_2)$, between two conformations q_1 and q_2 where n is the number of residues:

$$r_{\text{dist}}(q_1, q_2) = \sum_{0 \le i < j \le 2n} (r_{q_1}(i, j) \ne r_{q_2}(i, j)).$$

3.2   Automatic Roadmap Construction

Roadmap accuracy depends on the sampling density. Previously, this was user specified and difficult to tune. Here, we automate roadmap construction by building the roadmap incrementally [47]. We first build a roadmap with a low sampling density as described above. Then, we test the roadmap to see if it has stabilized as specified by a set of evaluation criteria. We continue to augment the roadmap with more samples and connections until it satisfies the evaluation criteria. This provides two key advantages over our previous work: (1) the roadmap is constructed automatically at the appropriate resolution, and (2) we reuse all previous computation reducing runtime cost by several factors.

For protein folding, we build a roadmap until the secondary structure formation order along its pathways stabilizes. A piece of secondary structure is 'formed' when the distance between its rigidity map (defined in Section 3.1) and that piece's rigidity map in the target state is within 0.8, normalized to the range [0,1]. The pathway's secondary structure formation order is then the order at which pieces are 'formed.' We examine every pathway in the roadmap from an unstructured conformation to the target state and group them by this ordering. We consider the roadmap stable when the percentage of each group does not vary from the previous roadmap by more than 10%.

3.3   Mapping Specified Transitions

We extend our PRM framework to study specific large-scale conformational changes by iteratively sampling around each target conformation and connecting samples together as described earlier in Section 3. Thus our roadmaps contain the target conformations, as well as the transitions between them, and approximate the energy landscape encompassing the transition under study.

We can study problems such as transitions between known folding intermediates, transitions between bound and unbound conformations to a ligand, misfolded proteins, and allostery interactions. For example, several devastating diseases such as scrapie in sheep and goats, bovine spongiform encephalopathy (Mad Cow disease), and Creutzfeldt-Jakob disease in humans are caused by misfolded proteins called prions [38]. Insight into how these proteins misfold could help develop better drugs.

To map specific large-scale transitions, we interleave sampling and connection to incrementally build a roadmap as in Section 3.2. The only difference here is we sample around each target conformation (as in Section 3.1) during each round of roadmap construction. Then we connect samples together and compute edge weights as before. We continue until the roadmap adequately represents the protein's energy landscape near the target conformations and between them. From this roadmap, we can extract multiple low energy transition pathways between target conformations and characterize the energy barriers between them.

We build the roadmap until the maximum network flow between each target conformation pair is above a threshold. For maximum network flow, edges are assigned a capacity, and the goal is to determine how much flow can be achieved between two points in the graph. Here, we define edge capacity as the inverse of the edge weight. Thus, the maximum network flow between two conformations approximates the transition rate between them [27].

4   Results and Discussion

We investigate the ability of our rigidity-based sampling strategy to efficiently sample the protein's conformation space. We also look at examples of large-scale conformational change between specific target states for several small proteins and compare our results with 'morphs' from the Database of Macromolecular Movements [15, 14]. In all experiments, we set P_{flex} to 0.8 and P_{rigid} to 0.2. We use a straight line local planner and attempt to connect each conformation with its 50 nearest neighbors. We measure distance between two conformations as the difference between their rigidity maps (see Section 3.1).

Improved Sampling. Rigidity analysis coupled with automatic roadmap construction greatly improves the efficiency of our PRM framework by restricting the sample space in a physically realistic way. We can build smaller roadmaps that better reflect the landscape. We built roadmaps for several previously studied proteins [2, 41]. For each protein, we compare our new automatic framework with rigidity-based sampling to our previous sampling technique with fixed sampling density. Table 2 shows the roadmap size and connectivity from both methods. Both methods give the same secondary structure formation order distribution. When available, these results also indicate the same dominant secondary structure formation order seen in experiment [31]. In all cases, the rigidity-based roadmaps produce equivalent folding pathways as the previous method with smaller, more efficient roadmaps and increases connectivity. Thus, we can study much larger proteins than before.

Case study of proteins G, L NuG1, and NuG2. Proteins G, L, and mutants of protein G, NuG1 and NuG2 [36], present a good test case for our technique because they are known to fold differently despite having similar structure. All proteins are composed of a central α-helix and a 4-stranded β-sheet: β strands 1 and 2 form the N-terminal hairpin (β1-2) and β strands 3 and 4 form the C-terminal hairpin (β3-4). Native state out-exchange experiments and pulse labeling/competition experiments for proteins G and L indicate that β1-2 forms first in protein L, and β3-4 forms first in protein G [31]. This is consistent with Φ-value analysis on G [34] and L [25]. In [36], protein G is mutated in both hairpins to increase the stability of β1-2 and decrease the stability of β3-4. Φ-value analysis indicates that the hairpin formation order for both NuG1 and NuG2 is switched from the wild type.

Our previous sampling strategy [42] was able to capture the folding differences between proteins G and L, but not between protein G and NuG1 or NuG2. Our new rigidity-based sampling and analysis is able to also capture the correct folding behavior of NuG1 and NuG2, see Table 3. In addition, our rigidity-based technique can also help to explain the stability shift in NuG1 and NuG2. For

Table 2. Comparison of rigidity-based sampling to previous work for several proteins. In all cases, rigidity-based sampling significantly reduces the required roadmap size ($N+E$) to produce equivalent pathways. It also increased roadmap connectivity (E/N).

PDB Identifier	Length	Structure	Gaussian Sampling				Rigidity Sampling			
			Nodes	Edges	$N+E$	E/N	Nodes	Edges	$N+E$	E/N
1AB1	46	$2\alpha + 2\beta$	24206	386974	411180	15.99	6000	158286	164286	26.38
1CCM	46	$1\alpha + 3\beta$	43646	728964	772610	16.70	10000	456080	466080	45.61
1RDV	52	$2\alpha + 3\beta$	33691	457392	491083	13.58	4000	166702	170702	41.68
1EGF	53	3β	27356	391146	418502	14.30	4000	164902	168902	41.23
1PRB	53	5α	44551	696708	741259	15.64	4000	126562	130562	31.64
1SMU	54	$3\alpha + 3\beta$	35501	557416	592917	15.70	4000	158852	162852	39.71
1FCA	55	$2\alpha + 4\beta$	38216	489840	528056	12.82	4000	162526	166526	40.63
1VGH	55	$1\alpha + 4\beta$	38216	631936	670152	16.54	4000	157454	161454	39.36
1GB1	56	$1\alpha + 4\beta$	34236	912908	947144	26.66	4000	160552	164552	40.14
1SHG	57	5β	24696	270232	294928	10.94	18000	654884	672884	36.38
1BPI	58	$2\alpha + 2\beta$	28426	399418	427844	14.05	4000	112010	116010	28.00
4PTI	58	$2\alpha + 2\beta$	39121	389468	428589	9.96	4000	160100	164100	40.03
1HCC	59	7β	33691	453628	487319	13.46	28000	1079904	1107904	38.57
1BDD	60	3α	58486	888298	946784	15.19	6000	195950	201950	32.66
1TCP	60	$2\alpha + 2\beta$	32786	354262	387048	10.81	4000	163692	167692	40.92
2ADR	60	$2\alpha + 2\beta$	42723	701942	744665	16.43	8000	339498	347498	42.44
2PTL	62	$1\alpha + 4\beta$	23921	281334	305255	11.76	4000	159728	163728	39.93
1COA	64	$1\alpha + 5\beta$	27746	403438	431184	14.54	4000	160838	164838	40.21
2CI2	65	$2\alpha + 5\beta$	27746	389670	417416	14.04	8000	228706	236706	28.59
1NYF	67	5β	23921	262376	286297	10.97	6000	249450	255450	41.58
1MJC	69	7β	23481	226942	250423	9.66	4000	153140	157140	38.29
1HOE	74	7β	30626	184012	214638	6.01	4000	103668	107668	25.92
1UBQ	76	$1\alpha + 5\beta$	25206	236216	261422	9.37	4000	154192	158192	38.55
1O6X	81	$2\alpha + 3\beta$	40931	342138	383069	8.36	4000	133544	137544	33.39
1PBA	81	$4\alpha + 3\beta$	26476	203974	230450	7.70	8000	282960	290960	35.37
2ABD	86	5α	27956	681796	709752	24.39	18000	953900	971900	52.99

Table 3. Comparison of secondary structure formation orders for proteins G, L, NuG1, and NuG2 with known experimental results: [1]hydrogen out-exchange experiments [31], [2]pulsed labeling/competition experiments [31], and [3]Φ-value analysis [36]. Brackets indicate no clear order. In all cases, our technique predicted the secondary structure formation order seen in experiment. Only formation orders greater than 1% are shown.

Protein	Experimental Formation Order	Rigidity Formation Order	%
G	$[\alpha,\beta1,\beta3,\beta4]$, $\beta2^1$ $[\alpha,\beta4]$, $[\beta1,\beta2,\beta3]^2$	α, $\beta3$-4, $\beta1$-2	99.4
L	$[\alpha,\beta1,\beta2,\beta4]$, $\beta3^1$ $[\alpha,\beta1]$, $[\beta2,\beta3,\beta4]^2$	$\beta1$-2, α, $\beta3$-4	100.0
NuG1	$\beta1$-2, $\beta3$-4^3	α, $\beta1$-2, $\beta3$-4	97.6
		$\beta1$-2, α, $\beta3$-4	1.6
NuG2	$\beta1$-2, $\beta3$-4^3	α, $\beta1$-2, $\beta3$-4	96.6
		$\beta1$-2, α, $\beta3$-4	1.1
		$\beta3$-4, $\beta1$-2, α	1.1

example, consider their native state rigidity maps shown in Figure 3. In all four proteins, the central alpha helix remains completely rigid. We also see increased rigidity in $\beta1$-2 from protein G to NuG1 and NuG2 as suggested in [36].

We can also use rigidity-based analysis to study dynamic changes along a folding pathway, see Figure 4. We see a distinction between the profiles for protein G where $\beta3$-4 forms first and the others where $\beta1$-2 forms first. For protein G, the rigidity profile (a) shows a plateau halfway along the folding

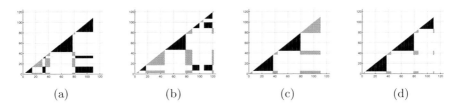

Fig. 3. Rigidity maps for the native states of proteins (a) G, (b) L, (c) NuG1, and (d) NuG2. Rigid clusters are black and dependent hinge sets are shaded/green.

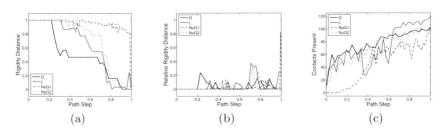

Fig. 4. Folding pathway profiles for proteins G, L, NuG1, and NuG2: (a) rigidity distance to the target state, (b) relative rigidity distance to the target state, and (c) contacts present. There is a distinction between the profiles for protein G where β3-4 forms first and the others where β1-2 forms first.

pathway, where the others do not. Protein G (b) also exhibits larger changes in rigidity earlier in the pathway while the others exhibit larger changes later.

Large-Scale Conformational Change. Calmodulin is a 148-residue signaling protein that binds to Ca^{2+} to regulate several processes in the cell. It is composed of 4 EF-hands joined by a flexible central α helix. When binding to Ca^{2+}, it undergoes two large-scale conformational changes: (1) the central α helix unravels to bring the protein from a dumbbell conformation to a more globular conformation (Figure 5a–b) and (2) the α helices in each domain reorganize (Figure 5c–d).

We built a roadmap biased towards both target states as outlined in Section 3.3. Figure 6 compares pathway profiles of the most energetically feasible transition between the two states in our roadmap and 'morphs' of various

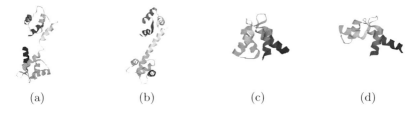

Fig. 5. Conformational changes of calmodulin: (a) calcium-free state (1CFD) to (b) bound state (1CLL) and of the N-terminal domain: (c) calcium-free to (d) bound

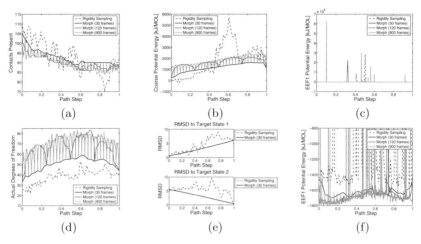

Fig. 6. Pathway profiles for the calmodulin N-terminal domain: (a) contacts present, (b) coarse potential energy, (c,f) all-atoms potential energy, (d) dof computed by rigidity analysis, and (e) RMSD to both target states. For RMSD, only the 30 frame 'morph' shown because all resolutions are nearly identical.

resolution obtained from the Morph server [15, 14]. We examined pathway profiles for energy, contacts present, dof computed by rigidity analysis, RMSD distance to the target states, and rigidity distance to the target states. Note that since the Morph server alters the original target conformations, their profile endpoints do not always align with our pathways. One striking observation is the regularity of the concavities for the 'morphs' corresponding to the various resolution levels across all the profiles except for the RMSD profiles in which the RMSD to the target states seems to change monotonically with the path step. These regularities in the 'morphs' would not be expected in actual transition pathways, e.g., one would not expect a monotonic increase in RMSD from 1CFD to 1CLL. In contrast, our roadmap pathways profiles are more plausible — they exhibit trends, but also have reasonable fluctuations. Indeed, this type of behavior has also been observed by other researchers, e.g., in [48], Monte Carlo simulations indicate a wide range of transition pathways and event durations.

Figure 6a,d shows the contacts present and dof computed by rigidity analysis along the pathway. Note that the protein does not completely unfold, but maintains a large number of contacts and loses few dof. Generally, the actual dof is inversely proportional to the number of contacts present. It is interesting to note, however, that we see a slight break in this relationship on the second half of the pathway where the peaks in dof do not match up with the peaks in number of contacts. Regions of the protein become stressed when the number of contacts increases without a corresponding decrease in dof.

We investigated several other protein transitions in a similar way, see Table 4. We measure % dof gained as the difference between the maximum dof along the pathway and the minimum dof of the starting/ending conformations, as a percentage of the total dof possible (2*length). Most transitions do not involve

Table 4. Pathway results for transitions studied. Most do not involve large unfolding.

Transition IDs		Length	% Dof Gained	# Barriers
2VGH	1VGH	55	21.8	2
1PRV	1PRU	56	5.4	0
1BMR	1FH3	67	32.8	0
1CFD	1CLL	72	18.1	1
1CMF	1CMG	73	24.7	2
1FOX	2FOW	76	3.9	0
1PFH	1HDN	85	43.5	1

a complete unfolding of the protein. In fact, several have % dof gain less than 10%. We also captured different types of transitions including smooth transitions without any significant energy barriers (i.e., 1PRV, 1BMR, and 1FOX) and those with multiple energy barriers (i.e., 2VGH and 1CMF).

We also compared 'morphs' of various resolutions to our transition pathways when possible. (The Morph server was not able to produce some higher resolution 'morphs' for transitions 1BMR–1FH3 and 1PRV–1PRU.) Across all transitions, we observed the same concavity pattern phenomenon for the 'morph' transitions as seen in calmodulin (Figure 6) for energy, contacts, degrees of freedom, and rigidity distance to the targets. Here also, the RMSD to the target states essentially changed monotonically with the path step. Again, our pathways did not exhibit these unrealistic regularities. Additional path profiles for all the transitions studied here can be found on our folding server: http://parasol.tamu.edu/foldingserver/

5   Conclusion

In this paper, we describe how to augment our PRM-based approach to study a broader range of motions for larger proteins. We proposed a method based on rigidity theory to sample more efficiently and to generate transitions between specified conformations. We also demonstrated that our approach yields more physically realistic transitions than those produced by the computationally less expensive Morph server. We invite the community to help enrich our publicly available motion database at http://parasol.tamu.edu/foldingserver/

Acknowledgments

We thank Ileana Streanu, Leslie Kuhn, and Walter Whitely for introducing us to rigidity theory and many insightful discussions. We also thank Luke Hunter, Kasia Leyk, and Dawen Xie for help with experiments.

References

1. E. Alm and D. Baker. Prediction of protein-folding mechanisms from free-energy landscapes derived from native structures. *Proc. Natl. Acad. Sci. USA*, 96(20):11305–11310, 1999.

2. N. M. Amato, K. A. Dill, and G. Song. Using motion planning to map protein folding landscapes and analyze folding kinetics of known native structures. *J. Comput. Biol.*, 10(3-4):239–256, 2003. Special issue of Int. Conf. Comput. Molecular Biology (RECOMB) 2002.

3. N. M. Amato and G. Song. Using motion planning to study protein folding pathways. *J. Comput. Biol.*, 9(2):149–168, 2002. Special issue of Int. Conf. Comput. Molecular Biology (RECOMB) 2001.

4. M. Apaydin, D. Brutlag, C. Guestrin, D. Hsu, and J.-C. Latombe. Stochastic roadmap simulation: An efficient representation and algorithm for analyzing molecular motion. In *Proc. Int. Conf. Comput. Molecular Biology (RECOMB)*, pages 12–21, 2002.

5. M. Apaydin, A. Singh, D. Brutlag, and J.-C. Latombe. Capturing molecular energy landscapes with probabilistic conformational roadmaps. In *Proc. IEEE Int. Conf. Robot. Autom. (ICRA)*, pages 932–939, 2001.

6. D. Baker. A surprising simplicity to protein folding. *Nature*, 405:39–42, 2000.

7. O. B. Bayazit, G. Song, and N. M. Amato. Ligand binding with OBPRM and haptic user input: Enhancing automatic motion planning with virtual touch. In *Proc. IEEE Int. Conf. Robot. Autom. (ICRA)*, pages 954–959, 2001. This work was also presented as a poster at *RECOMB 2001*.

8. H. Berman, J. Westbrook, Z. Feng, G. Gilliland, T. Bhat, H. Weissig, I. Shindyalov, and P. Bourne. The protein data bank. *Nucleic Acids Research*, 28(1):235–242, 2000.

9. N. Boutonnet, M. Rooman, and S. Wodak. Automatic analysis of protein conformational changes by multiple linkage clustering. *J. Mol. Biol.*, 253:633–647, 1995.

10. J. Bryngelson, J. Onuchic, N. Socci, and P. Wolynes. Funnels, pathways, and the energy landscape of protein folding: A synthesis. *Protein Struct. Funct. Genet*, 21:167–195, 1995.

11. D. Case. Molecular dynamics and normal mode analysis of biomolecular rigidity. In M. Thorpe and P. Duxbury, editors, *Rigidity theory and applications*, pages 329–344. Kluwer Academic/Plenum Publishers, 1999.

12. J. Cortes, T. Simeon, M. Remaud-Simeon, and V. Tran. Geometric algorithms for the conformational analysis of long protein loops. *J. Computat. Chem.*, 25, 2004.

13. D. Covell. Folding protein α-carbon chains into compact forms by Monte Carlo methods. *Proteins: Struct. Funct. Genet.*, 14(4):409–420, 1992.

14. N. Echols, D. Milburn, and M. Gerstein. Molmovdb: analysis and visualization of conformational change and structural flexibility. *Nucleic Acids Res.*, 31:478–482, 2003.

15. M. Gerstein and W. Krebs. A database of macromolecular motions. *Nucleic Acids Res.*, 26:4280–4290, 1998.

16. J. Haile. *Molecular Dynamics Simulation: elementary methods*. Wiley, New York, 1992.

17. B. M. Hespenheide, A. Rader, M. Thorpe, and L. A. Kuhn. Identifying protein folding cores from the evolution of flexible regious during unfolding. *J. Mol. Gra. Model.*, 21:195–207, 2002.

18. D. Jacobs. Generic rigidity in three-dimensional bond-bending networks. *J. Phys. A: Math. Gen.*, 31:6653–6668, 1998.

19. D. Jacobs and M. Thorpe. Generic rigidity percolation: The pebble game. *Phys. Rev. Lett.*, 75(22):4051–4054, 1995.

20. D. J. Jacobs, A. Rader, L. A. Kuhn, and M. Thorpe. Protein flexiblility predictions using graph theory. *Proteins Struct. Funct. Genet.*, 44:150–165, 2001.

21. J. Janin and S. Wodak. Structural domains in proteins and their role in the dynamics of protein function. *Prog. Biophys. Mol. Biol.*, 42:21–78, 1983.
22. P. Karplus and G. Schulz. Prediction of chain flexibility in proteins. *Naturwissenschaften*, 72:212–213, 1985.
23. L. E. Kavraki, P. Svestka, J. C. Latombe, and M. H. Overmars. Probabilistic roadmaps for path planning in high-dimensional configuration spaces. *IEEE Trans. Robot. Automat.*, 12(4):566–580, August 1996.
24. O. Keskin, R. Jernigan, and I. Bahar. Proteins with similar architecture exhibit similar large-scale dynamic behavior. *Biophys. J.*, 78:2093–2106, 2000.
25. D. E. Kim, C. Fisher, and D. Baker. A breakdown of symmetry in the folding transition state of protein l. *J. Mol. Biol.*, 298:971–984, 2000.
26. A. Kolinski and J. Skolnick. Monte Carlo simulations of protein folding. *Proteins Struct. Funct. Genet.*, 18(3):338–352, 1994.
27. S. V. Krivov and M. Karplus. Free energy disconnectivity graphs: Application to peptide models. *J. Chem. Phys*, 114(23):10894–10903, 2002.
28. M. Lei, M. I. Zavodszky, L. A. Kuhn, and M. F. Thorpe. Sampling protein conformations and pathways. *J. Comput. Chem.*, 25:1133–1148, 2004.
29. M. Levitt. Protein folding by restrained energy minimization and molecular dynamics. *J. Mol. Biol.*, 170:723–764, 1983.
30. M. Levitt and A. Warshel. Computer simulation of protein folding. *Nature*, 253:694–698, 1975.
31. R. Li and C. Woodward. The hydrogen exchange core and protein folding. *Protein Sci.*, 8(8):1571–1591, 1999.
32. J. Ma and M. Karplus. The allosteric mechanism of the chaperonin groel: a dynamic analysis. *Proc. Natl. Acad. Sci. USA*, 95:8502–8507, 1998.
33. V. Maiorov and R. Abagyan. A new method for modeling large-scale rearrangements of protein domains. *Proteins*, 27:410–424, 1997.
34. E. L. McCallister, E. Alm, and D. Baker. Critical role of β-hairpin formation in protein g folding. *Nat. Struct. Biol.*, 7(8):669–673, 2000.
35. V. Munoz, E. R. Henry, J. Hoferichter, and W. A. Eaton. A statistical mechanical model for β-hairpin kinetics. *Proc. Natl. Acad. Sci. USA*, 95:5872–5879, 1998.
36. S. Nauli, B. Kuhlman, and D. Baker. Computer-based redesign of a protein folding pathway. *Nature Struct. Biol.*, 8(7):602–605, 2001.
37. W. Nichols, G. Rose, L. T. Eyck, and B. Zimm. Rigid domains in proteins: an algorithmic approach to their identification. *Proteins*, 23:38–48, 1995.
38. S. Prusiner. Prions. *Proc. Natl. Acad. Sci. USA*, 95(23):13363–13383, 1998.
39. A. Rader, B. M. Hespenheide, L. A. Kuhn, and M. Thorpe. Protein unfolding: Rigidity lost. *Proc. Natl. Acad. Sci. USA*, 99(6):3540–3545, 2002.
40. A. Singh, J. Latombe, and D. Brutlag. A motion planning approach to flexible ligand binding. In *7th Int. Conf. on Intelligent Systems for Molecular Biology (ISMB)*, pages 252–261, 1999.
41. G. Song. *A Motion Planning Approach to Protein Folding*. Ph.D. dissertation, Dept. of Computer Science, Texas A&M University, December 2004.
42. G. Song, S. Thomas, K. Dill, J. Scholtz, and N. Amato. A path planning-based study of protein folding with a case study of hairpin formation in protein G and L. In *Proc. Pacific Symposium of Biocomputing (PSB)*, pages 240–251, 2003.
43. M. J. Sternberg. *Protein Structure Prediction*. OIRL Press at Oxford University Press, 1996.
44. S. Sun, P. D. Thomas, and K. A. Dill. A simple protein folding algorithm using a binary code and secondary structure constraints. *Protein Eng.*, 8(8):769–778, 1995.

45. X. Tang, B. Kirkpatrick, S. Thomas, G. Song, and N. M. Amato. Using motion planning to study rna folding kinetics. In *Proc. Int. Conf. Comput. Molecular Biology (RECOMB)*, pages 252–261, 2004.

46. W. Whiteley. Some matroids from discrete applied geometry. *Contemp. Math.*, 197:171–311, 1996.

47. D. Xie, S. Thomas, J.-M. Lien, and N. M. Amato. Incremental map generation. Technical Report TR05-006, Parasol Lab, Dept. of Computer Science, Texas A&M University, Sep 2005.

48. D. M. Zuckerman. Simulation of and ensemble of conformational transitions in a united-residue model of calmodulin. *J. Phys. Chem*, 108:5127–5137, 2004.

Predicting Experimental Quantities in Protein Folding Kinetics Using Stochastic Roadmap Simulation

Tsung-Han Chiang[1], Mehmet Serkan Apaydin[2], Douglas L. Brutlag[3],
David Hsu[1], and Jean-Claude Latombe[3]

[1] National University of Singapore, Singapore 117543, Singapore
[2] Dartmouth College, Hanover, NH 03755, USA
[3] Stanford University, Stanford, CA 94305, USA

Abstract. This paper presents a new method for studying protein folding kinetics. It uses the recently introduced Stochastic Roadmap Simulation (SRS) method to estimate the transition state ensemble (TSE) and predict the rates and Φ-values for protein folding. The new method was tested on 16 proteins. Comparison with experimental data shows that it estimates the TSE much more accurately than an existing method based on dynamic programming. This leads to better folding-rate predictions. The results on Φ-value predictions are mixed, possibly due to the simple energy model used in the tests. This is the first time that results obtained from SRS have been compared against a substantial amount of experimental data. The success further validates the SRS method and indicates its potential as a general tool for studying protein folding kinetics.

1 Introduction

Protein folding is a crucial biological process in nature. Starting out as a long, linear chain of amino acids, a protein molecule remarkably configures itself, or *folds*, into a unique three-dimensional structure, called the *native state*, in order to perform vital biological functions. There are two separate, but related problems in protein folding: structure prediction and folding kinetics. In the former problem, we are only interested in predicting the final three-dimensional structure, i.e., the native state, attained in the folding process. In the latter problem, we are interested in the folding process itself, e.g., the kinetics and the mechanism of folding. We have at least two important reasons for studying the folding process. First, better understanding of the folding process will help explain why and how proteins misfold and find therapies for debilitating diseases such as Alzheimer's disease or Creutzfeldt-Jakob ("mad cow") disease. Second, this will aid in the development of better algorithms for structure prediction.

In this work, we apply computational methods to study the kinetics of protein folding, specifically, to predict the folding rates and the Φ-values. The folding rate measures how fast a protein evolves from an unfolded state to the native state. The Φ-value measures the extent to which a residue of a protein attains its native conformation when the protein is in the transition state of the folding process. Performing such computational studies was once very difficult, due to a lack of good models of protein folding, a lack of efficient computational methods to predict experimental quantities based on theoretical

A. Apostolico et al. (Eds.): RECOMB 2006, LNBI 3909, pp. 410–424, 2006.

models, and a lack of detailed experimental results to validate the predictions. However, important advances have been made in recent years. On the theoretical side, the energy landscape theory [4, 7] offers a global view of protein folding in microscopic details based on statistical physics. It hypothesizes that proteins fold in a multi-dimensional energy funnel by following a myriad of pathways, all leading to the same native state. On the experimental side, residue-specific measurements of the folding process (see, e.g., [14]) provide detailed experimental data to validate theoretical predictions.

Our work takes advantage of these developments. To compute the folding rate and Φ-values of a protein, we first estimate the transition state ensemble (TSE), which is a set of high-energy protein conformations that limits the folding rate. We use the recently introduced *Stochastic Roadmap Simulation* (SRS) method [3] on a folding energy landscape proposed in [12]. SRS samples the protein conformational space and builds a directed graph, called the *stochastic conformational roadmap*. The nodes of the roadmap represent sampled protein conformations, and the edges represent transitions between the conformations. The roadmap compactly encodes a huge number of folding pathways and captures the stochastic nature of the folding process. Using the roadmap, we can efficiently compute the folding probability (P_{fold}) [8] for each sampled conformation in the roadmap and decide which conformations belong to the TSE. Finally, we estimate folding rates and Φ-values using the set of conformations in the TSE.

We tested our method on 16 proteins with sizes ranging from 56 to 128 residues and validated the results against experimental data. The results show that our method predicts folding rates with accuracy better than an existing method based on dynamic programming (DP) [12]. In the following, this existing method will be called the DP method, for lack of a better name. More importantly, our method provides a much more discriminating estimate of the TSE: our estimate of the TSE contains less than 10% of all sampled conformations, while the estimate by the DP method contains 85–90%. The more accurate estimate better reveals the composition of the TSE and makes our method more suitable for studying the mechanisms of protein folding. For Φ-value prediction, the accuracy of our method varies among the proteins tested. The results are comparable to those obtained from the DP method, but both methods need to be improved in accuracy to be useful in practice.

From a methodology point of view, this is the first time that results based on P_{fold} values computed by SRS were compared against substantial amount of experimental data. Earlier work on SRS compared it with Monte Carlo simulation and showed that SRS is faster by *several orders of magnitude* [3]. The comparison with experimental data serves as a test of the methodology, and the success further validates the SRS method and indicates its potential as a general tool for studying protein folding kinetics.

2   Related Work

There are many approaches for studying protein folding kinetics, including all-atom or lattice molecular dynamics simulation (see [9] for a survey), solving master equations [6, 21], and estimating the TSE [1, 12]. Recently, several related methods succeeded in predicting folding rates and Φ-values [1, 12, 15], using simplified energy functions that depend only on the topology of the native state of a protein. Our work also uses such an energy function, but instead of searching for rate-limiting "barriers" on the energy landscape as in [1, 12], we estimate the TSE by using SRS to compute P_{fold} values and then estimate the folding rates and Φ-values based on the energy of conformations in the TSE.

SRS is inspired by the probabilistic roadmap (PRM) methods for robot motion planning [5]. In motion planning, our goal is to find a path for a robot to move from an initial configuration to a goal configuration without colliding with any obstacles. The main idea of PRM methods is to sample at random the space of all robot configurations— a space conceptually similar to a protein conformation space—and construct a graph that captures the connectivity of this space. Methods derived from PRM have been applied to ligand-protein docking [17], protein folding [3, 2], and RNA folding [19]. In our earlier work, we used SRS to study protein folding, but the results were compared only with those obtained from Monte Carlo simulation. Here, we extend the work to compute folding rates and Φ-values and validate the results directly against experimental data. SRS has also been combined with molecular dynamics simulation to study protein folding rates and mechanisms [18].

3   Overview

The *conformation* of a protein is a set of parameters that specify uniquely the structure of the protein, e.g., the backbone torsional angles ϕ and ψ. The *conformational space* $\mathcal{C}$ contains all the conformations of a protein. If $\mathcal{C}$ is parametrized by d conformational parameters, then a conformation can be regarded as a point in a d-dimensional space.

Each conformation q of a protein has an associated energy value $E(q)$, determined by the interactions between the atoms of the protein and between the protein and the surrounding medium, e.g., the van der Waals and electrostatic forces. The energy E is a function defined over $\mathcal{C}$ and is often called the *energy landscape*. According to the energy landscape theory, proteins fold along many pathways over the energy landscape. These pathways start from unfolded conformations and all lead to the same native state.

To understand protein folding kinetics, we need to analyze the folding pathways and identify those conformations, called the *transition state ensemble* (TSE), that act as barriers on the energy landscape and limit the folding rate. For convenience, we also say that such conformations are in the transition state. In the simple case where there is a dominant folding pathway with a single major energy peak along the pathway, the TSE can be defined as the conformations with energy at or near the peak value. In general, there may be many pathways, and along every pathway, there may be multiple energy peaks. This makes the TSE more difficult to identify. To address this issue, Du et al. introduced the notion of P_{fold} [8]. In a folding process, the P_{fold} value of a conformation q is defined as the probability of a protein reaching the folded (native) state before reaching an unfolded state, starting from conformation q. P_{fold} measures the kinetic distance between q and the folded state. From any conformation q with P_{fold} value greater than 0.5, the protein is more likely to fold first than to unfold first. Thus q is kinetically closer to the folded state. The TSE is defined as the set of conformations with P_{fold} equal to 0.5. Defining the TSE using P_{fold} has many advantages. In particular,

P_{fold} is not determined by any specific pathway, but depends on all the pathways from unfolded states to the folded state. It thus captures the ensemble behavior of folding.

We can compute P_{fold} value for a conformation q by performing many folding simulation runs from q and count the number of times that they reach the folded state before an unfolded one. However, a large number of simulation runs are needed to estimate the P_{fold} value accurately, and doing so for many conformations incurs prohibitive computational cost. The SRS method approximates the P_{fold} values for many conformations simultaneously in a much more efficient way. In the following, we first describe the computation of the TSE using SRS (Sect. 4) and then the computation of folding rates (Sect. 5) and Φ-values (Sect. 6) based on the energy of conformations in the TSE.

4   Estimating the TSE Using Stochastic Roadmap Simulation

SRS is an efficient method for exploring protein folding kinetics by examining many folding pathways simultaneously. We use SRS to compute P_{fold} values and then determine the TSE based on the computed P_{fold} values.

4.1   A Simplified Folding Model

To study protein folding kinetics, we need an energy function that accurately models the interactions within a protein and the interactions between a protein and the surrounding medium at the atomic level. For this, we use the simple, but effective energy model developed by Garbuzynskiy et al. [12]. This model is based on the topology of a protein's native state. An important concept here is that of *native contact*. Two atoms are considered to be in contact if the distance between them is within a suitably chosen threshold. A native contact between two atoms of a protein is a contact that exists in the native state. Given a conformation q, we can obtain all the native contacts in q by computing the pairwise distances between the atoms of the protein.

The energy model that we use divides a protein into contiguous segments of five residues each. Each segment must be either folded or unfolded completely. In other words, atoms within a folded segment must gain all their native contacts with other atoms in the folded segments, while atoms within an unfolded segment are assumed to form a disordered loop and lose all their native contacts. We thus represent the conformation of a protein by a binary vector, with 1 representing a folded segment and 0 representing an unfolded segment. In particular, the folded (native) conformation is $(1, 1, \ldots, 1)$, and the unfolded conformation is $(0, 0, \ldots, 0)$.

Using this representation, a protein with N residues has $2^{\lceil N/5 \rceil}$ distinct conformations. To further reduce computation time, Garbuzynskiy et al. suggested a restriction which accepts only conformations with at most two unfolded regions in the middle of a protein plus two unfolded regions at the ends of the protein. With a maximum of four unfolded regions, we can capture the folding and unfolding of proteins with up to roughly 100 residues [11].

The free energy of a conformation q is calculated based on the number of native contacts and the length of unfolded segments in q:

$$E(q) = \varepsilon \cdot n(q) - T \cdot (2.3R \cdot \mu(q) + S(q)) . \tag{1}$$

In the formula above, $n(q)$ is the number of native contacts in the folded segments of q, $\mu(q)$ is the number of residues in the unfolded segments of q, and $S(q)$ is the entropy for closing the disordered loops. For the rest, which are all constants, ε is the energy of a single native contact, T is the absolute temperature, and R is the gas constant. A similar energy function has been used in the work of Alm and Baker [1].

Our model uses all the atoms of a protein, including the hydrogen atoms, to calculate the energy. For protein structures determined by X-ray crystallography, hydrogen atoms are missing and we added them using the Insight II program at pH level 7.0.

4.2   Constructing the Stochastic Conformational Roadmap

A stochastic conformational roadmap G is a directed graph. Each node of G represents a conformation of a protein. Each directed edge from a node q_i to a node q_j carries a weight P_{ij}, which represents the probability for a protein to transit from q_i to q_j. If there is no edge from q_i to q_j, the probability P_{ij} is 0; otherwise, P_{ij} depends on the energy difference between q_i and q_j, $\Delta E_{ij} = E(q_j) - E(q_i)$.

The transition probability P_{ij} is defined according to the Metropolis criterion, which is also used in Monte Carlo simulation:

$$P_{ij} = \begin{cases} (1/n_i)\exp(-\frac{\Delta E_{ij}}{k_\mathrm{B}T}) & \text{if } \Delta E_{ij} > 0 \\ 1/n_i & \text{otherwise} \end{cases},$$

where n_i is the number of outgoing edges of q_i, k_B is the Boltzmann constant, and T is the absolute temperature. The factor $1/n_i$ normalizes the effect that different nodes may have different numbers of outgoing edges. We also assign the self-transition probability:

$$P_{ii} = 1 - \sum_{j \neq i} P_{ij},$$

which ensures that the transition probabilities from any node sums to 1.

SRS views protein folding as a random walk on the roadmap graph. If q_F and q_U are the two roadmap nodes representing the folded and the unfolded conformation, respectively, every path in the roadmap from q_U to q_F represents a potential folding pathway. Thus, a roadmap compactly encodes an exponential number of folding pathways.

To construct the roadmap G using the folding model described in Sect. 4.1, we enumerate the set of all allowable conformations in the model (with the restriction of a maximum of four unfolded regions) and use them as the nodes of G. There is an edge between two nodes if the corresponding conformations differ by exactly one folded or unfolded segment.

4.3   Computing $\mathbf{P_{fold}}$

$\mathrm{P_{fold}}$ measures the kinetic distance between a conformation q and the native state q_F. The main advantage of using $\mathrm{P_{fold}}$ to measure the progress of protein folding is that it takes into account all folding pathways sampled from the protein conformation space and does not assume any particular pathway *a priori*.

Recall that the $\mathrm{P_{fold}}$ value τ of a conformation q is defined as the probability of a protein reaching the native state q_F before reaching the unfolded state q_U, starting from q.

Instead of computing τ by brute force through many Monte Carlo simulation runs, we construct a stochastic conformational roadmap and apply the first step analysis [20]. Let us consider what happens after a single step of transition:

- We may reach a node in the folded state, which, by definition, has P_{fold} value 1.
- We may reach a node in the unfolded state, which has P_{fold} value 0.
- Finally, we may reach an intermediate node q_j with P_{fold} value τ_j.

The first step analysis conditions on the first transition and gives the following relationship among the P_{fold} values:

$$\tau_i = \sum_{q_j \in \{q_F\}} P_{ij} \cdot 1 + \sum_{q_j \in \{q_U\}} P_{ij} \cdot 0 + \sum_{q_j \notin \{q_F, q_U\}} P_{ij} \cdot \tau_j, \qquad (2)$$

where τ_i is the P_{fold} value for node q_i. In our simple folding model, both the folded and the unfolded state contains only a single conformation, but in general, they may contain multiple conformations.

The relationship in (2) gives a linear equation for each unknown τ_i. The resulting linear system is sparse and can be solved efficiently using iterative methods [3].

The largest protein that we tested has 128 residues, resulting in a total of 314,000 allowable conformations. It took SRS only about a minute to compute P_{fold} values for all the conformations on a PC workstation with a 1.5GHz Itanium2 processor and 8GB of memory.

4.4   Estimating the TSE

After computing the P_{fold} value for each conformation, we identify the TSE by extracting all conformations with P_{fold} value 0.5. However, due to the simplification and discretization used in our folding model, we need to broaden our selection criteria slightly and identify the TSE as the set of conformations with P_{fold} values within a small range centered around 0.5. We found that the range between 0.45 to 0.55 is usually adequate to account for the model inaccuracy in our tests, and we used it in all the results reported below.

4.5   An Example on a Synthetic Energy Landscape

Consider a tiny fictitious protein with only two residues. Its conformation is specified by two backbone torsional angles ϕ and ψ. For the purpose of illustration, instead of using the simplified energy function described in Sect. 4.1, this example uses a saddle-shaped energy function over a two-dimensional conformation space (Fig. 1a) in which the two torsional angles vary continuously over their respective ranges. On this energy landscape, almost all intermediate conformations have energy levels at least as high as the unfolded conformation q_U and the native conformation q_F. This synthetic energy landscape is conceptually similar to more realistic energy models commonly used. Namely, to go from q_U to q_F, a protein must pass through energy barriers.

The computed P_{fold} values for this energy landscape is shown in Fig. 1b. A comparison of the two plots in Fig. 1 shows that the conformations with P_{fold} value 0.5 correspond well with the energy barrier that separates q_U and q_F.

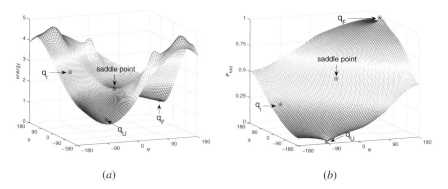

Fig. 1. P_{fold} values for a synthetic energy landscape. (*a*) A synthetic energy landscape. (*b*) The computed P_{fold} values.

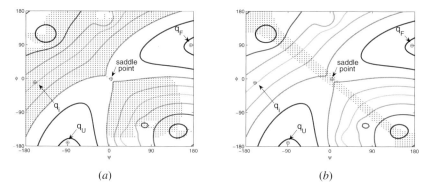

Fig. 2. Estimation of the TSE for the energy landscape shown in Fig. 1. The conformation-space region corresponding to the TSE is shaded and overlaid on the contour plot of the energy landscape. (*a*) The DP method. (*b*) The SRS method.

5   Predicting Folding Rates

The folding rate is an experimentally measurable quantity that determines how fast the protein proceeds from the unfolded state to the folded state. By observing how it varies under different experimental conditions, we can gain an understanding of the important factors that influence the folding process.

The speed at which a protein folds depends exponentially on the height of the energy barrier that must be overcome during the folding process. The higher the barrier, the harder it is for the unfolded protein to reach the folded state and the slower the process. Because of the exponential dependence, even a small difference in the height of the energy barrier has significant effect on the folding rate. Therefore, accurately identifying the TSE is crucial for predicting the folding rate.

5.1 Methods

After identifying the TSE using the SRS method described in the previous section, we compute the folding rate the same way as that in [12], for the purpose of easy comparison. First, we calculate E_{TSE}, the total energy of the TSE, according to the following relationship [12]:

$$\exp(-\frac{E_{\text{TSE}}}{RT}) = \sum_{q \in \text{TSE}} \exp(-\frac{E(q)}{RT}), \tag{3}$$

where the summation is taken over the set of all conformations in the TSE, R is the gas constant and T is the absolute temperature. We then compute the rate constant k_f according to the following theoretical dependence [12]:

$$\ln(k_f) = \ln(10^8) - (\frac{E_{\text{TSE}}}{RT} - \frac{E(q_U)}{RT}), \tag{4}$$

where $E(q_U)$ is the energy of the q_U.

5.2 Results

Using data from the Protein Data Bank (PDB), we computed folding rates for 16 proteins (see Appendix A for the list). The results are shown in Fig. 3. The horizontal axis of the chart corresponds to the experimentally measured folding rates (see [12] for the sources of data), and the vertical axis corresponds to the predicted values. The best-fit lines of the data are also shown. For comparison, we also computed the folding rates using the DP method [12] and show the results in the same chart. Note that since the chart plots $\ln k_f$, it basically compares the height of the energy barrier.

Fig. 3 shows that both methods can predict the trend reasonably well. The best-fit line of SRS is closer to the diagonal, indicating better predictions. This is confirmed by comparing the average error in $\ln k_f$ for the two methods.

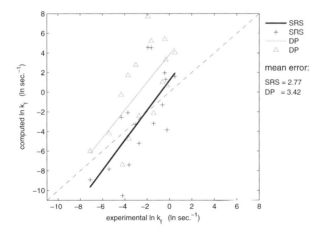

Fig. 3. Predicted folding rates versus the experimentally measured folding rates

It is interesting to note that DP consistently predicts higher k_f compared to SRS. Since a higher k_f corresponds to lower energy barrier along the folding pathway, the TSE identified by DP must have lower energy. This is significant in terms of the accuracy of folding rate prediction and suggests that an important difference exists between the TSE estimated by SRS and that estimated by DP.

5.3   Accuracy in Estimating the TSE

The difference between SRS and DP in estimating the TSE becomes more apparent when we compare the percentage of sampled conformations that are present in the TSE. Fig. 4 shows that the TSE estimated by SRS includes less than 10% of all allowable conformations. In contrast, the TSE estimated by DP includes, surprisingly, 85-90%. Closer inspection reveals that the TSE computed by SRS is mostly a subset of the TSE computed by DP. Combining this observation with the better prediction accuracy of SRS, we conclude that the additional 80% or so conformations identified by DP are not only unnecessary, but also negatively affect folding rate prediction.

Although it is difficult to know the true percentage of conformations that should belong to the TSE, careful examination of the DP method shows that it indeed may include in the TSE many conformations that are suspicious. This is best illustrated using the example in Fig. 1a. According to the DP method, a conformation q belongs to the TSE, if q has the highest energy along the folding pathway that has the lowest energy barrier among all pathways that go through q. This definition tries to capture the intuition that q is the location of minimum barrier on the energy landscape. For the energy landscape shown in Fig. 1, the globally lowest energy barrier is clearly the conformation q_s at the saddle point. So q_s belongs to the TSE. For any other conformation q, there are two possibilities. When $E(q) < E(q_s)$, any path through q must have a barrier higher than or equal to $E(q_s)$, and q cannot possibly achieve the highest energy along the path. Thus, q does not belong to the TSE. The problem arises when $E(q) \geq E(q_s)$. In this case, to place q in the TSE, all it takes is to find a path that goes through q and does not pass through any other conformation with energy higher than $E(q)$. This can be easily accomplished on the saddle-shaped energy landscape for most conformations with $E(q) \geq E(q_s)$, e.g., the conformation q_i indicated in Fig. 1. Including such

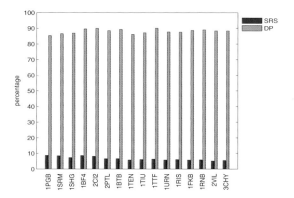

Fig. 4. The percentage of conformations in the TSE

conformations in the TSE seems counter-intuitive, as they do not constitute a barrier on the energy landscape.

As we have seen in Sect. 4.5, the SRS method includes in the TSE only those conformations near the barrier of the energy landscape , but the DP method includes many additional conformations, some of which are far below the energy of the barrier (see Fig. 2 for an illustration). Therefore, the TSE estimated by DP tend to have lower energy than the TSE estimated by SRS, resulting in over-estimated folding rates.

6 Predicting Φ-Values

Φ-value analysis is the only experimental method for determining the transition-state structure of a protein at the resolution of individual residues [10]. Its main idea is to mutate carefully selected residues of a protein, measure the resulting energy changes, and infer from them the structure of the protein in the transition state. Here, we would like to predict Φ-values computationally.

6.1 Methods

The Φ-value indicates the extent to which a residue has attained the native conformation when the protein is in the transition state of the folding process. More precisely, the Φ-value of a residue r is defined as:

$$\Phi_r = \frac{\Delta_r[E_{\text{TSE}} - E(q_{\text{U}})]}{\Delta_r[E(q_{\text{F}}) - E(q_{\text{U}})]},\tag{5}$$

where $\Delta_r[E_{\text{TSE}} - E(q_{\text{U}})]$ is the change in the energy difference between the TSE and the unfolded state q_{U} as a result of mutating r. Similarly, $\Delta_r[E(q_{\text{F}}) - E(q_{\text{U}})]$ is the mutation-induced change in the energy difference between the native state q_{F} and the unfolded state q_{U}. See Fig. 5 for an illustration. A Φ-value of 1 indicates that the mutation of residue r affects the energy of the transition state as much as the energy of the native state, relative to the energy of the unfolded state. So, in the transition state, r must have fully attained the native conformation, according to energy considerations.

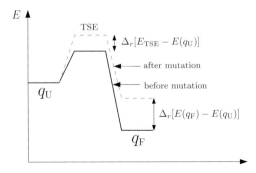

Fig. 5. Φ-value

Similarly, a Φ-value of 0 indicates that in the transition state, the residue remains unfolded. A fractional Φ-value value between 0 and 1 indicates that the residue has only partially attained its native conformation. By analyzing the Φ-value of each residue of a protein, we can elucidate the structure of the TSE.

Using (1) and (3), we can simplify (5) and obtain the following expression for the Φ-value of residue r:

$$\Phi_r = \frac{\sum_{q \in \text{TSE}} P(q) \cdot \Delta_r n(q)}{\Delta_r n(q_{\text{F}})}, \tag{6}$$

where $P(q)$ is the Boltzmann probability for conformation q and $\Delta_r n(q)$ is the change in the number of native contacts for conformation q as a result of mutating r.

6.2   Results on Φ-Value Prediction

The Φ-value is more difficult to predict than the folding rate, because it is a detailed experimental quantity and requires an accurate energy model for prediction. We computed Φ-values for 16 proteins listed in Appendix A, but got mixed results. Fig. 6

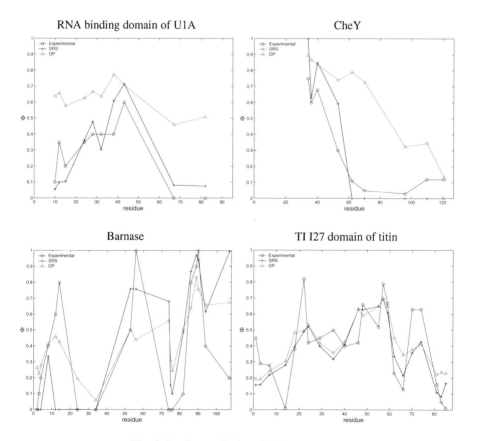

Fig. 6. Φ-value predictions for four proteins

Table 1. Performance of SRS and DP in Φ-value prediction. For each protein, the average error of computed Φ-values is calculated. The table reports the mean, the minimum, and the maximum of average errors over the 16 proteins tested.

Method	Mean	Min	Max
SRS	0.21	0.11	0.32
DP	0.24	0.13	0.35

shows a comparison of the Φ-values computed by SRS and DP and the Φ-values measured experimentally. The sources of the experimental data are available in [12]. In general, our Φ-value predictions based on X-ray crystallography structures are better than those based on NMR structures. When compared with DP, SRS is much better for some proteins, such as CheY and the RNA binding domain of U1A, both of which have X-ray crystallography structures. For the other proteins, the results are mixed. In some cases (e.g., barnase), our results are slightly better, and in others (e.g., TI I27 domain of titin), slightly worse. Table 1 shows the performance of SRS and DP over the 16 proteins tested. Since Φ-values range between 0 and 1, the errors are fairly large for both SRS and DP. To be useful in practice, more research is needed for both methods.

6.3   Results on the Order of Native Structure Formation

An important advantage of using P_{fold} as a measure of the progress of folding is that P_{fold} takes into account all sampled folding pathways and is not biased towards any specific one. We have seen how to use P_{fold} to estimate Φ-values, which give an indication of the progress of folding in the transition state only. We can extend this method to observe the details of the folding process, in particular, the order of native structure formation, by plotting the progression of each residue with respect to P_{fold}.

Each plot in Fig. 7 shows the frequency with which a residue achieves its native conformation in a Boltzmann weighted ensemble of conformations with approximately same P_{fold} values. For CheY, residues 1 to 40 gain their native conformation very early in the folding process. The coherent interactions between neighboring residues is consistent with the mainly helical secondary structure of these residues. Residues 50 to 80 are subsequently involved in the folding nucleus as folding progresses. The folding of barnase is more cooperative and involves many regions of the protein simultaneously. Residues 50 to 109 dominate the folding process early on, and the simultaneous progress of different regions corresponds to the formation of the β sheet. The helical residues 1 to 50 gain native conformation very late in the folding. The order of native structure formation that we observed is consistent with that obtained by Alm et al. [1].

The accuracy of Φ-value prediction gives an indication of the reliability of such plots. We made similar plots for the other proteins. Although we were able to see interesting trends for some of the other proteins, the plots are not shown here, because of the low correlation of their Φ-value predictions to experimental values. Verifying the accuracy of such plots directly is difficult, due to the limited observability of the protein folding

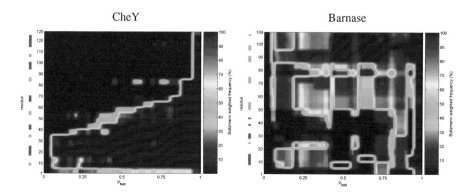

Fig. 7. Sequence of secondary structure formation. The colored bar on the left of each plot indicates secondary structures, red for helices and green for strands.

process and the limited experimental data available. The reliance on other simulation results for verification is almost inevitable.

7   Conclusion

This paper presents a new method for studying protein folding kinetics. It uses the Stochastic Roadmap Simulation method to compute the P_{fold} values for a set of sampled conformations of a protein and then estimate the TSE. The TSE is of great importance for understanding protein folding, because it gives insight into the main factors that influence folding rates and mechanisms. Knowledge of the structure of the TSE may be used to re-engineer folding in a principled way [16]. One main advantage of SRS is that it efficiently examines a huge number of folding pathways and captures the ensemble behavior of protein folding. Our method was tested on 16 proteins. The results show that our estimate of the TSE is much more discriminating than that of the DP method. This allows us to obtain better folding-rate predictions. We have mixed results in predicting Φ-values. One likely reason is that Φ-value prediction requires a more detailed model than the one that we used. The success of SRS on these difficult prediction problems further validates the SRS method and indicates its potential as a general tool for studying protein folding kinetics.

The 16 proteins that we studied fold via a relatively simple two-state transition mechanism. It would be interesting to further test our method on more complex proteins, such as those that fold via an intermediate. We also plan to improve Φ-value prediction by using a better energy model and to predict other experimental quantities, such as hydrogen-exchange protection factors [13].

Acknowledgements. M. S. Apaydin's work at Dartmouth was supported by the following grants to Bruce R. Donald: NIH grant R01-GM-65982 and NSF grant EIA-0305444. D. Hsu's research is partially supported by grant R252-000-145-112 from the National University of Singapore. J.C. Latombe's research is partially supported by NSF grants CCR-0086013 and DMS-0443939, and by a Stanford BioX Initiative grant.

References

1. E. Alm and D. Baker. Prediction of protein-folding mechanisms from free-energy landscapes derived from native structures. *Proc. Nat. Acad. Sci. USA*, 96:11305–11310, 1999.
2. N. M. Amato, K. A. Dill, and G. Song. Using motion planning to map protein folding landscapes and analyze folding kinetics of known native structures. In *Proc. ACM Int. Conf. on Computational Biology (RECOMB)*, pages 2–11, 2002.
3. M. S. Apaydin, D. L. Brutlag, C. Guestrin, D. Hsu, and J.-C. Latombe. Stochastic roadmap simulation: An efficient representation and algorithm for analyzing molecular motion. In *Proc. ACM Int. Conf. on Computational Biology (RECOMB)*, pages 12–21, 2002.
4. J. D. Bryngelson, J. N. Onuchic, N. D. Socci, and P. G. Wolynes. Funnels, pathways, and the energy landscape of protein folding: A synthesis. *Proteins: Structure, Function, and Genetics*, 21(3):167–195, 1995.
5. H. Choset, K. M. Lynch, S. Hutchinson, G. Kantor, W. Burgard, L. E. Kavraki, and S. Thrun. *Principles of Robot Motion: Theory, Algorithms, and Implementations*, chapter 7. The MIT Press, 2005.
6. M. Cieplak, M. Henkel, J. Karbowski, and J. R. Banavar. Master equation approach to protein folding and kinetic traps. *Phys. Rev. Let.*, 80:3654, 1998.
7. K. A. Dill and H. S. Chan. From Levinthal to pathways to funnels. *Nature Structural Biology*, 4(1):10–19, 1997.
8. R. Du, V. S. Pande, A. Y. Grosberg, T. Tanaka, and E. S. Shakhnovich. On the transition coordinate for protein folding. *J. Chem. Phys.*, 108(1):334–350, 1998.
9. Y. Duan and P. A. Kollman. Computational protein folding: From lattice to all-atom. *IBM Systems J.*, 40(2):297–309, 2001.
10. A. Fersht. *Structure and Mechanism in Protein Science: A Guide to Enzyme Catalysis and Protein Folding*. W.H. Freeman & Company, New York, 1999.
11. A. V. Finkelstein and A. Y. Badretdinov. Rate of protein folding near the point of thermodynamic equilibrium between the coil and the most stable chain fold. *Folding and Design*, 2(2):115–121, 1997.
12. S. O. Garbuzynskiy, A. V. Finkelstein, and O. V. Galzitskaya. Outlining folding nuclei in globular proteins. *J. Mol. Biol.*, 336:509–525, 2004.
13. V. J. Hilser and E. Freire. Structure-based calculation of the equilibrium folding pathway of proteins. Correlation with hydrogen exchange protection factors. *J. Mol. Biol.*, 262(5):756–772, 1996.
14. L. S. Itzhaki, D. E. Otzen, and A. R. Fersht. The structure of the transition state for folding of chymotrypsin inhibitor 2 analysed by protein engineering methods: evidence for a nucleation-condensation mechanism for protein folding. *J. Mol. Biol.*, 254(2):260–288, 1995.
15. V. Muñoz and William A. Eaton. A simple model for calculating the kinetics of protein folding from three-dimensional structures. *Proc. Nat. Acad. Sci. USA*, 96:11311–11316, 1999.
16. B. Nölting. *Protein Folding Kinetics: Biophysical Methods*. Springer, 1999.
17. A. P. Singh, J.-C. Latombe, and D. L. Brutlag. A motion planning approach to flexible ligand binding. In *Proc. Int. Conf. on Intelligent Systems for Molecular Biology*, pages 252–261, 1999.
18. N. Singhal, C. D. Snow, and V. S. Pande. Using path sampling to build better Markovian state models: Predicting the folodiing rate and mechanism of a tryptophan zipper beta hairpin. *J. Chemical Physics*, 121(1):415–425, 2004.
19. X. Tang, B. Kirkpatrick, S. Thomas, G. Song, and N. M. Amato. Using motion planning to study RNA folding kinetics. In *Proc. ACM Int. Conf. on Computational Biology (RECOMB)*, pages 252–261, 2004.

20. H. Taylor and S. Karlin. *An Introduction to Stochastic Modeling*. Academic Press, New York, 1994.
21. T. R. Weikl, M. Palassini, and K. A. Dill. Cooperativity in two-state protein folding kinetics. *Protein Sci.*, 13(3):822–829, 2004.

A   The List of Proteins Used for Testing

For each protein used in our test, the table below lists its name, PDB code, the number of residues, and the experimental method for structure determination.

Protein	PDB code	No. Res.	Exp. Meth.
B1 IgG-binding domain of protein G	1PGB	56	X-ray
Src SH3 domain	1SRM	56	NMR
Src-homology 3 (SH3) domain	1SHG	57	X-ray
Sso7d	1BF4	63	X-ray
CI-2	2CI2	65	X-ray
B1 IgG-binding domain of protein L	2PTL	78	NMR
Barstar	1BTB	89	NMR
Fibronectin type III domain from tenascin	1TEN	89	X-ray
TI I27 domain of titin	1TIU	89	NMR
Tenth type III module of fibronectin	1TTF	94	NMR
RNA binding domain of U1A	1URN	96	X-ray
S6	1RIS	97	X-ray
FKBP-12	1FKB	107	X-ray
Barnase	1RNB	109	X-ray
Villin 14T	2VIL	126	NMR
CheY	3CHY	128	X-ray

An Outsider's View of the Genome

Carl Zimmer

Guilford, CT, USA

Abstract. Genomics has transformed biology, including our understanding of evolution. Comparisons of the human genome to those of chimpanzees, rats, and other species have generated tremendous insights into the deep history of many evolutionary processes including gene duplication, neutral evolution, chromosome rearrangement, and changes in gene expression. These insights not only enrich our understanding of the history of life, but may also help guide research in medicine and conservation biology.

Yet genomics has not obliterated the deep tension between molecular and organismal biology that has existed for decades. In the 1960s, this tension was dramatically illustrated in the struggle between James Watson and E.O. Wilson over the direction of biological research at Harvard. Watson championed the reductionist methods of molecular biology, declaring, "There is only one science: physics. Everything else is social work." Wilson represented the social workers of biology—scientists who studied organisms. Their fight led to the biology department splitting in two.

Forty years later, this tension remains strong in the postgenomic era. Based solely on the analysis of genomes, scientists today frequently make sweeping claims about various aspects of evolution, such as the origins of complexity and patterns of biogeography. Rarely do these scientists consult a paleontologist about what the fossil record has to say on these matters. If they did, they would discover a far more intricate reality than reflected in their genome-based generalizations. In my talk, I will discuss some case studies in the perils of genomic myopia. I will also discuss examples of how computational biologists can work fruitfully with paleontologists and other organismal biologists to draw more reliable conclusions about evolution.

A. Apostolico et al. (Eds.): RECOMB 2006, LNBI 3909, p. 425, 2006.

Alignment Statistics for Long-Range Correlated Genomic Sequences

Philipp W. Messer[1], Ralf Bundschuh[2], Martin Vingron[1], and Peter F. Arndt[1]

[1] Max Planck Institute for Molecular Genetics, Ihnestr. 73, 14195 Berlin, Germany
[2] Department of Physics, Ohio State University, 191 W Woodruff Av.,
Columbus OH 43210-1117, USA

Abstract. It is well known that the base composition along eukaryotic genomes is long-range correlated. Here, we investigate the effect of such long-range correlations on alignment score statistics. We model the correlated score-landscape by means of a Gaussian approximation. In this framework, we can calculate the corrections to the scale parameter λ of the extreme value distribution of alignment scores. To evaluate our approximate analytic results, we perform a detailed numerical study based on a simple algorithm to efficiently generate long-range correlated random sequences. We find that the mean and the exponential tail of the score distribution are in fact influenced by the correlations along the sequences. Therefore, the significance of measured alignment scores in biological sequences will change upon incorporation of the correlations in the null model.

1 Introduction

Recent years have witnessed an impressive advance of bioinformatics sequence analysis tools, aiming at deeper insight to the functional organization and evolutionary dynamics of genomic DNA sequences. Popular examples include algorithms for genome annotation, homology detection between genomic regions of different organisms, or the prediction of transcription factor binding sites [1, 2].

Bioinformatics methods frequently yield probabilistic statements. Usually the statistical significance of a computational prediction is characterized by a p-value, specifying the likelihood that this prediction could have arisen by chance. The calculation of p-values requires an appropriate null model of DNA, which reflects our assumptions about the "background" statistical features of the sequence under consideration. The challenging task is to decide on the set of statistical features a suitable null model should obey. Ideally, one incorporates those features into the null model which describe the background "noise" of the DNA sequence, but still allow to discern the specific signal the computational analysis tries to detect.

The simplest DNA background model is an *iid* model, given by a random sequence with letters drawn independently from an identical distribution [2]. The iid model can incorporate the length and the average composition of the sequences under consideration, but it lacks any specific structure concerning the

A. Apostolico et al. (Eds.): RECOMB 2006, LNBI 3909, pp. 426–440, 2006.

arrangement of the nucleotides along the DNA. In particular, it is not capable of incorporating correlations in base composition along the sequences. Up to a certain degree, this additional complexity can be taken into account by an nth order Markov model, specifying the transition probabilities $P(a_{i+1}|a_{i-n+1}, \cdots, a_i)$ in a genomic sequence $\boldsymbol{a} = a_1, \ldots, a_N$ [2]. Assuming the sequences to be generated by Markov processes already allows to incorporate a multitude of spatial statistical features into the model, like e.g. the preferential occurrence of DNA motifs, local peculiarities in genomic composition, or specific dinucleotide frequencies. In contrast to iid sequences, where all letters are uncorrelated, Markov processes lead to, so called, *short-range correlations* in the nucleotide composition [3]. They are characterized by an exponential decay of the correlations between two different bases with increasing distance along the sequence.

A statistical measure of the correlations in genomic base composition is the autocorrelation function $C(r)$. It quantifies the deviations in the joint probability of finding equal bases at a distance of r basepairs along the DNA backbone compared to that in a random sequence of independent letters with the same nucleotide frequencies $p_{a \in \{A,C,T,G\}}$,

$$C(r) \equiv \sum_a \left[P(a_i = a_{i+r} = a) - p_a^2 \right]. \tag{1}$$

We have $C(r) = 0$ $(r > 0)$ for iid sequences, while $C(r) \propto \exp(-\beta r)$ for short-range correlated sequences, e.g. those generated by Markov processes.

With the rapidly growing availability of whole-genome sequence data the correlations along genomic DNA can nowadays be studied systematically over a wide range of scales and organisms. A striking observation in this field was the finding of *long-range correlations* in the base composition of genomes more than a decade ago [4, 3, 5]. They are characterized by a power-law decay of the correlation function for large r,

$$C(r) \propto r^{-\alpha}, \tag{2}$$

and therefore decay much slower compared to short-range correlations. By now it is well established that long-range correlations in base composition appear in the genomes of most eukaryotic species [6, 7, 8] with two examples shown in Fig. 1. Little is known about the origin of genomic long-range correlations, so far. However, their ubiquity among eukaryotic genomes points towards a universal mechanism. A likely dynamical scenario is that they are generated by the stochastic processes of molecular sequence evolution, as has been discussed in [9, 10, 11].

The widespread presence of long-range correlations in genomes raises the question if they need to be incorporated into an accurate null model of eukaryotic DNA and how that would change the p-value calculations. In this article, we address this question in the context of sequence alignment, which constitutes the most commonly used computational tool of molecular biology today [12, 13]. We tackle the problem of calculating sequence alignment significance values for null models with long-range sequence composition correlations with both, analytical, as well as numerical methods. On the analytical side, we introduce a novel

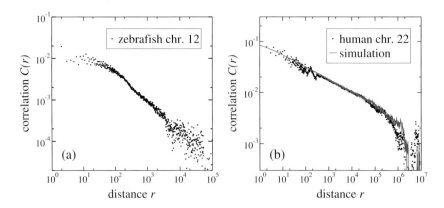

Fig. 1. Long-range correlations in the base composition of two eukaryotic chromosomes. In the double-logarithmic plots, power-law correlations $C(r) \propto r^{-\alpha}$ show up as straight lines with slope α. They extend over distances of several orders of magnitude. In (b) we demonstrate our capability of simulating long-range correlated sequences with similar amplitude and correlation exponent $\alpha \approx 0.232$, as measured in Human chr. 22.

approach, the Gaussian approximation, which allows us to calculate the corrections to the scale parameter λ of the alignment score distribution for correlated sequences. Long-range correlated sequences cannot be generated by an nth order Markov process with finite n [3]. The numerical approach therefore only recently has come within reach due to results derived in [10, 11], where we proposed a biologically motivated algorithm capable of efficiently generating long-range correlated sequences with arbitrary correlation parameters. As the main result of our analysis, it turns out that long-range correlations in the sequences lead to considerable deviations in the score statistics of sequence alignment.

After presenting a short review of sequence alignment in section 2, we analytically treat the alignment of long-range correlated sequences in section 3. A numerical evaluation of the approximative analytic results is presented in section 4. In section 5, we discuss the relevance of this effect for genomic sequence alignment by analyzing the magnitude of the corrections to the score significance values using correlation parameters, measured in eukaryotic genomes. The implications of our findings in a bioinformatics context are discussed at the end of this article.

2   Sequence Alignment and Significance Assessment

The goal of DNA sequence alignment is to assign to a given pair of genomic sequences $\boldsymbol{a} = a_1, \cdots, a_N$ and $\boldsymbol{b} = b_1, \cdots, b_M$ a measure of their similarity. The simplest version of sequence alignment is *gapless* alignment. A local gapless alignment $\mathcal{A}$ of the two sequences consists of a substring $a_{i-l+1} \cdots a_i$ of length l of sequence $\boldsymbol{a}$ and a substring $b_{j-l+1} \cdots b_j$ of sequence $\boldsymbol{b}$ of the same length. Each such alignment is assigned a score $S_{\mathcal{A}} = \sum_{k=0}^{l-1} s(a_{i-k}, b_{j-k})$, where $s(a, b)$

is some given scoring matrix measuring the mutual degree of similarity of the different letters of the alphabet. For DNA sequence comparison, one often uses the simple match-mismatch matrix [14]

$$s(a,b) = \begin{cases} 1 & : \quad a = b \\ -\mu & : \quad a \neq b \end{cases}.$$ (3)

The computational task is to find the alignment $\mathcal{A}$, which gives the highest total score

$$S \equiv \max S_{\mathcal{A}}.$$ (4)

For the purpose of detecting weak sequence homologies, alignment algorithms can also take into account insertions and deletions in either one of the two sequences during biological evolution [14]. For such *gapped* alignments, each gap contributes a (negative) gap cost γ to the total score of the alignment. Using affine gap costs, one additionally distinguishes between the gap initiation cost γ_i and the gap extension cost γ_e.

Since an alignment score S is assigned to any pair of sequences, also to biologically completely unrelated ones, it is helpful to know the distribution of S in an appropriate null model. The knowledge of this distribution gives the possibility to assign p-values to alignment results; they specify the probability that a high score could have arisen by chance in order to be able to distinguish true evolutionary relationship from random similarities. As already mentioned in the introduction, a frequently used null model for that purpose is the iid model. For ungapped alignment of long sequences $(M, N \gg 1)$, the distribution of S for the iid model has been worked out rigorously [15, 16, 17]; it is a Gumbel or extreme value distribution, with its probability density function given by

$$\text{pdf}(S) = KMN\lambda \exp\left(-\lambda S - KMNe^{-\lambda S}\right).$$ (5)

The distribution is characterized by the two parameters λ and K. In the iid case, the scale parameter λ is the unique positive solution of the equation

$$\langle \exp(\lambda s)\rangle = \sum_{a,b} p_a p_b \exp\left[\lambda s(a,b)\right] = 1.$$ (6)

The other parameter K then determines the mean of the distribution.

For gapped alignment, no rigorous theory for the distribution of S exists, so far. However, numerical evidence strongly suggests that the distribution is still of Gumbel form [18, 19, 20, 21]. Using this empirical applicability, it has been shown in [22, 23, 24] that λ for local gapped alignment in the iid model can be derived solely from studying the much simpler global alignment, where one is interested in the path with the highest score $h \equiv \max h_{\mathcal{A}}$, connecting the beginning (a_1, b_1) to the end (a_N, b_N) of a given pair of sequences $\boldsymbol{a}$ and $\boldsymbol{b}$ (we set $M = N$, from now on). One defines a *generating function*

$$Z_N(\lambda) \equiv \langle \exp(\lambda h)\rangle,$$ (7)

where the brackets $\langle \cdot \rangle$ denote an average over all possible pairs of random sequences $\boldsymbol{a}$ and $\boldsymbol{b}$ of length N. The *central conjecture* in [22] then states that λ is determined by the solution of the equation

$$\lim_{N \to \infty} \frac{1}{N} \log Z_N(\lambda) = 0. \tag{8}$$

Following the results of [25, 26], this allows for a very efficient computation of λ for gapped alignment in the iid model.

3 The Gaussian Approximation

In this section, we derive approximate analytical results for the parameter λ of the score distribution one obtains for alignment of random sequences with long-range correlations. We restrict ourselves to gapless alignment, since we expect qualitatively similar results for the gapped case. This will also be confirmed by the numerical data we present in section 5. For simplicity, we furthermore assume a uniform distribution of the four nucleotides; a generalization to sequences with biased composition is straightforward.

The approach employed in the following is based on the assumption that for local gapless alignment of correlated sequences the distribution of the maximal scores obeys Gumbel form, and λ is still determined by Eq. (8). The score of the global alignment is given by the sum over all elementary scores $s_i = s(a_i, b_i)$ along the diagonal of the alignment-lattice. Defining $\boldsymbol{s} = (s_1, \ldots, s_N)$, we have

$$h = \sum_{i=1}^{N} s_i = \mathbf{1}^{\mathrm{t}} \boldsymbol{s}. \tag{9}$$

The ensemble average of Eq. (7) over all realizations of the two sequences $\boldsymbol{a}$ and $\boldsymbol{b}$ can therefore be expressed in terms of an average over all score vectors $\boldsymbol{s}$. While the probability of a score vector factorizes in the iid model, $P(\boldsymbol{s}) = \prod_i P(s_i)$, this is no longer the case for correlated sequences. However, approximate values for the probabilities $P(\boldsymbol{s})$ in the correlated case can still be derived by a Gaussian approximation. The idea of this approach is to replace the discrete variables s_i by continuous Gaussian variables. More precisely, an individual discrete score $s_i = \{1, -\mu\}$ at position i along the diagonal of the alignment-lattice will now be allowed to take continuous values, distributed according to a normal distribution

$$\mathrm{pdf}(s_i) = \frac{1}{\sqrt{2\pi\sigma^2}} \exp \frac{-(s_i - \langle s \rangle)^2}{2\sigma^2}. \tag{10}$$

Mean and variance are chosen in accordance with the original discrete score distribution, i.e., $\langle s \rangle = 1/4 - 3\mu/4$, and $\sigma^2 = 3(1 + \mu)^2/16$.

The probability $P(\boldsymbol{s})$ of a score vector $\boldsymbol{s}$ is then determined by an N-dimensional Gaussian distribution

$$P(\boldsymbol{s}) = [(2\pi)^N \det \boldsymbol{\sigma}]^{-1/2}$$
$$\exp\left[-\frac{1}{2}(\boldsymbol{s} - \langle \boldsymbol{s} \rangle)^{\mathrm{t}} \boldsymbol{\sigma}^{-1}(\boldsymbol{s} - \langle \boldsymbol{s} \rangle)\right], \tag{11}$$

with $\langle s \rangle = (\langle s \rangle, \dots, \langle s \rangle)$ and the covariance matrix σ, defined by

$$\sigma_{ij} = \langle s(i)s(j) \rangle - \langle s(i) \rangle \langle s(j) \rangle. \tag{12}$$

The diagonal elements of σ are given by the variance of an individual score, $\sigma_{ii} = \sigma^2$. The non-diagonal elements $\sigma_{i \neq j}$ can be expressed in terms of the correlation function $C(r)$ of the sequences a and b,

$$\sigma_{ij} = \frac{1}{3}(1 + \mu)^2 C^2(|i - j|). \tag{13}$$

In this expression the correlation function $C(r)$ is squared, since (13) describes the correlations of the similarity scores which arise from a comparison of two sequences. The non-diagonal elements vanish for iid sequences.

Using the distribution (11), the calculation of the generating function (7) amounts to the evaluation of an N-dimensional Gaussian integral, which can be solved explicitly,

$$\begin{aligned} Z_N(\lambda) &= \int ds \, P(s) \exp\left(\lambda \mathbf{1}^{t} s \right) \\ &= [(2\pi)^N \det \sigma]^{-1/2} \\ &\quad \int ds \, e^{-\frac{1}{2}(s - \langle s \rangle)^{t} \sigma^{-1}(s - \langle s \rangle) + \lambda \mathbf{1}^{t} s} \\ &= \exp\left(\lambda \mathbf{1}^{t} \langle s \rangle + \frac{1}{2} \lambda^2 \mathbf{1}^{t} \sigma \mathbf{1} \right). \end{aligned} \tag{14}$$

The central conjecture (8) then implies

$$0 = \lim_{N \to \infty} \frac{1}{N}\left(\lambda \mathbf{1}^{t} \langle s \rangle + \frac{1}{2} \lambda^2 \mathbf{1}^{t} \sigma \mathbf{1} \right). \tag{15}$$

Notice that this expression coincides with the result obtained by applying the central conjecture to the Taylor series approximation of the generating function (7) up to second order. Using Eq. (13) yields

$$\lambda = \frac{-2\langle s \rangle}{\sigma^2 + \frac{2}{3}(1 + \mu)^2 \lim_{N \to \infty} \sum_{i=1}^{N} C^2(i)}. \tag{16}$$

The first term σ^2 in the denominator of (16) is related to the individual fluctuations of a single score element, irrespective of correlations along the sequences. The second term, on the other hand, vanishes for iid sequences and determines the corrections to λ due to correlations.

In case of long-range correlations, i.e., $C(r) = cr^{-\alpha}$, and assuming $\alpha > 1/2$, we obtain

$$\lambda = \frac{-2\langle s \rangle}{\sigma^2 + \frac{2}{3}(1 + \mu)^2 c^2 \zeta(2\alpha)}, \tag{17}$$

where $\zeta(x)$ is the Riemann zeta function. Consequently, the Gaussian approximation predicts deviations in λ for the alignment of long-range correlated sequences compared to iid sequences. A detailed numerical analysis of this analytic result will be performed in section 4. Notice that for $\alpha \leq 1/2$ the sum $\sum_{i=1}^{\infty} C^2(i)$ diverges, resulting in $\lambda = 0$. This might indicate a transition from local to global alignment in the Gaussian approximation, which will be discussed in section 4.3.

As a first evaluation of the Gaussian approximation, we investigate its predictions for sequences $\boldsymbol{a} = (a_1, \ldots, a_N)$ generated by a Markov process. We consider a first order process with four different states $A_i \in \{A, C, T, G\}$. Starting with a random nucleotide a_1, the transition probabilities are defined by

$$P(a_{i+1}|a_i) = \begin{cases} p & : \quad a_{i+1} = a_i \\ \frac{1}{3}(1-p) & : \quad a_{i+1} \neq a_i \end{cases}. \tag{18}$$

This process generates short-range correlations in the sequences of the form $C(r) = c \exp(-\beta r)$ with $\beta = -\log(4p/3 - 1/3)$ and $c = 3/4$. For this case, the Gaussian approximation (16) yields

$$\lambda = \frac{-2\langle s \rangle}{\sigma^2 + \frac{2}{3}(1 + \mu)^2 c^2/(\exp(2\beta) - 1)}. \tag{19}$$

This can be compared to an exact analytical result for λ obtained by equating the largest eigenvalue of a modified λ-dependent transition matrix of the underlying Markov process to one [16]. As is shown in Fig. 2, the Gaussian approximation (19) fits well to the exact resuls; deviations for large β vanish for decreasing β. Notice that the limit $\beta \to \infty$ corresponds to $p \to 1/4$, describing the asymptotics of an uncorrelated iid sequence. The deviations of the Gaussian

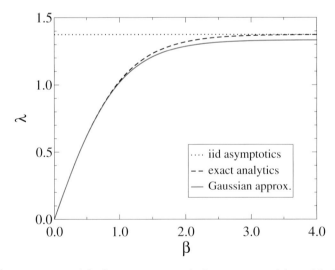

Fig. 2. λ for sequences with short-range correlations generated by a Markov process. The dashed line is the exact result [16] for the Markov process defined in (18), using $\mu = 3$. The solid line is the corresponding result of the Gaussian approximation, as derived in Eq. (19). Solving Eq. (6) yields the iid asymptotics $\lambda \approx 1.374$ (dotted line).

approximation for this regime result from the fact that the third and all higher cumulants of the distribution (10) vanish, which they do not for the discrete distribution.

4   Numerical Results

4.1   Generation of Long-Range Correlated Random Sequences

Numerical evaluation of the results obtained in the previous section hinges on the knowledge of the score distribution $\text{pdf}(S)$ for local gapless alignment of pairs of long-range correlated random sequences. However, the efficient generation of such sequences is quite intricate. In [10], we have proposed a biologically motivated model of sequence evolution which generates sequences with the desired statistical features. Furthermore, it has recently been shown [11] that there exists a much larger class of dynamical processes, so called, *expansion-randomization* processes, which allow for the efficient generation of sequences with arbitrary long-range correlations.

Based on [11], we use a single-site duplication-mutation algorithm to generate long-range correlated sequences. We start with a sequence of one random nucleotide a_1, and the dynamics of the model is defined by the following update rules:

1. A random position j of the sequence is chosen.
2. The nucleotide a_j is either mutated to a random but different nucleotide with probability P_{mut}, or duplicated with probability $P_{\text{dup}} = 1 - P_{\text{mut}}$. The duplication process inserts a copy of a_j at position $j + 1$, thereby increasing the sequence length by one.

This process generates sequences of arbitrary length N in a time $O[N \log (N)]$ with asymptotic long-range correlations in their nucleotide composition. The correlation function of the generated sequences is given in terms of the Euler beta function $B(x, y)$ by [10]

$$C(r) = \frac{3}{4}\alpha B(r + 1, \alpha). \qquad (20)$$

In the large r limit, this yields $C(r) \propto r^{-\alpha}$. By varying the mutation probability $0 < P_{\text{mut}} < 1$, the decay exponent α of the long-range correlations can be tuned to any desired positive value, as it is determined by

$$\alpha = \frac{8}{3}\frac{P_{\text{mut}}}{1 - P_{\text{mut}}}. \qquad (21)$$

Using this model, we are now in the position to efficiently generate large ensembles of long-range correlated sequences needed for an accurate measurement of the tail of the distribution $\text{pdf}(S)$. For the alignment, we use the standard Smith-Waterman dynamic programming algorithm [14] with scoring matrix (3) and $\mu = 3$.

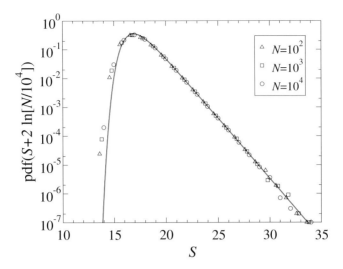

Fig. 3. Convergence of the distribution pdf(S) for long-range correlated sequences with $\alpha = 2.0$ to a Gumbel form. The solid line is a Gumbel distribution, as specified in Eq. (5) with $N = M = 10^4$ and fitted parameters $\lambda = 0.9614$ and $K = 0.119$. λ was obtained by fitting a linear function to log[pdf(S)] for $21 < S < 31$, K has then been estimated by fitting the data to (5) in the same interval. In order to be able to compare the shape of pdf(S) for different N, the distributions have to be rescaled by a transformation pdf(S) $\rightarrow$ pdf($S + 2 \ln [N/N_0]$) with reference length $N_0 = 10^4$.

4.2   The Gumbel Distribution of Alignment Scores

The Gaussian model is based on the assumption that the score distribution pdf(S) is of Gumbel form for long-range correlated sequences. Consequently, our first numerical analysis aims at a verification of this conjecture. In Fig. 3, we show the measured pdf(S) for long-range correlated sequences with $\alpha = 2.0$, estimated from ensembles of 10^7 pairs of random sequence realizations generated by the above specified algorithm. For large N, the distribution asymptotically approaches a Gumbel form. As is the case for the iid model, finite-size corrections come into play for short sequence lengths [20, 27, 28]. These deviations primarily show up in the small S regime, while the more relevant large S regime converges fast for increasing N.

   Now, that we have verified the shape of the score distribution to be of Gumbel form, we can test the accuracy of the analytic predictions for λ derived by the Gaussian approximation. Here we restrict ourselves to the discussion of the regime $\alpha > 1/2$, where the Gaussian approximation predicts finite values of λ; the regime $\alpha \leq 1/2$ will be investigated below.

   We compare our numerical data to Eq.(16), using correlations of the form (20). Results are shown in Fig. 4. The Gaussian approximation captures the qualitative behavior of the numerical data. Again, the right side of the plot reveals the deviations of the Gaussian approximation concerning its iid asymptotics given by $\alpha \rightarrow \infty$. With increasing correlation strength, i.e., smaller values of

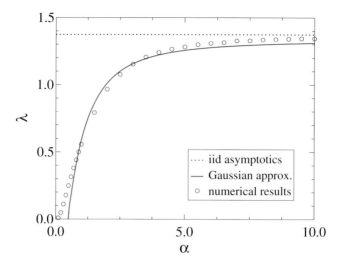

Fig. 4. λ for a null model with long-range correlated sequences in dependence of the correlation exponent α. The solid line is the analytic result of the Gaussian approximation, one obtains by estimating Eq. (16) using the correlations (20) of our simulated sequences. Numerically measured values of λ for different correlation parameters α are denoted by symbols. For our simulation, we use sequences of length $N = 10^3$ and average over ensembles of 10^8 pairs of sequences.

α, λ decreases, confirming that long-range correlations systematically raise the probability of measuring high alignment scores.

So far, our investigations of the alignment score distribution for long-range correlated sequences have focused on the exponential tail of $\text{pdf}(S)$. We now turn to the second parameter K. For that purpose, we recall that the mean of a Gumbel distribution (5) is determined by

$$\langle S \rangle = \frac{\Gamma + \log\left(KN^2\right)}{\lambda},\tag{22}$$

Table 1. Dependence of $\langle S \rangle$ and K on the exponent α. We use simulated sequences of length $N = 10^3$ and average over ensembles of 10^8 pairs of sequences for each value of α to obtain numerical values of λ and $\langle S \rangle$. The values of K have been calculated using Eq. (22).

α	λ	$\langle S \rangle$	K
(iid)	1.374	9.71	3.50×10^{-1}
4.0	1.240	10.61	2.90×10^{-1}
2.0	0.967	12.65	1.15×10^{-1}
1.0	0.556	18.07	1.30×10^{-2}

where $\Gamma \approx 0.5772$ is the Euler-Mascheroni constant. Thus, knowing λ, the parameter K can easily be calculated by measuring the mean $\langle S \rangle$ of the score distribution. As shown in Table 1, K is significantly affected by the presence of long-range correlations in the sequences to be aligned; it decreases with increasing correlation-strength. However, the mean of the distribution is, as expected, shifted to larger values of S for decreasing values of α, since K contributes only logarithmically in Eq. (22) and the change in $\langle S \rangle$ is dominated by the decrease of λ.

4.3   The Score Distribution for $\alpha \leq 1/2$

In the regime $\alpha > 1/2$, the score distribution is of Gumbel form and the Gaussian approximation suitably fits the numerical values of λ. For values of $\alpha \leq 1/2$, the Gaussian approximation yields $\lambda = 0$, which might indicate a transition from local to global alignment. For simulated sequences of finite length, on the other hand, one still measures finite values of λ (Fig. 4). The numerical investigation of this regime is complicated by a distinct finite size effect: according to the results derived in [11], an individual alignment of two finite sequences will have a systematic bias of $\langle s \rangle$ towards either $\langle s \rangle = 1$, or $\langle s \rangle = -\mu$, depending on whether by chance the two initial random letters a_1 and b_1 of our sequence generation algorithm were equal for the two sequences to be aligned, or not. This effect causes strong deviations of $\text{pdf}(S)$ from a Gumbel form for small S. However, the tail of the distribution is still exponential for finite sequences, and therefore allows for a measurement of λ. It is dominated by those realizations of the ensemble, where both sequences started with the same letter since they lead to systematically higher values of $\langle s \rangle$ and therefore also higher scores S.

As can be seen in Fig. 4, λ approaches zero for finite sequences not until the "infinite" correlation strength limit $\alpha \to 0$. Further analysis is needed to decide on whether there actually is a transition to global alignment for a particular $\alpha > 0$ in the limit $N \to \infty$, or not. If this is the case, then the rate of convergence for $\lambda \to 0$ is at most logarithmically.

However, for practical applications this transition is irrelevant. Finite sequences always have a positive λ, also in the regime $\alpha \leq 1/2$. For these particular choices of parameters, λ needs to be measured numerically.

5   Consequences for Alignments of Genomic Sequences

It has been shown that long-range correlations in base composition increase the probability of measuring high scores for pairwise sequence alignment. In a biological context, this raises the question whether the effect causes a significant change of the p-values for DNA alignment? In order to address this issue, we investigate the deviations of the score distribution for correlation parameters of genomic magnitude compared to iid sequences. As an example, we consider the measured correlation function of Human chromosome 22, shown in Fig. 1(b). Using the simulation algorithm introduced in section 4.1 we can generate long-range correlated random sequences with the corresponding exponent $\alpha \approx 0.232$.

By randomly mutating 85% of the sites after sequence build up, the correlation amplitude is reduced to the genomic value, while the exponent remains unchanged [11]. As can be seen in Fig. 1(b), this procedure allows us to generate random sequences featuring comparable correlations as Human chr. 22.

We perform ungapped, as well as gapped alignment with affine gap costs for 10^7 pairs of random sequences with length $N = 10^3$ from the above specified ensemble. Alignment parameters are chosen in accordance with the NCBI default values $\mu = 3$, gap initiation cost $\gamma_i = 5$, and gap extension cost $\gamma_e = 2$ [29]. In Fig. 5 we show the measured score distributions for the simulated chr. 22 sequences compared to iid sequences. The resulting parameters λ and $\langle S \rangle$ are presented in Table 2. It turns out that the difference in the score distributions between ungapped and gapped alignment is negligible for the parameters used.

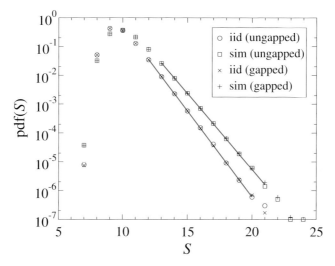

Fig. 5. The score distribution for ungapped and gapped alignment of simulated sequences with correlations comparable to those of Human chromosome 22. The straight lines are the fits to the exponential tails of the score distributions, obtained by fitting a linear function to log[pdf(S)] in the depicted intervals.

Table 2. Fitted parameters λ and $\langle S \rangle$ for the iid ensemble and simulated Human chr. 22 sequences of length $N = 10^3$. In the last column, exemplary p-values of a score $S' = 18$ are shown.

ensemble	λ	$\langle S \rangle$	$P(S \geq 18)$
iid (ungapped)	1.374	9.714	3.3×10^{-6}
sim. chr. 22 (ungapped)	1.191	10.164	2.8×10^{-5}
iid (gapped)	1.373	9.714	3.2×10^{-6}
sim. chr. 22 (gapped)	1.215	10.163	2.7×10^{-5}

The deviations in λ between the iid ensemble and the simulated Human chr. 22 sequences are approximately 15% in both cases, and the mean of the score distributions for the correlated sequences is significantly higher. In combination, both effects substantially change the p-values of high scores compared to the iid model, as can be seen in Table 2. The p-value of a specific score S' is thereby defined by the integral $P(S \geq S') = \int_{S'}^{\infty} \text{pdf}(S)dS$. For an exemplary score $S' = 18$, this p-value will be increased by almost one order of magnitude if one incorporates the genomic correlations into the null model.

6   Discussion

Long-range correlations are a widespread statistical feature of eukaryotic DNA. In this article, it has been shown that incorporation of this feature into the null model substantially influences the score statistics of sequence alignment. While the p-values of the scores are systematically increased, the ranking of hits will not be significantly changed. The effect is therefore relevant whenever one is actually interested in p-values, e.g., when specifying a cutoff in order to distinguish true evolutionary relationship from random similarities.

One has to keep in mind that genomic DNA is a highly heterogeneous environment: it consists of genes, noncoding regions, repetitive elements etc., and all of these substructures may imprint their signature on the amount of correlations found in a particular genomic region. Long-range correlations are by definition a feature on larger scales. Our findings are therefore naturally applicable to the alignment of larger genomic regions. This includes the identification of duplicated regions, or conserved syntenic segments between chromosomes of different species, which often extend over many kilobases up to several megabases. However, long-range correlations will also influence the statistics of search algorithms for short DNA motifs if the query sequences are large enough for long-range correlations to be measured.

Moreover, it will be interesting to analyze possible effects of long-range correlations on the statistics of other widely used sequence analysis tools, e.g., the prediction of transcription factor binding sites. Further investigation is needed to assess the relevance of long-range correlations for other statistical predictions. Finally, more accurate null models of DNA sequences utilizing quantitative correlation features will help to reduce the often encountered high false-positive rate of bioinformatics analysis tools.

References

1. Waterman, M.S.: Introduction to Computational Biology: Maps, Sequences, and Genomes. CRC Press (1995)
2. Durbin, R., Eddy, S., Krogh, A., Mitchison, G.: Biological Sequence Analysis. Cambridge University Press (1998)
3. Peng, C.K., Buldyrev, S.V., Goldberger, A.L., Havlin, S., Sciortino, F., Simons, M., Stanley, H.E.: Long-range correlations in nucleotide sequences. Nature **356** (1992) 168

4. Li, W., Kaneko, K.: Long-range correlation and partial $1/f^\alpha$ spectrum in a non-coding DNA sequence. Europhys. Lett. **17** (1992) 655
5. Voss, R.F.: Evolution of long-range fractal correlations and $1/f$ noise in DNA base sequences. Phys. Rev. Lett. **68** (1992) 3805
6. Arneodo, A., Bacry, E., Graves, P.V., Muzy, J.F.: Characterizing long-range correlations in DNA sequences from wavelet analysis. Phys. Rev. Lett. **74** (1995) 3293
7. Bernaola-Galvan, P., Carpena, P., Roman-Roldan, R., Oliver, J.L.: Study of statistical correlations in DNA sequences. Gene **300** (2002) 105
8. Li, W., Holste, D.: Universal $1/f$ noise, crossovers of scaling exponents, and chromosome-specific patterns of guanine-cytosine content in DNA sequences of the human genome. Phys. Rev. E **71** (2005) 041910
9. Li, W.: Expansion-modification systems: A model for spatial $1/f$ spectra. Phys. Rev. A **43** (1991) 5240
10. Messer, P.W., Arndt, P.F., Lässig, M.: Solvable sequence evolution models and genomic correlations. Phys. Rev. Lett. **94** (2005) 138103
11. Messer, P.W., Lässig, M., Arndt, P.F.: Universality of long-range correlations in expansion-randomization systems. J. Stat. Mech. (2005) P10004
12. Altschul, S.F., Gish, W., Miller, W., Myers, E.W., Lipman, D.J.: Basic local alignment search tool. J. Mol. Biol. **215** (1990) 403
13. Altschul, S.F., Madden, T.L., Schäffer, A.A., Zhang, J., Zhang, Z., Miller, W., Lipman, D.J.: Gapped blast and psi-blast: a new generation of protein database search programs. Nucleic Acids Research **25** (1997) 403
14. Smith, S.F., Waterman, M.S.: Comparison of biosequences. Adv. Appl. Math. **2** (1981) 482
15. Karlin, S., Altschul, S.F.: Methods for assessing the statistical significance of molecular sequence features by using general scoring schemes. Proc. Natl. Acad. Sci. U.S.A. **87** (1990) 2264
16. Karlin, S., Dembo, A.: Limit distribution of the maximal segmental score among Markov-dependent partial sums. Adv. Appl. Prob. **24** (1992) 113
17. Karlin, S., Altschul, S.F.: Applications and statistics for multiple high-scoring segments in molecular sequences. Proc. Natl. Acad. Sci. U.S.A. **90** (1993) 5873
18. Smith, T.F., Waterman, M.S., Burks, C.: The statistical distribution of nucleic acid similarities. Nucleic Acids Res **13** (1985) 645
19. Waterman, M.S., Vingron, M.: Rapid and accurate estimates of statistical significance for sequence data base searches. Proc. Natl. Acad. Sci.U.S.A. **91** (1994) 4625
20. Altschul, S.F., Gish, W.: Local alignment statistics. Methods Enzymol. **266** (1996) 460
21. Mott, R.: Maximum likelihood estimation of the statistical distribution of Smith-Waterman local sequence similarity scores. Bull. Math. Biol. **54** (1999) 59
22. Bundschuh, R.: An analytic approach to significance assessment in local sequence alignment with gaps. In: RECOMB2000. (2000) 86
23. Bundschuh, R.: Asymmetric exclusion process and extremal statistics of random sequences. Phys. Rev. E **65** (2002) 031911
24. Grossmann, S., Yakir, B.: Large deviations for global maxima of independent superadditive processes with negative drift and an application to optimal sequence alignments. Bernoulli **10** (2004) 829
25. Park, Y., Sheetlin, S., Spouge, J.L.: Accelerated convergence and robust asymptotic regression of the Gumbel scale parameter for gapped sequence alignment. Journal of Physics A **38** (2005) 97
26. Chia, N., Bundschuh, R.: A practical approach to significance assessment in alignment with gaps. In: RECOMB2005. (2005) 474

27. Altschul, S.F.: Amino acid substitution matrices from an information theoretic perspective. J. Mol. Biol. **219** (1991) 555
28. Yu, Y.K., Bundschuh, R., Hwa, T.: Statistical significance and extremal ensemble of gapped local hybrid alignment. LNP Vol. 585: Biological Evolution and Statistical Physics **585** (2002) 3
29. http://www.ncbi.nlm.nih.gov/BLAST

Simple and Fast Inverse Alignment[*]

John Kececioglu and Eagu Kim

Department of Computer Science,
The University of Arizona, Tucson, AZ 85721, USA
{kece, egkim}@cs.arizona.edu

Abstract. For as long as biologists have been computing alignments of
sequences, the question of what values to use for scoring substitutions
and gaps has persisted. While some choices for substitution scores are
now common, largely due to convention, there is no standard for choosing
gap penalties. An objective way to resolve this question is to learn the
appropriate values by solving the *Inverse String Alignment Problem*:
given examples of correct alignments, find parameter values that make
the examples be optimal-scoring alignments of their strings.

We present a new polynomial-time algorithm for Inverse String Alignment that is *simple* to implement, *fast* in practice, and for the first time
can learn hundreds of parameters simultaneously. The approach is also
flexible: minor modifications allow us to solve inverse *unique* alignment
(find parameter values that make the examples be the unique optimal
alignments of their strings), and inverse *near-optimal* alignment (find parameter values that make the example alignments be as close to optimal
as possible). Computational results with an implementation for global
alignment show that, for the first time, we can find best-possible values
for all 212 parameters of the standard protein-sequence scoring-model
from hundreds of alignments in a few minutes of computation.

Keywords: Sequence analysis, parametric sequence alignment, substitution score matrices, affine gap penalties, supervised learning, linear
programming, cutting plane algorithms.

1   Introduction

Perhaps the most studied problem in computational biology is the alignment of
biological sequences with substitutions, insertions, and deletions. The standard
formulations of string alignment optimize a sum of scores for each type of operation, often giving a penalty for a run of insertions or deletions, called a gap,
that is linear in the length of the gap. When performing sequence alignment in
practice, the question of what weights and penalties to use inevitably arises. An
interesting attack on this question is *parametric sequence alignment*, where for a
given pair of strings, the alignment problem is solved for effectively all possible
choices of the scoring parameters, thereby eliminating the need to specify *any*
weights and penalties. The problem with this approach is that it in effect defers

---

[*] Research supported by US National Science Foundation Grant DBI-0317498.

A. Apostolico et al. (Eds.): RECOMB 2006, LNBI 3909, pp. 441–455, 2006.

the question, since eventually a user must choose one of the solutions, and on what basis should this be done? Essentially one needs an example of a *correct alignment* to discriminate among the welter of parametric solutions. When one does solve the parametric problem and knows a biologically correct alignment of the sequences, this alignment is used to decide what region of the parameter space makes the correct alignment have optimal score.

This optimal parameter choice could be found much more directly, however, by solving *inverse parametric alignment*. In this problem, the input is an alignment of a pair of strings, and the output is a choice of parameters that makes the input alignment be an optimal-scoring alignment of its strings. We present a simple and fast algorithm for inverse parametric alignment that for the first time is capable of determining all substitution weights and gap penalties simultaneously. Such an algorithm, applied to a collection of benchmark protein-sequence alignments that are constructed by aligning their three-dimensional structures, could provide the first rigorous way of determining substitution scores and gap penalties for characterized classes of proteins.

Related Work. The first algorithms for *parametric* alignment of two sequences were discovered in the early 1990's by Waterman, Eggert and Lander [16] and Gusfield, Balasubramanian and Naor [7]. These algorithms handled *two* parameters, usually the gap open and extension penalties with fixed substitution scores. Zimmer and Lengauer [17] addressed numerical stability. Gusfield et al. [7] also bounded the number of regions in the decomposition of the parameter space, and constructed the decomposition with one optimal alignment computation per region. Fernández-Baca, Seppäläinen and Slutzki [4] showed these bounds are asymptotically tight, and initiated the study of parametric multiple-sequence alignment. Gusfield and Stelling [8] released a software implementation called XPARAL, and were the first to consider *inverse* parametric alignment, for which they gave a heuristic that attempted to avoid computing a decomposition of the entire parameter space. Pachter and Sturmfels [13] explored the relation of algebraic statistics to parametric sequence alignment.

Recently, Sun, Fernández-Baca and Yu [15] gave the first direct algorithm for inverse parametric alignment. While they consider three parameters, their solution effectively fixes one parameter value at zero. For two strings of length n, their algorithm runs in $O(n^2 \log n)$ time. Their approach is involved, and does not appear to have been implemented.

In contrast to prior work, our algorithm for inverse parametric alignment is simple to implement, does not compute a decomposition of the entire parameter space, solves both the near-optimal and unique-optimal inverse alignment problems, handles a set of input alignments, and for the first time can quickly solve problems with hundreds of free parameters.

Overview. In the next section we give a precise statement of the inverse parametric alignment problem and two variations: inverse near-optimal and unique-optimal alignment. Section 3 reduces all three variations to linear programming. Section 4 explains how the resulting linear programs, even though they have an exponential number of inequalities, can be solved in polynomial time. Section 5 then discusses a version of this approach, called a cutting-plane algorithm, that is highly effective in practice. Finally Section 6 presents experimental results with an implementation for inverse optimal and near-optimal global alignment.

2   Inverse Alignment and Its Variations

The standard string alignment problem is: given a pair of strings A and B and a function f that scores alignments, where f usually has several parameters that weight substitutions, insertions, deletions, and identities, find an alignment $\mathcal{A}$ of A and B that has optimal score under f. The *inverse alignment problem* turns this around: given an alignment $\mathcal{A}$, find values for the parameters of f that make $\mathcal{A}$ be an optimal alignment. (From the view of machine learning, this learns the parameters for optimal alignment from training examples of correct alignments.) Of course to find reliable values when f has many parameters may require several alignments. Formally, we define inverse alignment as follows.

Definition 1 (Inverse Optimal Alignment). The *Inverse String Alignment Problem* is the following. The input is a collection of alignments $\mathcal{A}_1, \ldots, \mathcal{A}_k$ of strings, and an alignment scoring function f_w with parameters $w = (w_1, \ldots, w_p)$. The output is values $x = (x_1, \ldots, x_p)$ for the parameters such that each $\mathcal{A}_i$ is an optimal alignment of its strings under f_x.   □

For example, the $\mathcal{A}_i$ might be structural global alignments of pairs of protein sequences, and f might score alignments using substitution scores σ_{ab} for all pairs of amino acids a, b together with a gap-open penalty γ and a gap-extension penalty λ. In this case, scoring function f has 212 parameters. For another example, the $\mathcal{A}_i$ might be local alignments of pairs of strings, also scored using substitutions and gap penalties, or the $\mathcal{A}_i$ might even be alignments of alignments [11] scored using the weighted sum-of-pairs measure.

Note that Inverse Optimal Alignment may have no solution: it may be that no choice x for the parameter values makes the $\mathcal{A}_i$ all be optimal alignments. An algorithm for inverse alignment must detect this situation, and report that no solution exists.

Given that it may be impossible to find parameters that make the alignments in a collection all be optimal, we might instead seek the next-best thing: parameters that make the alignments all be *near-optimal*. When the objective is to minimize scoring function f, we say an alignment $\mathcal{A}$ of a set of strings $\mathcal{S}$ is ϵ-*optimal* for some $\epsilon \geq 0$ if

$$f(\mathcal{A}) \ \leq \ (1{+}\epsilon)\, f(\mathcal{A}^*), \tag{1}$$

where $\mathcal{A}^*$ is an optimal alignment of $\mathcal{S}$ under f. Note that when $\epsilon = 0$, an ϵ-optimal alignment is optimal.

Definition 2 (Inverse Near-Optimal Alignment). The *Inverse Near-Optimal Alignment Problem* is: given a collection of alignments $\mathcal{A}_i$, scoring function f, and a real number $\epsilon \geq 0$, find parameter values x such that each alignment $\mathcal{A}_i$ is ϵ-optimal under f_x.   □

For large enough ϵ, Inverse Near-Optimal Alignment always has a solution. In practice though we might not know an appropriate value of ϵ in advance. Nevertheless an algorithm for Inverse Near-Optimal Alignment can be used to efficiently find the *smallest* value ϵ^* for which there is a solution, to any desired accuracy $\xi > 0$. First, by repeated doubling, find an upper bound b on ϵ^* by iteratively solving Inverse Near-Optimal Alignment with $\epsilon = 2^i \xi$ for an i of $1, 2, \ldots$ until a solution is found. Then, given upper bound b and lower bound $a = b/2$ on ϵ^*, perform binary search on the real values in interval $[a, b]$ that are spaced distance ξ apart. This finds the best possible ϵ to within accuracy ξ using $O(\log(\epsilon/\xi))$ calls to an algorithm for Inverse Near-Optimal Alignment. As we show in Section 6, such an approach is very fast in practice.

Finally, in some applications of inverse alignment we may need to find parameter values that make a given alignment be the *unique* optimal alignment of its strings. (For example, suppose we have alignment software that attempts to optimize a scoring function; when testing how well the software performs at recovering a benchmark alignment, the best parameter values to use would be those that make the benchmark be the unique optimal alignment.) To be the unique optimal alignment, every other alignment must score worse. We quantify how much worse as follows. When the objective is to minimize scoring function f, we say an alignment $\mathcal{A}$ of a set of strings $\mathcal{S}$ is δ-*unique* for some $\delta > 0$ if

$$f(\mathcal{B}) \;\geq\; f(\mathcal{A}) + \delta,$$

for every alignment $\mathcal{B}$ of $\mathcal{S}$ other than $\mathcal{A}$.

Definition 3 (Inverse Unique-Optimal Alignment). The *Inverse Unique-Optimal Alignment Problem* is: given a collection of alignments $\mathcal{A}_i$, scoring function f, and a real number $\delta > 0$, find parameter values x such that each alignment $\mathcal{A}_i$ is a δ-unique alignment of its strings under f_x.   □

Note that using the same doubling and binary-search idea described above to find the smallest ϵ for Inverse Near-Optimal Alignment, we can find the *largest* $\delta > 0$ for which the $\mathcal{A}_i$ are all δ-unique, to within accuracy $\xi > 0$, using $O(\log(\delta/\xi))$ calls to an algorithm for Inverse Unique-Optimal Alignment.

When the alignment scoring function f is *linear* in its parameters—as is the case for most forms of alignment used in practice (including the standard formulations of global and local alignment)—all three variations of inverse alignment can be solved using linear programming, as we show next.

3   Reduction to Linear Programming

For most standard forms of alignment, the alignment scoring function f is a linear function of its parameters. We make this precise as follows. In general suppose that f scores an alignment $\mathcal{A}$ by measuring $p+1$ features of $\mathcal{A}$ through functions $f_0, f_1, \ldots, f_p$, and combines these measures into one score through a weighted sum involving p parameters $w_1, \ldots, w_p$, by

$$f(\mathcal{A}) \;:=\; f_0(\mathcal{A}) + f_1(\mathcal{A})\, w_1 + \cdots + f_p(\mathcal{A})\, w_p. \tag{2}$$

Then we say f is *linear* in parameters $w_1, \ldots w_p$. (Note that f_0 is not weighted by a parameter.) When we want to indicate the dependence of function f on all its parameters $w = (w_1, \ldots, w_p)$, we write f_w.

For a concrete example, consider *standard global alignment* of two protein sequences with linear gap penalties, where a substitution of letter a by b has similarity value σ_{ab}, and a gap of length ℓ incurs a penalty of $\gamma + \lambda \ell$. (A *gap* in an alignment is a maximal run of either insertions or deletions; the *length* of the gap is the number of letters in the run. Here γ is the gap-open penalty and λ is the gap-extension penalty.) Suppose we have fixed all the similarity values σ_{ab} by choosing one of the standard substitution-score matrices (such as a PAM [2] or BLOSUM [9] matrix). If the only parameter values we want to find through inverse alignment are the gap open and extension penalties, we have $p = 2$ parameters: γ and λ. For an alignment $\mathcal{A}$, let

- $g(\mathcal{A})$ be the number of gaps in $\mathcal{A}$,
- $\ell(\mathcal{A})$ be the total length of all gaps in $\mathcal{A}$, and
- $s(\mathcal{A})$ be the total score of all substitutions (including identities) in $\mathcal{A}$.

Then the similarity score of alignment $\mathcal{A}$ is

$$f(\mathcal{A}) := s(\mathcal{A}) - g(\mathcal{A})\gamma - \ell(\mathcal{A})\lambda. \tag{3}$$

Here $(f_0, f_1, f_2) = (s, g, \ell)$ and $(w_1, w_2) = (-\gamma, -\lambda)$ in the notation of (2).

On the other hand, if no parameters are fixed and we want to find values for all the substitution scores σ_{ab} and gap penalties simultaneously, then the scoring function becomes

$$f(\mathcal{A}) := \left(\sum_{a,b} h_{ab}(\mathcal{A}) \sigma_{ab} \right) - g(\mathcal{A})\gamma - \ell(\mathcal{A})\lambda, \tag{4}$$

where a and b range over all letters in the alphabet, and the functions $h_{ab}(\mathcal{A})$ count the number of substitutions in $\mathcal{A}$ that replace a by b. For the protein alphabet of 20 amino acids, there are 210 substitution parameters σ_{ab}. These plus the two gap parameters gives $p = 212$ total parameters. Here $f_0(\mathcal{A}) = 0$ in the notation of equation (2).

When the scoring function f is linear in its parameters, we can solve Inverse Optimal, Near-Optimal, and Unique-Optimal Alignment using linear programming. Recall that the *Linear Programming Problem* is: given a collection of variables $x = (x_1, \ldots, x_n)$, a system of linear inequalities in the variables x, and a linear objective function in the variables x, find an assignment x^* of real values to the variables that satisfies all the inequalities and minimizes the objective function. In matrix notation, given a system of m inequalities in the n variables whose left-hand sides are specified by an $m \times n$ coefficient matrix A and whose right-hand sides are specified by an m-vector b, together with an n-vector c of coefficients for the objective function, Linear Programming finds

$$x^* := \underset{x \geq 0}{\operatorname{argmin}}\{cx : Ax \geq b\}.$$

Here x^* is an *optimal solution* to the linear program, and any $x \geq 0$ that satisfies $Ax \geq b$ is a *feasible solution*. In general a linear program may be *infeasible* (has no feasible solution), *bounded* (has an optimal feasible solution), or *unbounded* (has feasible solutions that are arbitrarily good under the objective).

Inverse Optimal Alignment. We can solve Inverse Optimal Alignment for a linear scoring function in a very natural way by linear programming. The *variables* $x = (x_1, \ldots, x_p)$ in the linear program correspond to the scoring-function parameters $w = (w_1, \ldots, w_p)$. Note that the condition $x \geq 0$ for linear programs is not a restriction. We can scale any linear scoring function by a positive amount (without changing the relative rank of alignments) so its parameters lie in the interval $[-1, 1]$. Then replacing every occurrence of parameter w_i by $x_i - 1$ yields variables satisfying $x \geq 0$.

We use the following system of *inequalities*. Let $\mathcal{S}_i$ be the set of strings that $\mathcal{A}_i$ aligns. For each $\mathcal{A}_i$ and *every* alignment $\mathcal{B}$ of $\mathcal{S}_i$, we have an inequality

$$f_x(\mathcal{B}) \;\; \geq \;\; f_x(\mathcal{A}_i). \tag{5}$$

These inequalities simply express that $\mathcal{A}_i$ is a minimum-score alignment of strings $\mathcal{S}_i$, and hence $\mathcal{A}_i$ under parameter values x is an optimal alignment. (Note that if the objective is to maximize scoring function f, the direction of inequality (5) should be reversed. Then negating all the inequalities puts them into the canonical form $Ax \geq b$ for the linear program.) Written in terms of $x_1, \ldots, x_p$, inequality (5) is by equation (2) equivalent to the linear inequality

$$\sum_{1 \leq j \leq p} \big(f_j(\mathcal{B}) - f_j(\mathcal{A}_i)\big)\, x_j \;\; \geq \;\; \big(f_0(\mathcal{A}_i) - f_0(\mathcal{B})\big). \tag{6}$$

Note that for any given alignments $\mathcal{A}_i$ and $\mathcal{B}$, the quantities $f_j(\mathcal{B}) - f_j(\mathcal{A}_i)$ in inequality (6) are constants that serve as the coefficients of the variables x.

Of course this yields a linear program with a huge number of inequalities. Suppose $\mathcal{A}_i$ aligns two strings of length n. The number of alignments of this pair of strings [5] is $\Theta((3 + \sqrt{2})^n / n^{1/2}) = \Omega(4^n)$, which is the number of alignments $\mathcal{B}$. Every such $\mathcal{B}$ generates an inequality in the linear program. So an inverse alignment problem with p parameters and k input alignments, each of which aligns two or more strings of length n or greater, generates a linear program with $\Omega(k\, 4^n)$ inequalities in p variables.

Surprisingly, for many forms of sequence alignment this linear program can be solved in *polynomial time*—even though it has an exponential number of inequalities—due to a deep result that we call the Separation Theorem. In Section 4 we discuss how this theorem guarantees that we can efficiently solve this linear programming formulation.

One advantage of this linear programming-based approach is that we may also specify any linear *objective function* that we wish for the linear program. While every feasible solution $x \geq 0$ that satisfies the above inequalities $Ax \geq b$ yields a choice of parameters that makes the $\mathcal{A}_i$ optimal, some choices may be more biologically desirable. For instance with linear gap penalties, biologists generally prefer a *large* gap-open penalty γ and a *small* gap-extension penalty λ, since real alignments typically consist of a few long gaps. We are free to use any objective function that is linear in x to pick out a feasible solution that is more desirable. Section 5 discusses some objective functions that are appropriate for standard global alignment.

Near-Optimal Alignment. To extend this to Inverse Near-Optimal Alignment simply involves modifying the inequalities (5). Given $\epsilon \geq 0$, we use the system

$$(1 + \epsilon)\, f_x(\mathcal{B}) \;\geq\; f_x(\mathcal{A}_i), \tag{7}$$

which for each $\mathcal{A}_i$ again has an inequality for every alignment $\mathcal{B}$ of strings $\mathcal{S}_i$. This is a linear inequality as well in the variables $x_1, \dots, x_p$. Note that if inequality (7) holds for every $\mathcal{B}$, then in particular it holds for $\mathcal{B} = \mathcal{A}^*$ where $\mathcal{A}^*$ is an optimal alignment of $\mathcal{S}_i$ under f_x—and vice versa. So by the definition given in inequality (1), the system ensures each $\mathcal{A}_i$ is ϵ-optimal.

Unique-Optimal Alignment. To solve Inverse Unique-Optimal Alignment for a given $\delta > 0$, the system simply has an inequality

$$f_x(\mathcal{B}) \;\geq\; f_x(\mathcal{A}_i) \,+\, \delta, \tag{8}$$

for each $\mathcal{A}_i$ and every alignment $\mathcal{B}$ of $\mathcal{S}_i$ with $\mathcal{B} \neq \mathcal{A}_i$, which is again a linear inequality in x.

We next explain how this linear programming formulation can be solved in polynomial time for most forms of sequence alignment.

4   Solving the Linear Program

One of the truly far-reaching results in linear programming is what we call the Separation Theorem. This result was discovered in the early 1980's by Grötschel, Lovász and Schrijver [6], Padberg and Rao [14], and Karp and Papadimitriou [10]. To explain it requires a few concepts. Linear programming optimizes a linear function of real variables over a domain given by linear inequalities. Geometrically this domain, which is an intersection of half-spaces, is a convex body called a *polyhedron*. If the inequalities have rational coefficients, the polyhedron is *rational*. A polyhedron that contains no infinite rays is *bounded*.

The *optimization problem* for a rational polyhedron $\mathcal{P} \subseteq \mathcal{R}^d$ is: Given rational coefficients c that specify the objective function, find a point $x \in \mathcal{P}$ that minimizes cx, or determine that $\mathcal{P}$ is empty. The *separation problem* for $\mathcal{P}$ is: Given a point $y \in \mathcal{R}^d$, either (1) find rational coefficients w and b that specify an inequality such that $wx \leq b$ for all $x \in \mathcal{P}$, but $wy > b$; or (2) determine that $y \in \mathcal{P}$. In other words, a *separation algorithm* that solves the separation problem for polyhedron $\mathcal{P}$ determines whether point y lies inside $\mathcal{P}$, and if it lies outside, finds an inequality that is satisfied by all points in $\mathcal{P}$ but is violated by y. Such a *violated inequality* gives a hyperplane that separates y from $\mathcal{P}$.

The Separation Theorem says that, remarkably, optimization and separation are equivalent: an efficient separation algorithm for a linear program yields an efficient algorithm for solving that linear program, and vice versa.

Theorem 1 (Equivalence of Separation and Optimization [6, 14, 10]).
The optimization problem on a bounded rational polyhedron can be solved in polynomial time if and only if the separation problem can be solved in polynomial time. □

The precise definition of polynomial time in the above is rather technical, but essentially means polynomial in the number n of *variables* in the system of inequalities describing the polyhedron (really, polynomial in its dimension and the number of digits in the rational coefficients). The import is that a linear program that is implicitly described by a list $\mathcal{L}$ of $2^{\Omega(n)}$ inequalities can be solved in $n^{O(1)}$ time if, for any candidate solution y, one can in $n^{O(1)}$ time determine that y satisfies all the inequalities in $\mathcal{L}$, or if it does not, report an inequality in $\mathcal{L}$ that y violates. Of course a separation algorithm that simply scans list $\mathcal{L}$ and tests each inequality will not achieve this time bound.

The proof of Theorem 1 exploits properties of the ellipsoid algorithm for linear programming. As a consequence, the polynomials bounding the running times have high degree, so the theorem does not directly yield algorithms for quickly solving exponentially-large linear programs in practice. Its main use is in proving that a polynomial-time algorithm exists.

To solve a linear program in practice using a separation algorithm, the following iterative approach is usually taken.

(1) Start with a small subset S of the inequalities in $\mathcal{L}$.
(2) Compute an optimal solution x to the linear program given by subset S.
(3) Call the separation algorithm for $\mathcal{L}$ on x. If the algorithm reports that x satisfies $\mathcal{L}$, output x and halt: x is an optimal solution for $\mathcal{L}$.
(4) Otherwise, add the violated inequality returned by the separation algorithm to S, and loop back to Step (2).

This kind of approach is known as a *cutting-plane algorithm*. Such algorithms often find optimal solutions very quickly in practice, even if they are not guaranteed to run in polynomial time. In Section 6 we show that the resulting cutting-plane algorithm for global alignment is indeed fast, solving instances with hundreds of parameters and alignments in a few minutes of computation.

In the remainder of this section we show that a polynomial-time *alignment algorithm* (in other words, an algorithm that computes an optimal alignment given fixed values for the parameters) yields a polynomial-time *separation algorithm* for our linear programming formulations of inverse alignment. Combined with Theorem 1, this proves our main result.

Theorem 2 (Complexity of Inverse Alignment). *Inverse Optimal and Near-Optimal Alignment can be solved in polynomial time for any form of alignment in which: (1) the alignment scoring-function is linear in its parameters, (2) the parameters values can be bounded, and (3) for any fixed parameter choice, an optimal alignment can be found in polynomial time. Inverse Unique-Optimal Alignment can be solved in polynomial time if in addition, for any fixed parameter choice, a next-best alignment can be found in polynomial time.* □

Optimal and Near-Optimal Alignment. We now give separation algorithms for each variation of inverse alignment. Recall that given an assignment of values x for the scoring-function parameters, a separation algorithm decides whether x satisfies all the inequalities in the linear program, and if it does not, identifies a violated inequality.

For inverse *optimal* alignment, the linear program consists of inequalities (5) for each input alignment $\mathcal{A}_i$. Conceptually we have a different separation algorithm for the inequalities associated with each $\mathcal{A}_i$. To separate all inequalities, run the separation algorithms for $\mathcal{A}_1, \ldots, \mathcal{A}_k$ consecutively. As soon as one algorithm finds a violated inequality, we halt and return the inequality. If no algorithm finds a violated inequality, we report x satisfies the linear program.

To separate the system of inequalities associated with a particular $\mathcal{A}_i$, simply compute an optimal alignment $\mathcal{B}^*$ under scoring function f_x over the strings $\mathcal{S}_i$ that $\mathcal{A}_i$ aligns. If $f_x(\mathcal{B}^*) \geq f_x(\mathcal{A}_i)$, then by transitivity inequality (5) holds for all $\mathcal{B}$, so x satisfies this system. On the other hand if $f_x(\mathcal{B}^*) < f_x(\mathcal{A}_i)$, this gives a violated inequality to report.

For inverse *near-optimal* alignment, we use an identical approach on inequalities (7). Note that this runs in $O(kt)$ time, where t is the time to compute an optimal alignment and k is the number of input alignments. So if t is polynomial, this separation algorithm runs in polynomial time.

Unique-Optimal Alignment. For inverse unique-optimal alignment, to separate the system of inequalities (8) associated with $\mathcal{A}_i$, we again compute an optimal alignment $\mathcal{B}^*$ of $\mathcal{S}_i$ under f_x. If $\mathcal{B}^* \neq \mathcal{A}_i$, then $f_x(\mathcal{B}^*) \geq f_x(\mathcal{A}_i) + \delta$ is a violated inequality. If $\mathcal{B}^* = \mathcal{A}_i$, compute a *next-best* alignment $\mathcal{C}^*$ of $\mathcal{S}_i$. If $f_x(\mathcal{C}^*) \geq f_x(\mathcal{A}_i) + \delta$, then by transitivity x satisfies the system; otherwise, this gives a violated inequality. Note that this runs in polynomial time if a next-best alignment can be computed in polynomial time (which is the case for standard string alignment [3]).

In the next section we use Theorem 2 to show that for *global alignment*, all variations of inverse alignment can be solved in polynomial time. The key point is showing how to bound the values of alignment parameters.

5   Application to Global Alignment

To obtain a cutting-plane algorithm for a particular form of alignment, such as global or local alignment of two strings, several details must be worked out to apply the general approach of Section 4. These include how to find an *initial subset* of the inequalities that yields a bounded linear program, and how to choose an appropriate *objective function*. Here we discuss these in the context of global alignment, but similar ideas apply to local alignment as well.

We use the definition of *standard global alignment* given at the beginning of Section 3, in which matches between pairs of letters are weighted by arbitrary substitution scores, and gaps are penalized using gap open and extension penalties. For inverse global alignment we separately consider two forms of the scoring function: when substitution scores are *varying* as given by equation (4), and when they are *fixed* as given by equation (3).

Initializing the Cutting-Plane Algorithm. Typically cutting-plane algorithms take as their initial set of inequalities just the trivial inequalities $x \geq 0$ of the linear program. For objective functions of biological interest, however, this trivial linear program is unbounded in the direction of the objective function. Consequently with this choice the first iteration of the cutting-plane algorithm would fail, as the first call to the linear programming solver to find an initial candidate solution x would report that the problem is unbounded, and return no solution.

When the substitution costs and gap penalties are all *varying*, we can set absolute upper and lower limits on the values of the parameters, and solve the linear program within the resulting bounding box as follows. By scaling the linear scoring-function by a positive factor (which does not change the relative rank of alignments), we can always make the largest parameter value hit 1. Then all parameters lie in the bounding box $0 \leq x \leq 1$, which we take as the initial set of inequalities for the cutting-plane algorithm.

When substitution costs are *fixed*, however, the bounding-box approach does not work (as the linear program may be unbounded). Instead we take the following approach. The linear programming problem is now a two-dimensional problem in the (γ, λ)-plane, where we associate γ with the vertical axis and λ with the horizontal axis. We say inequality I is a *bounding inequality* if the linear program consisting of I and the trivial inequalities $(\gamma, \lambda) \geq 0$ is bounded. In general, the linear program is bounded if and only if there exists (1) a bounding inequality, or (2) two inequalities where one is a downward halfspace, the other is an upward halfspace, and the slope of the downward inequality is less than the slope of the upward inequality. Furthermore, if they exist, these inequalities together with the trivial inequalities yield an initial set for the cutting-plane algorithm of at most four inequalities that give a bounded linear program.

We can find this set if it exists by identifying a downward inequality D of minimum slope and an upward inequality U of maximum slope. If D or U is a bounding inequality, or D's slope is less than U's, the linear program is bounded, and if not it is unbounded. For near-optimal inverse alignment, the general form of an inequality is $\gamma \Delta g + \lambda \Delta \ell \leq -\Delta s$, where $\Delta g := g(\mathcal{A}) - (1+\epsilon) g(\mathcal{B})$ for input alignment $\mathcal{A}$, and similarly for $\Delta \ell$ and Δs. This inequality is downward if $\Delta g > 0$, upward if $\Delta g < 0$, and its slope is $-\frac{\Delta \ell}{\Delta g}$. Thus for fixed $\mathcal{A}$ and ϵ, the direction and slope of an inequality is strictly a function of $g(\mathcal{B})$ and $\ell(\mathcal{B})$. For the two strings $\mathcal{A}$ aligns, functions g and ℓ range over a linear number of integer values, so the problem of finding a downward or upward inequality of optimal slope is certainly solvable. With further analysis, one can find the optimal inequalities in $O(1)$ time. Due to page limits we omit the details.

Choosing an Objective Function. As mentioned in Section 3, we are free to use any objective function we wish for the linear program, and we can exploit this freedom to pick a feasible solution that is biologically more desirable.

With *fixed* substitution scores, the parameters are γ and λ. Biologists generally prefer large γ and small λ, as in this regime optimal alignments tend to consist of a few long gaps, which is observed in biologically correct alignments. So one possibility for an objective is the linear combination $\max\{\gamma - \lambda\}$.

With *varying* substitution scores, the parameters are γ, λ, and all σ_{ab}. When the alignment problem seeks to minimize the alignment scoring function, so the σ_{ab} are treated as costs, we might want to maximize the separation between true substitution costs σ_{ab} (where $a \neq b$) and identity costs σ_{aa}. Then one possibility for the objective is to maximize the difference between the minimum true substitution cost and the maximum identity cost (so they are as far apart as possible). We can express this in our linear programming formulation by adding two new variables: s, which will equal the minimum true substitution cost, and i, which will equal the maximum identity cost. Using the objective $\max\{s-i\}$, and adding the inequalities $s \leq \sigma_{ab}$ for all $a \neq b$, and $i \geq \sigma_{aa}$ for all a, will achieve this goal. (Another possibility is to maximize the difference between the average true substitution cost and the average identity cost, which is also an objective that is linear in the parameters.) This objective on substitution scores can be combined with our objective on gap penalties by $\max\{s - i + \gamma - \lambda\}$.

Finally, note that for every objective, we can select two extreme solutions: x_{large}, which is the optimal solution under the objective, and x_{small}, which is the optimal solution in the direction opposite to the objective. Since the domain of feasible solutions for a linear program is convex, any convex combination of these two extremes, $x_\alpha := (1-\alpha)\, x_{\text{large}} + \alpha\, x_{\text{small}}$, where $0 \leq \alpha \leq 1$, is also a feasible solution. For example, $x_{1/2}$ may tend to be a more central parameter choice that generalizes to alignments outside the training set of input alignments $\mathcal{A}_i$ (which is borne out by our experiments of the next section).

6   Computational Results

We now present results from computational experiments on biological data with an implementation of our algorithms for inverse optimal and near-optimal global alignment. The implementation solves the problem both with *fixed* substitution scores (where $p = 2$ gap-penalty parameters are found), and with *varying* substitution scores (where for protein sequences all $p = 212$ parameters of the scoring function are simultaneously found). To solve linear programs we use the GNU Linear Programming Kit. For the linear programs we use the objective function $\max\{s - i + \gamma - \lambda\}$, where s and i are the minimum substitution and maximum identity costs, as described in Section 5. To find violated inequalities quickly, we maintain a queue Q of alignments $\mathcal{A}_i$ that generated a violated inequality the last time their separation algorithm was called. To find the next violated inequality, we remove an $\mathcal{A}_i$ from the front of Q, call its separation algorithm, and add it to the rear of Q if it generates another violated inequality. Figure 1 illustrates solving an instance with this implementation.

We ran several types of experiments on biological data. For the experiments, we chose six multiple sequence alignments from the PALI database [1] of structural protein alignments. (For each protein family in the SCOP protein classification database [12], PALI contains a multiple sequence alignment of the family based on aligning protein structures.) Table 1 describes the PALI families we chose, which are: T-boxes (box), NADH oxidoreductases (nad), Kunitz

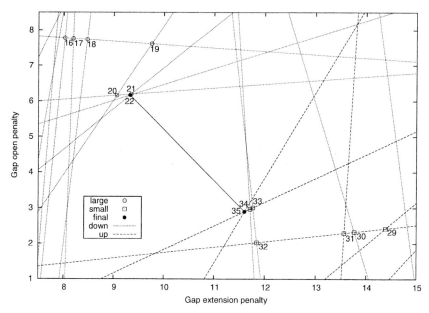

Fig. 1. Finding gap penalties by the cutting-plane algorithm. Labeled points are successive solutions for inverse near-optimal global alignment using PAM250 substitution scores on the pec dataset at the best possible $\epsilon = 0.06$ (see Tables 1 and 2). Solutions found when maximizing the linear program objective $\gamma - \lambda$ are *large* points; those found when minimizing this objective are *small* points. The optimal solutions for maximization and minimization are the two *final* points. Successive violated inequalities are also plotted, along with their half-space direction of *up* or *down*. Numbers labeling points give the order in which solutions were computed. The solid line-segment between the final large and small points is the *blend* line between these extremes.

inhibitors (kun), Sir2 transcription regulators (sir), apolipoproteins (apo), and pectin methylesterases (pec). Each family was reduced to 20 members by removing outlier sequences. For the training and testing experiments described next, these 20 members were partitioned into two groups of 10 members each, called the *training set* and the *test set*.

To investigate whether it is possible to learn scoring parameters from a training set of pairwise alignments that will apply to other alignments, we ran the following experiment. For each dataset with varying substitution scores, we found the smallest ϵ for the training set such that each induced pairwise alignment on the training set is ϵ-optimal; we call this smallest value ϵ_{train}. We also computed the same quantity for the test set, called ϵ_{test}. Then for each training set, we computed two extreme choices of parameters at ϵ_{train}, x_{large} and x_{small}, which are the parameter choices that respectively maximized and minimized the linear programming objective. We then searched for the convex combination between these two extremes that yielded a parameter choice for the test set with the smallest ϵ. For a given $0 \leq \alpha \leq 1$, the convex combination is $\alpha\, x_{\text{large}} + (1-\alpha)\, x_{\text{small}}$. The α that gave the smallest ϵ for the test set is called α_{blend}, and its corresponding ϵ is called ϵ_{blend}. These values are shown for the six datasets in Table 1. The table also gives the value of ϵ for the convex combinations $\alpha \in \{0, \frac{1}{2}, 1\}$. Notice that $\epsilon_{1/2}$ is surprisingly close to ϵ_{blend}, which is within roughly one percent of ϵ_{test}. This indicates that the best blend from the training set was nearly a best possible parameter choice for the test set. Notice also on this data the central parameter $\alpha = 1/2$ always yields a smaller ϵ than the extremes $\alpha = 0, 1$, indicating that central parameters generalize better.

To get a feel for how much closer to optimal we could make the induced pairwise alignments in a dataset by varying the substitution scores versus using a standard substitution matrix, we performed the following experiment. For all 20 members of each of the PALI datasets, we computed the minimum ϵ for the set of all 190 induced pairwise alignments, with varying substitution scores and the fixed PAM [2] and BLOSUM [9] substitution scores shown in Table 2. As might

Table 1. Generalizing from training to test sets. For each multiple sequence alignment dataset listed below, best possible parameters computed on a *training* subset are applied to a disjoint *test* subset. For each dataset the table lists the PALI accession number, the average sequence length, and the average percent-identity over all induced pairwise alignments. The meaning of the closeness entries is given in the text; values for ϵ are reported as percentages. All substitution scores and gap penalties are free parameters in these experiments.

Dataset	PALI number	Sequence length	Percent identity	Closeness ϵ						
				ϵ_{train}	ϵ_{test}	ϵ_{blend}	α_{blend}	ϵ_0	$\epsilon_{1/2}$	ϵ_1
box	333	183	14.3	1.7	**1.2**	1.5	0.47	1.7	**1.5**	2.6
nad	419	151	16.1	2.8	**2.8**	3.1	0.37	3.7	**3.1**	4.7
kun	409	172	15.0	3.9	**3.7**	4.2	0.49	4.4	**4.2**	4.8
sir	633	197	16.8	3.1	**2.2**	3.0	0.55	4.3	**3.0**	4.4
apo	99	143	17.8	1.9	**1.7**	3.2	0.22	3.6	**3.4**	4.6
pec	483	299	17.0	1.2	**2.0**	3.1	0.65	4.4	**3.1**	4.4

Table 2. Closeness to optimality for fixed and varying substitution scores. For each PALI dataset and for fixed or varying substitution scores, the smallest ϵ such that all induced pairwise alignments are ϵ-optimal is reported as a percentage.

Dataset	Closeness ϵ				
	varying	fixed			
		PAM250	PAM120	BL045	BL080
box	**2**	5	9	8	9
nad	**4**	9	15	12	18
kun	**5**	8	11	9	12
sir	**4**	11	16	12	16
apo	**3**	21	45	34	58
pec	**3**	6	9	8	11

be expected, the minimum ϵ when substitution scores are free parameters is smaller than the minimum ϵ for a fixed substitution matrix. Notice that on these datasets PAM250 gave the smallest ϵ of the four matrices considered. On the other hand, the optimal substitution matrix found when substitution scores were free parameters has roughly at most half the ϵ for PAM250, and is sometimes much better. Notice also that with varying substitution scores, one can consistently come very close to optimal on every dataset.

Finally, Table 3 gives running times and number of violated inequalities for the cutting-plane algorithm on the experiments of Table 2. Running times are wall-clock times in seconds to solve Inverse Near-Optimal Alignment for a given ϵ during the binary search for the smallest ϵ. Each binary search concluded in at most 8 iterations. Every iteration, while considering an input of 190 alignments, finished in roughly under a minute for fixed substitution scores, and under roughly 4 minutes for varying substitution scores while finding values for 212 parameters. Experiments were on a 3 GHz Pentium 4 with 1 GB of RAM.

Table 3. Running time and number of violated inequalities. For each PALI dataset, times and number of inequalities are reported for computing the ϵ of Table 2. Columns report the median and extreme values across the binary search iterations.

Dataset	Time (sec)				Violated inequalities					
	fixed		varying		fixed			varying		
	med	max	med	max	min	med	max	min	med	max
box	24	25	5	123	4	11	23	5	32	972
nad	17	18	7	46	3	12	21	5	118	590
kun	23	30	11	83	2	13	22	7	176	681
sir	29	31	13	96	2	11	20	5	196	832
apo	16	16	14	264	6	17	22	13	236	1398
pec	63	65	23	226	2	10	20	3	238	1087

7   Conclusion

We have presented a new approach to inverse parametric sequence alignment. The approach is actually quite general, and solves inverse parametric optimization in polynomial time for *any* optimization problem (not just sequence alignment) whose objective function is linear in its parameters, whose parameters can be bounded, and that can be solved in polynomial time when all parameters are fixed. Experiments on structural alignments from a protein family database show we can find all 212 parameters of the standard protein-sequence scoring-model from hundreds of pairwise alignments in a few minutes of computation.

Many lines of investigation remain open. Can parameter values be learned from both positive and *negative* examples? Can our algorithm aid the *evaluation* of alignment software by testing programs at parameter settings that make benchmark alignments score as close to optimal as possible? Can cutting planes be efficiently found for inverse alignment that are *facet-defining* inequalities?

References

1. Balaji, S., S. Sujatha, S.S.C. Kumar and N. Srinivasan. "PALI: a database of alignments and phylogeny of homologous protein structures." *Nucleic Acids Research* 29:1, 61–65, 2001.
2. Dayhoff, M.O., R.M. Schwartz and B.C. Orcutt. "A model of evolutionary change in proteins." In M.O. Dayhoff, editor, *Atlas of Protein Sequence and Structure* 5:3, National Biomedical Research Foundation, Washington DC, 345–352, 1978.
3. David Eppstein. "Finding the k shortest paths." *SIAM Journal on Computing* 28:2, 652–673, 1998.
4. Fernández-Baca, D., T. Seppäläinen and G. Slutzki. "Bounds for parametric sequence comparison." *Discrete Applied Mathematics* 118:3, 181–192, 2002.
5. Griggs, J.R., P. Hanlon, A.M. Odlyzko and M.S. Waterman. "On the number of alignments of k sequences." *Graphs and Combinatorics* 6, 133–146, 1990.
6. Grötschel, M., L. Lovász and A. Schrijver. "The ellipsoid method and its consequences in combinatorial optimization." *Combinatorica* 1, 169–197, 1981.
7. Gusfield, Dan, K. Balasubramanian and Dalit Naor. "Parametric optimization of sequence alignment." *Algorithmica* 12, 312–326, 1994.
8. Gusfield, Dan and Paul Stelling. "Parametric and inverse-parametric sequence alignment with XPARAL." *Methods in Enzymology* 266, 481–494, 1996.
9. Henikoff, S. and J.G. Henikoff. "Amino acid substitution matrices from protein blocks." *Proceedings of the National Academy of Sciences USA* 89, 10915–10919, 1992.
10. Karp, R.M. and C.H. Papadimitriou. "On linear characterization of combinatorial optimization problems." *SIAM Journal on Computing* 11, 620–632, 1982.
11. Kececioglu, John and Dean Starrett. "Aligning alignments exactly." Proceedings of the 8th ACM *Conference on Research in Computational Molecular Biology*, 85–96, 2004.
12. Murzin, A.G., S.E. Brenner, T. Hubbard and C. Chothia. "SCOP: a structural classification of proteins database for the investigation of sequences and structures." *Journal of Molecular Biology* 247, 536–540, 1995.
13. Pachter, Lior and Bernd Sturmfels. "Parametric inference for biological sequence analysis." *Proceedings of the National Academy of Sciences USA* 101:46, 16138–16143, 2004.
14. Padberg, M.W. and M.R. Rao. "The Russian method for linear programming III: bounded integer programming." Technical Report 81-39, Graduate School of Business and Administration, New York University, 1981.
15. Sun, Fangting, David Fernández-Baca and Wei Yu. "Inverse parametric sequence alignment." *Journal of Algorithms* 53, 36–54, 2004.
16. Waterman, M., M. Eggert and E. Lander. "Parametric sequence comparisons." *Proceedings of the National Academy of Sciences USA* 89, 6090–6093, 1992.
17. Zimmer, Ralf and Thomas Lengauer. "Fast and numerically stable parametric alignment of biosequences." Proceedings of the 1st ACM Conference on *Research in Computational Molecular Biology*, 344–353, 1997.

Revealing the Proteome Complexity by Mass Spectrometry

Roman A. Zubarev

Laboratory for Biological and Medical Mass Spectrometry,
Uppsala University, Sweden
roman.zubarev@bmms.uu.se

Abstract. The complexity of higher biological organisms is astounding, but the source of this complexity is far from obvious. With the emergence of epigenetics, the assumed main source of complexity has been shifted from the genome to pre- and post-translational modifications in proteins. There are estimated 100,000 different protein sequences in the human organism, and perhaps 10-100 times as many different protein forms. Analysis of the human proteome is a much more challenging task than that of the human genome. The challenge is to provide sufficient amount of information in experimental datasets to match the underlying complexity.

Mass spectrometry (MS) is one of the most informative techniques, widely used today for protein characterization. MS is the fastest growing spectroscopy area, which in 2005 has overtaken NMR as the prime research field. After a major revolution in the late 1980s (awarded by the Nobel prize in Chemistry in 2002), MS has continued to develop rapidly, showing amazing ability for innovation. Today, several different types of mass analyzers are competing with each other for the future. This diversity means that the field of MS, although a century old, is still in the fast evolving phase and is far from saturation.

Despite the rapid progress, today's MS tools are still largely insufficient. Mathematical models of the MS-based proteomics analysis as well as experimental assessments showed large disproportions between the information content of the experimental MS datasets and the underlying sample complexity. One of the most desired improvements would be the higher quality of ion fragmentation in tandem mass spectrometry (MS/MS). The latter parameter boils down to the ability to specifically fragment each of the chemical bonds (C-C, C-N and N-C) linking amino acid residues in a polypeptide sequence. This formidable physicochemical challenge is met by recently emerged techniques involving ion-electron reactions.

Characterization of primary polypeptide sequences of unmodified amino acids is a basic task in proteomics. Recent large-scale evaluation has shown that de novo sequencing by conventional MS/MS is insufficiently reliable. Fortunately, novel fragmentation techniques improved the situation and allowed the first proteomics-grade de novo sequencing routine to be developed.

Another group of challenges relates to the ability to extract maximum information from MS/MS data. The database search technologies

A. Apostolico et al. (Eds.): RECOMB 2006, LNBI 3909, pp. 456–457, 2006.

developed in the late 1990s are still the backbone of routine proteomics analyses, but they are rapidly becoming insufficient. Typically, only 5 to 15% of all MS/MS data produce "hits" in the database, with the bulk of the data being discarded. Research in that issue has led to the emergence of a quality factor for MS/MS data (S-score). S-score analysis has shown that only half of the data are discarded for a good reason, while another half could be utilized by improved algorithms. Such algorithms specially designed to deal with any mutation or modification have recently uncovered hundreds of new types of modifications in the human proteome. High mass accuracy reveals the elemental compositions of these modifications, and MS/MS determines their positions. The potential of such algorithms for unearthing the vast and previously invisible world of modifications and thus tackling proteome's enormous complexity will be discussed.

Motif Yggdrasil: Sampling from a Tree Mixture Model

Samuel A. Andersson and Jens Lagergren

Stockholm Bioinformatics Center and School of Computer Science and
Communication, Royal Institute of Technology, SE-100 44 Stockholm, Sweden
samme@nada.kth.se, jensl@nada.kth.se

Abstract. In phylogenetic foot-printing, putative regulatory elements
are found in upstream regions of orthologous genes by searching for
common motifs. Motifs in different upstream sequences are subject to
mutations along the edges of the corresponding phylogenetic tree, consequently taking advantage of the tree in the motif search is an appealing idea. We describe the Motif Yggdrasil sampler; the first Gibbs
sampler based on a general tree that uses unaligned sequences. Previous
tree-based Gibbs samplers have assumed a star-shaped tree or partially
aligned upstream regions. We give a probabilistic model describing upstream sequences with regulatory elements and build a Gibbs sampler
with respect to this model. We apply the collapsing technique to eliminate the need to sample nuisance parameters, and give a derivation of
the predictive update formula. The use of the tree achieves a substantial
increase in nucleotide level correlation coefficient both for synthetic data
and 37 bacterial *lexA* genes.

An important part of the regulation of genes in the cell is played by transcription
factors, that selectively bind DNA and affect the expression of a specific subset
of genes. The binding sites are typically situated upstream of the regulated gene
and are often degenerated, which makes identifying them difficult.

Phylogenetic foot-printing is one of the main approaches to *in silico* detection of regulatory elements in DNA. It uses orthologous genes of a gene family.
The homology, i.e., the common origin, makes it likely that regulatory elements
belonging to the upstream region of the common ancestor of the gene family
can also be found in the upstream regions of the extant orthologs. In principle, whereas the regulatory elements (REs) are functional and should be conserved during evolution, the parts of an upstream region not containing REs
are non-functional and should be less conserved; the difference in conservation
being dependent on the time and rate of evolution. It is important to notice
that conserved means having been affected by *fewer* mutations, i.e., nucleotides
substitutions. In phylogenetic foot-printing, evaluation of motif candidates is an
attempt to test the hypothesis that the candidate is a real motif that has evolved
from an ancestral RE, by measuring the number of mutations that have occurred
under this hypothesis.

A. Apostolico et al. (Eds.): RECOMB 2006, LNBI 3909, pp. 458–472, 2006.
© Springer-Verlag Berlin Heidelberg 2006

There are a number of methods for finding motifs in DNA-sequences [1, 2, 3, 4, 5, 6, 7, 8, 9, 10, 11], deterministic as well as probabilistic. One commonly used and successful probabilistic method is the Gibbs Motif sampler (GMS) [1, 2]. In the design of several phylogenetic footprinting methods, attempts have been made to take advantage of the phylogenetic tree for the gene family [7, 8, 9, 10, 11]. Using the tree S for the gene family, it is far easier to accurately measure the number of mutations that have occurred during the evolution of the genes. Consider the following example; exactly half the nucleotides of a column c of a motif are A and exactly half are T. If there is an edge of S that separates the genes associated with motifs having A in column c and those associated with motifs having T in column c, then the discrepancy can be explained with a single mutation.

Motivated by the realization that conservation is best measured given the phylogenetic tree for the orthologs, Blanchette and coworkers pioneered the area with a deterministic phylogenetic foot-printing algorithm based on the phylogenetic parsimony score [7, 8]. Their algorithm is rather time inefficient and can only be used for short motifs with low parsimony score. It also seems fairly common that some positions in a motif are relatively unconstrained or that some are constrained but to any of two specific nucleotides (say purines), toggling between the two, [12]. In both these cases, the parsimony score is high. The GMS, however, can model a few columns by fairly uniform distributions on all four nucleotides or some columns with a fairly uniform distribution on just two specific nucleotides, thereby managing these two problems.

There are also probabilistic algorithms designed to take advantage of the tree [9, 10, 11]. Some algorithms [9, 11], use standard models for sequence evolution. This requires that the upstream regions can be at least partially aligned, which is realistic only for organisms that are very closely related. The model in [11] also uses the same substitution matrices and rate of evolution for all positions in the motifs and background, except for a fixed multiplicative factor used to decrease the motif rate. This deviates from reality since the rate of evolution in motifs shows position-specific variation [13]. It is also commonly believed that the substitution matrices varies between motif positions.

We have designed a tree-based Gibbs sampler that, given upstream regions of orthologous genes and their phylogenetic tree, i.e., the corresponding species tree, finds regulatory motifs in those upstream regions. Each upstream region is associated with a leaf of the phylogenetic tree. The Motif Yggdrasil sampler (MY sampler) can handle any tree, as long as it is the true tree for the family, and does not require edge (a.k.a. branch) lengths or rates of evolution, both of which are often uncertain. Since the GMS represents well tested and functional technology, a key aspect in the design of the MY sampler was to make it as similar as possible to the GMS, e.g. to allow toggling, but to obtain higher sensitivity and specificity by utilizing the phylogenetic tree. Since our model is tree-based and share similarities with mixture models, we call it a tree mixture model.

A Gibbs sampler is based on a probabilistic model. The better the model describes the reality, the more likely the sampler is to generate the correct output.

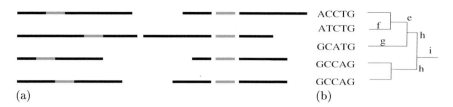

(a)   (b)

Fig. 1. (a) To the left upstream sequences with embedded motif instances (red) and to the right the motif alignment. (b) A tree and a motif alignment generated from it.

A good intuitive way of thinking of the models relevant to motif finding is as follows. First are a $k \times w$ motif alignment matrix M and k background sequences $b^1, \ldots, b^k$ generated. Thereafter, for each $1 \leq i \leq k$, the i:th row of M is inserted in b^i giving rise to the i:th upstream region s^i. A partition of any sequence into a background and a motif part is then possible. The motif parts can be extracted to form a motif alignment, as is illustrated in Fig. 1a. The problem of motif finding is to, given $s^1, \ldots, s^k$, recover the partition into a motif alignment and background. Motif loss and multiple motifs can be handled similar to BioProspector [14].

In the model underlying the GMS, each column of the motif is generated by a single distribution, where all positions of the column are treated in the same way. In our probabilistic model, each column of the motif is generated from a mutation on a single edge e of the phylogenetic tree S of the gene family (on each side of the edge, i.e., in each of the subtrees of $S \setminus e$, the positions, of the motif column, associated with the leaves are generated by a single distribution, the very same type of distribution used to generate the entire column in the model underlying the GMS). The mutation in the motif can, for instance, model a mutation in the associated transcription factor, that changed the preferential binding in this motif position, or model a change in affinity, hence altering the expression profile of the gene, with no changes in the transcription factor.

The choice of mutation edge for one column is independent of that of any other column, i.e., the mutation edge is likely to vary across the columns of the motif alignment. Also, among the possible choices of an edge is the empty edge, which corresponds to using one distribution for the entire column, equivalent to the model underlying the GMS.

Among the data, the start positions of the motifs and the edges used to generate the motifs are considered hidden data. We look for the motifs but not the edges. The edges are nuisance parameters and we use the collapsing technique [15] to avoid sampling them. The MY sampler proceeds by stepwise sampling a new motif start position in a randomly chosen upstream sequence.

We have chosen to compare our tree-based sampler to the GMS, as opposed to other motif finding methods, for mainly three reasons. Firstly, comparing motif finding algorithms is difficult. Actually, how such comparisons are done best is an interesting problem which is actively researched [16, 17, 18]. In our case the comparison is simplified by the similarity of the compared methods. Secondly, our main interest is the probabilistic model, and whether the incorporation of the phylogenetic tree improves the discriminatory power of the motif search.

Thirdly, the GMS is a commonly used and well-respected motif finding method, resting on a solid statistical foundation.

1 Method

The Motif Yggdrasil Sampler. MCMC is a technique that facilitates estimation of the stationary distribution of a Markov chain. It provides a uniform framework to design transition probabilities of a Markov chain so that a sought stationary probability distribution is obtained, which in Bayesian inference applications is the posterior distribution, (for details we refer to [19, 20]). A random walk is performed in the Markov chain according to the transition probabilities. As in the GMS, we seek, as opposed to the posterior distribution, merely the state with maximum posterior probability. This state also has maximum likelihood. Notice that the probabilistic models, w.r.t. which these probabilities are defined, differs for the MY sampler and the GMS. In a Gibbs sampler, a state in the Markov chain is a vector and in a state transition only one component of the vector is changed. In a state vector $(m_1, \ldots, m_k)$, the component m_i is a start position of a candidate motif instance in the i:th input string, which given the motif width uniquely determines the motif candidate. In each iteration of the MY sampler, a new state is chosen by altering one uniformly selected component, based on the so called *predictive update formula*. Below we derive an efficient algorithm for computing the predictive update formula for the MY model. Notice that the Gibbs sampling framework together with the capacity to compute the predictive update formula gives our MY sampling algorithm.

The Motif Yggdrasil Model. Below the generative probabilistic Motif Yggdrasil model (MY model) used in the Motif Yggdrasil sampler is introduced formally. The intuition behind the model has been described in the introduction. We begin with an example. Fig. 1b shows an example of a motif alignment that could have been obtained from the tree shown in the figure. In the figure, each distribution used, has probability 1 to generate one nucleotide and zero to generate the other nucleotides, which is not true in general. The edges used to generate the positions are e, f, g, h, and i, in order. The edge i represents the *empty edge*. The edges adjacent to the root, h, is counted only once in our model.

Definitions. Let $\mathcal{N} = \{A, C, G, T\}$, denote the alphabet of nucleotides. By *Bernoulli-DNA* with parameters $\theta = (\theta_A, \theta_C, \theta_G, \theta_T)$, where $\sum_{n \in \mathcal{N}} \theta_n = 1$, we mean a string or vector where a position is assigned the nucleotide $n \in \mathcal{N}$ with probability θ_n. The Dirichlet distribution is a conjugate prior for Bernoulli-DNA distributions. A tree S has vertex set $V(S)$, edge set $E(S)$, and leaf set $L(S)$. We extend the edge set of each tree S so that it includes a new edge, called the *empty edge*, denoted ε, such that $S \setminus \varepsilon$ contains one connected component only. Each tree considered will have leaves $[k]$, for some integer k. For a tree S

and an edge $e \neq \varepsilon$ of S, the forest $S \setminus e$ consists of two trees; the one containing the leaf 1 is called the *first component* and denoted $C_1(S \setminus e)$, analogously, the other tree is called the *second component* and denoted $C_2(S \setminus e)$, (e.g., in Fig. 1b the two bottom leaves belong to the same component when edge h is removed).

For given $k, n_1, \ldots, n_k$, w, and a given tree S with $L(S) = [k]$, the MY model generates: k sequences $s^1, \ldots, s^k$ where $s^i \in \mathcal{N}^{n_i}$, w edges $e_1, \ldots, e_w$ of $E(S) \cup \{\varepsilon\}$, k start positions $m_1, \ldots, m_k$ where $1 \leq m_i \leq n_i - w + 1$. The start position m_i is chosen uniformly from $1 \leq m_i \leq n_i - w + 1$. The edge e_i is chosen uniformly from $E(S) \cup \{\varepsilon\}$, where ε is the empty edge. For a sequence s^i, the positions $m_i, \ldots, m_i + w - 1$ are referred to as a *motif* instance. The positions $m_1 + l - 1, \ldots, m_k + l - 1$ are referred to as the *lth column of the motif*. The positions that do not belong to the motif are referred to as *background*.

For each $i \in [k]$, the background of s_i is multinomial-DNA distributed with nucleotide probabilities drawn from a Dirichlet distribution with parameters $\beta_A^i, \ldots, \beta_T^i$, for more information on Dirichlet distributions see, for instance, [21]. It is, however, possible to use a simplified variation where the entire background is described by a single multinomial-DNA distribution. The motif positions are drawn from $2w$ different Bernoulli-DNA distributions; the positions of the l:th motif column that belong to the same component of $S \setminus e_l$ constitute one Bernoulli-DNA. That is, if $i_1, \ldots, i_r \in [k]$ are the positions that belong to the first component of $S \setminus e_l$, then $s_{m_i+l-1}^{i_1}, \ldots, s_{m_i+l-1}^{i_r}$ is one Bernoulli-DNA with nucleotide probabilities $\theta_A^{l,1}, \ldots, \theta_T^{l,1}$, drawn from a Dirichlet distribution with parameters $\alpha_A^{l,1}, \ldots, \alpha_T^{l,1}$. Analogously, if $i_{r+1}, \ldots, i_k \in [k]$ are the positions that belong to the second component of $S \setminus e_l$, then $s_{m_i+l-1}^{i_{r+1}}, \ldots, s_{m_i+l-1}^{i_k}$ is one Bernoulli-DNA with nucleotide probabilities $\theta_A^{l,2}, \ldots, \theta_T^{l,2}$ drawn from a Dirichlet distribution with parameters $\alpha_A^{l,2}, \ldots, \alpha_T^{l,2}$. The hyper-parameters of our model are, for each $l \in [w]$, $\beta_A^i, \ldots, \beta_T^i$ and, for each $l \in [w]$ and $j \in [2]$, $\alpha_A^{l,j}, \ldots, \alpha_T^{l,j}$.

The MY model uses Bernoulli-DNA for each motif column because, compared to a multinomial-DNA, this promotes conserved motif columns. This is also what the GMS does. We use multinomial-DNA for the background, since there is no reason to promote conserved background. The reason why we do not use a distribution defined relative to the tree also for the background, is that we want the algorithm to be applicable to distantly related gene sequences.

The Predictive Update Formula. Below we describe the predictive update formula, for details see the Appendix. We will use m^i for the entire motif part of s^i and b^i for its background part, and not as previously merely the start position. The order is the natural one, although, this is not important, since nucleotide counts give a sufficient statistic. For notational convenience, we describe the predictive update formula for m^k, rather than for a general m^i. The following expression for the predictive update formula follows from the independence of background and motif

$$\Pr[m^k, b^k | m^{-k}, b^{-k}] = \Pr[m^k | m^{-k}]\Pr[b^k] \tag{1}$$

where $m^{-k} = \{m^i : i \neq k\}$ and $b^{-k} = \{b^j : j \neq k\}$. The expression (1) can be efficiently computed by breaking it down into efficiently computable parts. The motif part of (1) is, (for a complete derivation see the Appendix),

$$\Pr[m^k|m^{-k}] = \prod_{l\in[w]} \sum_{e\in E'(S)} \prod_{j\in[2]} \prod_{i\in(C_j(S\setminus e)\cap[k])} \int_\theta \Pr[m_l^i|\theta]\Pr[\theta|\alpha^{l,j}]\,d\theta \ . \quad (2)$$

The integral in (2) equals

$$\frac{\Gamma\left(\sum_{t\in\mathcal{N}}\alpha_t\right)\prod_{t\in\mathcal{N}}\Gamma\left(\alpha_t^k + c_t\right)}{\prod_{t\in\mathcal{N}}\Gamma\left(\alpha_t\right)\Gamma\left(\sum_{t\in\mathcal{N}}\alpha_t^k + c_t\right)} \quad (3)$$

where c_t is the count of nucleotide t in subtree $C_j(S \setminus e) \cap [k]$ and α_t is the pseudocount for nucleotide t from parameter vector $\alpha^{l,j}$. For the background part, a similar expression can be derived. These formulas can be calculated fast using an approximation [15] or by using a pre-computed look-up table.

Implementation. The MY sampler iteratively chooses a state in $(m^1, \dots, m^k)$, where m^i is the start position of the motif candidate in the i:th string, using the predictive update formula. For a window the size of the motif, slided along the i:th string, a score, up to a constant equal to the predictive update probability, is assigned to each valid position. A position is picked as the new m^i with a probability proportional to its score. To simplify the computation of the score, we use a slight modification of the MY model by modeling also the background by Bernoulli-DNA. This is also done in the GMS. The modification allows the scoring function to be represented by a positional weight matrix. In a modified version of the MY sampler, that takes a parameter u, any term in the predictive update formula is disregarded if it corresponds to an edge e such that a connected component of $S \setminus e$ has $\leq u$ leaves.

2   Results

To test the MY model and the the MY sampler, synthetic data were generated. The background, of length 200, was generated with a uniform distribution over the nucleotides and in each background sequence one motif was implanted at a randomly selected position. For Dataset A, we generated a random tree with 10 taxa, using a birth-death process and added exponential noise to the edge lengths. A motif alignment, 15 bp wide, was generated from the tree using the program Seq-Gen [22]. The Seq-Gen software evolved the motif according to the tree using the default F84 model of sequence evolution (for details see [22]).

The difficulty of finding the implanted motifs in a particular set of sequences is proportional to how much the motifs differ from each other. We therefore measure the difficulty of a dataset by the sum of Hamming distances, i.e., number of mismatches, of motif instances to their consensus motif sequence. Seq-Gen takes as an argument a branch length factor, which is multiplied to all branch lengths and hence effectively changes the size of the tree. This enabled us to

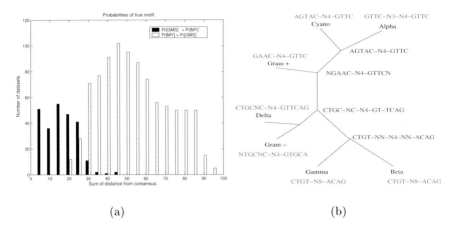

(a)   (b)

Fig. 2. (a) Probabilities of each motif model for Dataset A. The black bars indicate datasets where the GMS model assigns a higher probability than does the MY model. (b) Phylogenetic tree and binding site of LexA. Note that dashes are merely to enhance readability and do not indicate indels. Adapted from [23].

simulate datasets where the leaves have different evolutionary distances from each other although still related by a phylogenetic tree with the same topology. A total of 1162 individual datasets of differing Hamming distances to consensus, were made. We consider the MY model to describe the biological reality better than the model used to generate this first dataset. This dataset was chosen to test how well our model and sampler performs under circumstances deviating from the assumptions underlying our model.

We first evaluated the motif model of the GMS against our MY model, by calculating the probability of the true, i.e., planted, motif according to each model. This gives an indication of how well each model describes the actual motif. As is evident from Fig. 2a, for simple motifs that are relatively well conserved, the GMS model assigns the true motif a higher probability than does the MY model. This is expected since the MY model is developed specifically with more distant relatives in mind. If one edge alone is sufficient to explain the variation in a motif column, the weighted average that the MY model uses, will be lower than that edge's contribution. As stated previously, the GMS model corresponds to using the same probability distribution for the whole tree. This is a good approximation for well-conserved motifs. More distantly related, difficult motifs, are better described by the MY model.

We used both the MY sampler and our implementation of the GMS to be able to compare their performances. The results on Dataset A is summarized in Fig. 3, where the number of separate datasets is plotted against the sum of the Hamming distances from consensus for each motif instance in the dataset. For data with sum of distances < 20 both samplers found the correct motif and for datasets with sum of distances > 65 neither sampler found the correct motif. Note that these uninteresting extremes have been excluded in some figures.

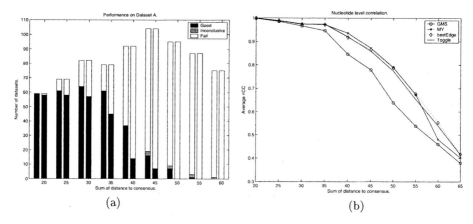

(a)   (b)

Fig. 3. (a) Performance of the MY sampler, (left bars), and the GMS, (right bars), on Dataset A, evaluated by probabilities. See text for details. (b) Performance of the GMS and the MY sampler on Dataset A, evaluated by average nCC in each bin.

In Fig. 3, the black bars indicate that the correct motif was found, i.e., the probability of the found motif was identical to the probability of the inserted motif. If the correct motif was not found, we evaluated the found motif and the true motif by the probability that it is a real motif according to the model, as calculated by the respective samplers. If the found motif had a higher probability than the correct motif, the sampler failed, white bars in Fig. 3. This reflects an imperfection in the probabilistic model, since a random motif was given a higher probability than the correct motif. If the found motif had a lower probability than the correct motif, indicated by grey bars in the figure, the run was inconclusive. It is well known [19, 20], that given enough time a Gibbs sampler will converge to its stationary (invariant) distribution, as long as the Markov chain is irreducible, aperiodic and has an invariant distribution. As can be seen in Fig. 3, the MY sampler consistently outperforms the GMS. For example, in bin 40 the MY sampler finds the correct motif in 37 datasets, while the GMS only find it in 14, i.e., the MY sampler outperforms the GMS by more than 250%.

A recent study [18] suggests using the nucleotide level correlation coefficient (nCC) for evaluating the performance of motif finding algorithms. An nCC of 1 corresponds to finding all correct nucleotides while not finding any false. A random assignment of nucleotides is expected to get an nCC of 0. The results, according to this criterion, of the GMS and the MY sampler can be seen in Fig. 3b. Notice that for datasets with a sum of Hamming distance to consensus in the range of 40 to 55, we have an average increase in performance of 10% to 25%, as compared to the GMS.

The MY model takes a sum of the probabilities over all edges in the tree. A natural variation is to consider only the edge with the highest probability, i.e., for each motif column, we choose the single edge that best explains that column given the observed nucleotides. This alternative model has also been evaluated and is called bestEdge in Fig. 3b. Another modified version picks one of the GMS and the MY model, based on their scores, and uses it for a user-defined number of iterations before making a new choice. This alternative model has also been evaluated and is called Toggle in Fig. 3b. For results on other synthetic data, see Fig. 4 and Fig. 5 in the Appendix.

The performance of each sampler was also tested on biological data from the *lexA* genes of several bacterial species. The autoregulatory *lexA* gene was chosen since a recent article [23] describes the evolution of its binding site. A PWM from Transfac [24] was used to scan *E. coli* for genes potentially regulated by the *lexA* transcription factor. The data was selected using the upstream sequences, ≤ 250 bp, of orthologs of the *E. coli* genes. The corresponding upstream sequences were scanned using the published consensus motif sequences [23, 25], and any hit was considered a true motif instance. The dataset consists of 37 upstream sequences with gapped motifs (motif width 12-16 bp), of several types. The protein sequences corresponding to upstream sequences with motif hits, were then used to construct a phylogenetic tree, based on 100 bootstrapped trees, using the PHYML software [26] with the JTT model for protein sequence evolution. The tree, with bootstrap support, is depicted in Fig. 6 in the Appendix.

The upstream sequences and a motif width of 12 bp, representing a core region shared by all species, were given as input to the GMS and to the MY sampler, which was also given the tree. The best found motif alignment according to the GMS model has nCC = 0.4968 (corrected for width and number of motifs) and probability 1.4×10^{-136}, which is higher than the true motif alignment that has probability 1.11×10^{-162}. The GMS sampler prefers a frameshifted motif alignment to the correct one. Under the same conditions the best motif alignment the MY sampler found, see Fig. 6, has a probability of 2.4×10^{-110}, lower than the true motif alignment, which the MY sampler gives a probability 1.39×10^{-104}. While this corresponds to an inconclusive run, its nCC of 0.8838 (corrected), is still in the order of two times that of the best GMS result.

3 Discussion

It is clear that there is useful information in the phylogeny of the regulated genes that may be advantageous in motif detection. From Dataset A it is obvious that the MY model does a better job modelling more evolutionary distant motifs. The observation that well-conserved motifs are well described by the GMS motif model is the basis for the bestEdge alteration of the MY model. In this version, the edge with highest probability, is greedily chosen instead of summing over all edges. In the case where the empty edge is the edge best explaining the data, the new model is the same as the Gibbs Motif sampler model. Results on Dataset A show that its performance is about the same as the original MY sampler. These are very interesting results and further investigation as to how this holds for biological data is warranted. The observation that binding sites often can be divided into subclasses, motivated [27] to develop a model that partitions the sequences and uses one PWM for each partition. Their approach, however, does not take the phylogenetic tree into account when partitioning the sequences.

Mazon *et al.* have shown how the consensus sequence of the *lexA* binding sites in Gram negative bacteria, Alpha proteobacteria, Delta proteobacteria, Gamma proteobacteria and Cyanobacteria, all can be derived from the consensus sequence of the Gram positive bacteria. This is shown in Fig. 2b, with the mutational events and sequences depicted in relation to the phylogenetic tree. Both the tree and all red sequences in the leaves are adapted from [23] and the blue sequences in the internal nodes are inferred, possible, ancestral sequences that could explain the observations.

The fact that the similarities of the *lexA* binding sites, found in as diverse species as all proteobacteria, Cyanobacteria, Gram positive bacteria and Gram negative bacteria, is consistent with the species tree, and that *in vitro* experiments [23] indicate that constraints on the transcription factor also follows the tree, give further support to both the MY sampling algorithm and the MY model.

The ideas presented can be generalized in many ways. By using gene trees, as with the *lexA* data, coregulated genes in the same species can be included in the analysis. This still includes evolutionary history, such as gene duplication events, which may help in finding the regulatory motif. The MY model can be generalized to use an arbitrary number of edges. Such a modification is likely to better describe a complex evolution with multiple mutational events per motif column. It would be very interesting to study whether this would give increased performance. The MY model can be extended to use several trees. This is particularly useful if the correct phylogeny of the genes of interest is unknown. One could then use many plausible trees at the same time, since any edge removal still corresponds to a partitioning of the leaves into two sets, each of which could be modelled by a Bernoulli-DNA model. In fact, any partitioning of the genes could be used, for instance splits trees or neighbour nets [28, 29]. In the MY model described in this paper, the prior $\Pr[E]$ is uniform, but it is quite possible to use other priors, e.g., an edge, e could be selected with a probability proportional to the length of edge e. This may be intuitive since the probability of a mutational event increases with the length of the edge, all other factors equal.

Another issue for future investigation is how to improve the convergence rate. The convergence rate for a Gibbs sampler is highly affected by the collinearity of the components in the state vector. It may, therefore, be possible to speed up the program by, e.g., grouping of the leaves and updating a group simultaneously. Since binding sites can have a constraint on distance from the gene, and therefore share similar start positions, this also makes sense in a biological setting.

Acknowledgment

We want to thank Lars Arvestad, Öjvind Johansson, Dannie Durand, Mathieu Blanchette and Wyeth W. Wasserman for fruitful discussions and two anonymous referees for helpful comments. We would also like to thank Wynand Alkema for generously providing us with the *lexA* dataset.

References

1. Lawrence, C.E., Altschul, S.F., Boguski, M.S., Liu, J.S., Neuwald, A.F., Wootton, J.C.: Detecting subtle sequence signals: a Gibbs sampling strategy for multiple alignment. Science **262** (1993) 208–14
2. Liu, J., Neuwald, A., Lawrence, C.: Bayesian models for multiple local sequence alignment and Gibbs sampling strategies. Journal of the American Statistical Association **90** (1995) 1156–70
3. Eskin, E., Pevzner, P.: Finding composite regulatory patterns in DNA sequences. Bioinformatics **18 Suppl 1** (2002) S354–63
4. Eskin, E.: From profiles to patterns and back again: a branch and bound algorithm for finding near optimal motif profiles. In: Proceedings of the eigth International Conference on Computational Molecular Biology (RECOMB-04), New York, NY, USA, ACM Press (2004) 115–124
5. Keich, U., Pevzner, P.: Subtle motifs: defining the limits of motif finding algorithms. Bioinformatics **18** (2002) 1382–1390
6. Buhler, J., Tompa, M.: Finding motifs using random projections. J Comput Biol **9** (2002) 225–242
7. Blanchette, M., Schwikowski, B., Tompa, M.: An exact algorithm to identify motifs in orthologous sequences from multiple species. Proc Int Conf Intell Syst Mol Biol **8** (2000) 37–45
8. Blanchette, M.: Algorithms for phylogenetic footprinting. In: Proceedings of the Fifth International Conference on Computational Molecular Biology (RECOMB-01), New York, ACMPress (2001) 49–58
9. Moses, A., Chiang, D., Eisen, M.: Phylogenetic motif detection by expectation-maximization on evolutionary mixtures. (2004) 324–335
10. Siddhartan, R., van Nimwegen, E., Siggia, E.D.: PhyloGibbs: A Gibbs sampler incorporating phylogenetic information. In E, E., (eds), W.C., eds.: RECOMB 2004 Satellite Workshop on Regulatory Genomics. (2005) 30–41
11. Li, X., Wong, W.: Sampling motifs on phylogenetic trees. Proc Natl Acad Sci USA **102** (2005) 9481–9486
12. Wray, G.A., Hahn, M.W., Abouheif, E., Balhoff, J.P., Pizer, M., Rockman, M.V., Romano, L.A.: The evolution of transcriptional regulation in eukaryotes. Mol Biol Evol **20** (2003) 1377–419
13. Moses, A.M., Chiang, D.Y., Kellis, M., Lander, E.S., Eisen, M.B.: Position specific variation in the rate of evolution in transcription factor binding sites. BMC Evol Biol **3** (2003) 19
14. Liu, X., Brutlag, D., Liu, J.: BioProspector: discovering conserved DNA motifs in upstream regulatory regions of co-expressed genes. (2001) 127–138
15. Liu, J.: The collapsed Gibbs sampler with applications to a gene regulation problem. Journal of the American Statistical Association **89** (1994)
16. Jensen, S.T., Liu, J.S.: Biooptimizer: a bayesian scoring function approach to motif discovery. Bioinformatics **20** (2004) 1557–64
17. Vavouri, T., Elgar, G.: Prediction of cis-regulatory elements using binding site matrices–the successes, the failures and the reasons for both. Curr Opin Genet Dev **15** (2005) 395–402
18. Tompa, M., Li, N., Bailey, T., Church, G., De Moor, B., Eskin, E., Favorov, A., Frith, M., Fu, Y., Kent, W., Makeev, V., Mironov, A., Noble, W., Pavesi, G., Pesole, G., Regnier, M., Simonis, N., Sinha, S., Thijs, G., van Helden, J., Vandenbogaert, M., Weng, Z., Workman, C., Ye, C., Zhu, Z.: Assessing computational tools for the discovery of transcription factor binding sites. Nat Biotechnol **23** (2005) 137–144

19. Gilks, W.R., Richardson, S., Spiegelhalter, D.J.: Markov Chain Monte Carlo in Practice. Chapman and Hall, Boca Raton, USA (1996)
20. Liu, J.S.: Monte Carlo strategies in Scientific Computing. Springer, New York (2003)
21. Durbin, R., Eddy, S., Krogh, A., Mitchison, G.: Biological Sequence Analysis: Probablistic Models of Proteins and Nucleic Acids. Cambridge University Press, Cambridge, UK (1998)
22. Rambaut, A., Grassly, N.: Seq-Gen: an application for the Monte Carlo simulation of DNA sequence evolution along phylogenetic trees. Comput Appl Biosci **13** (1997) 235–238
23. Mazon, G., Erill, I., Campoy, S., Cortes, P., Forano, E., Barbe, J.: Reconstruction of the evolutionary history of the LexA-binding sequence. Microbiology **150** (2004) 3783–3795
24. Wingender, E., Chen, X., Fricke, E., Geffers, R., Hehl, R., Liebich, I., Krull, M., Matys, V., Michael, H., Ohnhauser, R., Pruss, M., Schacherer, F., Thiele, S., Urbach, S.: The TRANSFAC system on gene expression regulation. Nucleic Acids Res **29** (2001) 281–283
25. Erill, I., Jara, M., Salvador, N., Escribano, M., Campoy, S., Barbe, J.: Differences in LexA regulon structure among Proteobacteria through in vivo assisted comparative genomics. Nucleic Acids Res **32** (2004) 6617–6626
26. Guindon, S., Gascuel, O.: A simple, fast, and accurate algorithm to estimate large phylogenies by maximum likelihood. Syst Biol **52** (2003) 696–704
27. Hannenhalli, S., Wang, L.: Enhanced position weight matrices using mixture models. Bioinformatics **21 Suppl 1** (2005) i204–i212
28. Huson, D.: Splitstree: analyzing and visualizing evolutionary data. Bioinformatics **14** (1998) 68–73
29. Bryant, D., Moulton, V.: Neighbor-net: an agglomerative method for the construction of phylogenetic networks. Mol Biol Evol **21** (2004) 255–265

Appendix

The Predictive Update Formula. We will now derive the predictive update formula, using collapsing [15]. We use the same notations as in the article. We will show that the predictive update formula (4), can be efficiently computed by breaking it down into efficiently computable parts as follows. We will first, in (5)-(6), derive an expression for $\Pr[m^k|m^{k-1}, ..., m^1]$. This expression, (6), is a sum of products which is trivial to compute given a procedure to compute the integral in (7). This is a common integral in applications with Dirichlet distributions. In (8), we give a standard expression for it, which for our application is efficiently computable, e.g., by using the approximation described in [15]. It is also possible to construct a table for all the values of c_t and α_t, for $t \in \mathcal{N}$, involved in our application, and then compute (8) in constant time by a table lookup. Finally in (9), we give an expression for $\Pr[b^k]$ which also is standard and its derivation is very similar to that of (8).

The following expression for the predictive update formula follows from the independence of background and motif and the independence between the background distributions

$$\Pr[m^k, b^k|m^{k-1}, ..., m^1, b^{k-1}, ..., b^1] = \Pr[m^k|m^{k-1}, ..., m^1]\Pr[b^k] \ . \quad (4)$$

This can be efficiently computed by breaking it down into efficiently computable parts. The motif part of the predictive update formula is

$$
\Pr[m^k|m^{k-1}, ..., m^1] = \int_\Theta \sum_{E \in E'(S)^w} \Pr[m^k, \Theta, E|m^{k-1}, ..., m^1] \, d\Theta
$$

$$
= \int_\Theta \sum_{E \in E'(S)^w} \frac{\Pr[m^k, ..., m^1, \Theta, E]}{\Pr[m^{k-1}, ..., m^1]} \, d\Theta
$$

$$
\propto \int_\Theta \sum_{E \in E'(S)^w} \Pr[m^n, ..., m^1|\Theta, E]\Pr[\theta, E] \, d\Theta
$$

$$
= \int_\Theta \sum_{E \in E'(S)^w} \prod_{i=1}^k \Pr[m^i|\Theta, E]\Pr[\Theta]\Pr[E] \, d\Theta \qquad (5)
$$

where $\Theta = \{\theta^{1,1}, \theta^{1,2}, ..., \theta^{w,1}\theta^{w,2}\}$ and $E'(S) = E(S) \cup \{\varepsilon\}$. In the MY model, the prior $\Pr[E]$ is uniform, i.e., each component e of E is selected uniformly from $E(S) \cup \{\varepsilon\}$. Disregarding the constant factor and expressing the probabilities $\Pr[m^i|\Theta, E]$ columnwise, we obtain

$$
\int_\Theta \prod_{l \in [w]} \sum_{e \in E'(S)} \prod_{j \in [2]} \prod_{i \in (C_j(S \backslash e) \cap [k])} \Pr[m_l^i|\theta^{l,j}, e]\Pr[\theta^{l,j}] \, d\Theta =
$$

$$
\prod_{l \in [w]} \sum_{e \in E'(S)} \prod_{j \in [2]} \prod_{i \in (C_j(S \backslash e) \cap [k])} \int_\theta \Pr[m_l^i|\theta]\Pr[\theta|\alpha^{l,j}] \, d\theta \ . \qquad (6)
$$

We will below drop the superscripts for parameters to the Dirichlet distribution to increase readability. Considering that m_l^i is Bernoulli-DNA with parameters θ such that $\Pr[\theta|\alpha^{l,j}] \in Dirichlet(\alpha^{l,j})$ and $\alpha^{l,j} = \{\alpha_A^{l,j}, ..., \alpha_T^{l,j}\}$, we obtain the following equality

$$
\prod_{i \in (C_j(S \backslash e) \cap [k])} \int_\theta \Pr[m_l^i|\theta]\Pr[\theta|\alpha] \, d\theta = \frac{\Gamma(\sum_{t \in \mathcal{N}} (\alpha_t))}{\prod_{t \in \mathcal{N}} \Gamma(\alpha_t)} \int_\theta \prod_{t \in \mathcal{N}} \theta^{c_t + \alpha_t - 1} \, d\theta \qquad (7)
$$

where, for each $t \in \mathcal{N}$, $c_t = |\{i \in (C_j(S \backslash e) \cap [k]) : m_l^i = t\}|$. The integral equals one over the normalizing constant for a Dirichlet distribution with parameters $c + \alpha$ giving

$$
\frac{\Gamma\left(\sum_{t \in \mathcal{N}} (\alpha_t)\right)}{\prod_{t \in \mathcal{N}} \Gamma(\alpha_t)} \frac{\prod_{t \in \mathcal{N}} \Gamma(c_t + \alpha_t)}{\Gamma\left(\sum_{t \in \mathcal{N}} (c_t + \alpha_t)\right)} \qquad (8)
$$

which thus describes the probability for one subtree, one column and one edge. Similarly, the following expression can be obtained for $\Pr[b^k]$

$$
\Pr[b^k] \propto \frac{\Gamma\left(\sum_{t \in \mathcal{N}} \beta_t^k\right) \prod_{t \in \mathcal{N}} \Gamma\left(\beta_t^k + c_t\right)}{\prod_{t \in \mathcal{N}} \Gamma\left(\beta_t^k\right) \Gamma\left(\sum_{t \in \mathcal{N}} \beta_t^k + c_t\right)} \qquad (9)
$$

where c_t is the number of occurrences of t in b^k. This way we can calculate (6) efficiently.

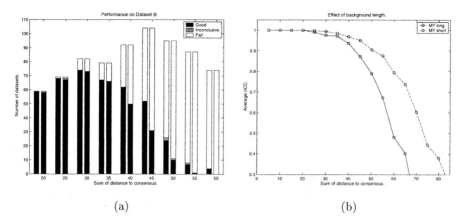

(a)   (b)

Fig. 4. Dataset B have the exact same motif instances as Dataset A, but with a shorter background, making the total length of each sequence 50 nucleotides. (a) Performance of the MY sampler, (left bars), and the GMS, (right bars). (b) Performance of the MY sampler on Dataset B, evaluated by average nCC in each bin. Performance on Dataset A has been included for comparison. The MY sampler still outperforms the GMS by approximately the same ratio.

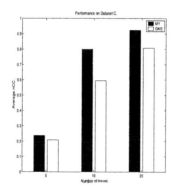

Fig. 5. Performance according to nCC of the GMS and the MY sampler on Dataset C. For each of three random trees, with 5, 10 and 20 leaves, 50 datasets were generated as follows. For each motif column an edge was randomly picked, and for subtree 1 of that edge, one nucleotide was uniformly selected, but with a 70% chance two nucleotides were picked, and randomly assigned to the leaves of the subtree. For subtree 2, a nucleotide was randomly selected among the ones not used in subtree 1. The motif length was 8 and background length was 42 nucleotides. Since half the edges are adjacent to a leaf, which would result in only one nucleotide generated from a different distribution, we required the smallest subtree to be bigger than a constant. For the trees with 10 and 20 leaves we used a constant of 2 and 4, respectively. A dataset with 20 leaves could still, have as many as 16 nucleotides in a motif column completely conserved. Both sampler will therefore benefit from larger trees.

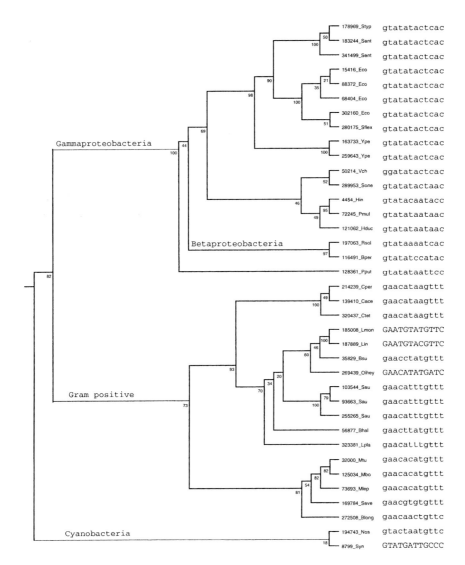

Fig. 6. Topology of the phylogenetic tree, with bootstrap support, of orthologs to *E. coli* LexA along with a motif alignment found by the MY sampler (MY model gives it probability 2.4×10^{-110}). Lower-case letters indicate true motif instances and upper-case letters indicate background. Notice that the almost perfect palindromic sequences it finds in two *Listeria* species, both are consistent with the two core parts, i.e., the first 4 and last 4 nucleotides, except for one C to T mutation. The probability of the true motif alignment is 1.11×10^{-162} for the GMS model and 1.39×10^{-104} for the MY model.

A Study of Accessible Motifs and RNA Folding Complexity

Ydo Wexler[1,*,**], Chaya Zilberstein[1,*], and Michal Ziv-Ukelson[2,*]

[1] Dept. of Computer Science, Technion - Israel Institute of Technology, Haifa 32000, Israel
{ywex, chaya}@cs.technion.ac.il
[2] School of Computer Science, Tel-Aviv University, Tel-Aviv 69978, Israel
michaluz@post.tau.ac.il

Abstract. mRNA molecules are folded in the cells and therefore many of their substrings may actually be inaccessible to protein and microRNA binding. The need to apply an accessability criterion to the task of genome-wide mRNA motif discovery raises the challenge of overcoming the core $O(n^3)$ factor imposed by the time complexity of the currently best known algorithms for RNA secondary structure prediction [24, 25, 43].

We speed up the dynamic programming algorithms that are standard for RNA folding prediction. Our new approach significantly reduces the computations without sacrificing the optimality of the results, yielding an expected time complexity of $O(n^2\psi(n))$, where $\psi(n)$ is shown to be constant on average under standard polymer folding models. Benchmark analysis confirms that in practice the runtime ratio between the previous approach and the new algorithm indeed grows linearly with increasing sequence size.

The fast new RNA folding algorithm is utilized for genome-wide discovery of accessible cis-regulatory motifs in data sets of ribosomal densities and decay rates of *S. cerevisiae* genes and to the mining of exposed binding sites of tissue-specific microRNAs in *A. Thaliana*.

Further details, including additional figures and proofs to all lemmas, can be found at: http://www.cs.tau.ac.il/~michaluz/ QuadraticRNAFold.pdf

1 Introduction

The brief "lives" of messenger RNAs (mRNAs) begin with transcription and ultimately end in degradation. During their "lives", mRNAs are translated into proteins. This whole process is regulated in a highly organized fashion to ensure that specific genes are expressed at the appropriate times and levels in response to various genetic and environmental stimuli [11, 35]. It is well-known that mRNA decay and translation are affected by cis-regulatory motifs within mRNAs. These motifs serve as binding sites for trans-regulatory proteins and microRNAs[1]. Several cis-regulatory RNA motifs were previously discovered experimentally, such as AREs (AU-Rich Elements) [28, 40], which

---

* These authors contributed equally to the paper.
** To whom correspondence should be addressed.
[1] microRNAs are single-stranded RNA molecules, typically 20-25 nucleotides long, that bind to a target mRNA and induce quick mRNA degradation or inhibit protein translation [21, 31].

A. Apostolico et al. (Eds.): RECOMB 2006, LNBI 3909, pp. 473–487, 2006.

are mRNA destabilizing elements involved in mRNA decay, and TOPs [13, 36], which control the translation of ribosomal proteins and elongation factors.

Recently, new and interesting data has become available which measures, on a genome-wide scale, the ribosomal densities of mRNAs which reflect translation rates [3], and additional data that measures mRNA decay rates [37]. The results of these measurements, if incorporated with genome-wide mRNA sequences, may reveal a wealth of novel cis-regulatory elements underlying both processes. However, since RNA elements are characterized by both *primary sequence* and higher order *structural constraints*, the identification of RNA elements is more complicated than identification of DNA elements. During the last decade, many computational efforts have been made to develop tools for the identification of RNA elements that are common to a group of functionally or evolutionarily related genes. Some of these methods rely on a first step that involves multiple alignment [2] and require that the sequences be highly similar to begin with, while other methods can detect locally conserved RNA sequence and structure elements in a subset of unaligned sequences [16, 26]. However, the complexity of these methods makes their application impractical for handling the large number of sequences involved in eukaryotic genome-wide analysis. Nevertheless, it turns out that most of the RNA regulatory motifs discovered so far are simple stem and loop structures with a consensus motif residing in the loop area (e.g. IRES) [13, 36].

Further note that the focus on *local* 2D structural conservation ignores the *global* consideration of whether or not the primary sequence sites are indeed accessible to protein binding. In order to allow the binding between the target cis-regulatory motif and the trans-regulatory proteins or microRNAs, the base pairs in the motif must be free of any other chemical bond. This is due to the fact that the chemical recognition is based on an interaction between amino acids residing in the protein and the corresponding nucleotides in the cis-regulatory motif residing in the mRNA [6], or on base pairing between the microRNA sequence and the motif nucleotides.

The above requirement for chemical availability of motifs to protein binding calls for the formalization of an accessability criterion:

Definition 1 ("accessible" substring). *Let S be a sequence and s a region i.e. substring in S. We say that s is **accessible** iff the following two conditions apply:*

1. *There exists a 2D structure of S with predicted free energy G_1 in which none of the nucleotides of s is engaged in base pairing.*
2. *$G_1 - G_0 \leq \delta$, where δ is a user defined threshold parameter, and G_0 is the optimal folding free energy of the full string S.*

In this paper we suggest a novel approach to the genome-wide discovery of RNA cis-regulatory motifs. In our framework, motifs are scored according to their statistical significance when applying the above accessibility criterion. In order to accommodate this, the input mRNA sequences are first filtered according to Definition 1. This is done in order to reduce the noise created by motifs which are not exposed to trans-regulatory binding (see Figure 1).

"Accessible site" criteria have been previously employed both in the context of microRNA target prediction [27] and in antisense oligonucleotide hybridization predictions [4, 15, 23, 30]. Nevertheless, in the antisense prediction application only a

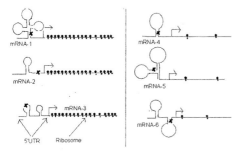

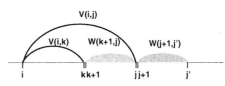

Fig. 2. The competition between candidate $V(i,k)$ and candidate $V(i,j)$ for the minimal $W(i,j')$. Candidate $V(i,k)$ has an advantage over candidate $V(i,j)$ in the additional potential cost for segment $s_{j+1}\ldots s_{j'}$ since it has a wider left-scope for combining this segment in a structure with $W(k+1,j)$. Therefore, if $V(i,k)+W(k+1,j) \leq V(i,j)$ then by triangle inequality $V(i,k)+W(k+1,j') \leq V(i,j)+W(j+1,j')$.

Fig. 1. Applying the accessibility criterion to genome-wide motif discovery in mRNA ribosomal density data. The motif X may be predicted to differentiate between the set of mRNAs with high density (left) and the set with low density (right) since its occurrences in mRNAs 1,4,5 and 6 are inaccessible.

single target, the mRNA corresponding to the gene which is targeted for "knock out", needs to be scanned for accessible sites. For this task, the current RNA folding prediction tools are sufficient. However, such tools could not be practically scaled up to serve whole genome motif discovery, where thousands of mRNAs need to be mined for accessible sites, without raising severe efficiency problems: the complexity of RNA structure prediction allowing multiple loops but no pseudoknots is $O(n^3)$ to begin with, where n is the size of an RNA sequence (typically ~ 2000). This complexity is further increased to $O(n^3 \cdot m)$ by the need to exhaustively run a sliding window across the input sequences, where $m = O(n)$ is the number of different starting positions of accessible regions that need to be considered in each gene. Note that the sliding window computational challenge is not addressed by Robins et al. [27], where the computation is simplified by the fact that only a single optimal folding is computed per gene. Thus, the task of mining accessible sites for genome-wide motif discovery creates a heavy $O(n^3 \cdot g \cdot m)$ bottleneck in terms of computational complexity, where g is the number of genes in the genome under study (typically in the thousands).

The practical considerations raised by such a complexity are exemplified as follows: suppose the genome under study contains 6000 mRNA sequences, of size ~ 2000 nucleotides each, in which we need to consider all potential sites obtained by sliding a window of size $k \ll 2000$. Given that the folding prediction computation for each sequence takes about twenty seconds[2]: the total time needed for the computation of all relevant accessible sites in this case would be $6000 \cdot 2000 \cdot 20$ seconds ≈ 7.61 years! Further note that even if we confine our search to ~ 300 windows in the UTR regions, the time needed still sums up to more than a year. This example demonstrates the need for efficient folding algorithms, especially when dealing with whole-genome scale data.

---

[2] The average folding time for this estimation was measured using the RNAFOLD program in Vienna package 1.4 on a 2G Hrz PC with 1G RAM and the average was taken over 100 random sequences of size 2000nt each.

Could the classical $O(n^3)$ algorithms for RNA secondary structure prediction [25, 43], which have been heavily used by the bioinformatics community in the last two decades, still be substantially sped up? Furthermore, could such a speed up be implemented via a practical, low-constant algorithm?

These important challenges are addressed in the rest of this paper, where we describe a new dynamic programming algorithm that exploits the combination of two properties to speed up RNA secondary structure prediction: one is the observed *triangle inequality* property of the matrices commonly used in RNA secondary structure prediction (Section 2.2), and the other is the *polymer-zeta* behavior of RNA folding with respect to increased sequence size (Section 2.4). These observations are utilized here via a simple candidate list algorithm, called Algorithm CANDIDATEFOLD (Section 2.3), which significantly reduces the computations without sacrificing the optimality of the results (no heuristics are used). The expected time complexity of Algorithm CANDIDATEFOLD is $O(n^2\psi(n))$ instead of the previously known $O(n^3)$, where $\psi(n)$ is shown to converge to a constant under models previously described for RNA folding and re-validated by our simulations (see Section 2.5). Furthermore, due to the simplicity of Algorithm CANDIDATEFOLD, it is indeed much faster than the classical algorithm in practice, as supported by experimental performance results in Section 3. Clearly, this new algorithm for speeding up RNA folding prediction is applicable to a wide range of additional biological applications, especially to those that require a substantial amount of RNA folding computations.

Based on the efficient new RNA folding algorithm CANDIDATEFOLD, we conducted a study which examines the contribution of the "accessible site" criterion to the discovery of RNA motifs that would otherwise be obscured by noise. The new approach was applied to quantitative data sets of ribosomal densities and decay rates of almost all (i.e. ~ 6000) *S. cerevisiae* genes. By applying our approach, some biologically interesting and statistically significant motifs were discovered (Section 5). For example, the p-value of the motif $AGCKTTA$ in the decay rates data was $5 \cdot 10^{-7}$. This p-value was due to the fact that the average half-life (i.e. $log(2)$/decay rate) of 24 genes that were found to contain this motif in an accessible substring was 26 days, while the half-life of the background population was 15 days. Relaxing the accessibility criterion lowered the significance of the motif by raising its p-value to 0.008.

We also employed the "accessible target" criterion to analyze microRNAs regulating tissue specific processes in *A. Thaliana*. Interesting tissue specific microRNAs were discovered (see Fig. 4).

2   The Accessible Site Prediction Engine

2.1   Preliminaries of RNA Folding Prediction Via Minimum Energy

RNA is typically produced as a single stranded molecule which then folds intramolecularly to form a number of short base-paired stems. This base-paired structure is called the *secondary structure* of the RNA. Base pairs almost always occur in a nested fashion in RNA secondary structure. Informally, this means that if we draw arcs over an RNA sequence connecting base pairs, none of the arcs cross each other. When nonnested base pairs occur, they are called *psuedoknots*. Most of the dynamic programming algorithms which are standard for RNA structure prediction do not deal with pseudo-knots. This is done mostly in order to simplify the problem and is justified by the fact that short pseudoknots do not contribute much to the overall energy and long pseudo-knots are kinetically difficult to form [20]. Therefore, in this paper we assume that no two arcs cross, however multiple loops are indeed allowed.

Under the above assumptions, a model was proposed in Tinoco et al. [32] to calculate the stability (in terms of free energy) of a folded RNA molecule by adding independent contributions from base pair stacking and loop-destabilizing terms from the secondary structure. This model has proven to be a good approximation of the forces governing RNA structure formation, thus allowing fair predictions of real structures by determining the most stable structures in the model of a given sequence. Based on this model, algorithms for computing the most stable structures have been proposed (Nussinov and Jacobson, 1980 [25]; Zuker and Steigler, 1981 [43]), and various tools for RNA secondary structure prediction were developed. The tools commonly used today are MFOLD [42], Vienna Package [14] and FOLDRNA [41].

The thermodynamic parameters used by our accessible site prediction engine are experimentally derived and are identical to those used by the RNA folding tools listed above, where the following four recursions are combined to model RNA secondary structure folding. Note that the recursions depend on the nature of the energy rules for loops, where $eh(i, j)$ is the energy of the hairpin loop closed by the base pair i, j, $es(i, j)$ is the energy of the stacked pair i, j and $i + 1, j - 1$ and $ebi(i, j, i', j')$ is the energy of a bulge or an interior loop closed by i, j with i', j' accessible from i, j. Also note the boundary conditions $W(i, j) = V(i, j) = +\infty$ if $j - i < 4$. More detailed recursions, based on the ones given here, take into consideration exterior base stacking [43]. These are not elaborated here for the sake of simplicity of presentation, however the same reasoning applies to this extension as well. The recursion equations are explicated below:

$$W(i, j) = \min\{V(i, j), W(i+1, j), W(i, j-1), \min_{i \leq k < j}\{W(i, k) + W(k+1, j)\} \quad (1)$$

Eq. 1 computes the optimal folding of substring $s_i, \ldots, s_j$, which is the value of the entry in row i and column j of the main, upper-triangular DP table W. The computation of this table involves the matrix V whose entries are computed via the following equations.

$$V(i, j) = \min\{eh(i, j), es(i, j) + V(i+1, j-1), VBI(i, j), VM(i, j)\} \quad (2)$$

Eq. 2 computes the optimal folding energy of a substring $s_i \ldots s_j$ in which s_i base pairs with s_j.

$$VBI(i, j) = \min_{i < i' < j' < j}\{ebi(i, j, i', j') + V(i, j)\} \quad (3)$$

Eq. 3 computes the score of an optimal folding of substring $s_i, \ldots, s_j$ given that there is an internal loop formed at indices (i, i', j', j).

$$VM(i, j) = \min_{i \leq k < j-1}\{W(i+1, k) + W(k+1, j-1)\} + a \quad (4)$$

where a is a constant multi-branch penalty.

Time Analysis of the Classical RNA Folding Prediction Engine. The above recursions are implemented by maintaining four tables of size $O(n^2)$ each. Eq. 1 is clearly $O(n^3)$. Given the values computed for Eq. 1, the values for Eq. 4 can be computed in $O(n^2)$ time and space via direct look-up of the minima values previously computed for Eq. 1. Eq. 2 is also $O(n^2)$.

Eq. 3 for the computation of internal loop size energies is naively $O(n^4)$. Practically, it is standard to assume that RNA interior loop size is bounded by a constant (15 nt in room temperature and up to 30 nt in extreme heat). The program RNAFOLD in Vienna package [14] as well as the MFOLD program [42] use constant gap size in both directions to reduce the complexity of Eq. 3 to $O(n^2)$. Lygnso *et. al.* [22] show how to reduce the complexity of this equation to $O(n^3)$ without binding the gap size. On the theoretical front, Waterman and Smith showed how to compute internal loops in $O(n^3)$, assuming that the loop penalty is a function of its size [34]. Eppstein, Galil and Giancarlo [7, 9] considered loop destabilizing functions satisfying certain convexity or concavity conditions, and developed an $O(n^2 \log^2 n)$ algorithm for this case. This was later improved to $O(n^2 \log n)$ [1], and finally to $O(n^2 \alpha(n))$ (where α is the inverse of Ackerman's function) for logarithmically growing destabilizing functions [19].

Conclusion 1. *The $O(n^3)$ bottleneck to RNA Folding Prediction complexity is based on the computation of the minimization term* $\min_{i \leq k < j} \{W(i, k) + W(k+1, j)\}$ *in Eq. 1.*

Note that the $O(n^3)$ bound applies to both the *worst case* and the *expected case* time complexities of the classical RNA folding algorithm, since Eq. 1 is called $O(n^2)$ times and each call involves the computation of the minimum over $O(n)$ elements on average.

2.2 Triangle Inequality in the Context of Dynamic Programming

In this section we formalize the *triangle inequality* property in the context of dynamic programming tables and show that the main matrix W, which is the final output of the RNA folding recursions given in the previous section, obeys this property. Let M be a $n \times n$ matrix in which each entry $M(i, j)$ $(i \leq j)$ is computed by the following formula:

$$M(i, j) = \min_{i < k \leq j} \{M(i, k) + M(k+1, j)\}$$

The well-known inverse quadrangle inequality property [10] is defined as follows.

Definition 2. *A matrix M obeys the* **inverse quadrangle inequality** *condition iff*

$$\forall \ i < i' < j < j' \qquad M(i, j') \leq M(i, j) + M(i', j') - M(j', j)$$

Both the quadrangle and the inverse quadrangle inequalities have previously been used to speed up dynamic programming [5, 10]. However, both the quadrangle inequality and the inverse quadrangle inequality are strong constraints on the input behavior, and do not apply to the matrix computed for RNA folding (see Eqs. 1- 4 above). However, a special weaker case of the inverse quadrangle inequality, the *triangle inequality* property, which is much more common in practice in various applications, will be used in this paper to speed up RNA folding prediction.

Definition 3. *A matrix M obeys the* **triangle inequality** *property iff*

$$\forall \ i < j < j' \qquad M(i, j') \leq M(i, j) + M(j + 1, j').$$

2.3   A Simple 1D Candidate List Approach to the Construction of W

Let $S = s_1 \ldots s_n$ denote a given RNA sequence. The next two definitions describe specific folding concepts that will be used in the description of the new algorithm.

Definition 4 (Structure). *A* **structure** *over a sequence* $s_i \ldots s_j$ *is a folding in which* s_i *base pairs with* s_j.

Definition 5 (Partition Point). *A* **partition point** *in a given folding of* $S = s_1 \ldots s_n$ *is an index* k, *such that there is no structure over* $s_i \ldots s_j$ *in this folding, where* $1 \leq i \leq k$ *and* $k < j \leq n$.

In this section we describe an alternative approach to the computation of W, which prunes Eq. 1. Similarly to the standard algorithm, the new algorithm computes the values of W row by row, in bottom-up order (decreasing row index). For each row i of W, the entry $W(i, j)$ is computed in left-to-right order (increasing column index). However, the suggested new algorithm, called CANDIDATEFOLD, differs from the original one in the application of Eq. 1 to the computation of $W(i, j)$. In a given row i, instead of considering $O(n)$ possible partition points for each column j in Eq. 1, the new algorithm only considers a list of candidate partition points, which are maintained in the form of a simple candidate list. In the following sections we show that the expected maximal size of this candidate list for an n-sized sequence, denoted $\psi(n)$, is constant.

In order to clearly define the properties that make a potential partition point a qualified candidate, we first need to simplify Eq. 1. Note that, if the main diagonal $W(r, r)$ was set to zero, then the two terms $W(i + 1, j)$ and $W(i, j - 1)$ in Eq. 1 could be embedded into the minimization term as special cases. $W(i + 1, j)$ would then be obtained as a special case $k = i$ to yield the sum $W(i, i) + W(i + 1, j)$ which is exactly $W(i + 1, j)$; similarly, $W(i, j - 1)$ would be obtained as the special case $k = j - 1$ to yield the sum $W(i, j - 1) + W(j, j)$ which is exactly $W(i, j - 1)$. However, the problem is that setting $W(r, r) = 0$ would contradict the boundary conditions set by Zuker and Stiegler [43], which assume that $W(r, r) = \infty$.

Therefore, we add two auxiliary matrices, denoted W' and V', computed via the recursions as given below, where Eq. 7 replaces the previous Eq. 1. Note that the matrix W' is added in order to get around the above boundary condition problem, while matrix V' serves to simplify the presentation of the algorithm which is described in the next section.

$$W(i, j) = W'(i, j) \ \forall j \geq i + 4 \tag{5}$$

$$V'(i, j) = V(i, j) \ \forall j \geq i + 4 \tag{6}$$

$$W'(i, j) = \min\{V'(i, j), \min_{i \leq k < j}\{W'(i, k) + W'(k + 1, j)\}\} \tag{7}$$

The matrices W' and V' are initialized as follows. $W'(i, j) = V'(i, j) = +\infty$ if $0 < j - i < 4$, and $W'(i, i) = V'(i, i) = 0$. In this formulation, the matrix W' preserves the minimum energy values of W everywhere except in the main diagonal entries. The correctness of this re-formulation is asserted via the following claim.

Claim. The values of $W(i,j)$ and $V(i,j)$, as computed via Eqs. 2-7, are identical to those obtained when using Eqs. 1-4.

The next claim is immediate from Definition 2 and Eq. 7.

Claim. The matrix W', as computed by Eq. 7, obeys the triangle inequality.

The above claim is used in the next lemma to show that any sum which yields the minimum of Eq. 7 can be reformulated as a corresponding, equal-scoring sum, in which the left term is a structure (see Definition 4).

Lemma 1. *Consider Eq. 7. For every entry $W'(i,j)$, if there exists an index k, $i \leq k < j$, such that $W'(i,j) = W'(i,k)+W'(k+1,j)$, then $W'(i,k') = V'(i,k')$ for some index $k' \leq k$.*

According to Lemma 1, Eq. 7 can be reformulated as follows.

$$W'(i,j) = \min\{V'(i,j), \min_{i \leq k < j}\{V'(i,k) + W'(k+1,j)\}\} \qquad (8)$$

Naively, after the transformation to Eq. 8, there are still n candidate partition points which compete for the optimal score in the minimization term. However, the next theorem exposes a dominance relationship between these candidates (see Figure 2).

Theorem 1. *If $V'(i,j) \geq V'(i,k) + W'(k+1,j)$ for some $i < k < j$. Then,*

$$\forall j' > j \qquad V'(i,j) + W'(j+1,j') \geq V'(i,k) + W'(k+1,j').$$

Theorem 1 exposes redundancies in the $O(n)$ computation of Eq. 8, which could be avoided by maintaining a list of only those candidates that are not dominated by others.

Definition 6 (candidate). *A column index j is a **candidate** in a row $i \leq j$ iff $V'(i,j) < W'(i,k) + W'(k+1,j) \ \forall i \leq k < j$.*

The above definition can be applied to speed up the computation of $W'(i,j)$, as follows: rather than considering all possible n partition point indices for the computation of Eq. 7, one could query the list that contains only partition points that satisfy the candidacy criterion according to Definition 6. This is formalized in the following equation,

$$W'(i,j) = \min\{V'(i,j), \min_{\forall k \in candidate\_list}\{V'(i,k) + W'(k+1,j)\} \qquad (9)$$

Eq. 9 is implemented via a candidate list that is empty at the start of each row and is extended throughout the left-to-right computation of row i by appending only those partition points which are candidates by Definition 6. Each partition point is considered for candidacy once per row, when its column is reached. The psuedo-code for the algorithm for computing Eq. 7, denoted *Algorithm* CANDIDATEFOLD, is given below.

Algorithm CANDIDATEFOLD:

```
0    for each row i := n to 1 do
1           candidate_list ← NULL
2           for each column j := i to n do
```

3        $W'(i, j) \leftarrow \min_{\forall k \in candidate\_list}\{V'(i, k) + W'(k+1, j)\}$

4   *if* $(V'(i, j) < W'(i, j))$ *then*

5                $W'(i, j) \leftarrow V'(i, j)$

6   Append j to the *candidate_list*

Expected Case Time Analysis of the Improved RNA Folding Prediction Engine.
Let $\psi(n)$ denote the expected maximal size of the candidate list in a sequence of size n.
Algorithm CANDIDATEFOLD computes each entry in the n^2-sized energy-matrix W'.
Each such calculation requires the computation of Eq. 9, where the major work is that
of computing the minimum among $O(\psi(n))$ candidates. All other recursions remain
unchanged. Therefore, the overall average time complexity is $O(n^2 \cdot \psi(n))$ if the standard bound on interior loop size is followed, or otherwise $O(n^2 \cdot \max\{\psi(n), \alpha(n)\})$,
where $\alpha(n)$ is the inverse ackerman function.

In the next sections we analyze the expected growth of the candidate list size with
respect to increasing sequence size and assert the surprising fact that $\psi(n)$ converges
to a constant. This leads to the conclusion that Algorithm CANDIDATEFOLD improves
the standard $O(n^3)$ classical algorithm (analyzed in section 2.1) by a linear factor on
average.

2.4 The Polymer-Zeta Property of RNA Folding

The *polymer-zeta* property is defined as follows.

Definition 7. *Let* $P(i, j)$ *denote the probability of a structure over the substring*
$s_i \ldots s_j$ *under a given set* Λ *of folding rules, where* $j - i = m$. *We say that* Λ *fol-*
lows the **polymer-zeta** *property if* $P(i, j) = b/m^c$ *for some constants* $b, c > 0$.

Previous work shows that RNA, which folds like other polymers, obeys the polymer-
zeta property, namely, the probability that a structure is formed over the subsequence
between two positions distant m monomers apart is $P(m) = b/m^c$ where $b = 1$ and
$c > 1$ [17, 18]. This fact is explained by modeling the 2D folding of a polymer chain
as a self-avoiding random walk (SAW) in a 2D lattice [33]. In this model the spacial
position of every nucleotide in the original polymer corresponds to a random step in
the lattice, where edges of the lattice represent possible transition directions. Since this
model of polymer folding also ignores pseudoknots, the walk is called "self avoiding",
i.e. an assumption is followed that the walk does not intersect the prefix of the chain. The
query of interest here is the probability that the mth step in the self avoiding random
walk occupies the same vertex in the lattice as the origin. The theoretical exponent
for the two dimensional SAW model is known to be $c = 1.5$ [8]. This is supported in
practice by simulations for collapsing polymers of sequence size up to 3200, as reported
in [17]. These simulations exhibited an exponent of 1.375 at low temperatures and 1.571
in higher temperatures.

    Our dynamic programming algorithm follows the thermodynamic rules defined by
Mathews *et al.* [24], which were derived experimentally to model RNA folding. We
ran our own simulations in order to assert that this model indeed follows the previously

analyzed single structure formation probabilities in polymer folding, which were found to obey the polymer-zeta property. We used 50,000 mRNA sequences with an average length of 1992 nucleotides from the NCBI databases and found that the probability that the optimal folding forms a structure over $s_i \ldots s_j$, where $m = j - i$, is estimated to be $2.11 \cdot m^{-1.47}$. The degree exponent c was estimated in our study to be ~ 1.47 by applying standard statistical procedures (approximating the MLE parameter followed by running "Kolmogorov-Smirnov" and "chi-square" goodness-of-fit tests, using the R statistical analysis package, *http://www.r-project.org*).

2.5   Bounds on $\psi(n)$

We next analyze $\psi(n)$ based on our findings. The following observation is immediate from Lemma 1.

Observation 1. *A new candidate j is added to the candidate list, in step 6 of Algorithm* CANDIDATEFOLD, *iff the optimal predicted folding of substring $s_i \ldots s_j$ forms a single structure from index i to index j. The only exception to this case is the boundary condition candidate i, which is always added as a "virtual" structure to the list.*

Given that the probability for a new candidate situated m bases away from the start of the sequence is $b \cdot m^{-c}$, the expected number of candidates in a sequence of length n is $\psi(n) = b \sum_{i=1}^{n} i^{-c}$. This summation could assume one of three values, according to the estimated c:

1. For values $c \geq 1$ this series is a partial sum of the *Riemann Zeta function* defined as $\sum_{i=1}^{\infty} i^{-c}$.
   (a) If $c > 1$, this series is known to converge and thus, $\psi(n) = O(1)$.
   (b) if $c = 1$, we get a partial sum of the first n elements of the Harmonic series, which is known to be less or equal to $1 + \ln(n)$ and thus $\psi(n) = O(\log n)$.
2. if $c < 1$, we use the power means inequality to obtain the bound $\psi(n) = O(n^{1-c} (\log n)^c)$.

Theorem 2. *Applying Algorithm* CANDIDATEFOLD *to the folding of a polymer chain of size n that obeys the polymer-zeta property with $c > 1$, requires an average of $O(n^2)$ operations.*

Recall that our simulations estimate c to be 1.47, which implies that $\psi(n) \sim 2.11 \cdot 2.74 \approx 5.7$, which is a constant. Therefore, applying Algorithm CANDIDATEFOLD to the folding of an RNA sequence of size n takes $O(n^2)$ time on average.

3   The Performance of the New RNA Folding Engine

To demonstrate the power of algorithm CANDIDATEFOLD in practice we ran it against a naive version of our folding program, which predicts the minimum free energy structure using the classical algorithm of Zuker and Stiegler [43]. The data set included 150,000 sequences: 300 sequences for every possible size in the range 500-1000.

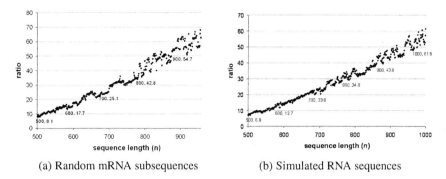

(a) Random mRNA subsequences   (b) Simulated RNA sequences

Fig. 3. The average measured run-time ratio of naive/CANDIDATEFOLD as a function of increasing sequence size

Figure 3 demonstrates that the average run time ratio (computed by dividing the run times of the classical algorithm with ours) is linear in the sequence length n, reconfirming our time complexity analysis. In Figure 3(a), the analysis was done for 100 sequences for each possible size in the range 500-1000, which were extracted as randomly chosen subsequences from 50,000 complete mRNA sequences taken from NCBI databases. The analysis shown in Figure 3(b) was done for 100 sequences of each size in the same range, which were generated using a Markov-model imitating software. This sequence-simulation program takes a set of sequences to imitate and a Markovian order as input, and generates an output of random sequences according to a Markov-model of the desired order. The input consisted of 50,000 complete mRNA sequences downloaded from the NCBI database and the Markovian order parameter was set to 6. The same results emerged when using the remaining 50,000 mRNA sequences as input for a zero order Markovian model simulator.

4   Methods for Mining Accessible Cis and Trans Regulatory Motifs

Our method for discovering novel cis-regulatory motifs incorporates large scale decay rate and ribosomal density measurements, combined with the information from mRNA sequences of the genome under study. It can be formulated as follows. Given a set of mRNAs $G = S_1 \ldots S_g$, a parameter k denoting motif window size (could be slightly longer than the motif residing in the window), and a pre-defined energy threshold δ, we apply the following simple two-stage approach:

Stage 1: Process the sequence set G to extract all "accessible" windows by running a sliding window of size k across the mRNA sequence and testing each window for compliance with Definition 1. For each shifted window this testing is conducted by masking the nucleotides inside the window in order to prevent their engagement in base pairing. Then, the minimal energy for folding the whole sequence with the masked window is computed and compared to the minimal folding energy of the original, unmasked sequence. The folding energies were computed via algorithm CANDIDATEFOLD.

Stage 2: This stage takes as input the accessible substrings, extracted in the first stage, and seeks regulatory motifs residing in the data. Two statistical techniques are applied here, depending on whether the sought motif is cis or trans regulatory:

Cis-regulatory motifs: Enumerate all motifs up to a given size k over the IUPAC alphabet [38]. For each motif use the new data created in stage 1 instead of the original genomic sequences, to compute a t-score [12] reflecting the functionality of that motif. If the p-value associated with the computed t-score is small enough, report the motif. This stage can be efficiently executed by using a variation of the algorithm of Sagot *et al.* [29] combined with the statistical computation of the t-score [38] and adapted to handle the new "accessible window" data.

Trans-Regulatory Signals (microRNAs). The search for microRNAs is similar to that of motif discovery, except for the following difference: instead of considering accessible mRNA motifs, we considered accessible sites that were predicted to hybridize well with the subject microRNAs, as described in [39].

5   A Biological Study of Accessible Regulatory RNA Elements

We conducted a study in order to test our novel approach, which applies the "accessibility" criterion to RNA motif discovery. Using various data sets, significant motifs were discovered, including some cis-regulatory degradation and translation motifs and tissue-specific microRNAs.

In each of the conducted experiments, two data sets were studied: a set containing only "accessible" substrings, according to Definition 1, and a "control" set which included the original complete mRNA sequences. A comparison of the results obtained for each of the two sets repeatedly confirms the contribution of the "accessibility" criterion as a powerful filter for masking out noise associated with inaccessible motifs and raising the significance score of otherwise invisible motifs.

Translation Related Motifs. Arava *et al.* [3] measured the ribosomal densities of almost all the mRNAs of the yeast *S. cerevisiae* under normal cell conditions, using the following method. First, mRNAs are extracted from the cells and separated by velocity sedimentation. Then, each fraction across the gradient is analyzed by microarray techniques for its mRNA content. Based on this, a fraction is assigned to each mRNA: the lower this fraction is, the higher the mRNA's ribosomal density is. We applied our approach to this data in order to detect translation cis-regulatory elements within 5' untranslated region (5'UTR)[3]. A few novel potential cis-regulatory elements were discovered that may affect translational efficiency (see Table 1). In particular, the average ribosomal density of the set of mRNAs containing the motif $AGSNNK$ in accessible substrings was low in comparison to the background. Thus, $AGSNNK$ seems to be a translation repressor.

Degradation Related Motifs. We applied our approach to the whole-genome mRNA decay rate data measured by Yang et al [37] in order to seek mRNA

---
[3] We used as 5'UTRs the regions spanning 150bp upstream to the translation start codons.

Table 1. Motifs potentially regulating mRNA translations. The accessible substring criterion was applied with window size 10 and $\delta = 2Kcal$. The average ribosomal density without the motif was computed based on ~ 5000 different genes.

Motif	Number of occurrences	Average density with the motif	Average density without the motif	p-value confined to accessible substrings	p-value in any substring	Hypothesized function
$ACASACT$	14	1.7	0.7	10^{-18}	10^{-4}	Translation enhancer
$AGSNNK$	1292	0.6	0.7	10^{-11}	10^{-3}	Translation repressor

Table 2. Motifs potentially regulating mRNA degradations. The first 3 columns refer to the case of accessible substring with window size 10 and $\delta = 2Kcal$. The average half life without the motif was computed based on ~ 5000 different genes.

Motif	Number of occurrences	Average half-life with the motif	Average half-life without the motif	p-value confined to accessible substrings	p-value in any substring	Hypothesized function
$AGCKTTA$	24	26.54	15.46	$4.83 \cdot 10^{-7}$	0.0083	Stabilizer
$GGGCYTR$	5	57.75	15.5	$2.76 \cdot 10^{-9}$	0.0081	Stabilizer
$ACMGCGT$	4	42.75	15.49	$4.84 \cdot 10^{-7}$	0.01198	Stabilizer

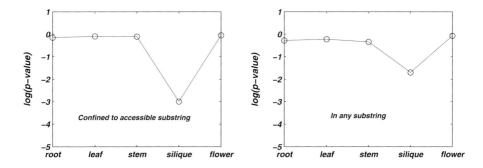

Fig. 4. miR-161 and it's p-values in different plant tissues. The accessible substring criterion was applied with window size 25 and $\delta = 6Kcal$.

stability-regulating elements within 3' UTRs[4]. We successfully identified some novel potential cis-regulatory motifs that may affect mRNA stability (see Table 2). For example, the average half-lives (i.e. $log(2)$/Decay rate) of the set of mRNAs containing the IUPAC motif $AGCKTTA$ in accessible substrings was high in comparison to the background. Thus, $AGCKTTA$ seems to be a strong mRNA stabilizer. Table 2 also demonstrates that, when relieving the accessibility criterion, the significance of the p-values substantially dropped.

Tissue Specific microRNAs. In order discover microRNAs, which are potential trans-factors influencing mRNA stabilities, we collected the genome-wide expression (measured using a microarray technique) profiles of 5 *A. Thaliana* tissues, including flowers, stems, siliques, leaf, and root. MicroRNAs with potential tissue-specific

---

[4] We used as 3'UTRs the regions spanning 150bp downstream to the stop codons of genes.

activity were discovered[5]. These microRNAs showed a significant p-value for binding in one of the tissues and non-significant p-values in the rest of the tissues. For example, the microRNA miR-161, represented in Figure 4, is specific to silique tissue. Interestingly, the figure demonstrates that in most of the tissues the $p - values$ corresponding to the first (accessible substring) and second (control) input sets are almost similar. However, in the silique tissue, where the microRNA miR-161 seems to be active, the difference between the two input sets becomes conspicuous.

Acknowledgments. Many thanks to Yoav Arava for inspiration and data, as well as for fruitful discussions. We thank Micheal Zuker for very helpful advice. The authors are also grateful to Ron Shamir, Ron Y. Pinter, Dan Geiger, Zohar Yakhini, Jeannette Schmidt, Christos Faloutsos, Eleazar Eskin and Firas Swidan for helpful discussions and comments. The research of Michal Ziv-Ukelson was supported in part by the Aly Kaufman Post Doctoral Fellowship.

References

1. A. Aggarawal and J. Park. Notes on searching in multidimensional monotone arrays. *Proc. 29th IEEE Symp. on Foundations of Computer Science*, 497–512, 1988.
2. V. Akmaev, S. Kelley, and G. Stormo. A phylogenetic approach to RNA structure prediction. *Proc Int Conf Intell Syst Mol Biol*, 235:10–17, 1999.
3. Y. Arava, Y. Wang, J. Storey, C. Liu, P. Brown, and D. Herschlag. Genome-wide analysis of mRNA translation profiles in saccharomyces cerevisiae. *PNAS*, 100:3889–3894, 2003.
4. R. Christofferson et al. Application of computational technologies to ribozyme biotechnology products. *J.Molecular Struct.(Theochem)*, 311:273, 1994.
5. M. Crochemore, G. Landau, B. Schieber, and M. Ziv-Ukelson. *Re-Use Dynamic Programming for Sequence Alignment:An Algorithmic Toolkit. String Algorithmics*. KCL Press, 2005.
6. D. Draper. Themes in RNA-protein recognition. *J Mol Biol*, 293(2):255–270, 1999.
7. D. Eppstein, Z. Galil, and R. Giancarlo. Speeding up dynamic programming. *Proc. 29th IEEE Symp. on Foundations of Computer Science*, 488–496, 1988.
8. M. Fisher. Shape of a self-avoiding walk or polymer chain. *JCP*, 44:616–622, 1966.
9. Z. Galil and R. Giancarlo. Speeding up dynamic programming with applications to molecular biology. *Theoretical Computer Science*, 64:107–118, 1989.
10. R. Giancarlo. *Dynamic Programming: Special Cases*. In Pattern Matching Algorithms, A. Apostolico and Z. Galil eds., Oxford University Press, 1997.
11. E. Goodwin, P. Okkema, T. C. Evans, and J. Kimble. Translational regulation of tra-2 by its 3' untranslated region controls sexual identity in c. elegans. *Cell*, 75:329–339, 1993.
12. C. Goulden. *Methods of Statistical Analysis*. New York: Wiley, 2 edition, 1956.
13. N. Gray and M. Wickens. *Annu Rev Cell Dev Biol*, 14:399–458, 1998.
14. I. L. Hofacker. Vienna RNA secondary structure server. *NAR*, (13):3429–3431, 2003.
15. A. Jayaraman and S.P.Walton. Rational selection and quantitative evaluation of antisense oligonucleotides. *Biochim.Biophys. Acta*, 1520:105, 2001.
16. Y. Ji, X. Xu, and G. Stormo. *Bioinformatics*, 20:1591–1602, 2004.
17. A. Kabakcioglu and A. Stella. A scale-free network hidden in the collapsing polymer. *ArXiv Condensed Matter e-prints*, Sept. 2004.

---

[5] We tested a set of ~ 100 previously discovered microRNAs retrieved from the the RFAM database http://www.sanger.ac.uk/Software/Rfam. The microarray data of the 5 tissues and the mRNA sequences were retrieved from the TAIR database http://www.Arabidopsis.org.

18. Y. Kafri, D. Mukamel, and L. Peliti. Why is the dna denaturation transition first order? *Physical Review Letters*, 85:4988–4991, 2000.

19. L. Larmore and B. Schieber. On-line dynamic programming with applications to the prediction of RNA secondary structure. *J. Algorithms*, 12(3):490–515, 1991.

20. T. Liu and R. Bundschuh. Quantification of the differences between quenched and annealed averaging for RNA secondary structures. *ArXiv Physics e-prints*, Apr. 2005.

21. C. Llave et al. Cleavage of scarecrow-like mRNA targets directed by a class of arabidopsis miRNA. *Science*, 297:2053–2056, 2002.

22. R. B. Lyngsø, M. Zuker, and C. N. S. Pedersen. An improved algorithm for RNA secondary structure prediction. Technical Report RS-99-15, brics, 1999.

23. D. Mathews et al. *RNA*, 5, 1458-1469, 1999.

24. D. Mathews, J. Sabina, M. Zuker, and D. Turner. *JMB*, 288:911, 1999.

25. R. Nussinov and A. Jacobson. Fast algorithm for predicting the secondary structure of single-stranded RNA. *PNAS*, 77(11):6309–6313, 1980.

26. G. Pavesi et al. an algorithm for finding conserved secondary structure motifs in unaligned RNA sequences. *NAR*, 32:3258–3269, 2004.

27. Robins et al. *PNAS*, 102:4006–4009, 2005.

28. J. Ross. mRNA stability in mammalian cells. *Microbiol Rev*, 59(3):423–450, 1995.

29. M. Sagot. Spelling approximate or repeated motifs using a suffix tree. *LNCS*,111-127, 1998.

30. L. Smith et al. *Eur. J. Pharm. Sci.*, 11:191, 2000.

31. G. Tang et al. framework for RNA silencing in plants. *Genes Dev*, 17:49–63, 2003.

32. I. Tinoco et al. *Nature New Biology*, 246:40–41, 1973.

33. C. Vanderzande. *Lattice Models of Polymers (Cambridge Lecture Notes in Physics 11)*. Cambridge University Press, 1998.

34. M. Waterman and T. Smith. Rapid dynamic programming algorithms for RNA secondary structure. *Adv. Appl. Math.*, 7:455–464, 1986.

35. M. Welsh, N. Scherberg, R. Gilmore, and D. Steiner. Translational control of insulin biosynthesis. *Biochem. J.*, 235:459–467, 1986.

36. G. Wilkie, K. Dickson, and N. Gray. Regulation of mRNA translation by 5'- and 3'-utr-binding factors. *Trends Biochem Sci*, 28:182–188, 2003.

37. E. Yang et al. Decay rates of human mRNAs: correlation with functional characteristics and sequence attributes. *Genome Res*, 13:1863–1872, 2003.

38. C. Zilberstein, E. Eskin, and Z. Yakhini. Sequence motifs in ranked expression data. In *The First RECOMB Satellite Workshop on Regulatory Genomics*, 2004.

39. C. Zilberstein, M. Ziv-Ukelson, R. Y. Pinter, and Z. Yakhini. A high-throughput approach for associating microRNAs with their activity conditions. In *RECOMB*, 133–151, 2005.

40. A. Zubiaga, J. Belasco, and M. Greenberg. The nonamer uuauuuauu is the key au-rich sequence motif that mediates mRNA degradation. *Mol.Cell.Biol.*, 15:2219–2230, 1995.

41. M. Zuker. Computer prediction of RNA structure. *Methods Enzymol.*, 180:262–288, 1989.

42. M. Zuker. *NAR*, (13):3406–15, 2003.

43. M. Zuker and P. Stiegler. Optimal computer folding of large RNA sequences using thermodynamics and auxiliary information. *NAR*, 9(1):133–148, 1981.

A Parameterized Algorithm for Protein Structure Alignment

Jinbo Xu[1,2,*], Feng Jiao[3], and Bonnie Berger[1,*]

[1] Department of Mathematics and Computer Science and AI Laboratory, MIT
bab@csail.mit.edu
[2] Toyota Technological Institute at Chicago, USA
j3xu@tti-c.org
[3] School of Computer Science, University of Waterloo, Canada
fjiao@cs.uwaterloo.ca

Abstract. This paper proposes a parameterized algorithm for aligning two protein structures, in the case where one protein structure is represented by a contact map graph and the other by a contact map graph or a distance matrix. If the sequential order of alignment is not required, the time complexity is polynomial in the protein size and exponential with respect to two parameters $\frac{D_u}{D_l}$ and $\frac{D_c}{D_l}$, which usually can be treated as constants. In particular, D_u is the distance threshold determining if two residues are in contact or not, D_c is the maximally allowed distance between two matched residues after two proteins are superimposed, and D_l is the minimum inter-residue distance in a typical protein. This result indicates that if both $\frac{D_u}{D_l}$ and $\frac{D_c}{D_l}$ are small enough, then there is a polynomial-time approximation scheme for the non-sequential protein structure alignment problem. Empirically, both $\frac{D_u}{D_l}$ and $\frac{D_c}{D_l}$ are very small and can be treated as constants. This result clearly demonstrates that the hardness of the contact-map based protein structure alignment problem is related not to protein size but to several parameters, which depend on how the protein structure alignment problem is modeled. The result is achieved by decomposing the protein structure using tree decomposition and discretizing the rigid-body transformation space. We have implemented our algorithm and preliminary experimental results indicate that on a Linux PC, it takes from ten minutes to one hour to align two proteins with approximately 100 residues.

1 Introduction

The structure of a protein plays an instrumental role in determining its functions. Two proteins with similar three-dimensional structure are more likely to have the same function than two without similar structure. Pairwise protein structure alignment tools routinely have been used to study the relationship between proteins. Many algorithms have been developed to solve this problem based on various alignment models [1, 2, 3, 4, 5, 6, 7, 8, 9]. Please refer to Lancia & Istrail's [10] and Lemmen & Lengauer's [11] papers for a survey on this problem. Though empirically many

---

* Corresponding authors.

A. Apostolico et al. (Eds.): RECOMB 2006, LNBI 3909, pp. 488–499, 2006.
© Springer-Verlag Berlin Heidelberg 2006

heuristic-based algorithms can generate a good alignment, there are few theoretical studies of the problem [12, 13].

In this paper, we consider only protein backbone alignment. There are two major methods to measure the similarity between two proteins: the coordinate distance-based method and inter-residue contact-based method. The first type of measure uses the Euclidean distance between two matched residues or atoms in the two proteins compared. Many programs such as STRUCTAL [4], 3dSearch [5], and VAST [6] belong to this category. To use this method, the optimal rigid-body transformation between two proteins must be determined. The other type of measure employs a contact map graph to describe the structure of a protein and compares the contact map graphs of two proteins under consideration [8, 9]. A contact in a protein is a pair of residues that are spatially close to each other. A contact map graph consists of all the residues (i.e, vertices) and their contacts (i.e., edges) and is inferred from crystal structures. Using this method, the protein structure alignment problem is often formulated as a maximum common subgraph problem. It is unnecessary to find the optimal rigid-body transformation before obtaining the best match between two proteins. Usually the rigid-body transformation is calculated after the best match is determined. A variant of a contact map representation of a protein structure is a distance matrix in which an element is the spatial distance between two residues. Two distance matrices are compared to render the best common submatrix. Several widely used protein structure alignment tools such as DALI [2], CE [14] and SARF [3] employ the distance matrix representation of a protein structure.

Previous studies show that contact map-based protein structure alignment is NP-hard and also hard to approximate [13, 15, 16], regardless of whether the alignment is sequential or non-sequential. A non-sequential alignment refers to one in which the sequential order of residues in a protein is ignored, and only the spatial proximity between two residues is taken into consideration. Many structure alignment tools support both sequential or non-sequential structure alignment [2, 17, 18, 19].

Many protein structure comparison programs such as DALI [2] use heuristic algorithms to find a good, but not the best, alignment. The advantage of these algorithms is that they are computationally efficient. While these algorithms have no performance guarantee, empirically they generate good alignment accuracy. There are also some globally optimal algorithms for this problem. For example, Lancia *et al.* [8] used a branch-and-cut method to find the optimal alignment between two proteins when a protein is modeled by a contact map. Later, Caprara & Lancia also developed a Lagrangian relaxation algorithm [9], which runs fast and sometimes can generate a globally optimal solution. The disadvantage of these algorithms is that they do not have good theoretical time complexity. Recently, Kolodny & Linial [12] proposed an interesting polynomial-time approximation scheme for this problem when a STRUCTAL-type objective function [4] (i.e., Gerstein & Levitt's coordinate distance based measurement) is used to measure the similarity between two proteins. However, there is still no good approximation algorithm in the case where the two proteins under consideration are modeled by a contact map. Instead, Goldman, Papadimitriou & Istrail have shown that, based on the maximum common subgraph formulation, the contact map-based protein structure alignment problem is hard to approximate [13]. The hardness of the protein structure alignment problem partially comes from the fact that when two contact maps are aligned, the geometric information in the protein structure is not taken into consideration.

Surprisingly, we show that this problem can be approximated within $(1 + \epsilon)$ times optimal in the case where the parameters the algorithm depends on are constant, which usually is the case. The major contribution of this paper is a parameterized algorithm for the protein structure alignment problem when one protein structure is modeled by a contact map graph and the other by a contact map or a distance matrix. Let $OPT(D_c)$ denote the optimal alignment score between two proteins where D_c is the maximally allowed distance between two matched residues after two proteins are superimposed. Our parameterized algorithm can generate a non-sequential alignment and its corresponding rigid-body transformation such that: i) the alignment score is at least $(1 - \frac{1}{k})OPT(D_c)$; ii) the distance between two matched residues is no more than $(1+\epsilon)D_c$ after two proteins are superimposed by the generated rigid-body transformation, where ϵ is a small positive number; and iii) the running time is $O(k^2 poly(n)2^{tw \lg \Delta}/(\epsilon D_c)^6)$, where $poly(n)$ is a polynomial in the protein size n, $tw = O(k^2 \frac{\max\{2D_c, D_u\}^3}{D_l^3})$, $\Delta = (1 + \frac{2D_c}{D_l})^3$, D_u is the distance threshold determining if two residues are in contact or not, and D_l is the minimum inter-residue distance in a protein. The same algorithm also works for the sequential protein structure alignment problem, although its theoretical time complexity is not as good as that of nonsequential alignment. We achieved this result by applying the following techniques: 1) instead of finding the best alignment first and then the rigid-body transformation, we simultaneously search for the best rigid-body transformation and the best alignment; 2) the whole rigid-body transformation space is discretized into a polynomial number of discrete transformations; 3) one protein structure is decomposed into small blocks and each block is aligned to another structure separately, using a tree-decomposition based method.

2   Preliminaries

Fixed-Parameter (Parameterized) Algorithm. Fixed-parameter algorithms are an approach to solving *NP*-hard problems. The time complexity of a fixed-parameter algorithm is polynomial in the problem size but exponential with respect to some parameters. If all these parameters are constants, then the fixed-parameter algorithm can terminate within polynomial time.

Polynomial Time Approximation Scheme. A polynomial-time approximation scheme (PTAS) is a type of approximation algorithm for optimization problems. For any given $\epsilon > 0$, this type of algorithm produces a solution of the optimization problem that is within an ϵ factor of the optimal. The running time of the algorithm is polynomial with respect to the problem size if ϵ is fixed. Usually, the smaller ϵ is, the greater the running time.

Protein Structure Alignment Problem. We use a contact map graph $G = (V, E)$ to model a protein structure in $\Re^3$. Each residue is represented by a vertex in V and associated with the 3D coordinates of its residue center. For each residue, we use its C_α atom as the residue center. There is a contact edge $(i, j) \in E$ between two residues i and j if and only if their spatial distance is within a given distance cutoff D_u. In a typical protein, two residues cannot be arbitrarily close, which is one of the underlying reasons why lattice models can be used to approximate protein folding. According to simple statistics on the PDB database [20], 99% of inter-residue distances are more than $3.5\mathring{A}$. Let the constant D_l $(D_l > 0)$ denote the minimum inter-residue distance in a protein. Therefore, it can be easily verified that any residue can be adjacent to at most $(1 + \frac{2D_u}{D_l})^3$ residues.

Given a protein chain A, let $G[A]$ denote its contact map graph. For a substructure P of A, let $G[P]$ denote the contact map subgraph induced by substructure P. Given two protein chains A and B, an alignment between A and B is a pair of substructures P and Q satisfying the following conditions:

- P is a substructure of A and Q of B;
- There is a one-to-one mapping between the residues in P and Q. One residue p in A is equivalent to residue q in B if and only if p is mapped to q. One contact edge in $G[P]$ is equivalent to one in $G[Q]$ if and only if their two end points are equivalent.

The optimal alignment between A and B is the alignment such that the number of equivalent contact edges is maximized. If we know the equivalent residues between A and B, then the rigid-body transformation between A and B can be calculated by the method described in Arun et al.'s paper [21]. After A and B are superimposed, the deviation between two equivalent residues cannot be too large. We use the distance parameter D_c to denote the maximum Euclidean distance between any two equivalent residues after superimposing these two proteins.

In doing protein structure alignment, we can choose to enforce the sequential order or not. If the sequential order is enforced, then for any two residues p_i and p_j in P and their equivalent residues q_i and q_j in Q, if p_i occurs before p_j along the primary sequence of A, then q_i also occurs before q_j along that of B. Some protein structure alignment tools can only generate sequential alignment [2], while some tools can generate non-sequential alignment [3, 17].

In this paper, we study the following problem.

Problem 1. Given two proteins A and B, each is represented by a contact map graph. There is a contact between two residues if their distance is no more than D_u. The optimal alignment between A and B is an alignment such that the number of equivalent contact edges is maximized and after the two proteins are superimposed, the Euclidean distance between two equivalent residues is no more than a threshold D_c.

Let $E[A]$ and $E[B]$ denote the set of contacts in proteins A and B, respectively. For any residue u in A, let $M(u)$ denote its equivalent residue in B. If there is no equivalent residue for u, then $M(u) = \phi$. The protein structure alignment problem is to maximize the following objective function:

$$\sum_{u,v \in V[A], u < v} f(u, v, M(u), M(v)) \tag{1}$$

where

$$f(u, v, M(u), M(v)) = \begin{cases} -\infty & M(u) = M(v) \neq \phi \\ 1 & (u, v) \in E[A], (M(u), M(v)) \in E[B]. \\ 0 & otherwise \end{cases} \quad (2)$$

Note that $f(u, v, M(u), M(v)) = -\infty$ is used to avoid two different residues u and v being aligned to the same residue in B.

We can further generalize the above problem to the case where protein A is represented as a contact graph and protein B as a distance matrix. That is, $f(u, v, M(u), M(v)) = h(|u - v|, |M(u) - M(v)|)$ when $(u, v) \in E[A]$ where $h(x, y)$ takes two contact distances and outputs a positive value. The closer these two contact distances, the higher the output.

The algorithm described in this paper can solve the protein structure alignment problem with Eq. 2 as the objective function. To enforce sequential order in the alignment, we can set $f(u, v, M(u), M(v))$ to be $-\infty$ if $u < v$ while $M(u) > M(v)$.

3   Structure Alignment with a Specific Transformation

In this section, we assume that the spatial positions of the two proteins are fixed and find the best mapping between them by maximizing Eq. 1.

3.1   An Exact Protein Structure Alignment Algorithm

Here we describe a tree-decomposition based algorithm for the optimal protein structure alignment problem, assuming that the positions of both proteins are fixed. This algorithm has an exponential time complexity and will be used as a subroutine of the final algorithm described in the following section. Please refer to Robertson and Seymour [22] for the definition of a tree decomposition.

In Eq. 2, in order to detect if two residues in A align to the same residue in B, we have to enumerate all the residue pairs in A. To be able to easily detect if two residues in protein A are aligned to the same residue in B or not, we extend the contact graph $G[A]$ to $G'[A] = (V[A], E'[A])$ by adding more edges to $G[A]$. Besides all the edges in $G[A]$, we add one extra edge (u, v) to $G'[A]$ if the distance between u and v is less than $2D_c$ but more than D_u. Therefore, for any two residues u and v in A, if there is no edge between them in $G'[A]$, then they cannot align to the same residue in B since the distance between two equivalent residues is no more than D_c. Using the extended graph, we can revise the objective function in Eq. 1 as follows:

$$\sum_{(u,v) \in E'[A]} f(u, v, M(u), M(v)) \quad (3)$$

where

$$f(u, v, M(u), M(v)) = \begin{cases} -\infty & M(u) = M(v) \neq \phi \\ 1 & (u, v) \in E[A], (M(u), M(v)) \in E[B] \\ 0 & otherwise \end{cases}$$

Since now we only need to enumerate all the edges in $G'[A]$ to calculate the objective function in Eq. 3, we can perform a tree-decomposition on graph $G'[A]$ and then use the same tree-decomposition based algorithm as described in the side chain packing paper [23] to maximize the objective function. Any two residues in A which might align to the same residue in B appear simultaneously in at least one tree decomposition component. So when doing calculations on this tree decomposition component, we can detect if these two residues are aligned to the same residue or not. Using the same proof technique as in paper [23], we can prove that the treewidth of $G'[A]$ is no more than $O(\frac{\max\{2D_c, D_u\}}{D_l} n^{2/3} \lg n)$. Since the distance between two matched residues is no more than D_c, each residue in A can be aligned to at most $O\left((1 + \frac{2D_c}{D_l})^3\right)$ residues in B. So we have the following theorem.

Theorem 1. *Let A and B be two protein structures in $\Re^3$. Assume that the spatial positions of A and B are fixed and the distance between two equivalent residues is no more than D_c. There is an algorithm with time complexity $O(n2^{tw \lg \Delta})$ generating the optimal non-sequential alignment between A and B, where n is the protein size, $\Delta = O\left((1 + \frac{2D_c}{D_l})^3\right)$, and $tw = O(\frac{\max\{2D_c, D_u\}}{D_l} n^{2/3} \lg n)$.*

Assume that protein A is inscribed in a minimal axis-parallel 3D rectangle and the widths along each dimension are W_x, W_y, and W_z respectively. The following lemma gives another upper bound on the running time of the tree-decomposition based algorithm. Please see the supplemental material at our website[1] for its proof.

Lemma 1. *Let A and B be two protein structures in $\Re^3$. Assume that the spatial positions of A and B are fixed and the distance between two equivalent residues is no more than D_c. There is an algorithm with time complexity $O(n2^{tw \lg \Delta})$ generating the optimal non-sequential alignment between A and B, where n is the protein size, $\Delta = O\left((1 + \frac{2D_c}{D_l})^3\right)$, and $tw = O(\frac{\max\{2D_c, D_u\}}{D_l^3} \min\{W_x W_y, W_x W_z, W_y W_z\})$.*

3.2   A PTAS for Protein Structure Alignment

In this subsection, we describe a polynomial-time approximation scheme (PTAS) for the protein structure alignment problem. The basic idea is to partition protein A into small blocks, align each block to B separately and then finally combine the alignment results. Assume that protein A is inscribed in a minimal axis-parallel 3D rectangle and the widths along each dimension are W_x, W_y, and W_z, respectively. We also use D to denote $\max\{2D_c, D_u\}$.

Theorem 2. *Let A and B be two protein structures in $\Re^3$. Assume that the spatial positions of A and B are fixed and the distance between two residues is no more than D_c. Then there is an algorithm with time complexity $O(nk2^{tw \lg \Delta})$ generating a non-sequential alignment between A and B with an alignment score at least $(1 - \frac{4}{k})$ times the best possible, where n is the protein size, k is a positive integer, $\Delta = O\left((1 + \frac{2D_c}{D_l})^3\right)$, and $tw = O(k\frac{\max\{2D_c, D_u\}^2}{D_l^3} \min\{W_x, W_y, W_z\})$.*

---

[1] http://ttic.uchicago.edu/~jinbo/StructureAlignment.htm

Proof. Without loss of generality, assume $W_x = \min\{W_x, W_y, W_z\}$. The intuition is to cut the protein structure A into non-overlapping blocks using k different partitioning schemes. Each block can be tree-decomposed into components containing no more than $O(k\frac{\max\{2D_c, D_u\}^2}{D_l^3}W_x)$ residues. Therefore, the structure alignment between each block and B can be done within time proportional to $O(\Delta^{O(k\frac{\max\{2D_c, D_u\}^2}{D_l^3}W_x)})$ where recall that Δ is the maximum number of residues in B that a residue in A can align to. We then prove that among k different partitioning schemes, at least one can give us a good structure alignment. Please see the supplemental material for an example of $k(= 3)$ different partition schemes.

Using a group of hyperplanes $y = y_j = jD$ ($j = 0, 1, ..., \frac{W_y}{D}$), we can partition the protein A into $\frac{W_y}{D}$ basic blocks along the y-axis, each of which has dimension $W_x \times D \times W_z$. Let T_j ($j = 1, 2..., \frac{W_y}{D}$) denote the set of residues contained in the basic block $\{(x, y, z)|0 \le x \le W_x, y_{j-1} \le y < y_j, 0 \le z \le W_z\}$. Let R_j denote the union of $T_{j+1}, T_{j+2}, ..., T_{j+k-1}$[2]. Let $G(R_j)$ denote the subgraph induced by R_j. Similarly, let $G(T_j)$ denote the subgraph induced by T_j plus the contact edges between T_j and its two adjacent blocks. We optimize the structure alignment using k different partition schemes and prove that at least one of them will give a good alignment. For a given partition scheme s ($0 \le s < k$), let $RS_s = \bigcup_{j:j\%k=s} G(R_j)$ [3] and $TS_s = \bigcup_{j:j\%k=s} G(T_j)$. RS_s refers to the shadowed areas and TS_s refers to the non-shadowed areas plus the edges connecting shadowed and non-shadowed areas. Each residue in protein A can only be aligned to residues in B which is no more than D_c away. Therefore, any two residues in different R_j will not be aligned to the same residue in B. We align the structure in RS_s to protein B first and then align the remaining residues to B, using our tree-decomposition based algorithm. Let $E(RS_s)$ and $E(TS_s)$ denote the optimal alignment score of RS_s and TS_s, respectively, and $E_s = E(RS_s) + E(TS_s)$. The union of RS_s and TS_s contains all the residues and inter-residue contact edges in the protein A. So the alignment score E_s is greater than or equal to the globally optimized alignment score E_{opt}.

$$E_s = E(RS_s) + E(TS_s) \ge E_{opt} \tag{4}$$

Summing over all values of s in Eq. 4, we have the following:

$$\sum_{0 \le s < k} E_s \ge kE_{opt} \tag{5}$$

Now we will prove that $\sum_{s=0}^{k-1} E(TS_s)$ is no more than $4E_{opt}$. Then, the sum of all the $E(RS_s)$ is at least $(k-4)E_{opt}$ and there is at least one s^* such that $E(RS_{s^*}) \ge (1-\frac{4}{k})E_{opt}$. Therefore, there is a structure alignment with score at least $(1-\frac{4}{k})E_{opt}$.

The union of all the TS_s is equal to $\bigcup_j G(T_j)$, which can be divided into four disjoint subsets $\bigcup_j G(T_{l+4j})$ ($0 \le l < 4$) such that for a given l, $G(T_{l+4j_1})$ and $G(T_{l+4j_2})$

---

[2] If the subscript of B is greater than $\frac{W_y}{D}$, then we replace the subscript with its modulus over $\frac{W_y}{D}$.

[3] In this paper, j%k represents j module k.

are disjoint if $j_1 \neq j_2$. So the whole alignment score between $\bigcup_j G(T_{l+4j})$ and B is no more than E_{opt} no matter how we do the alignment. Therefore, $\sum_{s=0}^{k-1} E(TS_s)$ is no more than $4E_{opt}$.

For each partition scheme s, the algorithm aligns the partial structure in RS_s to B. Based on Lemma 1, the structure alignment between the partial structure in R_j and B can be optimized by an algorithm with time complexity $O(|R_j|2^{tw \lg \Delta})$. Once the structure alignment between RS_s and B is fixed, the algorithm aligns the remaining structure to protein B. So the time complexity of structure alignment for each partition scheme is $O(n2^{tw \lg \Delta})$ and the time complexity of this algorithm is $O(kn2^{tw \lg \Delta})$.

In the proof of the above theorem, we partition a protein into small blocks along one dimension. Actually, we can further cut a protein into smaller blocks along two dimensions. Based on Theorem 2, we arrive at the following theorem, which is proved in the supplemental material.

Theorem 3. *Let A and B be two protein structures in $\Re^3$. Assume that the spatial positions of A and B are fixed and the distance between two equivalent residues is no more than D_c. Then there is an algorithm with time complexity $(nk^2 2^{tw \lg \Delta})$ generating a non-sequential alignment between A and B with an alignment score at least $\left(1 - \frac{8}{k}\right)$ times the best possible, where n is the protein size, k is a positive integer $\Delta = O\left((1 + \frac{2D_c}{D_l})^3\right)$, and $tw = O(k^2 \frac{\max\{2D_c, D_u\}^3}{D_l^3})$.*

4 Structure Alignment with All the Transformations

In this section, we assume that we can move protein A in any way and the position of protein B is fixed. We are going to find the best transformation of A such that the objective function in Eq. 1 is maximized. Kolodny and Linial [12] achieved a PTAS algorithm for the coordinate based structure alignment problem by discretizing the rigid-body transformation space into a polynomial number of discrete transformations. We will present a similar but more involved discretization technique for our problem.

A rigid-body transformation consists of two steps: rotation and translation. Mathematically, it can be represented by a triple (w, θ, t), where w is a normalized vector in $\Re^3$, θ the rotation angle and t the translation. The vector w and the angle θ form a quaternion, which is the classic representation for rotation. The normalized vector w is the unit axis around which an object is rotated by θ. Assume $\hat{v}$ to be the resultant vector for rotating a vector v by an angle of θ around a unit axis w. Then $\hat{v}$ can be calculated using the following formula:

$$\hat{v} = w(v \cdot w) + (v - w(v \cdot w))cos(\theta) + (v \times w)sin(\theta) \qquad (6)$$

where $\cdot$ is the dot product of two vectors and $\times$, the cross product. According to Eq. 6, if the unit axis w is changed by a small degree δw, then $|\hat{v}|$ will be changed by $O(|v||\delta w|)$. If the rotation angle θ is changed by $\delta\theta$, then $|\hat{v}|$ will be changed by $O(|v||\delta\theta|)$. Without loss of generality, we can assume that the unit axis w originates at the center point of a protein structure. Then $|v| \leq R$ where R is the radius of a protein structure. A small change in the unit axis w by ϵ/R or the rotation angle θ by ϵ/R will change $|\hat{v}|$ by at most ϵ. All the unit axes form the surface of a sphere with radius 1, and the rotation angle ranges from 0 to 2π.

For any given vector v, a translation t will lead to a new vector $\hat{v} = v+t$. Therefore, a small change in the translation t by $(\epsilon, \epsilon, \epsilon)$ will change $|\hat{v}|$ by at most $O(\epsilon)$. Assume that a protein structure A is enclosed in a rectangle with dimensions $W_x(A)$, $W_y(A)$ and $W_z(A)$. Then all the possible translations between proteins A and B are in a rectangle with dimensions $W_x(A) + W_x(B)$, $W_y(A) + W_y(B)$, and $W_z(A) + W_z(B)$.

Since a small change in the transformation will not greatly change the spatial position of protein A, we can discretize the whole transformation space into a polynomial number of possible transformations. By working on these possible discrete transformations, we can find an alignment between two proteins with an alignment score very close to the optimal. In fact, we can find all the possible transformations that lead to a near-optimal alignment.

Theorem 4. *Let $OPT(D_c)$ denote the optimal alignment score between two proteins A and B when the distance between two equivalent residues is no more than D_c after two proteins are superimposed. There is an algorithm to generate a non-sequential alignment between two proteins such that i) the time complexity of this algorithm is $O(k^2 n^3 \Delta^{tw}/(\epsilon D_c)^6)$ or $O(k^2 n^5 \Delta^{tw}/(\epsilon D_c)^6)$ where $\Delta = O((1 + \epsilon)^3 D_c^3/D_l^3)$ and $tw = O(k^2 \max\{2D_c, D_u\}^3/D_l^3)$; ii) the alignment score is no less than $(1 - \Theta(\frac{1}{k})) OPT(D_c)$; and iii) the distance between two equivalent residues is no more than $(1 + \epsilon)D_c$.*

Proof. Given two possible rigid transformations (w_1, θ_1, t_1) and (w_2, θ_2, t_2), assume they satisfy the following conditions:

$$|w_1 - w_2| \leq \epsilon D_c/3R, \tag{7}$$
$$|\theta_1 - \theta_2| \leq \epsilon D_c/3R, \tag{8}$$
$$|t_1 - t_2| \leq \epsilon D_c/3. \tag{9}$$

Let $\hat{A}_i$ denote the transformation of A by (w_i, θ_i, t_i) $(i = 1, 2)$. For any residue r in $\hat{A}_i$, let $\hat{r}_i$ denote the image of r in $\hat{A}_i$. It can be verified that $|\hat{r}_1 - \hat{r}_2| \leq \epsilon D_c$. Let $N_i(r, d)$ denote the set of residues in B such that the distance between $\hat{r}_i$ and any residue in $N_i(r, d)$ is no more than d. We can easily verify that $N_1(r, D_c) \subseteq N_2(r, D_c(1 + \epsilon))$ and $N_2(r, D_c) \subseteq N_1(r, D_c(1 + \epsilon))$. Let $OPT(d, w, \theta, t)$ denote the optimal alignment score (i.e., the objective function in Eq. 1) between A and B when A is transformed by (w, θ, t) and the deviation between two equivalent residues is no more than d. Then we have $OPT(D_c, w_1, \theta_1, t_1) \leq OPT(D_c(1 + \epsilon), w_2, \theta_2, t_2)$ since $N_1(r, D_c) \subseteq N_2(r, D_c(1 + \epsilon))$.

Given a small positive constant ϵ, we can discretize the unit axis with step size $\epsilon D_c/3R \times \epsilon D_c/3R$, the rotation angle with step size $\epsilon D_c/3R$ and the translation with step size $\epsilon D_c/3$. The whole transformation space is discretized into a set of $O\left(R^3 V/(\epsilon^6 D_c^6)\right)$ points where $V = (W_x(A) + W_x(B))(W_y(A) + W_y(B))(W_z(A) + W_z(B))$. Let $\sum$ denote this set of discrete transformations. For any possible transformation (w_1, θ_1, t_1), there is a discrete transformation $(w_2, \theta_2, t_2) \in \sum$ such that conditions (7)-(9) are satisfied. That is, $OPT(D_c, w_1, \theta_1, t_1) \leq OPT(D_c(1 +$

$\epsilon), w_2, \theta_2, t_2)$. So $OPT(D_c) \leq \max_{(w,\theta,t)\in\sum} OPT(D_c(1 + \epsilon), w, \theta, t)$. For each discrete transformation, according to Theorem 3, there is an algorithm with time complexity $O(k^2 n \Delta^{tw})$ to calculate $OPT(D_c(1 + \epsilon), w_2, \theta_2, t_2)$. This algorithm will generate an alignment with score at least $\left(1 - \Theta(\frac{1}{k})\right) OPT(D_c(1+\epsilon), w_2, \theta_2, t_2)$. Enumerating all the discrete transformations in $\sum$, we can generate an alignment with score at least $\left(1 - \Theta(\frac{1}{k})\right) OPT(D_c)$ and the deviation between two equivalent residues is no more than $(1 + \epsilon)D_c$. The running time of the above procedure is $O(k^2 n \Delta^{tw} R^3 V/(\epsilon D_c)^6)$.

According to paper [24], V is proportional to the protein size. For a globular protein, $R = O(\sqrt[3]{n})$, so the time complexity of the above algorithm is $O(k^2 n^3 \Delta^{tw}/(\epsilon D_c)^6)$. For other proteins, $R = O(n)$, so the time complexity is $O(k^2 n^5 \Delta^{tw}/(\epsilon D_c)^6)$.

This result indicates that as long as the ratio between $\max\{2D_c, D_u\}$ and D_l is small compared to the protein size, there is a polynomial-time approximation scheme for the non-sequential protein structure alignment problem. If $\max\{2D_c, D_u\}/D_l$, l, k, and ϵ are constants, then the time complexity is polynomial. Therefore, we can claim that there is fixed-parameter polynomial-time algorithm for the contact map-based protein structure alignment problem if the sequential order is not enforced.

Combining the exact algorithm described in Subsection 3.1 and the discretization technique in this section, we have the following theorem for the structure alignment problem.

Theorem 5. *There is an algorithm to generate a non-sequential alignment with a score at least $OPT(D_c)$ such that i) the time complexity of this algorithm is $O(n^3 \Delta^{tw}/ (\epsilon D_c)^6)$ for globular proteins or $O(n^5 \Delta^{tw}/(\epsilon D_c)^6)$ for others, where $\Delta = O((1 + \epsilon)^3 D_c^3/D_l^3)$ and $tw = O(\frac{\max\{2D_c, D_u\}}{D_l} n^{2/3} \lg n)$ and ii) the distance between two equivalent residues is no more than $(1 + \epsilon)D_c$.*

5 Experimental Results

We have implemented the exact tree-decomposition algorithm described in Subsection 3.1 and the discretization algorithm described in Section 4. The algorithm is implemented on a cluster of Linux PCs with 2.5 GHz CPU. In total, we used 15 proteins from two different folds in the test set described in [25] to test our algorithm. We set the contact distance cutoff D_u to 6.75 $\mathring{A}$ and the maximum distance between two matched residues D_c to 3.0 $\mathring{A}$.

In doing structure alignment, we always fix protein B and transform protein A. The space of unit rotation axis is discretized into a 36×18 longitude-latitude grid. The rotation angle is evenly discretized into 36 possible angles. The translation space is discretized into $35 \times 35 \times 35$ discrete points. That is, if we fix the center of protein B to the origin, then the possible center positions of protein A form a set $\{(x/2, y/2, z/2)| -17 \leq x \leq 17, -17 \leq y \leq 17, -17 \leq z \leq 17\}$. We start from $(0, 0, 0)$ and gradually increase the distance between two protein centers to search for the best translation position. In total, the rigid-body transformation space is discretized into $1,000,188,000$ discrete transformations.

Currently, only the non-sequential alignment result is tested. Please see the supplemental material for the detailed alignment results. The running time of aligning one protein pair ranges from ten minutes to one hour. According to Caprara *et. al.* [25], for the contact distance threshold 6.75 $\overset{\circ}{A}$, we can cluster two proteins into the same fold if the number of aligned contacts is at least 0.559 times $\min\{c_A, c_B\}$ where c_A and c_B are the numbers of contacts of both proteins, respectively. Our experimental results comply with this criterion very well. However, to achieve the maximum number of aligned contacts, $D_c = 3.0\overset{\circ}{A}$ may not be big enough for some protein pairs. For example, we need a bigger D_c to obtain more aligned contacts between 1b00a and 1dbwa although $D_c = 3.0\overset{\circ}{A}$ gives a very good alignment between 2pcy and 2plt. We plan to investigate the cutoff value of D_c further. While the sequential order in the alignment is not required, there are almost no sequential disorders in the generated alignment if two proteins are in the same class.

6 Conclusion

This paper presents a parametrized algorithm for the contact map-based protein structure alignment problem, which has been proven to be *NP*-hard. The time complexity is polynomial in the protein size and exponential with respect to several parameters, which usually can be treated as constants. However, the method proposed in this paper might not be useful for everyday structure alignment since while theoretically significant, the computational time complexity is still expensive. A tool based on this method can be used as a benchmark to evaluate the performance of other heuristic-based structure alignment algorithms.

References

1. M. Comin, C.Guerra, and G. Zanotti. PROuST: a comparison method of three-dimensional structures of proteins using indexing techniques. *Journal of Computational Biology*, 11(6):1061–1072, 2004.
2. L. Holm and C. Sander. Protein structure comparison by alignment of distance matrices. *Journal of Molecular Biology*, 233:123–138, 1993.
3. N.N. Alexandrov. SARFing the PDB. *Protein Engineering*, 9:727–732, 1996.
4. M. Gerstein and M. Levitt. Using iterative dynamic programming to obtain accurate pairwise and multiple alignments of protein structures. In *Proceedings of International Conference on Intelligent Systems in Molecular Biology*, pages 59–67, 1996.
5. A.P. Singh and D.L. Brutlag. Hierarchical protein structure superposition using both secondary structure and atomic representations. In *Proceedings of International Conference on Intelligent Systems in Molecular Biology*, pages 284–93, 1997.
6. J.F. Gibrat, T. Madej, and S.H. Bryant. Surprising similarities in structure comparison. *Current Opinion in Structural Biology*, (6):377–385, 1996.
7. T. Akutsu and H. Tashimo. Protein structure comparison using representation by line segment sequences. In *Proceedings of Pacific Symposium on Biocomputing '96 (PSB'96)*, pages 25–40, 1996.
8. G. Lancia, R. Carr, B. Walenz, and S. Istrail. 101 optimal PDB structure alignments: a branch-and-cut algorithm for the maximum contact map overlap problem. In *RECOMB 2001*, pages 193–202. ACM Press, 2001.
9. A. Caprara and G. Lancia. Structural alignment of largesize proteins via Lagrangian relaxation. In *RECOMB 2002*, pages 100–108. ACM Press, 2002.

10. G. Lancia and S. Istrail. Protein structure comparison: Algorithms and applications. In *Mathematical Methods for Protein Structure Analysis and Design*, volume 2666 of *Lecture Notes in Computer Science*, pages 1–33, 2003.
11. C. Lemmen and T. Lengauer. Computational methods for the structural alignment of molecules. *Journal of Computer-Aided Molecular Design*, 14:215–232, 2000.
12. R. Kolodny and N. Linial. Approximate protein structural alignment in polynomial time. *PNAS*, 101(33):12201–12206, 2004.
13. D. Goldman, C.H. Papadimitriou, and S. Istrail. Algorithmic aspects of protein structure similarity. In *FOCS 99: Proceedings of the 40th Annual Symposium on Foundations of Computer Science*, pages 512–522. IEEE Computer Society, 1999.
14. I.N. Shindyalov and P.E. Bourne. Protein structure alignment by incremental combinatorial extension (CE) of the optimal path. *Protein Engineering*, 11(9):739–747, 1998.
15. O. Verbitsky. On the largest common subgraph problem, 1994. Unpublished manuscript.
16. S. Jokisch and H. Müller. Inter-point-distance-dependent approximate point set matching. Technical Report Research Report No. 653, 1997.
17. X. Yuan and C. Bystroff. Non-sequential structure-based alignments reveal topology-independent core packing arrangements in proteins. *Bioinformatics*, 27:1010–1019, 2005.
18. O. Dror, H. Benyamini, R. Nussinov, and H. Wolfson. MASS: Multiple structural alignment by secondary structures. *Bioinformatics*, 19(Suppl. 1):95–104, 2003.
19. J. Zhu and Z. Weng. FAST: A novel protein structure alignment algorithm. *Proteins: Structure Function, and Bioinformatics*, 2004. In Press.
20. H.M. Berman, J. Westbrook, Z. Feng, G. Gilliland, T.N. Bhat, H. Weissig, I.N. Shindyalov, and P.E. Bourne. The protein data bank. *Nucleic Acids Research*, 28:235–242, 2000.
21. K.S. Arun, T.S. Huang, and S.D. Blostein. Least-square fitting of two 3-d point sets. *IEEE Trans. on Pattern Analysis and Machine Intelligence*, 9(5):698–700, 1987.
22. N. Robertson and P.D. Seymour. Graph minors. II. algorithmic aspects of tree-width. *Journal of Algorithms*, 7:309–322, 1986.
23. J. Xu. Rapid side-chain packing via tree decomposition. In *RECOMB 2005*, volume 3500 of *Lecture Notes in Bioinformatics*. Springer, May 2005.
24. M.H. Hao, S. Rackovsky, A. Liwo, M.R. Pincus, and H.A. Scheraga. *Proc. Natl. Acad. Sci. USA*, 89:6614–6618, 1992.
25. A. Caprara, R. Carr, S. Istrail, G. Lancia, and B. Walenz. 101 optimal PDB structure alignments: a branch-and-cut algorithm for the maximum contact map overlap problem. *Journal of Computational Biology*, 11(1):27–52, 2004.

Geometric Sieving: Automated Distributed Optimization of 3D Motifs for Protein Function Prediction

Brian Y. Chen[1,*], Viacheslav Y. Fofanov[2,*], Drew H. Bryant[5],
Bradley D. Dodson[1], David M. Kristensen[3,4], Andreas M. Lisewski[4],
Marek Kimmel[2], Olivier Lichtarge[3,4], and Lydia E. Kavraki[1,3,5,**]

[1] Department of Computer Science, Rice University, Houston, TX 77005, USA
[2] Department of Statistics, Rice University
[3] Structural and Computational Biology and Molecular Biophysics,
Baylor College of Medicine, Houston, TX 77005, USA
[4] Department of Molecular and Human Genetics, Baylor College of Medicine
[5] Department of Bioengineering, Rice University
kavraki@cs.rice.edu

Abstract. Determining the function of all proteins is a recurring theme in modern biology and medicine, but the sheer number of proteins makes experimental approaches impractical. For this reason, current efforts have considered in silico function prediction in order to guide and accelerate the function determination process. One approach to predicting protein function is to search functionally uncharacterized protein structures (*targets*), for substructures with geometric and chemical similarity (*matches*), to known active sites (*motifs*). Finding a match can imply that the target has an active site similar to the motif, suggesting functional homology.

An effective function predictor requires effective motifs - motifs whose geometric and chemical characteristics are detected by comparison algorithms within functionally homologous targets (*sensitive* motifs), which also are not detected within functionally unrelated targets (*specific* motifs). Designing effective motifs is a difficult open problem. Current approaches select and combine structural, physical, and evolutionary properties to design motifs that mirror functional characteristics of active sites.

We present a new approach, Geometric Sieving (GS), which refines candidate motifs into *optimized motifs* with maximal geometric and chemical dissimilarity from all known protein structures. The paper discusses both the usefulness and the efficiency of GS. We show that candidate motifs from six well-studied proteins, including α-Chymotrypsin, Dihydrofolate Reductase, and Lysozyme, can be optimized with GS to motifs that are among the most sensitive and specific motifs possible for the candidate motifs. For the same proteins, we also report results that relate evolutionarily important motifs with motifs that exhibit maximal geometric and chemical dissimilarity from all known protein structures.

---

[*] Equal Contribution.
[**] Corresponding author.

A. Apostolico et al. (Eds.): RECOMB 2006, LNBI 3909, pp. 500–515, 2006.
© Springer-Verlag Berlin Heidelberg 2006

Our current observations show that GS is a powerful tool that can complement existing work on motif design and protein function prediction.

1   Introduction

The determination of protein function is an important goal in biology, but experimental techniques for determining function are expensive and time consuming. One way to accelerate this process is to use computational techniques to search the structure of functionally uncharacterized proteins (*targets*), for *matches* of geometric and chemical similarity to known functional sites (*motifs*). To achieve this, algorithms like Geometric Hashing [1], JESS [2], and Match Augmentation [3] identify a subset of a target with the greatest geometric and chemical similarity to the motif. Typically, geometric similarity is measured by least root mean squared distance (LRMSD[1]) and chemical similarity is ensured by examining the chemical compatibility of corresponding matches. The identification of a match with statistically significant LRMSD can suggest that the target and motif have similar function [2, 3, 4].

Designing effective motifs is a two-sided open problem: The geometric configuration and chemical makeup of effective motifs must be similar to functionally related proteins (*sensitive*), as well as dissimilar to functionally unrelated proteins (*specific*). For this reason, it is difficult to select *motif points*, the points in space with chemical labels which comprise motifs, so that sensitivity and specificity are simultaneously maximized. Many methods for designing motifs exist, and we are only able to include a partial list here. Motifs have been designed using evolutionary significance and proximity to binding sites [5]. Motifs have also been designed using literature search and PSI-BLAST alignments of literature-defined motifs from the Catalytic Site Atlas [6, 7]. Still other motifs are designed using surface exposure, and algorithms for detecting conserved binding patterns [8]. The work presented in this paper complements these methods with a novel criteria for motif design and an algorithm that can be used to further improve existing motifs.

Contributions and Outline. We begin by describing the design and implementation of *Geometric Sieving* (GS), an algorithm for refining candidate motifs into *optimized motifs*. As input, GS accepts a selection of candidate motif points, chosen perhaps by another motif design algorithm, called the *input set*, and the number k of motif points desired in the optimized motif. GS outputs an optimized motif: a motif of k candidate motif points with the *greatest geometric and chemical dissimilarity* to all known protein structures. We refer this property as *Geometric Uniqueness*.

The motivation and inspiration for defining Geometric Uniqueness stems from several observations in our earlier work [3, 5] and the work of other researchers [2, 4], where it has been observed that motifs which are highly representative

---

[1] LRMSD is the root mean square distance (RMSD) between two sets of points in 3D, aligned with smallest RMSD.

of protein function do not occur in a large fraction of the known proteins. One question that we posed is whether geometric and chemical dissimilarity of a motif to all other known proteins (a.k.a. Geometric Uniqueness) can be computed in a reasonable amount of time and whether Geometric Uniqueness can be used to identify sensitive and specific motifs. After we obtained a positive answer to the above question for a limited but well-designed set of experiments, we proceeded to investigate a second question which is whether Geometric Uniqueness correlates with other characteristics of active sites. For example, evolutionarily significant amino acids, those most associated with important evolutionary divergences, as defined in [9, 10], are often related to active sites [5]. We observed, on our limited set of examples, a correlation between Geometric Uniqueness and evolutionary significance.

Measuring and optimizing Geometric Uniqueness is a nontrivial computational problem because numerous structural comparisons must be made between many motifs and many protein structures. In Section 2, we present recent advances in the field of motif comparison algorithms that enabled the development of GS. In Section 3, we detail the GS algorithm, a distributed algorithm coupled with on-line statistical optimization, which measures Geometric Uniqueness to optimize motifs. Our experimental results are shown in Section 4. Targeting our first question, we optimized input sets derived from six well-studied proteins. On these examples, optimized motifs computed by GS had among the highest sensitivity and specificity of every subset motif definable from the input sets. Using information from the Evolutionary Trace (ET) [5, 9] we observed, on our examples, that evolutionarily significant motifs exhibited higher Geometric Uniqueness.

This paper does not advocate that Geometric Uniqueness should be the sole criterion for defining effective motifs. It argues, rather, that Geometric Uniqueness is an interesting property that seems to be useful for refining existing motifs. It also argues that GS is a novel methodology which can be used to optimize motifs designed by human intuition, or by other motif design methods, such as the milestone algorithm MultiBind [8]. It finally argues that Geometric Uniqueness can be compared with other known criteria for selecting motifs in an effort to better understand and finally attack the difficult problem of protein function prediction.

2   Related Work

Motif Types. The many approaches to designing effective motifs have created different types of motifs: motifs have been composed of points on the Connolly surface [11] representing electrostatic potentials [12], of hinge-bending sets of points in space [13], of sets of "pseudo-centers" representing protein-ligand interactions [8], or of points taken from atom coordinates with evolutionary data [3, 9], to name a few. Depending on how motif points are defined, they have different labels associated with them and these labels need to be taken into account when comparing motifs. GS is orthogonal to the choice of motif type and could be applied with any of the motif types above.

In this work, a motif S is a set of m points $\{s_1, \ldots, s_m\}$ in three dimensions, whose coordinates are taken from backbone and side-chain atoms. Each *motif point* s_i in the motif has an associated *rank* $p(s_i)$, a measure of the functional significance of the motif point. Each s_i also has a set of alternate amino acid *labels* $l(s_i) \subset \{GLY, ALA, ...\}$, which represent residues this amino acid has mutated to during evolution. Labels permit our motifs to simultaneously represent many homologous active sites with slight mutations, not just a single active site. In this paper, we obtain labels and ranks using ET [9, 10].

Motif Comparison Algorithms. GS requires a geometric and chemical comparison algorithm to compare motifs to targets. Many such algorithms exist, but differ fundamentally in that they are optimized for comparing different types of motifs. There are algorithms for comparing graph-based motifs [14], algorithms for finding catalytic sites [2], and the seminal Geometric Hashing framework [1] which can search for many types of motifs, including motifs based on atom position [15], points on Connolly face centers [16], catalytic triads [17], and flexible protein models [13]. The comparison algorithm we use in this work is Match Augmentation (MA) [3], because of its availability and compatibility with our selected motif type. GS is independent of MA, and adapting another comparison algorithm to use our motifs could be equally successful.

MA compares a motif S to a target T, a protein structure encoded as n *target points*: $T = \{t_1, \ldots t_n\}$, where each t_i is taken from atom coordinates, and labeled $l(t_i)$ for the amino acid t_i belongs to. A match M, is a bijection correlating all motif points in S to a subset of T of the form $M = \{(s_{a_1}, t_{b_1}), (s_{a_2}, t_{b_2}) \ldots (s_{a_m}, t_{b_m})\}$. Referring to Euclidean distance between points a and b as $||a - b||$, an acceptable match requires:

Criterion 1. $\forall i$, s_{a_i} and t_{b_i} are biologically compatible: $l(t_{b_i}) \in l(s_{a_i})$.
Criterion 2. LRMSD alignment, via rigid transformation A of S, causes
$$\forall i, ||A(s_{a_i}) - t_{b_i}|| < \epsilon,$$ our threshold for geometric similarity.

MA takes as input a motif S and a target T. MA outputs the match with smallest LRMSD among all matches that fulfill the criteria. Partial matches correlating subsets of S to T are rejected. By establishing a threshold for acceptable geometric similarity, the second criterion causes MA to return match LRMSDs bounded by ϵ, even if the smallest LRMSD is not very low. This allows us to generate a spectrum of matches ranging from high to low geometric and chemical similarity, which we refer to as a motif profile.

Obtaining Motif Profiles. The basic object of comparison used by GS is the *motif profile*, a set of matches S_Ω between a single motif S and a very large set of targets, Ω. We compute these matches with MA. Motif profiles are best visualized as frequency distributions (see Figure 1(a)), which are essentially histograms that plot frequency (the number of matches with a particular LRMSD) versus LRMSD. We apply kernel density estimation procedures [18] to estimate population density from the motif profile, using Gaussian Kernel smoothing to interpo-

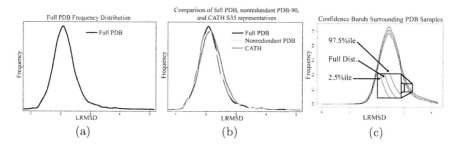

Fig. 1. (a) Typical frequency distribution of matches between a motif and the PDB [21]. (b) Comparison of PDB, sequentially nonredundant PDB, and CATH representatives. (c) Confidence band demonstrating the accuracy of samples of the PDB. This data computed using the motif C42, H57, C58, D102, D194, S195, S214 from α-Chymotrypsin (1acb).

late between data points, as in previous work [3]. Optimal bin-widths determined by Sheather-Jones method [19, 20] were used to avoid under- and over-smoothing.

The purpose of Ω is to represent the set of all known protein structures. We have found, however, that different representations of Ω tend not to have significant effect on the actual shape of motif profiles generated. For the 6 motifs optimized for this work, as well as 12 motifs used in previous work [3], we observed strong similarity between motif profiles calculated with the PDB (Ω_0), and Ω_{nr25} and Ω_{nr90}, two sets of sequentially nonredundant PDB structures having no more than 25% (resp. 90%) sequence identity. A similar comparison was true when using the CATH [22] database, a multi-level nested categorization of increasingly specific protein sequence and structure classifications. We selected a representative of every category at the three most specific levels: Topologies (Ω_T), Homologous Superfamiles (Ω_H), and Sequence Families Ω_S. In our experience, motif profiles on these representatives also resemble Ω_0, in increasing degrees of similarity corresponding to increasingly specific levels of CATH. The similarity between the Ω_0 (black), Ω_{nr25} (light grey) and Ω_S (dark grey) is plotted in Figure 1(b). Ω_{nr90}, Ω_T, and Ω_H were excluded for clarity, but are closely related.

We have also observed that motif profiles on Ω_0 are exceptionally robust to random sampling. Ω_5 is the random 5% sample of PDB structures in Ω_0, and motif profiles with this set are called S_{Ω_5}. In our experience, for any S, S_{Ω_5} resembles S_{Ω_0} with high accuracy. This can be seen in Figure 1(c), where we overlayed 5000 distinct S_{Ω_5} samples with a single S_{Ω_0}, the center line in Figure 1(c). 95% of the 5000 S_{Ω_5} fell within the upper and lower lines, demonstrating that motif profiles based on Ω_5 retain high similarity to motif profiles based on Ω_0. This is a result of sampling a largely unimodal distribution.

GS is not dependent on the selection of Ω, but because our observations suggest that motif profiles based on many logical representations of Ω, including Ω_S, Ω_H, Ω_T, Ω_{nr25}, and Ω_{nr90}, differ little from motif profiles based on Ω_5, this paper proceeds by using Ω_5. 5% sampling greatly reduces the number of matches necessary to compute a motif profile, while its simple definition promotes the reproducibility of this work. Other investigations could use alternate selections of Ω.

Motif profiles are especially useful for determining the statistical significance of matches with a given motif S. In previous work, we showed that nonparametric density estimation of motif profiles generated with S can be used to calculate p-values, which measure the statistical significance of any match of S [3]. Matches with low p-values, which correspond to high statistical significance, seem to correlate with functional similarity [3]. This result corroborates earlier work which applied parametric approaches [2, 4] to generate other measures of statistical significance which also correlate with functional homology.

3   Geometric Sieving

GS accepts an input set, a collection of candidate motif points which could be selected by another motif design method, or provided by a user seeking to improve a motif. GS also requires k, the number of candidate motif points expected in the output, and, as discussed in the previous section, a geometric comparison algorithm compatible with the motif type used. The output of GS is the subset motif with k points that has highest Geometric Uniqueness.

GS is a refinement process, not a motif discovery algorithm. If no subset motif of size k has geometric and chemical similarity to functionally homologous active sites, then GS cannot select one which does. For this reason, the input set is assumed to contain a subset motif of size k, which has basic geometric and chemical similarity to functional homologs of the input set. By this assumption, matches to functional homologs remain in the low-LRMSD tail of the motif profile for many subset motifs, while functionally unrelated proteins, the vast majority of matches in a motif profile, gravitate around the large mode near the median LRMSD. The difference in LRMSD between this low-LRMSD tail and the major mode of the distribution causes matches to functional homologs to be statistically significant relative to the distribution overall [3]. With many different combinations of motif points to choose from, in the form of varying subset motifs, we can select the motif profile which maximizes the LRMSD difference between the low-LRMSD tail and the major mode. As a result, matches to functional homologs will be maximally statistically significant for the input set considered. Geometric Sieving implements this task by analyzing motif profiles.

In this work, between two motif profiles, the motif profile with higher median LRMSD has higher Geometric Uniqueness. Medians are computed on kernel density smoothed motif profiles. While other statistics for quantitative comparison exist, such as the mode, our experimentation shows that comparing the medians of motif profiles is an elegant and effective approach for determining which motif is more Geometrically Unique. In addition, medians are not affected by extreme values at the tails of the distribution. Estimating the true median of the population from a sample is less prone to sampling errors and errors due to incorrect choice of smoothing parameters than mode estimation. Confidence bounds about the median, an integral part of our approach, are better studied than confidence bounds about the mode. Finally, in our results, we show the connection between medians and the actual distribution, demonstrating that motif profiles with higher medians are motif profiles with more and/or higher match LRMSDs.

The motif *size*, the number of motif points in a motif, is partially related to Geometric Uniqueness. Larger motifs specify more geometric constraints, and so tend to have higher LRMSD matches than smaller motifs [3]. Thus, we avoid comparing motif profiles from subset motifs of different sizes, ensuring that only the true geometric and chemical differences drive the motif profile comparison. This is why k, the size of the optimized motif, is an input. The operation and success of GS is not affected by k, and our results hold over varying k, as we will demonstrate later. Selecting an ideal k a priori remains an open problem, and the subject of continuing research.

3.1   The Geometric Sieving Algorithm

GS has two phases: GATHERING and ANALYSIS, which are described in Algorithms A1 and A2. Ignoring the ELIMINATION step (* in Algorithm A1) for now, the GATHERING phase uses MA to iteratively compute motif profiles (outer loop of Algorithm A1) for every subset motif of size k (inner loop of Algorithm A1). These motif profiles are passed to the ANALYSIS phase, which calculates the medians of each motif profile, and identifies the subset motif with the highest median LRMSD. This subset motif is returned as the optimized motif.

A1 GATHERING	**A2** ANALYSIS	**A3** ELIMINATION
Input: Input Motif S **Input:** Desired size k **for** each T_i in Ω_5 **do** **for** all subset motifs S' of size k **do** Run MA with S' and T_i MA returns match M Store M in profile S'_Ω **end for** ELIMINATION* **end for**	**Input:** all motif profiles S'_Ω from GATHERING phase Calculate $m(S'_\Omega)$ for all S'_Ω Find the motif profile S'_Ω with highest $m(S'_\Omega)$ **Output:** S', the optimized motif.	**Input:** all motif profiles S_Ω from GATHERING phase $\forall\, S'_\Omega$, compute $r(S_\Omega)$ $\forall\, r(S_\Omega)$, find l eliminate all $r(S'_\Omega)$ with $u < l$

The GATHERING phase is embarrassingly parallel. Given a set of c processors, we can obtain a $(c-1)$-times linear speedup by offloading the task of calculating each match between the current subset motif S', target T_i pair to another processor. This produces a client/server architecture where the server implements GATHERING, and offloads MA problems to the clients.

Further modifications to GS can increase performance. In particular, let us now consider the optimization procedure ELIMINATION (Algorithm A3) which is called from GATHERING. Note that when we call ELIMINATION during GATHERING, all motif profiles are only partially computed. Eventually ANALYSIS will identify the optimized motif by selecting the motif profile that has the highest median. A closer look at the computations happening during GATH-ERING revealed that some motif profiles have medians significantly lower than others. Since we are only interested in the motif profile with the highest median, we can stop computing matches for motif profiles that have significantly lower medians, saving computation time. For this reason, in Algorithm A1, we apply ELIMINATION (see outer loop of Algorithm A1), which determines for which motif profiles we can stop computing matches. These motif profiles will be *eliminated* in the next loop through GATHERING. ELIMINATION need not be applied at every iteration of the outer loop of GATHERING, as it will have a limited effect. Instead, we define a parameter called the *step size* and we call ELIMINATION after *step size* iterations of the outer loop of GATHERING.

As we pointed out above, when we call ELIMINATION during GATHER-ING (see Algorithm A3), all motif profiles are only partially computed. At this point in the algorithm, comparing the medians of these partial motif profiles can be affected by sampling error. For this reason, ELIMINATION computes a 95% Confidence Interval $r(S''_\Omega)$ (see method of Efron and Tibshirani [23, 24, 25]), which has 95% probability of containing the median $m(S'_\Omega)$ of S'_Ω. Therefore, for two partially computed motif profiles S'_Ω, S''_Ω, if $r(S'_\Omega) > r(S''_\Omega)$ do not overlap, there is low probability that $m(S'_\Omega) < m(S''_\Omega)$. Since we are interested only in the motif profile with highest median LRMSD, it is thus unnecessary to finish computing S''_Ω because S'' is not the optimized motif with high probability.

We apply this fact during ELIMINATION by finding l, the highest lower bound of all confidence intervals, and eliminate all subset motifs having confidence intervals with upper bound $u < l$. In the next loop through GATHERING, we do not calculate matches for eliminated subset motifs. If only one subset motif remains, or if GATHERING completes, we proceed to the ANALYSIS phase, which identifies the motif profile, that has not been eliminated, with the highest median. This is returned as the output of GS.

4   Experimental Results

We begin our experimentation by demonstrating that GS is a practical and efficient tool for motif optimization. Using input sets derived from 6 well-studied proteins, we show that different subset motifs derived from the same input set produce motif profiles which measurably vary in the median. We also demonstrate that estimating medians with a 95% confidence bound and eliminating subset motifs via ELIMINATE reduces the number of calculations necessary to correctly determine the motif profile with highest median. On our small data set, we made two key observations: First, sensitive and specific optimized motifs can be identified by Geometric Uniqueness. Second, evolutionary significant subset motifs tend to be more Geometrically Unique than evolutionarily insignificant amino acids. Full details can be found at:

http://www.cs.rice.edu/~brianyc/papers/RECOMB2006/.

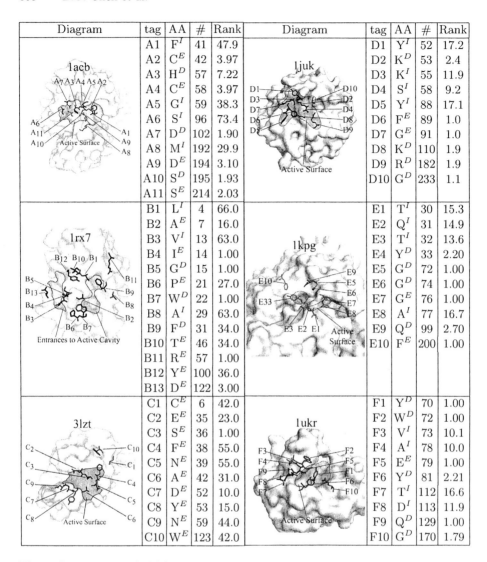

Diagram	tag	AA	#	Rank	Diagram	tag	AA	#	Rank
1acb	A1	F^I	41	47.9	1juk	D1	Y^I	52	17.2
	A2	C^E	42	3.97		D2	K^D	53	2.4
	A3	H^D	57	7.22		D3	K^I	55	11.9
	A4	C^E	58	3.97		D4	S^I	58	9.2
	A5	G^I	59	38.3		D5	Y^I	88	17.1
	A6	S^I	96	73.4		D6	F^E	89	1.0
	A7	D^D	102	1.90		D7	G^E	91	1.0
	A8	M^I	192	29.9		D8	K^D	110	1.9
	A9	D^E	194	3.10		D9	R^D	182	1.9
	A10	S^D	195	1.93		D10	G^D	233	1.1
	A11	S^E	214	2.03						
1rx7	B1	L^I	4	66.0	1kpg	E1	T^I	30	15.3
	B2	A^E	7	16.0		E2	Q^I	31	14.9
	B3	V^I	13	63.0		E3	T^I	32	13.6
	B4	I^E	14	1.00		E4	Y^D	33	2.20
	B5	G^D	15	1.00		E5	G^D	72	1.00
	B6	P^E	21	27.0		E6	G^D	74	1.00
	B7	W^D	22	1.00		E7	G^E	76	1.00
	B8	A^I	29	63.0		E8	A^I	77	16.7
	B9	F^D	31	34.0		E9	Q^D	99	2.70
	B10	T^E	46	34.0		E10	F^E	200	1.00
	B11	R^E	57	1.00						
	B12	Y^E	100	36.0						
	B13	D^E	122	3.00						
3lzt	C1	C^E	6	42.0	1ukr	F1	Y^D	70	1.00
	C2	E^E	35	23.0		F2	W^D	72	1.00
	C3	S^E	36	1.00		F3	V^I	73	10.1
	C4	F^E	38	55.0		F4	A^I	78	10.0
	C5	N^E	39	55.0		F5	E^E	79	1.00
	C6	A^E	42	31.0		F6	Y^D	81	2.21
	C7	D^E	52	10.0		F7	T^I	112	16.6
	C8	Y^E	53	15.0		F8	D^I	113	11.9
	C9	N^E	59	44.0		F9	Q^D	129	1.00
	C10	W^E	123	42.0		F10	G^D	170	1.79

Fig. 2. Input sets used. "AA": amino acid type; "#": PDB residue number; "Rank": ET rank.

4.1   Primary Data

Input Sets. Earlier work has produced examples of motifs designed with evolutionarily significant amino acids [3] and amino acids with documented function [6], which were sensitive and specific. Inspired by these approaches, we selected evolutionarily significant (^{E}, Figure 3) and functionally documented (^{D}, Figure 3) amino acids for each of our six input sets, except Lysozyme (3lzt). We also included evolutionarily insignificant amino acids (^{I}, Figure 3), chosen from the same region of the protein.

PDB Code	Amino Acids and Citations	EC class	size	k
1acb	S195 H57 D102 [26]	3.4.21.1	11	7
1rx7	W22 [27], G15, D27, F31, H45, I50, G96 [28]	1.5.1.3	13	10
3lzt	Control	3.2.1.17	10	8
1juk	Lys53, Lys110, Arg182, Gly233 [29]	4.1.1.48	10	6
1kpg	G72, G74, Q99, Y33 [30]	2.1.1.79	10	6
1ukr	Y70, W72, E79, Y81, Q129, E170 [31]	3.2.1.8	10	6

Fig. 3. Functionally documented amino acids used in our input sets (cited), with protein EC class, input set size ("size"), and subset motif size (k)

Having chosen evolutionarily significant and functionally documented amino acids as part of each input set, we postulated that these "motif-worthy" amino acids, and not the evolutionarily insignificant amino acids, would create the most sensitive and specific motifs. For this reason, k was chosen in each case as the total number of evolutionarily significant and functionally documented amino acids in each input set. This guarantees that one subset motif from each input set would contain only evolutionarily significant and functionally documented amino acids, while the other subset motifs must contain evolutionarily insignificant amino acids. As a control, the Lysozyme input set (3lzt) was composed entirely of evolutionarily significant amino acids.

Functional Homologs. Measuring sensitivity and specificity requires a benchmark set of functional homologs. We use the functional classification of the Enzyme Commission [32] (EC), which identifies families of functional homologs for each input set (see Figure 3). Structure fragments and mutants were removed.

The Protein Data Bank. In this paper, we use Ω_5, as mentioned in Section 2, which is sampled from the set of crystallographic protein structures in the Protein Data Bank on Sept 1, 2005. PDB entries with multiple chains were divided into separate structures, producing 79322 structures. While this could prevent the identification of matches to active sites that span multiple chains, it is not clear from the PDB file format how to determine which chains are intended to be in complex. Incorrectly combining chains can lead to searches within physically impossible colliding molecules. Since none of the active sites used in this study span multiple chains, separation was the most reproducible and well defined policy.

Implementation Specifics. GS uses the Message Passing Interface [33] (MPI) protocol for interprocess communication, and was tested on a 16-node Athlon 1900MP cluster. The Rice TeraCluster, a cluster of 272 800Mhz Intel Itanium2s, and Ada, a Cray XD1 with 672 2.2Ghz AMD Opteron cores, computed final data. ϵ (see Section 2) was set to 7Å.

4.2 Median LRMSD Differentiates Motif Profiles

As mentioned in Section 4.1, our input sets were defined on both evolutionarily significant and insignificant amino acids, as well amino acids with documented

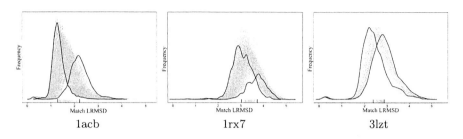

Fig. 4. Motif profiles generated using GS

function. Since GS calculates motif profiles for every possible subset motif, we hypothesized that the diversity of these input sets would present a spectrum of motif profile medians, and that medians within this spectrum would vary sufficiently to justify motif profile comparison by measuring median LRMSD.

Experiment. Each of our six input sets has between 10 and 13 motif points, and a specific k for each input set. GS computed motif profiles for every combination of k motif points in each input set. For example, α-Chymotrypsin and DHFR each contained, respectively, 7 and 10 amino acids which were either evolutionarily significant or functionally documented, out of the 11 and 13 amino acids total. Running GS with $k = 7$ and $k = 10$, respectively, GS exhaustively analyzed all combinations of 7 and 10 amino acids as the subset motifs considered. We expected the Lysozyme input set, a control composed entirely of evolutionarily significant amino acids, to have a narrower spectrum of median LRMSDs, relative to the other sets of motif profiles.

Observations. The medians of the motif profiles generated (vertical hashes on the x-axes in Figure 4), occurred in ranges of approximately 1 Å LRMSD. Motif profiles corresponding to the highest medians clearly had more matches at higher LRMSDs than motif profiles at the lowest medians, and thus higher Geometric Uniqueness. This is demonstrated by darkened hashes and darkened curves in Figure 4, where the biggest differences in medians (darkened hashes) correlated to obvious differences in motif profiles (darkened curves). Lysozyme, which did not contain a spectrum of evolutionarily insignificant and significant amino acids, had a smaller range of medians. Higher median LRMSD in this application is clearly directly associated with more and higher match LRMSDs, showing on these examples that medians can be used to measure Geometric Uniqueness.

4.3   Median Estimation Cuts Runtime, Minor Accuracy Loss

Our implementation of GS uses online estimation of motif profile medians, reducing the number of matches which need to be calculated before the optimized motif is identified. Using input sets from Section 4.2, we first generated matches without using the ELIMINATION optimization, mentioned in Section 3. Next, we repeated this calculation with the ELIMINATION optimization, with step sizes of 100 and 500, to stop sampling on motif profiles which clearly did

Input Set	Time-Full	Matches-Full	Time-500	Matches-500	Time-100	Matches-100
1acb*	12545:33:20	1,322,230	2683:07:40	186,883	1424:13:20	97,836
1rx7*	10826:50:00	1,211,266	915:20:40	203,356	554:56:40	107,657
3lzt*	1204:52:00	184,395	227:56:00	97,593	942:00:00	92,099
1juk	1059:06:40	1,100,452	100:33:20	183,086	22:13:20	87,098
1kpg	1224:53:20	1,092,748	80:26:40	179,721	22:46:40	78,014
1ukr	2030:26:40	1,063,797	150:13:20	110,043	35:40:00	74,613

Fig. 5. Speedups from Median Estimation: Execution time and number of matches computed, using step sizes of 100, 500, and exhaustive sampling. * = Run on the Rice TeraCluster. Remaining runs were done on Ada.

not have the highest median LRMSD, thereby reducing the number of matches necessary.

Observations. Median estimation substantially reduces running time necessary to determine the optimized motif. Operating at step sizes of 100, GS can identify the optimized motif an average of 10 times faster than GS without median estimation. This speedup follows directly from the early elimination of motifs which, with high probability, do not have the highest median. At step sizes of 100, GS can identify the optimized motif with an average of 10 times less matches than GS without median estimation. Figure 5 describes the precise number of matches and time consumed.

Median estimation is very accurate. In every case described in Figure 5, median estimation identified the same optimized motif as GS using full sampling. However, at step size 100, GS also identifies an alternative subset motif for 3lzt. GS was unable to eliminate the alternative subset motif because overlapping confidence intervals (see Section 3.1) did not separate by the time sampling was complete. The same was true at a step size of 500 for 3lzt, and 1ukr. This suggests that for some motifs, achieving certainty of the optimized motif beyond 95% confidence can require sampling more than 5% of the PDB. Median estimation strongly accelerates the determination of the optimized motif with minor sacrifices in accuracy.

4.4   Geometric Uniqueness Identifies Effective Motifs

GS was designed for the purpose of improving the sensitivity and specificity of motifs by identifying the subset motif with highest median LRMSD, our measure of Geometric Uniqueness. We demonstrate that optimized motifs, on our six input sets, are among the most sensitive and specific of all possible motifs definable from the input sets.

Experiment. For each input set, we computed a match between every possible subset motif and every functional homolog in the corresponding EC class, except for the identical structure. Then, for each match, we accessed the p-value, a measure of statistical significance determined using a method from previous work [3]. Using $\alpha = .02$, our standard of statistical significance, we determined

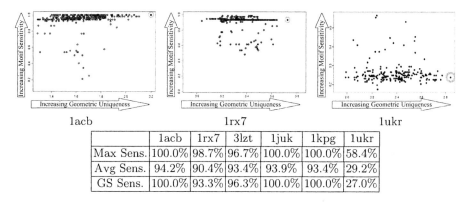

	1acb	1rx7	3lzt	1juk	1kpg	1ukr
Max Sens.	100.0%	98.7%	96.7%	100.0%	100.0%	58.4%
Avg Sens.	94.2%	90.4%	93.4%	93.9%	93.4%	29.2%
GS Sens.	100.0%	93.3%	96.3%	100.0%	100.0%	27.0%

Fig. 6. Sensitivity of 1acb, 1rx7, 1ukr vs median LRMSD (above), and sensitivity per input set: the most sensitive subset motif, the average sensitivity, and the sensitivity of the optimized motif from GS (table)

the number of matches with p-values below α - the true positives. The proportion of true positives relative to the total number of functional homologs is the sensitivity of the motif. With α at .02, specificity was always slightly above 98%.

Observations. In exhaustive comparison to all possible motifs definable from the input sets at their respective k values, GS identified optimized motifs which were quite sensitive, at a high level of specificity. From the 6 input motifs, GS produced 5 optimized motifs with greater sensitivity than the average subset motif from the same input set (see Figure 6). The exception, 1ukr, displayed no subset motifs with high sensitivity, even though it was created with the same criteria as the other input sets. Overall, Geometric Sieving performed well, identifying optimized motifs among the most sensitive of 5 out of 6 input sets, except where no effective motif could be found.

4.5 Geometric Uniqueness Correlates with Evolutionary Significance

Using the motif profiles calculated over Ω_5, we have the median LRMSD of every subset motif. Since we also have the evolutionary significance of every amino acid

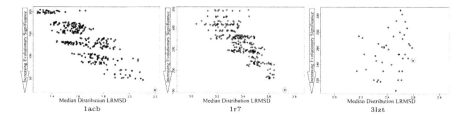

Fig. 7. Geometric Uniqueness vs. Evolutionary Significance

in our input sets, we can evaluate the evolutionary significance of every subset motif relative to its Geometric Uniqueness. In this experiment, we represented the total evolutionary significance of a subset motif as the sum of the ET ranks of its elements. Increasing sums relate to decreasing evolutionary significance, displayed on the vertical axis in Figure 7. Median LRMSD was plotted on the horizontal axis.

Observations. Motif profiles with high medians corresponded to subset motifs with evolutionarily significant amino acids (grey circles in Figure 7). In all cases but Lysozyme (3lzt), the input sets used demonstrate how evolutionary significance increases proportionately with increasing median LRMSD. In Lysozyme, a control set where every candidate motif point was evolutionarily significant, no apparent trend is visible. The existence of this apparent trend suggests that Geometric Uniqueness may be tied to evolutionary conservation.

5   Conclusions

We have presented GS, a novel distributed algorithm for exhaustively refining input sets of candidate motif points into optimized motifs. We have implemented GS with techniques and optimizations suitable for large scale distributed systems, and tested it on a cluster with more than 600 CPUs. By demonstrating refinement on 6 well-studied input sets, we show that, at a very high level of specificity, the optimized motifs from these examples were among the most sensitive of all motifs definable from these input sets. Using GS in conjunction with the Evolutionary Trace permitted us to demonstrate examples where amino acids that are evolutionarily significant are also Geometrically Unique. Our current observations show that GS is a powerful motif refinement algorithm which can be used in conjunction with other motif design techniques in an effort to create sensitive and specific motifs. In the future, we hope to accomplish larger-scale investigations to help clarify the problem of selecting the appropriate motif size, which remains an open problem, and also to understand how Geometric Uniqueness can be combined with other motif design principles to produce more effective motifs.

Acknowledgements. This work is supported by a grant from the National Science Foundation NSF DBI-0318415. Additional support is gratefully acknowledged from training fellowships the Gulf Coast Consortia (NLM Grant No. 5T15LM07093) to B.C. and D.K.; from March of Dimes Grant FY03-93 to O.L.; from a Whitaker Biomedical Engineering Grant and a Sloan Fellowship to L.K; and from a VIGRE Training in Bioinformatics Grant from NSF DMS 0240058 to V.F. Experiments were run on equipment funded by NSF EIA-0216467 and NSF CNS-0523908. Large production runs were done on equipment supported by NSF CNS-042119, Rice University, and partnership with AMD and Cray. D.B. has been partially supported by the W.M. Keck Undergraduate Research Training Program and by the Brown School of Engineering at Rice University.

B.D. has been partially supported by the Rice Century Scholar Program and by the W.M. Keck Center. The authors are exceptionally grateful for the assistance of Anand P. Dharan, Colleen Kenney, Amanda E. Cruess and Yi-Chieh J. Wu.

References

1. Wolfson H.J. and Rigoutsos I. Geometric hashing: An overview. *IEEE Comp. Sci. Eng.*, 4(4):10–21, Oct 1997.
2. Barker J.A. and Thornton J.M. An algorithm for constraint-based structural template matching: application to 3D templates with statistical analysis. *Bioinf.*, 19 (13):1644–1649, 2003.
3. Chen B.Y. et al. Algorithms for structural comparison and statistical analysis of 3d protein motifs. *Proceedings of Pacific Symposium on Biocomputing 2005*, pages 334–45, 2005.
4. Stark A., Sunyaev S., and Russell R.B. A model for statistical significance of local similarities in structure. *J. Mol. Biol.*, 326:1307–1316, 2003.
5. Yao H. et. al. An accurate, sensitive, and scalable method to identify functional sites in protein structures. *J. Mol. Biol.*, 326:255–261, 2003.
6. Laskowski R.A., Watson J.D., and Thornton J.M. Protein function prediction using local 3d templates. *Journal of Molecular Biology*, 351:614–626, 2005.
7. Porter C.T., Bartlett G.J., and Thornton J.M. The catalytic site atlas: a resource of catalytic sites and residues identified in enzymes using structural data. *Nucleic Acids Research*, 32:D129–D133, 2004.
8. Shatsky M., Shulman-Peleg A., Nussinov R., and Wolfson H.J. Recognition of binding patterns common to a set of protein structures. *Proceedings of RECOMB 2005*, pages 440–55, 2005.
9. Lichtarge O., Bourne H.R., and Cohen F.E. An evolutionary trace method defines binding surfaces common to protein families. *J. Mol. Biol.*, 257(2):342–358, 1996.
10. Lichtarge O., Yamamoto K.R., and Cohen F.E. Identification of functional surfaces of the zinc binding domains of intracellular receptors. *J.Mol.Biol.*, 274:325–7, 1997.
11. Connolly M.L. Solvent-accessible surfaces of proteins and nucleic acids. *Science*, 221:709–713, 1983.
12. Kinoshita K. and Nakamura H. Identification of protein biochemical functions by similarity search using the molecular surface database ef-site. *Protein Science*, 12: 15891595, 2003.
13. Shatsky M., Nussinov R., and Wolfson H.J. Flexprot: Alignment of flexible protein structures without a predefinition of hinge regions. *Journal of Computational Biology*, 11(1):83–106, 2004.
14. Artymuik P.J. et. al. A graph-theoretic approach to the identification of three dimensional patterns of amino acid side chains in protein structures. *J. Mol. Biol.*, 243:327–344, 1994.
15. Bachar O. et. al. A computer vision based technique for 3-d sequence independent structural comparison of proteins. *Prot. Eng.*, 6(3):279–288, 1993.
16. Rosen M. et. al. Molecular shape comparisons in searches for active sites and functional similarity. *Prot. Eng.*, 11(4):263–277, 1998.
17. Wallace A.C., Laskowski R.A., and Thornton J.M. Derivation of 3D coordinate templates for searching structural databases. *Prot. Sci.*, 5:1001–13, 1996.
18. Silverman B.W. *Density Estimation for Statistics and Data Analysis.* Chapman and Hall: London, 1986.

19. Jones M.C., Marron J.S., and Sheather S.J. A brief survey of bandwidth selection for density estimation. *J. Amer. Stat. Assoc.*, 91:401–407, Mar 1996.

20. Sheather S.J. and Jones M.C. A reliable data-based bandwidth selections method for kernel density estimation. *J. Roy. Stat. Soc.*, 53(3):683–690, 1991.

21. Berman H.M. et. al. The protein data bank. *Nucleic Acids Research*, 28:235–242, Sept 2000.

22. Orengo C.A., Michie A.D., Jones S., Jones D.T., Swindells M.B., and Thornton J.M. Cath- a hierarchic classification of protein domain structures. *Structure.*, 5 (8):1093–1108, 1997.

23. Efron B. and Tibshirani R. The bootstrap method for standard errors, confidence intervals, and other measures of statistical accuracy. *Statistical Science*, 1(1):1–35, 1986.

24. Efron B. Better bootstrap confidence intervals (with discussion). *J. Amer. Stat. Assoc.*, 82:171, 1987.

25. Efron B. and Tibshirani R.J. *An Introduction to the Bootstrap.* Chappman & Hall, London, 1993.

26. Blow D.M., Birktoft J.J., and Hartley B.S. Role of a buried acid group in the mechanism of action of chymotrypsin. *Nature*, 221(178):337–40, Jan 1969.

27. Reyes V. et. al. Isomorphous crystal structures of *Escherichia coli* dihydrofolate reductase complexed with folate, 5-deazafolate, and 5,10-dideazatetrahydrofolate: mechanistic implications. *Biochemistry*, 34:2710–2723, 1995.

28. Bystroff C. et. al. Crystal structures of *Escherichia coli* dihydrofolate reductase: the nadp$^+$ holoenzyme and the folate-nadp$^+$ ternary complex. substrate binding and a model for the transition state. *Biochemistry*, 29:3263–3277, 1990.

29. Knochel T.R. et al. The crystal structure of indole-3-glycerol phosphate synthase from the hyperthermophilic archaeon sulfolobus solfataricus in three different crystal forms: effects of ionic strength. *J. Mol. Biol.*, 262:502–515, 1996.

30. Huang C.-C. et al. Crystal structures of mycolic acid cyclopropane synthases from mycobacterium tuberculosis. *J. Biol. Chem.*, 277:11559–11569, 2002.

31. Krengel U. and Dijkstra B.W. Three-dimensional structure of endo-1,4-beta-xylanase i from aspergillus niger: Molecular basis for its low ph optimum. *J. Mol. Biol.*, 263:70–78, 1996.

32. International Union of Biochemistry. Nomenclature Committee. *Enzyme Nomenclature.* Academic Press: San Diego, California, 1992.

33. Snir M. and Gropp W. *MPI: The Complete Reference (2nd Edition).* The MIT Press, 1998.

A Branch-and-Reduce Algorithm for the Contact Map Overlap Problem

Wei Xie and Nikolaos V. Sahinidis*

Department of Chemical and Biomolecular Engineering,
University of Illinois at Urbana-Champaign,
600 South Mathews Avenue, Urbana, IL, 61801
nikos@uiuc.edu

Abstract. A fundamental problem in molecular biology is the comparison of 3-dimensional protein folds in order to develop similarity measures and exploit them for protein clustering, database searches, and drug design. Contact map overlap (CMO) is one of the most reliable and robust measures of protein structure similarity. Fold comparison can be done by aligning the amino acid residues of two proteins in a way that maximizes the number of common residue contacts. CMO maximization is gaining increasing attention because it results in protein clusterings in good agreement with classification by experts. However, CMO maximization is an $\mathcal{NP}$-hard problem and few exact algorithms exist for solving this problem to global optimality.

In this paper, we propose a branch-and-reduce exact algorithm for the CMO problem. Contrary to previous approaches, we do not transform CMO to other combinatorial optimization problems for solution. Instead, we address the problem directly in its natural form. By exploiting the problem's mathematical structure, we develop bounding and reduction procedures that lead to a very efficient algorithm. We present extensive computational results for over 36000 test problems from the literature. These results demonstrate that our algorithm is significantly faster and solves many more challenging test sets than the best previous algorithms for CMO. Furthermore, the algorithm results in protein clusters that are in excellent agreement with the SCOP database.

1 Introduction

Understanding the function of genes and proteins is a challenging problem in molecular biology and is gaining increasing attention as is suggested by the recent Protein Structure Initiative [1]. The task is now possible thanks to the large number of structures that have been deposited in databases [2, 3]. To facilitate high throughput gene function prediction, it is important to be able to automatically find sequences that have similar structures to a given sequence. This automatic similarity detection requires efficient algorithms for aligning the structures of different proteins in a way that highlights their similarities.

---

* Corresponding author.

A. Apostolico et al. (Eds.): RECOMB 2006, LNBI 3909, pp. 516–529, 2006.

Sequence alignment methods, together with numerous scoring matrices and gap penalty schemes, have been extensively studied and used in the molecular biology literature. Recent work in profile-based sequence alignment improves the sensitivity of finding remote homologs. However, evidence suggests that structure alignment methods are usually more accurate than sequence alignment [4]. As a result, interest in structure alignment methods has grown because of their "standards of truth" certificates [5, 6].

As many established databases are devoted to protein structure and function, various algorithms have been proposed to solve the structural alignment problem according to different criteria [7]. Among various approaches that have been studied over the years, the contact map overlap maximization problem [8, 9] has received special attention because it provides protein clusters that agree well with classification by experts. To compare proteins A and B using CMO, contact maps are first extracted from each structure separately by identifying geometrically close residues. Depending on how such spatial closeness is measured, several definitions of contact exist in the literature [10, 11, 12]. Then, the contact maps of proteins A and B are superimposed so as to maximize the common contacts between the aligned residues. Optimal mapping of two contact maps usually provides interesting insight into the structural similarity of the original protein structures.

CMO is known to be $\mathcal{NP}$-hard [13] and exact algorithms for CMO are exponential in the worst case. Carr et al. [14] and Lancia et al. [15] proposed to solve the CMO maximization problem as a mixed-integer linear program. Their method is the first approach that guarantees global optimality. They also proposed a number of strong cuts by showing that instances of CMO can be reduced to instances of the maximum independence set problem. Later, Caprara and Lancia [16] and Caprara et al. [12] proposed a Lagrangian relaxation procedure to yield an upper bound for CMO. Combined with several clever lower bounding methods, Caprara et al. managed to solve a large number of practically interesting CMO instances. Strickland et al. [17] observed that CMO instances can be reduced to instances of the maximum clique problem. In addition to this observation, they studied a number of reduction techniques, and showed that these methods dramatically reduce problem size. Their method guarantees a globally optimal alignment via branch-and-bound on a reduced maximum clique problem instance.

In this paper, we take a different approach to solve the CMO problem to global optimality. Instead of reducing CMO to other problems, we solve this problem *directly*. We feel it is easier to exploit the mathematical structure of the problem in its original setting, and use dynamic programming as our major tool to design an efficient branch-and-bound algorithm. Due to the extensive application of reduction techniques in this algorithm, we refer to this algorithm as a branch-and-*reduce* algorithm.

The remainder of this paper is organized as follows. In section 2, we introduce notation necessary to define CMO in mathematical terms. Then, we provide the description of our algorithm in Section 3. Due to space limitations, we omit all the proofs and include them in a forthcoming paper [18]. Several test sets are used in Section 4 to demonstrate the power of our algorithm. Finally, in Section 5, we discuss our experience of solving hard CMO instances and suggest some directions for future work.

2   Problem Statement

Consider proteins A and B with m and n residues, respectively. Let x, x', and x'' take values from $\{1, 2, \ldots, m\}$ denoting residues on protein A. Similarly, y, y', and y'' take values from $\{1, 2, \ldots, n\}$ and denote residues on protein B. The contact map of each protein is described via the *contact map matrices* E^A and E^B, of dimension $m \times m$ and $n \times n$, respectively. In particular, element $E^A_{xx'}$ (resp. $E^B_{yy'}$) has a value of 1 if and only if residues x and x' (resp. y and y') are in contact, and 0 otherwise. "In contact" here means that the two corresponding residues are within a prespecified distance. A contact map matrix can be thought of as the node-incidence matrix of an undirect *contact map graph*, where nodes correspond to amino acid residues and an edge between any two nodes exists if and only if the corresponding residues are in contact.

The contact map problem is to find the largest common substructure for any two given protein contact map graphs. For this, we must develop a correspondence (alignment) between the node sets (residues) of the two contact graphs and identify the number of common contacts (edges) for all corresponding node pairs. If, in this process, residue x aligns to residue y, we say that (x, y) forms a *pair*. If a common contact results by aligning residue x to residue y and residue x' to residue y', we say that pairs (x, y) and (x', y') lead to an *overlap*. If pairs (x, y) and (x', y') form an overlap, we will set $h(x, y, x', y')$ to 1; otherwise, $h(x, y, x', y')$ will be set to zero.

A prerequisite for two pairs to form an overlap is that the order of residues in the original proteins be preserved. In other words, either the pair of inequalities $x < x'$ and $y < y'$ or the pair of inequalities $x > x'$ and $y > y'$ must hold for pairs (x, y) and (x', y') to be considered for an overlap. This requirement is the *non-crossing* property in the CMO literature. Two non-crossing pairs are also called *parallel* pairs.

We use the interval product $[x, x'] \times [y, y']$ to denote the set of pairs $\{(x'', y'') : x \leq x'' \leq x', y \leq y'' \leq y'\}$. For any given set of residue pairs S, the set $Q(S)$ is defined as the set of subsets of S that contain only parallel pairs. The CMO problem is then to find a one-to-one mapping from a subset of residues in one protein to a subset of residues in another protein so that the resultant number of overlaps is maximized. In mathematical terms, the CMO problem for proteins A and B defined above is:

$$\max_{Q' \in Q([1,m] \times [1,n])} \frac{1}{2} \sum_{(x,y) \in Q'} \sum_{(x',y') \in Q'} h(x, y, x', y')$$

Figure 1 provides an example of a protein with seven residues and a protein with nine residues, together with their contact map graphs (solid edges). Optimal

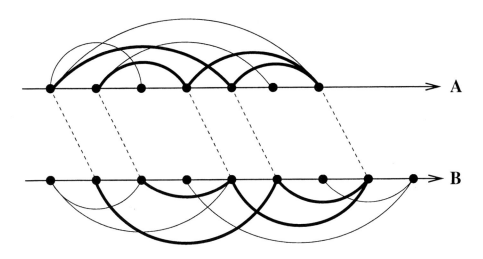

Fig. 1. Example of CMO problem

alignments of four overlaps are indicated by connecting corresponding residues with dashed lines, and contacts that result in overlaps are highlighted.

3   Branch-and-Reduce Algorithm

Our algorithm is based on the principles of branch-and-bound. We start by creating a root node that permits all possible residue pairs to be considered in the final solution. We estimate lower and upper bounds on the maximal overlap for this node. These bounds are bounds on the objective of the original CMO instance. If these lower and upper bounds coincide we terminate with an optimal solution. Otherwise, we choose a pair from all admissible pairs and construct two children nodes by enforcing this pair in one of the two children nodes while disallowing it in the second node. Lower and upper bounds for the optimal objective are calculated for each of the two children nodes and the process is repeated for each descendant while nodes are deleted whenever they are found to be inferior to mappings identified in the process. Application of this procedure in a tree search guarantees that the algorithm will identify a globally optimal solution in a finite number of steps.

In each node V of the branch-and-bound tree, we are dealing with a CMO instance with three different types of pairs:

1. All pairs $(x, y) \in \mathbb{C}(V)$ must be included in the solution. Here, $\mathbb{C}(V)$ denotes the set of *chosen* pairs for node V.
2. None of the pairs $(x, y) \in \mathbb{D}(V)$ can be included in the solution. Here, $\mathbb{D}(V)$ denotes the set of *disallowed* pairs for node V.
3. The remaining pairs have the freedom to enter or not enter the final solutions and form the set of *free* pairs, $\mathbb{F}(V)$, for node V.

Clearly, an instance of this restricted CMO problem reduces to an instance of CMO when $\mathbb{F}(V) = [1, m] \times [1, n]$ and $\mathbb{C}(V) = \mathbb{D}(V) = \emptyset$. We use $\mathrm{RCMO}(V)$ to denote the restricted CMO on node V.

An outline of our algorithm is shown in Algorithm 1. This algorithm emphasizes the use of reduction techniques. It also relies on a number of observations to simplify the problem. In the subsequent subsections, we discuss the details of these tools.

Algorithm 1. Branch-and-Reduce Algorithm

1: Create a root node V_0 with $\mathbb{C}(V_0) = \mathbb{D}(V_0) = \emptyset$, and $\mathbb{F}(V_0) = [1, m] \times [1, n]$. Set the upper bound $UB(V_0) = \infty$ and the lower bound $LB(V_0) = -\infty$. Initialize the list of open nodes $L = \{V_0\}$ and the value of the incumbent to $LB = -\infty$.
2: **while** $|L| > 0$ **do**
3:    Choose a node $V \in L$ with the highest $UB(V)$ and let $L = L \backslash \{V\}$.
4:    Apply *Simple Reduction I* (Proposition 4).
5:    Eliminate all $(x, y) \in \mathbb{C}(V)$ as described in Proposition 1.
6:    Apply *Simple Reduction II* (Proposition 5), *Reduction via Dominance* (Proposition 6), and *Reduction via Objective* (Proposition 7).
7:    Compute an upper bound for V. If $UB(V) \leq LB$, delete V and finish this iteration.
8:    Compute a lower bound for V. If $LB(V) > LB$, update LB and the best solution found.
9:    Choose a pair $(x, y) \in \mathbb{F}(V)$ to branch. Create two children nodes V_1 and V_2 and let $UB(V_1) = UB(V)$, $\mathbb{C}(V_1) = \mathbb{C}(V) \cup \{(x, y)\}$, $\mathbb{F}(V_1) = \mathbb{F}(V) \backslash \{(x, y)\}$, $\mathbb{D}(V_1) = \mathbb{D}(V)$, $UB(V_2) = UB(V)$, $\mathbb{C}(V_2) = \mathbb{C}(V)$, $\mathbb{F}(V_2) = \mathbb{F}(V) \backslash \{(x, y)\}$, and $\mathbb{D}(V_2) = \mathbb{F}(V) \cup \{(x, y)\}$. Update $L = L \cup \{V_1, V_2\} \backslash \{V\}$.
10: **end while**
11: Return LB and the best solution found.

3.1   Pair Elimination

Recall that an instance of RCMO in node V is defined over a set of chosen pairs $\mathbb{C}(V)$, a set of disallowed pairs $\mathbb{D}(V)$, and a set of free pairs $\mathbb{F}(V)$. The following observation suggests $\mathbb{C}(V)$ can be removed from consideration.

Proposition 1. *Problem* $\mathrm{RCMO}(V)$ *is equivalent to the optimization problem*

$$\max_{Q' \in Q(\mathbb{F}(V))} \sum_{(x', y') \in Q'} g(x', y') + \frac{1}{2} \sum_{(x', y') \in Q'} \sum_{(x'', y'') \in Q'} h(x', y', x'', y''). \quad (1)$$

where

$$g(x', y') := \sum_{(x'', y'') \in \mathbb{C}(V)} h(x', y', x'', y''), \forall (x', y') \in \mathbb{F}(V)$$

In particular:

1. *An optimal solution of (1) is also an optimal solution for* $\mathrm{RCMO}(V)$ *(excluding the pairs in* $\mathbb{C}(V)$*).*

2. *The optimal objective function value of (1) is obtained from the optimal objective function value of* $\mathrm{RCMO}(V)$ *by subtracting the quantity:*

$$\sigma(V) := \frac{1}{2} \sum_{(x',y') \in \mathbb{C}(V)} \sum_{(x'',y'') \in \mathbb{C}(V)} h(x',y',x'',y'').$$

For convenience of the discussion to follow, we define (1) as the *Generalized Contact Map Overlap problem* (GCMO), and use $\mathrm{GCMO}(V)$ to denote the GCMO instance on node V.

3.2 Interval Contribution for Residue Pair

We define $\bar{Q}(x,y,S)$ to be the set of all subsets of pairs in S that are mutually parallel and parallel to pair (x,y), and define array $p(x,y,S)$ to be the number of maximal overlaps with pair (x,y) for a subset of parallel pairs from a given set S, i.e.

$$p(x,y,S) := \max_{Q' \in \bar{Q}(x,y,S)} \sum_{(x',y') \in Q'} h(x,y,x',y').$$

We will use $p^-(x,y,x',y')$ to denote $p(x,y,S)$ when S is in the form of $[1,x'] \times [1,y'] \cap \mathbb{F}(V)$, and $p^+(x,y,x',y')$ when S is in the form of $[x',m] \times [y',n] \cap \mathbb{F}(V)$. In addition, we define

$$\bar{h}(x,y,x',y') := \begin{cases} h(x,y,x',y'), & \text{if } (x',y') \in \mathbb{F}(v), \\ 0, & \text{otherwise.} \end{cases}$$

Proposition 2 and 3 suggest an efficient algorithm to compute $p^+(x,y,x',y')$ and $p^-(x,y,x',y')$.

Proposition 2. *Computing all* $p^+(x,y,x',y')$ *and* $p^-(x,y,x',y')$ *for a given* $(x,y) \in \mathbb{F}(V)$ *requires* $\mathbb{O}(mn)$ *time complexity and* $\mathbb{O}(mn)$ *space complexity. Retrieval of* $p^+(x,y,x',y')$ *and* $p^-(x,y,x',y')$ *requires* $\mathbb{O}(1)$.

Proposition 3. *The time and space complexities for computing* $p^+(x,y,x',y')$ *and* $p^-(x,y,x',y')$ *in Proposition 2 can be improved to* $O(N_A N_B)$ *with preprocessing time complexity* $\mathbb{O}(m+n)$*, where* N_A *(resp.* N_B*) is the maximal node degree of the contact graph of protein A (resp. protein B). Retrieval of a single* $p^+(x,y,x',y')$ *or* $p^-(x,y,x',y')$ *still requires* $\mathbb{O}(1)$ *time.*

3.3 Reduction

The purpose of reduction is to develop techniques to eliminate inferior pairs early in the search process so as to reduce the computational effort in subsequent nodes. In this section, we discuss four types of reduction techniques that we have developed, namely, Simple Reduction I (Proposition 4), Simple Reduction II (Proposition 5), Reduction via Domination (Proposition 6), and Reduction via Objective (Proposition 7).

The simple reduction principles are observations that allow fast identification of some free pairs as disallowed via the next two propositions:

Proposition 4. *A pair* $(x, y) \in \mathbb{F}(V)$ *that crosses some* $(x', y') \in \mathbb{C}(V)$ *can be immediately eliminated by setting* $(x, y) \in \mathbb{D}(V)$.

Proposition 5. *If pair* $(x, y) \in \mathbb{F}(V)$ *satisfies* $g(x, y) = 0$ *and* $\bar{h}(x, y, x', y') = 0$ *for any* $(x', y') \in \mathbb{F}(V)$, *then there must exist an optimal solution* $Q^\dagger$ *for* GCMO(V) *that does not contain* (x, y). *Therefore, pair* (x, y) *can be eliminated.*

The general principle for reduction via domination is that an pair may be disallowed if we can show that, for any solution that contains this pair, substituting this pair with another one leads to a solution that is not worse. Direct application of this principle to RCMO(V) or the associated GCMO(V) is not straightforward, mainly due to the fact that the substituting pair may cross pairs parallel to the substituted pair. However, some special pairs permit this type of reduction as follows:

Proposition 6. *Define the* upper-left corner *to be the minimal residue* x_L *from amongst those* x *with some* $(x, y) \in \mathbb{F}(V)$. *For any two pairs* $(x_L, y_1) \in \mathbb{F}(V)$ *and* $(x_L, y_2) \in \mathbb{F}(V)$ *with* $y_1 < y_2$, *substituting* (x_L, y_1) *for* (x_L, y_2) *in any feasible solution for* RCMO(V) *always results in another feasible solution. Further,* (x_L, y_1) *dominates* (x_L, y_2) *if*

$$g(x_L, y_1) - g(x_L, y_2) + \sum_{y \in \{y_2+1, \dots, n\}} \min\{0, E^B_{y_1 y} - E^B_{y_2 y}\} \geq 0.$$

Similar claims hold for the *lower-left corner*, the *upper-right corner*, and the *lower-right corner*, which are defined in a similar manner as the upper-left corner.

Finally, we can eliminate a pair $(x, y) \in \mathbb{F}(V)$ if we can show that any solution containing it has no better objective than LB. Although direct application of this criterion is practically impossible because it is equivalent to solving a CMO instance of smaller size, we can over-estimate the objective in order to apply this rule as follows:

Proposition 7. *There must exist an optimal solution for* RCMO(V) *that does not contain pair* $(x, y) \in \mathbb{F}(V)$ *if the following holds:*

$$g(x, y) + \left\lfloor \max_{Q' \in \bar{Q}(x, y, \mathbb{F}(V))} \sum_{(x', y') \in Q'} \left(g(x', y') + h(x', y', x, y) + \tfrac{1}{2} s(x', y', x, y) \right) \right\rfloor$$
$$\leq LB - \sigma(V)$$

where

$$s(x', y', x, y) := p^-(x', y', x-1, y-1) + p^+(x', y', x+1, y+1).$$

3.4   Upper and Lower Bounding

Bounding RCMO(V) from below usually involves finding a good feasible solution, while bounding it from above requires constructing a relaxation that over-estimates the objective. In this subsection, we develop such bounds.

Proposition 8. *Define*

$$t(x,y) := \begin{cases} g(x,y), & \text{if } (x,y) \in \mathbb{F}(V); \ x=1 \text{ or } y=1, \\ \max\{g(x,y), g(x,y)+w(x,y)\}, & \text{if } (x,y) \in \mathbb{F}(V); \ x>1 \text{ and } y>1, \\ -\infty, & \text{otherwise}, \end{cases} \tag{2}$$

where

$$w(x,y) := \begin{cases} -\infty, & \\ \qquad \text{if } [1,x-1] \times [1,y-1] \cap \mathbb{F}(V) = \emptyset \\ \max_{(x',y')\in[1,x-1]\times[1,y-1]\cap\mathbb{F}(V)} \{t(x',y') + \bar{h}(x',y',x,y) \\ \qquad\qquad\qquad +\frac{1}{2}u(x',y',x,y)\}, \\ \qquad\text{otherwise}. \end{cases}$$

and

$$u(x',y',x,y) := p^+(x',y',x+1,y+1) + p^-(x,y,x'-1,y'-1).$$

Then

$$\lfloor \sigma(V) + \max_{(x,y)\in\mathbb{F}(V)} t(x,y) \rfloor \tag{3}$$

is an upper bound for RCMO(V).

While computing the above upper bound, we can keep track of the pair (x',y') that achieves the maximum in (2) for each (x,y). In particular, we define *previous pairs* as $\mathrm{pr}(x,y)$ to be equal to $(0,0)$ if $t(x,y) = g(x,y)$. Otherwise, we set:

$$\mathrm{pr}(x,y) = \mathrm{argmax}_{(x',y')\in[1,x-1]\times[1,y-1]\cap\mathbb{F}(V)} \{t(x',y') + \bar{h}(x',y',x,y) \\ +\frac{1}{2}u(x',y',x,y)\}.$$

Once we have determined the pair (x^*, y^*) that achieves the maximum in (3), we can backtrack and add recursively these previous pairs to set $Q^\S$ (see Algorithm 2).

Algorithm 2. Backtrack $Q^\S$

1: $Q^\S = \emptyset$; $(x,y) = (x^*, y^*)$
2: **while** $(x,y) \neq (0,0)$ **do**
3:    $Q^\S = Q^\S \cup \{(x,y)\}$; $(x,y) = \mathrm{pr}(x,y)$
4: **end while**
5: return $Q^\S$.

Proposition 9. *The pairs in set* $Q^\S := \{(x_i^\S, y_i^\S)\}$ *provide a feasible solution for* GCMO(V).

Therefore, we simply take

$$\sigma(V) + \sum_i \left(g(x_i^\S, y_i^\S) + \frac{1}{2}\sum_j h(x_i^\S, y_i^\S, x_j^\S, y_j^\S) \right)$$

as a lower bound for RCMO(V). In fact, computational experience indicates that this lower bound is likely to yield good solutions very quickly during the branch-and-reduce search.

3.5 Branching

As is shown in Algorithm 1, we need to choose a pair $(x, y) \in \mathbb{F}(V)$ to branch on if the current node V cannot be pruned. Although many alternatives exist, we choose the branching pair $(x_{\mathrm{br}}, y_{\mathrm{br}})$ as follows:

$$(x_{\mathrm{br}}, y_{\mathrm{br}}) \in \mathrm{argmax}_{(x_i^\S, y_i^\S) \in Q^\S} \left\{ g(x_i^\S, y_i^\S) + p^-(x_i^\S, y_i^\S, x_i^\S - 1, y_i^\S - 1) \right.$$

$$\left. + p^+(x_i^\S, y_i^\S, x_i^\S + 1, y_i^\S + 1) \right\}. \quad (4)$$

The philosophy behind such a choice is to induce a balanced branch-and-bound tree, which usually suggests an overall small tree size. The child node V_2 with the $(x_{\mathrm{br}}, y_{\mathrm{br}})$ set to $\mathbb{D}$ has more *degrees of freedom* than its brother node V_1 and is, therefore, more likely to result in a higher upper bound. By choosing a pair that maximizes (4), we hope that the upper bound for V_2 achieves maximal decrement compared to its parent and is, therefore, close to the upper bound of V_1.

4 Computational Studies

In this section, we use three sets of test problems to study the performance of the proposed algorithm.

The first test set is taken from [17] and contains 11 alignment instances of medium size proteins. Results for this test set are provided in Table 1. The first column of this table shows the pairs of proteins to be aligned. The second column shows the maximal contact overlaps. The remaining three columns of the table present CPU times for three algorithms. For the first two algorithms, we report the results earlier reported in [14] and [17]. Although the contact maps of the three algorithms depend on the thresholds used and may not be identical, they are likely to be very similar. Even when the different computing platform's LINPACK scores are taken into account, we can see that the proposed algorithm is faster than that of [14] by one to two orders of magnitude and at least as fast as that of [17] for all instances except one. More importantly, for the top five instances, which are difficult, the proposed algorithm is 3 to 9 times faster than [17].

Next, we test the branch-and-reduce algorithm on a set of 36046 pairwise alignment instances of 269 medium size proteins. This set was initially introduced by Lancia et al. [15]. We ran the proposed algorithm for ten days on three workstations with 3.0 GHz CPU and 1.0 G RAM each. Over this time, the algorithm managed to solve 2309 instances to global optimality. Only 1680 of these instances were solved to global optimality in the past [12].

Table 1. Results for the Sokol test set

Pair	Obj.	CPU sec		
		Carr et al. [14] [a]	Strickland et al. [17] [b]	Branch-and-Bound [c]
3ebx-1era	31	487	236	0.72
1bpi-2knt	29	423	182	0.46
1knt-1bpi	30	331	110	0.38
6ebx-1era	20	427	101	0.73
2knt-5pti	28	760	95	0.32
1bpi-1knt	31	331	19	0.24
1bpi-5pti	42	320	30	0.29
1knt-2knt	39	52	0	0.15
1knt-5pti	28	934	46	0.57
1vii-1cph	6	12	0	0.00
3ebx-6ebx	28	388	6	0.12

[a]Computing hardware is similar to Strickland et al. [17].
[b]SGI workstation with 200 MHz CPU and LINPACK score of 32 [19].
[c]Dell workstation with a 3.0 GHz P4 processor and LINPACK score of 1414 [19].

The third test set involves 780 pairwise alignment instances. Initially suggested by Skolnick and described in [15], this set contains 40 large protein domains: 1b00A (1), 1dbwA (2), 1nat (3), 1ntr (4), 1qmpA (5), 1qmpB (6), 1qmpC (7), 1qmpD (8), 1rn1A (9), 1rn1B (10), 1rn1C (11), 3chy(12), 4tmyA (13), 4tmyB (14), 1bawA (15), 1byoA (16), 1byoB (17), 1kdi (18), 1nin (19), 1pla (20), 2b3iA (21), 2pcy (22), 2plt (23), 1amk (24), 1aw2A (25), 1b9bA (26), 1btmA (27), 1htiA (28), 1tmhA (29), 1treA (30), 1tri (31), 3ypiA (32), 8timA (33), 1ydvA (34), 1b71A (35), 1bcfA (36), 1dpsA (37), 1fha (38), 1ier (39) and 1rcd (40). Each domain entry contains a name and its assigned index in parentheses. The domain name is the PDB code for the protein containing it. The chain index is appended to the PDB code whenever the protein has multiple chains. As shown in Table 2, these 40 protein domains are divided into four categories.

Table 2. Protein domains in the Skolnick test set

Categories	Residues	Sequence similarity	Domain indices
1	124	15–30%	1–14
2	99	35–90%	15–23
3	250	30–90%	24–34
4	170	7–70%	35–40

In order to generate contact maps, we consider two residues to be in contact if their C_α's are within 7Å, as suggested by [10]. Routines in the BALL package [20] were used to compute the distances between residues and generate the contact maps. The results are depicted in Fig. 2, where a • indicates instances solved by both [12] and the proposed algorithm, while a × indicates problems that were solved only by the proposed algorithm. It can be seen in this figure

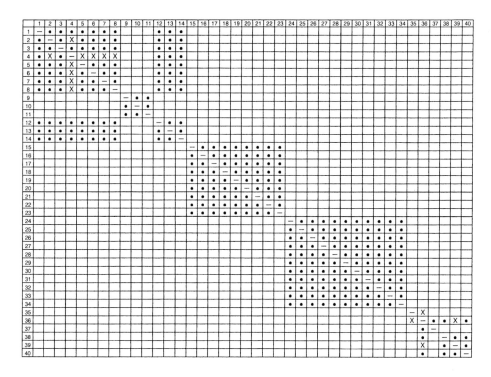

Fig. 2. Instances of the Skolnick test set solved by [12] and the proposed algorithm (●=solved by [12] and the proposed algorithm, ×=solved only by the proposed algorithm)

that the proposed algorithm solves 7 more instances in addition to the 150 instances solved earlier by [12]. However, neither [12] nor the proposed algorithm are able to solve a single instance that aligns two protein domains from different categories. This is due to the fact that such instances are much harder than instances within the same category, as is argued by [12].

In order to measure the structural similarity of two domains z_1 and z_2, we use the similarity index

$$\theta(z_1, z_2) := \frac{2\mathrm{LB}(z_1, z_2)}{\mathcal{E}(z_1) + \mathcal{E}(z_2)},$$

where $\mathrm{LB}(z_1, z_2)$ is the best objective found by aligning domain z_1 with z_2. We also introduce $\mathcal{E}(z_1)$ and $\mathcal{E}(z_2)$ to denote the total number of contacts for domains z_1 and z_2, respectively. As a result of this definition, the more similar two domains are, the closer the similarity index is to one.

In the left side of Fig. 3, we plot $\theta(z_1, z_2)$ against z_1 and z_2. It is easy to see that five clusters result from pairwise alignments by permuting 12 (3chy), 13 (4tmyA), and 14 (4tmyB), with 9 (1rn1A), 10 (1rn1B), and 11 (1rn1C), which is shown in the right side of Fig. 3. This result is in excellent agreement with the SCOP database (version 1.69) [21, 22], as is shown in Table 3.

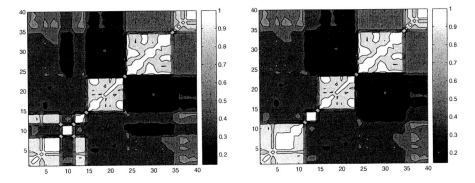

Fig. 3. Cluster analysis on the Skolnick test set

Table 3. Clusters of proteins in the Skolnick test set

Cluster	Domains	SCOP		
		Fold	Superfamily	Family
1	1–8,12–14	Flavodoxin-like	Che Y-like	Che Y-related
2	9–11	Microbial ribonucleases	Microbial ribonucleases	Fungi ribonucleases
3	15–23	Cuperdoxin-like	Cuperdoxins	Plastocyanin/azurin-like
4	24–34	TIM-beta/alpha-barrel	Triosephosphate isomerase (TIM)	Triosephosphate isomerase (TIM)
5	35–40	Ferritin-like	Ferritin-like	Ferritin

5 Conclusions

The main contribution of this work is the development of a branch-and-reduce algorithm for the CMO problem. It is quite interesting to note that the relatively weaker bound of (3) results in a more efficient algorithm than a stronger bound from a branch-and-cut method [14]. This is likely due to the fact that a strong bound may take too much time to compute to justify its merit. Perhaps more important in the context of hard CMO problems is the role of reduction. For large and difficult instances, even the relatively cheap upper bound (3) becomes expensive, as we observed by looking at the computational times at the root node and its immediate children. However, subsequent nodes turn out to be much easier because branch-and-bound manages to remove a large number of inferior pairs. Therefore, it is of particular interest to consider the development of even more powerful reduction techniques that can be computed efficiently. Since we have not thoroughly tested alternative branching schemes, another interesting future research direction would be to study whether better branching schemes can lead to significant improvement of the branch-and-reduce algorithm.

Aligning proteins with different structures still seems much harder than those with similar structures. Although we managed to solve several instances with medium size proteins (e.g., around 60 residues) in two hours of CPU time on a standard workstation, attempts on certain larger instances failed. However, alignments are usually performed on pairs with medium to significant level of similarity, where global optimization algorithms can solve fairly large instances and provide critical insights to molecular biologists.

Acknowledgement

The authors are grateful to Professors Giuseppe Lancia and Joel Sokol who kindly provided us their test problems for this work, and to three anonymous referees whose comments significantly improved the presentation of this work.

References

1. NIH: Protein structural initiative: Better tools and better knowledge for structural genomics. (Web) http://nigms.nih.gov/psi/.
2. Berman, H.M., Westbrook, J., Feng, Z., Gilliland, G., Bhat, T.N., Weissig, H., Shindyalov, I.N., Bourne, P.E.: The protein data bank. Nucleic Acids Research **28** (2000) 235–242
3. Hulo, N., Sigrist, C.J.A., Saux, V.L., Langendijk-Genevaux, P.S., Bordoli, L., Gattiker, A., De Castro, E., Bucher, P., Bairoch, A.: Recent improvements to the PROSITE database. Nucleic Acids Research **32** (2004) 134–137
4. Pearson, W.R., Sierk, M.L.: The limits of protein sequence comparison? Current opinion in structural biology **15** (2005) 254–260
5. Vogt, G., Etzold, T., Argos, P.: An assessment of amino acid exchange matrices in aligning protein sequences: The twilight zone revisited. Journal of Molecular Biology **249** (1995) 816–831
6. Orengo, C.A., Michie, A.D., Jones, S., Jones, D.T., Swindells, M.B., Thornton, J.M.: CATH-A hierarchic classification of protein domain structures. Structure **5** (1997) 1093–1108
7. Godzik, A.: The structural alignment between two proteins: Is there a unique answer? Protein science **5** (1996) 1325–1338
8. Godzik, A., Skolnick, J., Kolinski, A.: A topology fingerprint approach to inverse protein folding problem. Journal of Molecular Biology **227** (1992) 227–238
9. Godzik, A., Skolnick, J.: Flexible algorithm for direct multiple alignment of protein structures and sequences. Computer applications in biosciences: CABIOS **10** (1994) 587–596
10. Zaki, M.J., Jin, S., Bystroff, C.: Mining residue contacts in proteins using local structure predictions. In: Proceedings. IEEE Symposium on Bioinformatics and Biomedical Engineering, Los Alamitos, CA, IEEE Computer Society (2000) 168–175
11. Zhao, Y., Karypis, G.: Prediction of contact maps using support vector machines. In: Proceedings. Third IEEE International Symposium on Bioinformatics and Bioengineering, Los Alamitos, CA, IEEE Computer Society (2003) 26–36
12. Caprara, A., Carr, R., Istrail, S., Lancia, G., Walenz, B.: 1001 optimal PDB structure alignments: Integer programming methods for finding the maximum contact map overlap. Journal of Computational Biology **11** (2004) 27–52

13. Goldman, D.: Algorithmic aspects of protein folding and protein structure similarity. PhD thesis, University of California at Berkeley (2000)
14. Carr, R.D., Lancia, G., Istrail, S.: Branch-and-cut algorithms for independent set problems: Integrality gap and an application to protein structural alignment. Technical report, Sandia National laboratories (2000)
15. Lancia, G., Carr, R., Walenz, B., Istrail, S.: 101 optimal PDB structure alignments: A branch-and-cut algorithm for the maximum contact map overlap problem. In: Proceedings of Annual International Conference on Computational Biology (RECOMB). (2001) 193–202
16. Caprara, A., Lancia, G.: Structural alignment of large-size proteins via Lagrangian relaxation. In: Proceeding of Internation Conference on Computational Biology (RECOMB). (2002) 100–108
17. Strickland, D.M., Barnes, E., Sokol, J.S.: Optimal protein structure alignment using maximum cliques. Operations Research **53** (2005) 389–402
18. Xie, W., Sahinidis, N.V.: A reduction-based exact algorithm for the contact map overlap problem. In preparation (2005)
19. Dongarra, J.J.: Performance of various computers using standard linear equations software. Technical report, University of Tennessee, Knoxville, TN (2005) http://www.netlib.org/benchmark/performance.ps.
20. Kohlbacher, O., Lenhof, H.: BALL—Rapid software prototyping in computational molecular biology. Bioinformatics **16** (2000) 815–824
21. Murzin, A., Brenner, S.E., Hubbard, T., Chothia, C.: SCOP: A structural classification of protein database for the investigation of sequences and structures. Journal of Molecular Biology **247** (1995) 536–540
22. Andreeva, A., Howorth, D., Brenner, S.E., Hubbard, T.J.P., Chothia, C., Murzin, A.G.: SCOP database in 2004: Refinements integrate structure and sequence family data. Nucleic Acids Research **32** (2004) D226–D229

A Novel Minimized Dead-End Elimination Criterion and Its Application to Protein Redesign in a Hybrid Scoring and Search Algorithm for Computing Partition Functions over Molecular Ensembles

Ivelin Georgiev[1,*], Ryan H. Lilien[1,2,3,*], and Bruce R. Donald[1,3,4,5,**]

[1] Dartmouth Computer Science Department, Hanover, NH 03755, USA
[2] Dartmouth Medical School, Hanover, NH
[3] Dartmouth Center for Structural Biology and Computational Chemistry, Hanover, NH
[4] Dartmouth Department of Chemistry, Hanover, NH
[5] Dartmouth Department of Biological Sciences, Hanover, NH
Bruce.R.Donald@dartmouth.edu

Abstract. Novel molecular function can be achieved by redesigning an enzyme's active site so that it will perform its chemical reaction on a novel substrate. One of the main challenges for protein redesign is the efficient evaluation of a combinatorial number of candidate structures. The modeling of protein flexibility, typically by using a rotamer library of commonly-observed low-energy side-chain conformations, further increases the complexity of the redesign problem. A dominant algorithm for protein redesign is Dead-End Elimination (DEE), which prunes the majority of candidate conformations by eliminating *rigid* rotamers that provably are not part of the Global Minimum Energy Conformation (GMEC). The identified GMEC consists of rigid rotamers that have not been energy-minimized and is referred to as the *rigid-GMEC*. As a post-processing step, the conformations that survive DEE may be energy-minimized. When energy minimization is performed after pruning with DEE, the combined protein design process becomes heuristic, and is no longer provably accurate: That is, the rigid-GMEC and the conformation with the lowest energy among all energy-minimized conformations (the *minimized-GMEC*, or *minGMEC*) are likely to be different. While the traditional DEE algorithm succeeds in not pruning rotamers that are part of the rigid-GMEC, it makes no guarantees regarding the identification of the minGMEC. In this paper we derive a novel, provable, and efficient DEE-like algorithm, called *minimized-DEE (MinDEE)*, that guarantees that rotamers belonging to the minGMEC will not be pruned, while still pruning a combinatorial number of conformations. We show that MinDEE is useful not only in identifying the minGMEC, but also as a filter in an ensemble-based scoring and search algorithm for protein redesign that exploits energy-minimized conformations. We compare our results both to our previous computational predictions of protein designs and to biological activity assays of predicted protein mutants. Our provable and efficient minimized-DEE algorithm is applicable in protein redesign, protein-ligand binding prediction, and computer-aided drug design.

* These authors contributed equally to the work.
** This work is supported by grants to B.R.D. from the National Institutes of Health (R01 GM-65982), and the National Science Foundation (EIA-0305444).

A. Apostolico et al. (Eds.): RECOMB 2006, LNBI 3909, pp. 530–545, 2006.
© Springer-Verlag Berlin Heidelberg 2006

1   Introduction

Computational Protein Design. The ability to engineer proteins has many biomedical applications. Novel molecular function can be achieved by redesigning an enzyme's active site so that it will perform its chemical reaction on a novel substrate. A number of computational approaches to the protein redesign problem have been reported. To improve the accuracy of the redesign, protein flexibility has been incorporated into most previous structure-based algorithms for protein redesign [30, 14, 13, 12, 1, 18, 15]. A study of bound and unbound structures found that most structural changes involve only a small number of residues and that these changes are primarily side-chains, and not backbone [22]. Hence, many protein redesign algorithms use a rigid backbone and model side-chain flexibility with a rotamer library, containing a discrete set of low-energy commonly-observed side-chain conformations [19, 25]. The major challenge for redesign algorithms is the efficient evaluation of the exponential number of candidate conformations, resulting not only from mutating residues along the peptide chain, but also by employing rotamer libraries. The development of pruning conditions capable of eliminating the majority of mutation sequences and conformations in the early, and less costly, redesign stages has been crucial.

Non-ensemble-based algorithms for protein redesign are based on the assumption that protein folding and binding can be accurately predicted by examining the GMEC. Since identifying the GMEC using a model with a rigid backbone, a rotamer library, and a pairwise energy function is known to be NP-hard [24], different heuristic approaches (random sampling, neural network, and genetic algorithm) have been proposed [30, 14, 13, 12, 20]. A provable and efficient deterministic algorithm, which has become the dominant choice for non-ensemble-based protein design, is Dead-End Elimination (DEE) [6]. DEE reduces the size of the conformational search space by eliminating *rigid* rotamers that provably are not part of the GMEC. Most important, since no protein conformation containing a dead-ending rotamer is generated, DEE provides a combinatorial factor reduction in computational complexity.

When energy minimization is performed after pruning with DEE, the process becomes heuristic, and is no longer provably accurate: a conformation that is pruned using rigid-rotamer energies may subsequently minimize to a structure with lower energy than the rigid-GMEC. Therefore, the traditional DEE conditions are not valid for pruning rotamers when searching for the lowest-energy conformation among all energy-minimized rotameric conformations (the *minimized-GMEC*, or *minGMEC*).

NRPS Redesign and K^*. Traditional ribosomal peptide synthesis is complemented by non-ribosomal peptide synthetase (NRPS) enzymes in some bacteria and fungi. NRPS enzymes consist of several domains, each of which has a separate function. Substrate specificity is generally determined by the adenylation (A) domain [28, 3, 27]. Among the products of NRPS enzymes are natural antibiotics (penicillin, vancomycin), anti-fungals, antivirals, immunosuppressants, and antineoplastics. The redesign of NRPS enzymes can lead to the synthesis of novel NRPS products, such as new libraries of antibiotics [2]. The main techniques for NRPS enzyme redesign are *domain-swapping* [29, 26, 7, 21], *signature sequences* [28, 8, 3], and *active site manipulation from a structure-based mutation search utilizing ensemble docking* (the K^* method [17]).

The K^* algorithm [17] has been demonstrated for NRPS redesign, but is a general algorithm that is, in principle, capable of redesigning any protein. K^* is an ensemble-based scoring technique that uses a Boltzmann distribution to compute partition functions for the bound and unbound states of a protein. The ratio of the bound to the unbound partition function is used to compute a provably-good approximation (K^*) to the binding constant for a given sequence. A volume and a steric filter are applied in the initial stages of a redesign search to prune the majority of the conformations from more expensive evaluation. The number of evaluated conformations is further reduced by a provable ε-approximation algorithm. Protein flexibility is modeled for *both* the protein *and* the ligand using energy-minimization and rotamers [17].

Contributions of the Paper. Boltzmann probability implies that low-energy conformations are more likely to be assumed than high-energy conformations. The motivation behind energy minimization is therefore well-established and algorithms that incorporate energy minimization often lead to more accurate results. However, if energy minimization is performed *after* pruning with DEE, then the combined protein design process is heuristic, and not provable. We show that a conformation pruned using rigid-rotamer energies may subsequently minimize to surpass the putative rigid-GMEC.

We derive a novel, provable, and efficient DEE-like algorithm, called *minimized-DEE (MinDEE)*, that guarantees that no rotamers belonging to the minGMEC will be pruned. We show that our method is useful not only in (a) identifying the minGMEC (a non-ensemble-based method), but also (b) as a filter in an ensemble-based scoring and search algorithm for protein redesign that exploits energy-minimized conformations. We achieve (a) by implementing a MinDEE/A^* algorithm in a search to switch the binding affinity of the Phe-specific adenylation domain of the NRPS Gramicidin Synthetase A (GrsA-PheA) towards Leu. The latter goal (b) is achieved by implementing MinDEE as a combinatorial filter in a hybrid algorithm,[1] combining A^* search and our previous work on K^* [17]. The experimental results, based on a 2-point mutation search on the 9-residue active site of the GrsA-PheA enzyme, confirm that the new Hybrid MinDEE-K^* algorithm has a much higher pruning efficiency than the original K^* algorithm. Moreover, it takes only *30 seconds* for MinDEE to determine which rotamers can be provably pruned. We make the following contributions in this paper:

1. Derivation of MinDEE, a novel, provable, and efficient DEE-like algorithm that incorporates energy minimization, with applications in both non-ensemble- and ensemble-based protein design.
2. Introduction of a MinDEE/A^* algorithm that identifies the minGMEC and returns a set of low-energy conformations;
3. Introduction of a hybrid MinDEE-K^* ensemble-based scoring and search algorithm, improving on our previous work on K^* [17] by replacing a constant-factor with a combinatorial-factor provable pruning condition; and
4. The use of our novel algorithms in a redesign mutation search for switching the substrate specificity of the NRPS enzyme GrsA-PheA; we compare our results to previous computational predictions of protein designs and to biological activity assays of predicted protein mutants.

---

[1] For brevity, we will henceforth refer to this algorithm as the *Hybrid MinDEE-K^** algorithm.

2   Derivation of the Minimized-DEE Criterion

2.1   The Original DEE Criterion

In this section we briefly review the *traditional-DEE* theorem [6, 23, 11]. Traditional-DEE refers to the original DEE, which is not provably correct when used in a search for the minGMEC. The total energy, E_T, of a given rotameric-based conformation can be written as $E_T = E_{t'} + \sum_i E(i_r) + \sum_i \sum_{j>i} E(i_r, j_s)$, where $E_{t'}$ is the template self-energy (i.e., backbone energies or energies of rigid regions of the protein not subject to rotamer-based modeling), i_r denotes rotamer r at position i, $E(i_r)$ is the self energy of rotamer i_r (the intra-residue and residue-to-template energies), and $E(i_r, j_s)$ is the non-bonded pairwise interaction energy between rotamers i_r and j_s. The rotamers assumed in the rigid-GMEC are written with a subscript g. Therefore i_g is the rotamer assumed in the rigid-GMEC at position i. The following two bounds are then noted: for all i, j $(i \neq j)$, $\max_{s \in R_j} E(i_t, j_s) \geq E(i_t, j_g)$ and $\min_{s \in R_j} E(i_g, j_s) \leq E(i_g, j_g)$, where R_j is the set of allowed rotamers for residue j. For clarity, we will not include R_j in the limits of the max and min terms, since it will be clear from the notation from which set s must be drawn. The DEE criterion for rotamer i_r is defined as:

$$E(i_r) + \sum_{j \neq i} \min_s E(i_r, j_s) > E(i_t) + \sum_{j \neq i} \max_s E(i_t, j_s). \tag{1}$$

Any rotamer i_r satisfying the DEE criterion (Eq. 1) is provably not part of the rigid-GMEC $(i_r \neq i_g)$, and is considered 'dead-ending.' Extensions to this initial DEE criterion allow for additional pruning while maintaining correctness with respect to identifying the rigid-GMEC [6, 10, 11, 23].

2.2   DEE with Energy Minimization: MinDEE

We now derive generalized DEE pruning conditions which can be used when searching for the minGMEC. The fundamental difference between traditional-DEE and MinDEE

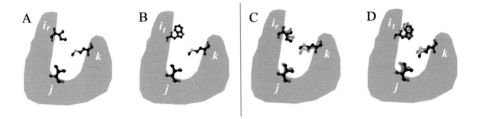

Fig. 1. Energy-Minimized DEE. Without energy minimization the swapping of rotamer i_r for i_t (A to B) leaves unchanged the conformations and self and pairwise energies of residues j and k. When energy minimization is allowed, the swapping of rotamer i_r for rotamer i_t (C to D) may cause the conformations of residues j and k to minimize (i.e., move) to form more energetically favorable interactions (from the faded to the solid conformations in C and D).

is that the former enjoys significant independence among multiple energy terms during a rotamer swap (Fig. 1). Therefore, to be provably correct, MinDEE must account for a range of possible energies. The conformation of a residue may change during energy minimization, however we constrain this movement to a region of conformation space called a *voxel* to keep one rotamer from minimizing into another. In this framework, the voxel $\mathcal{V}(i_r)$ for rotamer i_r is simply all conformations of residue i within a $\pm\theta$ range around each rotamer dihedral when starting from the rotamer[2] i_r. We similarly define the voxel $\mathcal{V}(i_r, j_s)$ for the pair of rotamers i_r and j_s to be the region of conformation space $\mathcal{V}(i_r) \times \mathcal{V}(j_s)$. Next, we can define the *maximum, minimum,* and *range* of voxel energies: $E_\oplus(i_r) = \max\limits_{z \in \mathcal{V}(i_r)} E(z)$; $E_\ominus(i_r) = \min\limits_{z \in \mathcal{V}(i_r)} E(z)$; and $E_\oslash(i_r) = E_\oplus(i_r) - E_\ominus(i_r)$. Analogous definitions exist for pairwise terms (Fig. 4 in [9, p. 25]). For a given protein, we define a *rotamer vector* $A = (A_1, A_2, \cdots, A_n)$ to specify the rotamer at each of the n residue positions; $A_i = r$ when rotamer r is assumed by residue i. We then define the *conformation vector* $A^\bullet = (A_1^\bullet, A_2^\bullet, \cdots, A_n^\bullet)$ such that $A_i^\bullet$ is the conformation of residue i in the voxel-constrained minimized conformation, i.e., $A_i^\bullet \in \mathcal{V}(A_i)$ and $A^\bullet = (A_1^\bullet, A_2^\bullet, \cdots, A_n^\bullet) = \underset{B=(B_1, B_2, \cdots, B_n) \in \prod_{i=1}^{n} \mathcal{V}(A_i)}{\operatorname{argmin}} E(B)$, where $E(B)$ is the energy of the system specified by conformation vector B. For the energy-minimized conformation starting from rotamer vector A, we define the self-energy of rotamer i_r as $E_\odot(i_r|A) = E(A_i^\bullet)$ and the pairwise interaction energy of the rotamer pair i_r, j_s as $E_\odot(i_r, j_s|A) = E(A_i^\bullet, A_j^\bullet)$ where $E(A_i^\bullet)$ is the self-energy of residue i in conformation $A_i^\bullet$ and $E(A_i^\bullet, A_j^\bullet)$ is the pairwise energy between residues i and j in conformations $A_i^\bullet$ and $A_j^\bullet$. We can then express the minimized energy of A, $E_T(A)$ as: $E_T(A) = E_{t'} + \sum_i E_\odot(i_r|A) + \sum_i \sum_{j>i} E_\odot(i_r, j_s|A)$. Let G represent the rotamer vector that minimizes into the minGMEC and $E_T(G)$ be the energy of the minGMEC. Let $G_{i_g \to i_t}$ be the rotamer vector G where rotamer i_g is replaced with i_t. We know that $E_T(G_{i_g \to i_t}) \geq E_T(G)$, so we can pull residue i out of the two summations, obtaining:

$$E_{t'} + E_\odot(i_t|G_{i_g \to i_t}) + \sum_{j \neq i} E_\odot(i_t, j_g|G_{i_g \to i_t}) + \sum_{j \neq i} E_\odot(j_g|G_{i_g \to i_t})$$

$$+ \sum_{j \neq i} \sum_{k \neq i, k > j} E_\odot(j_g, k_g|G_{i_g \to i_t}) \geq E_{t'} + E_\odot(i_g|G)$$

$$+ \sum_{j \neq i} E_\odot(i_g, j_g|G) + \sum_{j \neq i} E_\odot(j_g|G) + \sum_{j \neq i} \sum_{k \neq i, k > j} E_\odot(j_g, k_g|G). \tag{2}$$

The $E_{t'}$ terms (Sec. 2.1) are independent of rotamer choice, are equal, and can be canceled. We make the following trivial upper and lower-bound observations:

$$E_\odot(i_t|A) \leq E_\oplus(i_t); \quad E_\odot(i_t, j_g|A) \leq \max_{s \in R_j} E_\oplus(i_t, j_s); \tag{3}$$

$$E_\odot(j_g|A) \leq E_\oplus(j_g); \quad E_\odot(j_g, k_g|A) \leq E_\oplus(j_g, k_g); \tag{4}$$

$$E_\ominus(i_g) \leq E_\odot(i_g|A); \quad \min_{s \in R_j} E_\ominus(i_g, j_s) \leq E_\odot(i_g, j_g|A); \tag{5}$$

$$E_\ominus(j_g) \leq E_\odot(j_g|A); \quad E_\ominus(j_g, k_g) \leq E_\odot(j_g, k_g|A). \tag{6}$$

---

[2] The voxel for each rotamer can be multi-dimensional, depending on the number of dihedrals.

Substituting Eqs. (3-6) into Eq. (2), we obtain:

$$E_\oplus(i_t) + \sum_{j \neq i} \max_s E_\oplus(i_t, j_s) + \sum_{j \neq i} E_\oplus(j_g) + \sum_{j \neq i} \sum_{k \neq i, k > j} E_\oplus(j_g, k_g) \geq$$
$$E_\ominus(i_g) + \sum_{j \neq i} \min_s E_\ominus(i_g, j_s) + \sum_{j \neq i} E_\ominus(j_g) + \sum_{j \neq i} \sum_{k \neq i, k > j} E_\ominus(j_g, k_g). \quad (7)$$

We now define the MinDEE criterion for rotamer i_r to be:

$$E_\ominus(i_r) + \sum_{j \neq i} \min_s E_\ominus(i_r, j_s) - \sum_{j \neq i} \max_s E_\oslash(j_s) - \sum_{j \neq i} \sum_{k \neq i, k > j} \max_{s,u} E_\oslash(j_s, k_u) >$$
$$E_\oplus(i_t) + \sum_{j \neq i} \max_s E_\oplus(i_t, j_s). \quad (8)$$

Proposition 1. *When Eq. (8) holds, rotamer i_r is provably not part of the minGMEC.*

Proof. When Eq. (8) holds, we can substitute the left-hand side of Eq. (8) for the first two terms of Eq. (7), and simplify the resulting equation to:

$$E_\ominus(i_r) + \sum_{j \neq i} \min_s E_\ominus(i_r, j_s) - \sum_{j \neq i} \max_s E_\oslash(j_s) - \sum_{j \neq i} \sum_{k \neq i, k > j} \max_{s,u} E_\oslash(j_s, k_u)$$
$$+ \sum_{j \neq i} E_\oslash(j_g) + \sum_{j \neq i} \sum_{k \neq i, k > j} E_\oslash(j_g, k_g) > E_\ominus(i_g) + \sum_{j \neq i} \min_s E_\ominus(i_g, j_s). \quad (9)$$

We then substitute the following two bounds $\sum_{j \neq i} \max_s E_\oslash(j_s) \geq \sum_{j \neq i} E_\oslash(j_g)$ and $\sum_{j \neq i} \sum_{k \neq i, k > j} \max_{s,u} E_\oslash(j_s, k_u) \geq \sum_{j \neq i} \sum_{k \neq i, k > j} E_\oslash(j_g, k_g)$ into Eq. (9) and reduce: $E_\ominus(i_r) + \sum_{j \neq i} \min_s E_\ominus(i_r, j_s) > E_\ominus(i_g) + \sum_{j \neq i} \min_s E_\ominus(i_g, j_s)$. Thus, when the MinDEE pruning condition Eq. (8) holds, $i_r \neq i_g$ and we can provably eliminate rotamer i_r as not being part of the minGMEC. $\square$

MinDEE (unlike traditional-DEE) accounts for energy changes during minimization (addends 3-4 in Eq. 8). Using precomputed energy bounds, the MinDEE condition (Eq. 8) can be computed as efficiently as the traditional-DEE condition (Eq. 1). Here we presented a generalization of traditional-DEE, to obtain an initial MinDEE pruning criterion. Analogously to the traditional-DEE extensions [6, 10, 11, 23], we also derived extensions to MinDEE to improve its pruning efficiency (see Appendix A in [9]).

3   Minimized-DEE/A^* Search Algorithm (Non-ensemble-Based Redesign)

3.1   Traditional-DEE with A^*

In [16], an A^* branch and bound algorithm was developed to compute a number of low-energy conformations for a single mutation sequence (i.e., a single protein). In this algorithm, traditional-DEE was first used to reduce the number of side-chain conformations, and then surviving conformations were enumerated in order of conformation energy by expanding sorted nodes of a conformation tree.[3] The following derivation of the DEE/A^* combined search closely follows [16]. The A^* algorithm scores each node in a conformation tree using a scoring function $f = g + h$, where g is the cost of the path from the root to that node (the energy of all self and pairwise terms assigned through depth d) and h is an estimate (lower bound) of the path cost to a leaf node (a lower bound on the sum of energy terms involving unassigned residues). The value of g (at depth d) can be expressed as $g = \sum_{i=1}^{d}(E(i_r) + \sum_{j=i+1}^{d} E(i_r, j_s))$. The lower bound h can be written as $h = \sum_{j=d+1}^{n} E_j$, where n is the total number of flexible residues and $E_j = \min_s(E(j_s) + \sum_{i=1}^{d} E(i_r, j_s) + \sum_{k>j}^{n} \min_u E(j_s, k_u))$. The A^* algorithm maintains a list of nodes (sorted by f) and in each iteration replaces the node with the smallest f value by an expansion of the children of that node, until the node with the smallest f value is a leaf node, corresponding to a fully-assigned conformation. To reduce the branching factor of the conformation tree, the DEE algorithm is used to preprocess the set of allowed rotamers. If low-energy conformations within E_w of the GMEC are to be returned by the DEE/A^* search, then the DEE criterion (Eq. 1) must be modified to only eliminate rotamers that are provably not part of any conformation within E_w of the GMEC: $E(i_r) - E(i_t) + \sum_{j \neq i} \min_s E(i_r, j_s) - \sum_{j \neq i} \max_s E(i_t, j_s) > E_w$.

3.2   MinDEE with A^*

The traditional-DEE/A^* algorithm [16] can be extended to include energy minimization by substituting MinDEE for traditional-DEE. So that no conformations within E_w of the minGMEC are pruned, the MinDEE equation (Eq. 8) becomes:

$$E_\ominus(i_r) + \sum_{j \neq i} \min_s E_\ominus(i_r, j_s) - \sum_{j \neq i} \max_s E_\oslash(j_s) - \sum_{j \neq i} \sum_{k \neq i, k > j} \max_{s,u} E_\oslash(j_s, k_u)$$
$$-E_\oplus(i_t) - \sum_{j \neq i} \max_s E_\oplus(i_t, j_s) > E_w . \tag{10}$$

We modify the definition of the A^* functions g and h to use the minimum energy terms $E_\ominus(\cdot)$: $g = \sum_{i=1}^{d}(E_\ominus(i_r) + \sum_{j=i+1}^{d} E_\ominus(i_r, j_s))$, and $h = \sum_{j=d+1}^{n} E_j$, where $E_j = \min_s \left(E_\ominus(j_s) + \sum_{i=1}^{d} E_\ominus(i_r, j_s) + \sum_{k>j}^{n} \min_u E_\ominus(j_s, k_u) \right)$. A lower bound on the minimized energy of the partially-assigned conformation is given by g, while a lower bound on the minimized energy for the unassigned portion of the conformation is given by h. Thus, the MinDEE/A^* search generates conformations in order of increasing *lower bounds* on the conformation's *minimized* energy.

We combine our modified MinDEE criterion (Eq. 10) with the modified A^* functions g and h above in a provable search algorithm for identifying the minGMEC and obtaining a set of low-energy conformations. First, MinDEE prunes the majority of the conformations by eliminating rotamers that are provably not within E_w of the minGMEC. The remaining conformations are then generated in order of increasing *lower bounds*

---

[3] In a conformation tree, the rotamers of flexible residue i are represented by the branches at depth i. Internal nodes of a conformation tree represent partially-assigned conformations and each leaf node represents a fully-assigned conformation (see Fig. 3 in [17, p. 745]).

on their minimized energies. The generated conformations are energy-minimized and ranked in terms of increasing *actual* minimized energies.

MinDEE/A^* must guarantee that upon completion all conformations within E_w of the minGMEC are returned. However, the minGMEC may not be among the top A^* conformations if the lower bound on its energy does not rank high. We therefore derive the following condition for halting the MinDEE/A^* search. Let $B(s)$ be a lower bound on the energy of conformation s (see Appendix B in [9], which describes how lower energy bounds are precomputed for all rotamer pairs) and let E_m be the current minimum energy among the minimized conformations returned so far in the A^* search.

Proposition 2. *The MinDEE/A^* search can be halted once the lower bound $B(c)$ on the energy of the next conformation c returned by A^*, satisfies $B(c) > E_m + E_w$. The set of returned conformations is guaranteed to contain every conformation whose energy is within E_w of the energy of the minGMEC. Moreover, at that point in the search, the conformation with energy E_m is the minGMEC.*

The proof of Proposition 2 can be found in [9, Sec. 3.2]. Using both MinDEE and A^*, our algorithm obtains a combinatorial pruning factor by eliminating the majority of the conformations, which makes the search for the minGMEC computationally feasible. MinDEE/A^* incorporates energy minimization with provable guarantees, and is thus more capable of returning conformations with lower energy states than traditional-DEE.

4   Hybrid MinDEE-K^* Algorithm (Ensemble-Based Redesign)

We now present an extension and improvement to the original K^* algorithm [17] by using a version of the MinDEE criterion plus A^* branch-and-bound search. The K^* ensemble-based scoring function approximates the protein-ligand binding constant with the following quotient: $K^* = \frac{q_{PL}}{q_P q_L}$, where q_{PL}, q_P, and q_L are the partition functions for the protein-ligand complex, the free (unbound) protein, and the free ligand, respectively. A partition function q over a set (ensemble) of conformations S is defined as $q = \sum_{s \in S} \exp(-E_s/RT)$, where E_s is the energy of conformation s, T is the temperature in Kelvin, and R is the gas constant. In a naive K^* implementation, each partition function would be computed by a computationally-expensive energy minimization of all rotamer-based conformations. However, because the contribution to the partition function of each conformation is exponential in its energy, only a subset of the conformations contribute significantly. By identifying and energy-minimizing *only* the significantly-contributing conformations, a provably-accurate ε-approximation algorithm substantially improved the algorithm's efficiency [17]. The MinDEE criterion must be used in this algorithm because the K^* scoring function is based on *energy-minimized* conformations. Since pruned conformations never have to be examined, the Hybrid MinDEE-K^* algorithm provides a combinatorial improvement in runtime over the previously described constant-factor ε-approximation algorithm [17].

MinDEE (Eq. 8) can prune rotamers across mutation sequences.[4] By pruning *across* mutations with MinDEE, we risk pruning conformations that could otherwise contribute

---

[4] A *mutation sequence* specifies an assignment of amino-acid type to each residue in a protein.

substantially to the partition functions, thus violating our provably-good partition function approximation (Sec. 4.1). Hence, we derive a modified version of MinDEE, called *Single-Sequence* MinDEE (*SSMinDEE*), that is capable of pruning rotamers only within a single mutation sequence; the MinDEE criterion (Eq. 8) is valid for SSMinDEE.

4.1   Efficient Partition Function Computation Using A^* Search

Using A^* with SSMinDEE, we can generate the conformations of a rotamerically-based ensemble in order of increasing lower bounds on the conformation's minimized energy. As each conformation c is generated from the conformation tree, we compare a lower bound[5] $B(c)$ on its conformational energy to a moving *stop-threshold* and stop the A^* search once $B(c)$ becomes greater than the threshold, since all remaining conformations are guaranteed to have minimized energies above the stop-threshold. We now prove that a partial partition function q^* computed using only those conformations with energies below (i.e., better than) the stop-threshold will lie within a factor of ε of the true partition function q. Thus, q^* is an ε-approximation to q, i.e., $q^* \geq (1 - \varepsilon)q$.

During the application of the MinDEE criterion (Eq. 8), we can easily piggyback the computation of a lower bound B_{i_r} on the energy of all conformations that contain a pruned rotamer i_r: $B_{i_r} = E_{t'} + E_\Theta(i_r) + \sum_{j \neq i} \min_s E_\Theta(j_s) + \sum_{j \neq i} \min_s E_\Theta(i_r, j_s)$ $+ \sum_{j \neq i} \sum_{k \neq i, k > j} \min_{s,u} E_\Theta(j_s, k_u)$. Now, let E_0 be the minimum lower energy bound among all conformations containing at least one pruned rotamer, $E_0 = \min_{i_r \in S} B_{i_r}$, where S is the set of pruned *rotamers*. E_0 can be precomputed during the MinDEE stage and prior to the A^* search. Let p^* be the partition function computed over the set P of pruned *conformations*, so that $p^* \leq k \exp(-E_0/RT)$, where $|P| = k$. Also, let X be the set of conformations not pruned by MinDEE and let q^* be the partition function for the top m conformations already returned by A^*; let q' be the partition function for the n conformations that have not yet been generated, all of which have energies above E_t, so that $q' \leq n \exp(-E_t/RT)$; note that $|X| = m + n$. Finally, let $\rho = \frac{\varepsilon}{1-\varepsilon}$. We can then guarantee an ε-approximation to the full partition function q using:

Proposition 3. *If the lower bound $B(c)$ on the minimized energy of the $(m+1)^{\text{st}}$ conformation returned by A^* satisfies $B(c) \geq -RT (\ln(q^*\rho - k \exp(-E_0/RT)) - \ln n)$, then the partition function computation can be halted, with q^* guaranteed to be an ε-approximation to the true partition function q, that is, $q^* \geq (1 - \varepsilon)q$.*

Proof. The full partition function q is computed using all conformations in both P and X: $q = q^* + q' + p^*$. Thus, $q \leq q^* + n \exp(-E_t/RT) + k \exp(-E_0/RT)$. Hence, $q^* \geq (1 - \varepsilon)q$ holds if $q^* \geq (1 - \varepsilon)(q^* + n \exp(-E_t/RT) + k \exp(-E_0/RT))$. Solving for E_t, we obtain the desired stop-threshold:

$$- RT (\ln(q^*\rho - k \exp(-E_0/RT)) - \ln n) \leq E_t. \tag{11}$$

We can halt the search once a conformation's energy lower bound becomes greater than the stop-threshold (Eq. 11), since then q^* is already an ε-approximation to q.   □

---

[5] Efficiently computed as a sum of precomputed pairwise minimum energy terms (see Appendix B in [9]).

```
Initialize: n ← Number of Rotameric Conformations; q* ← 0
while (n > 0)
    c ← GetNextAStarConf()
    if B(c) ≤ −RT (ln(q*ρ − k exp(−E₀/RT)) − ln n)
        q* ← q* + exp (−ComputeMinEnergy(c)/RT)
        n ← n − 1
    else  Return q*
if q*ρ < k exp(−E₀/RT)
    RepeatSearch(q*, ρ, k, E₀)
else  Return q*
```

Fig. 2. Intra-Mutation Filter for Computing a Partition Function with Energy Minimization Using the A* Search. q^* is the running approximation to the partition function. The function $B(\cdot)$ computes the energy lower bound for the given conformation (see Appendix B in [9]). The function ComputeMinEnergy$(\cdot)$ returns a conformation's energy after energy minimization. The function GetNextAStarConf() returns the next conformation from the A* search. The function RepeatSearch$(\cdot)$ sets up and repeats the mutation search if an ε-approximation is not achieved after the generation of all A^* conformations; the search is repeated at most once. Upon completion, q^* represents an ε-approximation to the true partition function q, such that $q^* \geq (1 - \varepsilon)q$.

If at some point in the search, the stop-threshold condition has not been reached and there are no remaining conformations for A^* to extract ($n = 0$), then $q' = 0$ by definition, and $q = q^* + p^*$. Hence, if $q^*\rho \geq k\exp(-E_0/RT)$, then $q^* \geq (1 - \varepsilon)q$ is already an ε-approximation to q; otherwise, the set of pruned rotamers must be reduced to guarantee the desired approximation accuracy (see Fig. 2 and [9, p. 13] for details).

Proposition 3 represents an *intra-mutation* energy filter for pruning within a single mutation sequence (Fig. 2). For an analogous provable partition-function approximation for pruning *across* mutation sequences (so that conformations for a given sequence can be pruned based on the K^* scores computed for other sequences), see [9, Sec. 4.2].

Table 1. Conformational Pruning with Hybrid MinDEE-K^*. The initial number of conformations for the GrsA-PheA 2-residue Leu mutation search is shown with the number of conformations remaining after the application of volume, single-sequence minimized-DEE, steric, and energy (with A^*) pruning. The A^* energy filter is based on the ε-approximation algorithm in Sec. 4.1. The pruning factor represents the ratio of the number of conformations present before and after the given pruning stage. The pruning-% (in parentheses) represents the percentage of remaining conformations eliminated by the given pruning stage.

	Conf. Remaining	Pruning Factor (%)
Initial	6.8×10^8	-
Volume Filter	2.04×10^8	3.33 (70.0)
SSMinDEE Filter	8.83×10^6	23.12 (95.7)
Steric Filter	5.76×10^6	1.53 (34.7)
A^* Energy Filter	2.78×10^5	20.7 (95.2)

We now have all the necessary tools for our ensemble-based Hybrid MinDEE-K^* algorithm. The volume filter (Sec. 5) in the original K^* is applied first to eliminate under- and over-packed mutation sequences; this is followed by the combinatorial SS-MinDEE filter and the A^* energy filter using the ε-approximation algorithm above (see Table 1). A steric filter (Sec. 5), similar to the one in [17], prevents some high-energy conformations (corresponding to steric clashes) with good lower bounds from being returned by A^*, gaining an additional combinatorial speedup. Only the conformations that pass all of these filters are energy-minimized and used in the computation of the partition function for the conformational ensemble. Finally, the K^* score for a given mutation is computed as the ratio of the bound and unbound partition functions. Hybrid MinDEE-K^* efficiently prunes the majority of the mutation sequences and conformations from more expensive full energy-minimization (see Appendix B in [9]), while still giving provable guarantees on the accuracy of its score predictions.

5   Methods

Structural Model. Our structural model is the same as the one used in the original K^* [17]. In our experiments, the structural model consists of the 9 active site residues (D235, A236, W239, T278, I299, A301, A322, I330, C331) of GrsA-PheA (PDB id: 1AMU) [4], the steric shell (the 30 residues with at least one atom within 8 Å of a residue in the active site), the amino acid substrate, and the AMP cofactor. Flexible residues are represented by rotamers from the Lovell *et al.* rotamer library [19]. Each rotameric-based conformation is minimized using steepest-descent minimization and the AMBER energy function (electrostatic, vdW, and dihedral energy terms) [31,5]. For full details of our structural model, see [9, Sec. 5].

Energy Precomputation for Lower Bounds, B($\cdot$). The MinDEE criterion (Eq. 8) uses both min and max *precomputed* energy terms to determine which rotamers are not part of the minGMEC. There is no need to re-compute the min and max energies every time Eq. (8) is evaluated. See Appendix B in [9] for a detailed discussion.

Approximation Accuracy. We use $\varepsilon = 0.03$, thus guaranteeing that the computed partial partition functions will be at least 97% of the corresponding full partition functions.

Filters. *Volume filter:* Mutation sequences that are over- or under-packed by more than 30Å^3 compared to the wildtype PheA are pruned; *Steric filter:* Conformations in which a pair of atoms' vdW radii overlap by more than 1.5Å prior to minimization are pruned; *Sequence-space filter:* The active site residues are allowed to mutate to the set (GAVLIFYWM) of hydrophobic amino acids; *MinDEE:* We use an implementation of the MinDEE analog to the simple coupled Goldstein criterion ([10] and Fig. 4d in [9]).

6   Results and Discussion

In this section, we compare the results of GMEC-based protein redesign without (traditional-DEE/A^*) and with (MinDEE/A^*) energy minimization. We also compare the redesign results when energy minimization is used without (MinDEE/A^*) and with (Hybrid MinDEE-K^*) conformational ensembles. We further compare our ensemble-based redesign results both to our previous computational predictions of protein designs and to biological activity assays of predicted protein mutants.

Comparison to Biological Activity Assays. Similarly to [17], we simulated the biological activity assays of L-Phe and L-Leu against the wildtype PheA enzyme and the double mutant T278M/A301G [28]. In [28], T278M/A301G was shown to have the desired switch of specificity from Phe to Leu by performing activity assays. The activity for both the wildtype and the mutant protein sequences was normalized, so that the substrate with the larger activity was assigned a specificity of 100%, while the other substrate was assigned specificity relative to the first one. The wildtype PheA had a specificity of 100% for Phe and approximately 7% for Leu; the double mutant had a specificity of 100% for Leu and approximately 40% for Phe. The computed Hybrid MinDEE-K^* normalized scores qualitatively agreed with these results, showing the desired switch of specificity for T278M/A301G. The wildtype sequence had a normalized K^* score of 100% for Phe and 0.01% for Leu; the double mutant had a normalized score of 100% for Leu and 20% for Phe.

Comparison to Traditional-DEE. For comparison, the simple coupled Goldstein traditional-DEE criterion [10] was used in a redesign search for changing the specificity of the wildtype PheA enzyme from Phe to Leu, using the experimental setup in Sec. 5. A comparison to the rotamers in the minGMEC A236M/A322M (see MinDEE/A^* results below), revealed that 2 of these 9 rotamers were in fact *pruned* by traditional-DEE. As an example, the minGMEC was energy-minimized from a conformation that included rotamer 5 [19] of Met at residue 236. This particular rotamer (χ angles $-177°$, $180°$, and $75°$) was pruned by traditional-DEE. We then energy-minimized A236M/A301G, the rigid-GMEC obtained by traditional-DEE/A^*, and determined that its energy was higher (by appx. 5 kcal/mol) than the energy for the minGMEC obtained by MinDEE/A^*. Moreover, a total of 104 different conformations minimized to a lower energy than the rigid-GMEC. These results confirm our claim that traditional-DEE is not provably-accurate with energy-minimization; they also show that conformations pruned by traditional-DEE may minimize to a lower energy state than the rigid-GMEC.

Hybrid MinDEE-K^*. The experimental setup for Leu redesign with Hybrid MinDEE-K^* is as described in Sec 5. The 2-point mutation search took approximately 10 hours on a cluster of 24 processors. Only 30% of the mutation sequences passed the volume filter, while MinDEE pruned over 95% of the remaining conformations. The use of the ε-approximation algorithms reduced the number of conformations that had to be subsequently generated and energy-minimized by an additional factor of twenty (see Table 1). A brute-force version of Hybrid MinDEE-K^* that did not utilize any of the filters, would take approximately 2,450 times longer (appx. 1,023 days).

The two top-scoring sequences are A301G/I330W and A301G/I330F for both Hybrid MinDEE-K^* and the original K^* [17]. These novel mutation sequences were tested in the wetlab and were shown to have the desired switch of specificity from

Phe to Leu [17]. Moreover, the other known successful redesign T278M/A301G [28] is ranked 4th. Furthermore, all of the top 17 Hybrid MinDEE-K^* sequences contain the mutation A301G, which is found in all known native Leu adenylation domains [3]. These results show that our algorithm can give reasonable predictions for redesign.

To compare the efficiency of Hybrid MinDEE-K^* and the original K^*, we measured the number of fully-evaluated conformations. The original K^* (using the better minimizer of Hybrid MinDEE-K^*, see Appendix B in [9]) fully-evaluated approx. 30% more conformations than the 2.78×10^5 evaluated by Hybrid MinDEE-K^* (Table 1). Thus, Hybrid MinDEE-K^* is much more efficient at obtaining the desired results.

MinDEE/A^*. We now discuss results from our non-ensemble-based experiments using MinDEE/A^*. To redesign the wildtype PheA enzyme so that its substrate specificity is switched towards Leu, we used the experimental setup described in Sec. 5. The MinDEE filter on the bound protein:ligand complex pruned 206 out of the 421 possible rotamers for the active site residues, reducing the number of conformations that were subsequently supplied to A^* by a factor of 2,330. We then extracted and minimized all conformations over the 2-point mutation sequences using A^* until the halting condition defined in Proposition 2 was reached, for $E_w = 8.5$ kcal/mol. A total of 813 conformations, representing 45 unique mutation sequences, had actual minimized energies within 8.5 kcal/mol of the minGMEC energy. The top-ranked MinDEE/A^* sequence is A236M/A322M; the minGMEC is obtained from this sequence. The entire redesign process took approximately 14 days on a single processor, with more than $120,000$ extracted conformations before the search could be provably halted. Thus, the provable accuracy of the results comes at the cost of this computational overhead. Note, however, that a redesign effort without a MinDEE filter and a provable halting condition would be computationally infeasible.

Like A301G/I330W and A301G/I330F, the top 5 MinDEE/A^* sequences are unknown in nature. To assess the switch of specificity from Phe to Leu, we extracted the minimum-energy conformation for these top 5 Leu-binding sequences. Each of these 5 conformations was then energy-minimized when bound to Phe. Whereas the Leu-bound energies were negative and low, the corresponding Phe-bound energies were positive and high. Thus, the top sequences are predicted to bind more stably to Leu, as desired.

Only 9 of the 45 MinDEE/A^* sequences passed the Hybrid MinDEE-K^* volume filter. Moreover, only 5 of the MinDEE/A^* sequences were found in the top 40 Hybrid MinDEE-K^* sequences, indicating that ensemble-scoring yields substantially different predictions from single-structure scoring using the minGMEC, where only the minimized *bound* state of a *single* conformation is considered (see Fig. 3 in [9, p. 20]). We can conclude that, currently, MinDEE appears useful as a filter in the Hybrid MinDEE-K^* algorithm; however, the incorporation of additional information, such as a comparison to negative design (the energies to bind the wild-type substrate), may promote MinDEE as a valuable stand-alone non-ensemble-based algorithm for protein redesign.

7   Conclusions

When energy-minimization is required, the traditional-DEE criterion makes no guarantees about pruning rotamers belonging to the minGMEC. In contrast, a rotamer is only pruned by MinDEE if it is provably not part of the minGMEC. We showed experimentally that the minGMEC can minimize to lower energy states than the rigid-GMEC, confirming the feasibility and significance of our novel MinDEE criterion. When used as a filter in *ensemble-based* redesign, MinDEE efficiently reduced the conformational and sequence search spaces, leading both to predictions consistent with previous redesign efforts and novel sequences that are unknown in nature. Our Hybrid MinDEE-K^* algorithm showed a significant improvement in pruning efficiency, as compared to the original K^* algorithm. Redesign searches for two other substrates, Val and Tyr, have also been performed, confirming the generality of our algorithms.

Protein design using traditional-DEE uses neither ensembles nor rotamer minimization. In our experiments, we reported the relative benefits of incorporating ensembles and energy-minimization into a provable redesign algorithm. A major challenge for protein redesign algorithms is the balance between the efficiency and accuracy with which redesign is performed. While the ability to prune the majority of mutation/conformation search space is extremely important, increasing the accuracy of the model is a prerequisite for successful redesign. It would be interesting to implement finer rotamer sampling and more accurate (and hence more expensive) energy functions, remove bias in the rotamer library by factoring the Jacobian into the partition function over torsion-angle space, and incorporate backbone flexibility. An accurate and efficient algorithm for redesigning natural products should prove useful as a technique for drug design.

Acknowledgments. We thank Prof. A. Anderson, Dr. S. Apaydin, Mr. J. MacMaster, Mr. A. Yan, Mr. B. Stevens, and all members of the Donald Lab for helpful discussions and comments.

References

1. D. Bolon and S. Mayo. Enzyme-like proteins by computational design. *Proc. Natl. Acad. Sci. USA*, 98:14274–14279, 2001.
2. D. Cane, C. Walsh, and C. Khosla. Harnessing the biosynthetic code: combinations, permutations, and mutations. *Science*, 282:63–68, 1998.
3. G. Challis, J. Ravel, and C. Townsend. Predictive, structure-based model of amino acid recognition by nonribosomal peptide synthetase adenylation domains. *Chem. Biol.*, 7: 211–224, 2000.
4. E. Conti, T. Stachelhaus, M. Marahiel, and P. Brick. Structural basis for the activation of phenylalanine in the non-ribosomal biosynthesis of Gramicidin S. *EMBO J.*, 16:4174–4183, 1997.
5. W. Cornell, P. Cieplak, C. Bayly, I. Gould, K. Merz, D. Ferguson, D. Spellmeyer, T. Fox, J. Caldwell, and P. Kollman. A second generation force field for the simulation of proteins, nucleic acids and organic molecules. *J. Am. Chem. Soc.*, 117:5179–5197, 1995.
6. J. Desmet, M. Maeyer, B. Hazes, and I. Lasters. The dead-end elimination theorem and its use in protein side-chain positioning. *Nature*, 356:539–542, 1992.

7. S. Doekel and M. Marahiel. Dipeptide formation on engineered hybrid peptide synthetases. *Chem. Biol.*, 7:373–384, 2000.

8. K. Eppelmann, T. Stachelhaus, and M. Marahiel. Exploitation of the selectivity-conferring code of nonribosomal peptide synthetases for the rational design of novel peptide antibiotics. *Biochemistry*, 41:9718–9726, 2002.

9. I. Georgiev, R. Lilien, and B. R. Donald. A novel minimized dead-end elimination criterion and its application to protein redesign in a hybrid scoring and search algorithm for computing partition functions over molecular ensembles. Technical Report 570, Dartmouth Computer Science Dept., http://www.cs.dartmouth.edu/reports/abstracts/TR2006-570, 2006.

10. R. Goldstein. Efficient rotamer elimination applied to protein side-chains and related spin glasses. *Biophys. J.*, 66:1335–1340, 1994.

11. D. Gordon and S. Mayo. Radical performance enhancements for combinatorial optimization algorithms based on the dead-end elimination theorem. *J. Comput. Chem.*, 19:1505–1514, 1998.

12. H. Hellinga and F. Richards. Construction of new ligand binding sites in proteins of known structure: I. Computer-aided modeling of sites with pre-defined geometry. *J. Mol. Biol.*, 222:763–785, 1991.

13. A. Jaramillo, L. Wernisch, S. Héry, and S. Wodak. Automatic procedures for protein design. *Comb. Chem. High Throughput Screen.*, 4:643–659, 2001.

14. W. Jin, O. Kambara, H. Sasakawa, A. Tamura, and S. Takada. De novo design of foldable proteins with smooth folding funnel: Automated negative design and experimental verification. *Structure*, 11:581–591, 2003.

15. A. Keating, V. Malashkevich, B. Tidor, and P. Kim. Side-chain repacking calculations for predicting structures and stabilities of heterodimeric coiled coils. *Proc. Natl. Acad. Sci. USA*, 98:14825–14830, 2001.

16. A. Leach and A. Lemon. Exploring the conformational space of protein side chains using dead-end elimination and the A* algorithm. *Proteins*, 33:227–239, 1998.

17. R. Lilien, B. Stevens, A. Anderson, and B. R. Donald. A novel ensemble-based scoring and search algorithm for protein redesign, and its application to modify the substrate specificity of the Gramicidin Synthetase A phenylalanine adenylation enzyme. *Journal of Computational Biology*, 12(6–7):740–761, 2005.

18. L. Looger, M. Dwyer, J. Smith, and H. Hellinga. Computational design of receptor and sensor proteins with novel functions. *Nature*, 423:185–190, 2003.

19. S. Lovell, J. Word, J. Richardson, and D. Richardson. The penultimate rotamer library. *Proteins*, 40:389–408, 2000.

20. J. Marvin and H. Hellinga. Conversion of a maltose receptor into a zinc biosensor by computational design. *PNAS*, 98:4955–4960, 2001.

21. H. Mootz, D. Schwarzer, and M. Marahiel. Construction of hybrid peptide synthetases by module and domain fusions. *Proc. Natl. Acad. Sci. USA*, 97:5848–5853, 2000.

22. R. Najmanovich, J. Kuttner, V. Sobolev, and M. Edelman. Side-chain flexibility in proteins upon ligand binding. *Proteins*, 39(3):261–8, 2000.

23. N. Pierce, J. Spriet, J. Desmet, and S. Mayo. Conformational splitting: a more powerful criterion for dead-end elimination. *J. Comput. Chem.*, 21:999–1009, 2000.

24. N. Pierce and E. Winfree. Protein design is *NP*-hard. *Protein Eng.*, 15:779–782, 2002.

25. J. Ponder and F. Richards. Tertiary templates for proteins: use of packing criteria in the enumeration of allowed sequences for different structural classes. *J. Mol. Biol.*, 193:775–791, 1987.

26. A. Schneider, T. Stachelhaus, and M. Marahiel. Targeted alteration of the substrate specificity of peptide synthetases by rational module swapping. *Mol. Gen. Genet.*, 257:308–318, 1998.

27. D. Schwarzer, R. Finking, and M. Marahiel. Nonribosomal peptides: from genes to products. *Nat. Prod. Rep.*, 20:275–287, 2003.

28. T. Stachelhaus, H. Mootz, and M. Marahiel. The specificiy-conferring code of adenylation domains in nonribosomal peptide synthetases. *Chem. Biol.*, 6:493–505, 1999.
29. T. Stachelhaus, A. Schneider, and M. Marahiel. Rational design of peptide antibiotics by targeted replacement of bacterial and fungal domains. *Science*, 269:69–72, 1995.
30. A. Street and S. Mayo. Computational protein design. *Structure*, 7:R105–R109, 1999.
31. S. Weiner, P. Kollman, D. Case, U. Singh, C. Ghio, G. Alagona, S. Profeta, and P. Weiner. A new force field for molecular mechanical simulation of nucleic acids and proteins. *J. Am. Chem. Soc.*, 106:765–784, 1984.

10 Years of the International Conference on Research in Computational Molecular Biology (RECOMB)

Sarah J. Aerni and Eleazar Eskin

The RECOMB 10th Year Anniversary Committee

The tenth year of the annual International Conference on Research in Computational Biology (RECOMB) provides an opportunity to reflect on its history. RECOMB has been held across the world, including 6 different countries spanning 3 continents (Table 1). Over its 10 year history, RECOMB has published 373 papers and 170 individuals have served on its various committees. While there are many new faces in RECOMB each year, a significant number of researchers have participated over many years forming the core of the RECOMB community.

Over the past ten years, members of the RECOMB community were key players in many of the advances in Computational Biology during this period. These include the sequencing and assembly of the human genome, advances in sequence comparison, comparative genomics, genome rearrangements and the HapMap project among others.

Table 1. The locations and dates of each year of RECOMB. The program and conference chair are listed for each conference in the final two columns.

	Location	Dates	Program Chair	Conference Chair
1997	Santa Fe, USA	January 20-23	Michael Waterman	Sorin Istrail
1998	New York, USA	March 22-25	Pavel Pevzner	Gary Benson
1999	Lyon, France	April 11-14	Sorin Istrail	Mireille Régnier
2000	Tokyo, Japan	April 8-11	Ron Shamir	Satoru Miyano
2001	Montreal, Canada	April 22-25	Thomas Lengauer	David Sankoff
2002	Washington, USA	April 18-21	Eugene Myers	Sridhar Hannenhalli
2003	Berlin, Germany	April 10-13	Webb Miller	Martin Vingron
2004	San Diego, USA	March 27-31	Dan Gusfield	Philip Bourne
2005	Boston, USA	May 14-18	Satoru Miyano	Jill Mesirov, Simon Kasif
2006	Venice, Italy	April 2-5	Alberto Apostolico	Concettina Guerra

10 Years of RECOMB Papers

Over RECOMB's 10 year history, 731 authors have published a total of 373 papers in the conference proceedings. These papers span the diversity of research areas in

A. Apostolico et al. (Eds.): RECOMB 2006, LNBI 3909, pp. 546–562, 2006.

Computational Biology and present many new computational techniques for the analysis of biological data.

It should be noted that some authors have variances in how names appear throughout the years, including differing first names, initials, and middle names. While every effort was made to normalize the names, any such error could lead to the skewing of data and there may be small errors in the reporting of individual participation throughout the paper.

As a preliminary analysis, we consider the number of papers for each researcher that has appeared throughout the 10 years of RECOMB in the proceedings. In such a measure, Richard Karp who has authored 12 different papers in RECOMB throughout the 10 years would be the top participant.

Using the graph in Figure 1, we can identify the most collaborative members of the RECOMB community (hubs in a protein network). The most collaborative authors are the individuals that have the most number of co-authors. Ron Shamir is the most collaborative RECOMB author with 22 co-authors (Table 3).

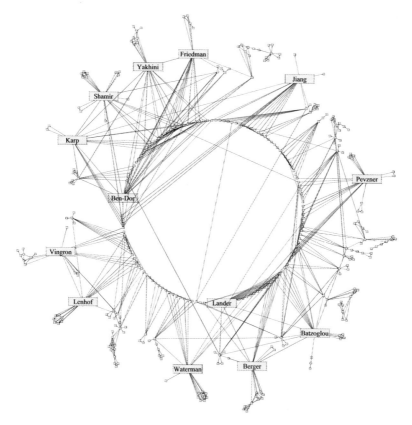

Fig. 1. Graphical view of interactions between RECOMB authors represented as a "protein interaction network" (giant component). Vertices of the graph represent authors while edges connect vertices corresponding to co-authors. Authors whose names are displayed are authors who have at least 16 coauthors.

Table 2. RECOMB's most prolific authors. The table identifies authors who have published at least 4 papers in RECOMB.

Author	Number of Papers	Author	Number of Papers
Richard Karp	12	Serafim Batzoglou	6
Ron Shamir	11	Dan Gusfield	6
Pavel Pevzner	11	Webb Miller	5
Bonnie Berger	10	Fengzhu Sun	5
Amir Ben-Dor	10	Ralf Bundschuh	5
Nir Friedman	9	Jeremy Buhler	5
Eugene Myers	9	Jens Lagergren	5
Zohar Yakhini	9	Roded Sharan	4
Tao Jiang	9	Benno Schwikowski	4
Benny Chor	8	Nancy Amato	4
Michael Waterman	8	Eran Halperin	4
David Sankoff	8	Zheng Zhang	4
Martin Vingron	7	Martin Farach-Colton	4
Ting Chen	7	Sorin Istrail	4
Steven Skiena	7	Vlado Dancík	4
Eric Lander	7	Golan Yona	4
Hans-Peter Lenhof	6	Dannie Durand	4
John Kececioglu	6	Mathieu Blanchette	4
Vineet Bafna	6	Adam Siepel	4
Bruce Donald	6	Tatsuya Akutsu	4
David Haussler	6	Eran Segal	4
Lior Pachter	6	Thomas Lengauer	4

Table 3. RECOMB contributors with more than 10 co-authors. For each author the number of individuals with whom they have coauthored papers is listed.

Author Name	Num ofCoauthors	Author Name	Num of Coauthors
Ron Shamir	22	Hans-Peter Lenhof	16
Serafim Batzoglou	20	Vlado Dancik	14
Bonnie Berger	20	Steven Skiena	14
Pavel Pevzner	20	Benny Chor	14
Michael Waterman	19	Lydia Kavraki	13
Zohar Yakhini	19	Bruce Donald	13
Tao Jiang	18	Martin Farach-Colton	12
Richard Karp	18	Sorin Istrail	12
Eric Lander	18	Lior Pachter	12
Nir Friedman	18	Eugene Myers	11
Amir Ben-Dor	17	David Sankoff	11
Martin Vingron	17	Vineet Bafna	11

Similarly, we can identify which groups of authors have had the most success working together (complexes in protein networks). The team of Eric S. Lander, Bonnie Berger and Serafim Batzoglou have published 3 papers together and are the only group of three authors which have published more than two papers. The most prolific pair of authors is Amir Ben-Dor and Zohar Yakhini who have published 7 papers together. 21 pairs of authors have published at least 3 papers as shown in Table 4.

Relationships between individual authors can be established in other ways as well. In Figure 2 we analyze the relationships between the most prolific authors (Table 2). By examining the relationships between individuals as advisors in both PhD and postdoctoral positions, the connections between the most prolific authors can be seen as a phylogeny. In addition, the individuals are shown on a timeline indicating the times at which they first began publishing in the field of Computational Biology.

We manually classified each paper into one of 16 categories: Protein structure analysis, Molecular Evolution, Sequence Comparison, Motif Finding, Sequence analysis, Population genetics/SNP/Haplotyping, Physical and Genetic Mapping, Gene Expression, Systems Biology, RNA Analysis, Genome rearrangements, Computational

Table 4. Coauthor Pairs. All pairs of authors who have written 3 or more papers accepted by RECOMB throughout the 10 year history of the conference are listed in the table.

Author Names		Number of Papers
Amir Ben-Dor	Zohar Yakhini	7
Bonnie Berger	Eric Lander	4
Zheng Zhang	Webb Miller	4
Serafim Batzoglou	Bonnie Berger	3
Serafim Batzoglou	Eric Lander	3
Amir Ben-Dor	Benny Chor	3
Amir Ben-Dor	Richard Karp	3
Amir Ben-Dor	Benno Schwikowski	3
Benny Chor	Tamir Tuller	3
Tao Jiang	Richard Karp	3
Richard Karp	Ron Shamir	3
David Haussler	Adam Siepel	3
Eric Lander	Jill Mesirov	3
Fengzhu Sun	Ting Chen	3
Ralf Zimmer	Thomas Lengauer	3
Bruce Donald	Christopher Langmead	3
Bruce Donald	Ryan Lilien	3
Nir Friedman	Yoseph Barash	3
Michael Hallett	Jens Lagergren	3
Guang Song	Nancy Amato	3
Eran Segal	Daphne Koller	3

Proteomics, Recognition of Genes, Microarray design, DNA computing and Other. Using these classifications, we can observe which authors have written the most about a single topic and which authors have written about the most topics. Both Bonnie Berger and Benny Chor have contributed the most papers (6) on a single topic, Protein Structure Analysis and Molecular Evolution respectively. Table 5 shows the top contributors in a single area.

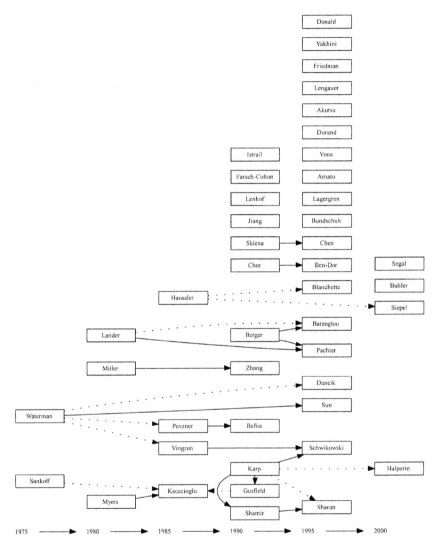

Fig. 2. Phylogeny of Authors. In this figure authors are organized across a timeline representing their earliest publications in the field of Computational Biology. Solid lines indicate PhD advisors, while dotted lines represent postdoctoral advisors. While we attempted to accurately link the timeline and RECOMB authors/genealogy, the figure represents only approximate time estimates and approximate topology of the RECOMB tree.

Table 5. Most consistent authors. For each author in the table, a subject is indicated for which he or she has written at least 3 papers. The number of papers in the 10 years of RECOMB by the author on the given subject is indicated.

Author Name	Author Name	Num of papers
Benny Chor	Molecular Evolution	6
Bonnie Berger	Protein structure analysis	6
David Sankoff	Genome rearrangements	5
Ralf Bundschuh	Sequence Comparison	5
Bruce Donald	Protein structure analysis	5
Nir Friedman	Gene Expression	5
Jens Lagergren	Molecular Evolution	5
Amir Ben-Dor	Gene Expression	4
Richard Karp	Physical and Genetic Mapping	4
Webb Miller	Sequence Comparison	4
David Haussler	Molecular Evolution	4
Zohar Yakhini	Gene Expression	4
Lior Pachter	Recognition of Genes	4
Dannie Durand	Molecular Evolution	4
Eugene Myers	Sequence Comparison	3
John Kececioglu	Sequence Comparison	3
Tao Jiang	Physical and Genetic Mapping	3
Ron Shamir	Sequence analysis	3
Michael Waterman	Physical and Genetic Mapping	3
Hans-Peter Lenhof	Protein structure analysis	3
Zheng Zhang	Sequence Comparison	3
Dan Gusfield	Population genetics/SNP/Haplotyping	3
Tandy Warnow	Molecular Evolution	3
Douglas Brutlag	Protein structure analysis	3
Jon Kleinberg	Protein structure analysis	3
Franco Preparata	Sequence analysis	3
Chris Bailey-Kellogg	Protein structure analysis	3
Michael Hallett	Molecular Evolution	3
Jonathan King	Protein structure analysis	3
Jeremy Buhler	Sequence Comparison	3
Kaizhong Zhang	RNA Analysis	3
Nancy Amato	Protein structure analysis	3
Eran Halperin	Population genetics/SNP/Haplotyping	3
Ryan Lilien	Protein structure analysis	3
Tamir Tuller	Molecular Evolution	3

Table 6. Most Diverse Authors. These are authors spanning the largest number of subjects. Authors are given who have papers in RECOMB in more than 4 subjects.

# of Subjects	Author Name	# of Subjects	Author Name
8	Pavel A. Pevzner	5	Fengzhu Sun
7	Steven S. Skiena	5	Ting Chen
7	Richard M. Karp	4	Tatsuya Akutsu
7	Ron Shamir	4	Bonnie Berger
6	Amir Ben-Dor	4	Hans-Peter Lenhof
6	Tao Jiang	4	Benno Schwikowski
6	Martin Vingron	4	Dan Gusfield
6	Eric S. Lander	4	Thomas Lengauer
6	Zohar Yakhini	4	Vineet Bafna
5	Serafim Batzoglou	4	Nir Friedman
5	Eugene W. Myers	4	Eran Segal
5	Michael S. Waterman	4	Roded Sharan

On the opposite end of the spectrum are the authors who contributed papers on different topics (Table 6).

For each author we create a topic profile which is a 16 dimensional vector containing the number of papers of each topic that an individual has published in RECOMB normalized by dividing by the total number of papers published. Intuitively, an author's topic profile represents the areas of research in which the author works on. Not surprisingly, co-authors tend to work on the same topics. The average pairwise Euclidean distance between any two authors topic profile is 1.19 while the average distance between co-authors is only 0.61. Similarly, papers written by the same author tend to be on the same topic. The chances that any two papers are on the same topic are 0.09 while the chance that two papers that share one author is on the same topic is 0.21.

Trends in RECOMB Authors over Time

The number of authors contributing to the conferences has fluctuated with the largest number in 2006 at 134. 1998 represents the year in which the fewest number of authors submitted multiple papers, that is, most authors had a single paper that was accepted to the conference (Table 7)

2006 had the lowest proportion of single-authored papers with only one of the 40 accepted papers showing a single author (Table 8 and Figure 3).

It appears that over the years there is a trend in an increase in the number of authors per paper with a slight decrease in papers per author. This indicates that while there are more authors on any one single paper, authors are less likely to have multiple papers in any given year.

Table 7. "Authors per paper" and "papers per author" statistics

Year	Papers	Authors	Averages	
			Author per Paper	Paper per Author
1997	42	101	2.8	1.2
1998	38	96	2.6	1.0
1999	35	106	3.3	1.0
2000	36	122	3.8	1.1
2001	35	92	2.8	1.1
2002	35	87	2.7	1.1
2003	35	88	2.8	1.1
2004	38	111	3.1	1.1
2005	39	121	3.4	1.1
2006	40	134	3.4	1.1

Table 8. Author Numbers in Papers. The table shows the percent of papers in each that had the given number of authors determined by counting the number of papers with the indicated number of authors and dividing it by the total number of papers in RECOMB in that year.

Year	Percent of papers with given number of authors										
	1	2	3	4	5	6	7	8	9	10	11
1997	19.0	42.9	14.3	9.5	9.5	0.0	0.0	2.4	0.0	0.0	2.4
1998	18.4	31.6	28.9	15.8	5.3	0.0	0.0	0.0	0.0	0.0	0.0
1999	8.6	37.1	22.9	8.6	11.4	5.7	5.7	0.0	0.0	0.0	0.0
2000	2.8	30.6	19.4	22.2	13.9	5.6	0.0	0.0	0.0	0.0	5.6
2001	17.1	31.4	28.6	8.6	8.6	2.9	2.9	0.0	0.0	0.0	0.0
2002	14.3	34.3	25.7	17.1	8.6	0.0	0.0	0.0	0.0	0.0	0.0
2003	8.6	42.9	25.7	5.7	17.1	0.0	0.0	0.0	0.0	0.0	0.0
2004	7.9	39.5	21.1	13.2	10.5	2.6	2.6	2.6	0.0	0.0	0.0
2005	2.6	28.2	28.2	25.6	7.7	2.6	2.6	2.6	0.0	0.0	0.0
2006	2.5	30.0	32.5	12.5	12.5	5.0	2.5	0.0	2.5	0.0	0.0

There are multiple ways to gauge the participation of individuals in the conference. One such measure might be to determine the span of years over which individuals have papers appearing in the proceedings. This was measured by determining the years of the first followed by the most recent papers of individual authors, and determining the span of years over which they had participated. Using such a measure, ten authors have papers published over a span of all ten years of the conference listed in the table. These authors are Benny Chor, Bonnie Berger, Sampath Kannan, John Kececioglu, Martin Vingron, David Haussler, Pavel Pevzner, Serafim Batzoglou, Dan Gusfield and Tao Jiang.

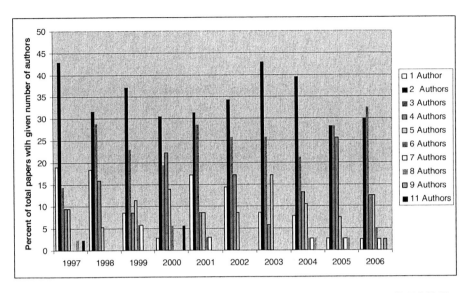

Fig. 3. Distribution of papers with given number of authors over the 10 years of RECOMB

Table 9. Authors with RECOMB papers in most number of years

Author Name	Num of Years	Author Name	Num of Years
Bonnie Berger	9	Tao Jiang	5
Pavel Pevzner	9	Benno Schwikowski	4
Ron Shamir	8	Nancy Amato	4
Amir Ben-Dor	8	Eran Halperin	4
Richard Karp	8	Vineet Bafna	4
Benny Chor	7	Sorin Istrail	4
Zohar Yakhini	7	Webb Miller	4
David Sankoff	7	Vlado Dancík	4
Eugene Myers	6	David Haussler	4
Bruce Donald	6	Mathieu Blanchette	4
Lior Pachter	6	Fengzhu Sun	4
Hans-Peter Lenhof	5	Adam Siepel	4
John Kececioglu	5	Serafim Batzoglou	4
Nir Friedman	5	Dan Gusfield	4
Martin Vingron	5	Jens Lagergren	4
Ting Chen	5	Steven Skiena	4
Ralf Bundschuh	5	Eric Lander	4
Jeremy Buhler	5	Thomas Lengauer	4
Michael Waterman	5		

However, such a measure may not be completely representative of a researcher's participation in the conference. Over the 10 years of RECOMB, no author has contributed to every year of the conference (Table 9).

Trends in RECOMB Paper Topics

As bioinformatics has grown and changed over the 10 years since RECOMB's inception, so have the subjects which comprise the papers accepted at each conference (Table 10). Some subjects, such as Protein Structure Analysis has remained a stronghold in the papers throughout the 10 years of RECOMB. Not only is it the most represented subject over the course of time, at 72 total papers in this field, with a steady portion of the total papers in each year in this field, it entails nearly 30 percent of the accepted papers in 2006.

Table 10. Distribution of topics of RECOMB papers. "Other" category includes more specific subjects such as drug design, DNA denaturization, etc.

Subject	Total	1997	1998	1999	2000	2001	2002	2003	2004	2005	2006
Protein structure analysis	72	16.7	18.4	25.7	27.8	17.1	17.1	11.4	15.8	12.8	30.0
Molecular Evolution	52	4.8	15.8	14.3	13.9	11.4	8.6	11.4	13.2	20.5	25.0
Sequence Comparison	40	28.6	21.1	2.9	8.3	5.7	8.6	2.9	7.9	7.7	10.0
Motif Finding	32	0.0	15.8	5.7	8.3	14.3	8.6	11.4	15.8	7.7	0.0
Sequence analysis	22	0.0	0.0	5.7	5.6	22.9	11.4	8.6	5.3	0.0	2.5
Population genetics/ SNP/ Haplotyping	21	2.4	2.6	0.0	0.0	0.0	11.4	20.0	7.9	7.7	5.0
Physical and Genetic Mapping	20	23.8	7.9	8.6	5.6	0.0	0.0	2.9	0.0	2.6	0.0
Gene Expression	20	0.0	0.0	8.6	11.1	8.6	17.1	5.7	2.6	2.6	0.0
Systems Biology	20	0.0	0.0	5.7	2.8	2.9	2.9	11.4	5.3	12.8	10.0
RNA Analysis	18	0.0	2.6	2.9	2.8	2.9	5.7	2.9	10.5	7.7	10.0
Genome rearrangements	15	9.5	5.3	2.9	2.8	0.0	2.9	5.7	2.6	5.1	2.5
Computational Proteomics	14	0.0	0.0	2.9	2.8	8.6	0.0	2.9	5.3	10.3	5.0
Recognition of Genes	10	7.1	0.0	2.9	2.8	5.7	0.0	0.0	5.3	2.6	0.0
Other	10	0.0	10.5	11.4	2.8	0.0	0.0	0.0	2.6	0.0	0.0
Microarray design	5	2.4	0.0	0.0	2.8	0.0	5.7	2.9	0.0	0.0	0.0
DNA computing	2	4.8	0.0	0.0	0.0	0.0	0.0	0.0	0.0	0.0	0.0

While protein structure remained a consistent part of the RECOMB content, other subjects have fluctuated, disappeared, or gained strength over time. Sequence comparison, which composed well over 25 percent of all papers in the first year of RECOMB, fell to 10 percent of the total content of the 2006 conference. Similarly, Physical and Genetic Mapping which exceeded protein structure analysis in 1997 has completely disappeared in 2006. RNA analysis and Systems Biology have also been growing in popularity since the first papers were accepted in the subjects in 1998 and 1999 respectively.

Computational Proteomics and Population Genetics each represented five percent of the total number of accepted papers. While neither was very abundant in the first four years of the conference, they seem to be gaining momentum over time. Genome rearrangement has maintained a consistent presence throughout the 10 years of RECOMB. Most notably, however, is the area of molecular evolution which has evolved from a small presence of 4.8 percent of all accepted papers in 1997 to 25 percent of the total accepted papers in 2006.

Table 11. Paper Acceptance Rates. The table gives the paper acceptance rates based on the number of papers submitted and accepted over the 10 years of RECOMB.

Year	Number Submitted	Number Accepted	Rate
1997	117	43	37%
1998	123	38	31%
1999	147	35	24%
2000	110	36	33%
2001	128	35	27%
2002	118	35	30%
2003	175	35	20%
2004	215	38	18%
2005	217	38	18%
2006	215	40	19%

Table 12. Proportion of USA/Non-USA RECOMB papers

Year	USA	Non-USA
1997	67%	33%
1998	66%	34%
1999	66%	34%
2000	54%	46%
2001	69%	31%
2002	86%	14%
2003	71%	29%
2004	74%	26%
2005	54%	46%
2006	65%	35%

RECOMB has grown more competitive over time, with an increase in submissions to over 200 in the last three years (Table 11). The number of submissions in 2006 has nearly doubled over the first year of the conference.

Origins of RECOMB Papers

The first authors of the papers have spanned the globe, representing 25 countries. While US first authors regularly contributed over 60 percent of the papers accepted to the conference, in 2000 and 2005, held in Tokyo and Boston respectively, the split neared 50 percent (Table 12). Most strikingly, over 85 percent of the papers the 2002 conference held in Washington, DC had first authors from US institutions.

Israel, Germany and Canada had first authors contributing papers to nearly every conference (Figure 4). Israel became the second most represented country during 5 years, including 2003 when the conference was held in Germany where 80 percent of non-US authors were from Israel. Canada, Germany and Italy represented the runner-up position during 2 years each. Italy contributed the largest proportion of first authored papers during 2002 when 40% of non-USA first authors were from Italian institutions, which is the second largest percentage in any year.

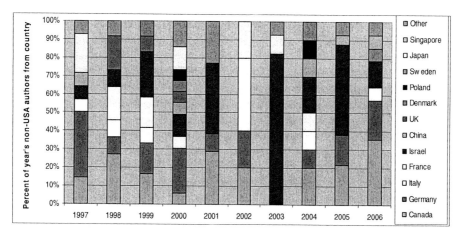

Fig. 4. Distribution of countries of origin of non-US first authors[1]

Throughout RECOMB's history, over 90% of the first authors were involved in the public sector with the exception of a brief interruption in 2001 when just over 11% of first authors were from Industry. In 2002, the conference was hosted by Celera, during which nearly 9% of first authors were involved in the private sector, the second largest amount during the conference's history. However, the contributions from industry have steadily declined since 2002.

---

[1] Category Other includes Chile, Belgium, Australia, Spain, Netherlands, Finland, Switzerland, New Zealand, Austria, and Taiwan.

RECOMB's Most Cited Papers

Several of the papers published in RECOMB have had a significant influence on research in Computational Biology and have been widely cited. Table 13 contains a list of the most cited RECOMB papers as of January 2006 according to Google Scholar. A difficulty in obtaining this list is that many of the RECOMB papers are later published in journals and the citations are split between the original RECOMB version and the journal version which may lead to some inaccuracies in calculating the number of citations.

Table 13. RECOMB's most cited papers. The number of citations given in the final column is based on the journal in which they were published, and are accurate as of January 1, 2006 when the citations were last confirmed.

Paper Title	RECOMB Year	Journal	# Citations
Nir Friedman, Michal Linial, Iftach Nachman, Dana Pe'er. "Using Bayesian networks to analyze expression data"	2000	J Comp Biol 2000:7	506
Manolis Kamvysselis, Nick Patterson, Bruce Birren, Bonnie Berger, Eric S. Lander. "Whole-genome comparative annotation and regulatory motif discovery in multiple yeast species"	2003	Nature 2003: 423	385
Amir Ben-Dor, Zohar Yakhini. "Clustering gene expression patterns"	1999	J Comp Biol 1999:6	355
Harmen J. Bussemaker, Hao Li, Eric D. Siggia. "Regulatory element detection using correlation with expression (abstract only)"	2001	Nat Genet 2001:27	265
Amir Ben-Dor, Laurakay Bruhn, Nir Friedman, Iftach Nachman, Michèl Schummer, Zohar Yakhini. "Tissue classification with gene expression profiles"	2000	J Comp Biol 2000:7	245
Serafim Batzoglou, Lior Pachter, Jill P. Mesirov, Bonnie Berger, Eric S. Lander. "Human and mouse gene structure: comparative analysis and application to exon prediction"	2000	Genome Res 2000:10	190
Isidore Rigoutsos, Aris Floratos. "Motif discovery without alignment or enumeration"	1998	Bioinformatics 2000:14	150
Jeremy Buhler, Martin Tompa. "Finding motifs using random projections"	2001	J Comp Biol 2002:9	138
Martin G. Reese, Frank H. Eeckman, David Kulp, David Haussler. "Improved splice site detection in Genie"	1997	J Comp Biol 1997:4	131
Haim Kaplan, Ron Shamir, Robert E. Tarjan. "Faster and simpler algorithm for sorting signed permutations by reversals"	1997	SIAM J Comput 1999:29	127

Table 13. (*Continue*)

Alberto Caprara. "Sorting by reversals is difficult"	1997	RECOMB 1997	111
Vlado Dancík, Theresa A. Addona, Karl R. Clauser, James E. Vath, Pavel A. Pevzner. "De Novo Peptide Sequencing via Tandem Mass Spectrometry"	1999	J Comp Biol 1999:6	109
Pierluigi Crescenzi, Deborah Goldman, Christos Papadimitriou, Antonio Piccolboni, Mihalis Yannakakis. "On the complexity of protein folding"	1998	J Comp Biol 1998:5	104
Bonnie Berger, Tom Leighton. "Protein folding in the hydrophobic-hydrophilic (HP) is NP-complete"	1998	J Comp Biol 1998:5	89
Donna K. Slonim, Pablo Tamayo, Jill P. Mesirov, Todd R. Golub, Eric S. Lander. "Class prediction and discovery using gene expression data"	2000	RECOMB 2000	87
Mathieu Blanchette. "Algorithms for phylogenetic footprinting"	2001	J Comp Biol 2002:9	84
David Sankoff, Mathieu Blanchette. "Multiple genome rearrangements"	1998	J Comp Biol 1998:5	84
Donna K. Slonim, Leonid Kruglyak, Lincoln Stein, Eric S. Lander. "Building human genome maps with radiation hybrids"	1997	J Comp Biol 1997:4	81

Table 14. Invited speakers over the 10 years of RECOMB

Year	Speaker names
1997	David Botstein, Sam Karlin, Martin Karplus, Eric Lander, Robert Lipshutz, Jonathan King, Rich Roberts, Temple Smith, Terry Speed
1998	Ruben Abagyan, Charles Cantor, David Cox, Ron Davis, Klaus Guberna tor, Joshua Lederberg, Michael Levitt, David Schwartz, John Yates
1999	Peer Bork, Cyrus Chothia, Gene Myers, John Moult, Pitor Slonimsky, Ed Southern, Peter Willett, John Wooley
2000	Eric Davidson, Walter Gilbert, Takashi Gojobori, Leroy Hood, Minoru Kanehisa, Hans Lehrach, Yvonne Martin, Yusuke Nakamura, Svante Paabo
2001	Mark Adams, Roger Brent, George Church, Franz Lang, Klaus Lindpaintner, Yvonne Martin, Mark Ptashne, Philip Sharp, Matthias Wilm
2002	Ruben Abagyan, Ali Brivanlou, Evan Eichler, Harold Garner, David Ho, Gerry Rubin, Craig Venter, Marc Vidal
2003	Edward Trifonov, Christiane Nüsslein-Volhard, Árpád Furka, Andrew Clark, David Haussler, Arthur Lesk, Dieter Oesterhelt, Terry Speed, Kari Stefansson
2004	Carlos Bustamante, Russell Doolittle, Andrew Fire, Richard Karp, William McGinnis, Deborah Nickerson, Martin Nowak, Christine Orengo, Elizabeth Winzeler
2005	David Altshuler, Wolfgang Baumeister, James Collins, Charles DeLisi, Jonathan King, Eric Lander, Michael Levine, Susan Lindquist
2006	Anne-Claude Gavin, David Haussler, Ajay Royyuru, David Sankoff, Michael Waterman, Carl Zimmer, Roman Zubarev

RECOMB Keynote Speaker Series

The conference has been honored to have many excellent speakers throughout the 10 years of the conference. Every year between 7 and 9 distinguished individuals were invited to deliver lectures at the conference in a variety of fields (Table 14).

RECOMB includes a distinguished lecture series which consists of the Stanislaw Ulam Memorial Computational Biology lecture, the Distinguished Biology lecture, and New Technologies lectures delivered by a different set of individuals every year (Table 15) with the exception of 1999 and 2005. In 199 There was no Biology lecture, while in 2005 no distinguished lectures were delivered on new technologies. 2004 included an additional address in which Richard Karp delivered the lecture awarded the Fred Howes Distinguished Service Award.

Table 15. Distinguished lecture series in Computational Biology, Biology, and New Technologies

Year	Stanislaw Ulam Memorial Computational Biology Lecture	Distinguished Biology Lecture	Distinguished New Technologies Lecture
1997	Eric Lander	Rich Roberts	Robert Lipshutz
1998	Joshua Lederberg	Ron Davis	David Cox
1999	Pitor Slonimsky		Ed Southern
2000	Minoru Kanehisa	Walter Gilbert	Leroy Hood
2001	George Church	Philip Sharp	Mark Adams
2002	Craig Venter	David Ho	Harold Garner
2003	Edward Trifonov	Christiane Nüsslein-Volhard	Árpád Furka
2004	Russell Doolittle	Andrew Fire	Carlos Bustamante
2005	Charles DeLisi	Jonathan King	
2006	Michael Waterman	Anne-Claude Gavin	Roman Zubarev

The RECOMB Organizers

Since its inception in 1997, many scientists have participated in the conference in many fashions. While the committees have enjoyed the membership of over 170 different individuals between 1997 and 2006, many have participated over multiple years. The Steering Committee had consistent presence of 5 scientists between 1997 and 2005, including Michael Waterman, Pavel Pevzner, Ron Shamir, Sorin Istrail and Thomas Lengauer. The steering committee included 6 members throughout the first 8 years of the conference, with Richard Karp rounding out the group through 2003, and passing the position on to Terry Speed in 2004. In 2005 Michal Linial joined the Steering Committee to increase its size to 7.

Table 16. RECOMB Committee Membership. Each year shows the number of members in each committee.

Year	Number of Members		
	Steering	Organizing	Program
1997	6	5	23
1998	6	4	21
1999	6	6	29
2000	6	8	27
2001	6	9	23
2002	6	11	28
2003	6	5	31
2004	6	9	42
2005	7	17	43
2006	7	9	38

Table 17. RECOMB Program Committee Membership

Name	Years
Michael Waterman	10
Pavel Pevzner	10
Ron Shamir	10
Thomas Lengauer	10
Sorin Istrail	10
Martin Vingron	9
Richard Karp	9
Terry Speed	7
David Sankoff	6
Satoru Miyano	6
Gene Myers	5
Tandy Warnow	5
Dan Gusfield	5
Gordon Crippen	5
Sridhar Hannenhalli	5

The organizing committee has had a far more variable composition. Between 1997 and 2006, a total of 81 individuals have comprised the committee. The program committee has grown in size throughout the years of the conference (Table 16). While the size of the organizing and program committees do not correlate perfectly, the trend toward an increasing number of members per year has been exhibited in both. Numerous individuals have served on program committees in multiple years (Table 17).

RECOMB Funding

RECOMB has received support from a variety of sources. The US Department of Energy, US National Science Foundation and the SLOAN Foundation have been 3 major sponsors over the 10 years. Many other sponsors have significantly contributed to the conference, including IBM, International Society for Computational Biology (ISCB), SmithKline Beecham, Apple, Applied Biosystems, Celera, Compaq, Compugen, CRC Press, Glaxo-SmithKline, Hewlett-Packard, The MIT Press and the Broad Institute, Accelerys. Affymetrix, Agilent Technologies, Aventis, Berlin Center for Genome Based Bioinfornatics-BCB, Biogen, Boston University's Center for Advanced Genomic Technology, Centre de recherche en calcul applique (CERCA), CNRS, Conseil Regional Rhone-Alpes, Eurogentec-Seraing, Geneart GmbH, Genome Therapeutics, IMGT, INRA, LION Biosceince, LIPHA, Mairie de Lyon, Mathworks, Millennium Pharmaceuticals, Max Planck Institute for Molecular Genetics, Microsoft Research, NetApp, Novartis, Paracel, Partek Incorporated, Pfizer, Rosetta Biosoftware, Schering AG, Sun Microsystems, Technologiestiftung Berlin, The European Commission, High-level Scientific Conferences, The German Federal Ministry for Education and Research, The San Diego Supercomputer Center, The University of California-San Diego, Timelogic, Wyeth, Universitat degli Studi di Padova, Italy, DEI and AICA.

Conclusion

The approach of the 10th RECOMB conference held in Venice Italy provides us an opportunity to reflect on RECOMB's history. The landscape of computational biology has changed drastically since the first RECOMB Conference was held in Santa Fe, New Mexico. Today's conference contains papers covering research topics that did not exist 10 years ago. Over this period, many individuals have made significant research contributions through published papers. Many of the original founders of the RECOMB conference are still active, and many new faces are becoming active in the community each year.

Sorting by Weighted Reversals, Transpositions, and Inverted Transpositions

Martin Bader and Enno Ohlebusch

Computer Science Faculty,
University of Ulm, 89069 Ulm, Germany
martin.bader@uni-ulm.de, enno.ohlebusch@uni-ulm.de

Abstract. During evolution, genomes are subject to genome rearrangements that alter the ordering and orientation of genes on the chromosomes. If a genome consists of a single chromosome (like mitochondrial, chloroplast or bacterial genomes), the biologically relevant genome rearrangements are (1) *inversions*—also called *reversals*—where a section of the genome is excised, reversed in orientation, and reinserted and (2) *transpositions*, where a section of the genome is excised and reinserted at a new position in the genome; if this also involves an inversion, one speaks of an *inverted transposition*. To reconstruct ancient events in the evolutionary history of organisms, one is interested in finding an optimal sequence of genome rearrangements that transforms a given genome into another genome. It is well known that this problem is equivalent to the problem of "sorting" a signed permutation into the identity permutation. The complexity of the problem is still unknown. The best polynomial-time approximation algorithm, recently devised by Hartman and Sharan, has a 1.5 performance ratio. However, it applies only to the case in which reversals and transpositions are weighted equally. Because in most organisms reversals occur more often than transpositions, it is desirable to have the possibility of weighting reversals and transpositions differently. In this paper, we provide a 1.5-approximation algorithm for sorting by weighted reversals, transpositions and inverted transpositions for biologically realistic weights.

1   Introduction

During evolution, genomes are subject to genome rearrangements that alter the ordering and orientation (strandedness) of genes on the chromosomes. Because these events are rare compared to point mutations, they can give us valuable information about ancient events in the evolutionary history of organisms. For this reason, one is interested in the most "plausible" genome rearrangement scenario between two (or multiple) species. More precisely, given two genomes, one wants to find an optimal (shortest) sequence of rearrangement operations that transforms one into the other. Here we will focus on genomes that consists of a single (circular) molecule of DNA such as mitochondrial, chloroplast or bacterial genomes. As usual, the genomes are represented by a signed permutation,

A. Apostolico et al. (Eds.): RECOMB 2006, LNBI 3909, pp. 563–577, 2006.

i.e., an ordering of signed genes where the sign indicates the orientation (the strand). In this paper we do not consider unsigned permutations. In the single chromosome case, the relevant genome rearrangements are *inversions* (where a section of the genome is excised, reversed in orientation, and reinserted) and *transpositions* (where a section of the genome is excised and reinserted at a new position in the genome; if this also involves an inversion, one speaks of an *inverted transposition*). As is usually done in bioinformatics, we will use the terms "reversal" and "transreversal" as synonyms for "inversion" and "inverted transposition." It is well known that the problem of finding an optimal sequence of rearrangement operations that transforms a permutation into another permutation is equivalent to the problem of "sorting" a permutation by the same set of operations into the identity permutation. Let us briefly recall what is known for various sets of operations. In a seminal paper, Hannenhalli and Pevzner showed that the problem of sorting by reversals can be solved in polynomial time [13]. The Hannenhalli-Pevzner theory was simplified [5] and the running time of their algorithm was improved several times. To date, a subquadratic time algorithm [19] is available, and the reversal distance problem (which asks solely for the minimum number of required reversals, but not for the sequence of reversals) is solvable in linear time [1,6]. It is also worth mentioning that the problem of sorting an *unsigned* permutation by reversals is NP-hard [9] and the currently best approximation algorithm has the performance ratio 1.375 [7].

If one restricts the set of operations to transpositions (T), to transpositions and reversals (T + R), or to transpositions, reversals, and transreversals (T + R + TR), the complexity of the problem is still unknown. There exist polynomial-time approximation algorithms, and the best of them are listed in the table below.

operations	T	T + R	T + R + TR
performance ratio	1.375	2	1.5
references	[10]	[20, 17]	[15]

The biologically most relevant scenario is the T + R + TR case because in reality genomes are reorganized by all three kinds of operations. A drawback of Hartman and Sharan's [15] 1.5-approximation algorithm is that it applies only to the case in which reversals and transpositions are weighted equally (called the unweighted case in this paper). Because a transposition can create two cycles in the reality-desire diagram while a reversal can create at most one cycle (see below), the algorithm generally favors transpositions. Consequently, the sequence of rearrangement operations returned by that algorithm will often significantly deviate from the "true" evolutionary history because in most organisms transpositions are observed much less frequently than reversals. Thus, it is desirable to have the possibility of weighting reversals and transpositions differently. Given such weights, the weighted genome rearrangement problem asks for a sorting sequence of rearrangement operations such that the sum of the weights of the operations in the sequence is minimal. That is, a shortest sequence is not necessarily optimal. However, this problem is poorly studied. To our knowledge, there are only two algorithms that tackle it. The first is a $(1+\varepsilon)$-approximation algorithm devised by Eriksen [11]. It uses a weight proportion 2:1 (transposition:reversal) and has the tendency to use as much reversals as possible. The second algorithm is implemented in the software tool DERANGE II [8]. It is a greedy algorithm that works on the breakpoint distance and can only guarantee an approximation ratio of 3. In this paper, we will present a 1.5-approximation algorithm for any weight proportion between 1:1 and 2:1. Hence, our result closes the gap between the result of Hartman and Sharan [15] for the 1:1 proportion and that of Eriksen [11] for the 2:1 proportion. As the previous state of the art approximation algorithms for this problem, our algorithm proceeds by case analysis. In contrast to them, however, it is based on a (nontrivial) lower bound on the weighted rearrangement distance that is based on the number of odd *and* the number of even cycles. The running time of our algorithm is $O(n^2)$ in the naive implementation, but the time complexity can be improved to $O(n^{3/2} \log n)$.

2   Preliminaries

A *signed circular permutation* $\pi = (\pi_1 \ldots \pi_n)$ is a permutation of $(1 \ldots n)$, in which the indices are cyclic (i.e., n is followed by 1) and each element is labeled by plus or minus. We will use the term "permutation" as short hand for signed circular permutation. The *reflection* of a permutation π is the permutation $(-\pi_n \cdots - \pi_1)$. It is considered to be equivalent to π. Two consecutive elements π_i, π_{i+1} form an *adjacency* if $\pi_i = +x$ and $\pi_{i+1} = +(x + 1)$, or if $\pi_i = -x$ and $\pi_{i+1} = -(x - 1)$. Otherwise, they form a *breakpoint*. A *segment* $\pi_i \ldots \pi_j$ (with $j \geq i$) of a permutation π is a consecutive sequence of elements in π, with π_i as first element and π_j as last element. There are three possible rearrangement operations on a permutation π. A *transposition* $t(i, j, k)$ (with $i < j$ and $k < i$ or $k > j$) is an operation that cuts the segment $\pi_i \ldots \pi_{j-1}$ out of π, and reinserts it before the element π_k. A *reversal* $r(i, j)$ (with $i < j$) is an operation that inverts the order of the elements of the segment $\pi_i \ldots \pi_{j-1}$. Additionally, the sign of every element in the segment is flipped. A *transreversal* $tr(i, j, k)$ (with $i < j$ and $k < i$ or $k > j$) is the composition $t(i, j, k) \circ r(i, j)$ of a reversal and a transposition. In other words, the segment $\pi_i \ldots \pi_{j-1}$ will be cut out of π, inverted, and reinserted before π_k. A *sequence* of operations $op_1, op_2, \ldots, op_k$ applied to a permutation π yields the permutation $op_k \circ op_{k-1} \circ \cdots \circ op_1(\pi)$. In the following, reversals have weight w_r and transpositions as well as transreversals have weight w_t. As reversals usually occur much more frequently than transpositions and transreversals, we assume that $w_r \leq w_t$. The weight of a sequence is the sum of the weights of the operations in it. The problem of *sorting by weighted reversals, transpositions, and inverted transpositions* is defined as follows: Given a permutation π, find a sequence (of these operations) of minimum weight that transforms π into the identity permutation. This minimum weight will be denoted by $w(\pi)$.

In practice, it is also of interest to sort linear permutations. It has been proven by Hartman and Sharan [15] that sorting circular permutations is linearly equivalent to sorting linear permutations if yet another operation *revrev* is used that inverts two consecutive segments of the permutation. As long as the weights for transreversals and revrevs are the same, the proof also holds for sorting with weighted operations. Hence, our algorithm for circular permutations can be adapted to an algorithm for linear permutations that also uses revrevs.

2.1   The Reality-Desire Diagram

The reality-desire diagram [18] is a graph that helps us analysing the permutation; see Fig. 1. It is a variation of the breakpoint graph first described in [3]. The reality-desire diagram of a permutation $\pi = (\pi_1 \ldots \pi_n)$ can be constructed as follows. First, the elements of π are placed counterclockwise on a circle. Second, each element x of π labeled by plus is replaced with the two nodes $-x$ and $+x$, while each element x labeled by minus is replaced with $+x$ and $-x$. We call the first of these nodes the *left node* of x and the other the *right node* of x. Third, *reality-edges* are drawn from the right node of π_i to the left node of π_{i+1} for each index i (indices are cyclic). Fourth, *desire-edges* or *chords* are drawn from node $+x$ to node $-(x + 1)$ for each element x of π. We can interpret reality-edges as the actual neighborhood relations in the permutation, and desire-edges as the desired neighborhood relations. As each node is assigned exactly one reality-edge and one desire-edge, the reality-desire diagram decomposes into cycles. The *length* of a cycle is the number of chords in it. A k-cycle is a cycle of length k. If k is odd (even), we speak of an odd (even) cycle. The number of odd (even) cycles in π is denoted by $c_{odd}(\pi)$ ($c_{even}(\pi)$). It is easy to see that a 1-cycle corresponds to an adjacency and vice versa. A reversal cuts the permutation at two positions, while a transposition (transreversal) cuts it at three positions. Hence each of the operations cuts two or three reality-edges and moves the nodes. We say that the operation *acts* on these edges. Desire-edges are never changed by an operation.

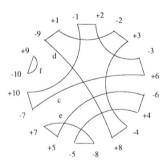
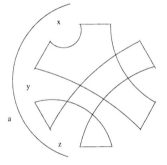

Fig. 1. Left: The reality-desire diagram of $\pi = (+1 + 9 + 10 + 7 - 5 + 8 + 4 + 6 + 3 + 2)$ contains the cycles $c, d, e,$ and f. Cycles d and e are intersecting, cycles c and d are interleaving, and all other pairs of cycles do not intersect. Right: The configuration that consists of the cycles d and e. Labels x, y, and z mark three positions in the configuration and the arc a consists of these positions.

2.2   Some Observations About Cycles

The following notions are illustrated in Fig. 1. A *configuration* is a subset of the cycles of the reality-desire diagram of a permutation. Configurations help us to focus on a few cycles in the reality-desire diagram instead of examining the whole diagram. A *position* in a configuration is the position between two consecutive reality-edges in the configuration. An *arc* a is a series of consecutive positions of a configuration, bounded by two reality-edges r_1 and r_2. Two chords d_1 and d_2 are *intersecting* if they intersect in the reality-desire diagram. More precisely, the endpoints of the chords must alternate along the circle in the configuration. Two cycles are intersecting if a pair of their chords is intersecting. Two cycles are *interleaving* if their reality-edges alternate along the circle. A rearrangement operation is called x_y-*move* if it increases the number of cycles by x and the operation is of type y (where r stands for a reversal, t for a transposition, and tr for a transreversal). For example, a transposition that splits one cycle into three is a 2_t-move. A reversal that merges two cycles is a -1_r-move. An $m_1m_2 \ldots m_n$-sequence is a sequence of n operations in which the first is an m_1-move, the second an m_2-move and so on. A cycle c is called r-*oriented* if there is a 1_r-move that acts on two of the reality-edges of c. Otherwise, the cycle is called r-unoriented. A cycle c is called t-*oriented* if there is a 2_t-move or a 2_{tr}-move that acts on three of the reality-edges of c. Otherwise, the cycle is called t-unoriented. A reality-edge is called *twisted* if its adjacent chords are intersecting; see Fig. 2. A chord is called twisted if it is adjacent to a twisted reality-edge; otherwise, it is called nontwisted. A cycle is called k-*twisted* if k of its reality-edges are twisted. If $k = 0$, we also say that the cycle is nontwisted.

Lemma 1. *A 2-cycle is r-oriented if and only if it is 2-twisted.*

Proof. There are only two possible configurations for a 2-cycle. If the cycle is 2-twisted, a reversal that acts on its reality-edges splits the cycle into two 1-cycles (adjacencies). Otherwise, no such move is possible.

Lemma 2 (proven in [14]). *A 3-cycle is t-oriented if and only if it is 2- or 3-twisted.*

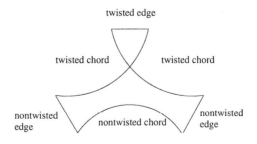

Fig. 2. An example for twisted reality-edges and twisted chords

Lemma 3 (*proven in [12]*). *If a cycle c of length ≥ 2 has a nontwisted chord, then there is another cycle d that intersects with this nontwisted chord of c.*

3   The Algorithm

We begin by introducing a new scoring function that allows us to show a very good lower bound for sorting by weighted reversals, transpositions, and inverted transpositions. Then, we will use the fact that a permutation can be transformed into an equivalent *simple permutation* without violating this lower bound. Because the sorting of the original permutation can be mimicked by the sorting of the simple permutation, we merely have to take care of simple permutations.

3.1   A Lower Bound

It has been proven by Gu et al. [12] that every operation changes the number of odd cycles by at most two. This fact leads to the following lower bound on $d(\pi)$.

Theorem 4 (*goes back to [4, 12, 15]*). *For any permutation $\pi = (\pi_1 \ldots \pi_n)$, the inequality $d(\pi) \geq (n - c_{odd}(\pi))/2$ holds, where $d(\pi)$ denotes the minimum number of reversals, transpositions, and inverted transpositions required to sort π into the identity permutation.*

For sorting by *weighted* reversals, transpositions, and inverted transpositions, this bound is not good enough because it does not distinguish between the weights of the operations. More precisely, adapting the bound to the weighted case would lead to the bound $w(\pi) \geq (n - c_{odd}(\pi))w_r/2$ because $w_r \leq w_t$. However, the only way how a reversal can increase c_{odd} by two is to split an even cycle into two odd cycles. We will now define a scoring function that treats such a reversal and a transposition splitting one odd cycle into three odd cycles equally.

Definition 5. *The score $\sigma(\pi)$ of a permutation π is defined by*

$$\sigma(\pi) = c_{odd}(\pi) + \left(2 - \frac{2w_r}{w_t}\right) c_{even}(\pi)$$

Let op_i be a rearrangement operation. The weight w_i of op_i is defined to be w_r if op_i is a reversal and w_t otherwise. Furthermore, we define $\Delta\sigma_i = \sigma(op_i(\pi)) - \sigma(\pi)$ to be the gain in score after the application of op_i to the permutation π (a negative gain is possible). It is not difficult to verify that for each operation op_i, the inequality $\Delta\sigma_i/w_i \leq 2/w_t$ holds provided that $w_r \leq w_t \leq 2w_r$. Moreover, for the two operations discussed immediately before Definition 5, the inequality becomes an equality.

Lemma 6. *For any permutation $\pi = (\pi_1 \ldots \pi_n)$ and weights w_r, w_t with $w_r \leq w_t \leq 2w_r$:*

- $\sigma(\pi) = n$ if π is the identity permutation
- $\sigma(\pi) \leq n - 1$ if π is not the identity permutation

Proof. If π is the identity permutation, the reality-desire diagram consists of n 1-cycles (adjacencies), so $\sigma(\pi) = c_{odd}(\pi) = n$. Otherwise, the diagram has at least one cycle of length ≥ 2. Therefore, it has at most $n - 1$ cycles. An odd cycle adds 1 to the score, while an even cycle adds $2 - \frac{2w_r}{w_t}$. With $w_t \leq 2w_r$ it follows that $2 - \frac{2w_r}{w_t} \leq 1$. Thus, $\sigma(\pi) \leq n - 1$.

Theorem 7. *For any permutation π and weights w_r, w_t with $w_r \leq w_t \leq 2w_r$, we have*

$$w(\pi) \geq lb(\pi) \ where \ lb(\pi) = c_{even}(\pi)w_r + \left(\frac{n - c_{odd}(\pi)}{2} - c_{even}(\pi) \right) w_t$$

Proof. Let $op_1, op_2, \ldots, op_k$ be an optimal sorting sequence of π, i.e., $w(\pi) = \sum_{i=1}^{k} w_i$. We have $\sigma(\pi) + \sum_{i=1}^{k} \Delta\sigma_i = n$ because π is transformed into the identity permutation, which has score n. It follows from $\Delta\sigma_i \leq w_i \frac{2}{w_t}$ that $n \leq \sigma(\pi) + \sum_{i=1}^{k} w_i \frac{2}{w_t} = \sigma(\pi) + w(\pi)\frac{2}{w_t}$. Hence $w(\pi) \geq (n - \sigma(\pi))\frac{w_t}{2} = lb(\pi)$.

3.2 Transformation into Simple Permutations

The analysis of cycles of arbitrary length is rather complicated. For this reason, a permutation will be transformed into a so-called simple permutation. A cycle is called *long* if its length is greater than 3. A permutation is called *simple* if it contains no long cycles. According to [13, 17, 14, 15], there is a padding algorithm that transforms any permutation π into a simple permutation $\tilde{\pi}$. Each transformation step increases n and c_{odd} by 1, and leaves c_{even} unchanged. Hence $lb(\tilde{\pi}) = lb(\pi)$. As the padding algorithm just adds elements to π, π can be sorted by using a sorting sequence of $\tilde{\pi}$ in which the added elements are ignored. Consequently, the resulting sorting sequence of π has the same or a smaller weight than the sorting sequence of $\tilde{\pi}$. In the next subsection, we will present an algorithm that takes a simple permutation $\tilde{\pi}$ as input and outputs a sorting sequence $op_1, op_2, \ldots, op_k$ of $\tilde{\pi}$ such that $\sum_{i=1}^{k} w_i \leq 1.5 \, lb(\tilde{\pi})$. Altogether, this yields a 1.5-approximation for sorting by weighted reversals, transpositions, and inverted transpositions because $w(\pi) \leq \sum_{i=1}^{k} w_i \leq 1.5 \, lb(\tilde{\pi}) = 1.5 \, lb(\pi) \leq 1.5 \, w(\pi)$.

Note that it is not possible to transform 2-cycles into 3-cycles as done in [15] because these transformations would change the score and the lower bound.

3.3 The Algorithm for Simple Permutations

Given a simple permutation π, the overall goal is to find a sorting sequence $op_1, op_2, \ldots, op_k$ of π such that $\sum_{i=1}^{k} \Delta\sigma_i \geq \sum_{i=1}^{k} w_i \frac{4}{3w_t}$. By a reasoning similar to the proof of Theorem 7, it then follows $\sum_{i=1}^{k} w_i \leq 1.5 \, lb(\pi)$. To achieve this goal, we search for a "starting sequence" $op_1, \ldots, op_j$ of at most four operations

Table 1. The algorithm's decision tree if it begins with an r-unoriented 2-cycle c. All cycles are considered to be r-unoriented 2-cycles or t-unoriented 3-cycles because r-oriented 2-cycles or t-oriented 3-cycles can directly be eliminated. Cross-references α and β can be found in this table, γ and δ in Table 2, while ε, ζ, and η are in Table 3.

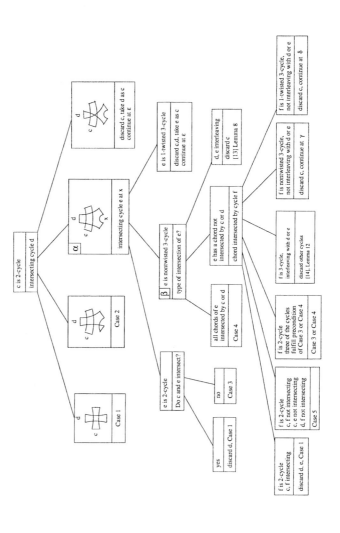

Table 2. The algorithm's decision tree if it begins with a nontwisted 3-cycle c. All cycles are considered to be r-unoriented 2-cycles or t-unoriented 3-cycles because r-oriented 2-cycles or t-oriented 3-cycles can directly be eliminated. Cross-references α and β can be found in Table 1, γ and δ in this table, while ε, ζ, and η are in Table 3.

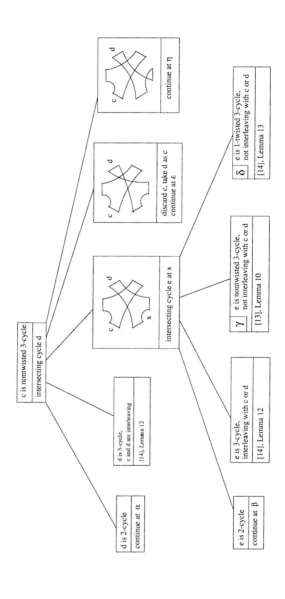

Table 3. The algorithm's decision tree if it begins with a 1-twisted 3-cycle c. Again, all cycles are r-unoriented 2-cycles or t-unoriented 3-cycles. Cross-references α and β can be found in Table 1, γ and δ in Table 2, while ε, ζ and η are in this table.

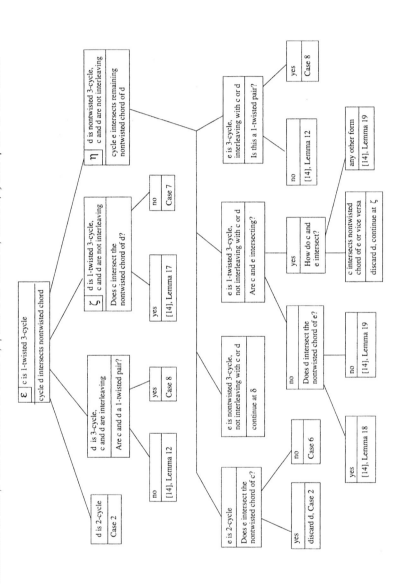

(i.e., $1 \leq j \leq 4$) such that $\sum_{i=1}^{j} \Delta\sigma_i \geq \sum_{i=1}^{j} w_i \frac{4}{3w_t}$. This procedure is iterated (i.e., we next search for a starting sequence of $op_j \circ \cdots \circ op_1(\pi)$ etc.) until the identity permutation is reached.

The algorithm starts by searching for an arbitrary cycle c of length ≥ 2 in the reality-desire diagram of π. If the cycle is an r-oriented 2-cycle or a t-oriented 3-cycle, the starting sequence can consist solely of the operation op_1 that eliminates this cycle (i.e., op_1 is a 1_r, 2_t or 2_{tr} move that cuts the cycle into 1-cycles). This is because $\Delta\sigma_1/w_1 = 2/w_t \geq 4/3w_t$. Otherwise, according to Lemma 3, c must have a nontwisted chord that is intersected by another cycle d. The algorithm now searches for this cycle and examines the configuration of the cycles c and d. Depending on the configuration found, the algorithm either directly outputs a starting sequence that meets the requirements or, again by Lemma 3, there must be a chord in the configuration that is intersected by a cycle e that is not yet in the configuration. Consequently, the algorithm searches for this cycle and adds it to the configuration. This goes on until a configuration is found for which a starting sequence can be provided. The algorithm is based on a descision tree that can be found in Tables 1, 2, and 3. Note that every configuration consists of at most four cycles.

A careful inspection of the starting sequences described in [14] and [15] for configurations that do not contain 2-cycles reveals that these sequences also work in our case. Therefore, we merely have to consider configurations with at least one r-unoriented 2-cycle (recall that r-oriented 2-cycles can immediately be eliminated). These cases are listed below and example configurations can be found in Fig. 3.

Case 1. *c and d are two intersecting 2-cycles (Fig. 3a).*

Case 2. *A 2-cycle c intersects the nontwisted chord of a 1-twisted 3-cycle d (Fig. 3b).*

Case 3. *c and e are 2-cycles, whereas d is a nontwisted 3-cycle. c and e are not intersecting, and each nontwisted chord of d is intersected by c or e (Fig. 3c).*

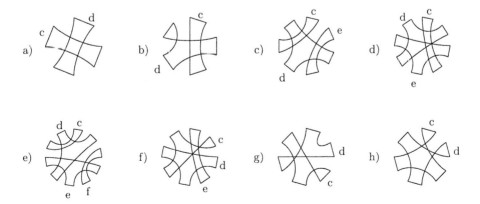

Fig. 3. Example configurations for the new cases to be taken into account

Case 4. *c is a 2-cycle, whereas d and e are intersecting nontwisted 3-cycles. c intersects the nontwisted chords of d and e that are not intersected by the other 3-cycle (Fig. 3d).*

Case 5. *d and e are two intersecting nontwisted 3-cycles, whereas c and f are 2-cycles. c intersects with the nontwisted chord of d that is not intersected by e, and f intersects with the nontwisted chord of e that is not intersected by d. c and d do not intersect with f, and e does not intersect with c (Fig. 3e).*

Case 6. *c is a 1-twisted 3-cycle and d is a nontwisted 3-cycle that intersects the nontwisted chord of c. The remaining chord of d (the one not intersected by c) is intersected by a 2-cycle e that does not intersect the nontwisted chord of c (Fig. 3f).*

Case 7. *c and d are two intersecting 1-twisted 3-cycles. d intersects the non-twisted chord of c, but c does not intersect the nontwisted chord of d (Fig. 3g).*

Case 8. *Two 1-twisted 3-cycles c and d form a 1-twisted pair (Fig. 3h).*

Although the last two cases do not contain a 2-cycle, they have to be taken into account because in these cases we need a further intersecting cycle, which may be a 2-cycle.

To exemplify our method, we will give the starting sequences for Cases 4 and 6. Figs. 4 and 5 depict the configurations before and after the application of an operation in the sequence. In each configuration, the reality-edges on which the next operation acts are marked with x or *. If three edges are marked with *, the operation is a transposition. If two edges are marked with x and one is marked with *, the operation is a transreversal, and the segment between the two x will be inverted. If two edges are marked with x and none is marked with *, the operation is a reversal.

A full listing of the starting sequences can be found in [2]. In the following Δc_{odd} (Δc_{even}) denotes the change in the number of odd (even) cycles after the application of the starting sequence.

Lemma 8. *For Case 4, there is a $0_r 1_r 2_{tr}$-sequence with $\Delta c_{odd} = 4$ and $\Delta c_{even} = -1$.*

Proof. The sequence is described in Fig. 4. We have $\sum \Delta \sigma_i / \sum w_i = 2(w_r + w_t)/w_t(2w_r + w_t)$. This value varies from $4/3w_t$ (for $w_t : w_r = 1 : 1$) to $3/2w_t$ (for $w_t : w_r = 2 : 1$).

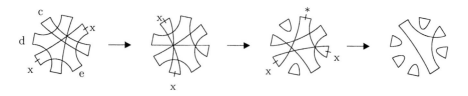

Fig. 4. Sequence for Case 4

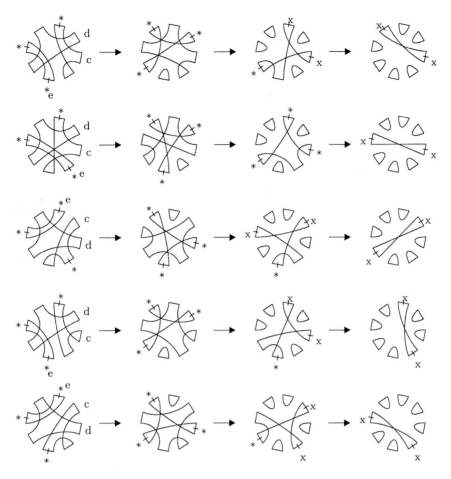

Fig. 5. The sequences for Case 6

Lemma 9. *For Case 6, there is a $0_t 2_t 2_t 1_r$-sequence or a $0_t 2_t 2_{tr} 1_r$-sequence with $\Delta c_{odd} = 6$ and $\Delta c_{even} = -1$.*

Proof. There are five possible configurations. For all of them, a sequence is described in Fig. 5. The last operation of each sequence is a reversal that splits the last 2-cycle into adjacencies (the resulting configurations are not shown in the figure). Note that for these sequences, we have $\sum \Delta \sigma_i / \sum w_i = (4w_t + 2w_r)/w_t(3w_t + w_r)$. This value varies from $10/7w_t$ (for $w_t : w_r = 2 : 1$) to $3/2w_t$ (for $w_t : w_r = 1 : 1$).

4   Further Improvements

Our algorithm is implemented in $C++$ and it has time complexity $O(n^2)$. There are several possible improvements to the basic algorithm that can decrease its running time or its approximation ratio. Some of these improvements are:

- Using a special data structure described in [16], it is possible to find the different cases in sublinear time. The running time improves to $O(n^{3/2}\sqrt{\log n})$; cf. [15].
- Examining configurations with more cycles could improve the approximation ratio. Using this strategy, Elias and Hartman [10] recently succeeded in improving the performance ratio for sorting by transpositions from 1.5 to 1.375. It is highly expected that this strategy can also improve the performance ratio of sorting by weighted reversals, transpositions, and inverted transpositions.
- The algorithm can be combined with a greedy strategy: Instead of beginning with the first cycle in the reality-desire diagram, we start the search at each cycle in the diagram, and use a sequence with the best gain in score. This increases the running time by a factor of n, but the algorithm will find better sorting sequences, and changes in the weight function result in different sorting sequences.

References

1. D.A. Bader, B.M.E. Moret, and M. Yan. A linear-time algorithm for computing inversion distance between signed permutations with an experimental study. *Journal of Computational Biology*, 8:483–491, 2001.
2. M. Bader. Sorting by weighted transpositions and reversals. Master's thesis, University of Ulm, December 2005.
3. V. Bafna and P.A. Pevzner. Genome rearrangements and sorting by reversals. *SIAM Journal on Computing*, 25(2):272–289, 1996.
4. V. Bafna and P.A. Pevzner. Sorting by transpositions. *SIAM Journal on Discrete Mathematics*, 11(2):224–240, 1998.
5. A. Bergeron. A very elementary presentation of the Hannenhalli-Pevzner theory. *Discrete Applied Mathematics*, 146(2):134–145, 2005.
6. A. Bergeron, J. Mixtacki, and J. Stoye. Reversal distance without hurdles and fortresses. In *Proc. 15th Annual Symposium on Combinatorial Pattern Matching*, volume 3109 of *Lecture Notes in Computer Science*, pages 388–399. Springer-Verlag, 2004.
7. P. Berman, S. Hannenhalli, and M. Karpinski. 1.375-approximation algorithm for sorting by reversals. In *Proc. of the 10th Annual European Symposium on Algorithms*, volume 2461 of *Lecture Notes in Computer Science*, pages 200–210. Springer-Verlag, 2002.
8. M. Blanchette, T. Kunisawa, and D. Sankoff. Parametric genome rearrangement. *Gene*, 172:GC11–17, 1996.
9. A. Caprara. Sorting permutations by reversals and Eulerian cycle decompositions. *Journal on Discrete Mathematics*, 12:91–110, 1999.
10. I. Elias and T. Hartman. A 1.375-approximation algorithm for sorting by transpositions. In *Proc. of 5th International Workshop on Algorithms in Bioinformatics*, volume 3692 of *Lecture Notes in Bioinformatics*, pages 204–215. Springer-Verlag, 2005.
11. N. Eriksen. $(1 + \epsilon)$-approximation of sorting by reversals and transpositions. *Theoretical Computer Science*, 289(1):517–529, 2002.

12. Q.-P. Gu, S. Peng, and I.H. Sudborough. A 2-approximation algorithm for genome rearrangements by reversals and transpositions. *Theoretical Computer Science*, 210(2):327–339, 1999.
13. S. Hannenhalli and P.A. Pevzner. Transforming cabbage into turnip (polynomial algorithm for sorting signed permutations by reversals). *Journal of the ACM*, 48:1–27, 1999.
14. T. Hartman. A simpler 1.5-approximation algorithm for sorting by transpositions. In *Proc. of the 14th Annual Symposium on Combinatorial Pattern Matching*, volume 2676 of *Lecture Notes in Computer Science*, pages 156–169. Springer-Verlag, 2003.
15. T. Hartman and R. Sharan. A 1.5-approximation algorithm for sorting by transpositions and transreversals. In *Proc. of 4th International Workshop on Algorithms in Bioinformatics*, volume 3240 of *Lecture Notes in Bioinformatics*, pages 50–61. Springer-Verlag, 2004.
16. H. Kaplan and E. Verbin. Efficient data structures and a new randomized approach for sorting signed permutations by reversals. In *Proc. of 14th Symposium on Combinatorial Pattern Matching*, volume 2676 of *Lecture Notes in Computer Science*, pages 170–185. Springer-Verlag, 2003.
17. G.-H. Lin and G. Xue. Signed genome rearrangement by reversals and transpositions: Models and approximations. *Theoretical Computer Science*, 259(1-2): 513–531, 2001.
18. J. Setubal and J. Meidanis. *Introduction to Computational Molecular Biology*. PWS Publishing, Boston, M.A., 1997.
19. E. Tannier and M.-F. Sagot. Sorting by reversals in subquadratic time. In *Proc. of the 15th Annual Symposium on Combinatorial Pattern Matching*, volume 3109 of *Lecture Notes in Computer Science*, pages 1–13. Springer-Verlag, 2004.
20. M.E.T. Walter, Z. Dias, and J. Meidanis. Reversal and transposition distance of linear chromosomes. In *Proc. of the Symposium on String Processing and Information Retrieval*, pages 96–102. IEEE Computer Society, 1998.

A Parsimony Approach to Genome-Wide Ortholog Assignment

Zheng Fu[1], Xin Chen[2], Vladimir Vacic[1], Peng Nan[3],
Yang Zhong[1], and Tao Jiang[1,4]

[1] Computer Science Department, University of California - Riverside
[2] School of Physical and Mathematical Sci., Nanyang Tech. Univ., Singapore
[3] Shanghai Center for Bioinformatics Technology, Shanghai, China
[4] Tsinghua University, Beijing, China

Abstract. The assignment of orthologous genes between a pair of
genomes is a fundamental and challenging problem in comparative genomics, since many computational methods for solving various biological
problems critically rely on *bona fide* orthologs as input. While it is usually done using sequence similarity search, we recently proposed a new
combinatorial approach that combines sequence similarity and genome
rearrangement. This paper continues the development of the approach
and unites genome rearrangement events and (post-speciation) duplication events in a single framework under the parsimony principle. In this
framework, orthologous genes are assumed to correspond to each other
in the most parsimonious evolutionary scenario involving both genome
rearrangement and (post-speciation) gene duplication. Besides several
original algorithmic contributions, the enhanced method allows for the
detection of inparalogs. Following this approach, we have implemented
a high-throughput system for ortholog assignment on a genome scale,
called MSOAR, and applied it to the genomes of human and mouse.
As the result will show, MSOAR is able to find 99 more true orthologs
than the INPARANOID program did. We have also compared MSOAR
with the iterated exemplar algorithm on simulated data and found that
MSOAR performed very well in terms of assignment accuracy. These
test results indiate that our approach is very promising for genome-wide
ortholog assignment.

1   Introduction

Orthologs and *paralogs*, originally defined in [6], refer to two fundamentally different types of homologous genes. They differ in the way that they arose: orthologs
are genes that evolved by speciation, while paralogs are genes that evolved by
duplication. To better describe the evolutionary process and functional diver-
sification of genes, paralogs are further divided into two subtypes: *outparalogs*,
which evolved via an ancient duplication preceding a given speciation event under consideration, and *inparalogs*, which evolved more recently, subsequent to
the speciation event [16][10]. For a given set of inparalogs on a genome, there

A. Apostolico et al. (Eds.): RECOMB 2006, LNBI 3909, pp. 578–594, 2006.

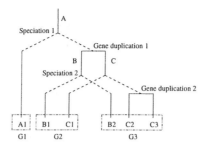

Fig. 1. An illustration of orthologous and paralogous relationships. After two speciation events and two gene duplications, three present genomes, $G_1 = (A_1)$, $G_2 = (B_1, C_1)$ and $G_3 = (B_2, C_2, C_3)$ are obtained. In this scenario, all genes in G_2 and G_3 are co-orthologous to gene A_1. Genes B_1 and C_1 are outparalogs w.r.t. G_3 (*i.e.*, the 2nd speciation), and are inparalogs w.r.t. G_1 (*i.e.*, the 1st speciation). Gene C_2 is the direct descendant (*i.e.*, true exemplar) of the ancestral gene C while C_3 is not, if C_3 is duplicated from C_2. C_1 and C_2 are said to form a pair of main orthologs.

commonly exists a gene that is the direct descendant of the ancestral gene of the set, namely the one that best reflects the original position of the ancestral gene in the ancestral genome. Sankoff [17] called such a gene the *true exemplar* of the inparalogous set. Given two genomes, two sets of inparalogous genes (one from each genome) are *co-orthologous* if they are descendants of the same ancestral gene at the time of speciation. These concepts are illustrated in Figure 1.

Clearly, orthologs are evolutionary and, typically, functional counterparts in different species. Therefore, many existing computational methods for solving various biological problems, *e.g.*, the inference of functions of new genes and the analysis of phylogenetic relationship between different species, use orthologs in a critical way. A major complication with the use of orthologs in these methods, however, is that orthology is not necessarily a one-to-one relationship because a single gene in one phylogenetic lineage may correspond to a whole family of inparalogs in another lineage. More caution should be taken while such one-to-many and many-to-many relationships are applied to the transfer of functional assignments because inparalogs could have acquired new functions during the course of evolution. As a consequence, the identification of orthologs and inparalogs, especially those one-to-one orthology relationships, is critical for evolutionary and functional genomics, and thus a fundamental problem in computational biology.

It follows from the definition of orthologs and paralogs that the best way to identify orthologs is to measure the divergence time between homologous genes in two different genomes. As the divergence time could be estimated by comparing the DNA or protein sequences of genes, most of the existing algorithms for ortholog assignment, such as the well-known COG system [21][23] and IN-PARANOID program [16], rely mainly on sequence similarity (usually measured via BLAST scores [1]). An implicit, but often questionable, assumption behind these methods is that the evolutionary rates of all genes in a homologous family are equal. Incorrect ortholog assignments might be obtained if the real rates of evolution vary significantly between paralogs. On the other hand, we observe that molecular evolution proceeds in two different forms: local mutation and global rearrangement. Local mutations include base substitution, insertion and deletion, and global rearrangements include reversal (*i.e.* inversion), translocation, fusion, fission and so on. Apparently, the sequence similarity-based methods for ortholog assignment make use of local mutations only and neglect genome rearrangement events that might contain a lot of valuable information.

In our recent papers [4][5], we initiated the study of ortholog assignment via genome rearrangement and proposed an approach that takes advantage of evolutionary evidences from both local mutations and global rearrangements. It begins by identifying homologous gene families on each genome and the correspondence between families on both genomes using homology search. The homologs are then treated as copies of the same genes, and ortholog assignment is formulated as a natural optimization problem of rearranging one genome consisting of a sequence of (possibly duplicated) genes into the other with the minimum number of reversals, where the most parsimonious rearrangement process should suggest orthologous gene pairs in a straightforward way. A high-throughput system, called SOAR, was implemented based on this approach. Though our preliminary experiments on simulated data and real data (the X chromosomes of human and mouse) have demonstrated that SOAR is very promising as an ortholog assignment method, it has the drawback of ignoring the issue of inparalogs. In fact, it assumed that there were no gene duplications after the speciation event considered. As a consequence, it only outputs one-to-one orthology relationships and every gene is forced to form an orthologous pair. Moreover, it is only able to deal with unichromosomal genomes. In this paper, we present several improvements that are crucial for more accurate ortholog assignment. In particular, the method will be extended to deal with inparalogs explicitly by incorporating a more realistic evolutionary model that allows duplication events after the speciation event. In summary, our main contributions in this study include

- We introduce a subtype of orthologs, called *main orthologs*, to delineate sets of co-orthologous genes. For two inparalogous sets of co-orthologous genes, the main ortholog pair is simply defined as the two true exemplar genes of each set (see Figure 1 for an example).[1] Since a true exemplar is a gene that best reflects the original position of the ancestral gene in the ancestral genome, main orthologs are therefore the *positional counterpart* of orthologs in different species. By definition, main orthologs form a one-to-one correspondence, thus allowing for the possibility of direct transfer of functional assignments. We believe that, compared with other types of ortholog pairs, main orthologs are more likely to be functional counterparts in different species, since they are both evolutionary and positional counterparts.
- In our previous study, the evolutionary model assumes that there is no gene duplication subsequent to the given speciation event. Thus, no inparalogs are assumed to be present in the compared genomes, which is clearly inappropriate for nuclear genomes. In this paper, we propose a parsimony approach

---

[1] Note that, our definition of a main ortholog pair is different from the one in [16], where it refers to a mutually best hit in an orthologous group.

based on a more realistic evolutionary model that combines both rearrangement events (including reversal, translocation, gene fusion and gene fission) and gene duplication events. This will allow us to treat inparalogs explicitly. More specifically, in order to assign orthologs, we reconstruct an evolutionary scenario since the splitting of the two input genomes, by minimizing the (total) number of reversals, translocations, fusions, fissions and duplication events necessary to transform one genome into the other (*i.e.*, by computing the *rearrangement/duplication* distance between two genomes). Such a most parsimonious evolutionary scenario should reveal main ortholog pairs and inparalogs in a straightforward way.

– Computing the rearrangement/duplication distance between two genomes is known to be very hard. We have developed an efficient heuristic algorithm that works well on large multichromosomal genomes like human and mouse. We strengthen and extend some of the algorithmic techniques developed in [4][5], including (sub)optimal assignment rules, minimum common partition, and maximum graph decomposition, as well as a new post-processing step that removes "noise" gene pairs that are most likely to consist of inparalogs.
– Based on the above heuristic algorithm, we have implemented a high-through put system for automatic assignment of (main) orthologs and the detection of inparalogs on a genome scale, called MSOAR. By testing it on simulated data and human and mouse genomes, the MSOAR system is shown to be quite effective for ortholog assignment. For example, it is able to find 99 more true ortholog pairs between human and mouse than INPARANOID [16].

Related work. In the past decade, many computational methods for ortholog assignment have been proposed, most of which are based primarily on sequence similarity. These methods include the COG system [21][23], EGO (previously called TOGA)[11], INPARANOID [16], and OrthoMCL [12], just to name a few. Some of these methods combine sequence similarity and a parsimony principle, such as the reconciled tree method [25] and the bootstrap tree method [20], or make use of synteny information, such as OrthoParaMap [3] and the recent method proposed by Zheng *et al.* [26]. However, none of these papers use genome rearrangement. On the other hand, there have been a few papers in the literature that study rearrangement between genomes with duplicated genes, which is closely related to ortholog assignment. Sankoff [17] proposed an approach to identify the true exemplar gene of each gene family, by minimizing the breakpoint/reversal distance between two reduced genomes that consist of only true exemplar genes. El-Mabrouk [14] developed an approach to reconstruct an ancestor of a modern genome by minimizing the number of duplication transpositions and reversals. The work in [13][18] attempts to find a one-to-one gene correspondence between gene families based on conserved segments. Very recently, Swenson *et al.* [19] presented some algorithmic results on the cycle splitting problem in a combinatorial framework similar to the one introduced in [4][5].

The rest of the paper is organized as follows. We first discuss the parsimony principle employed in our ortholog assignment approach in Section 2. Section 3 describes the heuristic algorithm implemented in MSOAR. Section 4 will present our experiments on simulated data and on the whole genome data of human and mouse. Finally, some concluding remarks are given in Section 5.

2   Assigning Orthologs Under Maximum Parsimony

The two genomes to be compared, denoted as Π and Γ, have typically undergone series of genome rearrangement and gene duplication events since they split from their last common ancestral genome. Clearly, we could easily identify main orthologs and inparalogs if given such an evolutionary scenario. Based on this observation, we propose an approach to reconstruct the evolutionary scenario on the basis of the parsimony principle, *i.e.*, we postulate the minimal possible number of rearrangement events and duplication events in the evolution of two genomes since their splitting so as to assign orthologs. Equivalently, it can be formulated as a problem of finding a most parsimonious transformation from one genome into the other by genome rearrangements and gene duplications, without explicitly inferring their ancestral genome. Let $R(\Pi, \Gamma)$ and $D(\Pi, \Gamma)$ denote the number of rearrangement events and the number of gene duplications in a most parsimonious transformation, respectively, and $RD(\Pi, \Gamma)$ denotes the *rearrangement/duplication (RD) distance* between Π and Γ satisfying $RD(\Pi, \Gamma) = R(\Pi, \Gamma) + D(\Pi, \Gamma)$. Most genome rearrangement events will be considered in this study, including reversal, translocation, fusion and fission.

In practice, we will impose two constraints on this optimization problem, based on some biological considerations. First, we require that at least one member of each family that appears in the other genome be assigned orthology, because each family should provide an essential function and the gene(s) retaining this function is more likely conserved during the evolution. Second, observe that the assignment of orthologs that leads to the minimum rearrangement/duplication distance is not necessarily unique. Therefore, among all assignments with the minimum rearrangement/duplication distance, we attempt to find one that also minimizes $R(\Pi, \Gamma)$, in order to avoid introducing unnecessary false orthologous pairs.

Figure 2 presents a simple example to illustrate the basic idea behind our parsimony approach. Consider two genomes, $\Pi = -b - a_1 + c + a_2 + d + a_4 + e + f + g$ and $\Gamma = +a_1 + b + c + a_2 + d + e + a_5 + f + a_3 + g$, sharing a gene family a with multiple copies. As shown in Figure 2, both genomes evolved from the same ancestral genome $+a + b + c + d + e + f + g$, Π by one inversion and one gene duplication and Γ by two gene duplications, respectively. By computing the rearrangement/duplication distance $RD(\Pi, \Gamma) = 4$, the true evolutionary scenario can be reconstructed, which then suggests that the two genes a_1 form a pair of main orthologs, as well as the two genes a_2. Meanwhile, a_3, a_4, and a_5 are inferred as inparalogs that were derived from duplications after the speciation event. It is interesting to see that here a_4 is not assigned orthology to a_3 or a_5 greedily. (Note that they are orthologs, but not main orthologs, by our definition.) This simple example illustrates that, by minimizing the reversal/duplication distance, our approach is able to pick correct main orthologs out of sets of inparalogs.

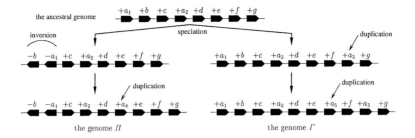

Fig. 2. An evolutionary history of two genomes Π and Γ. Π evolved from the ancestor by one inversion and one gene duplication, and Γ by two duplications.

Note that, although gene loss may occur in the course of evolution, it actually has no impact on the capability of assigning ortholog by our method. If an inparalog is lost, the gene loss event can be simply ignored and this will not affect ortholog assignment. If some gene of a main ortholog pair is lost, our approach attempts to identify the other gene as an inparalog rather than to assign it a wrong orthology, which also makes some sense especially when considering the transfer of functional assignment.

3   The MSOAR System

Following the parsimony principle discussed in the previous section, we have implemented a high-throughput system for ortholog assignment, called MSOAR. It employs a heuristic to calculate the rearrangement/duplication distance between two genomes, which can be used to reconstruct a most parsimonious evolutionary scenario. In this section, we discuss in detail the heuristic algorithm.

We represent a gene by a symbol of some finite alphabet $\mathcal{A}$, and its orientation by the sign $+$ or $-$. A chromosome is a sequence of genes, while a genome is a set of chromosomes. Usually, a genome is represented as a set $\Pi = \{\pi(1), \cdots, \pi(N)\}$, where $\pi(i) = \langle \pi(i)_1 \cdots \pi(i)_{n_i} \rangle$ is a sequence of oriented genes in the ith chromosome. Recall the genome rearrangement problem between two genomes with distinct oriented genes. Hannenhalli and Pevzner developed algorithms for calculating genome rearrangement distance on both unichromosomal [7] and multichromosomal genomes [8] in polynomial time. The rearrangement distance between multichromosomal genomes is the minimum number of *reversals, translocations, fissions* and *fusions* that would transform one genome into the other. Given two multichromosomal genomes Π and Γ, Hannenhalli and Pevzner [8] gave a formula for calculating the genome rearrangement distance (called the HP formula in this paper). Tesler [22], and Ozery-Flato and Shamir [15] then suggested some corrections to the formula (called the revised HP formula):

$$d(\Pi,\Gamma) = b(\Pi,\Gamma) - c(\Pi,\Gamma) + p_{\Gamma\Gamma}(\Pi,\Gamma) + r(\Pi,\Gamma) + \lceil \tfrac{s'(\Pi,\Gamma) - gr'(\Pi,\Gamma) + fr'(\Pi,\Gamma)}{2} \rceil$$

where $b(\Pi,\Gamma)$ is the number of black edges in the breakpoint graph $G(\Pi,\Gamma)$, $c(\Pi,\Gamma)$ is the overall number of cycles and paths, $p_{\Gamma\Gamma}(\Pi,\Gamma)$ is the number of

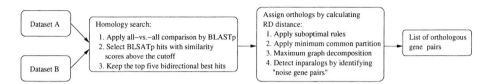

Fig. 3. An outline of MSOAR

the $\Gamma\Gamma$-paths, and r, s', gr' and fr' are some parameters in terms of real-knots [15]. In practice, the dominant parts of the formula are the first three terms.

When the genomes Π and Γ contain duplicated genes, however, the rearrangement/duplication distance problem (*i.e.* $RD(\Pi, \Gamma)$) cannot be directly solved by the revised HP formula. In fact, we can prove that computing the rearrangement/duplication distance is NP-hard by a reduction similar to the one employed in the proof of Theorem 4.2 of [5]. Note that, once the main ortholog pairs are assigned and the inparalogs are identified, the rearrangement/duplication distance can be easily computed as follows. The number of duplications is determined by the number of inparalogs. After removing all the inparalogs, the rearrangement distance between the reduced genomes, which now have equal gene content, can be computed using the above formula since every gene can be regarded as unique. An outline of MSOAR is illustrated in Figure 3.

3.1   Homology Search and Gene Family Construction

MSOAR starts by calculating the pairwise similarity scores between all gene sequences of the two input genomes. An all-*versus*-all gene sequence comparison by BLASTp is used to accomplish this. As in [16], two cutoffs are applied to each pair of BLASTp hits. Two genes are considered homologous if (1) the bit score is no less than 50 and (2) the matching segment spans above 50% of each gene in length. In order to eliminate potential false main ortholog pairs, we take the top five bidirectional best hits of each gene as its potential main orthologs if their logarithmic E-value is less than the 80% of the best logarithmic E-value. By clustering homologous genes using the standard single linkage method, we obtain gene families. A gene family is said to be *trivial* if it has cardinality exactly 2, *i.e.* with one occurrence in each genome. Otherwise it is said to be non-trivial. A gene belonging to a trivial (or non-trivial) family is said to be trivial (or non-trivial, resp.). We use a *hit graph* (denoted as $\mathcal{H}$) to describe the relationship between genes within each family. A hit graph is a bipartite graph illustrating the BLASTp hits between two genomes. Each vertex represents a gene and an edge connects two vertices from different genomes if they are potential main orthologs. Figure 4 gives an example of the hit graph. Adjacent genes in the hit graph are regarded as candidates for main ortholog pairs.

3.2   (Sub)Optimal Assignment Rules

We presented three assignment rules for identifying individual ortholog assignments that are (nearly) optimal in SOAR [4][5]. In MSOAR, we will add two

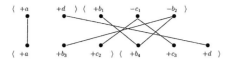

Fig. 4. A hit graph for genomes $\Pi = \{\langle +a, +d\rangle, \langle +b_1, -c_1, -b_2\rangle\}$ and $\Gamma = \{\langle +a, +b_3, +c_2\rangle, \langle +b_4, +c_3, +d\rangle\}$, each having two chromosomes

more assignment rules, which could make the system more efficient. The four rearrangement operations (reversal, translocation, fission and fusion) can be mimicked by reversals when we represent a multichromosomal genome by a *con-catenate* [8][22]. This approach reduces the problem of computing rearrangement distance between two multichromosomal genomes to the problem of computing the reversal distance between two *optimal concatenates*. Since these two new rules are only concerned with segments of consecutive genes within a single chromosome, which also form gene segments in an optimal concatenate, the unichromosomal HP formula [7] can be used to prove their (sub)optimality. Let G and H be two chromosomes in genomes Π and Γ, respectively. A *chromosome segment* is defined as a substring of some chromosome (*i.e.* a consecutive sequence of genes). A chromosome segment $(g_{i_1} g_{i_2} \cdots g_{i_n})$ in G *matches* a chromosome segment $(h_{j_1} h_{j_2} \cdots h_{j_n})$ in H if g_{i_t} and h_{j_t} are connected by an edge in the hit graph and have the same orientations for all $1 \leq t \leq n$.

Theorem 1. *Assume that a chromosome segment $(g_{i_1} g_{i_2} \cdots g_{i_n})$ in G, matches a chromosome segment $(h_{j_1} h_{j_2} \cdots h_{j_n})$ in H or its reversal, where g_{i_1} ,g_{i_n}, h_{j_1} and h_{j_n} are trivial but the other genes are not. Define two new genomes Π' and Γ' by assigning orthology between g_{i_t} and h_{j_t} or g_{i_t} and $g_{j_{n+1-t}}$ (in the case of matching by a reversal), for all $1 \leq t \leq n$. Then, $RD(\Pi, \Gamma) \leq RD(\Pi', \Gamma') \leq RD(\Pi, \Gamma) + 2$.*

Theorem 2. *Assume that for a chromosome segment $(g_{i_1} g_{i_2} \cdots g_{i_n})$ in G and a chromosome segment $(h_{j_1} h_{j_2} \cdots h_{j_n})$ in H, g_{i_1} matches h_{j_1}, g_{i_n} matches h_{j_n}, and $g_{i_2} \cdots g_{i_{n-1}}$ matches the reversal of $h_{j_2} \cdots h_{j_{n-1}}$, where g_{i_1}, g_{i_n}, h_{j_1} and h_{j_n} are trivial but the other genes are not. Define two new genomes Π' and Γ' by assigning orthology between g_{i_t} and $g_{j_{n+1-t}}$, for all $1 < t < n$. Then, $RD(\Pi, \Gamma) \leq RD(\Pi', \Gamma') \leq RD(\Pi, \Gamma) + 2$.*

3.3   Minimum Common Partition

We extend the *minimum common partition* (MCP) problem, which was first introduced in [4][5] to reduce the number of duplicates of each gene in ortholog assignment, to multichromosomal genomes. Use $\overline{\pi(i)}_j$ to represent a chromosome segment or its reversal in chromosome i of genome Π. A *chromosome partition* is a list $\{\overline{\pi(i)}_1, \overline{\pi(i)}_2, \cdots, \overline{\pi(i)}_n\}$ of chromosome segments such that the concatenation of the segments (or their reversals) in some order results in the chromosome i. A *genome partition* is the union of some partitions of all the chromosomes. A list of chromosome segments is called a *common partition* of two genomes Π and Γ if it is a partition of both Π and Γ. Furthermore, a *minimum common partition* is a partition with the minimum cardinality (denoted as $L(\Pi, \Gamma)$) over all possible common partitions of Π and Γ. The MCP problem is the problem of finding the minimum common partition between two given genomes. Two genomes have a common partition if and only if they have equal gene content (*i.e.* they have the same number of duplications for each gene).

We can further extend MCP to an arbitrary pair of genomes that might have unequal gene contents. A *gene matching* $\mathcal{M}$ between genomes Π and Γ is a matching between the genes of Π and Γ, which can be defined by a maximum matching in their hit graph $\mathcal{H}$. Given a gene mathing $\mathcal{M}$, two reduced genomes (denoted as $\tilde{\Pi}_{\mathcal{M}}$ and $\tilde{\Gamma}_{\mathcal{M}}$) with equal gene content can be obtained by removing all the unmatched genes. The minimum common partition of Π and Γ is defined as the minimum $L(\tilde{\Pi}_{\mathcal{M}}, \tilde{\Gamma}_{\mathcal{M}})$ among all gene matchings $\mathcal{M}$.

Given two genomes Π and Γ, recall that $RD(\Pi, \Gamma)$ is the rearrangement/duplication distance between them. Let N_u be the number of unmatched genes introduced by a gene matching and N_c be the number of chromosomes. Based on the fact that inserting two genes into the two genomes under consideration, one for each genome, will increase the rearrangement distance by at most three, the following theorem can be obtained to establish the relationship between the minimum common partition and the rearrangement/duplication distance.

Theorem 3. *For any two genomes Π and Γ, $(L(\Pi, \Gamma) - N_c - 2)/3 + N_u \leq RD(\Pi, \Gamma) \leq L(\Pi, \Gamma) + 2N_c + N_u + 1$.*

An efficient heuristic algorithm for MCP on unichromosomal genomes was given in [4][5]. The algorithm constructs the so called "pair-match" graphs and then attempts to find a large independent set. We extend the method to multichromosomal genomes in a straightforward way.

3.4 Maximum Graph Decomposition

After minimum common partition, the genomes Π and Γ may still contain duplicates, although the number of duplicates is expected to be small. In order to match all the genes, we define another problem, called *maximum graph decomposition* (MGD). The MGD problem is: among all pairs of reduced genomes of Π and Γ obtained by all possible gene matchings, find one with the maximum value of $c(\Pi, \Gamma) - p_{\Gamma\Gamma}(\Pi, \Gamma)$.

Using the basic framework developed in [4][5], we design a greedy algorithm in MSOAR to solve MGD using a new graph, called the *complete-breakpoint graph*. The complete-breakpoint graph associated with Π and Γ is denoted as $\mathcal{G}$, which is adapted from the breakpoint graph of multichromosomal genomes of equal gene content consisting of only singletons [8]. The prefix "complete" is added here to differentiate from the partial graphs in [4][5]. If Π and Γ have different numbers of chromosomes, add null chromosomes to the genome with fewer chromosomes to make them both have N_c chromosomes. As defined in [8], a *cap* is used as a marker that serves as a chromosomal end delimiter when we convert a multichromosomal genome into a unichromosomal genome. A *capping*

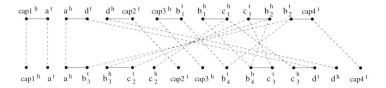

Fig. 5. The complete-breakpoint graph of two genomes with unequal gene contents $\Pi = \{\langle +a, +d \rangle, \langle +b_1, -c_1, -b_2 \rangle\}$ and $\Gamma = \{\langle +a, +b_3, +c_2 \rangle, \langle +b_4, +c_3, +d \rangle\}$. The hit graph of these two genomes is shown in Figure 4.

of a chromosome $\pi(i)$ is $\pi(i) = \langle \pi(i)_0 \pi(i)_1 \cdots \pi(i)_{n_i} \pi(i)_{n_i+1} \rangle$, where $\pi(i)_0$ is the left cap of $\pi(i)$, called *lcap*, and $\pi(i)_{n_i+1}$ is the right cap, called *rcap*. Choose any capping and an arbitrary concatenation to transform Π and Γ into unichromsomal genomes $\hat{\pi}$ and $\hat{\gamma}$. In the complete-breakpoint graph $\mathcal{G}$, every gene or cap g from each genome is represented by two ordered vertices $\alpha^t \, \alpha^h$ if α is positive or $\alpha^h \, \alpha^t$ if it is negative. Insert black edges between vertices that correspond to adjacent genes (or caps) in the same genome except the pairs of the form α_i^h and α_i^t from the same gene (or cap) α_i. Insert cross-genome grey edges $(\hat{\pi}_i^t, \hat{\gamma}_j^t)$ and $(\hat{\pi}_i^h, \hat{\gamma}_j^h)$ if gene $\hat{\pi}_i$ and gene $\hat{\gamma}_j$ are connected by an edge in the hit graph or they are the same caps. Next, we delete the left vertex of every lcap, the right vertex of every rcap and all the edges incident on them. The calculation of RD distance using the resulting graph no longer depends on the actual concatenations. Finally, we make the complete-breakpoint graph independent on the capping of Γ, by deleting the $2N_c$ black edges incident on the $\hat{\gamma}$ cap vertices. These cap vertices are called Π-caps. The vertex on the other end of a deleted black edge is called a Γ-tail unless the black edge arises from a null chromosome, in which case both of its ends are Π-caps. An example of the complete-breakpoint graph is shown in Figure 5. The complete-breakpoint graph contains both cycles and paths. Depending on whether the end points are both Π-caps, both Γ-tails or one of each, a path could be classified as a $\Pi\Pi$-path, $\Gamma\Gamma$-path or $\Pi\Gamma$-path.

After the complete-breakpoint graph is constructed, we try to find small cycles and short $\Pi\Gamma$-paths first, and then finish the decomposition by finding the rest of $\Pi\Pi$-paths and $\Gamma\Gamma$-paths. The decomposition has to satisfy the following three conditions: (1) every vertex belongs to at most one cycle or path (2) the two vertices representing each gene must be connected respectively to the two vertices of a single gene in the other genome by edges of the cycles or paths, otherwise both must be removed, *i.e.*, the connections satisfy a pairing condition; and (3) the edges within a genome and across genomes alternate in a cycle or a path. Intuitively, small cycles may lead to large cycle decompositions, although it is not always the case. Moreover, the more $\Pi\Gamma$-paths, the fewer $\Gamma\Gamma$-paths, because the number of Γ-tail vertices is fixed and each vertex can only belong to at most one path. Note that during the cycle decomposition, some gene vertices might have all of their cross-genome edges removed since Π and Γ may have unequal gene contents and these genes are regarded as inparalogs. If two gene vertices α_i^t and α_i^h have no cross genome edges incident on them during the cycle decomposition, they need to be removed from the complete-breakpoint graph right away and a black edge need to be inserted between two endpoints of the deleted black edges arising from α_i^t and α_i^h.

Any feasible solution of the MCD problem gives a maximal matching between the genes of Π and the genes of Γ. The genes that have not been matched will be assigned as inparalogs of the matched ones in the same family. The matched genes suggest possible main ortholog pairs and a rearrangement scenario to transform Π to Γ by the operations reversal, translocation, fusion and fission.

3.5   "Noise" Gene Pairs Detection

The maximum graph decomposition of a complete-breakpoint graph $\mathcal{G}$ determines a one-to-one gene matching between two genomes. Unmatched genes are removed since either they are potential inparalogs or their orthology counterparts were lost during the evolution. However, some individual paralogs might be forced to be assigned as main ortholog pairs because the maximum graph decomposition always gives a maximal matching between all the genes. Therefore, it is necessary to remove these "noise" gene pairs so that the output main ortholog pairs are more reliable.

After removing the unmatched genes, we obtain two reduced genomes with equal gene content. Remove all the gene pairs whose deletion would decrease the rearrangement distance of reduced genomes by at least two. Note that, in this case, the rearrangement/duplication distance will never increase since the deletion of a gene pair may only increase the number of duplications required in an optimal scenario by two. As mentioned before, we require that at least one main ortholog pair of each gene family be kept during this post-processing.

MSOAR combines the suboptimal ortholog assignment rules, heuristic MCP algorithm, heuristic MGD algorithm, and "noise" gene pair detection step to find all the potential main ortholog pairs and detect inparalogs.

4   Experiments

In order to test the performance of MSOAR as a tool of assigning orthologs, we have applied it to both simulated and real genome sequence data, and compared its results with two well-known algorithms in the literature, namely, an iterated version of the exemplar algorithm and INPARANOID.

4.1   Simulated Data

In order to assess the validity of our parsimony principle as a means of distinguishing main orthologs from inparalogs, we conduct two simple experiments to estimate the probability of inparalogs that may incorrectly be assigned orthology by transforming one genome into another with the minimum number of rearrangement and duplication events. The first experiment is done as follows. First, we simulate a genome G with 100 distinct genes, and then randomly perform k reversals on G to obtain another genome H. The boundaries of these reversals are uniformly distributed within the genome. Next, make a copy of

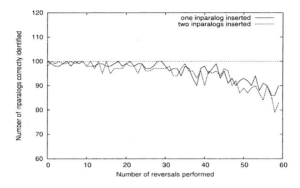

Fig. 6. Distribution of the number of inparalogs correctly identified by the parsimony principle

some gene that is randomly selected from H and insert it back into H as a duplicate. Clearly, the inserted gene is an inparalog, by definition. In this case, there are only two possible ortholog assignments between G and H. Therefore, we can easily calculate the rearrangement/duplicate distance between G and H, and know the ortholog assignment that will be made by the parsimony principle. We repeat the above procedure on 100 random instances for each k, $0 \leq k < 60$, and count the number of instances for which the inparalogs are correctly identified. The distribution of these inparalogs against the number k of reversals is plotted in Figure 6 (the curve marked "one inparalog inserted"). The result shows that with a very high probability ($> 90\%$, for each $k < 55$) the main ortholog and inparalog can be correctly identified. This suggests that an orthology assignment between two genes that are the positional counterparts between two genomes tends to result in the smaller rearrangement/duplication distance, compared to the distance given by an assignment involving inparalogs.

The second experiment is conducted to estimate the probability that two inparalogs, one from each genome, would be identified as a main ortholog pair, instead of as two individual inparalogs. Its data is generated similarly as the first experiment, except that a same copy of the gene is also inserted into genome G, resulting in a non-trivial gene family of size four in both genomes. As before, we count the number of instances for which two inparalogs are correctly identified, and plot its distribution in Figure 6 (the curve marked "two inparalogs inserted"). The result shows that it is very unlikely that two inparalogs from different genomes are assigned as a (main) ortholog pair. This and the above findings provide some basic support for the validity of using the parsimony approach to identify main orthologs.

We further use simulated data to assess the performance of our heuristic algorithm for ortholog assignment. In order to make a comparison test, we implemented the exemplar algorithm [17] and extended it into a tool for ortholog assignment as described in [5], called the iterated exemplar algorithm. The simulated data is generated as follows. Start from a genome G with n distinct symbols whose signs are generated randomly. Each symbol defines a single gene family.

Then randomly combine two gene families into a new family until r singletons are left in the genome G. Perform k reversals on G to obtain a genome H as in the previous experiments. Finally, randomly insert c inparalogs (each is a copy of some gene randomly selected) into the two genomes. Note that some singletons may be duplicated during this step so that more non-trivial gene families could be generated. The quadruple (n, r, k, c) specifies the parameters for generating two genomes as test data.

We run the iterated exemplar algorithm [17][5] and our heuristic algorithm on 20 random instances for each combination of parameters. The average performance of both algorithms is shown in Figure 7, in terms of the number of incorrectly assigned orthologs (*i.e.*, genes in a genome that are not assigned orthology to their positional counterparts in the other genome) and inparalogs. As we can see, our heuristic algorithm is quite reliable in assigning orthologs and identifying inparalogs. On average, the number of incorrect assignments generally increases as the number of reversals k increases. While both algorithms perform equally well for inparalogous gene identification, our heuristic algorithm produces fewer incorrect ortholog assignments than the iterated exemplar algorithm, especially for the instances generated using parameters $n = 100$, $r = 80$, and $c = 5$ (see Figure 7).

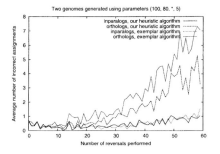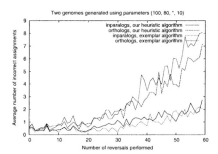

Fig. 7. Comparison of our heuristic and the exemplar algorithm on simulated data

4.2   Real Data

We consider two model genomes: Human (*Homo sapiens*) and Mouse (*Mus musculus*). Gene positions, transcripts and translations were downloaded from the UCSC Genome Browser [9] web site (http://genome.ucsc.edu). We used the canonical splice variants from the Build 35 human genome assembly (UCSC hg17, May 2004) and the Build 34 assembly of the mouse genome (UCSC mm6, March 2005). There are 20181 protein sequences in human genome assembly hg17 and 17858 sequences in mouse genome assembly mm6. Due to assembly errors and other reasons, 220 human and 114 mouse genes were mapped to more than one location in the respective genomes. For such a gene, we kept the first transcription start position which is closest to the 5$'$ end as its start coordinate. A homology search was then performed and a hit graph between human and mouse built as described in Section 3.1.

As shown in Table 1, before removing "noise" gene pairs, MSOAR assigned 13395 main orthologs pairs between human and mouse. Then MSOAR removed 177 "noise"gene pairs and output 13218 main orthologs pairs. The distribution of the number of orthologs assigned by MSOAR between human chromosomes and mouse chromosomes is illustrated in Fig 8. It shows that the top 3 chromosome pairs between human and mouse with the largest numbers of orthologs are human chromosome 17 vs. mouse chromosome 11, human chromosome 1 vs. mouse chromosome 4, and human chromosome X vs. mouse chromosome X, which are consistent with the Mouse Human synteny alignments.   (http://www.sanger.ac.uk/Projects/M_musculus/publications/fpcmap-2002/mouse-s.shtml).

We validate our assignments by using gene annotation, in particular, gene names. To obtain the most accurate list of gene names, we have cross-linked database tables from the UCSC Genome Browser with gene names extracted from UniProt [2] release 6.0 (September 2005). The official name of a gene is usually given to convey the character or function of the gene [24]. Genes with identical names are most likely to be an orthologous pair, although we should keep in mind that many names were given mostly based on sequence similarity and erroneous/inconsistent names are known to exist in the annotation. Some genes have names beginning with "LOC" or ending with "Rik" or even have no names, implying that these genes have not yet been assigned official names or their functions have not been validated. If a pair of genes output by MSOAR have completely identical gene symbol, we count them as a true positive pair; if they have different names without substring "LOC"or "Rik", it is a false positive pair; otherwise, it is counted as an unknown pair. We also calculate the total number of *assignable* pairs of orthologs, *i.e.* the total number of pairs of genes

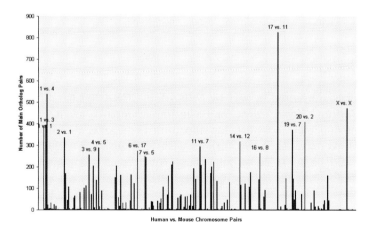

Fig. 8. Distribution of the number of ortholog pairs assigned by MSOAR across all pairs of the human and mouse chromosomes. The chromosome pairs with more than 250 main ortholog pairs are labeled. *E.g.*, the highest bar is human chromosome 17 vs. mouse chromosome 11, between which 825 main ortholog pairs were assigned.

with identical names. For example, there are 9891 assignable orthologous gene pairs between human and mouse. Before removing "noise" gene pairs, MSOAR predicted 13395 ortholog pairs, among which 9263 are true positives, 2171 are unknown pairs and 1961 are false positives, resulting in a sensitivity of 93.65% and a specificity of 81.01%. After removing "noise" gene pairs, MSOAR predicted 13218 ortholog pairs, among which 9214 are true positives, 2126 are unknown pairs and 1878 are false positives, resulting in a sensitivity of 93.16% and a specificity of 81.25%. It is interesting to note that the last step of MSOAR identified 177 "noise" gene pairs, among which 72.32% were false positives. This result shows that the identification of "noise" gene pairs effectively detects false positives and could provide more reliable ortholog assignment.

The comparison result between MSOAR and INPARANOID [16] is shown in Table 1. MSOAR was able to identify 99 more true ortholog pairs than IN-PARANOID, although it also reported more false positives.

Table 1. Comparison of ortholog assignments between MSOAR and INPARANOID

	assignable	assigned	true positive	unkown
MSOAR (before removing "noise" pairs)	9891	13395	9263	2171
MSOAR (after removing "noise" pairs)	9891	13218	9214	2126
INPARANOID	9891	12758	9115	2034

5   Concluding Remarks

Although we anticipate that the system MSOAR will be a very useful tool for ortholog assignment, more systematics tests will be needed to reveal the true potential of this parsimony approach. Our immediate future work includes the incorporation of transpositions into the system and consideration of weighing the evolutionary events.

Acknowledgment

This project is supported in part by NSF grant CCR-0309902, a DoE GtL sub-contract, National Key Project for Basic Research (973) grant 2002CB512801, NSFC grant 60528001, and a fellowship from the Center for Advanced Study, Tsinghua University. Also, the authors wish it to be known that the first two authors should be regarded as joint First Authors of this paper. The contact email addresses are {zfu,jiang}@cs.ucr.edu

References

1. S. Altschul *et al.* Gapped BLAST and PSI-BLAST: a new generation of protein database search programs. *Nucleic Acids Research*, vol. 25, no. 17, pp 3389-3402, 1997.
2. A. Bairoch *et al.* The Universal Protein Resource (UniProt). *Nuc. Acids Res.* 33:D154-D159, 2005.

3. S.B. Cannon and N.D. Young. OrthoParaMap: distinguishing orthologs from paralogs by integrating comparative genome data and gene phylogenies. *BMC Bioinformatics* 4(1):35, 2003.

4. X. Chen, J. Zheng, Z. Fu, P. Nan, Y. Zhong, S. Lonardi, and T. Jiang. Computing the assignment of orthologous genes via genome rearrangement. In *Proc. 3rd Asia Pacific Bioinformatics Conf. (APBC'05)*, pp. 363-378, 2005.

5. X. Chen, J. Zheng, Z. Fu, P. Nan, Y. Zhong, S. Lonardi, and T. Jiang The assignment of orthologous genes via genome rearrangement. *IEEE/ACM Transactions on Computational Biology and Bioinformatics*, vol. 2, no. 4, pp. 302-315, 2005.

6. W.M. Fitch. Distinguishing homologous from analogous proteins. *Syst. Zool.* 19: 99-113, 1970.

7. S. Hannenhalli and P. Pevzner. Transforming cabbage into turnip (polynomial algorithm for sorting signed permutations by reversals). *Proc. 27th Ann. ACM Symp. Theory of Comput.* (STOC'95), pp. 178-189, 1995.

8. S. Hannenhalli, P. Pevzner. Transforming men into mice (polynomial algorithm for genomic distance problem). *Proc. IEEE 36th Symp. Found. of Comp. Sci.*, 581-592, 1995.

9. D. Karolchik, K.M. Roskin, M. Schwartz, C.W. Sugnet, D.J. Thomas, R.J. Weber, D. Haussler and W.J. Kent. The UCSC Genome Browser Database. *Nucleic Acids Res.*, vol. 31, no. 1, pp. 51-54, 2003.

10. E. Koonin. Orthologs, paralogs, and evolutionary genomics. *Annu. Rev. Genet.*, 2005.

11. Y. Lee *et al.* Cross-referencing eukaryotic genomes: TIGR orthologous gene alignments (TOGA). *Genome Res.*, vol. 12, pp. 493-502, 2002.

12. L. Li, C. Stoeckert, D. Roos. OrthoMCL: identification of ortholog groups for eukaryotic genomes. *Genome Res.*, vol. 13, pp. 2178-2189, 2003.

13. M. Marron, K. Swenson, and B. Moret. Genomic distances under deletions and insertions. *Theoretic Computer Science*, vol. 325, no. 3, pp. 347-360, 2004.

14. N. El-Mabrouk. Reconstructing an ancestral genome using minimum segments duplications and reversals. *Journal of Computer and System Sciences*, vol. 65, pp. 442-464, 2002.

15. M. Ozery-Flato and Ron Shamir. Two notes on genome rearragnements. *Journal of Bioinformatics and Computational Biology*, Vol. 1, No. 1, pp. 71-94, 2003.

16. M. Remm, C. Storm, and E. Sonnhammer. Automatic clustering of orthologs and in-paralogs from pairwise species comparisons. *J. Mol. Biol.* 314, 1041-1052, 2001.

17. D. Sankoff. Genome rearrangement with gene families. *Bioinformatics* 15(11): 909-917, 1999.

18. K. Swenson, M. Marron, J. Earnest-DeYoung, and B. Moret. Approximating the true evolutionary distance between two genomes. *Proc. 7th SIA Workshop on Algorithm Engineering & Experiments*, pp. 121-125, 2005.

19. K. Swenson, N. Pattengale, and B. Moret. A framework for orthology assignment from gene rearrangement data. *Proc. 3rd RECOMB Workshop on Comparative Genomics*, Dublin, Ireland, LNCS 3678, pp. 153-166, 2005.

20. C. Storm and E. Sonnhammer. Automated ortholog inference from phylogenetic trees and calculation of orthology reliability. *Bioinformatics*, vol. 18, no. 1, 2002.

21. R.L. Tatusov, M.Y. Galperin, D.A. Natale, and E. Koonin. The COG database: A tool for genome-scale analysis of protein functions and evolution. *Nucleic Acids Res.* 28:33-36, 2000.

22. G. Tesler. Efficient algorithms for multichromosomal genome rearrangements. *Journal of Computer and System Sciences*, vol. 65, no. 3, pp. 587-609, 2002.

23. R.L. Tatusov, E. Koonin, and D.J. Lipman. A genomic perspective on protein families. *Science*, vol. 278, pp. 631-637, 1997.
24. H.M. Wain, E.A. Bruford, R.C. Lovering, M.J. Lush, M.W. Wright and S. Povey. Guidelines for human gene nomenclature. *Genomics* 79(4), 464-470, 2002.
25. Y.P. Yuan, O. Eulenstein, M. Vingron, and P. Bork. Towards detection of orthologues in sequence databases. *Bioinformatics*, vol. 14, no. 3, pp. 285-289, 1998.
26. X. Zheng *et al.* Using shared genomic synteny and shared protein functions to enhance the identification of orthologous gene pairs. *Bioinformatics* 21(6): 703-710, 2005.

Detecting the Dependent Evolution of Biosequences

Jeremy Darot[2,3,*], Chen-Hsiang Yeang[1,*], and David Haussler[1]

[1] Center for Biomolecular Science and Engineering, UC Santa Cruz
[2] Department of Applied Mathematics and Theoretical Physics,
University of Cambridge
[3] EMBL - European Bioinformatics Institute

Abstract. A probabilistic graphical model is developed in order to detect the dependent evolution between different sites in biological sequences. Given a multiple sequence alignment for each molecule of interest and a phylogenetic tree, the model can predict potential interactions within or between nucleic acids and proteins. Initial validation of the model is carried out using tRNA sequence data. The model is able to accurately identify the secondary structure of tRNA as well as several known tertiary interactions.

1 Introduction

Recent advances in systems biology and comparative genomics are providing new tools to study evolution from a systems perspective. Selective constraints often operate on a system composed of multiple components, such that these components evolve in a coordinated way. We use the term dependent evolution to denote the dependency of sequence evolution between multiple molecular entities. A molecular entity can be a protein, a non-coding RNA, a DNA promoter, or a single nucleotide or residue. Dependent evolution is prevalent in many biomolecular systems. Instances include neo-functionalization and pseudogene formation [1, 2], co-evolution of ligand-receptor pairs [3, 4], protein-protein interactions [5], residues contributing to the tertiary structure of proteins [6], and RNA secondary structure [7]. Understanding dependent evolution helps to predict the physical interactions and functions of biomolecules, reconstruct their evolutionary history, and further understand the relation between evolution and function.

In this work, we develop a computational method for detecting and characterizing dependent evolution in orthologous sequences of multiple species. Continuous-time Markov models of sequence substitutions encoding the dependent or independent evolution of two molecular entities are constructed. The spatial dependency of adjacent sites in the sequence is captured by a hidden Markov model (HMM) specifying the interaction states of sites. As a proof-of-concept demonstration, we apply the model to tRNA sequences and show that the method can identify their secondary and tertiary structure.

---

[*] Contributed equally to this work.

A. Apostolico et al. (Eds.): RECOMB 2006, LNBI 3909, pp. 595–609, 2006.

Models of co-evolution have been investigated in many previous studies. Some of these have demonstrated that the sequence substitution rates of proteins are correlated with their function [8] and relationships with other proteins, such as the number of interactions [5], their interacting partners [5], and their co-expressed genes [9]. The compensatory substitutions of RNA sequences have been used to predict RNA secondary structure [10, 11, 12, 13, 14, 15, 16, 7, 17]. Other studies have attempted to predict protein-protein interactions at the residue or whole-protein levels by using co-evolutionary models [3, 4, 6, 18]. We use a framework of continuous-time Markov models resembling those in [6, 18], although the assumptions and mathematical approaches are significantly different.

2   Methods

In this study we use both general and specific evolutionary models to detect the secondary and tertiary structure of tRNAs. These are well suited to a proof-of-concept demonstration since nucleotide pairs have fewer joint states than residue pairs ($4 \times 4 = 16$ compared to $20 \times 20 = 400$), their interaction rules are relatively simple (primarily Watson-Crick base pairing), the secondary and tertiary interactions of tRNAs are already mapped, and a large number of aligned tRNA sequences across many species are available.

The typical structure of the tRNA encoding methionine is shown in Fig. 1. It comprises four stems, three major loops and one variable loop. Each stem contains several nucleotide pairs forming hydrogen bonds (black bars in Fig. 1). Those base pairs typically conform with the Watson-Crick complementary rule

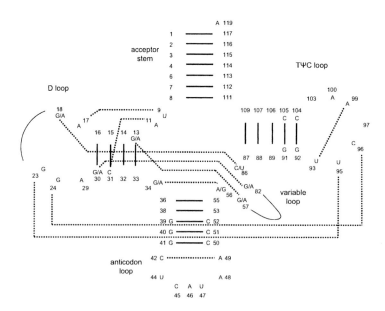

Fig. 1. tRNA secondary and tertiary structure

(AU or GC). Several GU pairs also form weaker hydrogen bonds (GU wobble). In addition, nucleotide pairs that are distant in the secondary structure may also form tertiary interactions (dotted lines in Fig. 1). Unlike secondary interactions, tertiary interactions do not necessarily conform with the Watson-Crick rules or GU wobble.

The co-evolutionary model that we developed is a probabilistic graphical model, operating on a given alignment of families of sequences for two molecular entities, along two orthogonal dimensions. The first dimension is time, with a continuous-time Markov process modeling the potentially coupled evolution of the two entities considered. This model operates at each position in the alignment, along a given phylogenetic tree. The second dimension is space, with an HMM operating along the sequence alignment and determining which regions of the two entities are co-evolving. Such graphical models were introduced by [19, 20] and have been recently adopted for instance by [21] to model the evolution of single molecular entities.

Consider first the sequence evolution model of a single nucleotide. It is a continuous-time Markov process with a substitution rate matrix $\mathbf{Q}$:

$$\frac{d\mathbf{P}(x(t))}{dt} = \mathbf{P}(x(t))\mathbf{Q}. \tag{1}$$

where $x(t)$ denotes the sequence at time t and $\mathbf{P}(x(t))$ a 1×4 probability vector of $x(t)$ being each nucleotide. $\mathbf{Q}$ is a 4×4 matrix with each row summed to zero. Different rate matrices have been developed in the literature of molecular evolution. In this work we use the HKY model [22], which characterizes $\mathbf{Q}$ by a stationary distribution $\boldsymbol{\pi}$ and a transition/transversion ratio κ:

$$\mathbf{Q} = \begin{pmatrix} - & \pi_C & \kappa\pi_G & \pi_T \\ \pi_A & - & \pi_G & \kappa\pi_T \\ \kappa\pi_A & \pi_C & - & \pi_T \\ \pi_A & \kappa\pi_C & \pi_G & - \end{pmatrix} \tag{2}$$

Each diagonal entry is the opposite of the sum of the other entries in the same row. The transition probability $P(x(t)|x(0))$ is an entry of the matrix exponential of $\mathbf{Q}t$:

$$P(x(t) = b|x(0) = a) = e^{\mathbf{Q}t}[a, b]. \tag{3}$$

Given a phylogenetic tree and the length of its branches, the marginal likelihood of the observed sequence data at the leaves is the joint likelihood summed over all possible states of internal (ancestral) nodes. This marginal likelihood can be efficiently calculated using a dynamic programming algorithm [23]. Briefly, let u be a node in the tree, v and w its children, and t_v, t_w the branch lengths of $(u, v), (u, w)$. Define $P(L_u|a)$ as the probability of all the leaves below u given that the base assigned to u is a. The algorithm is then defined by the recursion:

$$P(L_u|a) = \begin{cases} I(x_u = a) & \text{if } u \text{ is a leaf,} \\ \sum_b e^{\mathbf{Q}t_v}[a, b]P(L_v|b) \sum_c e^{\mathbf{Q}t_w}[a, c]P(L_w|c) & \text{otherwise.} \end{cases} \tag{4}$$

where $I(.)$ is the indicator function.

Now consider the sequence evolution model of a nucleotide pair. Define $\mathbf{x}(t) = (x_1(t), x_2(t))$ as the joint state of the nucleotide pair at time t. There are 16 possible joint states. The null model assumes that each nucleotide evolves independently with an identical substitution rate matrix. Therefore, the transition probability matrix is:

$$P(\mathbf{x}(t)|\mathbf{x}(0)) = (e^{\mathbf{Q}} \otimes e^{\mathbf{Q}})^t. \tag{5}$$

where $e^{\mathbf{Q}} \otimes e^{\mathbf{Q}}$ is the tensor product of two identical 4×4 matrices $e^{\mathbf{Q}}$. The outcome is a 16×16 matrix, specifying the transition probability of the joint state in a unit time. Each entry is the product of the corresponding entries in the single nucleotide substitution matrices. For instance,

$$\begin{aligned}
P(\mathbf{x}(1) &= (C, G)|\mathbf{x}(0) = (A, U)) \\
&= P(x_1(1) = C|x_1(0) = A)P(x_2(1) = G|x_2(0) = U) \\
&= e^{\mathbf{Q}}[A, C] \cdot e^{\mathbf{Q}}[U, G].
\end{aligned} \tag{6}$$

The substitution rate $\mathbf{Q}_2 = log(e^{\mathbf{Q}} \otimes e^{\mathbf{Q}})$ of the nucleotide pair transitions in (5) is a 16×16 matrix, with the rates of single nucleotide changes identical to those in (2) and zero rates on double nucleotide changes. More precisely, if we denote (a, b) the joint state of a nucleotide pair:

$$\mathbf{Q}_2((a_1, a_2), (b_1, b_2)) = \begin{cases}
\mathbf{Q}(a_1, b_1) & \text{if } a_2 = b_2, \\
\mathbf{Q}(a_2, b_2) & \text{if } a_1 = b_1, \\
-\mathbf{Q}(a_1, b_1) - \mathbf{Q}(a_2, b_2) & \text{if } a_1 = b_1, a_2 = b_2, \\
0 & \text{otherwise.}
\end{cases} \tag{7}$$

Equations (5) and (7) are equivalent, and the latter is discussed in [24]. Intuitively, if two nucleotides evolve independently, then during an infinitesimal time only one nucleotide can change, and the rate is identical to the single nucleotide transition rate.

The alternative model assumes that the evolution of the two nucleotides is coupled. One way to express their dependent evolution is to "reweight" the entries of the substitution rate matrix by a potential term ψ:

$$\mathbf{Q}_2^a = \mathbf{Q}_2 \circ \psi. \tag{8}$$

where ψ is a 16×16 matrix and $\circ$ denotes the following operation:

$$\mathbf{Q}_2(a, b) \circ \psi(a, b) = \begin{cases}
\mathbf{Q}_2(a, b) \cdot \psi(a, b) & \text{if } a \neq b, \mathbf{Q}_2(a, b) > 0, \\
\psi(a, b) & \text{if } a \neq b, \mathbf{Q}_2(a, b) = 0, \\
-\sum_{b' \neq b} \mathbf{Q}_2(a, b') \circ \psi(a, b') & \text{if } a = b.
\end{cases} \tag{9}$$

It multiplies an off-diagonal, nonzero entry $\mathbf{Q}_2(a, b)$ by $\psi(a, b)$, sets the value of a zero entry $\mathbf{Q}_2(a, b)$ as $\psi(a, b)$, and normalizes a diagonal entry as the opposite of the sum of the other entries in the same row. $\mathbf{Q}_2^a$ is a valid substitution rate matrix, thus its exponential induces a valid transition probability matrix.

We give (8) a mechanistic interpretation. The sequence substitution pattern of a co-evolving pair is the composite effect of neutral mutations, which occur independently at each nucleotide, and a selective constraint, which operates on the joint state. The potential term ψ rewards the state transitions that denote co-evolution and penalizes the others. We set the ratio between penalty and neutrality at ϵ, and the reward for simultaneous changes as r.

The choice of rewarded and penalized states is crucial. Here, we apply three different criteria to reweight the joint states. The first criterion rewards the state transitions that establish Watson-Crick base pairing from non-interacting pairs, penalizes the state transitions which break it, and is neutral for all other state transitions. We call it the "Watson-Crick co-evolution" or WC model. Specifically, the potential term is:

$$\psi(\mathbf{x}(0), \mathbf{x}(1)) = \begin{cases} \frac{1}{\epsilon} & \text{if } \mathbf{x}(0) \text{ is not WC and } \mathbf{x}(1) \text{ is WC,} \\ \epsilon & \text{if } \mathbf{x}(0) \text{ is WC and } \mathbf{x}(1) \text{ is not WC,} \\ 0 & \text{if } x_1(1) \neq x_1(0) \text{ and } x_2(1) \neq x_2(0), \\ 1 & \text{otherwise.} \end{cases} \quad (10)$$

The second criterion includes the GU/UG pairs (denoted GU since the order does not matter here) in the rewarded states. It thus rewards the state transitions that establish Watson-Crick or GU wobble base pairs, penalizes the state transitions which break the extended rule, and is neutral for all other state transitions. We call it the "Watson-Crick co-evolution with GU wobble" or WCW model. Specifically,

$$\psi(\mathbf{x}(0), \mathbf{x}(1)) = \begin{cases} \frac{1}{\epsilon} & \text{if } \mathbf{x}(0) \text{ is not WC or GU and } \mathbf{x}(1) \text{ is WC or GU,} \\ \epsilon & \text{if } \mathbf{x}(0) \text{ is WC or GU and } \mathbf{x}(1) \text{ is not WC or GU,} \\ 0 & \text{if } x_1(1) \neq x_1(0) \text{ and } x_2(1) \neq x_2(0), \\ 1 & \text{otherwise.} \end{cases} \quad (11)$$

Note that both the WC and the WCW model have zero rates on simultaneous nucleotide changes.

The third criterion does not use prior knowledge of Watson-Crick base pairing and GU wobble and only considers the simultaneous changes of the two nucleotides ("simple co-evolution" or CO model). It rewards the state transitions where both nucleotides change, and penalizes the state transitions where only one nucleotide changes. Recall that the rates of simultaneous changes in the independent model are zero. Therefore, we reward these transitions not by reweighting their entries in $\mathbf{Q}_2$, but by giving them a positive rate r. Specifically,

$$\psi(\mathbf{x}(0), \mathbf{x}(1)) = \begin{cases} r & \text{if } x_1(1) \neq x_1(0) \text{ and } x_2(1) \neq x_2(0), \\ \epsilon & \text{if either } x_1(1) = x_1(0) \text{ or } x_2(1) = x_2(0), \\ 1 & \text{otherwise.} \end{cases} \quad (12)$$

The CO model assumes that the interacting nucleotide pairs maintain stable states. In order to transition from one stable state to another, both nucleotides must change. We introduce this general model in order to capture tertiary interactions for which pairing rules are complex or unknown. Moreover, since this general model incorporates no knowledge about nucleotide interactions and has only two extra free parameters (ϵ and r), it can be directly extended to more complicated problems such as protein-protein interactions or multi-way interactions.

We apply the dynamic programming algorithm described in (4) to evaluate the marginal likelihood of the nucleotide pair data. Specifically, a, b and c are the joint states of nucleotide pairs and e^{Qt} is defined as in (5) for the null model and as the exponential of (8) times t for the alternative model.

In order to incorporate the spatial dimension of the nucleotide sequence into the model, we define an HMM for the "interaction states" of the aligned sequences. Suppose that the sequences of two molecular entities are aligned (e.g., a tRNA sequence is aligned with itself in the opposite direction) across all species. We define the "interaction state" $y(s)$ of the sequence pair at alignment position s as a binary random variable, indicating whether co-evolution occurs at position s (i.e., $y(s) = 1$) or not ($y(s) = 0$). The $y(s)$'s are the hidden variables of the HMM. Their transitions are specified by a homogeneous Markov chain with transition probability $P(y(s+1) = 1|y(s) = 0) = P(y(s+1) = 0|y(s) = 1) = \alpha$. The observed variable $X(s)$ comprises the sequences at position s across all species. The emission probability $P(X(s)|y(s))$ corresponds to the likelihood of the sequence data, conditioned on the null model of independent evolution or the alternative model of co-evolution. The likelihoods are evaluated by the aforementioned dynamic programming algorithm. Given the transition and emission probabilities, we apply the Viterbi algorithm to identify the interacting regions of the two sequences.

Issues arise when there are gaps in the aligned sequences. If "sparse" gaps appear at scattered positions in a few species, we treat them as missing data, by giving an equal probability to each nucleotide. If there are consistent gaps appearing in consecutive regions over many species, we ignore those regions when calculating the likelihood scores.

In order to quantify the confidence of the inferred interaction states, we used the log-likelihood ratio (LLR) between the co-evolutionary model and the null model, at each position within the Viterbi algorithm. Pollock et al. [6] have pointed out that a χ^2 distribution is not appropriate for such co-evolutionary models. For this reason, we have not reported the p-values that might have otherwise been calculated from a χ^2 distribution with one (WC and WCW models) or two (CO model) extra degrees of freedom.

3   Results

We applied our model to the methionine tRNA sequences of 60 species covering the three superkingdoms of life. Three different criteria were used to reward and penalize the joint state transitions in the model of dependent evolution: Watson-Crick base pairing, Watson-Crick base pairing with GU wobble, simultaneous changes. We compared the performance of each model in detecting secondary and tertiary interactions, and further investigated false positives and false negatives.

3.1   Data and Pre-processing

Aligned tRNA sequences were downloaded from the Rfam database [25]. Unique sequences for the methionine tRNA (tRNA-Met, ATG codon) were extracted for 60 species, including archea, bacteria, eukaryotes and their organelles (mitochondria and chloroplast). The length of the complete sequence alignment including gaps was $lseq = 119$ nucleotides. A phylogenetic tree was derived from these sequences using a Metropolis-coupled Markov chain Monte-Carlo (MC^3) simulation implemented in the *MrBayes* program [26]. The resulting tree was found to be robust and consistent with the tree topologies obtained by parsimony using the *DNAPARS* program of the *PHYLIP* package [27]. The phylogenetic tree of the tRNA data is reported in the supplementary materials.

The tRNA sequence was then paired with itself in the opposite direction in order to evaluate potential co-evolution between all possible nucleotide pairs. The first entity in the model was the tRNA sequence itself, and the second entity was the reversed sequence, shifted by a number of nucleotides varying from 1 to $lseq$, and "rolled over" to match the length of the first entity. The co-evolutionary signal, which is the Viterbi path of the HMM, was then plotted as a $lseq \times lseq$ matrix, where the x-axis represents the position in the sequence, and the y-axis the offset. As an example, the expected signal for the structure depicted in Fig.1 is shown in Fig. 2. The figure comprises four symmetric patterns, which correspond to the four stems of the tRNA secondary structure (in yellow): acceptor

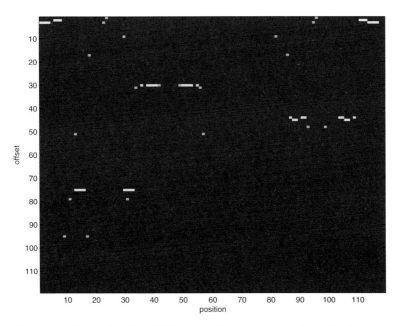

Fig. 2. Expected signal for the tRNA secondary structure (in yellow) and tertiary structure (in green)

stem at offset 2 and 3, anticodon stem at offset 30, TΨC stem at offset 44 and 45, and D stem at offset 75. The tertiary structure appears as symmetric isolated nucleotide pairs (in green). The patterns are not symmetric with respect to the diagonal line due to a gap between positions 60 and 80 covering padding to the variable loop.

3.2   Sensitivity Analysis

A sensitivity analysis was carried out, varying ϵ from 10^{-3} to 0.90, r from 0 to 0.5, and α from 0.05 to 0.45 (results not shown). It was found that the performance of the different methods depends on a reasonable choice of parameter values. Indeed, the co-evolutionary models merge with the independent model for $\epsilon = 1$ and $r = 0$, therefore no signal can be detected for these parameter values. Conversely, excessively small values for ϵ and large values for r compromise the performance of the analysis. The parameter α can be seen as a spatial "smoothing" factor, which tends to eliminate isolated hits as its value decreases. This can help to eliminate isolated false positives from the contiguous secondary structure signal, but can also prevent the identification of isolated tertiary interactions. We henceforth report the results for $\epsilon = 0.5$, $r = 0.05$ and $\alpha = 0.2$.

3.3   Watson-Crick Co-evolution

The co-evolutionary signal detected by the WC model is shown as a ROC curve and at a particular cutoff LLR value of 5.0 in Fig. 3. At this level of significance, 20 out of 21 secondary interactions were identified (in orange), and 4 out of 10 tertiary interactions (in red), resulting in 22 false positives (in light blue). The "missing" secondary interaction, between nucleotides 36 and 55, shows evidence of GU wobble, which can be contrasted with the purely Watson-Crick base pairing of the true positive pair 39-52 (Table 1). The WC model is not suited to the detection of such an interaction, though it is eventually picked up at a much lower significance level (Fig. 3(a)).

As expected, the four tertiary interactions identified by the WC model (Table 2) are mainly Watson-Crick, even though pairs 24-96 and 93-99, which

Table 1. Dinucleotide composition of one (a) true positive (b) false negative secondary interaction, WC model

39-52	A	C	G	U	36-55	A	C	G	U
A	0	0	0	14	A	1	0	0	2
C	0	0	5	0	C	0	0	18	3
G	0	39	0	0	G	0	1	0	2
U	2	0	0	0	U	25	0	8	0

Table 2. Dinucleotide composition of detected tertiary interactions, WC model

9-17	A	C	G	U	18-86	A	C	G	U	24-96	A	C	G	U	93-99	A	C	G	U
A	0	0	0	0	A	0	1	1	24	A	2	0	0	0	A	11	0	0	0
C	0	0	0	0	C	0	0	0	0	C	2	0	0	0	C	0	0	0	0
G	0	0	0	2	G	0	30	0	1	G	0	44	0	0	G	0	0	0	0
U	58	0	0	0	U	1	0	0	0	U	6	0	0	4	U	48	0	0	1

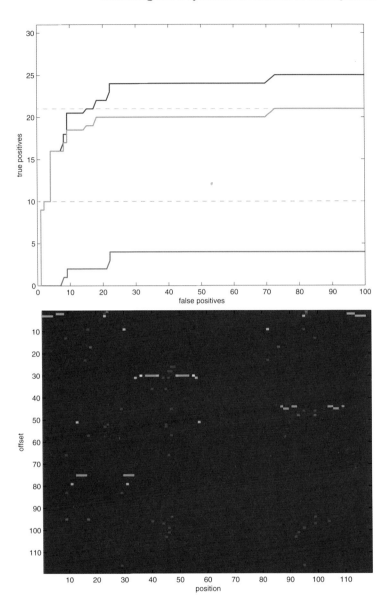

Fig. 3. Results from the WC model (a) ROC curves (b) signal at a $LLR = 5.0$ cut-off. The three ROC curves refer to the secondary structure (in orange), the tertiary structure (in red) and the complete structure (in blue) respectively.

are detected at a comparatively lower significance level, have some non-negligible terms off the second diagonal.

Many of the false positives seem to be vertically aligned in Fig. 3(b). A closer examination reveals that these are composed of nucleotides which are highly

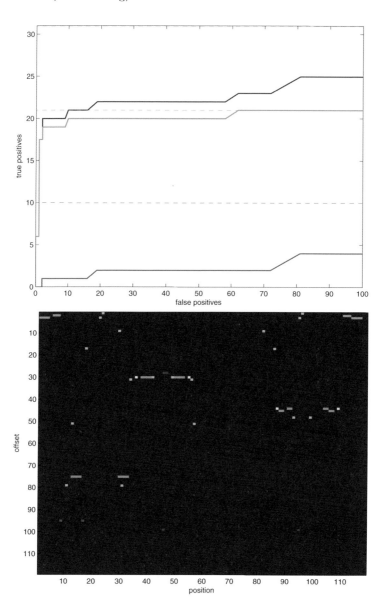

Fig. 4. Results from the WCW model (a) ROC curves (b) signal at a $LLR = 5.8$ cutoff. The three ROC curves refer to the secondary structure (in orange), the tertiary structure (in red) and the complete structure (in blue) respectively.

conserved individually, and appear to form a Watson-Crick pair without physically interacting. In particular, the constant nucleotides of the CAU anticodon at positions 45-47 form spurious Watson-Crick base pairs with other highly conserved nucleotides in the different loops of the tRNA structure.

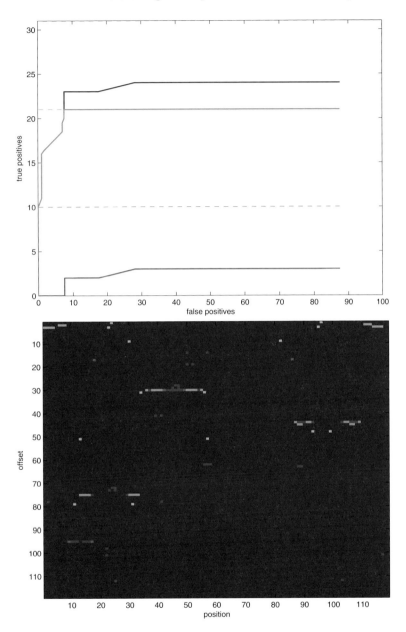

Fig. 5. Results from the CO model (a) ROC curves (b) signal at a $LLR = 0.8$ cut-off. The three ROC curves refer to the secondary structure (in orange), the tertiary structure (in red) and the complete structure (in blue) respectively.

3.4   Watson-Crick Co-evolution with GU Wobble

The co-evolutionary signal detected by the WCW model is shown as a ROC curve and at a particular cutoff LLR value of 5.8 in Fig. 4. At this level of significance, 19 out of 21 secondary interactions were identified, and 1 out of 10 tertiary interactions, for only 2 false positives. The much steeper ROC curve for secondary interactions demonstrates the benefit of incorporating additional biochemical knowledge into the model. Indeed, as many secondary interactions involve some degree of GU wobble, they are detected earlier by the WCW model than they were by the WC model. In contrast, the identification of tertiary interactions does not benefit from the refined model, because those rarely involve GU wobble. The only exception is the 23-95 pair, which involves GU wobble, but it is only detected for more than 200 false positives (beyond the boundaries of Fig. 4(a)).

3.5   Simple Co-evolution Model

The co-evolutionary signal detected by the CO model is shown as a ROC curve and at a particular cutoff LLR value of 0.8 in Fig. 5. At this level of significance, all secondary interactions were identified, and 3 out of 10 tertiary interactions, yielding 25 false positives. The tertiary interactions detected by the CO model include the pairs 9-17 and 18-86, which were also identified by the WC and WCW models. However, neither the 24-96, 93-99 nor 23-95 interactions were identified, as for those pairs one nucleotide often varies while the other remains constant (Table 2). Additionally, the 42-49 interaction was identified by the CO

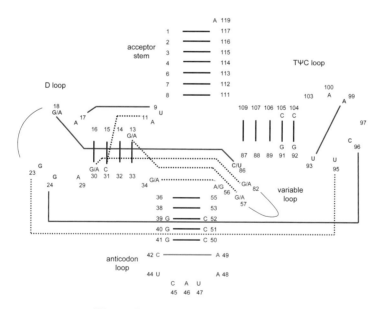

Fig. 6. Detection of tertiary interactions

model, which had not been detected by the WC and WCW models because it consists mainly of C-A and U-C pairs.

3.6   Detection of Tertiary Interactions: Summary

Figure 6 highlights the tRNA-Met tertiary interactions that have been detected using one of the three co-evolutionary models. Among the ten annotated tertiary interactions, six were identified by at least one of the models: 9-17 and 18-86 (all three models, solid blue), 24-96 and 93-99 (WC and WCW models, solid blue), 23-95 (WCW model, dotted blue), 42-49 (CO model, solid red). The four remaining interactions (dotted black) were not detected by any model. With the possible exception of 30-82, none of those pairs shows a particular bias towards Watson-Crick base pairing or simultaneous evolution in their dinucleotide composition, so the failure to detect them using the aforementioned models is not surprising.

4   Discussion

We have shown that a probabilistic graphical model incorporating neutral mutations, selective constraints and sequence adjacency can successfully identify the secondary and tertiary interactions in a tRNA structure.

The comparison of the results of the WC and WCW models indicates a trade-off between generality and performance. Indeed, increasing the specificity of the model by incorporating more biological knowledge significantly improves the detection of the secondary structure. However, the increased specificity of the WC and WCW models causes them to miss a non-Watson Crick tertiary interaction, which is detected by the much more general CO model. Given this trade-off, the performance of the CO model turns out to be surprisingly good for both secondary and tertiary interactions, and suggests that rewarding non-specific simultaneous changes is a simple, yet powerful approach. This result is encouraging when one considers using such probabilistic graphical models to investigate the co-evolution of more complex molecular systems, for which the interaction rules are not well characterized and the number of joint states is much larger, e.g., between proteins and nucleic acids.

Currently the parameters of the models – ϵ, r, α and the LLR cutoff – are set empirically. A more systematic way of estimating them from the data and testing the model in cross-validation would be a useful extension of this work.

Some scenarios beyond the co-evolution of physically interacting molecules may also be captured by this model. For instance, instead of rewarding simultaneous changes and penalizing unilateral changes, we can invert the potential term to reward unilateral changes and penalize simultaneous changes. A possible interpretation of this scenario is that the two entities are complementary in function, such as paralogous genes after their duplication. The conservation of one gene allows the evolution of the other, which can acquire a new function. A change in both genes, however, is likely to be detrimental to their original functions and thereby reduces the fitness. The modeling approach presented here could thus provide a general framework to study the dependent evolution of biosequences.

Supplementary Materials

The phylogenetic tree of the tRNA data across 60 species and the statistics of dinucleotide composition of all the tertiary interactions are reported in http://www.soe.ucsc.edu/~chyeang/RECOMB06/.

Acknowledgements

We thank Harry Noller for helpful discussions and providing information about tRNA secondary and tertiary interactions. CHY is sponsored by an NIH/NHGRI grant of UCSC Center for Genomic Science (1 P41 HG02371-02) and JD was sponsored by a Microsoft Research scholarship.

References

1. Ohno, S.: Evolution by gene duplication. Springer-Verlag, Heidelberg, Germany, 1970
2. Lynch, M., Conery, J.S.: The evolutionary fate and consequences of duplicated genes. Science **290** (2000) 1151–1155
3. Goh, C.S., Bogan, A.A., Joachmiak, M., Walther, D., Cohen, F.E.: Co-evolution of proteins with their interaction partners. J. Mol. Biol. **299** (2000) 283–293
4. Ramani, A.K., Marcotte, E.M.: Exploiting the co-evolution of interacting proteins to discover interaction specificity. J. Mol. Biol. **327** (2003) 273–284
5. Fraser, H.B., Hirsh, A.E., Steinmetz, L.M., Scharfe, C., Feldman, M.W.: Evolutionary fate in the protein interaction network. Science **296** (2002) 750–752
6. Pollock, D.D., Taylor, W.R., Goldman, N.: Coevolving protein residues: maximum likelihood identification and relationship to structure. J. Mol. Biol. **287** (1999) 187–198
7. Washietl, S., Hofacker, I.L., Stadler, P.F.: Fast and reliable prediction of noncoding RNAs. PNAS **102** (2005) 2454–2459
8. Wall, D.P., Hirsh, A.E., Fraser, H.B., Kumm, J., Giaver, G., Eisen, M., Feldman, M.W.: Functional genomic analysis of the rates of protein evolution. PNAS **102** (2005) 5483–5488
9. Jordan, I.K., Marino-Ramrez, L., Wolf, Y.I., Koonin, E.V.: Conservation and co-evolution in the scale-free human gene coexpression network. Mol. Biol. Evol. **21** (2004) 2058–2070
10. Noller, H.F., Woese, C.R.: Secondary structure of 16S ribosomal RNA. Science **212** (1981) 403–411
11. Hofacker, I.L., Fekete M., Flamm, C., Huynen, M.A., Rauscher, S., Stolorz, P.E., Stadler, P.F: Automatic detection of conserved RNA structure elements in complete RNA virus genomes. Nucleic Acids Res. **26** (1998) 3825-3836
12. Eddy, S.R.: Non-coding RNA genes and the modern RNA world. Nat. Rev. Genet. **2 (2001)** 919–929
13. Rivas, E., Klein, R.J., Jones, T.A., Eddy, S.R.: Computational identification of non-coding RNAs in E. coli by comparative genomics. Curr. Biol. **11** (2001) 1369-1373
14. di Bernardo, D., Down T., Hubbard, T.: ddbRNA: detection of conserved secondary structures in multiple alignments. Bioinformatics **19** (2003) 1606-1611

15. Coventry, A., Kleitman D.J., Berger, B.: MSARI: multiple sequence alignments for statistical detection of RNA secondary structure. PNAS **101** (2004) 12102-12107

16. Pedersen, J.S., Meyer, I.M., Forsberg, R., Simmonds, P., Hein, J.: A comparative method for finding and folding RNA secondary structures within protein-coding regions. Nucleic Acids Res. **32** (2004) 4925-4936

17. Washietl, S., Hofacker I.L., Lukasser, M., Huttenhofer, A., Stadler, P.F.: Mapping of conserved RNA secondary structures predicts thousands of functional noncoding RNAs in the human genome. Nat. Biotechnol. **23** (2005) 1383-1390

18. Barker, D., Pagel, M.: Predicting functional gene links from phylogenetic-statistical analyses of whole genomes. PLoS Comp. Biol. **1** (2005) 24-31

19. Yang, Z.: A space-time process model for the evolution of DNA sequences. Genetics **139** (1995) 993-1005

20. Felsenstein, J., Churchill, G.: A hidden Markov model approach to variation among sites in rate of evolution. Mol. Biol. Evol. **13** (1996) 93-104

21. Siepel, A., Haussler, D.: Combining phylogenetic and hidden Markov models in biosequence analysis. JCB **11** (2004) 413-428

22. Hasegawa, M., Kishino, H., Yano, T.: Dating the human-ape splitting by a molecular clock of mitochondrial DNA. J. Mol. Evol. **22** (1985) 160-174

23. Felsenstein, J.: Evolutionary trees from DNA sequences: a maximum likelihood approach. J. Mol. Evol. **17** (1981) 368-376

24. Pagel, M.: Detecting correlated evolution on phylogenies: a general method for the comparative analysis of discrete characters. Proceedings of the Royal Society in London, series B, **255** (1994) 37-45.

25. RNA families database. http://www.sanger.ac.uk/cgi-bin/Rfam/getacc?RF00005

26. MrBayes: Bayesian inference of phylogeny. http://mrbayes.csit.fsu.edu/index.php

27. http://evolution.genetics.washington.edu/phylip.html

Author Index